Yulei Shenjing Neifenmixue

鱼类神经内分泌学

［加］N.J. 伯尼尔　　［加］G. Van Der 克拉克
［加］A.P. 法雷尔　　［加］C.J. 布劳纳　　编著

林浩然　译

中山大学出版社
·广州·

版权所有　翻印必究

图书在版编目（CIP）数据

鱼类神经内分泌学/（加）伯尼尔（Nicholas J. Bernier）等编著；林浩然译. —广州：中山大学出版社，2017.11

书名原文：Fish Neuroendocrinology

ISBN 978-7-306-06211-6

Ⅰ. ①鱼… Ⅱ. ①伯… ②林… Ⅲ. ①鱼类—神经递体—研究 Ⅳ. ①Q959.405

中国版本图书馆 CIP 数据核字（2017）第 250950 号

出 版 人：	徐 劲
策划编辑：	廖丽玲
责任编辑：	廖丽玲
封面设计：	林绵华
责任校对：	杨文泉
责任技编：	何雅涛
出版发行：	中山大学出版社
电　　话：	编辑部 020 - 84110283，84113349
	发行部 020 - 84111998，84111981，84111160
地　　址：	广州市新港西路 135 号
邮　　编：	510275　传　真：020 - 84036565
网　　址：	http://www.zsup.com.cn　E-mail: zdcbs@mail.sysu.edu.cn
印 刷 者：	广州家联印刷有限公司
规　　格：	787mm×1092mm　1/16　33 印张　8 彩插　1020 千字
版次印次：	2017 年 11 月第 1 版　2007 年 11 月第 1 次印刷
定　　价：	160.00 元

如发现本书因印装质量问题影响阅读，请与出版社发行部联系调换

本书作者一览表

R. J. Balment（R. J. 巴尔门特），英国，曼彻斯特大学，生命科学系

N. J. Bernier（N. J. 伯尼尔），加拿大，圭尔夫大学，整合生物学系

L. F. Canosa（L. F. 坎诺沙），阿根廷，布宜诺斯艾利斯省，Chascomus 生物技术研究所

J. M. Cerdá-Reverter（J. M. 谢达-里维特），西班牙，卡斯特利翁，Torre 农业技术研究所

M. Chadzinska（M. 查德金斯卡），荷兰，瓦格宁根大学，细胞生物学和免疫学学科群

J. P. Chang（J. P. 张），加拿大，阿尔伯塔大学，生物科学系

G. Flik（G. 弗利克），荷兰，拉德堡德大学，动物生理学系

S. Holmgren（S. 霍尔姆格伦），瑞典，哥德堡大学，动物生理学系

O. Kah（O. 凯），法国，雷恩大学，生殖内分泌分子生物学实验室

H. Kawauchi（川内浩司），日本，北里大学，水产学系

S. P. Kelly（S. P. 凯利），加拿大，约克大学，生物学系

B. M. V. -van Kemenade（B. M. V. -Van 凯门纳德），荷兰，瓦格宁根大学，细胞生物学和免疫学学科群

P. H. M. Klaren（P. H. M. 克拉伦），荷兰，拉德堡德大学，动物生理学系

G. Van Der Kraak（G. Van Der 克拉克），加拿大，圭尔夫大学，整合生物学系

J. R. Metz（J. R. 梅特兹），荷兰，拉德堡德大学，动物生理学系

S. Moriyama（森山俊介），日本，北里大学，渔业科学学院

C. Olsson（C. 奥尔逊），瑞典，哥德堡大学，动物生理学系

S. A. Sower（S. A. 索沃），美国，新罕布什尔大学，分子与比较内分泌学中心

E. H. Stolte（E. H. 斯托尔特），荷兰，瓦格宁根大学，细胞生物学和免疫学学科群

Y. Takei（竹井祥郎），日本，东京大学，海洋研究所

S. Unniappan（S. 安尼阿潘），加拿大，约克大学，生物学系

H. Volkoff（H. 沃尔科夫），加拿大，纽芬兰纪念大学，生物学与生物化学系

A. O. L. Wong（A. O. L. 王），中国，香港大学，生物科学学院

内 容 简 介

本书为美国科学出版社出版的"鱼类生理学"系列专著中的第28卷《鱼类神经内分泌学》。它全面收集和总结了近数十年来在鱼类神经内分泌学方面发表的科学著作和研究成果，包括解剖和功能两部分。解剖部分介绍鱼类神经内分泌系统的解剖结构和下丘脑与脑垂体的靶标。功能部分着重在分子、细胞和系统水平阐述主要的神经激素在调控脑垂体激素和调节重要生理活动过程如液体平衡、食物摄取、消化道作用以及免疫反应等方面所起的功能与作用机理。

本书内容充实、系统全面、概念新颖、论述清晰，是一部学术水平很高的专著。

本书可供鱼类学、鱼类生理学、鱼类养殖生物学、鱼类内分泌学、鱼病学、鱼类免疫学、鱼类分子生物学以及比较生理学、比较内分泌学等相关学科与研究领域的科学技术工作者和高等院校有关专业的师生学习参考。

前　言

鱼类神经内分泌学的学术内容对我们深刻了解各种神经化学信使和系统的功能作用与进化发展正在持续不断地起着重要的促进作用。鱼类不仅具有一些独特的神经内分泌特征，对于新神经肽的发现，它们还一直是重要而易于研究处理的脊椎动物模型。近50年以来，神经内分泌学者们已经充分阐明了许多复杂而形式多样的激素和神经构造之间的相互作用。这些研究成果使我们深刻认识到一些特定的神经激素通路和信使的作用能够在水环境中保持生命稳态，而且这个功能系统的调控作用使鱼类能够具备高度多样性的生活史和生殖模式。

目前已有大量而又迅速增长的关于鱼类神经内分泌调控作用机理的学术资料，但还没有一本介绍鱼类神经内分泌学的专门教科书。事实上，除了少数关于哺乳类神经内分泌学的教科书之外，比较神经内分泌学的重要参考书系列亦很少见到。

因此，本卷"鱼类生理学"系列丛书提供了第一本《鱼类神经内分泌学》参考书。该书包括解剖和功能两部分。第一部分共两章，叙述鱼类脑神经内分泌系统的解剖构造和下丘脑与脑垂体的内分泌靶标。第二部分的各章着重在分子、细胞和系统水平阐述神经激素的作用机理。为此，作者们综述了主要的神经激素在调控脑垂体激素和调节关键生理活动过程如液体平衡、食物摄取、消化道作用等方面所起的功能作用。本书还有一章介绍了神经内分泌对免疫系统的调控作用。

我们非常感谢本书的作者们给予的支持和贡献。我们亦感谢所有审稿人提出的富有建设性意义的评议。最后，我们感谢Elsevier出版社的成员们为本书的顺利出版所做出的努力。

我们将这本书奉献给Richard E.（Dick）Peter博士。Peter博士在很多方面都是鱼类神经内分泌学的开拓者和奠基人。他在鱼类脑部定位图像研究中建立的一些关键技术所做出的贡献，使得一代神经内分泌学者们能够获得有益的手段，以深入研究鱼类脑的神经内分泌区域。Peter教授在神经内分泌调控鱼类生长、生殖和摄食等各个方面取得的研究成果都体现在本书的各个章节中。Peter教授是一位伟大的导师，对鱼类神经内分泌学的建立和发展做出了巨大的贡献！

N. J. 伯尼尔

G. Van Der 克拉克

A. P. 法雷尔

C. J. 布劳纳

术 语 缩 写

下列为本书中使用的术语缩写。标准的缩写在文中没有解释。非标准的缩写在每章中有说明，下面列出供读者参考。

标 准 缩 写

AMP，ADP，ATP	adenosine 5′-mono-, di-, and triphosphates	腺苷5′-一，二，和三磷酸盐
cDNA	complementary DNA	互补 DNA
cAMP	3′,5′-cyclic AMP	3,5′-环 AMP
DNA	deoxyribonucleic acid	脱氧核糖核酸
GDP，GMP，GTP	guanosine 5′-mono-, di-, and triphosphates	鸟苷5′-一，二，和三磷酸盐
EC_{50}	median effective concentration	半数有效浓度
ED_{50}	median effective dose	有效中量
IC_{50}	median inhibitory concentration	抑制中浓度
LD_{50}	median lethal dose	半数致死量
mRNA	messenger RNA	信使 RNA
NMR	nuclear magnetic resonance	核磁共振
PCR	polymerase chain reaction	聚合酶链反应
RIA	radioimmunoassay	放射免疫测定
RT-PCR	reverse transcription-PCR	反转录多聚合酶链反应
RNA	ribonucleic acid	核糖核酸

非标准缩写

A

A	anterior thalamic nucleus	前丘脑核
AA	arachidonic acid	花生四烯酸
AC	adenylate cyclase	腺苷酸环化酶
ACE	angiotensin-converting enzyme	血管紧张肽转变酶
ACh	acetylcholine	乙酰胆碱
ACTH	adrenocorticotropic hormone	促肾上腺皮质激素
AD	adrenaline	肾上腺素
AD	epinephrine	肾上腺素
AgRP	agouti-related protein	野灰蛋白相关蛋白
AM	adrenomedullin	肾上腺髓质素
ANG II	angiotensin II	血管紧张素II
ANP	atrial natriuretic peptide	心房钠尿肽
AP	area postrema	最后区
AR	androgen receptors	雄激素受体
ARC	arcuate nucleus	弓状核
ASP	agouti-signaling peptide	野灰蛋白信号传递蛋白
AVP	arginine-vasopressin	精氨酸升压素
AVT	arginine-vasotocin	精氨酸催产素

B

BBB	blood-brain barrier	血-脑屏障
BBS	bombesin	铃蟾肽
BNP	B-type natriuretic peptide	B-型钠尿肽

C

CAM	calmodulin	钙调蛋白
CART	cocaine-and amphetamine-regulated transcriptt	可卡因-与

	安非他明-调节转录体
CB1&CB2	cannabinoid receptor subtypes　大麻素受体亚型
CCK	cholecystokinin　缩胆囊素
CE	cerebellum　小脑
CFTR	cystic fibrosis transmembrane regulator　囊性纤维化跨膜调节蛋白
CGRP	calcitonin gene-related peptide　降钙素基因相关肽
ChAT	choline acetyl transferase　胆碱乙酰转移酶
CLIP	corticotropin-like intermediate lobe peptide　中间叶促皮质样肽
CLR	calcitonin receptor-like receptor　降钙素受体样受体
CNP	C-type natriuretic peptide　C-型钠尿肽
CNS	central nervous system　中枢神经系统
CNSS	caudal neurosecretory system　尾神经分泌系统
COX	cyclooxygenase　环加氧酶
CP	central posterior thalamic nucleus　中央后丘脑核
CR	corticosteroid receptor　皮质类固醇受体
CRF-BP	CRF binding protein　CRF结合蛋白
CRF	corticotropin-releasing factor　促肾上腺皮质激素释放因子
CRF_1 and CRF_2	CRF receptor subtypes　CRF受体亚型
CSF	cerebrospinal fluid　脑脊液
CVO	circumventricular　室周器

D

DA	dopamine　多巴胺
Dc	dorsal telencephalon pars centralis　背端脑中央部
Dd	dorsal telencephalon pars dorsalis　背端脑背部
DHT	dihydrotestosterone　双氢睾酮
Dl	dorsal telencephalon pars lateralis　背端脑外侧部
DL	dorsolateral thalamic nucleus　背外侧丘脑核
Dm	dorsal telencephalon pars medialis　背端脑内侧部
DOC	deoxycorticosterone　脱氧皮质酮
DOPAC	3,4 dihydroxyphenylacetic acid　3,4二羟苯乙酸
Dp	dorsal telencephalon pars posterior　背端脑后部
Dp	dorsal posterior thalamic nucleus　背后丘脑核
DYN	dynorphin　强啡肽

17, 20βP	17, 20βdihydroxy-4-pregnen-3-one	17, 20β 双羟黄体酮

E

E2	17β-estradiol	17β-雌二醇
E	ventral telencephalon entopeduncular nuclei	腹端脑内大脑脚核
EC	endocannabinoids	内大麻素
EDC	endocrine disrupting compound	破坏内分泌的化合物
EE2	ethinylestradiol	乙炔基雌二醇
END	endorphin	内啡肽
ENS	enteric nervous system	肠神经系统
EPO	erythropoietin	红细胞生成素
ER	estrogen receptor	雌激素受体
ERK1/2	extracellular signal-regulated kinase1/2	胞外信号调节激酶1/2

F

FR	fasciculus retroflexus	神经束反折
FSH	follicle-stimulating hormone	促卵泡激素
FSH	gonadotropin Ⅰ	促性腺激素 Ⅰ

G

GABA	γ-amino butyric acid	γ-氨基丁酸
GAD	glutamic acid decarboxylase	谷氨酸脱羧酶
GAL	galanin	甘丙肽
GALP	galanin-like peptide	甘丙肽样肽
GEP	gastroenteropancreatic	胃肠胰的
GFR	glomerular filtration rate	肾小球滤过率
GH	growth hormone (somatotropin)	生长激素
GHR	growth hormone receptor	生长激素受体
GHRH	growth hormone-releasing hormone	生长激素释放激素
GHS-R	growth hormone secretagogue receptor	生长激素促分泌受体
GI	gastrointestinal	胃肠的
GIP	gastric inhibitory peptide	肠抑胃肽
GLP	glucagon-like peptide	胰高血糖素样肽

GnIH	gonadotropin inhibitor hormone	促性腺激素抑制激素
GnRH-R	gonadotropin-releasing hormone receptor	促性腺激素释放激素受体
GnRH	gonadotropin-releasing hormone	促性腺激素释放激素
GPCR	G protein-coupled receptor	G蛋白偶联受体
GR	glucocorticoid receptor	糖皮质素受体
GRE	glucocorticoid responsive element	糖皮质激素应答元件
GRK	G-protein-coupled receptor kinase	G蛋白偶联受体激酶
GRL-R	ghrelin receptor	生长素释放肽受体
GRL	ghrelin	生长素释放肽
GRP	gastrin-releasing peptide	胃泌素释放肽
GTH	gonadotropins	促性腺激素
GVC	glossopharyngeal-vagal motor complex	吞咽-迷走运动复合体

H

Ha	anterior hypothalamus	前下丘脑
Hc	caudal hypothalamus	尾下丘脑
HCo	horizontal commissure	水平连合
Hd	dorsal hypothalamus	背下丘脑
HPA	hypothalamic-pituitary adrenal	下丘脑-脑垂体肾上腺的
HPG	hypothalamic-pituitary gonadal	下丘脑-脑垂体性腺的
HPI	hypothalamic-pituitary interrenal	下丘脑-脑垂体肾间腺的
HPT	hypothalamic-pituitary thyroid	下丘脑-脑垂体甲状腺
HRP	horseradish peroxidase	辣根过氧化物酶
Hsp	heat shock protein	热休克蛋白
Hv	ventral hypothalamus	腹下丘脑
HYP	hypothalamus	下丘脑
5-HIAA	5-hydroxy indole acetic acid	5-羟基吲哚乙酸
5-HT	5-hydroxytryptamine（serotonin）	5-羟色胺

I

ICCs	interstitial cells of Cajal	Cajal间质细胞
ICL	internal cell layer of the olfactory bulb	嗅球内细胞层
icv	intracerebroventricular	脑内室

IFN	interferon	干扰素
Ig	immunoglobulin	免疫球蛋白
IGF-I /IGF-II	insulin-like growth factor I / II	胰岛素样生长因子 I / II
IL (e.g., IL-1, IL-6)	interleukin	白细胞介素
IL	inferior lobe of the hypothalamus	下丘脑下叶
im	intramuscular (-ly)	肌肉内的
IP3	inositol 1,4,5-trisphosphate	肌醇 1,4,5-三磷酸
ip	intraperitoneal (-ly)	腹腔内的
ir	immunoreactive (-ity)	免疫反应的
IST	isotocin	硬骨鱼催产素
iv	intravenous (-ly)	静脉内的

J

JAK	janus kinase	詹纳斯激酶

K

KKS	kallikrein-kinin system	激肽释放酶-激肽系统
11KA	11-ketoandrostenedione	11-酮雄烯二酮
11KT	11-ketotestosterone	11-酮睾酮

L

LE	leu-enkephalin	亮-脑啡肽
LFB	lateral forebrain bundle	外侧前脑束
LH	gonadotropin II	促性腺激素 II
LH	luteinizing hormone	促黄体素
LH	lateral hypothalamic nucleus	下丘脑外侧核
LIF	leukemia inhibitory factor	白血病抑制因子
LPH	lipotropin	促脂解素
LPS	lipopolysaccharide	脂多糖
LR	lateral recess	外侧隐窝
LT	lymphotoxin	淋巴毒素

M

α2M	macroglobulin alpha-2	巨球蛋白 α-2
MAPK	mitogen-activated protein kinase	促分裂原活化蛋白激酶
MC	melanocortin	黑皮质素
MCH	melanin-concentrating hormone	黑色素浓集激素
MCR（MCIR-MC5R）	melanocortin receptor	黑皮质素受体
ME	met-enkephalin	蛋-脑啡肽
MHC	major histocompatibility complex	主要组织相容性复合体
MPO	median preoptic nucleus	正中视前核
MR	mineralocorticoid receptor	盐皮质激素受体
MRCs	mitochondria-rich cells	富线粒体细胞
MSH	melanocyte-stimulating hormone	促黑色素细胞激素

N

NA	noradrenaline	去甲肾上腺素
NA	nucleus ambiguous	疑核
NAPv	anterior periventricular nucleus	前室周核
NAT	anterior tuberal nucleus	前结节核
NCC	non specific cytotoxic cells	非特异性细胞毒素细胞
NCC	commissural nucleus of Cajal	Cajal 连合核
NCLI	central nucleus of the inferior lobe	下叶中央核
NDLI	diffuse nucleus of the inferior lobe	下叶弥散核
NE	entopeduncular nucleus	内大脑脚核
NEO	neoendorphin	新内啡肽
NH	neurohypophysial	神经脑垂体的
NK	natural killer	天然杀伤细胞
NKA	neurokinin A	神经激肽 A
NLT	lateral tuberal nucleus	外侧结节核
NMDA	N-methyl-D-aspartate	N-甲基-D-天冬氨酸
NMLI	medial nucleus of the inferior lobe	下叶正中核
NO	nitric oxide	一氧化氮
NOR	nucleus olfactoretinalis	嗅觉视网膜核
NOS	nitric oxide synthase	一氧化氮合酶
NP	natriuretic peptide	利尿钠肽

NP	paracommissural nucleus	旁连合核
NpCp	copeptine	混合肽素
NPO	nucleus preopticus	视前核
NPP	N-terminal peptide of pro-opiomelanocortin	N-端阿黑皮素原肽
NPP	preoptic parvocellular nucleus	视前小细胞核
NPPv	posterior periventricular nucleus	后室周核
NPR	natriuretic peptide receptor	利尿钠肽受体
NPT	posterior tuberal nucleus	后结节核
NPY	neuropeptide Y	神经肽 Y
NRL	lateral recess nucleus	外侧隐窝核
NRP	posterior recess nucleus	后隐窝核
NSC	suprachiasmatic nucleus	交叉上核
NSV	nucleus of the saccus vasculosus	血管囊核
NTe	nucleus of the thalamic eminentia	丘脑隆起核
NTS	nucleus tractus solitarius	孤束核

O

OB	olfactory bulb	嗅球
OpN	optic nerve	视神经
OT	optic tectum	视顶盖
OVLT	organon vasculosum of the lamina terminalis	终板的血管器
OX_1 and OX_2	orexin (hypocretin) receptors	食欲肽（下丘泌素）受体

P

$P450_{c11}$	11β-hydroxylase enzyme	11β-羟化酶
$P450_{scc}$	cytochrome P450 side chain cleavage enzyme	细胞色素 P450 侧链剪切酶
P	pituitary	脑垂体
PACAP	pituitary adenylate cyclase-activating polypeptide	脑垂体腺苷酸环化酶激活多肽
PAMPs	pathogen-associated molecular patterns	致病相关分子型式
PC	prohormone convertase	激素原转变酶
PD	pituitary pars distalis	脑垂体远侧部
PDYN	pro-dynorphin	强啡肽原

PENK	pro-enkephalin	脑啡肽原
PI3K	phosphatidylinositol 3-kinase	磷脂酰肌醇3-激酶
PI	pituitary pars intermedia	脑垂体中间部
PKA	protein kinase A	蛋白激酶A
PKC	protein kinase C	蛋白激酶C
PLC	phospholipase C	磷脂酶C
PM	magnocellular preoptic nucleus	大细胞视前核
PN	pituitary pars nervosa	脑垂体神经部
POA	preoptic area	视前区
POMC	pro-opiomelanocortin	阿黑皮素原
PP	pancreatic polypeptide	胰多肽
PP	parvocellular preoptic nucleus	小细胞视前核
PPa	parvocellular preoptic nucleus anterior pars	小细胞视前核前部
PPd	dorsal periventricular pretectal nucleus	背室周顶盖前核
PPD	pituitary proximal pars distalis	脑垂体近端远侧部
PPp	parvocellular preoptic nucleus posterior pars	小细胞视前核后部
PPv	ventral periventricular pretectal nucleus	腹室周顶盖前核
PR	progesterone receptor	黄体酮受体
PRL	prolactin	催乳激素
PRLR	prolactin receptor	催乳激素受体
PRP	PACAP-related peptide	PACAP-相关肽
PRR	pathogen recognition receptors	致病识别受体
PrRP	prolactin-releasing peptide	催乳激素释放肽
PSS	preprosomatostatin	前生长抑素原
PTN	posterior tuberal nucleu	后结节核
PVN	paraventricular nucleus	室旁核
PVO	paraventricular nucleus organ	室旁器
PY	peptide Y	肽Y
PYY	peptide tyrosine-tyrosine peptide YY	肽酪氨酸-酪氨酸肽YY

R

RAMP	receptor activity-modifying protein	受体活化-修饰蛋白
RAS	renin angiotensin system	肾素血管紧张素系统
RFa	arginyl-phenylalanyl-amide peptides	精氨酰-苯丙氨酰-酰

		胺肽
RFRP	RFamide-related peptide	RF酰胺相关肽
ROS	reactive oxygen species	活性氧种类
RPD	pituitary rostral pars distalis	脑垂体吻端远侧部
rT$_3$	3,3′,5′-triiodothyronine	3,3′,5′-三碘甲腺原氨酸

S

SAA	serum amyloid A	血清类淀粉A
SCO	subcommissural organ	连合下器
SFO	subfornical organ	穹窿下器
SL	somatolactin	生长乳素
SLR	somatolactin receptor	生长乳素受体
SOCS	suppressor of cytokine signaling	细胞因子信号抑制剂
SON	supraoptic nucleus	视上核
SS	somatostatin (SRIF, somatotropin release inhibiting factor)	生长抑素
SST$_{1-5}$	somatostatin receptors 1-5	生长抑素受体1-5
StAR	steroidogenic acute regulatory protein	类固醇生成急性调节蛋白
STAT	signal transducer and activator of transcription	信号转导及转录激活蛋白

T

T	testosterone	睾酮
T$_3$	3,5,3′-triiodothyronine	3,5,3′-三碘甲腺原氨酸
T$_4$	thyroxine	甲状腺素
TCR	T-cellreceptor	T-细胞受体
TEL	telencephalon	端脑
TGF	transforming growth factor	转化生长因子
TH	tyrosine hydroxylase	酪氨酸羟化酶
TK	tachykinin	速激肽
TLR	toll-like receptor	toll-样受体
TN	tuberal nucleus	结节核
TNF	tumor necrosis factor	肿瘤坏死因子
TR	thyroid hormone receptor	甲状腺激素受体

TRH	thyrotropin-releasing hormone	促甲状腺素释放激素
TSH	thyrotropin (thyroid-stimulating hormone)	促甲状腺素
TTX	tetrodotoxin	河鲀毒素
TV	ventral tuberal nucleus	腹结节核

U

UCN	urocortin	尾皮质素
U I	urotensin I	硬骨鱼紧张肽 I
U II	urotensin II	硬骨鱼紧张肽 II
URP	urotensin II -related peptide	硬骨鱼紧张肽 II -相关肽

V

Vc	ventral telencephalon central nuclei	腹端脑中央核
Vd	ventral telencephalon dorsal nuclei	腹端脑背核
Vi	ventral telencephalon intermediate nuclei	腹端脑中间核
VIP	vasoactive intestinal polypeptide	血管活性肠多肽
Vl	ventral telencephalon lateral nuclei	腹端脑外侧核
VM	ventromedial thalamic nucleus	腹中丘脑核
VNP	ventricular natriuretic peptide	心室利尿钠肽
Vp	ventral telencephalon postcommissural nuclei	腹端脑后连合核
Vs	ventral telencephalon supracommissural nuclei	腹端脑连合上核
VSCC	voltage-sensitive calcium channel	电压敏感钙通道
Vv	ventral telencephalon ventral nuclei	腹端脑腹核

Z

ZL	zona limitans diencephali	间脑界膜带

目 录

第一部分 鱼类神经内分泌系统的解剖构造

第1章 鱼类脑的神经内分泌系统 /3
  1.1 导言 /3
  1.2 下丘脑-脑垂体主要区域的细胞结构 /7
  1.3 鱼类脑的下丘脑-脑垂体分区：神经束追踪研究 /15
  1.4 中央神经激素 /17
  1.5 下丘脑调控脑垂体的肽类 /23
  1.6 下丘脑的神经递质 /39
  1.7 结束语 /41

第2章 下丘脑和脑垂体的内分泌靶标 /69
  2.1 导言 /69
  2.2 性类固醇激素 /70
  2.3 肾上腺皮质激素受体 /81
  2.4 代谢激素 /84
  2.5 结束语 /87

第二部分 鱼类神经内分泌系统的功能

第3章 GnRH系统和生殖的神经内分泌调控 /107
  3.1 导言 /107
  3.2 促性腺激素释放激素（GnRH） /108
  3.3 GnRH、FSH和LH的类固醇反馈调控 /111
  3.4 单胺和氨基酸神经递质 /113
  3.5 神经肽类 /117
  3.6 蛋白质激素 /119
  3.7 性腺发育的综合神经内分泌调控 /121
  3.8 展望 /123

第4章 鱼类生长激素的调控：一个包括下丘脑、外周和局部自分泌/旁分泌信号的多因子模式 /139

4.1 导言 /139
4.2 生长激素和生长激素受体 /139
4.3 鱼类生长激素的生物学功能 /140
4.4 生长激素分泌与合成的调控 /142
4.5 在下丘脑和脑垂水平 GH 调节剂的功能相互作用 /160
4.6 结束语 /163

第5章 神经内分泌调控鱼类催乳激素和生长乳素的分泌活动 /180

5.1 导言 /180
5.2 催乳激素分泌活动的神经内分泌调控 /181
5.3 生长乳素分泌活动的神经内分泌调控 /194
5.4 结束语 /198

第6章 促肾上腺皮质激素轴、促黑色素激素轴和促甲状腺激素轴对鱼类应激反应的调控和作用 /218

6.1 导言 /218
6.2 脑垂体细胞的下丘脑调控 /219
6.3 促肾上腺皮质激素细胞、促黑色素激素细胞和促甲状腺素细胞分泌的靶标和功能 /239
6.4 皮质醇和甲状腺激素的靶标、功能和反馈作用 /245
6.5 促肾上腺皮质激素轴、促黑色素激素轴和促甲状腺素激素轴对应激反应的作用 /252
6.6 展望 /257

第7章 硬骨鱼类神经内分泌-免疫相互作用 /289

7.1 神经内分泌-免疫相互作用 /289
7.2 神经内分泌因子对免疫的调节 /296
7.3 细胞因子对神经内分泌的调节 /310
7.4 结论和展望 /313

第8章 神经内分泌调控液体摄入和液体平衡 /338

8.1 导言 /338
8.2 液体摄入的调控 /343
8.3 液体平衡的调控：肾的和肾外的作用机理 /356
8.4 展望 /366

第9章 内分泌调控食物摄取 /388
    9.1 导言 /388
    9.2 内分泌调控 /395
    9.3 内在因子的影响 /406
    9.4 外界因子的影响 /408
    9.5 参与鱼类摄食的内分泌线路模型和结束语 /409

第10章 神经和内分泌调控消化道功能 /433
    10.1 导言 /433
    10.2 消化道神经和内分泌系统的解剖学 /433
    10.3 消化道的神经递质和激素 /436
    10.4 消化道神经分布和神经内分泌系统的发育 /441
    10.5 消化道活动性的调控 /443
    10.6 分泌作用和消化作用的调控 /450
    10.7 营养物吸收的调控 /453
    10.8 水分和离子运输的调控 /454
    10.9 内脏血液循环的调控 /454
    10.10 结束语 /457

索 引 /477

译后记 /507

第一部分 鱼类神经内分泌学

鱼类神经内分泌系统的解剖构造

第1章 鱼类脑的神经内分泌系统

研究神经元系统在脑垂体的神经分布对于了解一系列生命活动过程的调控作用至关重要。本章介绍鱼类神经内分泌区域的研究进展。首先，描述硬骨鱼类脑的主要神经内分泌区域的解剖学，并且规范了学者们提供的不同学术术语。硬骨鱼类缺乏一个正规的正中隆起（median eminence），下丘脑的神经元非常紧密地终止于腺脑垂体细胞，或者和它们形成突触联系。这种解剖学特征使得人们可以采用逆向追踪（retrograde tracing）实验方法来研究下丘脑-脑垂体系统。神经束追踪可以确证早期研究所表明的下丘脑视前区和结节下丘脑（tuberal hypothalamic）作为神经元细胞体的位置，它们的轴突沿着界限清晰的纤维神经束达到神经脑垂体和腺脑垂体。其次，对产生下丘脑释放性或抑制性肽类和神经递质以及分布到脑垂体的各种不同的神经元系统做了综述。最后，还介绍脑垂体各个不同区域的肽神经分布及其和各种不同分泌细胞的联系。

1.1 导　　言

Ernest Scharrer 的开拓性研究证明硬骨鱼类的脑存在着分泌细胞（Scharrer，1928）。这些研究表明腺型神经细胞的主要作用是分泌激素进入脑垂体的神经部（pars nervosa）。Scharrer 的观点完全是以形态学标准为基础，曾受到强烈的反对，并且没有立即为致力于建立一门新的神经科学——神经内分泌学的人们所接受。随后的染色研究在腹前脑（ventral forebrain）鉴别出两个区域，即视前区和下丘脑，它们是神经元细胞体所在的位置，它们的轴突沿着界限清晰的纤维神经束、脑垂体束和视前-脑垂体束到达神经脑垂体（Palay，1945）。这些开拓性研究揭示了存在着一个连接中枢神经（CNS）的下丘脑-脑垂体系统，正是这个神经内分泌系统调控着一系列生命活动过程。

鱼类下丘脑-脑垂体可以分为三个主要的区域：下丘脑，它是间脑的一部分；神经脑垂体，它由腹间脑衍生而来，是脑垂体的神经区；腺脑垂体，它是脑垂体的非神经元部分（见图1.1；Pogoda 和 Hammerschmidt，2007）。神经脑垂体由神经末梢组成，其神经细胞体主要处在视前区。这些细胞具有神经胶质细胞的特点，或者具有支持功能的垂体细胞（pituicyte）的特征。有些鱼类如斑光蟾鱼（*Porichthys notatus*），其神经脑垂体由神经组织形成一个柄而和脑分离（Sathyanesan，1965）；其他鱼类没有漏斗状的柄，或者神经脑垂体处在很短的漏斗状柄的末端（Gorbman 等，1983）。在板鳃类（Van de Kamer 和 Zandbergen，1981）和一些软骨硬鳞鱼类（Lagios，1968）中，神经脑垂体可分为正中隆起和神经部。正中隆起包括门脉系统，下丘脑神经元通过毛细血管网释放其分泌物进入血管系统，然后输送到腺脑垂体。

无颌类包括七鳃鳗和盲鳗，没有典型的正中隆起，但具有一个前神经血管区，分泌物能通过它由脑迅速扩散到腺脑垂体内（神经-扩散系统，Nozaki 等，1994）。相反，硬骨鱼类没有正中隆起，亦没有门脉系统，下丘脑神经元非常紧密地终止于腺脑垂体细胞中，减少了扩散的距离，或者和腺脑垂体细胞形成突触连接。这表明下丘脑能够通过对分泌细胞的直接作用实行对腺脑垂体的调控。图 1.2 表示在鱼类中观察到的腺脑垂体神经分布的几种模式。目前还有争论的是关于中枢神经分泌能否通过神经脑垂体的血管丛而间接到达腺脑垂体。这表明硬骨鱼类可能存在着双重的调控机理，即直接的（神经）和间接的（血管）通路（Hill 和 Henderson，1968）。但是，非常明显的是硬骨鱼类并不具有血管模式的通路，完全不同于四足类的正中隆起，由初级血管丛进入门静脉，进而引入次级的毛细血管丛（Gorbman 等，1983）。

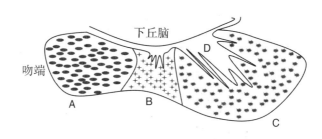

图 1.1　鱼类脑垂体构造

A. 吻端远侧部，或前腺脑垂体。B. 近端远侧部，或中腺脑垂体。C. 中间部，或后腺脑垂体。A、B、C 三区属于腺脑垂体。D. 神经垂体的神经部。

　　腺脑垂体细胞合成它们的激素并有序地保留在各自不同的区域（Pogoda 和 Hammerschmidt，2007）。和四足类不同，鱼类腺脑垂体缺少结节部（pars tuberis）。对鱼类腺脑垂体的分区有两种不同的命名术语（Green，1951；Pickford 和 Atz，1957）。这两种术语都把腺脑垂体分为三个区：最吻端的部分称为吻端远侧部（rostral pars distalis，RPD）（Green，1951），或称为前腺脑垂体（pro-adenohypophysis）（Pickford 和 Azt，1957）；接着是远侧部（psrs distalis，PD），称为近端远侧部（proximal pars distalis，PPD），或称为中腺脑垂体（meso-adenohypophysis）；然后是中间部（pars intermedia，PI）或后腺脑垂体（meta-adenohypophysis）。和哺乳类不同，鱼类的 PD 和 PI 之间没有明显的形态结构分隔。神经部（pars nervosa，PN）或神经脑垂体的后部和腺脑垂体的 PI 相嵌交错而形成神经中间叶（neurointermediate lobe，NIL）。神经脑垂体亦呈舌状侵入 PD。PN 包含来自视前区神经元的神经末梢，它们显现神经内分泌物的染色。神经脑垂体前部和 PD 相嵌交错并不显现神经内分泌物的染色，而是包含来自外侧结节核以及其他一些前脑腹区（亦可以列为脑垂体区）的神经纤维末梢（Gorbman 等，1983）。

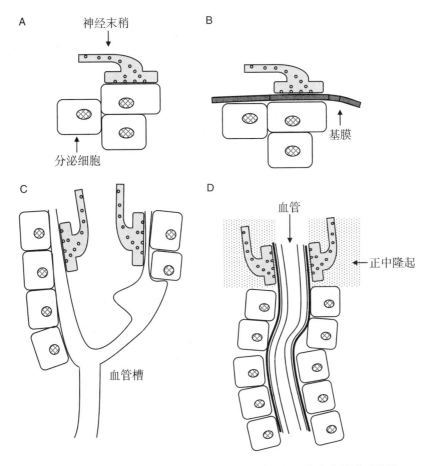

图 1.2 中枢神经系统和腺脑垂体内分泌细胞之间可能存在的联系类型

A、B、C 为硬骨鱼类，D 为非硬骨鱼类（即除硬骨鱼类之外的鱼类）。A. 神经内分泌末梢和内分泌细胞之间直接的突触连接，如海马（*Hippocampus* sp.）。B. 神经内分泌末梢和内分泌细胞之间存在着基膜，如丁鱥（*Tinca* sp.）。C. 神经内分泌末梢分布于内分泌细胞当中的血管槽内，如鳗鲡（*Anguilla anguilla*）。D. 神经内分泌末梢和腺脑垂体分泌细胞之间完全分隔，下丘脑神经元释放它的分泌物进入下丘脑血管系统（正中隆起），然后输送到腺脑垂体。在四足类脊椎动物亦观察到这种神经分布类型。（修改自 Vollrath，1967）

鱼类腺脑垂体至少合成 8 种激素（见图 1.3），它们可以分为 3 种类型（Pogoda 和 Hammerschmidt，2007）。第一种类型包括属于生长激素（GH）/催乳激素（PRL）/生长乳素（somatolactin，SL）家族的生长激素细胞产生的生长激素（GH）、催乳素细胞产生的催乳激素（PRL）和生长乳素细胞产生的生长乳素（SL）（Kawauchi 和 Sower，2006）。第二种类型包括甲状腺刺激激素或促甲状腺素（thyrotropin，TSH），和两种促性腺激素（gonadotropin，GTH），即卵泡刺激激素（follicle-stimulating hormone，FSH/GTH-Ⅰ）和促黄体激素（luteinizing hormone，LH/GTH-Ⅱ）。这 3 种激素都是异二聚体的（heterodimeric）糖蛋白，由一个共同的 α-亚基和一个不同基因编码的激素特异性 β-亚基组成。α-亚基基因可在促性腺激素细胞和促甲状腺素细胞中共同表达，而 β-亚基

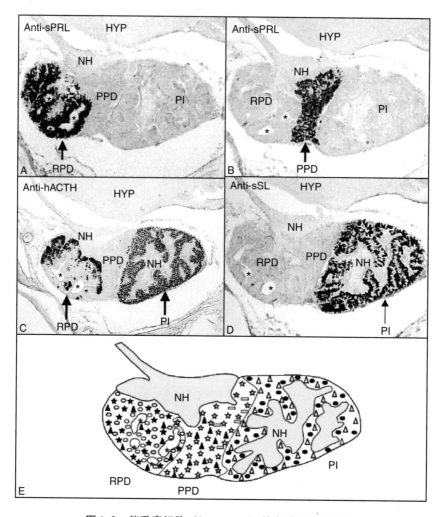

图 1.3 催乳素细胞（lactotropes）的免疫反应示范图

A. 抗大麻哈鱼催乳激素（sPRL）-免疫反应（ir）细胞，生长激素细胞；B. 抗鲷鱼生长激素（GH）-ir 细胞，促肾上腺皮质细胞；C. 抗人体促肾上腺皮质素细胞（ACTH）-ir 细胞和生长乳素细胞；D. 抗大麻哈鱼生长乳素（SL）-ir 细胞分布在美国鲱鱼（*Alosa sapidissima*）的腺脑垂体。E. 脑垂体矢切面图，表示腺脑垂体各种分泌细胞的分布情况。PRL（★），ACTH（○），GH（☆），GTH（▲），TSH（□），SL（●）和 MSH（△）。"*"表示空腔。HYP，下丘脑；NH，神经脑垂体；PL，中间部；PPD，近端远侧部；RPD，吻端远侧部。（修改自 Laiz-Carrión 等，2003）

表达则是细胞特异性的（Querat 等，2000）。第三种类型是由一个共同前体阿黑皮素原（proopiomelanocortin，POMC）衍生的肽类。它们由三个主要区域组成：N-端前-γ-黑色素细胞刺激激素（MSH）、中央肾上腺皮质激素（ACTH）和 C-端 β-促脂解素（lipotropin，LPH）。每个区都含有一个 MSH 肽，即 γ-MSH 在前-γ-MSH，α-MSH 在 ACTH 的 N-端序列，而 β-MSH 在 β-促脂解素。最后的区还包含 C-端 β-内腓肽。硬骨鱼类缺少 γ-MSH 区，但保留 α-MSH、β-MSH 区和一个 β-内腓肽。板鳃鱼类有第四个 MSH 区域，即 δ-MSH（Takahashi 等，2001）。硬骨鱼类常见的腺脑垂体分泌细胞分布型式（Laiz-

Carrión 等，2003）是：PRL 分泌细胞和 ACTH 分泌细胞位于 RPD，PRL 细胞位于 ACTH 细胞的腹方。GH 分泌细胞和 TSH 分泌细胞集中分布在 PPD，而 SL 分泌细胞分布在 PI 的最尾端区域，和黑色素分泌细胞（α-MSH 细胞）混合在一起。最后，FSHβ 和 LHβ 由分布于 PPD 的不同 GTH 分泌细胞产生，其中有些 LHβ 细胞排列在 PI 的外边缘。在 NIL 的后部和下丘脑下叶的中线，大多数硬骨鱼类和软骨鱼类具有血管囊（saccus vasculosus），它是一个由高度血管化的神经上皮组成的室周器（circumventricular organ），而这种神经上皮是由一种独特的称为冠细胞（coronet cell）以及接连脑脊液的神经元和支持细胞等所组成（Sueiro 等，2007）。

1.2 下丘脑-脑垂体主要区域的细胞结构

可以采用特异性组织染色技术研究中枢神经系统（CNS）的结构。Nissl 氏染色法采用碱性苯胺染剂，仍保留研究 CNS 细胞结构的常规技术。这种技术是分析 CNS 必不可少的第一步，是神经解剖学最有成效的方法之一。这种方法使 RNA 染成蓝色，显示所谓"Nissl 氏体"的定位；它是在神经元细胞质内发现的颗粒状嗜碱性小体，由粗面内质网和游离多核糖体（polyribosome）组成。这种染色法能显示独特的神经元核周体（neuronal perikarya），而树突、轴突和神经末梢都不着色。它还能使被称为"核"的成群神经元体划分定位。这些"核"能按照几种形态学指标如大小、形状、核周体染色强度、排列密度、细胞体分布形式、神经毡（neuropil）围绕细胞群和细胞群的密度等进行鉴别。

本节将介绍鱼类脑的下丘脑、脑垂体主要区域 Nissl 氏的细胞结构，并着重简要描述硬骨鱼类前脑（包括端脑和间脑）的形态构造（见图 1.4）。我们采用 Braford 和 Northcutt（1983）的命名，并附上 Cerdá-Reverter（2001a、b）的详细解释。对于进一步深入的比较研究，读者还需要查阅 Braford 和 Northcutt（1983）、Nieuwenhuys 等（1998）以及 Butler 和 Hodos（2005）等的著作。

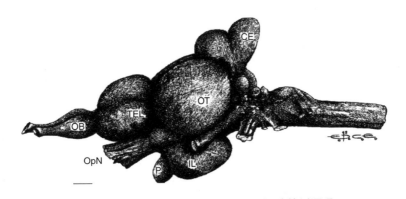

图 1.4　舌齿鲈（*Dicentrarchus labrax*）脑的侧面观

CE，小脑；IL，下丘脑的下叶；OB，嗅球；OpN，视神经；OT，视顶盖；P，脑垂体；TEL，端脑；标准尺度 = 1 mm。（修改自 Cerdá-Reverter 等，2000a）

1.2.1 端脑

对于硬骨鱼类，端脑（telencephalon）的局部解剖学是特殊的。大多数脊椎动物（如七鳃鳗、软骨鱼类、两栖类和羊膜动物）在繁育的初级阶段，端脑发育是通过所谓的"外凸"（evagination）过程实现的。在这个过程中，神经管腔增大而形成端脑室，使端脑突出和增大。相反，辐鳍鱼类的端脑发育是经过一个"外翻"（eversion）过程，神经管顶部向两侧延伸，从而使成对的背部向侧腹方伸展；外翻的结果使端脑为薄的组织覆盖住（见图1.5；Northcutt，1995）。这种不同的脑发育型式使我们难于指明辐鳍鱼类和其他脊椎动物端脑结构之间的同源性，而且还要提出它们各自不同的命名（Northcutt和Davis，1983；Meek和Nieuwenhuys，1998）。但是，采用神经元回路（neuronal circuitry）型式和化学递质与肽类分布的研究方法，已经确立了它们之间一些可能的同源性。

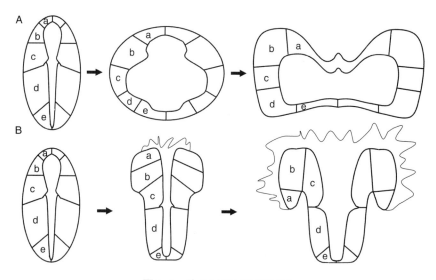

图1.5　外凸过程与外翻过程

A. 大多数脊椎动物在早期发育时前脑的外凸过程；B. 辐鳍鱼类前脑发育的外翻过程。比对图中英文字母所处的位置可以很容易地比较外凸过程和外翻过程之间的差别。

端脑由端脑半球和嗅球组成（见图1.4）。端脑半球可分为两个主要区，即一个背端脑区（大脑皮质，pallium）和一个腹端脑区（下大脑皮质，subpallium）。背端脑区是端脑的外翻部分，呈现出一些组织结构明显不同的区，其大多数细胞分布在远离脑室表面。相反，腹端脑区组成核团，它们大多靠近脑室（Wullimann和Mueller，2004）。

端脑背区（或大脑皮质）是外翻部分，亦是硬骨鱼类端脑的最大区。按照细胞结构的标准可以把它区分为三个室周区：一个外区包括中部（pars medialis，DM）、背部（pars dorsalis，Dd）和侧部（pars lateralis，Dl）；一个中区，即中部（pars centralis，Dc）；一个后区，即后部（pars posterior，Dp）。端脑背区的神经元并不延伸到脑垂体。端脑腹区是硬骨鱼类端脑的非外翻部分，位于视前区吻端。这个区分裂为前连合核

（precommissural nucleus）和后连合核（postcommissural nucleus）。前连合核是内折点，背核（Vd）、腹核（Vv）和侧核（Vl）是主要的前连合核；而后连合核包含中核（Vc）、上连合核（Vs）、后连合核（Vp）、中连合核（Vi）和内大脑脚核（见图1.6）。

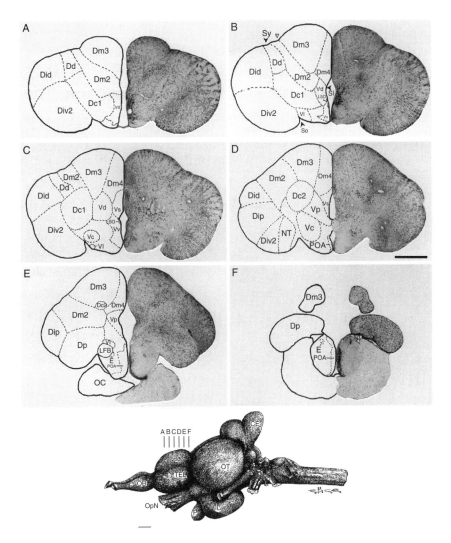

图1.6 舌齿鲈（*Dicentrarchus labrax*）脑经甲酚紫（cresyl violet）染色后横切面的显微图像，表示各种不同的端脑细胞团

Dc1，背端脑区，中部，再分区1；Dc2，背端脑区，中部，再分区2；Dd，背端脑区，背部；Dld，背端脑区，侧部；Dlp，背端脑区，侧后部；Dlv1，背端脑区，侧腹部，再分区1；Dlv2，背端脑区，侧腹部，再分区2；Dm2，背端脑区，中部，再分区2；Dm3，背端脑区，中部，再分区3；Dm4，背端脑区，中部，再分区4；Dp，背端脑区，后部；E，内大脑脚核（entopeduncular nucleus）；LSO，侧隔膜；NT，核带（nucleus taenia）；OB，嗅球；OC，视交叉；OpN，视神经；POA，视前区；Sy，Y字形沟（sulcus ypsiliformis）；Vc，腹端脑区，中央部；Vd，腹端脑区，背部；Vi，腹端脑区，中间部；Vl，腹端脑区，外侧部；Vp，腹端脑区，后连合部；Vs，腹端脑区，上连合部；Vv，腹端脑区，腹部。标准尺度 = 100 μm。（引自 Cerdá-Reverter 等，2001b）

根据神经内分泌的研究，一些硬骨鱼类都曾被报道 Vv 和 Vc 延伸到脑垂体。如舌齿鲈（*Dicentrarchus labrax*）（Cerdá-Reverter 等，2001a），其腹核（Vv）的特点是具有一个室管膜柱，由小的染色致密的细胞排列在脑室周围并和核的背方连接。中核（Vc）含有大而着色深的和侧前脑束（lateral forebrain bundle，LFB）纤维混合一起的核周体。神经元回路型式和化学递质与肽类的分布研究表明，端脑腹区的 Vd 和 Vc 相当于哺乳类髓纹形成（striatal formation），而 Vl 和 Vv 核相当于隔膜形成（septal formation）。学者们曾经为说明 Vl-Vv 和隔膜形成之间的同源性是否大量下行输出到达中线下丘脑而引起争论，这亦是羊膜类动物隔膜形成的特征（Wullimann 和 Mueller，2004）。

1.2.2 视前区

视前区（preoptic area）成为下丘脑结构和功能的连续统一体。可以恰当地认为视前区和下丘脑已形成一个单一的复合体，并且把视前区-下丘脑连续体划归到间脑内。当讨论到下丘脑调控脑垂体（hypophysiotropic）的活动时，内分泌学者们经常把视前区包括在下丘脑之内。

视前区围绕视前-室周隐窝（preoptic-periventricular recess），位于前连合（anterior commissure）和视交叉之间。根据 Braford 和 Northcutt（1983）与 Meek 和 Nieuwenhuys（1998）的研究，视前区可划分为大细胞的（magnocellular，PM）和小细胞的（parvocellular）视前（PP）核（见图 1.7）。大细胞视前核含有大的神经分泌细胞，它们可以再划分为小细胞（PMpc）、大细胞（PMmc）和巨大细胞（PMgc）部分。大细胞视前核具有一定的形状，在纵向平面，形似一个倒转的 L 字，其垂直杆位于吻端，而水平杆朝向背尾端延伸。吻腹端较小的细胞组成 PM 的小细胞部分，而背尾端较大的细胞组成大细胞部分和巨大细胞部分。小细胞的视前核可以进一步划分为前部（PPa）和后部（PPp），它们分别位于大细胞视前核的前方和尾腹方。按照彼得教授对金鱼（*Carassius auratus*）、底鳉（*Fundulus heteroclitus*）和两种鲑鱼类的命名（Peter 等，1975；Peter 和 Gill，1975；Billard 和 Peter，1982；Peter 等，1991），PPa 称为视前小细胞核（preoptic parvocellular nucleus，NPP），大细胞部分称为视前核（preoptic nucleus，NPO），而 PPp 称为室周核（periventricular nucleus），室周核又分为前室周核（anterior periventricular nucleus，NAPv）和后室周核（posterior periventricular nucleus，NPPv）。但是，NPPv 的尾区相当于 Braford 和 Northcutt（1983）所说的间脑界膜带（zona limitans，ZL）。按照 Cerdá-Reverter 对舌齿鲈的命名，NPP（Peter 和 Gill，1975）或者 PPa（Braford 和 Northcutt，1983）应称为小细胞视前核，并再划分为前腹部（anteroventral，NPOav）和小细胞部（NPOpc）。对于胡子鲶（*Clarias batrachus*）（Prasada Rao 等，1993），NPOav 被称为视前核（NPOs）的视上部（supraoptic division）。按照 Maler 对电鳗（*Apteronotus leptorhynchus*）（Maler 等，1991）的命名，PM（Braford 和 Northcutt，1983）曾被称为前小丘脑核（anterior hypothalamic nucleus，Ha），而 PPp（Braford 和 Northcutt，1983）再划分为 PPp 本身和前室周核（anterior periventricular nucleus，nAPv）。最后，所有的其他命名包括有些鱼类在视前区的大部分尾腹区称为上交叉核（suprachiasmatic nucleus，NSC）的，都不认为它是一个分离的核（Gómez-Segade 和 Anadón，1988）。

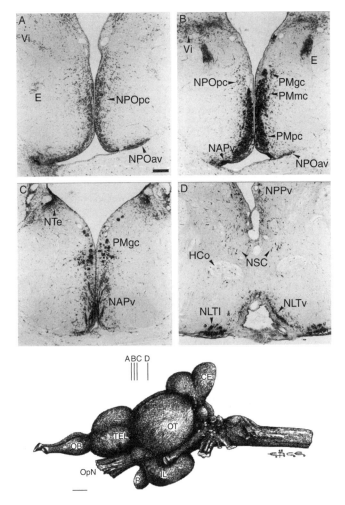

图1.7 舌齿鲈（*Dicentrarchus labrax*）脑以甲酚紫染色后横切面显微摄影图，表示视前不同的细胞团

E，内大脑脚核（entopeduncular nucleus）；HCo，水平连合；NAPv，前室周核；NLTl，侧结节核的外侧部；NLTv，侧结节核的腹部；NPOav，小细胞视前核的前腹部；NPOpc，小细胞视前核的小细胞部；NPPv，后室周核；NSC，上交叉核；NTe，丘脑隆起核（nucleus of the thalamic eminentia）；PMgc，大细胞视前核的巨细胞部分；PMmc，大细胞视前核的大细胞部分；PMpc，大细胞视前核的小细胞部分；Vi，腹端脑的中间部分。标准尺度 = 100 μm。（引自 Cerdá-Reverter 等，2001b）

1.2.3 下丘脑

下丘脑位于丘脑下方，视前区尾方。它在视交叉嵴（chiasmatic ridge）尾方分离，由于其围绕初生漏斗隐窝（infundibular recess）的排列稠密又深着色的细胞而很容易和视前区辨别（见图1.8）。下丘脑是最大的间脑脑区，通过含有下丘脑和视前神经内分泌纤维的柄和脑垂体连接（见第1.1节）。有些硬骨鱼类的漏斗隐窝或第三脑室的下丘脑部分向外侧伸展而形成成对的尾向壶腹，称为外侧隐窝（lateral recess）。在尾区，第三脑室衍生后隐窝（posterior recess），它亦形成小的向背侧方的壶腹（见图1.8；Braford 和 Northcutt）。硬骨鱼类下丘脑的细胞结构研究显示出很大的差别（Peter 等，1975；Braford 和 Northcutt，1983；Gómez-Segade 和 Anadón，1998；Striedter，1990；Maler 等，1991；Wullimann 等，1996；Cerdá-Reverter 等，2001b），然而，不管是同一个再分区在不同的硬骨鱼类中是相类似的，或者是再分区只是反映再分区的不同标准，它们都是不确定的（Meek 和 Nieuwenhuys，1998）。

根据 Meek 和 Nieuwenhuys（1998）的描述，硬骨鱼类的下丘脑可以划分为三个主要区，即室周区、正中位的结节区和由深的腹沟（ventral sulcus）从结节下丘脑分离出来的成对下叶。但是，有些学者把下丘脑划分为背区（Hd）、腹区（Hv）和尾区（Hc）（Braford 和 Northcutt，1983；Striedter，1990；Wullimann 等，1996）。腹区和尾区主要由下丘脑正中结节区组成。所有三个再分区都展现以侧移核为边界的室周细胞群体。按照 Braford 和 Northcutt（1983）的描述，下丘脑的吻区主要包含两个室周细胞群体，即背（Hd）和腹（Hv）室周下丘脑，以及两个侧移的细胞群体，即前结节核（NAT）和侧下丘脑核（LH）。两个室周的再分区之间的边界是弥散的。第三脑室的腹方由 Hv 的细胞形成边界，它们沿着脑室壁密致地排列，形成有1～10个细胞厚的层。Hd 的细胞稍大些，较为稀疏排列而形成一个1～4个细胞厚的层（Cerdá-Reverter 等，2001b）。前结节核（NAT）由位于 Hd 外侧的大而分散、呈纺锤形的细胞组成。所有的命名术语都和一个 NAT 的名称相一致，但它们所处的位置有所不同（Peter 和 Gill，1975；Braford 和 Northcutt，1983；Gómez-Segade 和 Anadón，1988；Striedter，1990；Maler 等，1991；Wullimann 等，1996）。有些学者认为 NAT 延伸到室周壁（Peter 和 Gill，1975；Striedter，1990；Maler 等，1991），而另一些学者把它进一步划分为吻区和腹区（Striedter，1990）。在斑马鱼（*Danio rerio*）的脑图中，Wullimann 等（1996）认为 Hv 比 Hd 向吻端（前方）延伸得更远些，所以，Hd 的吻端就被称为 Hv，因而，NAT 就位于 Hv 的外侧。

第三脑室的侧孔（lateral aperture）以 Hd 所在的吻-尾水平定界。正是在下丘脑脑室侧孔之前，LH 的小细胞向背侧方延伸。这个核位于 Hv 尾区的外侧，它的细胞似乎是包被着外侧隐窝（LR）的腹方。实际上，这个第三脑室的侧孔可以看作导向 Hc 的一个过渡区。在这个水平上，Hd 的细胞包被着 LR，而第三脑室的背区为大量后结节核（posterior tuberal nucleus，NPT）的细胞所占据。和 LR 从中央下丘脑室完全分离的情况大体相似并稍靠向尾方，Hc 展现出来。它是一个小区，边缘是后隐窝的尾部，背方止于 NPT。Hc 由厚的室管膜细胞（ependymal cell）层组成，并朝尾部的两侧延伸而形成一个柱状排列位置，沿着侧方直到后隐窝的小囊突（diverticula）。Hc 包含一个小细胞群，称为血管囊核（NSV）以及 Peter 和 Gill（1975）提到的后隐窝核（posterior recess nucleus，NRP）。

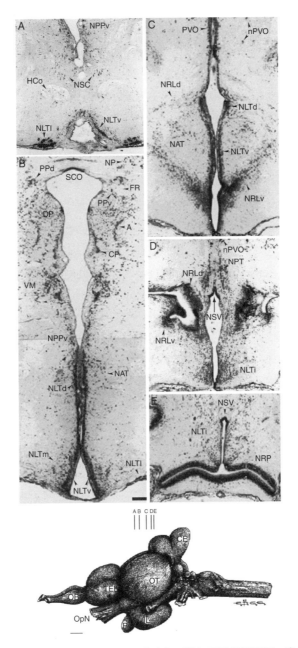

图1.8 舌齿鲈（*Dicentrarchus labrax*）脑以甲酚紫染色后横切面显微摄影图，表示不同的间脑细胞团

A，前丘脑核；CP，中央后丘脑核；DP，背后丘脑核；FR，反折的神经束（fasciculus retroflexus）；NAT，前结节核；NLTd，外侧结节核的背部；NLTi，外侧结节核的下部；NLTm，外侧结节核的中部；NLTl，外侧结节核的侧部；NLTv，外侧结节核的腹部；NP，副连合核；NPPv，后室周核；NPT，后结节核；nPVO，室旁核器；NRLd，外侧隐窝核的背部；NRLv，外侧隐窝核的腹部；NRP，后隐窝核；NSC，上视交叉核；NSV，血管囊核（nucleus of the saccus vasculosus）；PPd，背室周顶盖前核（dorsal periventricular pretectal nucleus）；PPv，腹室周顶盖前核；PVO，室旁器（paraventricular organ）；SCO，连合下器（subcommissural organ）；VM，腹中下丘脑核。标准尺度 = 100 μm。（引自 Cerdá-Reverter 等，2001b）

按照部分以 Peter 和 Gill（1975）的理论为基础的 Cerdá-Reverter 等（2001b）学者的命名法，吻室周区（rostral periventricular region）可以一致称为外侧结节核（lateral tuberal nucleus，NLT）。在横向平面，这个核可以分为两个不同的亚区，即背区（NLTd）和腹区（NLTv），它们分别相当于 Braford 和 Northcutt（1983）所提到的 Hd 和 Hv 的吻区。Cerdá-Reverter 等（2001b）曾描述在舌齿鲈下丘脑吻区内，有两个附加的侧移细胞群，即 NLT 的外侧部（NLTl）和中央部（NLTm）。NLTl 细胞是下丘脑横切面最早出现的细胞。舌齿鲈的 NLTl 由很大的染成黑色的细胞组成，它们排列在水平连合（horizontal commissure）的腹侧方并侧向初期的漏斗隐窝（Cerdá-Reverter 等，2001b）。在尾端，这些神经元比较小，位于脑的腹侧外表。吻端大的神经元亦为 Braford 和 Northcutt（1983）所识别并标明为腹结节核（ventral tuberal nucleus，TV），而较小的尾神经元称为 LH。NLTl 神经元可以采用化学方法区分，因为它们能合成硬骨鱼类脑的一种神经激素，即黑色素浓集激素（melanin-concentrating hormone，MCH）（Baker 和 Bird，2002）。NTLm 由散布在 NLT 腹方和侧方之间的细胞组成。这种再区分最初报道于非洲丽鱼（*Haplochromis burtoni*）（Fernald 和 Shelton，1985）。舌齿鲈的 NLTm 部分相当于 Peter 和 Gill（1975）区分的 NLT 前部和后部以及 Braford 和 Northcutt（1983）提到的腹下丘脑。

两种细胞群将稍后围绕 LR，即外侧隐窝核的背部（NRLd）和腹部（NRLv），在脑室侧孔形成之前就可以看到。NRLv 位于 NLTv 的腹方，这两个核一直共存到腹室周区被后隐窝核和外侧结节核侧下部（NLTi）在尾端所侵占。NRLv 相当于 Braford 和 Northcutt（1983）、Striedter（1990）和 Wullimann 等（1996）提到的 LH，亦包含 Hv 尾方的脑室细胞。NRLd 和室旁器（paraventricular organ，PVO）的细胞可能侧移，包被初生外侧隐窝（nascent lateral recess）的背方，而该隐窝的腹方则为 NRLv 的细胞所包被。根据 Braford 和 Northcutt（1983）的描述，NRLd 和 Hd 的尾区相匹配。

下叶是硬骨鱼类下丘脑最大的部分，能在脑的腹方表面清楚地看到。迁移自背室周的细胞群体可能"填满"下叶，因此，可以认为它们是 Hd 向外侧延伸的一部分（Wullimann 等，1996）。根据 Braford 和 Northcutt（1983）和 Wullimann 等（1996）的研究，可以清晰地区分三个细胞群体：Hd、下叶的弥散核（NDLI）和下叶的中央核（NCLI）。Hd 包被 LR，在下叶内占据中央位置；NDLI 由很小、着色淡而分散的细胞组成；而 NCLI 处在朝向尾外侧隆凸（lateral torus）的腹中位置，由大的纺锤形和卵圆形细胞组成。在舌齿鲈中，NDLI 可进一步划分为中间部（NDLIm）和外侧部（NDLIl）（Cerdá-Reverter 等，2001b），而包被 LR 的细胞群体是外侧隐窝核（NRLl）的外侧部分。此外，根据细胞结构的标准，我们阐明下叶的中间核（NMLI）是位于 NRLl 附近，小球核（nucleus glomerulosus）的腹中方和乳头体（corpus mamillare）的侧方。这个核稍后将在其他硬骨鱼类神经元回路的研究中说明。

1.3 鱼类脑的下丘脑-脑垂体分区：神经束追踪研究

硬骨鱼类由于缺少正中隆起和门脉系统而便于采用神经束追踪（tract-tracing）技术研究它们的下丘脑脑垂体神经元。这项研究以标志物（主要是凝集素和葡聚糖胺）的逆行性运送为基础。碳花色素苷荧光染料（carbocyanine fluorescent dyes）如 DiI（1-1′-二十八烷-3，3，3′，3′-四甲基吲哚碳花色素苷）（1-1′-dioctadecyl-3，3，3′，3′-tetramethylindocarbocyanin）比其他技术如放射自显影（autoradiography）、辣根过氧化物酶（horseradish peroxidase，HRP）或钴追踪等具有更多个优点。这些优点包括可以在低聚甲醛（paraformaldehyde）固定的组织中使用碳花色素苷做示踪物，而且它与其他荧光示踪物以及免疫组织化学技术等具有同样的效果（Holmqvist 等，1992）。这项技术是使用昆虫针或玻璃电极把小小的染料结晶埋植在组织内，经过孵育后进行组织切片。研究人员曾采用追踪技术研究胡子鲇（Rama Krishna 和 Subhedar，1989；Prasada Rao 等，1993）、金鱼（Fryer 和 Maler，1981；Anglade 等，1993）、褐电鳗（Johnston 和 Maler，1992；Zupanc 等，1999；Corrêa，2004）和大西洋鲑（*Salmo salar*；Holmqvist 等，1992；Holmqvist 和 Ekström，1995）的下丘脑脑垂体神经元。如上所述（见第 1.1 节），神经束追踪研究曾揭示神经元伸向脑垂体的两个主要脑区：视前区和结节下丘脑。但是，有时由于对鱼类脑的不同核采用另外一些名称，就难于对不同鱼类进行比较。表 1.1 总结了对硬骨鱼类采用神经束技术揭示的下丘脑-脑垂体各部分构造的分布情况。

表 1.1 硬骨鱼类脑的主要下丘脑-脑垂体区域

	电鳗	大西洋鲑	鲫鱼	胡子鲶	Bradford 和 Northcutt（1983）
技术	DiI 葡聚糖 生物胞素	DiI	HRP DiI	HRP 钴-赖氨酸	
端脑					
嗅球（OB）			X		OB
腹端脑腹部（Vv）	X	X	X		Vv
腹端脑上连合部（Vs）			X		Vs
腹端脑中部（Vc）			X	X	Vc
视前区					
室周视前核，前部（PPa）	X	X			PPa
室周视前核，后部（PPp）	X	X			PPp
上交叉核（NSC）	X	X	X		NSC
室周视前核（NPP）			X		PPa

续表1.1

	电鳗	大西洋鲑	鲫鱼	胡子鲶	Bradford 和 Northcutt（1983）
视前核，上视分区（NPOs 或 SO）				X	PPa
视前核，室旁分区（NPOp 或 PV）				X	PM
视前核小细胞部分（NPOpc）		X			PMpc
			X		PPa
视前核大细胞部分（NPOmc）		X			PMmc
			X		PMmc-PMgc
视前核巨大细胞部分（NPOgc）		X			PMgc
前室周核（NAPv）		X	X		PPp
后室周核（NPPv）			X		PPp
前下丘脑（Ha）	X				PM
结节下丘脑					
背下丘脑（Hd）	X				Hd
腹下丘脑（Hv）	X				Hv
尾下丘脑（Hc）	X				Hc
侧下丘脑（Hl）	X				LH, partially
前结节核（NAT）	X		X		NAT
侧结节核（NLT）		X		X	Hd/Hv/LH
侧结节核，前部（NLTa）	X				Hv and Hc
侧结节核，后部（NLTp）	X				Hc
侧结节核，侧部（NLTl）		X			LH
侧结节核，前部（NLTa）			X	X	Hv
侧结节核，后部（NLTp）			X		Hd/Hv
侧结节核，下部（NLTi）			X		Hv
侧隐窝核（NRL）		X	X	X	Hd
后隐窝核（NRP）		X	X	X	Hc
血管囊核（NSV）		X			PTN
下叶中央核（NCLI）	X				NCLI
后结节					
室旁器（PVO）		X		X	PVO
丘脑					
中央后核（CP）	X				CP
背侧丘脑核（DL）			X		Vm/DP/I/A
网状结构	X				R

在端脑的研究中曾证明一些硬骨鱼类的腹区（Vv）和中区（Vc）都发现逆行性标志的细胞和神经纤维。此外，一些下丘脑-脑垂体细胞亦曾经被发现分布在金鱼的嗅球和腹下丘脑的上连合部（Vs）。

视前区是神经分布到硬骨鱼类脑垂体的主要脑区。几种鱼类的研究证明前室周核（PPa）、后室周核（PPp）和上视交叉核（NSC）都伸向脑垂体（见表1.1）。此外，在视前核的三个再分区，即 NPOpc、NPOgc、NPOmc 中都发现逆行性神经细胞体。对大西洋鲑鱼的研究亦报道 DiI 标志的细胞分布在视前区吻端，它们的大小和形态都和 NPOgc 相似，但和 NPO 的主要细胞群明显不同。研究者曾建议把这个细胞群称为吻视前区（NPOr）。

结节下丘脑是神经分布于硬骨鱼类脑垂体的第二个主要下丘脑区。对几种鱼类的研究都曾报道在背、腹侧和尾下丘脑出现标志的结节下丘脑细胞。在前结节核（NAT）内出现下丘脑脑垂体细胞（hypophysiotropic cell）似乎是种类依存性的（species dependent）（见表1.1），而且只有一项研究曾报道在下叶出现逆行性标志细胞（Corrêa, 2004）。此外，采用 DiI 标志的金鱼，在腹结节核（TV）曾发现标志的细胞。

1.4　中央神经激素

1.4.1　精氨酸-加压催产素

精氨酸-加压催产素（arginine-vasotocin，AVT）肽属于神经九肽的家族，包括硬骨鱼类催产素（isotocin，IST）以及哺乳类同源的精氨酸血管加压素（arginine-vasopressin，AVP）和催产素（oxytocin，OXT）（Acher, 1996）。AVT 在内分泌调控盐分/水分稳态和血管机能方面起关键作用（McCormick 和 Bradshaw, 2006），还对一系列生理活动过程（Balment 等, 2006）和社会性行为（Goodson 和 Bass, 2001）起调节作用。AVT 是低等脊椎动物主要的神经脑垂体激素之一，由视前神经元合成，运行于视前-脑垂体神经束（preoptico-hypophysial tract），然后通过神经脑垂体轴突末梢而释放到血管系统内。对一些鱼类的研究都证明 AVT-免疫（ir）反应神经元完全属于视前核小细胞神经元和大细胞神经元（见表1.2；Goossens 等, 1977；Batten 等, 1990a；Duarte 等, 2001；Lema 和 Nevitt, 2004；Saito 等, 2004；Mukuda 等, 2005；Bond 等, 2007；Maruska 等, 2007）。然而，对光蟾鱼的免疫染色研究表明在腹下丘脑还有一个 AVT 细胞群体。这第二个神经元群由一些小而圆形的 AVT-免疫反应神经元组成，它们通常包埋在视前-脑垂体束（preoptico-hypophysial tract）内（Goodson 和 Bass, 2001）。这种鱼类的大多数 AVT 神经元都分布在 PPa 内，只有少数微弱标志的细胞位于 PPp 内。在 PM，大多数 AVT 细胞都处在大细胞分区内，而这些神经元受标志的程度要比分布于 PPa 与 PPp 的神经元强烈得多。AVT-免疫细胞的这种分布情况在大多数被研究的鱼类中是保守的，尽管也有不同的报道。对内华达鳉（*Cyprinodon nevadensis*）的研究曾报道类似的结果，但在 PPp 内没有发现 AVT-免疫反应细胞，而在 PM 的所有三个再分区，即

PMpc、PMmc 和 PMgc 中都有 AVT-免疫反应的细胞（Lema 和 Nevitt，2004）。对虹鳟（*Oncorhynchus mykiss*）的研究结果亦有类似的报道，AVT 神经元几乎占了 PM 巨大细胞部分神经分泌细胞的 50%，而在 PPa 中只占 15%。采用神经元神经束追踪结合 AVT 免疫组织化学技术对大西洋鲑的研究表明，在 PM 内大多数为 DiI 标志的神经元是 AVT-免疫反应的。然而，并非所有 AVT-免疫反应神经元都在脑垂体，因为有些神经细胞体，特别是在 PMgc 的，并未为 DiI 标志。在 PPa 内的 AVT 神经元并不显示逆行性标志，因而可以认为其是具有 AVT-免疫反应的非下丘脑视前区神经元。该研究报道的下丘脑和外下丘脑凸出体的视前神经分泌 AVT-免疫反应细胞（Holmqvist 和 Ekström，1995）亦曾在对光蟾鱼的研究中报道过（Goodson 和 Bass，2001）。采用单细胞染色技术对虹鳟的深入研究支持上述的研究结果，并且进一步表明，在所有的研究实例中，AVT-免疫反应神经元都同时伸向脑垂体和外下丘脑区，包括腹端脑、丘脑和中脑投射（project）。单个神经分泌细胞的多重伸出还未曾在其他脊椎动物的研究中报道。这种多重伸出适合于通过由不同的生理需求引起电活动的同步性，进而协同调控外周与中枢的分泌物（Saito 等，2004）。在金鱼、虹鳟和鳗鱼（*Anguilla* sp.）的 PM 中，AVT 神经元常依靠邻近脑室壁的体细胞（soma-somatic）的并列而聚集一起（Cumming 等，1982；Saito 等，2004）。这种解剖学的细胞并列的功能还不清楚，但可能参与交换局部场电位（field potential）的沟通，或者参与体-树状突（somatodedtritic）肽类的释放（Saito 等，2004）。在 PM 内 AVT 神经元的明显特征是频繁在它们基部突起之间接触，而这基部突起来自同一个或者不同的细胞群，它们可能亦参与 AVT 神经元当中的沟通（Saito 等，2004）。金鱼的超显微构造研究证明 AVT-免疫细胞轴突末梢在 PM 内形成轴-树突触（axo-dendritic synapse）。这表明在 AVT 细胞之间存在着突触（synaptic）联系，而在 PM 内具有局部的神经元回路（neuronal circuitry）（Cumming 等，1982），它们可能作用于神经分泌细胞的同步性，使分泌活动符合生理需求（Saito 和 Urano，2001；Saito 等，2004）。在 PM 内的 AVT 神经元进一步被证明和鳗鱼的体循环直接接触（Mukuda 等，2005），从外周接受化学信息。其实，在对牙鲆（*Platichthys flesus*）的研究中发现，AVT 神经元在 PM 的三个再分区中都表达糖皮质激素受体（glucocorticoid receptor），并为限制性应激反应所调控（Bond 等，2007）。对几种鱼类的研究还表明，AVT 神经元在 PM 和 PPa 中分别和 CRF 与甘丙肽（galanin）共存（colocalize）。对一些鱼类的研究还表明，AVT 神经元系统的另一个显著特点是两性异形（sexual dimorphism）（Grober 和 Sunobe，1996；Maruska 等，2007），并且在显示不同生殖特性（reproductive tactics）的雄鱼之间亦观察到差别（Foran 和 Bass，1998；Goodson 和 Bass，2001）。此外，黄点拟矶塘鳢（*Trimma okinawae*）的性反转和 AVT-免疫细胞的大小亦有联系，但细胞的数量没有变化（Grober 和 Sunobe，1996）。

表 1.2　包含神经肽的主要下丘脑-脑垂体区域

肽类	端脑	视前区	下丘脑
AgRP	—	—	Hv
AVT	—	PPa, PPp PMmc[1]	—
CART	腹端脑	PMmc	Hv, Hc, NCLI
CCK	腹端脑	PPa, PMpc, PMmc	Hv, Hc, Hd, PVO
CRF	OB, 腹端脑	PPa, PPp, PM, NSC	Hv, Hc
甘丙肽	OB, Vs	PPa, PM	Hv, LH, Hc
GHRH	—	PPa, PPp, PMpc, PMmc	Hv, LH, DF (NDLI)
GnRH	OB, Vv	PPa	LH
GRP	腹 TEL	PPa, PM, PPp	Hv, NAT, Hd, PVO
IST	—	PPa, PMmc	
MCH	—	—	LH, PVO, Hv
NPY	OB, Vd, Vv, Vs, Vl, Vc	PPa, PPp, PMgc	Hv, LH, NAT, Hc
食欲肽	—	PMmc, PMgc, PPp, NSC	PVO, Hc
PACAP	—	PPa, PPp, PMgc	Hv
RF-酰胺	OB	—	Hv (PrRP)
SS	OB, Vl, NE	PPa, PMpc, PMmc, PPp, NSC	Hv, Hd, Hc, PVO
TRH	OB, 腹 TEL	PPa, PMmc, PPp, NSC	Hv, NAT, NPT, Hc
UI	Vv, Vd, Vp	PPa, PPp,	Hv, Hc
α-MSH	—	Hv, LH	—

注：各个核的命名参照 Bradford 和 Northcutt（1983）。下丘脑-脑垂体各区域的缩写见表 1.1。AgRP，野灰蛋白相关肽；AVT，精氨酸-加压催产素；CART，可卡因-和安非他明-相关转录体；CCK，缩胆囊肽；CRF，促肾上腺皮质激素释放因子；GHRH，生长激素释放激素；GnRH，促性腺激素释放激素；GRP，促胃液素释放肽；IST，硬骨鱼催产素；MCH，黑色素浓集激素；MSH，黑色素刺激激素；NPY，神经肽 Y；PACAP，脑垂体腺苷酸环化酶激活多肽；SS，生长抑素；TRH，促甲状腺素释放激素；UI，硬骨鱼紧张素 I。

如上所述，视前 AVT 神经元主要伸向神经脑垂体。但是，在对舌齿鲈的研究中发现，神经脑垂体深入交错到 RPD 内，可以看到在 AVT 末梢和促皮质激素细胞之间的紧密联系（见表 1.3）。在脑垂体近端远侧部（PPD），AVT 纤维和 TSH 与 GH 细胞混合一起，偶尔亦和 GTH 细胞混合。在 PI 内，大量 AVT 纤维和 MSH 与生长乳素细胞接触（Moon 等，1989）。对帆鳍花鳉（*Poecilia latipina*）的超显微结构研究证明，AVT 纤维主要局限于神经脑垂体的毛细血管区（pericapillary area），但有些免疫反应轮廓紧靠基底层（basement lamina），且有的和基底膜中断而直接在 AVT 纤维和 ACTH 细胞之间接触，很少在 AVT 纤维和 PRL 细胞之间接触。在脑垂体中部，基底膜中断使得 AVT 纤维和 GH 细胞与 TSH 细胞混合一起，但很少和 GTH 细胞混合。和舌齿鲈相似，在 PI 内的 AVT 纤维和 MSH 细胞及生长乳素细胞接触（Batten，1986）。

表 1.3　鱼类脑垂体的肽能神经分布和靶细胞的表现型

肽	靶细胞类型	观察结果	参考文献
AVT	MSH, ACTH, SL, GH, TSH	偶尔在 GTH, PRL。大多数 AVT 免疫反应末梢围绕毛细血管	Batten (1986); Moons et al. (1989); Batten et al. (1999)
CART	GH, TSH	在 PI 中只有少量神经纤维	Singru et al. (2007)
CCK	GH, 非特异性 RPD 细胞	在金鱼中很少看到 CCK 神经分布	Moons et al. (1989); Himick et al. (1993); Batten et al. (1999)
CRF	MSH, TSH, ACTH	没有接触 GH 和 GTH 细胞	Moons et al. (1989); Batten et al. (1999); Matz and Hofeldt (1999); Duarte et al. (2001); Pepels et al. (2002)
Galanin	SL, GH, TSH, GTH, ACTH, PRL	在比目鱼和虹鳟中看到 SL 的神经分布, 在四眼鱼中则没有看到	Moons et al. (1989); Magliulo-Cepriano et al. (1993); Anglade et al. (1994); Power et al. (1996); Jadhao and Pinelli (2001)
GHRH	MSH, LS, GH, 作用较小的 GTH, ACTH (?)	金鱼、长颌姬鰕虎鱼、杜父鱼和银汉鱼有神经分布到 PI, 但鳕鱼、鳗鱼和鳟鱼没有。海鲈、鲻鱼、杜父鱼有神经分布到 RPD, 但鳗鱼、鲤鱼、金鱼、鳕鱼和鲑鳟鱼类没有	Marivoet et al. (1988); Moons et al. (1988, 1989); Pan et al. (1985); Olivereau et al. (1990); Miranda et al. (2002)
GnRH	SL, GTH, GH, PRL (?)	银汉鱼和尖吻鲈有神经分布到 RPD, 但其他鱼类没有	Parhar and Iwata (1994); Parhar et al. (1995); Parhar (1997); Batten et al. (1999); Vissio et al. (1999); Mousa and Mousa (2003); Pandolfi et al. (2005)
GRP	GH (?), 非特异性 PI 细胞	没有直接神经分布到 PPD; 可能从 PI 扩散过来	Himick and Peter (1995)
IST	MSH, ACTH, SL, GH, TSH	偶尔在 GTH 和 PRL, 实际上是 AVT 神经重叠分布。大多数 IST 免疫反应末梢围绕毛细血管	Batten (1986); Moons et al. (1989); Batten et al. (1999)
MCH	毛细血管, MSH, SL, ACTH	MCH 免疫反应末梢围绕毛细血管	Batten and Baker (1988); Powell and Baker (1988); Batten et al. (1999); Amano et al. (2003); Pandolfi et al. (2003)

续表 1.3

肽	靶细胞类型	观察结果	参考文献
NPY	MSH, GH, GTH	神经分布种类依赖性的,见第 1.5.9 节	Moons *et al.* (1989); Pontet *et al.* (1989); Zandbergen *et al.* (1994); Marchetti *et al.* (2000); Gaidwad *et al.* (2004)
PACAP	MSH, SL, β-内啡肽, GH, PRL, ACTH		Montero *et al.* (1998); Wong *et al.* (1998); Matsuda *et al.* (2005a、b)
RF-酰胺	非特异性的 PPD 细胞, PRL (PrRP)	在 RPD 局部产生(潜在的副分泌作用)。PrRP 神经分布	Rama Krishna *et al.* (1992); Magliuo-Cepriano *et al.* (1993); Amano *et al.* (2007)
SS	MSH, SL, GH, 作用较小的 GTH, PRL, ACTH	罗非鱼、海鲈、鲶鱼有 RPD 神经分布,但鳑鱼、长颌姬鰕虎鱼和鲷鱼没有	Olivereau *et al.* (1984a); Batten *et al.* (1985); Grau *et al.* (1985); Moons *et al.* (1989); Power *et al.* (1996); Batten *et al.* (1999)
TRH	MSH, GH, TSH(?)	TSH 未确实鉴别	Batten *et al.* (1999); Díaz *et al.* (2001, 2002)

ACTH, 促肾上腺皮质激素; AVT, 精氨酸催产素; CART, 可卡因-与安他非明-调节转录体; CCK, 缩胆囊素; CRF, 促肾上腺皮质激素释放因子; GH, 生长激素; GHRH, 生长激素释放激素; GnRH, 促性腺激素释放激素; GRP, 胃泌素释放肽; GTH, 促性腺激素; IST, 硬骨鱼催产素; MCH, 黑色素浓集激素; MSH, 促黑色素细胞激素; NPY, 神经肽 Y; PACAP, 脑垂体腺苷酸环化酶激活多肽; PI, 脑垂体中间部; PPD, 脑垂体近端远侧部; PRL, 催乳激素; PrRP, 催乳激素释放肽; RPD, 脑垂体吻端远侧部; SL, 生长乳素; SS, 生长抑素; TRH, 促甲状腺素释放激素; TSH, 促甲状腺素。

1.4.2 硬骨鱼类催产素

硬骨鱼类的催产素 (isotocin, IST) 是和哺乳类的催产素 (oxytocin, OXT) 同源的肽。和 AVT 神经元系统一样,对一些硬骨鱼类的研究表明它们的 IST-免疫反应细胞在脑的分布情况都十分相似,这些硬骨鱼类包括金鱼、欧洲鲽鱼 (*Pleuronectes platessa*; Goossens 等, 1977)、帆鳍花鳉 (Batten 等, 1990a)、光蟾鱼、海湾豹胆鱼 (*Opsanus beta*; Goodson 等, 2003)、虹鳟 (Saito 等, 2004) 和异带重牙鲷 (*Diplodus sargus*; Duarte 等, 2001) 等。在所有鱼类中, IST 都在 PPa 和 PM 内产生 (见表 1.2)。但是, 对光蟾鱼的研究报道在 PPp 亦有少量 IST-免疫反应的核周体。IST 的凸出物在硬骨鱼脑中广泛散布, 表明 IST 在生理、行为和感觉运动等方面起着广泛的调节作用。对虹鳟的研究显示, IST-免疫反应神经元处于 PM 的最外侧, 而 AVT 神经元分布在正中位置 (Saito 等, 2004)。在异带重牙鲷 (*Diplodus sargus*) 中, AVT 神经元比 IST 神经元更明显地向尾端延伸，但在金鱼、欧洲鲽鱼或帆鳍花鳉中，AVT/IST 免疫反应核周体都不表现出占优势的地位（Goossens 等，1977；Batten 等，1990a）。和 AVT 神经元一样，虹鳟的 IST 神经元聚集在一起，并和它们在 PM 区的突起之间接触。但是，细胞内染色证明在虹鳟 PM 同一个或者不同细胞群中，染料偶联（dye coupling）的影响出现在 IST 神经元当中，而 AVT 神经元没有。这表明电偶合（electrical coupling）包含在 IST 神经元之间的沟通当中。IST 神经元能为神经元回路而利用电的和/或化学的联会（synapse）。总之，研究表明 AVT 和 IST 神经元在 PM 内形成细胞型的特异性神经元，由它们产生神经脑垂体的和外下丘脑的凸出物（Saito 等，2004）。

对帆鳍花鳉的超显微构造研究证明，在后神经脑垂体，IST 纤维在数量上超过 AVT 纤维（2∶1），但在吻区，这两种神经纤维的数量相近（Batten，1986）。少量 IST 纤维和 RPD 的基底膜接触，这和 AVT 纤维、ACTH 细胞及 PRL 细胞的关系相似，但它们大多位于中部并包围着毛细血管（见表 1.3）。在神经脑垂体和腺脑垂体之间的分界面（interface），IST 纤维分布于 PPD 和 PI 的所有类型细胞的周围，特别是 TSH、GH 和 MSH 细胞。在海鲈脑垂体中亦观察到类似的神经纤维分布的情况（Moons 等，1989）。

1.4.3　黑色素浓集激素

从进化的观点看，黑色素浓集激素（melanin-concentrating hormone，MCH）可能是脊椎动物中枢神经系统内最引人注目的肽类之一。它起初从大麻哈鱼（*Oncorhynchus keta*）的脑垂体中分离纯化，鉴定为循环-环七十肽（circulating-cyclic heptadecapeptide），介导硬骨鱼类的体色变化（Kawauchi 等，1983），诱导黑色素体（melanosome）通过载黑色素细胞（melanophore）树状突进行可逆的传入聚集（centripetal aggregation）。这种色素细胞的中央结集引起折射率（refractive index）的变化，导致鳞片显得较为苍白，从而产生隐蔽性伪装的效果（Kawauchi 和 Baker，2004）。在硬骨鱼类中，当鱼体游到白色背景时，MCH 作为神经脑垂体激素就会释放到血液循环中。MCH 的大部分轴突进入脑垂体并中止于神经叶的后区。此外，阳性神经纤维通常分布在神经脑垂体 PI 的指状突上，即 MSH 和生长乳素细胞分布的位置（见表 1.3）。有些神经纤维亦插入腺脑垂体远侧部，表明它们参与腺脑垂体激素合成/释放的调控（Batten 和 Baker，1988；Powell 和 Baker，1988；Gröneveld 等，1995；Baker 和 Bird，2002；Amano 等，2003；Pandolfi 等，2003）。对鳗鱼（Powell 和 Baker，1988）和花鳉（Batten 和 Baker，1988）的电子显微镜研究表明，类 MCH 的神经末梢分布于将 PI 从神经组织隔开的厚的基底膜附近。MCH 末梢主要位于 α-MSH 细胞对面，通常很少中止于毛细血管周围。相反，MCH 纤维和围绕毛细血管的基底膜之间的接触明显地处在神经脑垂体的吻部和中部内。所以，MCH 末梢是处在很好的位置以便于释放激素沿着基底膜扩散到血管系统以及 PI 内（Powell 和 Baker，1988）。对绿花鳉的电子显微镜研究进一步表明有些 MCH 纤维能通过基底膜中断而和 PI 的内分泌细胞直接接触，通常大多是 MSH 细胞。这些研究亦报道 MCH 纤维和神经脑垂体的脑垂体细胞之间的突触联系，但这种接触的功能意义还不清楚。Gröneveld 等（1995）报道罗非鱼（*Oreochromis mossambicus*）

神经脑垂体通过原位杂交（*in situ* hybridization）进行 MCH 表达，并设想脑垂体细胞的亚群（subpopulation）有可能产生 MCH mRNA。

产生 MCH 的神经元细胞体通常位于鱼类 LH 或 NLTl 内（见表 1.2）。然而，MCH 产生的细胞群数量、细胞的形态以及相关脑室表面的位置等是变化不定的（Baker 和 Bird，2002）。在多鳍鱼类（polypteriform）和软骨硬鳞鱼类（chondrostean）中，MCH 神经元主要位于下丘脑室周表面。第一个神经元群于第三脑室背面和 PVO 联系，而第二个神经元群处于第三脑室外侧面周围。在全骨鱼类（holostean）中观察到的转移到 LH 的 MCH 分泌细胞群相当于多鳍鱼类与软骨硬鳞鱼类在室周的细胞群。硬骨鱼类大部分 MCH 神经元都从室周区（periventricular area）转移出来，而在真骨鱼类（euteleost）中，LH 的下丘脑脑垂体大细胞神经元就相当于最主要的 MCH 产生部位（Baker 和 Bird，2002）。分布在侧下丘脑的这类神经元是脑垂体神经分布的主要神经元，它们存在于所有曾被研究过的真骨鱼类之中。靠近尾端，正好在第三脑室下丘脑区的外侧孔处，有一个小核周体群在背方和 PVO 联系。在同样水平上，有一些产生 MCH 的散布细胞体出现在 Hv，靠近第三脑室外侧孔腹面。这些尾下丘脑 MCH 神经元的功能还不清楚，但是，在罗非鱼受到重复干扰后能观察到 MCH mRNA 水平的特异性升高。这些尾下丘脑细胞的 MCH 基因表达水平亦受到盐水刺激的影响（Gröneveld 等，1995）。

1.5　下丘脑调控脑垂体的肽类

1.5.1　缩胆囊肽

缩胆囊肽（cholecystokin，CCK）和促胃液素（gastrin）组成具有共同 C-端色-蛋-天冬-苯丙-NH_2 的肽类家族（综述见 Chandra 和 Liddle，2007；Rchfeld 等，2007）。CCK 的 C-端八肽的结构在进化过程中相当保守，在哺乳类、鸡、龟和蛙中都是同一性的，而鱼类只取代一个氨基酸（Johnsen，1998；Peyon 等，1998）。这个八肽是在神经系统产生的 CCK 主要肽，而较长的肽如 CCK58、CCK33 和 CCK22 亦在周围组织和血液循环中发现（Rehfeld 等，2007）。

CCK/促胃液素的核周体和神经纤维广泛分布在金鱼的前脑、中脑和后脑（Himick 和 Peter，1994）。高度浓集与众多的 CCK/促胃液素免疫反应核周体和神经纤维系统是在金鱼的后腹中下丘脑、腹侧下丘脑和下丘脑下叶（Himick 和 Peter，1994）。所以，CCK/促胃液素免疫反应细胞体散布在视前区、室周和外侧 Hv 和 Hc 以及背与腹丘脑（见表 1.2）。在对绿花鳉和虹鳟的研究结果中亦描述 CCK 免疫反应在脑部的类似分布情况（Notenboom 等，1981；Batten 等，1990a）。对虹鳟类的研究显示，CCK/促胃液素的细胞体位于 Hv 尾区的腹中下丘脑，由此处发出神经分布脑垂体。对花鳉的研究显示，CCK 免疫反应出现在脑的各个部分。显然，在腹端脑、视前区、结节下丘脑和室周下丘脑都检测到 CCK 免疫反应细胞体（Batten 等，1990a）。对金鱼的研究进一步显示，CCK mRNA 和 CCK 免疫反应一样在脑内广泛表达，而下丘脑是 CCK 表达的主要脑区（Peyon 等，1999）。原位杂交分析表明 CCK 是在腹后下丘脑表达（Peyon 等，1999）。在虹鳟体内曾分离出 CCK 三个不同的 mRNA（CCK-N、CCK-L、CCK-T），它们之间的差别是一个氨基酸的替换（Jensen 等，2001）。每个变异体都在脑和外周组织呈现特异性的表达型式，CCK-N 和 CCK-L 分别在视前区和腹后下丘脑出现（Jensen 等，2001）。

CCK/促胃液素免疫反应纤维由腹下丘脑发出，成束的靠近神经脑垂体而进入脑垂体中（见表 1.3）。在 PPD，神经束分为较小的束以及单一的神经纤维。这些免疫反应的神经纤维中止于将神经脑垂体和腺脑垂体分隔开的基底膜（Notenboom 等，1981；Batten 等，1999）。研究表明，CCK 免疫反应纤维和 GH 细胞关系密切（Moons 等，1989；Himick 等，1993；Batten 等，1999），有力地为 CCK 能够通过神经内分泌调控 GH 释放提供证据（Canasa 等，2007）。CCK 免疫反应纤维没有分布到 PI（Batten 等，1999；Himick 等，1993），亦不在绿花鲈、非洲鲶鱼和海鲈的 RPD 中出现（Batten 等，1999），只罕见地在金鱼的 RPD 中看到（Himick 等，1993）。

1.5.2 可卡因-和安非他明-调控的转录体

可卡因-和安非他明-调控的转录体（cocaine- and amphetamine-regulated transcript，CART）肽最初从牛下丘脑中被分离出来，并被认为是一个类 SS 的肽（Spiess 等，1981）。随后，在采用心理刺激药物如可卡因和安非他明处理后，发现 CART mRNA 表达能在大鼠的纹状体（striatum）内增强（Douglas 等，1995）。在对大鼠的研究中，CART 基因选择性剪接以产生包含 27 个氨基酸信号肽的 129 个或 116 个氨基酸组成的肽，于是就可得到 102 个氨基酸残基（长型）或者 89 个氨基酸残基（短型）组成的前肽（pro-peptide）（Douglas 等，1995）。这些前肽进一步加工以便释放主要的分泌产物，如大鼠长 CART［55～102 个氨基酸残基］或大鼠短 CART［42～89 个氨基酸残基］，这取决于前体的长度（Dylag 等，2006）。

在鱼类的研究中，CART mRNA 已在金鱼（Volkoff 和 Peter，2001）和鳕鱼（*Gadus morhua*）中鉴别出来，而 DNA 序列亦在斑马鱼和红鳍东方鲀（*Takifugu rubripes*）（Kehoe 和 Volkoff，2006）中完成测定。鱼类脑的 CART 肽免疫组织化学定位只在鲶鱼中采用抗鼠 CART［55-102］的抗体进行（Singru 等，2007）。研究结果证明，CART 免疫反应广泛分布在鲶鱼脑内，主要定位在神经内分泌区域（见表 1.2）。CART 细胞曾被报道分布在腹端脑内，包括 Vv、V、Vc 和 Vs 核。在 PMgc 检测到强的免疫反应。在神经脑垂体观察到 CART 免疫反应末梢，在整个 PPD 都检测到一些长的神经纤维，这是生长激素细胞和促甲状腺素细胞集中的部位（见表 1.3）。只有很少的神经纤维出现在鲶鱼的 PI 内。这种脑垂体 CART 的神经末梢亦来源于结节下丘脑，它的一些神经元对 CART 肽显示出免疫反应。学者们把这个区称为弓状核（arcuate nucleus），它相当于 Hv。此外，有些学者报道有 CART 免疫反应神经元分布在 Hc 或 NRP 和 NCLI，它们分别是鲶鱼和电鳗下丘脑调控脑垂体的脑区。

1.5.3 促肾上腺皮质激素-释放因子和相关肽类

鱼类的促肾上腺皮质激素-释放因子（corticotropin-releasing factor，CRF）系统包括四个神经肽，即 CRF，硬骨鱼尾紧张肽（urotensin I，UI）以及哺乳类直向同源物（orthologue），硬骨鱼尾紧张肽 2 和 3（UⅡ，UⅢ），两个 G 蛋白偶联受体（CRF_1 和 CRF_2）和一个结合蛋白（CRF-BP）（Bernier，2006）。CRF 系统对协调应激反应起关键作用。这个系统通过调节腺脑垂体 ACTH 和 MSH 细胞的活动性而调控脑垂体-肾上腺轴。但它们亦参与其他一些生理活动的调控（Flik 等，2006；详见本书第 6 章）。

原位杂交已经被用于研究白亚口鱼（*Catostomus commersoni*）（Okawara 等，1992）和斑马鱼（Alderman 和 Bernier，2007）CRF-表达神经元的分布。在斑马鱼中，CRF 在脑的几个下丘脑-脑垂体相关脑区，包括腹嗅球、端脑、视前区和结节下丘脑广泛表达（见表 1.2）。此外，在背端脑、背和腹丘脑、视顶盖（optic tectum，OT）和大脑脚盖（tegmentum）都有一定程度的表达。CRF 在腹端脑的 Vv 和 Vc 内表达（见表 1.2），它们都伸向鲑鱼和金鱼的脑垂体（见表 1.1）。罗非鱼免疫染色反应研究把 CRF-免疫反应定位在 Vl 和 Vc 内（Pepels 等，2002），而对绿花鲈（Batten 等，1990a）的研究表明 CRF 细胞群只分布在它们的外侧区。但是，对几种鱼类的研究都报道在端脑并没有 CRF 的免疫反应细胞体（Olivereau 等，1984b；Yulis 和 Lederis，1987；Olivereau 和 Olivereau，1988；Mancera 和 Fernández-Llébrez，1995；Matz 和 Hofeldt，1999；Zupanc 等，1999；Duarte 等，2001）。

视前区是 CRF 产生的主要脑区。斑马鱼视前区有四个 CRF 表达神经元群体，分别定位于 PPa、PPp、PM 和 NSC（Alderman 和 Bernier，2007）。先前已经报道 CRF 在亚口鱼的 PPa 和 PM 内表达（Yulis 和 Lederis，1987；Okawara 等，1992）。此外，将虹鳟以"禁闭"的方法诱导应激反应能刺激视前区合成 CRF（Ando 等，1999）。所有关于 CRF 在鱼脑分布研究的报道都证明 CRF 的免疫反应是在视前区，虽然 CRF 神经元回路在不同鱼类中有所差别。在绿花鲈的 PM 中只看到少量 CRF 细胞（Batten 等，1990a）。在鲑鳟鱼类、鳗鱼、电鳗和鲤科鱼类中，PM 和 PPa 的细胞群都显现 CRF-免疫反应（Olivereau 等，1984b；Olivereau 和 Olivereau，1988；Zupanc 等，1999）。对鲷科鱼类如异带重牙鲷（*Diplodus sargus*）（Duarte 等，2001）和金头鲷（*Sparus aurata*）（Mancera 和 Fernández-Llébrez，1995）进行的研究发现，在整个吻视前区没有 CRF 神经元，只在正对结节隐窝（ZL）外侧孔吻向的 PPp 有一些散布的核周体。对鳗鱼（Olivereau 等，1988）、绿花鲈（Batten 等，1990a）、鲤鱼（Huising 等，2004）和罗非鱼（Pepels 等，2002）的研究表明，在 PM 的一些 CRF 神经元含有 AVT；而在白亚口鱼中，所有的 CRF 神经元都出现 AVT-免疫反应（Yulis 和 Lederis，1988）。这种共定位亦出现在形成视前-下丘脑束的神经纤维中和伸向神经中叶（neurointermediate lobe）的神经脑垂体相嵌交错中。实际上，在白亚口鱼 PM 的 CRF 细胞似乎是神经分布到神经中叶，而在腹下丘脑的细胞则伸向腺脑垂体的 RPD（见表 1.3）。肽共定位亦表现在虹鳟的视前区（Ando 等，1999）。对电鳗的神经元束追踪结合 CRF 免疫组织化学研究证明，在 PPa 和

PM 中大约有 6% 的逆行追踪细胞呈现类 CRF-免疫反应。

在结节下丘脑内，CRF 表达曾定位在斑马鱼的 Hv 和 Hc 内。此外，在 PTN 中亦曾检测到一些 CRF mRNA 表达的神经元（Alderman 和 Bernier，2007）。但是，白亚口鱼的结节下丘脑没有杂交信号（Okawara 等，1992），而在 Hv 内却检测到 CRF-肽免疫反应（Yulis 和 Lederis，1987）。同样，在对鲷科鱼类（Mancera 和 Fernández-Llébrez，1995；Duarte 等，2001）、罗非鱼（Pepels 等，2002）和绿花鳉（Batten 等，1990a）的研究中都曾报道在 Hv 内有 CRF-免疫反应。绿花鳉的 CRF 神经元细胞体靠近垂体柄，表明神经分布脑垂体。几种硬骨鱼类包括电鳗、鲑鳟鱼类和鳗鱼，在结节下丘脑没有 CRF-免疫反应（Olivereau 和 Olivereau，1988；Zupanc 等，1999），只有 CRF 系统伸向金头鲷和异带重牙鲷的脑垂体（Mancera 和 Fernández-Llébrez，1995；Duarte 等，2001）。

学者们曾研究几种鱼类的 CRF 纤维在脑垂体的分布情况（见表 1.3）。早期对舌齿鲈的研究证明，大量 CRF 纤维进入后神经脑垂体和 PI（Moons 等，1989），但亦有一部分 CRF 阳性纤维到达 RPD 和 PPD。在对一些硬骨鱼类的研究中有类似的报道（Yulis 和 Lederis，1987；Matz 和 Hofeldt，1999），但是，鳗鱼、鲑鱼（Olivereau 和 Olivereau，1988）、异带重牙鲷（Duarte 等，2001）、罗非鱼（Pepels 等，2002）和鲤鱼（Huising 等，2004）的 PPD 没有 CRF-免疫反应。ACTH 细胞所在的 RPD，似乎是有些鱼类 CRF 纤维分布的主要部位（Olivereau 等，1984b），但在其他一些鱼类中，RPD 没有 CRF 免疫反应的纤维（Mancera 和 Fernández-Llébrez，1995）。在对有些鱼类的研究中，学者们曾认为 CRF 可能通过血管或者旁分泌形式到达 ACTH 细胞。在海鲈的 RPD，CRF 纤维分布在 ACTH 细胞附近，这和 ACTH 释放因子明确的作用相符合；而在 PPD，CRF 纤维和 TSH 细胞紧密并列，但不和生长激素细胞与促性腺激素细胞接触。在对大鳞大麻哈（*Oncorhynchus tshawytscha*）的研究中报道了类似的研究结果，用 CRF 和 TSH 抗体免疫染色后邻近的切片证明它们是部分重叠的（Matz 和 Hofeldt，1999）。先前对大麻哈鱼的离体研究证明 CRF 和 UI 具有促甲状腺激素释放的活性（Larsen 等，1998）。双重免疫染色法表明，CRF 纤维分布紧靠 ACTH 细胞，而神经垂体的免疫反应纤维终止于鲑鳟鱼类、鳗鱼（Olivereau 和 Olivereau，1988）的 MSH 细胞以及鲤鱼（Huising 等，2004）的腺脑垂体。异带重牙鲷的研究亦有类似报道，这表明 CRF 调控脑垂体 MSH 系统（Duarte 等，2001）。事实上，早已充分证明 CRF 刺激硬骨鱼类的 MSH 分泌（详见本书第 6 章）。

早期对鱼脑的研究证明类 UI 免疫反应存在于白亚口鱼的间脑（Yulis 等，1986）。免疫反应部位是在 Vp、Hv 和 Hc 内。对白亚口鱼的进一步研究表明，UI 免疫反应出现在 Vv 和顶盖前区（pretectal area）（Yulis 和 Lederis，1986）。在脑垂体中，类 UI 的纤维局限在 PDP，紧密靠近腺脑垂体细胞（见表 1.3）。同样，在对鳗鱼的研究中报道 UI 转录体存在于 PVO 的背与腹再分区（Kawauchi 等，2003）。根据 Braford 和 Northcutt（1983）的描述，PVO 的背与腹再分区分别相当于背下丘脑和腹下丘脑的尾区。最近的表达研究表明 UI mRNA 在斑马鱼脑广泛表达（Alderman 和 Bernier，2007）。UI 转录体定位在背端脑和腹端脑、视前区和结节下丘脑、中脑顶盖和被盖区（mesencephalic tectal and tegmental area）。在端脑，UI mRNA 表达细胞体位于 Vd 内以及可能是 Vp 的尾区。在视前区的表达局限于小细胞区，在 PPa 的标志要比 PPp 强。在结节下丘脑内，在 Hv 和 He 中检测到一些表达。

到目前为止，尿皮质激素（urocortin）2 和 3 的分布以及这些肽在参与鱼类神经内分泌系统的情况如何，都还不清楚。

1.5.4 甘丙肽

甘丙肽（galanin）是一个 N-端 29 个氨基酸的肽，最初分离自猪肠。它是由前肽原（prepropeptide）和被称为甘丙肽信使的联系肽（galanin message-associated peptide）一起经过蛋白酶解加工（proteolytic process）而成。这个神经肽和三个不同的 G-蛋白偶联受体结合，这些受体都展示它们在功能偶联方面和随后的信号活动方面的实质性差别。甘丙肽在许多哺乳动物的中枢和外周神经系统中广泛分布并且具有多种生物学效应，包括摄食和代谢、渗透压调节和水分摄入、学习和记忆增强、焦虑和相关行为、苏醒和入睡调节、生殖和伤害感受（nociception）等（Lang 等，2007）。已报道甘丙肽能神经纤维分布在几个脊椎动物类群包括鱼类的脑垂体内，表明甘丙肽能参与脑垂体分泌活动的调控（Moons 等，1989；Batten 等，1990a；Olivereau 和 Olivereau，1991；Mogliulo-Cepriano 等，1993；Anglade 等，1994；Rodríguez-Gómez 等，2000a；Jadhao 和 Pinelli，2001）。对鱼类的超显微结构研究进一步证明在神经脑垂体的中区有甘丙肽-免疫反应的神经末梢，有些神经纤维随着神经组织的交错衔接进入近端远侧部（PPD），终止于和 PRL、ACTH、GH、TSH 及 GTH 细胞相对的基底膜（见表 1.3），但并不对 PPD 任何一种内分泌细胞类型表现特别突出的联系（Batten 等，1990c，1999）。对海鲈的免疫细胞化学研究表明，甘丙肽神经纤维直接分布到腺脑垂体的 ACTH 和 PRL 细胞。同样，对斑剑尾鱼（*Xiphophorus maculates*）的研究证明，甘丙肽免疫反应纤维和 SL、PRL 及 GH 细胞共同定位。斑剑尾鱼的受体结合研究表明，甘丙肽受体局限于腺脑垂体吻部催乳激素细胞所占据的部位（Moons 等，1991）。

甘丙肽-免疫反应细胞群主要分布在视前区和结节下丘脑（见表 1.2），但在对一些鱼类的研究中曾报道有两性异形（sexual dimorphism）现象（Prasada Rao 等，1996；Jadhao 和 Meyer，2000；Rodríguez 等，2003）。然而在对四眼鱼（*Anableps anableps*）的研究中没有发现中枢甘丙肽能系统出现中间性异形（inter-sex dimorphism）现象（Jadhao 和 Pinelli，2001）。此外，甘丙肽-免疫反应神经元群体曾被报道出现在河鳟（*Salmo trutta fario*）（Rodríguez 等，2003）的嗅球、金鱼（Prasada Rao 等，1996）的 Vs 和虹鳟（Anglade 等，1994）的脑内。在大多数硬骨鱼类中，甘丙肽-免疫反应和/或基因表达细胞通常都是分布在 PPa。雄性金鱼（Prasada Rao 等，1996）和河鳟（Rodríguez 等，2003）的 PPa 免疫反应细胞的数量都大于雌鱼，但在红大麻哈鱼（*Oncorhynchus nerka*）（Jadhao 和 Meyer，2000）和绿花鳉（Cornbrooks 和 Parsons，1991）的 PPa 则报道没有性别的差异。对几种硬骨鱼类的研究曾报道在 PM 存在甘丙肽-免疫反应神经元的第二个群体（Batten 等，1990c；Magliulo-Cepriano 等，1993；Prasada Rao 等，2003；Adrio 等，2005）。在雌性金鱼和花鳉中没有这个神经元群体，但雄性则明显存在着（Cornbrooks 和 Parsons，1991；Prasada Rao 等，1996）。对红大麻哈鱼的研究报道的则是相反情况（Jadhao 和 Meyer，2000）。我们对金鱼进行的表达研究表明甘丙肽 mRNA 存在于 PPa 而不存在于 PM（Unniappan 等，2004）。双重标志研究表明在花鳉和青鳉的 PM 内有一些甘丙肽和 CRH 免疫反应共存的神经元（Batten 等，1990c）。在虹鳟（Anglade 等，1994）和塞内加尔鳎（*Solea senegalensis*）（Rodríguez-Gómez 等，2000a）的 PM 内没有发现甘丙肽免疫反应细胞。

产生甘丙肽的第三个细胞群体位于结节下丘脑的后区和尾区。对金鱼的原位杂交研究把甘丙肽 mRNA 定位在 LH 和 Hv 内（Unniappan 等，2004）。在对另外几种鱼类的研究中亦曾报道在 Hv 的细胞群体（Power 等，1996；Rodríguez-Gómezz 等，2000a），但这个部位有时候被称为侧结节核的后部（Batten 等，1999c；Magliulo-Cepriano 等，1993；Anglade 等，1994；Prasada Rao 等，1996；Jadhao 和 Meyer，2000；Jadhao 和 Pinelli，2001）或者 NAT（Rodríguez 等，2003）。这个存在于结节下丘脑的细胞群在对其他一些鱼类的研究中被描述为似乎延续到 Hc 和 PTN（Anglade 等，1994；Power 等，1996；Prasada Rao 等，1996；Jadhao 和 Pinelli，2001；Rodríguez 等，2003）。结节下丘脑甘丙肽细胞分布的两性异形现象只在河鳟的有关研究中有过报道，雄鱼的甘丙肽免疫反应细胞要比雌鱼的大些。

1.5.5 胃泌素释放肽

胃泌素释放肽（gastrin-releasing peptide，GRP）由 27 个氨基酸组成，是铃蟾肽（bombesin，BBS）和神经调节肽（neuromedin）B 家族肽的一部分（Martínez 和 Taché，2000）。这个肽曾从几种哺乳动物中分离出来，亦在其他脊椎动物，包括鱼类中发现（McDonald 等，1979；Holmgren 和 Jensen，1994）。这类肽的特征是 C-端高度保守并具有重要的生物学功能（Martínezz 和 Teché，2000）。由于 BBS 和 GRP 之间结构相似，外源 BBS 的效应能够很好地反映内源 GRP 的功能。BBS/GRP 肽广泛分布在胃肠消化道和 CNS（McCoy 和 Avery，1990），将它们通过腹腔或中枢施给哺乳类和鱼类时表现为强有力的厌食性物质（Gibbs 等，1979；Flynn，1991；Himick 和 Peter，1995；Volkoff 等，2005）。此外，一种类 BBS 肽亦参与鱼类消化道活动和内脏活动的调节（Holmgren 和 Jonsson，1988；见本书第 10 章）。BBS-免疫反应的纤维分布在腹端脑、视前区、结节下丘脑、后下丘脑和被公认为鱼类摄食中心相联系的脑区以及一些丘脑核（Himick 和 Peter，1995）。此外，BBS-免疫反应的核周体分布在视前区和室周区，包括 NAT、Hv（细胞群包被侧隐窝）和 Hc（Himick 和 Peter，1995）。虹鳟的 BBS-免疫反应细胞体出现在 PPp、Hv（包被侧隐窝）和 Hc（Cuadrado 等，1994）。BBS-免疫反应纤维广泛分布在间脑、中脑和后脑，而下丘脑是标志得最稠密的（Cuadrado 等，1994）。编码 GRP 的 mRNA 已在金鱼中被鉴别（Volkoff 等，2000）。RT-PCR 分析表明 GRP mRNA 在金鱼的脑以及卵巢、鳃、皮肤、消化道和脑垂体广泛表达（Volkoff 等，2000）。在脑垂体内，BBS/GRP 纤维主要分布在 PI。

1.5.6 促性腺激素释放激素

脊椎动物的促性腺激素释放激素（gonadotropin-releasing hormone，GnRH）是一个环状结构的十肽，氨基端有焦谷氨酸的修饰，而在羧基端有酰胺功能。至今已经鉴别多达 24 个 GnRH 的分子同种型（molecular isoform），其中 8 个变异体（variant）在硬骨鱼类的脑中发现，它们中的 6 个是硬骨鱼类特有的（Kah 等，2007）。White 及其同事（1998）提出把 GnRH 多样性划分为三个类型，即：GnRH1 型是脑垂体 GnRH 的各种变异体，如 mGnRH、cfGnRH、pjGnRH 和 sbGnRH；GnRH2 型包括存在于所有研究过的种类中脑的 cGnHR-Ⅱ；而 GnRH3 型包括几种鱼类具有的 sGnRH（Lethimonier 等，2004）。明显的是 1 型 GnRH 具有调控脑垂体激素分泌活动的功能，而 2 型和 3 型 GnRH 的作用还未完全确定。有些证据表明，2 型 GnRH 参与生殖行为调控（Volkoff 和 Peter，1999；Canosa 等，2008），而 3 型 GnRH 的作用则较难以捉摸。在一些硬骨鱼类中，1 型和 3 型 GnRH 在腹前脑共存，3 型 GnRH 的纤维亦分布到脑垂体（Kah 等，2007；Gonzále-Martínezz 等，2002；Pandolfi 等，2005），表明 3 型 GnRH 保留某些下丘脑调控脑垂体分泌活动的功能。

GnRH 肽在前脑的分布情况是从嗅球、端神经（terminal nerve）的神经节细胞（TNgc）延伸，沿着腹端脑而到视前区，有些例子是到达前腹下丘脑（Lepretre 等，1993；Montero 等，1994；Gonzále-Martínezz 等，2002；Mohamed 等，2005；Pandolfi 等，2005）。然而，硬骨鱼类 GnRH 的下丘脑调控脑垂体分泌活动的区域主要局限于小细胞视前区，并有少量细胞分布在中基下丘脑（mediobasal hypothalamus）（Lepretre 等，1993；Montero 等，1994；Yamamoto 等，1998；Gonzále-Martínezz 等，2002；Mousa 和 Mousa，2003）。只有两种类型 GnRH 的鱼类（1 型或 3 型在前脑，2 型在中脑），相同的 GnRH 都是沿着前脑的分布区分布。另一方面，具有三种类型 GnRH 的鱼类，1 型和 3 型主要重叠分布在前脑分布区，并且神经分布到脑垂体（Gonzále-Martínezz 等，2002；Pandolfi 等，2005）。通常，1 型（sGnRH）主要分布于嗅觉区和 TNgc，3 型（hgGnRH、sbGnRH、pjGnRH）主要在神经分布到脑垂体的视前区神经元表达（Gonzále-Martínezz 等，2002；Pandolfi 等，2005）。

关于 GnRH 在脑部的分布情况，金鱼是 cGnRH-Ⅱ 在前脑表达的唯一一种鱼类（Yu 等，1988）。免疫组织化学研究表明，sGnRH-免疫反应细胞定位在 TNgc、腹端脑、视前区和下丘脑（Kim 等，1995）。在前脑内，cGnRH-Ⅱ 表达的神经元数量较少，但和 sGnRH 的情况相似。此外，sGnRH-免疫反应和 cGnRH-Ⅱ-免疫反应的纤维都分布在下丘脑和脑垂体（Kim 等，1995）。最近对金鱼在排卵和产卵时 GnRH mRNA 水平和血清 LH 水平的联合分析研究结果表明，其前脑的 cGnRH-Ⅱ 并未参与启动排卵时 LH 的大量分泌活动（Canosa 等，2008）。

对鲑鳟鱼类和尼罗尖吻鲈（*Lates niloticus*）的研究已证明 GnRH 纤维分布到 PI，分别紧密靠近 SL 细胞（Parhar 和 Iwata，1994；Parhar 等，1995）和 SL 与 MSH 细胞（Mousa

和Mousa，2003）。此外，南美丽体鱼（*Cichlasoma dimerus*）的PI存在大量sbGnRH纤维，它们沿着边界分布并且可能和PI的内分泌细胞共同作用（Pandolfi等，2005）。对银汉鱼GnRH和SL的双重免疫染色研究表明，在PI的pjGnRH纤维和SL细胞之间紧密靠近（Vissio等，1999）。进一步的研究检测到GnRH的结合点是在银汉鱼脑垂体散布的SL细胞上（Stefano等，1999）。对绿花鳉、海鲈和印度鲶鱼的研究发现GnRH-免疫反应都深入到PPD GTH区的组织中（Moons等，1989；Batten等，1999）。对罗非鱼的研究亦发现GnRH纤维进入PPD并神经分布到GTH和GH细胞（Melamed等，1995；Parhar，1997）。在对欧洲鳗鱼进行的研究中观察到大量的mGnRH神经分布到脑垂体，但只检测到少量cGnRH-II-免疫反应纤维（Montero等，1994）。所以，在不同的鱼类，下丘脑调节脑垂体的GnRH类型都将神经分布到PPD，紧密靠近GH分泌细胞和GTH分泌细胞（Vissio等，1999；Mousa和Mousa，2003；Pandolfi等，2005）。

只有极少数硬骨鱼类的RPD存在GnRH纤维并影响RPD的激素分泌活动。在对罗非鱼的研究中发现，GnRH刺激离体的PRL释放（Weber等，1997）。在银汉鱼的散布PRL细胞发现GnRH的结合位点（Stefano等，1999），还发现pjGnRH-免疫反应纤维进入RPD（Vissio等，1999）。但是，在南美丽体鱼的RPD没有检测到GnRH纤维（Pandolfi等，2005）。

1.5.7　生长激素释放激素和脑垂体腺苷酸环化酶激活多肽

生长激素释放激素（growth hormone-releasing hormone，GHRH）和脑垂体腺苷酸环化酶激活多肽（pituitary adenyl cyclase activating polypeptide，PACAP）属于胰高血糖素（glucagon）/血管活性肠肽（vasoactive intestinal peptide，VIP）/促胰液素（secretin）的超级家族肽类，具有一系列结构与功能的相似性（综述见Sherwood等，2000；Vaudry等，2000）。显然，PACAP是家族中最保守的成员，亦是一个祖分子，通过外显子串联重复（tandem exon duplication）和基因复制而产生其他的成员（Sherwood等，2000）。在整个脊椎动物中，PACAP氨基酸序列的同一性变动于88%~97%之间，而在人类和非哺乳类脊椎动物之间，GHRH氨基酸序列只有32%~45%的同一性（Sherwood等，2000；Vaudry等，2000）。哺乳类各个的基因可以编码每个肽。此外，一个未知功能的C-肽可以和GHRH编码在一起，而PACAP的相关肽（PRP）只存在于PACAP相同的转录体中（Mayo等，1985；Hosoya等，1992）。可以认为被囊动物（tunicates）、硬骨鱼类、两栖类和鸟类的PACAP基因包括一个上游外显子编码GHRH（综述见Sherwood等，2000；Vaudry等，2000），然而，最近对鱼类的研究证明存在着编码GHRH的另一个不同的基因以及GHRH的特异性受体（Lee等，2007）。这个GHRH基因和哺乳动物的配对物（counterpart）应该是同源的，因为它衍生出来的肽刺激GH释放的能力要比鱼类原先的GHRH肽强（Lee等，2007）。所以，原先的GHRH肽和哺乳动物的PACAP相关肽（PRP）是同源的。这个发现亦可以说明为何这个原先的GHRH肽刺激GH分泌活动的能力要比PACAP弱些（Canosa等，2007；本书第4章）。无论如何，在鱼类中，PACAP对脑垂体具有明显的调控作用（Montero等，1998；Wong等，

1998；Kong 等，2007；Canosa 等，2008；Mitchell 等，2008）。

学者们对几种鱼类的脑研究了 GHRH 的免疫组织化学分布情况（Pan 等，1985；Marivoet 等，1988；Luo 和 Mckeown，1989；Batten 等，1990a；Olivereau 等，1990；Rao 等，1996；Miranda 等，2002）。早期研究对鳕鱼采用抗人 GHRH 中部和 C-端的抗血清，在 PM、LH 和 Hv 鉴别出两个主要的免疫反应细胞群。在 PM，小细胞和大细胞的神经元核周体都呈 GHRH 阳性，而在 Hv，只有大细胞类型才有免疫反应（Pan 等，1985）。海鲈阳性反应的神经元位于 PMpc、PMmc 和 Hv（Marivoet 等，1988）。虹鳟 GHRH-免疫反应核周体主要在 LH 和 Hv，在 PM 的尾区亦出现一小群 GHRH-免疫反应的细胞体（Luo 和 McKeown，1989）。金鱼、鲤鱼、鳗鱼、鲑鳟鱼类和多刺床杜父鱼（*Myoxocephalus octodecimspinosus*）等的 PMpc 和 PMmc 有 GHRH-免疫反应核周体，在 Hv 亦偶尔有（Olicereau 等，1990）。绿花鳉 GHRH-免疫反应细胞体只在 PMgc 和 PMpc 发现，而且这些细胞亦表达 AVT（Batten 等，1990a）。在金鱼视前区、Hv、松果核（pineal nucleus）和中脑盖（midbrain tegmentum）的外侧丘系（lateral lemniscus）都出现 GHRH-免疫反应核周体（Rao 等，1996）。金鱼的腹端脑、视前区、脑垂体、中盖脑和下丘脑下叶都有 GHRH-免疫反应纤维。在对银汉鱼的研究中还观察了 GHRH-免疫反应的个体发生过程（Miranda 等，2002）。

学者对欧洲鳗鱼（*Anguilla anguilla*）（Montero 等，1998）和日本䲢（*Uranoscopus japonicus*；Matsuda 等，2005a）研究了 PACAP 在脑部的分布情况。PACAP 的分布和 GHRH 相似，主要在视前区和下丘脑，亦出现在后脑的一些神经元中（Matsuda 等，2005a）。欧洲鳗鱼的 PACAP-免疫反应核周体出现在视前区、PPa、PPp 和 PM。此外，成群的免疫反应核周体亦分布到丘脑，在腹丘脑核和背丘脑核内（Montero 等，1998）。没有其他脑区出现 PACAP 阳性反应。免疫反应纤维在欧洲鳗鱼的脑内广泛分布（Montero 等，1998）。日本䲢的 PACAP-免疫反应细胞体分布于 PMpc 和 PMgc，以及 Hv、小脑体（corpus cerebelli）和脊髓的腹角（Matsuda 等，2005a）。

GHRH 和 PACAP 纤维分布在腺脑垂体所有三个 PD 再分区的混合内分泌细胞当中（见表 1.3）。海鲈、金鱼、鲤鱼、杜父鱼和银汉鱼的 GHRH-免疫反应纤维沿着血管进入相嵌交错的 PN 而伸进 PI 中（Marivoet 等，1988；Moons 等，1989；Olivereau 等，1990；Miranda 等，2002）。在脑垂体内，GHRH 纤维和 MSH 与 SL（或 PAS 阳性）细胞接触（Marivoet 等，1988；Moons 等，1989）。相反，在鳕鱼、鳗鱼和鳟鱼的脑内都没有发现 GHRH-免疫反应纤维（Pan 等，1985；Olivereau 等，1990）。GHRH-免疫反应纤维来自下丘脑，依不同鱼类以变化不定的数量伸进 PPD（Pan 等，1985；Olivereau 等，1990）。这些神经纤维穿行在 GH 细胞当中，末梢紧密靠近 GH 细胞，明显很少接近 GTH 细胞（Marivoet 等，1988；Moons 等，1988，1989）。此外，对海鲈的研究曾报道 GHRH 纤维和 TSH 细胞之间有紧密联系。最近，个体发生的研究证明孵化后一周的银汉鱼，视前区和脑垂体的 PPD 出现 GHRH-免疫组织反应的核周体和纤维（Miranda 等，2002），表明鱼类早期发育过程中 GHRH 可能参与调节 GH 的释放。海鲈、杜父鱼和鲻鱼（*Mugil cephalus*）的 GHRH-免疫反应神经纤维来源于下丘脑而延伸到 RPD

（Marivoet 等，1988；Moons 等，1989；Olivereau 等，1990），而鳗鱼、鲤鱼、金鱼、鲑鳟鱼类和鳕鱼的 GHRH-免疫反应神经纤维并不进入 RPD（Pan 等，1985；Olivereau 等，1990）。双重染色研究表明，在 GHRH-免疫反应神经纤维和 ACTH 细胞之间有紧密的联系（Marivoet 等，1988；Moons 等，1989）。

可以观察到 PACAP-免疫反应纤维穿过 PN 和日本鳉（Matsuda 等，2005a、b）与欧洲鳗鱼（Montero 等，1998）PI 内的 SL-、POMC(MSH)- 以及内啡肽免疫反应内分泌细胞紧密靠近。此外，在欧洲鳗鱼（Montero 等，1998）、金鱼（Wong 等，1998）和日本鳉（Matsuda 等，2005b）的 PPD 都曾发现 PACAP-免疫反应纤维。PACAP-免疫反应纤维亦在金鱼（Wong 等，1998）以及日本鳉幼鱼而不是成鱼（Matsuda 等，2005b）的生长激素细胞附近观察到。相反，没有看到 PACAP-免疫反应纤维分布到 GTH 细胞。在 RPD 内，PACAP-免疫反应纤维中止于紧密靠近的 PRL 细胞和 ACTH 细胞（Montero 等，1998；Matsuda 等，2005a、b）。

1.5.8　黑皮质素系统

中央黑皮质素系统（melanocortin system）包括 POMC-表达神经元、表达内源性黑皮质素拮抗物的神经元、野灰蛋白相关蛋白（agouti-related protein，AgRP），以及这些神经元表达的下游靶标和中央黑皮质素受体。POMC 基因编码一个复杂的前体，它展现三个主要区，每区都含有一个 MSH 肽（见第 1.1 节）。α-MSH 处于成为 ACTH N-端序列的第二区内（Cerdá-Reverter 等，2003a）。POMC 主要在脑垂体产生，其后转录的加工（post-transcriptional processing）则是组织特异性（tissue-specific）的。在 RPD 的促皮质激素细胞内由前转变酶（proconvertase 1，PC1）引起的蛋白酶剪切（proteolytic cleavage）产生 ACTH 和 β-促酯解素（β-lipotropin），而在 PI 的黑色素细胞内 PC1 和 PC2 裂解产生 α-MSH 和 β-内啡肽。POMC 亦在 CNS 的一些神经元群体中表达，主要是加工 α-MSH 和 β-内啡肽（综述见 Castro 和 Morrison，1997）。

黑皮质素 ACTH 和 MSH 参与一系列广泛的生理功能，并且是通过和 G-蛋白偶联受体家族的结合而发挥其作用。在四足类克隆了黑皮质素受体的 5 个亚型（MC1R—MC5R）（综述见 Schiöth 等，2005）。斑马鱼的基因组包含 6 个 MCR 亚型，这是因为 MC5R 的复制；而红鳍东方鲀（*Fugu rubripes*）的基因组只有 MC1R、MC2R、MC4R 和 MC5R 4 个亚型（Logan 等，2003）。

采用原位杂交技术进行的表达研究曾报道 POMC 主要在金鱼和鲑鱼的外侧结节核或 Hv 中产生（Salbert 等，1992；Cerdá-Reverter 等，2003a）。同样，采用抗 α-MSH 或 ACTH（1-39）的抗体进行的免疫染色研究报道黑皮质素-免疫反应主要出现在几种鱼类的结节下丘脑（Kishida 等，1998；Bird 等，1989；Olivereau 和 Olivereau，1990；Amano 等，2005；Forlano 和 Cone，2007；见表 1.2）。α-MSH 免疫反应核周体通常位于腹下丘脑后部，紧靠着下丘脑底面。鲤鱼的 MSH-免疫反应亦出现在 LH 和 MCH 神经元（Kishida 等，1988）。只有一项研究报道在鱼脑中存在 ACTH-免疫反应，即采用直接抗 α-MSH-促肾上腺皮质激素类中间肽（α-MSH-corticotropin-like intermediate peptide,

CLIP）的抗血清转变鲤鱼 ACTH（^{10}甘 – ^{23}酪），并不和 α-MSH 或 CLIP 产生交叉反应。ACTH-免疫反应神经元位于 PMgc。含有 ACTH 的细胞体对 AVT 是阴性的，脑垂体没有发现 ACTH 纤维。ACTH 在鱼脑内的功能还不清楚，但应激反应时，用 PCR 测定 POMC 在 PM 的表达是上调的（Metz 等，2004）。然而，对鲤鱼采用抗合成的 α-MSH，合成的 ACTH（11-24）或鲑鱼 N-端 POMC 肽的抗血清进行免疫染色后在 PM 并没有发现黑皮质素的免疫反应（Kishida 等，1988；Bird 等，1989）。α-MSH 纤维广泛分布在硬骨鱼类的脑内，但黑皮质素轴突末梢分布到脑垂体的并不多（见表 1.3）。对金鱼的研究表明，MSH 纤维并不深插入神经脑垂体。条斑星鲽没有 MSH 凸出物从下丘脑神经元进入脑垂体（Amano 等，2005），但在虹鳟的相关研究中可观察到一支稠密的神经束从腹方经过垂体柄中止于神经脑垂体。有些神经纤维亦混合在远侧部内，表明 α-MSH 具有神经内分泌作用（Vallarino 等，1989）。此外，鲤鱼在 LH 内的 MSH 细胞似乎伸到脑垂体（Kishida 等，1988）。

非典型的是，黑皮质素信号并不是唯一的为内源性同功物结合后所调节，因为自然存在的拮抗物野灰蛋白（agouti）和野灰蛋白相关蛋白（agouti-related protein，AgRp）通过与 MCRs 结合而和黑皮质素进行竞争。哺乳类的 AgRp 主要在下丘脑弓状核为 NPY 表达的同一神经元内产生，并能够和 MC3R 和 MC4R 结合而抑制黑皮质素的信号（Cerdá-Reverter 和 Peter，2003）。对斑马鱼进行的受体结合研究证明，AgRp 在 MC3R、MC4R 和 MC5R 起着竞争拮抗物的作用，这三个受体都在脑内表达（Song and Cone，2007）。至今只有一项研究精确地在硬骨鱼类脑内对 AgRp mRNA 进行定位（Cerdá-Reverter 和 Peter，2003），其他的研究报道在鱼脑内 AgRP 的免疫反应情况（Forlano 和 Cone，2007）。海鲈和斑马鱼的 AgRP 在腹下丘脑的后区产生。然而，没有关于 AgRP 纤维在脑垂体分布的报道。斑马鱼 AgRP 和 α-MSH 的凸出物明显地和金鱼表达 MC4R 的核（Cerdá-Reverter 等，2003b）以及 MC5R mRNA（Cerdá-Reverter，2003c）相一致。

1.5.9 神经肽 Y 家族肽类

神经肽酪氨酸（neuropeptide tyrosine，NPY）家族的肽类由 36 个氨基酸的肽组成，其羧基端（C-端）酰胺化（amidation）（Cerdá-Reverter 和 Larhammar，2000）。这个家族包括三个不同的肽：NPY、酪氨酸-酪氨酸肽（PYY）和胰多肽（pancreatic polypeptide，PP）。四足类脊椎动物产生所有这三种肽，而非四足类脊椎动物只有两种类型，即 NPY 和 PYY。硬骨鱼类在进化过程中曾发生第三轮基因复制，每个类型肽（即 NPY 和 PYY）合成两种不同的变体（Sundström 等，2008）。

对鱼类的研究证明，NPY 参与腺脑垂体细胞内分泌活动的调控（见本书第 4、5、6 章），包括 GH 和 LH（Kah 等，1989；Peng 等，1993；Cerdá-Reverter 等，1999）。

神经解剖学研究进一步确定在一些硬骨鱼类的神经脑垂体存在 NPY 的终端（NPY terminal）（Pontet 等，1989；Batten 等，1990a；Danger 等，1991；Cepriano 和 Schreibman，1993；Zandbergen 等，1994；Chiba 等，1996；Subhedar 等，1996；Rodríguez-Gómez 等，2001；Gaikwad 等，2004；Chiba，2005）。NPY 神经纤维到达鱼类腺脑垂

体，但依不同种类而分布在不同部位（见表1.3）。对鲶鱼（Gaikwad等，2004）、剑尾鱼（Cepriano和Schreibman，1993）和斑马鱼（Mathieu等，2002）的研究证明，NPY纤维分布于腺脑垂体的所有三个部分。电子显微镜研究揭示NPY-免疫反应颗粒和鲶鱼腺脑垂体三个部分分泌细胞的细胞质囊（cytoplasmic vesicle）相结合（Gaikwad等，2004）。对金鱼的研究报道了同样的研究结果，NPY-免疫反应神经分泌小囊和腺脑垂体大多数类型的分泌细胞直接接触，包括LH-FSH、GH和黑色素细胞（Pontet等，1989）。免疫反应研究亦报道，在塞内加尔鳎的PPD和PI存在NPY的终端（Rodríguez-Gómez等，2001）。对罗非鱼的研究表明，PI有大量NPY-免疫反应神经分布，而在远侧部（Sakharkar等，2005）则很少。胡子鲶（*Clarias gariepinus*）（Zandbergen等，1994）和鲤鱼（Marchetti等，2000）的NPY-免疫反应只局限于PI。

对香鱼（*Plecoglossus altivelis*）的研究证明神经脑垂体的NPY神经分布呈现季节性变化，表明它们参与生殖轴的调节。双重免疫染色研究亦证明NPY的终端在视前区和GnRH细胞紧密靠近。然而，尽管它们在神经脑垂体中混合在一起，但NPY和GnRH的终端是明显分开的（Chiba等，1996）。同样的情况亦出现在雀鳝（*Lepisosteus oculatus*）（Chiba等，2005）和鲶鱼（Gaikwad等，2005）中。对鲶鱼的超薄切片观察证明携带NPY-免疫反应金颗粒的神经内分泌轴突在PPD和含有GnRH的细胞紧密联系，表明NPY可能参与GnRH分泌活动的调控。事实上，在中枢施给NPY后能够增加嗅球、中嗅束、端脑/视前区和脑垂体的GnRH含量（Gaikwad等，2005）。

免疫化学和原位杂交研究已将NPY神经元细胞体定位于硬骨鱼类脑的几个下丘脑-脑垂体区（hypophysiotropic areas）。在硬骨鱼类脑最吻端的NPY-免疫反应定位在嗅球。海鲈的内细胞层（internal cell layer，ICL）（Cerdá-Reverter等，2000）有明显的NPY mRNA表达细胞群体。关于NPY是否产生在嗅觉视网膜核（nucleus olfactoretinalis）或TNgc有一些争论。免疫染色研究报道在鲶鱼（Gaikwad等，2004）、香鱼（Chiba等，1996）、雀鳝（Chiba等，2005）、罗非鱼（Sakharkar等，2005）、斑马鱼（Mathieu等，2002）、剑尾鱼（Cepriano和Schreibman，1993）和青鳉（Subhedar等，1996）的TNgc内出现NPY-免疫反应，但是，在虹鳟（Danger等，1991）、鲤鱼（Marchetti等，2000）、塞内加尔鳎（Rodríguez-Gómez等，2001）、金鱼（Pontet等，1989）和鲶鱼（Zandbergen等，1994）的TNgc则检测不到免疫反应。对海鲈（Cerdá-Reverter等，2000）、金鱼（Peng等，1994；Vecino等，1994）和鲑鱼（Silverstein等，1998）进行的表达研究从未报道NPY基因在TNgc表达。腹端脑被认为是鱼类（硬骨鱼类和非硬骨鱼类）NPY能系统的主要组成部分（Cerdá-Reverter和Larhammar，2000）。NPY基因表达或免疫反应曾报道出现在腹端脑的Vd、Vv、Vl、Vs、Vc（或NE）核。Vc可能是硬骨鱼脑NPY产生水平最高的部位，亦曾报道其是所有研究过的鱼类的NPY能核（NPYergic nucleus）。在Vs、Vv和Vc的神经元可以参与脑垂体分泌活动的调控，因为它们伸入到达脑垂体（见表1.1）。对雄性罗非鱼和金鱼的研究表明，Vc的NPY神经元对于性类固醇激素信号的加工处理可能起着重要中心的作用（Peng等，1994；Sakharkar等，2005）。事实上，将雄鱼阉割后显著增强NPY-免疫反应，而睾酮能逆转这个反应（Sakharkar 等，2005）。

硬骨鱼类的视前区存在着 NPY-免疫反应或 mRNA 转录体，但细胞数量要比在端脑看到的少得多。然而，对鲶鱼（Zandbergen 等，1994）和鲤鱼（Marchetti 等，2000）的研究未能证明 NPY-免疫反应存在于整个视前区，而对金鱼（Pontet 等，1989）和塞内加尔鲷（Rodríguez-Gómez 等，2001）的研究发现 NPY-免疫反应局限于 PPp 的最尾区。NPY 细胞体在视前区内的准确定位在不同种鱼类当中有所不同。对海鲈（Cerdá-Reverter 等，2000）、香鱼（Chiba 等，1996）、青鳉（Subhedar 等，1996）、雄罗非鱼（Sakharkar 等，2005）、斑马鱼（Mathieu 等，2002）和鲶鱼（Gaikwad 等，2004）等的研究曾报道最吻端 NPY 细胞群是在 PPa 内。对于斑马鱼和鲶鱼，PPa 是视前区内唯一产生 NPY 的核。NPY-免疫反应或基因表达亦曾报道出现在金鱼（Peng 等，1994）、青鳉（Subhedar 等，1996）、罗非鱼（Sakharkar 等，2005）和鲑鳟鱼类（Danger 等，1991；Silverstein 等，1998）的 PM 内。用性类固醇激素处理金鱼能增强 NPY 在 PM 的表达（Peng 等，1994），但对雄罗非鱼没有这种作用（Sakharkar 等，2005），在香鱼的视前区亦没有 NPY-免疫反应的季节性变化（Chiba 等，1996）。最后，在一些鱼类的研究中亦曾报道 NPY-免疫反应和/或表达出现在结节下丘脑，包括香鱼（Chiba 等，1996）、海鲈（Cerdá-Reverter 等，2000）、塞内加尔鲷（Rodríguez-Gómez 等，2001）、斑马鱼（Mathieu 等，2002）和雄罗非鱼（Sakharkar 等，2005）。海鲈的 NPY-表达细胞位于 Hv、NAT 和 Hc。雄罗非鱼（Sakharkar 等，2005）的 NPY-免疫反应水平并不受阉割以及随后用睾酮处理的影响，但在香鱼中出现季节性变化。NPY-免疫反应活动的季节性伴随着 NPY 能纤维在神经脑垂体的分布，表明 NPY 参与脑垂体分泌活动的调控。

1.5.10 食欲肽

食欲肽（Orexin）亦称为下丘泌素（hypocretin），是由肠降血糖素（incretin）家族一个共同前体产生的兴奋性神经调节肽（de Lecea 等，1998）。下丘泌素 1 和 2 通过和两个 G-蛋白偶联受体结合而发挥它们的生理功能；两个 G-蛋白偶联受体对下丘泌素显示不同的亲和力以及在 CNS 的不同分布情况（Sutcliffe 和 de Lecea，2000）。

鱼类的食欲肽最初是通过检索公共 DNA 序列数据库而在河鲀中被鉴别出来（Alvarez 和 Sutcliffe，2002）。和哺乳类不同，在斑马鱼、河鲀和三棘刺鱼中只鉴别出一个受体（Yokogawa 等，2007）。对几种鱼类的研究证明食欲肽能刺激食物摄取（Nakamachi 等，2006），而禁食能提高下丘脑食欲肽 mRNA 水平（Novak 等，2005；Nakamachi 等，2006；见本书第 9 章）。此外，食欲肽能增强金鱼的活动能力（Nakamachi 等，2006）。食欲肽过度表达能诱导斑马鱼出现一个类似失眠患者的表型（insomnia-like phenotype）（Prober 等，2006）。如同哺乳类，斑马鱼缺少一个有功能的食欲肽受体，表现为睡眠/清醒行为的稳定性受到破坏，但斑马鱼表型的食欲肽受体突变型只有睡眠的染色体断裂（fragmentation），因而减少了其在黑暗中的睡眠（Yokogawa 等，2007；见本书第 9 章）。对金鱼的研究证明含有食欲肽的神经元出现在和 MCH 神经元分布相同的脑区（Huesa 等，2005；Nakamachi 等，2006），即 LH、ZL 和 PTN。在 ZL 的神经元相当于和 PVO 联系而表达 MCH 的神经元群体。这亦是在对青鳉的研究观察到的仅有的食欲肽神经元群体（Amiya 等，2007）。在斑马鱼的视前区和结节下丘脑已鉴别两个突出的食欲肽-免疫反应神经元群体（见表 1.2）。在视前区内，食欲肽神经元分布在 PMmc、PMgc、PPp 和 NSC。第二个细胞群出现在侧隐窝的背方，和在金鱼与青鳉的研究中描述的细胞群体一样。只有结节下丘脑的细胞群才可以被认为是合成食欲肽的神经元，因为它们含有食欲肽-mRNA 和已经加工的肽。相反，在视前区的神经元群体缺乏前食欲肽原（preprohypocretin）mRNA（Kaslin 等，2004）。大量的食欲肽纤维从视前区发出，经过视前-脑垂体束（preoptico-hypophysial tract），向腹方转向视交叉后面，穿过下丘脑腹表面而到达脑垂体。在青鳉的脑垂体中没有发现食欲肽-免疫反应纤维（Amiya 等，2007），但在花鲈（*Lateolabrax japonicus*）的神经脑垂体观察到纤维染色图像（Suzuki 等，2007）。曾报道食欲肽-免疫反应细胞存在于青鳉（Amiya 等，2007）和花鲈（Suzuki 等，2007）的脑垂体内。花鲈的食欲肽-免疫反应定位于 GH 细胞，食欲肽和 GH 共存于分泌颗粒当中。

1.5.11　RF-酰胺肽

RF-酰胺肽（RF-amide peptides）的第一个成员由 Price 和 Greenberg（1997）在软体动物中发现。此后，一系列免疫组织化学的研究结果表明，类似于软体动物 FMRF-酰胺肽类存在于脊椎动物的脑内。例如，FMRF-酰胺-免疫反应出现在金鱼的 TNgc 和 GnRH 共表达（Stell 等，1984）。同样，剑尾鱼（*Xyphophorus helleii*）的 FMRF-酰胺-免疫反应位于 TNgc 的细胞体内（Magliulo-Cepriano 等，1993），有时和 GnRH-免疫反应共定位（见表 1.2）。这些细胞的突出体可以追踪到结节下丘脑的吻端，并且经过视前区。然而，含有 FMRF-酰胺-免疫反应的纤维要比 GnRH 纤维具有更广阔的分布范围，还出现在背前脑、OT 和 Hc（Magliulo-Cepriano 等，1993）。特别值得注意的是，到达结节下丘脑的神经纤维束是在达到性成熟时才完成它们的发育。此外，鲶鱼和剑尾鱼源自 TNgc 的 FMRF-酰胺-免疫反应纤维分布在脑垂体的三个分部（Rama Krishna 等，1992；Magliulo-Cepriano 等，1993）。但是，目前对和这些免疫反应相关的内源肽还不是很了解（见表 1.3）。具有 RF-酰胺模体（motif）的一些神经肽，例如，催乳激素释放肽（prolactin-releasing peptide，PrRP）（Moriyama 等，2002；Seale 等，2002；Sakamoto 等，2003b；Montefusco-Siegmund 等，2006；Amano 等，2007）、RF-酰胺相关肽（RF-amide-related peptide，RFRP）（Hinuma 等，2000；Fukusumi 等，2001；Ukena 等，2002）、LPXRF-酰胺（Tsutsui 和 Ukena，2006）和 Kisspeptin（van Aerle 等，2008）等都已确认存在于鱼类。

PrRP 从牛下丘脑提取物中被发现，是前脑垂体孤独 G-蛋白偶联受体（hGR3）的配体（Hinuma 等，1998）。这个肽能促进 PRL 释放（Hinuma 等，1998；Fujimoto 等，2006）。从日本鲫鱼（*Carassius cuvieri*）（Fujimoto 等，1998）、金鱼（Kelly 和 Peter，2006）、罗非鱼（Seale 等，2002）、大麻哈鱼（Moriyama 等，2002）和大西洋鲑（Montefusco-Siegmund 等，2006）的脑提取物中已经分离到 PrRP 的同源物。在硬骨鱼类中，PrPP 能促进离体和在体的 PRL 分泌活动（Moriyama 等，2002；Seale 等，2002；

Sakamoto 等, 2003a、b), 增强 PRL 基因表达 (Sakamoto 等, 2003a), 并且能调节金鱼的渗透压平衡和食物摄取 (Fujimoto 等, 2006; Kelly 和 Peter, 2006)。对虹鳟进行的免疫细胞化学染色研究表明, PrPP-免疫反应细胞体位于下丘脑的后部 (Moriyama 等, 2002)。此外, 在成年的网纹花鳉的 Hv 检测到 PrRP-免疫反应的细胞体 (Amano 等, 2007)。但大西洋鲑合成 PrRP 肽的免疫反应活动主要发生在小脑, 在 Hc 有少量微弱反应的细胞体, 而在脑垂体柄几乎看不到有阳性反应的纤维 (Montefusco-Siegmund 等, 2006)。在七鳃鳗的脑内鉴别了两种和硬骨鱼类 PrRP 同源的 RF-酰胺肽 (A 和 B) (Moriyama 等, 2007)。在这项研究中, 采用抗-鲑鱼 PrRP 的血清, 观察到 PrRP-免疫反应细胞体分布于下丘脑室周弓状核 (periventricular arcuate nucleus) 的腹部 (很可能相当于 Hv)。在网纹花鳉的脑垂体内可以看到 PRL 细胞附近有少量 PrRP-免疫反应的纤维 (Amano 等, 2007)。

Kisspeptin, 对孤独 G-蛋白偶联受体 54 (GPCR54) 展现同等物作用的内源肽, 是从人体胎盘中分离出来的 (Ohtaki 等, 2001)。这些 54-、14-、13-和 10-氨基酸的肽类具有共同的 RF-酰胺 C-端, 都是从黑素瘤的癌转移抑制基因 (metastasis suppressor gene) KISS-1 的产物衍生而来。最近的研究证明, Kisspeptin 是哺乳类启动青春期的重要因子 (Tena-Sempere, 2006)。GPCR54 的神经元数量以及表达 GPCR54 的水平都随着性腺成熟而增长 (Parhar 等, 2004)。一个 Kisspeptin 的负体 (counterpart) 最近在鱼类中通过计算机基因组分析而鉴别出来 (van Aerle 等, 2008), 但是还没有对 Kisspeptin 特异性的脑和/或脑垂体作图。在剑尾鱼研究中的 FMRF-免疫反应物质 (Magliulo-Cipriano 等, 1993) 可能是 kisspeptin, 因为如上所述, 它是在成年和性成熟时达到充分的发育, 将神经分布到基部下丘脑 (basal hypothalamus) 的核内。

1.5.12 生长抑素

生长抑素 (促生长素抑制素, somatostatin, SS) 是硬骨鱼类基础的和受刺激的 GH 分泌活动强有力的抑制剂 (Canosa 等, 2007)。这个肽属于多功能的肽类家族, 在硬骨鱼类包括由不同基因编码的三个不同的 SS 前体 (综述见 Nelson 和 Sheridan, 2005; Canosa 等, 2007)。PSS-I 是最保守的类型, 在 C-端编码 SS-14, 其氨基酸序列在所有研究过的脊椎动物中都是相同的。第二个 cDNA 称为 PSS-II, 只从硬骨鱼类中分离出来, 编码长度在 22 个氨基酸和 28 个氨基酸之间不同长度的 SS。这些肽的特点是, 它们的 C-端为 [酪^7, 甘^{10}] SS-14。从几种鱼类和其他脊椎动物中分离出 PSS-III (Nelson 和 Sheridan, 2005; Canosa 等, 2007)。PSS-III 编码肽的特点是 14-氨基酸 C-端的第二位是脯氨酸。目前对这些肽的生理功能尚未充分了解, 而大部分有关脑部定位和脑垂体神经分布的信息都是以 SS-14 免疫染色的研究结果为基础。尽管抗 SS-14 的抗体对其他 SS 肽类可以产生交叉反应 (Canosa 等, 2004), 但所有的数据都认为是以 SS-14 免疫染色进行研究得到的。

早期对金鱼的研究表明, SS-免疫反应存在于视前区的 PPa、PM、PPp、结节下丘脑的 Hv 和腹丘脑 (Kah 等, 1982) (见表 1.2)。Olivereau 等 (1984b) 研究 SS 在几种硬骨鱼类脑内的分布，发现 SS-免疫反应存在于 Vc、PPa、PPp、Hv 和 Hc。罗非鱼的 SS-免疫反应神经元定位于端脑 D1 和 Vc 的前部、视前区和结节下丘脑（Grau 等，1985）。对绿花鳉 SS 核周体的脑定位研究表明，SS 细胞分布在端脑的 Vl 和 Vc、PPa、PM 和视前区 PPp 的吻部、Hv、NAT 以及结节下丘脑的 Hc（Batten 等，1985）。欧洲鳉鱼和欧洲鳗鱼 SS-免疫反应的细胞体亦有类似的分布情况（Vigh-Teichmann 等，1983）。许多 SS-免疫反应细胞体都是 CSF-接触的神经元，特别是分布在前、侧和后隐窝如 Hv 和 Hc 的。电鳗、金鱼和海鲈 SS-免疫反应在脑的分布情况都和上述的相似（Sas 和 Maler，1991；Pickavance 等，1992；Power 等，1996）。在海鲈的下丘脑，SS-免疫反应分布于 PM、Hv 和 Hc，还有一个膨大的纤维网穿过视前-脑垂体束（preoptico-hypophysial tract）突向脑垂体（Power 等，1996）。

金鱼是唯一曾经被描述具有三个 SS 基因（Lin 等，1999）的硬骨鱼类。原位杂交研究表明，这三个 mRNA 在脑的分布有重叠，但在同一个细胞中则未必出现共表达情况（Canosa 等，2004）。SS-阳性细胞体的总体分布情况在不同的鱼类以及其他脊椎动物中都和上述的一致。PSS-I 和 PSS-III 在腹端脑、结节下丘脑、腹和背丘脑、腹后下丘脑的大部分以及中脑和后脑都有重叠。另一方面，PSS-II 只限于分布在下丘脑和腹丘脑的一些核：Hv、NCLI、VM 和 NTP（Canosa 等，2004）。PSS-I 和 PSS-III 表达型式的最明显差别出现在视前区，PSS-I 在 PM 而不在 PPp 的前部表达，而 PSS-III 的表达部位正好和 PSS-I 相反（Canosa 等，2004）。

PD 的三个区都有 SS-免疫反应的神经纤维分布（见表 1.3）。杜父鱼、绿花鳉和鲶鱼 PN 尾部的 SS-免疫纤维到达 PI 并终止于 MSH 和 SL 细胞之间（Olivereau 等，1984a；Batten 等，1999）。绿花鳉的 SS 和 MSH 双重染色研究表明 SS 纤维分布在 PI 的 MSH 细胞群中（Batten 等，1995）。SS-免疫纤维还被发现插入几种鱼类的 PPD 内（Kah 等，1982；Olivereau 等，1984a、b；Batten 等，1985；Grau 等，1985；Moons 等，1989；Sas 和 Maler，1991；Power 等，1996）。SS-免疫反应纤维还和 GH 细胞接触（Moons 等，1989；Power 等，1996），或者和 GH 及 GTH 细胞群（GH 细胞分布于周围）的基底膜接触（Olivereau 等，1984a、b；Batten 等，1999）。罗非鱼的 SS-免疫反应纤维可以在 RPD 清晰地看到，它们分布到 PRL 细胞内（Grau 等，1985）。罗非鱼 SS 的这种分布型式似乎是特异性的，因为青鳉或长颌姬鰕虎鱼（*Gillichthys mirabilis*）的 SS-免疫反应都不存在于 RPD（Grau 等，1985）。此外，海鲈的 SS 纤维很稀少甚至没有，并不分布到 PRL 或 ACTH 细胞中（Power 等，1996），亦从未在鲑鳟鱼类中看到 SS 纤维插入 RPD（Olivereau 等，1984a）。然而，对海鲈进行双重染色表明 SS 纤维和 ACTH 细胞紧密接触（Moons 等，1989）。此外，海鲈和印度鲶鱼的 SS-免疫反应末梢和 RPD 的基底膜形成突触联系（Batten 等，1999）。

1.5.13 促甲状腺素释放激素

促甲状腺素释放激素（thyrotropin-releasing hormone，TRH）是从猪和羊的下丘脑分离出来并进行化学鉴定的第一个下丘脑释放因子（Schally 等，1969；Guillemin，

1970）。对四足类（不包括两栖类）的研究已经充分证实 TRH 的作用是调控促甲状腺素（TSH）释放（Fliers 等，2006）。硬骨鱼类的 TRH 刺激几种脑垂体激素，例如 PRL（Batten 和 Wigham，1984；Wigham 和 Batten，1984；Barry 和 Grau，1986）、GH（Canosa 等，2007）和 MSH（Tran 等，1989；Lamers 等，1991），但关于它对脑垂体-甲状腺轴的作用却有不少争论（见本书第 6 章），因为已发表了一些互相矛盾的研究成果，表明 TRH 既有刺激性的作用（Tsuneki 和 Fernholm，1975；Eales 和 Himick，1988），又可能没有作用（Peter 和 McKeown，1975），甚至具有抑制性作用（Bromage，1975）。只有一项研究报告表明 TRH 神经分布到鲶鱼的 PPD（Batten 等，1999）（见表 1.3）。小的 TRH-免疫反应细胞体只在海鲈的 NAT 和包被侧隐窝（lateral recess）的 Hv 区（Batten 等，1990b）以及绿花鲷的 PPa（Batten 等，1990a）中被发现（见表 1.2）。在鲤鱼的相关研究中观察到类似的结果（Hamano 等，1990）。另一方面，鲑鳟鱼类的 TRH 细胞的分布似乎比较广些。大麻哈鱼的 TRH-阳性细胞体分布在嗅球的内细胞层、腹端脑（Vs）的上连合核和视前区。在视前区，TRH-免疫反应细胞体出现于 PMmc（Matz 和 Takahashi，1994）。用原位杂交技术对红大麻哈鱼进行研究观察到类似的结果（Ando 等，1998）。最近，在鳟鱼和斑马鱼的 OB 和腹端脑的几个区、视前区和上交叉区（suprachiasmatic region）、腹中下丘脑和一些丘脑核都观察到 TRH-免疫反应的核周体（Díaz 等，2001，2002）（见表 1.2）。

1.6　下丘脑的神经递质

1.6.1　氨基酸神经递质：谷氨酸盐和 γ-氨基丁酸

谷氨酸盐（glutamate）是脊椎动物 CNS 主要的兴奋性神经递质（Nakanishi，1992；Trudeau 等，2000b）。谷氨酸盐亦是重要的下丘脑-脑垂体（hypophysiotropic）调节剂，参与 LH、GH、PRL（Trudeau 等，1996；Holloway 和 Leatherland，1997；Trudeau 等，2000b；Bellinger 等，2006）以及 MSH 或 SL 分泌活动的调控，因为谷氨酸盐-免疫反应纤维出现在 NIL（Trudeau 等，1996）。再者，免疫组织化学研究揭示谷氨酸盐纤维分布在金鱼脑垂体（Trudeau 等，1996）。但对于硬骨鱼类，没有研究报道表明这些谷氨酸神经分布是来自中枢神经系统（见表 1.3）。

γ-氨基丁酸（gamma-amino butyric acid，GABA）是 CNS 主要的抑制性神经递质，尽管亦有关于 GABA 兴奋性作用的报道（Wagner 等，1997）。此外，GABA 对脑垂体激素的分泌活动起着重要的调控作用（Khan 和 Thomas，1999；Mañanos 等，1999；Trudeau 等，2000a、b；Martyniuk 等，2007b）。GABA 主要通过谷氨酸脱羧酶（glutamic acid decarboxylase，GAD）分解代谢的简单酶促作用从谷氨酸盐合成而来。因此，GAD 酶可以用作 GABA 能细胞体和纤维的标志物（Kah 等，1987；Martinoli 等，1990；Medina 等，1994；Anglade 等，1999）。在金鱼的嗅球和端脑可检测到 GAD-免疫反应细胞体。在间脑，含有 GABA 的细胞体出现在下丘脑，特别是视前区和结节区。此外，在腹后下丘脑、背和腹中丘脑以及顶盖前区（pretectal area）出现大量 GABA-阳性核周体（Martinoli 等，1990）。鳗鱼（Medina 等，1994）和虹鳟（Anglade 等，1999）的 GAD-免疫反应细胞体在前脑出现类似的分布情况。采用非放射的原位杂交分析金鱼脑部 GABA 能细胞体的分布，结果表明，GAD65、GAD67 的转录体和 GABA 代谢酶 GABA-T一起，主要分布于端脑的中区和腹区、视前区的 PM 和 Hv 以及 Hc 的侧方。一些研究报道表明 GABA 对 LH 的分泌活动起调节作用（Kah 等，1992；Sloley 等，1992；Trudeau 等，1993；Anglade 等，1998；Trudeau 等，2000b）。此外，GABA 能神经纤维分布到金鱼脑垂体的各个部分（Kah 等，1987）。

1.6.2 多巴胺

多巴胺（dopamine，DA）是脊椎动物 CNS 中一个重要的神经递质，具有重要的下丘脑-脑垂体（hypophysiotropic）调控功能。采用抗体或者核糖核酸探针（riboprobes）检测参与 DA 合成的酪氨酸羧化酶（tyrosine hydroxylase），已经广泛研究多巴胺能系统在鱼类神经系统中所起的作用（Ma，1997；Smeets 和 Gonzále，2000；Rink 和 Wullimann，2001）。此外，还采用 DA 本身的抗血清进行各项研究（Meek 等，1989；Roberts 等，1989；Pierre 等，1997）。

鱼类最密集的多巴胺神经元位于后结节（posterior tuberculum）和附近的下丘脑区（Ma，2003；Ma 和 Lopez，2003），特别是大量含有多巴胺的神经元出现在靠近脑室及其隐窝的核中（见表1.4）。此外，多巴胺能神经元分布在嗅球、视前区的腹区和结节下丘脑（Meek 等，1989；Roberts 等，1989；Pierre 等，1997）。早期研究已证明，多巴胺能神经元分布在金鱼的视前区和后下丘脑，并伸向脑垂体将神经纤维分布到 PD 的所有三个区（Kah 等，1984；Fryer 等，1985；Kah 等，1986）。大西洋鲑分布于脑垂体的主要多巴胺能神经纤维都来自 PPp 前区的亚群和交叉上核（Holmqvist 和 Ekström，1995）。斑马鱼的多巴胺能神经元主要分布于视前区的 PPa、PM、NSC、后结节和腹后下丘脑（Ma，2003）。欧洲鳗鱼将神经纤维分布于促性腺激素细胞的多巴胺能神经元主要来自 PPa 的吻区，而这些神经细胞能对类固醇激素起反应（Weltzien 等，2006）。

表1.4　含有神经递质的主要下丘脑-脑垂体（hypophysiotropic）区

神经递质	端脑	视前区	下丘脑
DA	OB	PPa, PMmc, PMgc, PPp, NSC	Hv, Hc, PVO
GABA	OB，腹端脑	PPa, PMpc, PMmc, PPp, NSC	Hv, LH, Hc, NAT, PVO, NDLI
S-羟色胺	OB, Vv	PMpc, PPp, NSC	Hv, Hd, Hc, PVO

注：各种核按照 Braford 和 Northcutt（1983）命名法命名。
下丘脑和脑垂体各个区的缩写见表1.1。

1.6.3 5-羟色胺

5-羟色胺（serotonin 或 5-hydroxytryptamine，5-HT）是一种吲哚胺（indoleamine）的神经递质，和儿茶酚胺神经递质 DA 和 NA 一起组成单胺群（monoamine group）。已证明 5-HT 具有行为方面（De Pedro 等，1998；Lin 等，2000；Johansson 等，2004）和神经内分泌方面（Khan 和 Thomas，1992；Trudeau，1997；Winberg 等，1997；Canosa 等，2007）的功能。对一些脊椎动物，包括鱼类，已经研究 5-羟色胺能细胞和神经纤维的分布情况（Kah 和 Chambolle，1983；Margolis-Kazan 等，1985；Ekström 和 Ebbesson，1989；Meek 和 Joosten，1989；Johnston 等，1990；Khan 和 Thomas，1993；Rodríguez-Gómez 等，2000b）。总而言之，5-HT 系统在鱼脑中有两个主要的定位：一个在前部，包括和尾下丘脑的脑室及其隐窝相联系的核；另一个在后部，在脑干（brainstem）内（见表1.4）。在吻部的 5-HT 系统主要由在 PVO 的含有 CSF 的细胞体组成，它们延伸到腹中下丘脑和视前区（Johnston 等，1990；Khan 和 Thomas，1993）。虽然已经证实 5-HT 对脑垂体 LH 分泌活动有直接的作用（Somoza 和 Peter，1991；Khan 和 Thomas，1992），但对一些硬骨鱼类的研究表明，在脑垂体只有很少量，甚至检测不到 5-HT 或者它的主要代谢物 5-羟基吲哚乙酸（5-hydroxyindoleacetic acid，5-HIAA）（Sloley 等，1992；Hernández-Rauda 等，1996）。Hernández-Rauda 和 Aldegunde（2002）分析了神经递质在纹眼笛鲷（*Lutjanus argentiventris*）性腺发育过程中的作用，发现在性腺发育的任何时期，在其脑垂体都检测不到 5-HT 或者 5-HIAA。同样，5-羟色胺能的神经纤维很少分布在海鲈的脑垂体（Batten 等，1993）。在塞内加尔鲷的脑垂体没有发现 5-羟色胺-免疫反应纤维（Rodríguez-Gómez 等，2000b）。然而，金鱼的 PPD 有 5-羟色胺免疫反应（Kah 和 Chambolle，1983）；在鲶鱼的脑垂体柄和脑垂体的各个部位都可以观察到 5-羟色胺能的纤维（Corio 等，1991）。在剑尾鱼（Margolis-Kazan 等，1985）和大西洋绒须石首鱼（Khan 和 Thomas，1993）的脑内，5-羟色胺能纤维出现在 PPD 和 PI。在大西洋绒须石首鱼（*Micropogonias undulatus*）的脑垂体，最强烈的 5-HT-免疫反应出现在 PPD，而一些散布在 PI 周边的细胞和它们的纤维则和 PPD 的内分泌细胞接触（Khan 和 Thomas，1993）（见表1.3）。

1.7 结 束 语

硬骨鱼类的下丘脑-脑垂体复合体，在正中隆起大大退化或者不存在的情况下，表现得尤为特化（sepecialization），其结果是下丘脑通过神经组织的相嵌交错而插入腺脑垂体内，分布神经纤维，实施对脑垂体的调控。通过适宜组合的激素受体表达以及神经元之间的沟通，这些神经元回路（neuronal circuitry）整合外部和体内环境的信息（见本书第2章）。有些信息是关于这些神经元回路如何受到体内因子的调控以及感觉信息如何转送到"调节作用的脑"中的。然而，关于下丘脑调控系统如何整合体内和外部的信息以及它们如何精准地产生一个协调一致的最后反应，目前学术界研究与了解得还很少。这些神经元调控系统究竟是如何相互作用以协调各种不同的生理功能，例如营养和生殖？

最近的逆行神经束追踪（retrograde tract-tracing）研究已经证实早期研究所表明的视前区和结节下丘脑是神经元细胞体所在的位置，它们的轴突沿着清晰的纤维束到达神经脑垂体和腺脑垂体（见第1.3节）。这些神经元释放激素到血管系统（见第1.4节）或者采用不同的释放肽（见第1.5节）和神经递质（见第1.6节），并通过它们和脑垂体结构与功能的联系而调控许多生理功能。采用双重标志、神经束追踪和脑垂体细胞孵育等技术的研究，已经对调控脑垂体各种不同分泌细胞的神经元进行定位，并鉴别这些神经元系统使用的释放肽和/或神经递质。直接神经分布到 RPD 内，邻近 ACTH 或 PRL 细胞的是含有 AVT、IST、MCH、GHRH、PACAP、SS、CRF、甘丙肽、CCK 和 PrRP 的神经元。PPD 的内分泌细胞（GH、LH、FSH 和 TSH）是由含有 AVT、IST、NPY、GHRH、PACAP、SS、CRF、甘丙肽、GnRH 和 CCK 的神经元分布神经纤维的，还在较小程度上包括含有 TRH 的神经元。最后，PI 是由含有 AVT、IST、MCH、NPY、GHRH、PACAP、SS、CRF、甘丙肽、GnRH、CCK、BBS 和 TRH 的神经元纤维相嵌交错的，并且和 MSH 或 SL 细胞紧密联系。在 PN 内的神经内分泌活动主要是由含有 AVT、IST 和 MCH 的神经元参与的。

J. M. 谢达-里维特

L. F. 坎诺沙

参考文献

Acher R. 1996. Molecular evolution of fish neurohypophysial hormones: Neutral and selective evolutionary mechanisms. *Gen. Comp. Endocrinol*, 102: 157-172.

Adrio F, Rodríguez M Á, Rodríguez-Moldes I. 2005. Distribution of galanin like immunoreactivity in the brain of the Siberian sturgeon (*Acipenser baeri*). *J. Comp. Neurol*, 487: 54-74.

Ahrens K, Wullimann M F. 2002. Hypothalamic inferior lobe and lateral torus connections in a percomorph teleost, the red cichlid (*Hemichromis lifalili*). *J. Comp. Neurol*, 449: 43-64.

Alderman S L, Bernier N J. 2007. Localization of corticotropin-releasing factor, urotensin I, and CRF-binding protein gene expression in the brain of the zebrafish, *Danio rerio*. *J. Comp. Neurol*, 502: 783-793.

Alvarez C E, Sutcliffe J G. 2002. Hypocretin is an early member of the incretin gene family. *Neurosci. Lett*, 324: 169-172.

Amano M, Takahashi A, Oka Y, Yamanome T, Kawauchi H, Yamamori K. 2003. Immunocytochemical localization and ontogenic development of melaninconcentrating hormone in the brain of a pleuronectiform fish, the barfin flounder. *Cell Tissue Res*,

311: 71-77.

Amano M, Takahashi A, Yamanome T, Oka Y, Amiya N, Kawauchi H, Yamamori K. 2005. Immunocytochemical localization and ontogenic development of α-melanocyte-stimulating hormone (α-MSH) in the brain of a pleuronectiform fish, barfin flounder. *Cell Tissue Res*, 320: 127-134.

Amano M, Oka Y, Amiya N, Yamamori K. 2007. Immunohistochemical localization and ontogenic development of prolactin-releasing peptide in the brain of the ovoviviparous fish species *Poecilia reticulata* (guppy). *Neurosci. Lett*, 413: 206-209.

Amiya N, Amano M, Oka Y, Iigo M, Takahashi A, Yamamori K. 2007. Immunohistochemical localization of orexin/hypocretin-like immunoreactive peptides and melanin-concentrating hormone in the brain and pituitary of medaka. *Neurosci. Lett*, 427: 16-21.

Ando H, Ando J, Urano A. 1998. Localization of mRNA encoding thyrotropin-releasing hormone precursor in the brain of sockeye salmon. *Zool. Sci*, 15: 945-953.

Ando H, Hasegawa M, Ando J, Urano A. 1999. Expression of salmon corticotropin-releasing hormone precursor gene in the preoptic nucleus in stressed rainbow trout. *Gen. Comp. Endocrinol*, 113: 87-95.

Anglade I, Zanbergen T, Kah O. 1993. Origin of the pituitary innervation in the goldfish. *Cell Tissue Res*, 273: 345-355.

Anglade I, Wang Y, Jensen J, Tramu G, Kah O, Conlon J M. 1994. Characterization of trout galanin and its distribution in trout brain and pituitary. *J. Comp. Neurol*, 350: 63-74.

Anglade I, Douard V, Le Jossic-Corcos C, Mañanos E L, Mazurais D, Michel D, Kah O. 1998. The GABAergic system: A possible component of estrogenic feedback on gonadotropin secretion in rainbow trout (*Oncorhynchus mykiss*). *BFPP-Bulletin Francais de la Peche et de la Protection des Milieux Aquatiques*, 71: 647-654.

Anglade I, Mazurais D, Douard V, Le Jossic-Corcos C, Mañanos E L, Michel D, Kah O. 1999. Distribution of glutamic acid decarboxylase mRNA in the forebrain of the rainbow trout as studied by *in situ* hybridization. *J. Comp. Neurol*, 410: 277-289.

Baker B I, Bird D J. 2002. Neuronal organization of melanin concentrating hormone system in primitive actinopterygians: Evolutionary changes leading to teleost. *J. Comp. Neurol*, 442: 99-114.

Balment R J, Lu W, Weybourne E, Warne J M. 2006. Arginine vasotocin a key hormone in fish physiology and behaviour: A review with insights from mammalian models. *Gen. Comp. Endocrinol*, 147: 9-16.

Barry T P, Grau E G. 1986. Estradiol-17b and thyrotropin-releasing hormone stimulate prolactin release from the pituitary gland of a teleost fish *in vitro*. *Gen. Comp. Endo-

crinol, 62: 306-314.

Batten T F C. 1986. Ultrastructural characterization of neurosecretory fibres immunoreactive for vasotocin, isotocin, somatostatin, LHRH and CRF in the pituitary of a teleost fish, *Poecilia latipinna. Cell Tissue Res*, 244: 661-672.

Batten T F C, Baker B I. 1988. Melanin-concentrating hormone (MCH) immunoreactive hypophysial neurosecretory system in the teleost *Poecilia latipinna*: Light and electron microscopic study. *Gen. Comp. Endocrinol*, 70: 193-205.

Batten T F C, Wigham T. 1984. Effects of TRH and somatostatin on releases of prolactin and growth hormone in vitro by the pituitary of *Poecilia latipinna*. II. Electron-microscopic morphometry using automatic image analysis. *Cell Tissue Res*, 237: 595-603.

Batten T F C, Groves D J, Ball JN. 1985. Immunocytochemical investigation of forebrain control by somatostatin of the pituitary in the teleost *Poecilia latipinna. Cell Tissue Res*, 242: 115-125.

Batten T F C, Cambre M L, Moons L, Vandesande F. 1990. Comparative distribution of neuropeptide-immunoreactive systems in the brain of the green molly, *Poecilia latipinna. J. Comp. Neurol*, 302: 893-919.

Batten T F C, Moons L, Cambre M L, Vandesande F, Seki T, Suzuki M. 1990. Thyrotropin-releasing hormone-immunoreactive system in the brain and pituitary gland of the sea bass (*Dicentrarchus labrax*, teleostei). *Gen. Comp. Endocrinol*, 79: 385-392.

Batten T F C, Moons L, Cambre M, Vandesande F. 1990. Anatomical distribution of galanin-like immunoreactivity in the brain and pituitary of teleost fishes. *Neurosci. Lett*, 111: 12-17.

Batten T F C, Berry P A, Maqbool A, Moons L, Vandesande F. 1993. Immunolocalization of catecholamine enzymes, serotonin, dopamine and L-dopa in the brain of *Dicentrarchus labrax* (Teleostei). *Brain Res. Bull*, 31: 233-252.

Batten T F C, Moons L, Vandesande F. 1999. Innervation and control of the adenohypophysis by hypothalamic peptidergic neurons in teleost fishes: EM immunohistochemical evidence. *Microsc. Res. Techniq*, 44: 19-35.

Bellinger F P, Fox B K, Wing Y C, Davis L K, Andres M A, Hirano T, Grau E G, Cooke I M. 2006. Ionotropic glutamate receptor activation increases intracellular calcium in prolactin-releasing cells of the adenohypophysis. *Am. J. Physiol*, 291: E1188E1196.

Bernier N J. 2006. The corticotropin-releasing factor system as a mediator of the appetite-suppressing effects of stress in fish. *Gen. Comp. Endocrinol*, 146: 45-55.

Billard R, Peter R E. 1982. A stereotaxic atlas and technique for the nuclei of the diencephalon of rainbow trout (*Salmo gairdneri*). *Reprod. Nutr. Develop*, 22: 1-25.

Bird D J, Baker B I, Kawauchi H. 1989. Immunocytochemical demonstration of melanin-concentrating hormone and proopiomelanocortin-like products in the brain of the trout and carp. *Gen. Comp. Endocrinol*, 74: 442-450.

Bond H, Warne J M, Balment R J. 2007. Effect of acute restraint on hypothalamic pro-vasotocin mRNA expression in flounder, *Platichthys flesus*. *Gen. Comp. Endocrinol*, 153: 221-227.

BrafordJr M R, Northcutt R G. 1983. Organization of the diencephalon and pretectum of the ray-finned fishes. In: Northcutt R G, Davis R E, Ed. *Fish Neurobiology*. Vol. 2. University of Michigan Press: Ann Arbor, 117-163.

Bromage N R. 1975. The effects of mammalian thyrotropin releasing hormone on the pituitary thyroid axis of teleost fish. *Gen. Comp. Endocrinol*, 25: 292-297.

Butler A B, Hodos W. 2005. *Comparative Vertebrate Neuroanatomy*. Wiley-Interscience: New Jersey.

Canosa L F, Cerdá-Reverter J M, Peter R E. 2004. Brain mapping of three somatostatin encoding genes in the goldfish. *J. Comp. Neurol*, 474: 43-57.

Canosa L F, Chang J P, Peter R E. 2007. Neuroendocrine control of growth hormone in fish. *Gen. Comp. Endocrinol*, 151: 1-26.

Canosa L F, Stacey N, Peter R E. 2008. Changes in brain mRNA levels of gonadotropin-releasing hormone, pituitary adenylate cyclase activating polypeptide, and somatostatin during ovulatory luteinizing hormone and growth hormone surges in goldfish. *Am. J. Physiol*, 295: R1815-R1821.

Castro M G, Morrison E. 1997. Post-translational processing of proopiomelanocortin in the pituitary and the brain. *Crit. Rev. Neurobiol*, 11: 35-57.

Cepriano K M, Schreibman M P. 1993. The distribution of neuropeptide Y and dynorphin immunoreactivity in the brain and pituitary gland of the platyfish, *Xiphophorus maculatus*, from birth to sexual maturity. *Cell Tissue Res*, 271: 87-92.

Cerdá-Reverter J M, Larhammar D. 2000. Neuropeptide Y family of peptides: structure, anatomical expression, function and molecular evolution. *Biochem. Cell Biol*, 78: 371-392.

Cerdá-Reverter J M, Peter R E. 2003. Endogenous melanocortin antagonist in fish. Structure, brain mapping and regulation by fasting of the goldfish agouti-related protein gene. *Endocrinology*, 144: 4552-4561.

Cerdá-Reverter J M, Sorbera L, Carrillo M, Zanuy S. 1999. Energetic dependence of NPY-induced LH secretion in a teleost fish (*Dicentrarchus labrax*). *Am. J. Physiol*, 46: R1627-R1634.

Cerdá-Reverter J M, Anglade I, Martínez-Rodríguez G, Mazurais D, Muñoz-Cueto J A, Carrillo M, Kah O, Zanuy S. 2000. Characterization of neuropeptide Y expression in the brain of a perciform fish, the sea bass (*Dicentrarchus labrax*). *J. Chem. Neuroanat*, 19: 197-210.

Cerdá-Reverter J M, Zanuy S, Munoz-Cueto J A. 2001. Cytoarchitectonic study of the brain of a perciform species, the sea bass (*Dicentrarchus labrax*). I. The telencephalon. *J. Morphol*, 247: 217-228.

Cerdá-Reverter J M, Zanuy S, Munoz-Cueto J A. 2001. Cytoarchitectonic study of the brain of a perciform species, the sea bass (*Dicentrarchus labrax*). II. The diencephalon. *J. Morphol*, 247: 229-251.

Cerdá-Reverter J M, Schiöth H B, Peter R E. 2003. The central melanocortin system regulates food intake in goldfish. *Regul. Pept*, 115: 101-113.

Cerdá-Reverter J M, Ringholm A, Schiöth H B, Peter R E. 2003. Molecular cloning, pharmacological characterization and brain mapping of the melanocortin 4 receptor in the goldfish: Involvement in the control of food intake. *Endocrinology*, 144: 2336-2349.

Cerdá-Reverter J M, Ling M, Schiöth H B, Peter R E. 2003. Molecular cloning, pharmacological characterization and brain mapping of the melanocortin 5 receptor in the goldfish. *J. Neurochem*, 87: 1354-1367.

Chandra R, Liddle R A. 2007. Cholecystokinin. *Curr. Opin. Endocrinol. Diabetes Obes*. 14: 63-67.

Chiba A. 2005. Neuropeptide Y-immunoreactive (NPY-ir) structures in the brain of the gar *Lepisosteus oculatus* (Lepisosteiformes, Osteichthyes) with special regard to their anatomical relations to gonadotropin-releasing hormone (GnRH) -ir structures in the hypothalamus and the terminal nerve. *Gen. Comp. Endocrinol*, 142: 336-346.

Chiba A, Chang Y, Honma Y. 1996. Distribution of neuropeptide Y and gonadotropin-releasing hormone immunoreactivities in the brain and hypophysis of the ayu, *Plecoglossus altivelis* (Teleostei). *Arch. Histol. Cytol*, 59: 137-148.

Corio M, Peute J, Steinbusch H W M. 1991. Distribution of serotonin- and dopamine-immunoreactivity in the brain of the teleost *Clarias gariepinus*. *J. Chem. Neuroanat*, 4: 79-95.

Cornbrooks E B, Parsons R L, 1991. Sexually dimorphic distribution of a galanin-like peptide in the central nervous system of the teleost fish *Poecilia latipinna*. *J. Comp. Neurol*, 304: 639-657.

Corrêa S. 2004. Re-evaluation of the afferent connections of the pituitary in the weakly electric fish *Apteronutus leptorhynchus*: An *in vitro* tract-tracing study. *J. Comp. Neurol*, 470: 39-49.

Cuadrado M I, Coveñas R, Tramu G. 1994. Distribution of gastrin-releasing peptide/bombesin-like immunoreactivity in the rainbow trout brain. *Peptides*, 15: 1027-1032.

Cumming R, ReavesJr T A, Hayward J N. 1982. Ultrastructural immunocytochemical characterization of isotocin, vasotocin and neurophysin neurons in the magnocellular preoptic nucleus of the goldfish. *Cell Tissue Res*, 223: 685-694.

Danger J M, Breton B, Vallarino M, Fournier A, Pelletier G, Vaudry H. 1991. Neuropeptide-Y in the trout brain and pituitary: Localization, characterization, and action on gonadotropin release. *Endocrinology*, 128: 2360-2368.

De Lecea L, Kilduff T S, Peyron C, Gao X B, Foye P E, Danielson P E, Fukuhara C, Battenberg E L F, Gautvik V T, BartlettII F S, Frankel W N, Van Den Pol A N, Bloom F E, Gautvik K M, Sutcliffe J G. 1998. The hypocretins: Hypothalamus-specific peptides with neuroexcitatory activity. *Proc. Natl. Acad. Sci. USA*, 95: 322-327.

De Pedro N, Pinillos M L, Valenciano A I, Alonso-Bedate M, Delgado M J. 1998. Inhibitory effect of serotonin on feeding behavior in goldfish: Involvement of CRF. *Peptides*, 19: 505-511.

Díaz M L, Becerra M, Manso M J, Anadón R. 2001. Development of thyrotropin-releasing hormone immunoreactivity in the brain of the brown trout *Salmo trutta fario*. *J. Comp. Neurol*, 429: 299-320.

Díaz M L, Becerra M, Manso M J, Anadón R. 2002. Distribution of thyrotropin-releasing hormone (TRH) immunoreactivity in the brain of the zebrafish (*Danio rerio*). *J. Comp. Neurol*, 450: 45-60.

Douglas J, McKinzie A A, Couceyro P. 1995. PCR differential display identifies a rat brain mRNA that is transcriptionally regulated by cocaine and amphetamine. *J. Neuroscim*, 15: 2471-2481.

Duarte G, Segura-Noguera M M, Martín del Río M P, Mancera J M. 2001. The hypothalamo-hypophyseal system of the white seabream *Diplodus sargus*: Immunocytochemical identification of arginine-vasotocin, isotocin, melanin-concentrating hormone and corticotropin-releasing factor. *Histochem. J*, 33: 569-578.

Dylag T, Kotlinska J, Rafalski P, Pachuta A, Siberring J. 2006. The activity of CART peptide fragments. *Peptides*, 27: 1926-1933.

Eales J G, Himick B A. 1988. The effect of TRH on plasma thyroid hormone levels of rainbow trout (*Salmo gairdneri*) and arctic charr (*Salvelinus alpinus*). *Gen. Comp. Endocrinol*, 72: 333-339.

Ekström P, Ebbesson S O. 1989. Distribution of serotonin-immunoreactive neurons in the brain of sockeye salmon fry. *J. Chem. Neuroanat*, 2: 201-213.

Fernald R D, Shelton L C. 1985. The organization of the diencephalon and pretectum in the cichlid fish, *Haplochromis burtoni*. *J. Comp. Neurol*, 238: 202-217.

Fliers E, Unmehopa U A, Alkemade A. 2006. Functional neuroanatomy of thyroid hormone feedback in the human hypothalamus and pituitary gland. *Mol. Cell. Endocrinol*, 251: 1-8.

Flik G, Klaren P H M, Van Den Burg E H, Metz J R, Huising M O. 2006. CRF and stress in fish. *Gen. Comp. Endocrinol*, 146: 36-44.

Flynn F W. 1991. Effects of fourth ventricle bombesin injection on meal-related parameters and grooming behavior. *Peptides*, 12: 761-765.

Foran C M, Bass A H. 1998. Preoptic AVT immunoreactive neurons of a teleost fish with alternative reproductive tactics. *Gen. Comp. Endocrinol*, 111: 271-282.

Forlano P M, Cone R D. 2007. Conserved neurochemical pathways involved in hypothalamic control of energy homeostasis. *J. Comp. Neurol*, 505: 235-248.

Fryer J N, Maler L. 1981. Hypophysiotrophic neurons in the goldfish hypothalamus demonstrated by retrograde transport of horseradish peroxidase. *Cell Tissue Res*, 218: 93-102.

Fryer J N, Boudreault-Chateauvert C, Kirby R P. 1985. Pituitary afferents originating in the paraventricular organ (PVO) of the goldfish hypothalamus. *J. Comp. Neurol*, 242: 475-484.

Fujimoto M, Takeshita K I, Wang X, Takabatake I, Fujisawa Y, Teranishi H, Ohtani M, Muneoka Y, Ohta S. 1998. Isolation and characterization of a novel bioactive peptide, *Carassius*RFamide (C-RFa), from the brain of the Japanese crucian carp. *Biochem. Bioph. Res. Co*, 242: 436-440.

Fujimoto M, Sakamoto T, Kanetoh T, Osaka M, Moriyama S. 2006. Prolactin-releasing peptide is essential to maintain the prolactin level and osmotic balance in freshwater teleost fish. *Peptides*, 27: 1104-1109.

Fukusumi S, Habata Y, Yoshida H, Iijima N, Kawamata Y, Hosoya M, Fujii R, Hinuma S, Kitada C, Shintani Y, Suenaga M, Onda H, et al. 2001. Characteristics and distribution of endogenous RFamide-related peptide-1. *Biochim. Biophys. Acta Mol. Cell Res*, 1540: 221-232.

Gaikwad A, Biju K C, Saha S G, Subhedar N. 2004. Neuropeptide Y in the olfactory system, forebrain and pituitary of the teleost, *Clarias batrachus*. *J. Chem. Neuroanat*, 27: 55-70.

Gaikwad A, Biju K C, Muthal P L, Saha S, Subhedar N. 2005. Role of neuropeptide Y in the regulation of gonadotropin releasing hormone system in the forebrain of *Clarias batrachus* (Linn.): Immunocytochemistry and high performance liquid chromatography-electrospray ionization-mass spectrometric analysis. *Neuroscience*, 133: 267-279.

Gibbs J, Fauser D J, Rowe E A. 1979. Bombesin suppresses feeding in rats. *Nature*, 282: 208-210.

Gómez-Segade P, Anadón R. 1988. Specialization in the diencephalon of advanced teleost. *J. Morphol*, 197: 71-103.

González-Martínez D, Zmora N, Mañanos E, Saligaut D, Zanuy S, Zohar Y, Elizur A, Kah O, Muñoz-Cueto J A. 2002. Immunohistochemical localization of three different prepro-GnRHs in the brain and pituitary of the European sea bass (*Dicentrarchus labrax*) using antibodies to the corresponding GnRH-associated peptides. *J. Comp. Neurol*, 446: 95-113.

Goodson J L, Bass A H. 2001. Social behavior functions and related anatomical characteristics of vasotocin/vasopressin systems in vertebrates. *Brain Res. Rev*, 35: 246-265.

Goodson J L, Evans A K, Bass A H. 2003. Putative isotocin distributions in sonic fish: Relation to vasotocin and vocal-acoustic circuitry. *J. Comp. Neurol*, 462: 1-14.

Goossens N, Dierickx K, Vandesande F. 1977. Immunocytochemical localization of vasotocin and isotocin in the preopticohypophysial neurosecretory system of teleosts. *Gen. Comp. Endocrinol*, 32: 371-375.

Gorbman A, Dickhoff W W, Vigna S R, Clark N B, Ralph C L. 1983. *Comparative Endocrinology*. Wiley-Interscience: New York.

Grau E G, Nishioka R S, Young G, Bern H A. 1985. Somatostatin-like immunoreactivity in the pituitary and brain of three teleost fish species-somatostatin as a potential regulator of prolactin cell-function. *Gen. Comp. Endocrinol*, 59: 350-357.

Green J D. 1951. The comparative anatomy of the hypophysis with special reference to its blood supply and innervation. *Am. J. Anat*, 88: 225-311.

Grober M S, Sunobe T. 1996. Serial adult sex change involves rapid and reversible changes in forebrain neurochemistry. *Neuroreport*, 7: 2945-2949.

Gröneveld D, Balm P H M, Martens G J M, Wenderlaar Bonga S E. 1995. Differential melanin-concentrating hormone expression in two hypothalamic nuclei of the teleost tilapia in response to environmental changes. *J. Neuroendocrinol*, 7: 527-533.

Guillemin R. 1970. Hormones secreted by the brain. Isolation, molecular structure and synthesis of the first hypophysiotropic hypothalamic hormone (to be discovered), TRF (thyrotropin-releasing factor). *Science*, 68: 64-67.

Hamano K, Inoue K, Yanagisawa T. 1990. Immunohistochemical localization of thyrotropin-releasing hormone in the brain of carp, *Cyprinus carpio*. *Gen. Comp. Endocrinol*, 80: 85-94.

Hernández-Rauda R, Aldegunde M. 2002. Changes in dopamine, norepinephrine and serotonin levels in the pituitary, telencephalon and hypothalamus during gonadal development of male *Lutjanus argentiventris* (Teleostei). *Mar. Biol*, 141: 209-216.

Hernández-Rauda R, Otero J, Rey P, Rozas G, Aldegunde M. 1996. Dopamine and serotonin in the trout (*Oncorhynchus mykiss*) pituitary: Main metabolites and changes during gonadal recrudescence. *Gen. Comp. Endocrinol*, 103: 13-23.

Hill J J, Henderson N E. 1968. The vascularization of the hypothalamichypophysial region of the eastern brook trout, *Salvelinus fontinalis*. *Am. J. Anat*, 122: 301-316.

Himick B A, Peter R E. 1994. CCK/gastrin-like immunoreactivity in brain and gut, and CCK suppression of feeding in goldfish. *Am. J. Physiol*, 267: R841-R851.

Himick B A, Peter R E. 1995. Bombesin-like immunoreactivity in the forebrain and pituitary and regulation of anterior pituitary hormone-release by bombesin in goldfish. *Neuroendocrinology*, 61: 365-376.

Himick B A, Golosinski A A, Jonsson A C, Peter R E. 1993. CCK/gastrinlike immunoreactivity in the goldfish pituitary - Regulation of pituitary hormone secretion by CCK-like peptides *in vitro*. *Gen. Comp. Endocrinol*, 92: 88-103.

Hinuma S, Habata Y, Fujii R, Kawamata Y, Hosoya M, Fukusumi S, Kitada C, Masuo Y, Asano T, Matsumoto H, Sekiguchi M, Kurokawa T, et al. 1998. A prolactin-releasing peptide in the brain. *Nature*, 393: 272-276.

Hinuma S, Shintani Y, Fukusumi S, Iijima N, Matsumoto Y, Hosoya M, Fujii R, Watanabe T, Kikuchi K, Terao Y, Yano T, Yamamoto T, et al. 2000. New neuropeptides containing carboxyterminal RFamide and their receptor in mammals. *Nat. Cell Biol*, 2: 703-708.

Holloway A C, Leatherland J F. 1997. The effects of N-methyl-D, L-aspartate and gonadotropin-releasing hormone on *in vitro* growth hormone release in steroidprimed immature rainbow trout, *Oncorhynchus mykiss*. *Gen. Comp. Endocrinol*, 107: 32-43.

Holmgren S, Jensen J. 1994. Comparative aspects on the biochemical identity of neurotransmitters of autonomic neurons. In: Nilsson S, Holmgren S, Ed. *Comparative physiology and evolution of the autonomic nervous system*. Harwood Academic: Chur, Switzerland, 69-95.

Holmgren S, Jonsson A C. 1988. Occurrence and effects on motility of bombesin related peptides in the gastrointestinal tract of the Atlantic cod, *Gadus morhua*. *Comp. Biochem. Physiol*, 89C: 249-256.

Holmqvist B I, Ekström P. 1995. Hypophysiotrophic systems in the brain of the Atlantic salmon: Neuronal innervation of the pituitary and the origin of pituitary dopamine and nonapeptides identified by means of combined carbocyanine tract tracing and immunocytochemistry. *J. Chem. Neuroanat*, 8: 125-145.

Holmqvist B I, Östholm T, Ekström P. 1992. DiI tracing in combination with the immunocytochemistry for analysis of connectivities and chemoarchitectonics of specific neuronal systems in a teleost, the Atlantic salmon. *J. Neurosci. Meth*, 42: 45-63.

Hosoya M, Kimura C, Ogi K, Ohkubo S, Miyamoto Y, Kugoh H, Shimizu M, Onda

H, Oshimura M, Arimura A, Fujino M. 1992. Structure of the human pituitary adenylate cyclase activating polypeptide (PACAP) gene. *Biochim. Biophys. Acta-Gene Struct. Expression*, 1129: 199-206.

Huesa G, van den Pol A N, Finger T E. 2005. Differential distribution of hypocretin (orexin) and melanin-concentrating hormone in the goldfish brain. *J. Comp. Neurol*, 488: 476-491.

Huising M O, Metz J R, van Schooten C, Taverne-Thiele A J, Hermsen T, Verburg-van Kemenade B M L, Flik G. 2004. Structural characterisation of a cyprinid (*Cyprinus carpio* L.) CRH, CRH-BP and CRH-R1, and the role of these proteins in the acute stress response. *J. Mol. Endocrinol*, 32: 627-648.

Jadhao A G, Meyer D L. 2000. Sexually dimorphic distribution of galanin in the preoptic area of red salmon, *Oncorhynchus nerka*. *Cell Tissue Res*, 302: 199-203.

Jadhao A, Pinelli C. 2001. Galanin-like immunoreactivity in the brain and pituitary of the "four-eyed" fish, *Anableps anableps*. *Cell Tissue Res*, 306: 309-318.

Jensen H, Rourke I J, Moller M, Jonson L, Johnsen A H. 2001. Identification and distribution of CCK-related peptides and mRNAs in the rainbow trout, *Oncorhynchus mykiss*. *Biochim. Biophys. Acta-Gene Struct. Expression*, 1517: 190-201.

Johansson V, Winberg S, Jonsson E, Hall D, Bjornsson B T. 2004. Peripherally administered growth hormone increases brain dopaminergic activity and swimming in rainbow trout. *Horm. Behav*, 46: 436-443.

Johnsen A H. 1998. Phylogeny of the cholecystokinin/gastrin family. *Front. Neuroendocrinol*, 19: 73-99.

Johnston S A, Maler L. 1992. Anatomical organization of the hypophysiotrophic systems in the electric fish, *Apteronotus leptorhynchus*. *J. Comp. Neurol*, 317: 421-437.

Johnston S A, Maler L, Tinner B. 1990. The distribution of serotonin in the brain of *Apteronotus leptorhynchus*: An immunohistochemical study. *J. Chem. Neuroanat*, 3: 429-465.

Kah O, Chambolle P. 1983. Serotonin in the brain of the goldfish, *Carassius auratus*. An immunocytochemical study. *Cell Tissue Res*, 234: 319-333.

Kah O, Chambolle P, Dubourg P, Dubois M P. 1982. Localisation immunocytochimique de la somatostatine dans le cerveau anterieur et l'hypophyse de deux teleosteens, le cyprin (*Carassius auratus*) et *Gambusia sp. Comptes Rendus des Seances de l'Academie des Sciences*, 294: 519-524.

Kah O, Chambolle P, Thibault J, Geffard M. 1984. Existence of dopaminergic neurons in the preoptic region of the goldfish. *Neurosci. Lett*, 48: 293-298.

Kah O, Dubourg P, Onteniente B. 1986. The dopaminergic innervation of the goldfish pituitary. An immunocytochemical study at the electron-microscope level using antibodies against dopamine. *Cell Tissue Res*, 244: 577-582.

Kah O, Dubourg P, Martinoli M G, Rabhi M, Gonnet F, Geffard M, Calas A. 1987. Central GABAergic innervation of the pituitary in goldfish: A radioautographic and immunocytochemical study at the electron microscope level. *Gen. Comp. Endocrinol*, 67: 324-332.

Kah O, Pontet A, Danger J M, Dubourg P, Pelletier G, Vaudry H, Calas A. 1989. Characterization, cerebral distribution and gonadotropin release activity of neuropeptide Y (NPY) in the goldfish. *Fish Physiol. Biochem*, 7: 69-76.

Kah O, Trudeau V L, Sloley B D, Chang J P, Dubourg P, Yu K L, Peter R E. 1992. Influence of GABA on gonadotrophin release in the goldfish. *Neuroendocrinology*, 55: 396-404.

Kah O, Lethimonier C, Somoza G, Guilgur L G, Vaillant C, Lareyre J J. 2007. GnRH and GnRH receptors in metazoa: A historical, comparative, and evolutive perspective. *Gen. Comp. Endocrinol*, 153: 346-364.

Kaslin J, Nystedt J M, Östergård M, Peitsaro N, Panula P. 2004. The orexin/hypocretin system in zebrafish is connected to the aminergic and cholinergic systems. *J. Neurosci*, 24: 2678-2689.

Kawauchi H, Baker B I. 2004. Melanin-concentrating hormone signaling systems in fish. *Peptides*, 25: 1577-1584.

Kawauchi H, Sower S A. 2006. The dawn and evolution of hormones in the adenohypophysis. *Gen. Comp. Endocrinol*, 148: 3-14.

Kawauchi H, Kawazoe I, Tsubokawa M, Kishida M, Baker B I. 1983. Characterization of melanin-concentrating hormone in chum salmon pituitaries. *Nature*, 305: 321-323.

Kawauchi N, Okubo K, Aida K. 2003. The expression and localization of corticotropin-releasing hormone and urotensin I transcripts in the Japanese eel, *Anguilla japonica*. *Fish Physiol. Biochem*, 28: 43-44.

Kehoe A S, Volkoff H. 2006. Cloning and characterization of neuropeptide Y (NPY) and cocaine and amphetamine regulated transcript (CART) in Atlantic cod (*Gadus morhua*). *Comp. Biochem. Physiol*, 146A: 451-461.

Kelly S P, Peter R E. 2006. Prolactin-releasing peptide, food intake, and hydromineral balance in goldfish. *Am. J. Physiol*, 291: R1474-R1481.

Khan I A, Thomas P. 1992. Stimulatory effects of serotonin on maturational gonadotropin release in the Atlantic croaker, *Micropogonias undulatus*. *Gen. Comp. Endocrinol*, 88: 388-396.

Khan I A, Thomas P. 1993. Immunocytochemical localization of serotonin and gonadotropin-releasing hormone in the brain and pituitary gland of the Atlantic croaker *Micro-*

pogonias undulatus. Gen. Comp. Endocrinol, 91: 167-180.

Khan I A, Thomas P. 1999. GABA exerts stimulatory and inhibitory influences on gonadotropin II secretion in the Atlantic croaker (*Micropogonias undulatus*). *Neuroendocrinology*, 69: 261-268.

Kim M H, Oka Y, Amano M, Kobayashi M, Okuzawa K, Hasegawa Y, Kawashima S, Suzuki Y, Aida K. 1995. Immunocytochemical localization of sGnRH and cGnRH-II in the brain of goldfish, *Carassius auratus. J. Comp. Neurol*, 356: 72-82.

Kishida M, Baker B I, Bird D J. 1988. Localisation and identification of melanocyte-stimulating hormones in the fish brain. *Gen. Comp. Endocrinol*, 71: 229-242.

Kong H S, Zhou H, Yang Y, He M, Jiang Y, Wong A O L. 2007. Pituitary Adenylate Cyclase-Activating Polypeptide (PACAP) as a Growth Hormone (GH)-releasing factor in grass carp: II. Solution structure of a brainspecific PACAP by nuclear magnetic resonance spectroscopy and functional studies on GH release and gene expression. *Endocrinology*, 148: 5042-5059.

Lagios M D. 1968. Tetrapod-like organization of the pituitary gland of the polypteriformid fishes, *Calamoichthys calabaricus* and *Polypterus palmas. Gen. Comp. Endocrinol*, 11: 300-315.

Laiz-Carrión R, Segura-Noguera M M, Martín del Río M P, Mancera J M. 2003. Ontogeny of adenohypophyseal cells in the pituitary of American shad (*Alosa sapidissima*). *Gen. Comp. Endocrinol*, 132: 454-464.

Lamers A E, Balm P H M, Haenen H E M G, Jenks B G, Wendelaar Bonga S E. 1991. Regulation of differential release of α-melanocyte-stimulating hormone forms from the pituitary of a teleost fish, *Oreochromis mossambicus. J. Endocrinol*, 129: 179-187.

Lang R, Gundlach A L, Kofler B. 2007. The galanin peptide family: Receptor pharmacology, pleiotropic biological actions, and implications in health and disease. *Pharmacol. Therapeut*, 115: 177-207.

Larsen D A, Swanson P, Dickey J T, Rivier J, Dickhoff W W. 1998. *In vitro* thyrotropin-releasing activity of corticotropin-releasing hormone-family peptides in coho salmon, *Oncorhynchus kisutch. Gen. Comp. Endocrinol*, 109: 276-285.

Lee L T O, Siu F K Y, Tam J K V, Lau I T Y, Wong A O L, Lin M C M, Vaudry H, Chow B K C. 2007. Discovery of growth hormone releasing hormones and receptors in nonmammalian vertebrates. *Proc. Natl. Acad. Sci. USA*, 104: 2133

Lema S C, Nevitt G A. 2004. Variation in vasotocin immunoreactivity in the brain of recently isolated populations of a death valley pupfish, *Cyprinodon nevadensis. Gen. Comp. Endocrinol*, 135: 300-309.

Lepretre E, Anglade I, Williot P, Vandesande F, Tramu G, Kah O. 1993. Comparative distribution of mammalian GnRH (gonadotrophin-releasing hormone) and chicken GnRH-II in the brain of the immature Siberian sturgeon (*Acipenser baeri*). *J. Comp. Neurol*, 337: 568-583.

Lethimonier C, Madigou T, Munoz-Cueto J A, Lareyre J J, Kah O. 2004. Evolutionary aspects of GnRHs, GnRH neuronal systems and GnRH receptors in teleost fish. *Gen. Comp. Endocrinol*, 135: 1-16.

Lin X W, Otto C J, Peter R E. 1999. Expression of three distinct somatostatin messenger ribonucleic acids (mRNAs) in goldfish brain: Characterization of the complementary deoxyribonucleic acids, distribution and seasonal variation of the mRNAs, and action of a somatostatin-14 variant. *Endocrinology*, 140: 2089-2099.

Lin X W, Volkoff H, Narnaware Y, Bernier N J, Peyon P, Peter R E. 2000. Brain regulation of feeding behavior and food intake in fish. *Comp. Biochem. Physiol*, 126A: 415-434.

Logan D W, Bryson-Richardson R J, Pagán K E, Taylor M S, Currie P D, Jackson I J. 2003. The structure and evolution of the melanocortin and MCH receptors in fish and mammals. *Genomics*, 81: 184-191.

Luo D, McKeown B A. 1989. Immunohistochemical detection of a substance resembling growth hormone-releasing factor in the brain of the rainbow trout (*Salmo gairdneri*). *Experientia*, 45: 577-580.

Ma P M. 1997. Catecholaminergic systems in the zebrafish. III. Organization and projection pattern of medullary dopaminergic and noradrenergic neurons. *J. Comp. Neurol*, 381: 411.

Ma P M. 2003. Catecholaminergic systems in the zebrafish. IV. Organization and projection pattern of dopaminergic neurons in the diencephalon. *J. Comp. Neurol*, 460: 13-37.

Ma P M, Lopez M. 2003. Consistency in the number of dopaminergic paraventricular organ-accompanying neurons in the posterior tuberculum of the zebrafish brain. *Brain Res*, 967: 267-272.

Magliulo-Cepriano L, Schreibman M P, Blum V. 1993. The distribution of immunoreactive FMRF-amide, neurotensin, and galanin in the brain and pituitary gland of three species of *Xiphophorus* from birth to sexual maturity. *Gen. Comp. Endocrinol*, 92: 269-280.

Maler L, Sas E, Johnston S, Ellis W. 1991. An atlas of the brain of the electric fish *Apteronotus leptorhynchus*. *J. Chem. Neuroana*, 4: 1-38.

Mañanos E L, Anglade I, Chyb J, Saligaut C, Breton B, Kah O. 1999. Involvement of γ-aminobutyric acid in the control of GTH-1 and GTH-2 secretion in male and female rainbow trout. *Neuroendocrinology*, 69: 269-280.

Mancera J M, Fernández-Llébrez P. 1995. Localization of corticotropin-releasing factor immunoreactivity in the brain of the teleost *Sparus aurata*. *Cell Tissue Res*, 281: 569-572.

Marchetti G, Cozzi B, Tavanti M, Russo V, Pellegrini S, Fabiani O. 2000. The distribution of Neuropeptide Y-immunoreactive neurons and nerve fibers in the forebrain of the carp *Cyprinus carpio* L. *J. Chem. Neuroanat*, 20: 129-139.

Margolis-Kazan H, Halpern-Sebold L R, Schreibman M P. 1985. Immunocytochemical localization of serotonin in the brain and pituitary gland of the platyfish, *Xiphophorus maculatus*. *Cell Tissue Res*, 240: 311-314.

Marivoet S, Moons L, Vandesande F. 1988. Localization of growth hormone releasing factor-like immunoreactivity in the hypothalamo-hypophyseal system of the frog (*Rana temporaria*) and the sea bass (*Dicentrarchus labrax*). *Gen. Comp. Endocrinol*, 72: 72-79.

Martínez V, Tache Y. 2000. Bombesin and the brain-gut axis. *Peptides*, 21: 1617-1625.

Martinoli M G, Dubourg P, Geffard M, Calas A, Kah O. 1990. Distribution of GABA-immunoreactive neurons in the forebrain of the goldfish, *Carassius auratus*. *Cell Tissue Res*, 260: 77-84.

Martyniuk C J, Awad R, Hurley R, Finger T E, Trudeau V L. 2007. Glutamic acid decarboxylase 65, 67, and GABA-transaminase mRNA expression and total enzyme activity in the goldfish (*Carassius auratus*) brain. *Brain Res*, 1147: 154-166.

Martyniuk C J, Chang J P, Trudeau V L. 2007. The effects of GABA agonists on glutamic acid decarboxylase, GABA-transaminase, activin, salmon gonadotrophin-releasing hormone and tyrosine hydroxylase mRNA in the goldfish (*Carassius auratus*) neuroendocrine brain. *J. Neuroendocrinol*, 19: 390-396.

Maruska K P, Mizobe M H, Tricas T C. 2007. Sex and seasonal co-variation of arginine vasotocin (AVT) and gonadotropin-releasing hormone (GnRH) neurons in the brain of the halfspotted goby. *Comp. Biochem. Physiol*, 147A: 129-144.

Mathieu M, Tagliafierro G, Bruzzone F, Vallarino M. 2002. Neuropeptide tyrosine-like immunoreactive system in the brain, olfactory organ and retina of the zebrafish, *Danio rerio*, during development. *Dev. Brain Res*, 139: 255-265.

Matsuda K, Nagano Y, Uchiyama M, Onoue S, Takahashi A, Kawauchi H, Shioda S. 2005. Pituitary adenylate cyclase-activating polypeptide (PACAP) -like immunoreactivity in the brain of a teleost, *Uranoscopus japonicus*: Immunohistochemical relationship between PACAP and adenohypophysial hormones. *Regul. Pept*, 126: 129-136.

Matsuda K, Nagano Y, Uchiyama M, Takahashi A, Kawauchi H. 2005. Immunohistochemical observation of pituitary adenylate cyclase-activating polypeptide (PACAP) and adenohypophysial hormones in the pituitary of a teleost, *Uranoscopus japonicus*. *Zool. Sci*, 22: 71-76.

Matz S P, Hofeldt G T. 1999. Immunohistochemical localization of corticotropin-releasing factor in the brain and corticotropin-releasing factor and thyrotropin-stimulating hormone in the pituitary of Chinook salmon (*Oncorhynchus tshawytscha*). *Gen. Comp. Endocrinol*, 114: 151-160.

Matz S P, Takahashi T T. 1994. Immunohistochemical localization of thyrotropin-releasing hormone in the brain of chinook salmon (*Oncorhynchus tshawytscha*). *J. Comp. Neurol*, 345: 214-223.

Mayo K E, Cerelli G M, Lebo R V. 1985. Gene encoding human growth hormone-releasing factor precursor: Structure, sequence, and chromosomal assignment. *P. Natl. Acad. Sci. USA*, 82: 63-67.

McCormick S D, Bradshaw D. 2006. Hormonal control of salt and water balance in vertebrates. *Gen. Comp. Endocrinol*, 147: 3-8.

McCoy J G, Avery D D. 1990. Bombesin: Potential integrative peptide for feeding and satiety. *Peptides*, 11: 595-607.

McDonald T J, Jornvall H, Nilsson G. 1979. Characterization of a gastrin releasing peptide from porcine non-antral gastric tissue. *Biochem. Biophys. Res. Commun*, 90: 227-233.

Medina M, Reperant J, Dufour S, Ward R, Le Belle N, Miceli D. 1994. The distribution of GABA-immunoreactive neurons in the brain of the silver eel (*Anguilla anguilla*L.). *Anat. Embryol*, 189: 25-39.

Meek J, Joosten H W J. 1989. Distribution of serotonin in the brain of the mormyrid teleost *Gnathonemus petersii*. *J. Comp. Neurol*, 281: 206-224.

Meek J, Nieuwenhuys R D. 1998. Holosteans and teleost. In: R. Nieuwenhuys, H. J. Ten Donkelaar, C. Nicholson, Ed. *The Central Nervous System of Vertebrates*. Vol. 1 Springer-Verlag: Heidelberg, 759-937.

Meek J, Joosten H W J, Steinbusch H W M. 1989. Distribution of dopamine immunoreactivity in the brain of the mormyrid teleost *Gnathonemus petersii*. *J. Comp. Neurol*, 2 (81): 362-383.

Melamed P, Eliahu N, Levavi-Sivan B, Ofir M, Farchi-Pisanty O, Rentier-Delrue F, Smal J, Yaron Z, Naor Z. 1995. Hypothalamic and thyroidal regulation of growth hormone in tilapia. *Gen. Comp. Endocrinol*, 97: 13-30.

Metz J R, Huising M O, Meek J, Taverne-Thiele A J, Bonga S E W, Flik G. 2004. Localization, expression and control of adrenocorticotropic hormone in the nucleus preopticus and pituitary gland of common carp (*Cyprinus carpio*L.). *J. Endocrinol*,

182: 23-31.

Miranda L A, Strobl-Mazzulla P H, Somoza G M. 2002. Ontogenetic development and neuroanatomical localization of growth hormone-releasing hormone (GHRH) in the brain and pituitary gland of pejerrey fish *Odontesthes bonariensis*. *Int. J. Dev. Neurosci*, 20: 503-510.

Mitchell G, Sawisky G R, Grey C L, Wong C J, Uretsky A D, Chang J P. 2008. Differential involvement of nitric oxide signaling in dopamine and PACAP stimulation of growth hormone release in goldfish. *Gen. Comp. Endocrinol*, 155: 318-327.

Mohamed J S, Thomas P, Khan I A. 2005. Isolation, cloning, and expression of three prepro-GnRH mRNAs in Atlantic croaker brain and pituitary. *J. Comp. Neurol*, 488: 384.

Montefusco-Siegmund R A, Romero A, Kausel G, Muller M, Fujimoto M, Figueroa J. 2006. Cloning of the prepro C-RFa gene and brain localization of the active peptide in *Salmo salar*. *Cell Tissue Res*, 325: 277.

Montero M, Vidal B, King J A, Tramu G, Vandesande F, Dufour S, Kah O. 1994. Immunocytochemical localization of mammalian GnRH (gonadotropin-releasing hormone) and chicken GnRH-II in the brain of the European silver eel (*Anguilla anguilla*L.). *J. Chem. Neuroanat*, 7: 227-241.

Montero M, Yon L, Rousseau K, Arimura A, Fournier A, Dufour S, Vaudry H. 1998. Distribution, characterization, and growth hormone-releasing activity of pituitary adenylate cyclase-activating polypeptide in the European eel, *Anguilla anguilla*. *Endocrinology*, 139: 4300-4310.

Moons L, Cambre M, Marivoet S, Batten T F, Vanderhaeghen J J, Ollevier F, Vandesande F. 1988. Peptidergic innervation of the adrenocorticotropic hormone (ACTH)-and growth hormone(GH)-producing cells in the pars distalis of the sea bass (*Dicentrarchus labrax*). *Gen. Comp. Endocrinol*, 72: 171-180.

Moons L, CambréM, Ollevier F, Vandesande F. 1989. Immunocytochemical demonstration of close relationships between neuropeptidergic nerve fibers and hormone-producing cell types in the adenohypophysis of the sea bass (*Dicentrarchus labrax*). *Gen. Comp. Endocrinol*, 73: 270-283.

Moons L, Batten T F C, Vandesande F. 1991. Autoradiographic distribution of galanin binding sites in the brain and pituitary of the sea bass (*Dicentrarchus labrax*). *Neurosci. Lett*, 123: 49-52.

Moriyama S, Toshihiro I T O, Takahashi A, Amano M, Sower S A, Hirano T, Yamamori K, Kawauchi H. 2002. A homolog of mammalian PRL-releasing peptide (fish arginyl-phenylalanyl-amide peptide) is a major hypothalamic peptide of prl release in teleost fish. *Endocrinology*, 143: 2071-2079.

Moriyama S, Kasahara M, Amiya N, Takahashi A, Amano M, Sower S A, Yamamori K, Kawauchi H. 2007. RFamide peptides inhibit the expression of melanotropin and growth hormone genes in the pituitary of an agnathan, the sea lamprey, *Petromyzon marinus*. *Endocrinology*, 148: 3740-3749.

Mousa M A, Mousa S A. 2003. Immunohistochemical localization of gonadotropin releasing hormones in the brain and pituitary gland of the Nile perch, *Lates niloticus* (Teleostei, Centropomidae). *Gen. Comp. Endocrinol*, 130: 245-255.

Mukuda T, Matsunaga Y, Kawamoto K, Yamaguchi K I, Ando M. 2005. "Blood-contacting neurons" in the brain of the Japanese eel *Anguilla japonica*. *J. Exp. Zool*, 303A: 366-376.

Nakamachi T, Matsuda K, Maruyama K, Miura T, Uchiyama M, Funahashi H, Sakurai T, Shioda S. 2006. Regulation by orexin of feeding behaviour and locomotor activity in the goldfish. *J. Neuroendocrinol*, 18: 290-297.

Nakanishi S. 1992. Molecular diversity of glutamate receptors and implications for brain function. *Science*, 258: 597-603.

Nelson L E, Sheridan MA. 2005. Regulation of somatostatins and their receptors in fish. *Gen. Comp. Endocrinol*, 142: 117-133.

Nieuwenhuys R, Ten Donkelaar H J, Nicholson C. 1998. *The Central Nervous System of Vertebrates*. Springer-Verlag: Heidelberg.

Northcutt R G. 1995. The forebrain of gnathostomes: in search of a morphotype. *Brain Behav. Evol*, 46: 275-318.

Northcutt R G, Davis R E. 1983. Telencephalic organization in ray-finned fishes. In: Northcutt R G, Davis R E, Ed. *Fish Neurobiology*. Vol. 2. University of Michigan Press: Ann Arbor, 203-236.

Notenboom C D, Garaud J C, Doerr-Schott J, Terlou M. 1981. Localization by immunofluorescence of a gastrin-like substance in the brain of the rainbow trout, *Salmo gairdneri*. *Cell Tissue Res*, 214: 247-255.

Novak C M, Jiang X, Wang C, Teske J A, Kotz C M, Levine J A. 2005. Caloric restriction and physical activity in zebrafish (*Danio rerio*). *Neurosci. Lett*, 383: 99-104.

Nozaki M, Gorbman A, Sower S A. 1994. Diffusion between the neurohypophysis and the adenohypophysis of lampreys, *Petromyzon marinus*. *Gen. Comp. Endocrinol*, 96: 385-391.

Ohtaki T, Shintani Y, Honda S, Matsumoto H, Hori A, Kanehashi K, Terao Y, Kumano S, Takatsu Y, Masuda Y, Ishibashi Y, Watanabe T T, et al. 2001. Metastasis suppressor gene KiSS-1 encodes peptide ligand of a G-protein-coupled receptor. *Nature*, 411: 613-617.

Okawara Y, Ko D, Morley S D, Richter D, Lederis K P. 1992. *In situ* hybridization of corticotropin-releasing factor-encoding messenger RNA in the hypothalamus of the white sucker, *Catostomus commersoni*. *Cell Tissue Res*, 267: 545-549.

Olivereau M, Olivereau J. 1988. Localization of CRF-like immunoreactivity in the brain and pituitary of teleost fish. *Peptides*, 9: 13-21.

Olivereau M, Olivereau J M. 1990. Corticotropin-like immunoreactivity in the brain and pituitary of three teleost species (goldfish, trout and eel). *Cell Tissue Res*, 262: 115-123.

Olivereau M, Olivereau J M. 1991. Immunocytochemical localization of a galanin-like peptidergic system in the brain and pituitary of some teleost fish. *Histochemistry*, 96: 343-354.

Olivereau M, Ollevier F, Vandesande F, Olivereau J. 1984. Somatostatin in the brain and the pituitary of some teleosts. Immunocytochemical identification and the effect of starvation. *Cell Tissue Res*, 238: 289-296.

Olivereau M, Ollevier F, Vandesande F, Verdonck W. 1984. Immunocytochemical identification of CRF-like and SRIF-like peptides in the brain and the pituitary of cyprinid fish. *Cell Tissue Res*, 237: 379-382.

Olivereau M, Moons L, Olivereau J, Vandesande F. 1988. Coexistence of corticotropin-releasing factor-like immunoreactivity and vasotocin in perikarya of the preoptic nucleus in the eel. *Gen. Comp. Endocrinol*, 70: 41-48.

Olivereau M, Olivereau J, Vandesande F. 1990. Localization of growth hormone-releasing factor-like immunoreactivity in the hypothalamo-hypophysial system of some teleost species. *Cell Tissue Res*, 259: 73-80.

Palay S L. 1945. Neurosecretion. VII. The preoptico-hypophyseal pathway in fishes. *J. Comp. Neurol*, 82: 129-143.

Pan J X, Lechan R M, Lin H D, Jackson I M D. 1985. Immunoreactive neuronal pathways of growth hormone-releasing hormone (GRH) in the brain and pituitary of the teleost *Gadus morhua*. *Cell Tissue Res*, 241: 487-493.

Pandolfi M, Cánepa M M, Ravaglia M A, Maggese M C, Paz D A, Vissio P G. 2003. Melanin-concentrating hormone system in the brain and skin of the cichlid fish *Cichlasoma dimerus*: Anatomical localization ontogeny and distribution in comparison to α-melanocyte-stimulating hormone-expressing cells. *Cell Tissue Res*, 311: 61.

Pandolfi M, Cueto J A M, Lo Nostro F L, Downs J L, Paz D A, Maggese M C, Urbanski H F. 2005. GnRH systems of *Cichlasoma dimerus* (Perciformes, Cichlidae) revisited: A localization study with antibodies and riboprobes to GnRH-associated peptides. *Cell Tissue Res*, 321: 219-232.

Parhar I S. 1997. GnRH in tilapia: three genes, three origins and their roles. In:

Parhar I S, Sakuma Y, Ed. *GnRH Neurons*, *Gene to behavior*. Brain Shuppan: Tokyo, 99-122.

Parhar I S, Iwata M. 1994. Gonadotropin releasing hormone (GnRH) neurons project to growth hormone and somatolactin cells in the Steelhead trout. *Histochemistry*, 102: 195-203.

Parhar I S, Iwata M, Pfaff D W, Schwanzel-Fukuda M. 1995. Embryonic development of gonadotropin-releasing hormone neurons in the Sockeye salmon. *J. Comp. Neurol*, 362: 256-270.

Parhar I S, Ogawa S, Sakuma Y. 2004. Laser-captured single digoxigenin-labeled neurons of gonadotropin-releasing hormone types reveal a novel G protein-coupled receptor (GPR54) during maturation in cichlid fish. *Endocrinology*, 145: 3613-3618.

Peng C, Chang J P, Yu K L, Wong A O L, Van Goor F, Peter R E, Rivier J E. 1993. Neuropeptide-Y stimulates growth hormone and gonadotropin-II secretion in the goldfish pituitary: Involvement of both presynaptic and pituitary cell actions. *Endocrinology*, 132: 1820-1829.

Peng C, Gallin W, Peter R E, Blomqvist A G, Larhammar D. 1994. Neuropeptide-Y gene expression in the goldfish brain: distribution and regulation by ovarian steroids. *Endocrinology*, 134: 1095-1103.

Pepels P P L M, Meek J, Wendelaar Bonga S E, Balm P H M. 2002. Distribution and quantification of corticotropin-releasing hormone (CRH) in the brain of the teleost fish *Oreochromis mossambicus* (tilapia). *J. Comp. Neurol*, 453: 247-268.

Peter R E, Gill V E. 1975. A stereotaxical atlas and technique for forebrain nuclei of the goldfish, *Carassius auratus*. *J. Comp. Neurol*, 159: 69-101.

Peter R E, McKeown B A. 1975. Hypothalamic control of prolactin and thyrotropin secretion in teleosts, with special reference to recent studies on the goldfish. *Gen. Comp. Endocrinol*, 25: 153-165.

Peter R E, Macey M J, Gill V E. 1975. A stereotaxic atlas and technique for forebrain nuclei of the killifish *Fundulus heteroclitus*. *J. Comp. Neurol*, 159: 103-127.

Peter R E, Crim L W, Billard R. 1991. A stereotaxical atlas and implantation technique for the nuclei of the diencephalon of Atlantic salmon (*Salmo salar*) parr. *Reprod. Nutr. Dev.* 31: 167-186.

Peyon P, Lin X W, Himick B A, Peter R E. 1998. Molecular cloning and expression of cDNA encoding brain preprocholecystokinin in goldfish. *Peptides*, 19: 199-210.

Peyon P, Saied H, Lin X, Peter R E. 1999. Postprandial, seasonal and sexual variations in cholecystokinin gene expression in goldfish brain. *Mol. Brain Res*, 74: 190-196.

Pickavance L C, Staines W A, Fryer J N. 1992. Distributions and colocalization of neuropeptide-Y and somatostatin in the goldfish brain. *J. Chem. Neuroanat*, 5: 221-233.

Pickford G E, Atz J W. 1957. *The Physiology of the Pituitary Gland of Fish*. New York Zoological Society: New York.

Pierre J, Mahouche M, Suderevskaya E I, Repérant J, Ward R. 1997. Immunocytochemical localization of dopamine and its synthetic enzymes in the central nervous system of the lamprey *Lampetra fluviatilis*. *J. Comp. Neurol*, 380: 119-135.

Pogoda H M, Hammerschmidt M. 2007. Molecular genetics of the pituitary development in zebrafish. *Semin. Cell Dev. Biol*, 18: 543-558.

Pontet A, Danger J M, Dubourg P, Pelletier G, Vaudry H, Calas A, Kah O. 1989. Distribution and characterization of neuropeptide Y-like immunoreactivity in the brain and pituitary of the goldfish. *Cell Tissue Res*, 255: 529-538.

Powell K A, Baker B I. 1988. Structural studies of nerve terminals containing melanin-concentrating hormone in the eel, *Anguilla anguilla*. *Cell Tissue Res*, 251: 433-439.

Power D M, Canario A V M, Ingleton P M. 1996. Somatotropin release-inhibiting factor and galanin innervation in the hypothalamus and pituitary of seabream (*Sparus aurata*). *Gen. Comp. Endocrinol*, 101: 264-274.

Prasada Rao P D, Job T C, Screibman M P. 1993. Hypophysiotrophic neurons in the hypothalamus of the catfish *Clarias batrachus*: A cobaltous lysine and HRP study. *Brain Behav. Evol.* 42: 24-38.

Prasada Rao P D, Murthy C K, Cook H, Peter R E. 1996. Sexual dimorphism of galanin-like immunoreactivity in the brain and pituitary of goldfish, *Carassius auratus*. *J. Chem. Neuroanat*, 10: 119-135.

Price D A, Greenberg M J. 1977. Structure of a molluscan cardioexcitatory neuropeptide. *Science*, 197: 670-671.

Prober D A, Rihel J, Onah A A, Sung R J, Schier A F. 2006. Hypocretin/orexin overexpression induces an insomnia-like phenotype in zebrafish. *J. Neurosci*, 26: 13400-13410.

Querat B, Sellouk A, Salmon C. 2000. Phylogenetic analysis of the vertebrate glycoprotein hormone family including new sequences of sturgeon (*Acipenserbaeri*) beta subunits of the two gonadotropins and the thyroid-stimulating hormone. *Biol. Reprod*, 63: 222-228.

RamaKrishna N S, Subhedar N. 1989. Hypothalamic innervation of the pituitary in the catfish, *Clarias batrachus* (L.): a retrograde horseradish peroxidase study. *Neurosci. Lett*, 107: 39-44.

RamaKrishna N S, Subhedar N, Schreibman M P. 1992. FMRFamide-like immunoreactive nervus terminalis innervation to the pituitary in the catfish, *Clarias batrachus* (Linn.): Demonstration by lesion and immunocytochemical techniques. *Gen. Comp. Endocrinol*, 85: 111-117.

Rao S D, Prasada Rao P D, Peter R E. 1996. Growth hormone-releasing hormone immunoreactivity in the brain, pituitary, and pineal of the goldfish, *Carassius auratus*. *Gen. Comp. Endocrinol*, 102: 210-220.

Rehfeld J F, Lennart F H, Goetze J P, Hansen T V O. 2007. The biology of cholecystokinin and gastrin peptides. *Curr. Top. Med. Chem*, 7: 1154-1165.

Rink E, Wullimann M F. 2001. The teleostean (zebrafish) dopaminergic system ascending to the subpallium (striatum) is located in the basal diencephalon (posterior tuberculum). *Brain Res*, 889: 316-330.

Roberts B L, Meredith G E, Maslam S. 1989. Immunocytochemical analysis of the dopamine system in the brain and spinal cord of the European eel, *Anguilla anguilla*. *Anat. Embryol*, 180: 401-412.

Rodríguez-Gómez F J, Rendón-Unceta M C, Sarasquete C, Muñoz-Cueto J A. 2000. Localization of galanin-like immunoreactive structures in the brain of the Senegalese sole, *Solea senegalensis*. *Histochem. J.* 32: 123-131.

Rodríguez-Gómez F J, Rendón-Unceta M C, Sarasquete C, Muñoz-Cueto J A. 2000. Distribution of serotonin in the brain of the Senegalese sole, *Solea senegalensis*: An immunohistochemical study. *J. Chem. Neuroanat*, 18: 103-118.

Rodríguez-Gómez F J, Rendón-Unceta C, Sarasquete C, Muñoz-Cueto J A. 2001. Distribution of neuropeptide Y-like immunoreactivity in the brain of the Senegalese sole (*Solea senegalensis*). *Anat. Rec*, 262: 227-237.

Rodríguez M A, Anadón R, Rodríguez-Moldes I. 2003. Development of galanin-like immunoreactivity in the brain of the brown trout (*Salmo trutta fario*), with some observations on sexual dimorphism. *J. Com. Neurol*, 465: 263-285.

Saito D, Urano A. 2001. Synchronized periodic Ca^{2+} pulses define neurosecretory activities in magnocellular vasotocin and isotocin neurons. *J. Neurosci*, 21: RC178.

Saito D, Komatsuda M, Urano A. 2004. Functional organization of preoptic vasotocin and isotocin neurons in the brain of rainbow trout: Central and neurohypophysial projections of single neurons. *Neuroscience*, 124: 973-984.

Sakamoto T, Agustsson T, Moriyama S, Itoh T, Takahashi A, Kawauchi H, Björnsson B T, Ando M. 2003. Intra-arterial injection of prolactin-releasing peptide elevates prolactin gene expression and plasma prolactin levels in rainbow trout. *J. Comp. Physiol. B*, 173B: 333-337.

Sakamoto T, Fujimoto M, Ando M. 2003. Fishy tales of prolactin-releasing peptide. *Int. Rev. Cytol*, 225: 91.

Sakharkar A J, Singru P S, Sarkar K, Subhedar N K. 2005. Neuropeptide Y in the forebrain of the adult male cichlid fish *Oreochromis mossambicus*: Distribution, effects of castration and testosterone replacement. *J. Comp. Neurol*, 489: 148-165.

Salbert G, Chauveau I, Bonnec G, Valotaire Y, Jego P. 1992. One of the two trout proopiomelanocortin messenger RNAs potentially encodes new peptides. *Mol. Endocrinol*, 6: 1605-1613.

Sas E, Maler L. 1991. Somatostatin-like immunoreactivity in the brain of an electric fish (*Apteronotus leptorhynchus*) identified with monoclonal antibodies. *J. Chem. Neuroanat*, 4: 155-186.

Sathyanesan A G. 1965. Hypothalamo-neurohypohyseal system in the normal and hypophysectomized teleost *Porichthys notatus* Girard and its response to continuous light. *J. Morphol*, 117: 25-48.

Schally A V, Redding T W, Bowers C Y, Barrett J F. 1969. Isolation and properties of porcine thyrotropin-releasing hormone. *J. Biol. Chem*, 244: 4077-4088.

Scharrer E. 1928. Untersuchungen über das Zwischenhirn der fische. I. *Z. Vergleich. Physiol*, 7: 1-38.

Schiöth H B, Haitina T, Ling M K, Ringholm A, Fredriksson R, Cerdá-Reverter J M, Klovins J J. 2005. Evolutionary conservation of the structural, pharmacological and genomic characteristics of the melanocortin receptors subtypes. *Peptides*, 26: 1886-1900.

Seale A P, Itoh T, Moriyama S, Takahashi A, Kawauchi H, Sakamoto T, Fujimoto M, Riley L G, Hirano T, Grau E G. 2002. Isolation and characterization of a homologue of mammalian prolactin releasing peptide from the tilapia brain and its effect on prolactin release from the tilapia pituitary. *Gen. Comp. Endocrinol*, 125: 328-339.

Sherwood N M, Krueckl S L, McRory J E. 2000. The origin and function of the pituitary adenylate cyclase-activating polypeptide (PACAP)/glucagon superfamily. *Endocr. Rev*, 21: 619-670.

Silverstein J T, Breininger J, Baskin D G, Plisetskaya E M. 1998. Neuropeptide Y-like gene expression in the salmon brain increases with fasting. *Gen. Comp. Endocrinol*, 110: 157-165.

Singru P S, Mazumdar M, Sakharkar A J, Lechan R M, Thim L, Clausen J T, Subhedar N K. 2007. Immunohistochemical localization of cocaine- and amphetamine-regulated transcript peptide in the brain of the catfish, *Clarias batrachus* (Linn.). *J. Comp. Neurol*, 502: 215-235.

Sloley B D, Kah O, Trudeau V L, Dulka J G, Peter R E. 1992. Amino acid neurotransmitters and dopamine in brain and pituitary of the goldfish: involvement in the regulation of gonadotropin secretion. *J. Neurochem*, 58: 2254-2262.

Smeets W J A J, González A. 2000. Catecholamine systems in the brain of vertebrates: new perspectives through a comparative approach. *Brain Res. Rev*, 33: 308-379.

Somoza G M, Peter R E. 1991. Effects of serotonin on gonadotropin and growth hormone release from *in vitro* perfused goldfish pituitary fragments. *Gen. Comp. Endocrinol*, 82: 103-110.

Song Y, Cone R D. 2007. Creation of a genetic model of obesity in a teleost. *FASEB J.* 21: 2042-2049.

Spiess J, Villarreal J, Vale W. 1981. Isolation and sequence analysis of a somatostatin-like polypeptide from ovine hypothalamus. *Biochemistry*, 20: 1982-1988.

Stefano A V, Vissio P G, Paz D A, Somoza G M, Maggese M C, Barrantes G E. 1999. Colocalization of GnRH binding sites with gonadotropin-, somatotropin-, somatolactin-, and prolactin-expressing pituitary cells of the pejerrey, *Odontesthes bonariensis*, *in vitro*. *Gen. Comp. Endocrinol*, 116: 133-139.

Stell W K, Walker S E, Chohan K S, Ball A K. 1984. The goldfish nervus terminalis: a luteinizing hormone-releasing hormone and molluscan cardioexcitatory peptide immunoreactive olfactoretinal pathway. *Proc. Natl. Acad. Sci. USA*, 81: 940-944.

Striedter G F. 1990. The diencephalon of the channel catfish, *Ictalurus punctatus*. I Nuclear organization. *Brain Behav. Evol*, 36: 329-354.

Subhedar N, Cerdá J, Wallace R A. 1996. Neuropeptide Y in the forebrain and retina of the killifish, *Fundulus heteroclitus*. *Cell Tissue Res*, 283: 313-323.

Sueiro C, Carrera I, Ferreiro S, Molist P, Adrio F, Anadón R, Rodríguez-Moldes I. 2007. New insights on saccus vasculosus evolution: A developmental and immunohistochemical study in elasmabranchs. *Brain Behav. Evol*, 70: 187-204.

Sundström G, Larsson T A, Brenner S, Venkatesh B, Larhammar D. 2008. Evolution of neuropeptide y family: New genes by chromosome duplications in early vertebrates and in teleost fishes. *Gen. Comp. Endocrinol*, 155: 705-716.

Sutcliffe J G, de Lecea L. 2000. The hypocretins: excitatory neuromodulatory peptides for multiple homeostatic systems, including sleep and feeding. *J. Neurosci. Res*, 62: 161-168.

Suzuki H, Miyoshi Y, Yamamoto T. 2007. Orexin-A (hypocretin 1)-like immunoreactivity in growth hormone-containing cells of the Japanese seaperch (*Lateolabrax japonicus*) pituitary. *Gen. Comp. Endocrinol*, 150: 205-211.

Takahashi A, Amemiya Y, Nozaki M, Sower S A, Kawauchi H. 2001. Evolutionary significance of proopiomelanocortin in agnatha and chondrichthyes. *Comp. Biochem. Physiol*, 29B: 283-289.

Tena-Sempere M. 2006. The roles of kisspeptins and G protein-coupled receptor-54 in pubertal development. *Curr. Opin. Pediatr*, 18: 442-447.

Tran T N, Fryer J N, Bennett H P J, Tonon M C, Vaudry H. 1989. TRH stimulates the release of POMC-derived peptides from goldfish melanotropes. *Peptides*, 10: 835-841.

Trudeau V L. 1997. Neuroendocrine regulation of gonadotrophin II release and gonadal growth in the goldfish, *Carassius auratus*. *Rev. Reprod*, 2: 55-68.

Trudeau V L, Sloley B D, Peter R E. 1993. GABA stimulation of gonadotropin-II release in goldfish: involvement of GABAA receptors, dopamine, and sex steroids. *Am. J. Physiol*, 265: R348-55.

Trudeau V L, Sloley B D, Kah O, Mons N, Dulka J G, Peter R E. 1996. Regulation of growth hormone secretion by amino acid neurotransmitters in the goldfish (I): Inhibition by N-methyl-D, L-aspartic acid. *Gen. Comp. Endocrinol*, 103: 129-137.

Trudeau V L, Kah O, Chang J P, Sloley B D, Dubourg P, Fraser E J, Peter R E. 2000. The inhibitory effects of (gamma) -aminobutyric acid (GABA) on growth hormone secretion in the goldfish are modulated by sex steroids. *J. Exp. Biol*, 203: 1477-1485.

Trudeau V L, Spanswick D, Fraser E J, Lariviere K, Crump D, Chiu S, MacMillan M, Schulz R W. 2000. The role of amino acid neurotransmitters in the regulation of pituitary gonadotropin release in fish. *Biochem. Cell Biol*, 78: 241-259.

Tsuneki K, Fernholm B. 1975. Effect of thyrotropin-releasing hormone on the thyroid of a teleost, *Chasmichthys dolicognathus*, and a hagfish, *Eptatretus burgeri*. *Acta Zool*, 56: 61-65.

Tsutsui K, Ukena K. 2006. Hypothalamic LPXRF-amide peptides in vertebrates: Identification, localization and hypophysiotropic activity. *Peptides*, 27: 1121-1129.

Ukena K, Iwakoshi E, Minakata H, Tsutsui K. 2002. A novel rat hypothalamic RFamide-related peptide identified by immunoaffinity chromatography and mass spectrometry. *FEBS Lett*, 512: 255-258.

Unniappan S, Cerdá-Reverter J M, Peter RE. 2004. *In situ* localization of preprogalanin mRNA in the goldfish brain and changes in its expression during feeding and starvation. *Gen. Comp. Endocrinol*, 136: 200-207.

Vallarino M, Delbende C, Ottonello I, Tranchand-Bunel D, Jegou S, Vaudry H. 1989. Immunocytochemical localization and biochemical characterization of α-melanocyte-stimulating hormone in the brain of the rainbow trout, *Salmo gairdneri*. *J. Neuroendocrinol*, 1: 53-60.

van Aerle R, Kille P, Lange A, Tyler C R. 2008. Evidence for the existence of a functional Kiss1/Kiss1 receptor pathway in fish. *Peptides*, 29: 57-64.

van de Kamer J C, Zandbergen M A. 1981. The hypothalamic-hypophyseal system and its evolutionary aspects in *Scyliorhinus caniculus*. *Cell Tissue Res*, 214: 575-582.

Vaudry D, Gonzalez B J, Basille M, Yon L, Fournier A, Vaudry H. 2000. Pituitary adenylate cyclase-activating polypeptide and its receptors: from structure to functions. *Pharmacol. Rev*, 52: 269-324.

Vecino E, Perez M T R, Ekstrom P. 1994. *In situ* hybridization of neuropeptide Y (NPY) mRNA in the goldfish brain. *NeuroReport*, 6: 127-131.

Vigh-Teichmann I, Vigh B, Korf H W, Oksche A. 1983. CSF-contacting and other somatostatin-immunoreactive neurons in the brains of *Anguilla anguilla*, *Phoxinus phoxinus*, and *Salmo gairdneri* (Teleostei). *Cell Tissue Res*, 233: 319-334.

Vissio P G, Stefano A V, Somoza G M, Maggese M C, Paz D A. 1999. Close association of gonadotropin-releasing hormone fibers and gonadotropin, growth hormone, somatolactin and prolactin expressing cells in pejerrey, *Odontesthes bonariensis*. *Fish Physiol. Biochem*, 21: 121-127.

Volkoff H, Peter R E. 1999. Actions of two forms of gonadotropin releasing hormone and a GnRH antagonist on spawning behavior of the goldfish *Carassius auratus*. *Gen. Comp. Endocrinol*, 116: 347-355.

Volkoff H, Peter R E. 2001. Characterization of two forms of cocaine- and amphetamine-regulated transcript (CART) peptide precursors in goldfish: molecular cloning and distribution, modulation of expression by nutritional status, and interactions with leptin. *Endocrinology*, 142: 5076-5088.

Volkoff H, Peyon P, Lin X, Peter R E. 2000. Molecular cloning and expression of cDNA encoding a brain bombesin/gastrin-releasing peptide-like peptide in goldfish. *Peptides*, 21: 639-648.

Volkoff H, Canosa L F, Unniappan S, Cerda-Reverter J M, Bernier N J, Kelly S P, Peter R E. 2005. Neuropeptides and the control of food intake in fish. *Gen. Comp. Endocrinol*, 142: 3-19.

Vollrath L. 1967. On neurosecretory innervation of the adenohypophysis in teleost fishes, especially in the Hippocampus cuda and Tinca tinca. *Z. Zellforsch. Mikrosk. Anat*, 78: 234-260.

Wagner S, Castel M, Gainer H, Yarom Y. 1997. GABA in the mammalian suprachiasmatic nucleus and its role in diurnal rhythmicity. *Nature*, 387: 598-603.

Weber G M, Powell J F F, Park M, Fischer W H, Craig A G, Rivier J E, Nanakorn U, Parhar I S, Ngamvongchon S, Grau E G, Sherwood N M. 1997. Evidence that gonadotropin-releasing hormone (GnRH) functions as a prolactin-releasing factor in a teleost fish (*Oreochromis mossambicus*) and primary structures for three native GnRH molecules. *J. Endocrinol*, 155: 121-132.

Weltzien F A, Pasqualini C, Sébert M E, Vidal B, Le Belle N, Kah O, Vernier P, Dufour S. 2006. Androgen-dependent stimulation of brain dopaminergic systems in the female European eel (*Anguilla anguilla*). *Endocrinology*, 147: 2964-2973.

White R B, Eisen J A, Kasten T L, Fernald R D. 1998. Second gene for gonadotropin-releasing hormone in humans. *Proc. Natl. Acad. Sci. USA*, 95: 305-309.

Wigham T, Batten T F C. 1984. *In vitro* effects of thyrotropin-releasing hormone and somatostatin on prolactin and growth hormone release by the pituitary of *Poecilia latipinna*. I. An electrophoretic study. *Gen. Comp. Endocrinol*, 55: 444-449.

Winberg S, Nilsson A, Hylland P, Söderstöm V, Nilsson G E. 1997. Serotonin as a regulator of hypothalamic-pituitary-interrenal activity in teleost fish. *Neurosci. Lett*, 230: 113-116.

Wong A O L, Leung M Y, Shea W L C, Tse L Y, Chang J P, Chow B K C. 1998. Hypophysiotropic action of pituitary adenylate cyclase-activating polypeptide (PACAP) in the goldfish: Immunohistochemical demonstration of PACAP in the pituitary, PACAP stimulation of growth hormone release from pituitary cells, and molecular cloning of pituitary type I PACAP receptor. *Endocrinology*, 139: 3465-3479.

Wullimann M, Mueller T. 2004. Teleostean and mammalian forebrains contrasted: Evidence from genes to behaviour. *J. Comp Neurol*, 475: 143-162.

Wullimann M F, Rupp B, Reichert H. 1996. *Neuroanatomy of Zebrafish Brain: A Topological Atlas*. Birkhaeuser Verlag: Switzerland.

Yamamoto N, Parhar I S, Sawai N, Oka Y, Ito H. 1998. Preoptic gonadotropin-releasing hormone (GnRH) neurons innervate the pituitary in teleosts. *Neurosci. Res*, 31: 31-38.

Yokogawa T, Marin W, Faraco J, Pézeron G, Appelbaum L, Zhang J, Rosa F, Mourrain P, Mignot E. 2007. Characterization of sleep in zebrafish and insomnia in hypocretin receptor mutants. *PLoS Biol*, 5: 2379-2397.

Yu K L, Sherwood N M, Peter R E. 1988. Differential distribution of two molecular forms of gonadotropin-releasing hormone in discrete brain areas of goldfish (*Carassius auratus*). *Peptides*, 9: 625-630.

Yulis C R, Lederis K. 1986. The distribution of "extraurophyseal" urotensin I-immunoreactivity in the central nervous system of *Catostomus commersoni* after urophysectomy. *Neurosci. Lett*, 70: 75-80.

Yulis C R, Lederis K. 1987. Co-localization of the immunoreactivities of corticotropin-releasing factor and arginine vasotocin in the brain and pituitary system of the teleost *Catostomus commersoni*. *Cell Tissue Res*, 247: 267-273.

Yulis C R, Lederis K. 1988. Occurrence of an anterior spinal, cerebrospinal fluid-contacting, urotensin II neuronal system in various fish species. *Gen. Comp. Endocri-*

nol, 70: 301-311.

Yulis C R, Lederis K, Wong K L, Fisher A W F. 1986. Localization of urotensin I- and corticotropin-releasing factor-like immunoreactivity in the central nervous system of *Catostomus commersoni*. *Peptides*, 7: 79-86.

Zandbergen M A, Voormolen A H T, Kah O, Goos H J T. 1994. Immunohistochemical localizatin of neuropeptide Y positive cell bodies and fibres in forebrain and pituitary of the African catfish, *Clarias gariepinus*. *Neth. J. Zool*, 44: 43-54.

Zupanc G K, Horschke I, Lovejoy D A. 1999. Corticotropin releasing factor in the brain of the gymnotiform fish, *Apteronotus leptorhynchus*: immunohistochemical studies combined with neuronal tract tracing. *Gen. Comp. Endocrinol*, 114: 349-364.

第 2 章 下丘脑和脑垂体的内分泌靶标

内环境的维持，亦称为体内稳态（homeostasis），要求生物有机体适应多种多样的环境状况。脊椎动物包括鱼类内环境的维持，是由多种反馈作用机理使激素变化保持在一定范围内，并通过精准调整内分泌的反应而实现的。对这种反馈作用机理阐述最清楚的事例是由一些作用于下丘脑-脑垂体复合体的外周激素实施的。例如，性类固醇激素对生殖轴的影响，皮质醇对促肾上腺皮质激素-释放因子（CRF）/促肾上腺皮质激素（ACTH）轴的作用，或者类胰岛素生长因子Ⅰ（IGF-I）对生长激素（GH）产生的影响，等等，这些都是通过脑和脑垂体的受体对这些外周激素的表达而实现的。这些作用机理了解得特别清楚的是生殖周期的来龙去脉，包括性类固醇激素对调控生殖轴的神经内分泌线路的反馈影响。然而，同样至关重要的是一些代谢性激素所产生的作用，包括胰岛素、IGF-I，瘦素（leptin）和皮质醇对较高层次调节中心的影响。这些复杂的调控作用机理以及它们的交叉作用对于确保在基础代谢、生殖活动和应激反应之间的能量分配是必不可少的，而所有这些对于个体和种族的生存都起着决定性的作用。本章的目的是综合介绍有关性类固醇激素、肾上腺皮质激素、瘦素、胰岛素、IGF-I 和甲状腺素等在鱼类脑表达部位的相关资料，同时亦提供有关这些受体和调控脑垂体的神经内分泌线路之间相互关系的信息。

2.1 导　　言

在 1865 年，Claude Bernard 提出了体内稳态（homeostasis）的概念，并强调"内部环境的稳定是自由与独立生存的条件"。这显然意味着任何重大的改变内部环境的趋向都会由于作用机理抗拒这种变化而被取消。这适合于许多生理作用机理，特别重要的是它们依靠持续而准确地调整中枢神经系统即下丘脑和外周内分泌腺体之间的相互作用而保持能量的稳态。下丘脑控制内分泌腺体的分泌活动，而内分泌腺体又通过反馈作用影响下丘脑，使生命有机体能在一定范围内持续地适应这些变动。这就是神经内分泌学的全部内容。正如一位本领域的开拓者——法国的 Claude Kordon 所说的"神经内分泌学是一门关注脑和激素之间相互交叉作用的科学"。

许多的相互作用发生在下丘脑，即脑的中心，一方面，整合代谢的、环境的和激素的信息；另一方面，激活特异性的神经元通道以调整行为的和生殖的反应。这个过程包含了特异性神经元群体之间精细的相互作用，以便能够监控输入的外周信号，进而使能量平衡、代谢、生长和生殖等生命活动能够协调进行。在整个进化过程中，这些作用机理得以选择性地发展以严格地保持着体内稳态，即"自由与独立生存的条件"，使能量的分配保证生殖活动持续顺利进行，从而使种族不断繁衍生存。

这些相互作用发生在所有的脊椎动物当中，以哺乳动物的报道与记载最多，最近的一些综述可供读者们查阅（Gamba 和 Pralong，2006；Tena-Sempere，2006；Navarro 等，2007；Gao 和 Horvath，2008；Goulis 和 Tarlatzis，2008；Popa 等，2008）。在这方面对鱼类研究得最少。但是，毫无疑问的是这些脑/激素的相互关系都出现在进化早期的脊椎动物当中，包括脑垂体的形成以及调控脑垂体激素的合成与释放，等等。由于这些脑垂体激素保守的功能，人们没有理由认为调控这些激素分泌活动的作用机理在鱼类和哺乳类之间有很大的差别。因而，正如比较内分泌学所阐明的，许多在哺乳类建立的概念和观点都可以适用于鱼类。本章的内容是综述鱼类脑和脑垂体激素的作用靶标，特别着重介绍性激素、糖皮质激素（glucocorticoids）和代谢激素（见图2.1）。

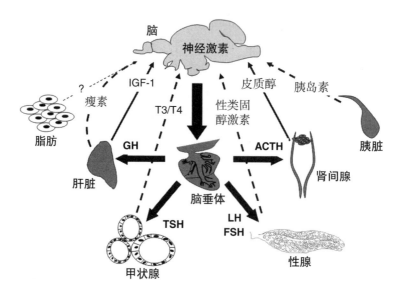

图2.1　鱼类保证维持激素环境的主要脑/激素关系
通过神经激素的合成，脑调控脑垂体激素的释放；这些脑垂体激素刺激外周激素的分泌活动。外周激素转而对脑/脑垂体复合体产生反馈作用，以调节神经内分泌系统的活动。

2.2　性类固醇激素

性类固醇激素在调控生殖活动中起着关键作用，它们亦参与能量平衡的调控（Gómez，2007；Lovejoy 和 Sainsbury，2009；O'Sullivan，2009）。在阉割/激素置换试验后都报道性类固醇激素负的和正的反馈作用机理（Donaldson 和 McBride，1967；Olivereau 和 Olivereau，1979；Crim 和 Evans，1983；Trudeau，1997）。性类固醇激素这种正的和负的反馈作用经常是矛盾的，往往取决于类固醇种类、鱼的种类、生理状态、靶组织（脑或脑垂体）以及研究参数（合成或释放）等等。另一些难以说明事实真相的原因是许多硬骨鱼类的生殖发育往往是不同步的（asynchronous），导致复杂的类固醇激素分泌型式，因而产生混乱的解析。

2.2.1 类固醇激素在鱼脑内产生

传统的观点认为类固醇激素影响下丘脑-脑垂体复合体，而这些类固醇激素是通过脑垂体激素，即促性腺激素（GTH）刺激外周腺体，主要是（但不是唯一的）性腺而产生的。这些类固醇激素转而反馈作用于神经内分泌系统，调控下丘脑-脑垂体激素的分泌活动，以便持续地使它们的活动适合于身体的生理需求。但是，一个新提出的观念是，脑本身就是类固醇激素生成的器官，它能表达几种类固醇激素生成的酶类，因而产生一系列类固醇激素，其功能作用还不是很清楚。神经类固醇激素的观点是 E. E. Baulieu 的实验室提出的，他们发现哺乳动物脑内孕烯醇酮（pregnenolone）和脱氢表雄酮（dehydroepiandrosterone）的浓度要比血浆高。此外，切除肾上腺和阉割后，脑的孕烯醇酮和脱氢表雄酮含量并不减少（Corpechot 等，1981；Robel 等，1995）。现在已经清楚地知道哺乳动物至少有几个脑区，主要是海马（hippocampus）区，能够利用胆固醇重新（de novo）产生雌激素和雄激素（Hojo 等，2004，2008）。

在鱼类中，脑能表达类固醇生成酶类的证据主要涉及 $cyp19a1b$ 基因的底物——P450 芳化酶 B（P450aromatase B，AroB）和 5α-还原酶（5α-reductase），这两种酶将睾酮分别转化为雌二醇和双氢睾酮（dihydrotestosterone）（Callard 等，1978，1981；Pasmanik 和 Callard，1985；Pasmanik 等，1988）。这些开拓性研究表明睾酮能够有效地在鱼脑内进行代谢活动。然而，到目前为止还缺乏有关类固醇产生所必需的其他一些类固醇生成酶类的研究报道。分解胆固醇的侧链需要 P450 scc 以产生孕烯醇酮。孕烯醇酮进一步代谢是由两个关键的类固醇生成酶，即 3β-羟基类固醇脱氢酶（3β-hydroxysteroid dehydrogenase，3β-HSD）/D4-D5 异构酶（isomerase）和细胞色素 P450c17（CYP17）实行的。3β-HSD 使孕烯醇酮脱氢与异构化而成为黄体酮（progesterone）。CYP17 使 C21 类固醇（17α-羟化酶活性）羧基化，接着两-碳侧链（C17，C20 裂解酶活性）分解。这将分别产生 C19 类固醇雄烷二酮（androstenedione）或脱氢表雄酮（dehydroepiandrosterone）。所有这些酶类在不同种鱼类脑中的表达都有文献报道，但还需要采用辅助的技术做精确的研究以进一步确认鱼类脑产生一系列有活性类固醇的能力。

采用整体杂交（in toto hybridization），P450 scc 能在成年斑马鱼（Danio rerio）脑中高度表达，主要是在前脑（Hsu 等，2002）。只有采用 PCR 能证明 CYP17 在鱼脑中表达（Halm 等，2003；Wang 和 Ge，2004；Tomy 等，2007），虽然初步的原位杂交结果已表明 CYP17 信息和 3β-HSD 转录体一起广泛出现在前脑中（Tong S. K.，Kah O.，Diotel N. 和 Chung B. C.，未发表结果）。3β-HSD 在成年斑马鱼脑中的表达和活性亦有报道（Sakamoto 等，2001）。采用抗纯化的牛肾上腺 3β-HSD 的抗体进行研究，发现成群的免疫反应细胞体定位于背端脑区、中后丘脑核、视前核、后结节核、旁室器（paraventricular organ）和中纵束（medial longitudinal fascicle）（Sakamoto 等，2001）。类-3β-HSD 的免疫反应亦在小脑浦肯野神经元（cerebellar Purkinje neurons）的细胞体观察到。这些研究结果有力地证明鱼脑能够产生黄体酮或者脱氢表雄酮。关于 17β-HSD，用 PCR 得到的研究结果表明，三个候选同源物中有 HSD17β1 和 HSD17β3 两个存在于斑马鱼中，并在其脑内表达（Mindnich 等，2004）。

值得指出的是，最近对黑棘鲷（*Acanthopagrus schlegeli*）的研究，在孵化后仅 60 天性腺性分化之前的幼鱼脑中存在着类固醇生成的关键酶类。这表明脑的类固醇生物合成能力要比性腺组织学分化早些出现（Tomy 等，2007）。在孵化后 120 天的黑棘鲷中，这些基因的 mRNA 出现同步性高峰，表明雌二醇可能在这期间在前脑和中脑局部形成（Tomy 等，2007）。在虹鳟（*Oncorhynchus mykiss*）全雄的群体中亦得到类似的研究结果（Vizziano 等，未发表结果），再次表明在发育的鱼类脑内类固醇的产生和性腺是没有关系的。最近对斑马鱼的研究表明，雌激素在脑内的产生可能和神经发生（neurogenesis）有联系（Pellegrini 等，2007；Mouriec 等，2008）。

和其他酶类不同，关于鱼脑内的芳化酶已有大量的研究资料。一些研究报道在发育的和成年的鱼体内，芳化酶 B 的表达严格地局限于放射状胶质细胞（radial glial cell）（见图 2.2）。这些细胞特征是核小，靠近脑室，长的放射状突起以尾足形式终止于脑表面（Rakic，1978；Bentivoglio 和 Mazzarello，1999）。放射状胶质细胞参与胚胎的神经发生；这些细胞在哺乳类神经发生后都消失掉，而不同的是它们大多继续存在于非哺乳类特别是鱼类成年的脑内。对光蟾鱼（*Porichthys notatus*）的研究首次报道芳化酶 B 在放射状胶质细胞的强烈表达（Forlano 等，2005）；随后在对虹鳟、斑马鱼、银汉鱼（*Odontesthes bonariensis*）和双带锦鱼（*Thalassoma bifasciatum*）（Menuet 等，2003，2005；Pellegrini 等，2005，Strobl-Mazzulla 等，2005；Marsh 等，2006；Kallivretaki 等，2007；Pellegrini 等，2007；Strobl-Mazzulla 等，2008）的研究中都有类似报道。这些芳化酶表达细胞大量处在前脑，主要分布在嗅球、端脑、视前区和中基下丘脑，特别是沿着侧隐窝和后隐窝。然而，除了和信息的分布一致之外，芳化酶 B 阳性细胞亦可以在视顶盖边沿的室周层（periventricular layer）、半规隆凸（torus semicircularis）以及沿着第四脑室观察到。总之，现在已研究清楚芳化酶在鱼脑的表达只限于放射状胶质细胞。芳化酶在鱼脑的高表达是由于 *cyp19a1b* 基因强烈地调控雌激素与可芳化的雄激素所致（Menuet 等，2005）。是否有其他的类固醇酶类亦在放射状胶质细胞表达，目前还不清楚。

芳化酶的细胞和它们的突起经常都和神经元紧密联系。明显的例子是促性腺激素释放激素（GnRH）神经元，它们在一些脑区有时候完全被放射状突起包围住（见图 2.2C）。虽然 GnRH 神经元并不表达雌激素受体（ER），但亦不能排除这些细胞产生的雌激素通过膜受体影响 GnRH 神经元的活动。在脑垂体，芳化酶 B 在近端远侧部（PPD）的非常类似促性腺激素细胞的细胞中以及神经垂体的细胞中强烈表达（见图 2.2D）。

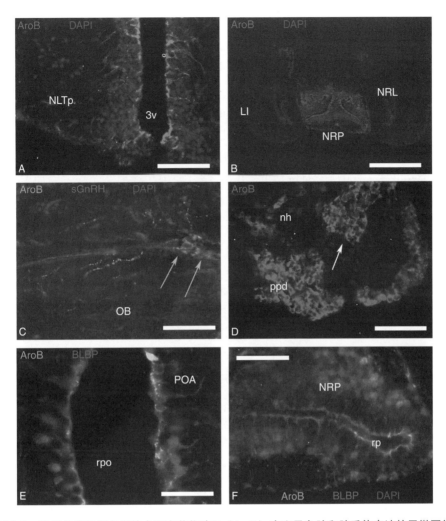

图 2.2 采用免疫组织化学技术研究芳化酶 B（AroB）在斑马鱼脑和脑垂体表达的显微图像

（A）在斑马鱼的后侧结节核（NLTp），芳化酶 B（AroB）的免疫组织化学示范图。红色信息只在第三脑室（3v）边沿的细胞检测到，并发出长的放射状突起朝向脑的腹面。标准尺度 = 50 μm。（B）在斑马鱼的后隐窝核（NRP），AroB 的免疫组织化学示范图。红色信息只在第三脑室（3v）边沿的细胞检测到，发出短的突起朝向后隐窝，发出长的放射状突起朝向脑的腹面。注意一较弱但仍明显的信号出现在下叶（LI）外侧隐窝核（NRL）的尾部。标准尺度 = 500 μm。（C）斑马鱼嗅球（OB）的矢切面，表示 AroB（绿色）的放射状突起围绕着一群 sGnRH 的细胞体（红色）。目前还不清楚这些 AroB 表达细胞产生的雌激素是否会影响到 GnRH 神经元的活动性。标准尺度 = 50 μm。（D）斑马鱼近端远侧部（PPD）AroB 的免疫组织化学示范图。在疑似的促性腺激素细胞观察到很强的信号，而在神经脑垂体（nh）的较小细胞检测到很弱的信号。标准尺度 = 50 μm。（E）和（F）表示斑马鱼 AroB 细胞的共表达情况。视前区（POA）细胞与后隐窝核（NRP）的细胞和脑脂结合蛋白（BLBP）进行 AroB 共表达。注意 BLBP（红色）出现于细胞的细胞质和细胞核，而 AroB 只出现于细胞质内。由于 BLBP 是哺乳类放射状胶质细胞的标志物，这就进一步证实 AroB 表达细胞的放射状特点。（E）标准尺度 = 25 μm；（F）标准尺度 = 50 μm。（见书后彩图）

如同芳化酶在放射状胶质细胞中表达的作用一样，最近的研究表明其存在着和神经原性（neurogenic）的联系（Mouriec 等，2008）。采用 BrdU 免疫组织化学技术和芳化酶 B 作为放射状胶质细胞的标记物进行研究，发现在短的存活时间（12h 和 24h）内，大量细胞呈现 BrdU 标志，它们相当于 AroB-阳性胶质细胞（Adolf 等，2006；Pellegrini 等，2007）。采用脑脂结合蛋白（brain lipid binding protein）的抗体亦能够表现在间脑和端脑增生细胞的放射性特点（见图 2.2E 和图 2.2F）。此外，用双重的 BrdU/PCNA 染色，在超过时间后，能表现新生细胞明显地从室周的增生区移走（Adolf 等，2006；Pellegrini 等，2007）。对斑马鱼，BrdU 结合使用几种神经元标志物如 Hu 或乙酰化的微管蛋白进行研究，结果表明许多新生细胞分化为神经元（Zupanc 等，2005；Adolf 等，2006；Pellegrini 等，2007）。

雌激素能通过雌激素 2-羟化酶（2-hydroxylase）的作用代谢为儿茶酚-雌激素（catechol-estrogen）。这些氧化的雌激素，例如 2-羟化雌激素（2-hydroxyestrogen），是潜在的双功能分子，能介导雌激素对神经内分泌和行为的影响。在对鱼类的研究中曾报道，儿茶酚-雌激素能在非竞争的条件下通过抑制酪氨酸羟化酶（tyrosinehydroxylase）的活性而降低儿茶酚胺的代谢。对于儿茶酚胺降解酶之一的儿茶酚-O-甲基转移酶（catechol-O-methyltransferase），儿茶酚-雌激素亦是竞争性底物（de Leeuw 等，1985；Goos 等，1985；Timmers 等，1988；Joy 和 Senthilkumaran，1998；Joy 等，1998；Chaube 和 Joy，2003）。

2.2.2 类固醇受体

雌激素、雄激素和黄体酮的作用都是通过属于核受体（nuclear receptor）家族的细胞内受体（intracellular receptors）介导的。这些家族的受体调控基本的生理功能，从生长发育直到精确协调体内稳态。已经清楚地知道核受体起着配体-活化转录因子（ligand-activated transcription factor）的作用，它们都共同使用在不同功能区的相同模块组织（modular organization）。在不同的结构和功能区当中有一个含锌指状 DNA 结合区（zinc finger DNA-binding domain）和一个配体-结合区。核受体家族有 48 个成员，包括糖皮质激素受体、黄体酮受体、雄激素受体、维生素 D 受体、甲状腺素受体、视黄酸受体（retinoid acid receptor），还有大量没有确定相应配体的孤独受体（orphan receptor）（Robinson-Rechavi 等，2003；Germain 等，2006）。因此，可以设想在进化过程中核受体是较早出现的；在后生动物（metazoans）进化期间，祖核受体（ancestral nuclear receptor）由一个孤独受体形成并获得配体结合能力。核受体通过和特异性 DNA 序列的直接相互作用而调控靶基因的表达（Wuertz 等，2007）。一些核受体三维（3D）结构的分析证明和配体结合时，配体结合区（LBD）经历构象变化（conformational change），以便补充辅激活蛋白（coactivator protein）。这些辅因子（cofactor）充实到一系列顺序安排的受体-辅激活蛋白中，其相互作用具有多种多样的功能区，主要是乙酰转移酶（acetyltransferase）、甲基转移酶（methyltransferase）或者泛素连接酶的活性。通过松弛染色质结构，这些辅因子对启动子（promoter）的循环补充，使得转录机（transcription machinery）能比较容易进入 DNA 中（Metivier 等，2003，2008）。

最近，新出现的类固醇膜受体家族使我们必须以当今的技术水平去进一步检测这个新生事物（Thomas 等，2006；Pang 等，2008；Pang 和 Thomas，2009）。本章不涉及这类受体，将在别的章节中对其进行综述。

2.2.2.1 雌激素受体

在哺乳类中发现 ERβ 之后，所有的脊椎动物都迅速确定有两个雌激素受体（estrogen receptors，ERs），即 ERα（esr1）和 ERβ（esr2），它们由两个不同的基因产生，具有部分不同的表达型式和对一些配体的结合能力（Kuiper 等，1998；Katzenellenbogen 等，2000）。在鱼类中，推想由于一个附加的基因组复制（Steinke 等，2006）而产生两个 ERβ，即 ERβ1（esr2b）和 ERβ2（esr2a）。在 ER 的 C-端，A/B 区含有一个配位-不依赖性（ligand-independent）的激活功能（AF1），并且在两个 ERs 之间是很不保守的。相反，在 C 区，即 DNA 结合区，在两个 ERs 之间是高度保守的（同一性高于95%），而在 NR 超级家族的成员之间亦是一样（同一性为40%～50%）。这个区含有两个锌指结构（zinc finger），以便能够识别特异性激素反应的 DNA 序列。在 ERs 中，称为雌激素-应答元件（estrogen-responsive element，ERE）的，是在瓜蟾（Xenopus）卵黄蛋白元基因确定的一个回文序列（palindromic sequence）带着契合序列（consensus sequence）（AGGTCAnnnTGACCT）（Klein-Hitpass 等，1986）。不太保守的 D 区参与 ERs 的三维结构，以保持 DNA 结合的稳定性。E/F 区是多功能的，包括和激素-依赖性反式激活（transactivation）功能（AF2）一起的 LBD。它是一种保守的 12 个螺旋（helice）（H1—H12）排列结构，由一个反平行 β-折叠（antiparallel β-sheet）和螺旋 H12 形成一个疏水性的（hydrophobic）配体袋（ligand pocket）。ERs 的配体结合区（LBD）是高度混杂的（promiscuous），能和各种各样的化合物结合，它们的大小和化学性质呈现明显的多样性（Thomas，2000；Singleton 和 Khan，2003）。这种特性对许多合成的化学物质都能起着异-雌激素的应答作用。和同类物结合时，H12 进行重定位（reposition），以参与 AF2 的反式激活功能，并且能够和辅因子相互作用。

有关鱼类无配体受体（unliganded receptor）亚细胞定位的研究资料不多。在哺乳类中，虽然它们主要是在核内发现，但亦有一小部分在细胞溶胶（cytosol）内。在核内，ERs 调节基因转录，翻译蛋白质以决定雌激素的生理功能。虽然对鱼类还没有相关文献报道，但不断增长的研究结果表明 ER 而不是 ERE 能和 DNA 中的调节序列结合（Carroll 等，2006）。特别是 ERα，能通过无须转录因子如 AP-1 或 Sp1 复合体参与的促进蛋白质-蛋白质相互作用而调节它们的转录（Nilsson 等，2001；Safe 和 Kim，2008）。

除了传统"教条"式的关于 ER 和转录机相互作用的分子作用机理之外，共激活因子/协阻抑物（co-activator/co-repressor）和其他的转录因子、ERs 和细胞内其他信息通道之间多层次的交流都已经迅速展现。值得注意的是，和膜联系的 ERs 能够激活几种信息的转导级联（transduction cascades），如促细胞分裂原-活化蛋白激酶（mitogen-activated protein kinase，MAPK）、蛋白激酶 C（protein kinase C，PKC）和磷脂酰肌醇 3-激酶（phosphatidylinositol3-kinase，PI3K）等，并可能引起所谓的雌激素"非基因组"效应。再者，AF1、ER 的独立激活-配体为生长因子的磷酸化作用而被激活，而生长因子能激活下游细胞激酶（MAPKs、PI3K、PKC）和其他信号通路。尽管 ER 类似的非传统的作用机理在鱼类的研究中还没有文献报道，但它们的存在不应被忽视。

雌激素受体是在鱼类体内鉴定得最清楚的类固醇受体，这主要是因为它们对调节卵黄蛋白元在肝脏中的表达起着关键作用。这个过程严格地取决于雌激素的作用，为我们提供一个研究基本的分子作用机理的良好模型。第一个 ER 是在虹鳟肝脏文库中克隆的（Pakdel 等，1990，1991）。随后，在脊椎动物中发现有两个 ER 基因，即 ERα 和 ERβ，因而把虹鳟克隆的 ER 定名为 ERα（esr1）。然而，由于硬骨鱼类特殊地出现第三次基因复制（3R）（Siegel 等，2007），在许多鱼类中都克隆到两个 ERβ（Hawkins 等，2000；Menuet 等，2002；Nagler 等，2007；Muriach 等，2008a、b）。大西洋绒须石首鱼（*Micropogonias undulates*）原先定名的 ERβ 和 ERα（Hawkins 等，2000），在明确了两个 ERβ 是产生于第三次基因复制（3R）之后，这些 ER 受体已经重新定名为 ERβ1（esr2b）和 ERβ2（esr2a）。目前雌激素受体已经在许多种鱼类中被克隆和鉴别（Pakdel 等，1991；Munoz-Cueto 等，1999；Xia 等，1999；Ma 等，2000；Pakdel 等，2000；Socorro 等，2000；Menuet 等，2002；Andreassen 等，2003；Teves 等，2003；Urushitani 等，2003；Halm 等，2004；Filby 和 Tyler 等，2005；Caviola 等，2007；Greytak 和 Callard，2007；Fu 等，2008）。但是，三种 ER 的序列都被研究清楚的鱼类还不多。

1. 雌激素受体在脑中的分布

对许多鱼类已经研究了三种雌激素受体的分布，主要是在信使（messenger）的水平，而有关这三种受体亚型的表达还有待研究。然而，很久以前采用注射含氚激素（tritiated hormone）和放射自显影（autoradiography）技术已经研究鱼类 ER 的分子鉴别和雌激素浓集细胞的分布（Davis 等，1977；Kim 等，1978，1979a、b）。例如，金鱼（*Carassius auratus*）和剑尾鱼（*Xiphophorus maculatus*）的雌激素浓集细胞主要分布于端脑的腹前连合和背上连合区、间脑的视前区、中部下丘脑和丘脑区（Kim 等，1978，1979b）。这些学者还证明脑垂体内 80% 的促性腺激素细胞吸收了含氚的雌激素（Kim 等，1979a）。这些早期的研究结果后来为原位杂交的研究所确认。虹鳟 ERα 的表达是被报道得最完整的，因为这是唯一一种在蛋白质和信使水平上被研究的鱼类（Anglade 等，1994；Pakdel 等，1994；Navas 等，1995；Linard 等，1996；Menuet 等，2001；Salbert 等，1991，1993）。这些结果首次指出早期学者采用放射自显影技术获得的研究记录是十分准确的（Kim 等，1978，1979b），同时亦表明采用免疫组织化学和原位杂交（见图2.3A和图2.3B）得到的研究结果都是互相一致的（Salbert等，1991；Anglade等，1994；Pakdel等，1994）。

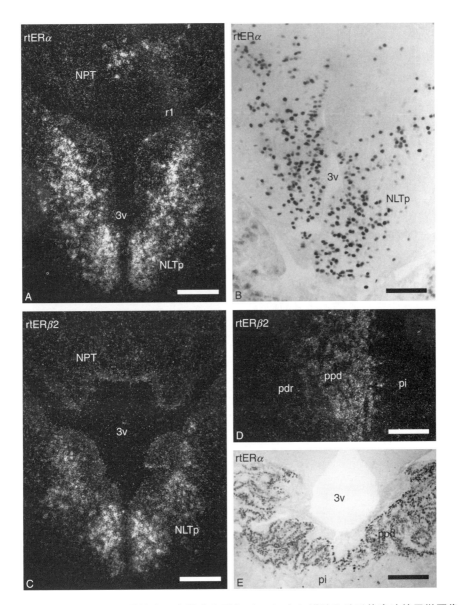

图2.3　采用免疫组织化学技术研究雌激素受体（ER）在虹鳟脑和脑垂体表达的显微图像

（A）和（B）是显微结构图，表示原位杂交（A）或免疫组织化学（B）证明雌激素受体α（rtERα）存在于虹鳟的脑内。标准尺度 = 60 μm。（C）为（A）附近的切片，表示 rtERβ2 在后外侧结节核（NLTp）的表达。注意 rtERβ2 在 NLT 的表达较弱，在后结节核（nucleus posterioris tuberis，NPT）没有表达。标准尺度 = 60 μm。（D）为雄性虹鳟脑垂体矢切面，rtERβ2 的分布情况。表示只在近端远侧部（PPD）有表达，而在吻端远侧部（RPD）和中间部（PI）没有表达。标准尺度 = 300 μm。（E）为免疫组织化学研究证明 rtERα 在虹鳟脑垂体 PPD 的表达情况。PI 的免疫反应很微弱。标准尺度 = 250 μm。rl，侧隐窝；3v，第三脑室。

此后，对一些鱼类主要采用原位杂交获得了许多研究结果（Salbert 等，1991，1993；Anglade 等，1994；Pakdel 等，1994；Navas 等，1995；Menuet 等，2001，2002；Andreassen 等，2003；Menuet 等，2003）。大部分研究结果都指出腹端脑、视前区（包括大细胞视前核）和中基下丘脑（mediobasal hypothalamus）是 ERα 表达最强的脑区。ERβ 表达的情况亦一样，但它只在少数鱼类研究中报道。通常，ERα 和 ERβ 大多是重叠的，但在斑马鱼（Hawkins 等，2000）和虹鳟（Lethimonier C. 和 Kah O.，未发表结果）中出现不同的分布情况（见图 2.2A 和图 2.3C）。在斑马鱼中，ERβ1 和 ERβ2 比 ERα 有更广的分布。它们都明显地出现在端脑和间脑的室周区，并参与雌激素对 AroB 表达的调控。在哺乳类，ERα 和 ERβ 经常在相同的细胞中发现（Adrio F. 和 Kah O.，未发表结果）。据关于舌齿鲈（*Dicentrachus labrax*）的研究报道，ERα 最大的分布区是在中脑顶盖（mesencephalic tectum）、大脑脚盖（tegmentum）和菱脑（rhombencephalon）（Muriach 等，2008b）。但是，这些研究结果还没有得到技术交叉效度分析（technique cross-validation）的确认，因而有待于进一步证实。采用 RNA 印迹法（northern blot）在虹鳟的视网膜和松果体检测到 ERα 的 mRNA，表明雌二醇能调控褪黑激素的分泌活动（Begay 等，1994）。

2. 雌激素受体表达细胞在鱼脑的特性

ERα-表达细胞的表型还很少有文献报道，仅对虹鳟研究了它们和生殖的神经内分泌调控的关系。

（1）KISS 神经元。Kisspeptins（KISS）在哺乳动物中已经成为调控 GnRH 神经元活动的关键因子（Popa 等，2008；Roa 等，2008）。KISS1 神经元似乎能够整合大量外来的和体内的信号，例如性腺类固醇、代谢因子、光周期和季节等。最近的研究表明，在弓状核的 KISS1 神经元能够调控雌雄性类固醇对 GnRH 和 GTH 分泌活动的负反馈作用。相反，至少在啮齿类动物中，KISS1 在前腹室周核（anteroventral periventricular nucleus，AVPV）的表达是性两态的，在 AVPV 的 KISS1 神经元能够介导 E_2 的正反馈作用以启动雌鱼排卵前 GnRH/促黄体激素（LH）的分泌活动（Popa 等，2008；Roa 等，2008）。已有充分的证据表明 KISS1 神经元在青春期激活 GnRH 系统起很大作用。对羊的研究表明 100% 的 KISS1 神经元在中视前区表达 ERα（Franceschini 等，2006）。

Kisspeptins 亦存在于鱼类，如青鳉（*Oryzias melastigma*）、斑马鱼和海鲈（Felip 等，2009；Kitahashi 等，2009）等，而且有两个 KISS 基因，即 KISS1 和 KISS2。对于两个 KISS 基因在青春期调控方面的意义如何，还很少报道。最近对青鳉的研究表明，KISS-表达神经元是性两态的，对类固醇颇为敏感。确实，鱼类 KISS1 神经元的数量在生殖时期和非生殖时期是不同的（Kanda 等，2008）。

（2）GnRH 神经元。在哺乳类和鱼类中，GnRH 神经元的活动性都强烈地受到雌激素和可芳化的雄激素的影响。双重免疫染色研究未能证明 ERα 在虹鳟雄成鱼的 550 GnRH 神经元中表达（Navas 等，1995）。这和在哺乳类脑中的研究结果一致，即 ERα 并不在 GnRH 神经元中表达（Wintermantel 等，2006）。但是，有学者曾认为 ERβ 膜受体的潜在表达能够介导迅速的反应（Abraham 等，2003，2004）。最近对小鼠采用转基因技术的研究证实 ERα 神经元表达是重要的，但排卵是由于雌激素作用于 ERα-表达神经元再传入到 GnRH 神经元而间接诱导的（Wintermantel 等，2006）。所以，尽管有研究报道类固醇影响鱼类的 GnRH 神经元（Trudeau 等，1992；Montero 等，1995；Breton 和 Sambroni，1996；Dubois 等，1998；Parhar 等，2001；Vetillard 等，2006），但它们似乎是由中间神经元，可能是 KISS 神经元介导的，和啮齿类动物的情况相似。

（3）多巴胺能神经元。已有一些文献报道多巴胺（DA）抑制许多硬骨鱼类的促性腺激素释放，如金鱼（Chang 和 Peter，1983；Chang 等，1990）、鳗鱼（*Anguilla anguilla*）（Weltzien 等，2006）和虹鳟（Linard 等，1995；Saligaut 等，1998）。联合采用免疫组织化学、原位杂交和损伤等技术的研究表明参与这些作用机理的多巴胺神经元位于前腹视前区（Peter 和 Paulencu，1980；Kah 等，1987b；Vetillard 等，2002；Weltzien 等，2006）。对虹鳟采用双重免疫组织化学技术，发现在前腹视前区的所有多巴胺神经元都强烈地表达 ERα（Linard 等，1996）。

（4）GABA 神经元。在鱼类中，最后通过化学鉴别的含有 ERα 的细胞是 GABA 神经元（Anglade I. 和 Kah O.，未发表结果）。GABA（γ-氨基丁酸）在鱼类的脑和脑垂体中强烈表达，并且明确表明它们参与前脑垂体功能的调控，特别是 GTH 的释放（Kah 等，1987a，1992；Trudeau，1997；Anglade 等，1999；Khan 和 Thomas，1999；Mananos 等，1999；Trudeau 等，2000；Fraser 等，2002）。对虹鳟的双重免疫组织化学研究表明在视前区和中基下丘脑的大部分 ERα-表达细胞都表达 ERα（Anglade I. 和 Kah O.，未发布结果）。这些研究结果充分说明，性类固醇激素深刻地影响着促性腺激素轴对 GABA 同等物或拮抗物的反应。

（5）黑色素浓集激素。睾酮和雌二醇能以性别依存的方式刺激黑色素浓集激素（MCH）mRNA 在下丘脑的表达，雌性表现出最大的反应。此外，离体试验证明梯度剂量的鲑鱼 MCH 能够刺激离散的脑垂体细胞分泌 LH，但并不刺激 GH 分泌。这个研究结果表明下丘脑的 MCH 可能参与类固醇的正反馈作用环对脑垂体 LH 分泌活动的影响（Cerda-Reverter 等，2006）。

3. 雌激素受体在脑垂体

最早是采用组织化学技术研究报道雌激素-浓集细胞的存在，荧光类固醇-激素结合物定位于脑垂体尾端远侧部（caudal pars distalis，CPD）的促性腺激素细胞，以及先前已证明含有 GTH 免疫反应的中间部（PI）细胞的细胞质和细胞核内（Schreibman 等，1982）。在克隆 ERs 之后，已证明许多鱼类包括虹鳟（见图 2.3D 和图 2.3E）和绵鳚（*Zoarces viviparous*）的脑垂体中存在 ER mRNAs（Andreassen 等，2003）。

2.2.2.2 **雄激素受体**

对硬骨鱼类，有关雄激素受体（andorogen receptor，AR）以及它们在脑与脑垂体复合体中表达的研究信息还相当欠缺。最早关于 ARs 在鱼类脑中存在的资料是来自采用含氚睾酮（T）或双氢睾酮（DHT）的开拓性研究（Davis 等，1977；Fine 等，1982；

Bsss 等，1986；Fine 等，1996）。然而，必须注意的是这两种雄激素都可以转化为雌激素化合物，T 转化为 E_2，而 DHT 可以转化为 5α-雄烯二酮（5α-androstan-3β, 17β-diol），一种具有雌激素活性的 DHT 代谢物（Kuiper 等，1998；Mouriec 等，2009）。所以，含氚的雄激素能转化为和雌激素受体结合的物质，当说明研究结果能证实在端脑和间脑含有大量吸收示踪物的细胞时就必须留意到这种情况。例如，在毒棘豹蟾鱼（Opsanus tau）中，吸收示踪物的细胞出现于背端脑，腹端脑的上连合核，视前小细胞前核（nucleus preopticus parvocellularis）和其他视前核，腹、背和尾下丘脑（Fine 等，1996）。此外，标志的细胞还定位在视顶盖、半规隆凸（torus semicircularis）、外侧瓣核（nucleus lateralis valvulae）和下网状结构（inferior reticular formation）（Fine 等，1996）。在非洲的长颌鱼中，浓集 ^{3}H-DHT 的细胞分布在靠近髓传递核（medullary relay nucleus）的网状结构内并神经分布到脊髓的电动神经元（electromotorneurons）（Bass 等，1986）。在对金鱼的研究中曾报道金鱼体内存在高密度的 ARs，性成熟的鱼比不成熟的鱼更能达到高峰（Pasmanik 和 Callard，1988）。尽管 AR 已在许多鱼类中克隆出来（Takeo 和 Yamashita，1999，2000；Todo 等，1999；Touhata 等，1999；Blazquez 和 Piferrer，2005；Olsson 等，2005；Hossain 等，2008；Liu 等，2009），但用原位杂交得到的研究结果还仅限于一两种鱼类（Burmeister 等，2007；Harbott 等，2007）。

在妊丽鱼（Astatotilapia burtoni）中克隆了两个 ARs，即 ARα 和 ARβ，并用原位杂交技术研究它们的分布情况。表现 ARα 和 ARβ 重叠分布的主要脑区是视前区和腹下丘脑。正是这些脑区依靠雄激素的信号，在调控鱼类 GTH 释放、性别分化和生殖行为等方面起着重要作用。在脑垂体，ARα 和 ARβ 以相似的型式表达，但 ARα mRNA 表现出较高的水平（Harbott 等，2007）。这些结果和对虹鳟的研究结果完全一致（Menuet 等，1999）。虹鳟的 ARα 和 ARβ（Takeo 和 Yamashita，1999）在皮质区（pallial region）、视前区、中基下丘脑、视顶盖和脑垂体都检测到；而在脑垂体中，ARα 的信使大多和促性腺激素细胞重叠一起（Menuet 等，1999）。ARα 在促性腺激素细胞的表达是和虹鳟的睾酮而不是雌二醇强烈促进 LH 释放的事实完全一致的，表明促性腺激素细胞存在着明显的雄激素效应（Breton 和 Sambroni，1996）。斑马鱼受精后 3～5 天，雄激素受体出现在脑、松果体原基和视网膜。在成鱼的脑内，AR 在端脑离散的区、视前区和室周下丘脑（periventricular hypothalamus）表达（Gorelick 等，2008），和虹鳟与妊丽鱼（Astatotilapia burtoni）的情况相似。

对大西洋绒须石首鱼曾用生物化学方法鉴别出两种不同的核受体 AR，定名为 AR1 和 AR2。值得注意的是 AR1 只出现在脑组织，而 AR2 出现在性腺和脑组织中（Sperry 和 Thomas，1999a，2000）。此外，两种 AR 具有不同的类固醇-结合特异性，AR1 对睾酮有很强的亲和力；AR2 对雄激素有较广的亲和力，且对结构不同的雄激素亦有较强的亲和力，包括 5α-还原的类固醇（5α-reduced steroid）（Sperry 和 Thomas，1999a、b，2000）。按照 Thomas 及其同事们的看法，存在着两种 ARs 可以说明 T 和 11KT 在雄性硬骨鱼类中具有不同的生理作用。睾酮能通过两种受体起作用，而雌鱼的主要雄激素睾酮可以特异地和 AR1 结合。他们还认为 5α-还原酶（5α-reductase）在鱼脑中大量表达（Pasmanik 和

Callard，1985），对于激活那一种 AR 可能起重要的决定作用。但是，关于 DHT 对于硬骨鱼类的生理重要性还没有深入研究。在斑马鱼中，DHT 转化为 5α-雄烯二酮（一种具有雌激素活性的 DHT 代谢物）之后能够激活芳化酶 B（cyp19a1b）基因（Mouriec 等，2009）。对于了解一些类固醇生成酶类特别是 5α-还原酶和 3β-羟基类固醇脱氢酶的作用来说，进一步阐明和掌握这两种 AR 受体亚型在鱼类脑中的表达是非常必要的。

目前学界对于 AR 在鱼脑中表达的调控还了解得很少。在妊丽鱼中，ER 和 AR 表达的优势度和睾丸大小、性类固醇水平之间存在正相关关系。这可能表示这些受体的表达程度受到性类固醇的上调，而雄鱼脑对性类固醇是相当敏感的（Burmeister 等，2007）。对于锯齿倒刺鲃（*Spinibarbus denticulatus*），在性腺充分发育的雄鱼脑垂体中，AR mRNA 显著增加，而在性腺发育成熟后期，脑中的 AR mRNA 增加（Liu 等，2009）。

2.2.2.3 黄体酮受体（PRs）

除了传统的黄体酮核受体（PR）之外，在鱼类和其他脊椎动物体内还有黄体酮膜受体（mPR）。这些膜受体介导黄体酮非传统的作用以诱导鱼类卵母细胞成熟（Zhu 等，2003；Thomas 等，2004；Hanna 等，2006；Thomas 等，2007）。在云纹犬牙石首鱼（*Cynoscion nebulosus*）（Zhu 等，2003）和其他鱼类（Hanna 等，2006；Mourot 等，2006）中，mPRα 在卵巢表达，亦在一些生殖组织和非生殖组织中，包括脑和脑垂体中广泛表达（Thomas，2008）。至今还没有对这类受体在鱼类脑和脑垂体表达部位的深入研究。但是，目前必须考虑到的是通过对 mPRs 的研究阐明黄体酮对鱼类脑-脑垂体复合体的潜在作用（Thomas，2008）。

关于黄体酮核受体 PRs 的研究资料亦很有限。对斑马鱼的研究表明 PRs 在鱼脑中表达的范围很广（Zhu 等，未发表资料）。

2.3 肾上腺皮质激素受体

皮质类固醇（corticosteroid）是由四足类动物的肾上腺皮质和鱼类的肾间腺产生的类固醇激素。皮质类固醇参与许多生理活动的调控，如应激反应、免疫反应、碳水化合物代谢、血液电解质水平以及行为等。糖皮质素（glucocorticoids）因作用于葡萄糖转移（一种应激之后恢复稳态的反应）而得名，但皮质醇（cortisol）对代谢、发育和免疫功能亦起着重要的调控作用。盐皮质素（mineralocorticoids）如醛固酮（aldosterone）（在哺乳类）调节电解质和水含量，主要是通过促进钠保留在肾脏内。显然，皮质醇是鱼类血液循环中最大量的糖皮质素，但对于由肾间腺分泌的皮质类固醇的确切性质曾经有过长期争论。目前的看法是鱼类的肾上腺缺少必要的酶类去完成醛固酮生物合成的最后一个步骤（Jiang 等，1998）。但是，克隆了虹鳟的类盐皮质素受体（MR）之后（Colombe 等，2000），最近的研究表明 11-脱氧皮质酮（11-deoxycorticosterone，DOC）是鱼类盐皮质素受体（MR）的内源性配体（Sturm 等，2005；Prunet 等，2006）。同样，在丽鱼类中克隆到一个 MR（Greenwood 等，2003），在虹鳟（Ducouret

等，1995；Bury 等，2003）和鲤鱼（*Cyprinus carpio*）（Stolte 等，2008）中克隆到两个糖皮质素受体（GR）。

2.3.1 盐皮质素受体（MR）

直到现在，有关盐皮质素受体在鱼类脑和脑垂体中表达的精确研究报道还很少。不过，已经在对虹鳟和鲤鱼的研究中证明脑是 MR 表达的主要部位（Sturm，2005；Stolte 等，2008）。

2.3.2 糖皮质素受体（GR）

在鲑鳟鱼类中最先通过结合研究鉴定脑的皮质醇受体，证明它是高亲和力、低容量的，在结合部位合成糖皮质素丙酮缩去类松（triamcinolone acetonide）。这个结合可以被皮质醇、地塞米松（dexamethasone）替换，亦可以在较小程度上被 RU38486 替换（Lee 等，1992；Knoebl 等，1996）。此后，在虹鳟中克隆到第一个糖皮质素受体（Ducouret 等，1995）。这个受体定名为 rtGR1，具有典型的类固醇核受体的结构，不同的是在两个锌指结构（zinc finger）之间插入一个 9-氨基酸。rtGR1 的其他序列给予受体对单一的糖皮质素应答元件（glucocorticoid-responsive element，GRE）以较强的结合亲和力，正如用 GST-DBDrtGR 融合蛋白质进行凝胶移位试验去除或不去除 9 个附加氨基酸所显示的。较强的亲和力是和受体在一个由单一 GRE 起动的报道基因上较强的组成转录活性相联系的，但和配体诱导的转录活性没有关系（Ducouret 等，1995；Lethimonier 等，2002b）。

2.3.2.1 糖皮质素受体的分布

Teitsma 和他的同事（1997）采用原位杂交技术进行受体在脑内分布的早期研究，结果表明大面积的信号出现在前脑，即皮质（pallial）和亚皮质（subpallial）区，视前核的所有亚分区（见图 2.4A 和图 2.4C）和侧结节核，这些都是主要的下丘脑-脑垂体区（Anglade 等，1993）。特别强烈的杂交信号出现在视前核的大细胞神经元，这是产生加压催产素（vasotocin）、硬骨鱼类催产素（isotocin）和 CRF 的部位（Teitsma 等，1997）。较弱的信号可在前室周核、上交叉核和丘脑区检测到。

直到现在，鱼类的 GR 抗体还只是从抗 rtGR1 受体 NH_2 端 A/B 区的前 165 个氨基酸得到的抗体（Tujague 等，1998）。由于 rtGR1 和 rtGR2 序列的相似性（Bury 等，2003），这个抗体似乎亦可以和 rtGR2 产生交叉反应。不过，免疫组织化学研究能很好地证实原位杂交的研究结果（Teitsma 等，1997，1998，1999），它们都一致确认鲤鱼的 GR1 和 GR2 信使呈现相同的分布型式（Stolte 等，2008），和所报道的 rtGR1 完全相同（Teitsma 等，1997）。免疫组织化学的染色反应从端脑延伸到脊髓，最高的密度是在脑的神经内分泌作用区，即视前区和中基下丘脑，视顶盖的室周区。同样的抗体亦应用于罗非鱼（*Oreochromis mossambicus*），结果稍有不同（Pepels 等，2004）。但是，由于罗非鱼和虹鳟 GRs 之间 A/B 区的低保守性，这项结果还需要采用其他技术进行交叉确认后才能证实。

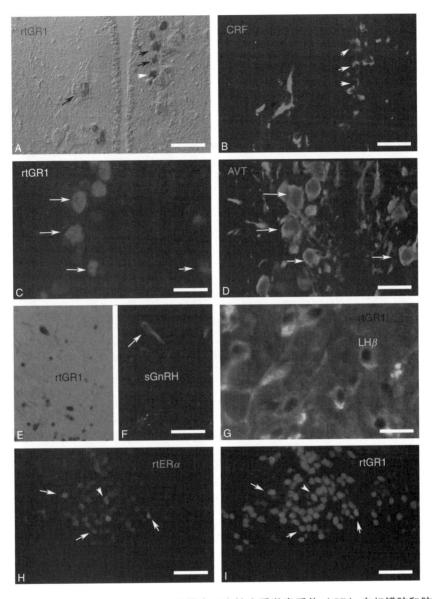

图 2.4 采用双重染色技术和共焦显微技术研究糖皮质激素受体（GR）在虹鳟脑和脑垂体表达的显微图像

（A）和（B）是采用 rtGR1（A）和促肾上腺皮质激素释放因子（CRF）（B）抗体对虹鳟大细胞视前核同一切片上进行双重染色的显微图像。注意（A）的 rtGR1 染色（箭头所示）经常都相应于（B）的 CRF-表达神经元的核。标准尺度 = 50 μm。（C）和（D）是采用 rtGR1（C）和加压催产素（AVT）（D）抗体对虹鳟大细胞视前核同一切片上进行双重染色的显微图像。注意（C）rtGR1 染色（箭头所示）经常都相应于（D）的 AVT-表达神经元的核。标准尺度 = 30 μm。（E）和（F）是采用 rtGR1（E）和鲑鱼 GnRH（sGnRH）（F）抗体对虹鳟腹端脑同一切片上进行双重染色的显微图像。注意（E）rtGR1 染色（箭头所示）相应于（F）的 sGnRH 神经元的核。标准尺度 = 50 μm。（G）是虹鳟腺脑垂体近端远侧部（PPD）的共焦显微图像，表示大部分 LHβ-阳性细胞（绿色）表达 rtGR1 受体（黑色）。标准尺度 = 30 μm。（H）和（I）是采用 rtERα（H）和 rtGR1（I）抗体对虹鳟腹视前区同一切片上进行双重染色的显微图像。注意两个染色的相应细胞核经常是重叠的（箭头所示）。标准尺度 = 100 μm。（见书后彩图）

2.3.2.2 化学鉴定糖皮质素受体-表达细胞

rtGR-免疫反应细胞高度密集在最重要的神经内分泌区和脑垂体，表明皮质醇对调控鱼类的适应性与应激性反应起着重要作用。共定位（colocalization）试验表明100%的CRF-免疫反应神经元在小细胞和大细胞的视前区强烈表达糖皮质素受体（见图2.4A和图2.4B），表明这是皮质醇对自身分泌活动调控神经元的反馈作用环（Teitsma等，1998）。对鲤鱼和罗非鱼的研究亦表明大细胞视前核是GR1和GR2的主要表达部位之一（Pepels等，2004；Stolte等，2008）。对虹鳟采用双重免疫染色技术，观察到rtGR1在许多大细胞神经元表达加压催产素（见图2.4C和图2.4D）（Teitsma等，1998）。

在脑垂体，rtGR-免疫反应都一致出现在吻端远侧部（RPD），主要是促皮质激素细胞（corticotropes）（Teitsma等，1998；Stolte等，2008），而催乳激素细胞是例外；在鳟鱼的近端远侧部（PPD）还包括促甲状腺素细胞、促性腺激素细胞（见图2.4G）和生长激素细胞（Teitsma等，1998，1999）。

应激反应和皮质醇能在不同的部位包括肝脏、脑和脑垂体影响生殖活动。对妊丽鱼（*Astatotilapia burtoni*）的研究表明GnRH系统是皮质醇重要的潜在靶标（Fox等，1997；Greenwood和Fernald，2004）。卵黄生成期的虹鳟，在尾腹端脑和前视前区有大量GnRH神经元表达rtGR1受体（见图2.4E和图2.4F）（Teitsma等，1999）。值得注意的是rtGR1亦在前腹视前区的多巴胺能神经元中强烈表达，而多巴胺可以抑制许多鱼类释放促性腺激素（Teitsma等，1999）。在这些神经元中，rtGR1和ERα之间很可能产生相互作用。在虹鳟的肝脏中，皮质醇通过干扰雌二醇影响ERα自身调节的分子作用机理而减少卵黄蛋白元生成（Lethimonier等，2002a）。这表明当GRs和ERα共表达时，可能会妨碍ERα-介导的活动。在虹鳟的脑内可以观察到许多ERα和rtGR1共表达的情况（见图2.4H和图2.4I）。

2.4 代谢激素

鱼类和其他动物一样，存在着能量用于生长和用于繁殖之间的冲突。这种冲突在卵生动物中由于子代低的成活率和配子生成高的能量耗费而尤为明显。所以，任何个体遭受代谢应激反应或者减少能量贮存都会导致青春期延缓和/或生育力下降。通过一系列神经肽类（如NPY、CART、KISS或GnRH等）组成的神经内分泌线路对代谢信息的感受，能量状态和生殖轴之间的相互关系得以保持，而且，直到近年来学界才对它们的特性和作用机理有所阐明。本文提到的由KISS-1基因编码的kisspeptins及其受体GPR54，作为正常青春期性成熟与性腺功能的标志性信息，使KISS-1/GPR54系统可能成为参与能量状态和繁殖的偶联关系。对于啮齿类，大量的证据表明KISS-1系统通过公认的瘦素-kisspeptin-GnRH（leptin-kisspeptin-GnRH）通道整合介导的代谢信息并输送到控制生殖活动的中心（Roa等，2008）。

KISS/GPR54系统在鱼类的研究中才刚开始有所阐明，尚欠缺详细的形态学基础。

目前已知道鱼类有两个 KISS 基因，编码 KISS1 和 KISS2（Felip 等，2009；Kitahashi 等，2009；Lee 等，2009），所以，还需要研究它们各自的表达部位。目前对青鳉 KISS1（Kanda 等，2008）和斑马鱼与青鳉的 KISS2 的初步研究（Kitahashi 等，2009）结果还很有限。然而，可以推测如果通过 KISS 神经元给 GnRH 系统以负的输入，就可以解释鱼类和哺乳类一样，已经阐明的生育力的变动和能量平衡的破坏是休戚相关的。

2.4.1 瘦素和瘦素受体

瘦素（leptin）于 1994 年被发现，从哺乳类的脂肪组织产生，瘦素的含量和体内脂肪贮存量大致呈正比（Robertson 等，2008）。瘦素通过作用于下丘脑腹中核（ventromedial nucleus）的食欲中心（appetite center）来调节食物的摄取。瘦素抑制 NPY 和野灰蛋白相关肽（AgRP），刺激 α-黑色素细胞刺激激素（α-MSH）。对于生殖，瘦素是脑活动的参数之一，用以确定能量水平以及给生殖功能的高度需求开"绿灯"。哺乳类有六个瘦素的受体（LepRa—LepRf），其中 LepRb 是唯一含有活动细胞内信号区的同等型（isoform）；相应的，下丘脑有高水平的 LepRb mRNA 表达。瘦素强烈地影响哺乳类的生殖功能，例如瘦素能使不育的 ob/ob 小鼠恢复生殖机能。一些研究表明瘦素能激活 GnRH 神经元，但 GnRH 神经元并不表达 LepRb（Finn 等，1998；Gamba 和 Pralong，2006）。最近的研究表明瘦素能调节 KISS1 在下丘脑的表达，还有一些证据表明 LepRb 存在于 KISS1 神经元（Smith 等，2006）。

在哺乳类中，LepRb 能够在表达食欲性神经肽 AgRP 与 NPY 和厌食性 POMC-激素原的神经元中表达（Robertson 等，2008）。

鱼类瘦素的主要来源是肝脏（Kurokawa 等，2005；Huising 等，2006），有关瘦素对摄食和生殖调控作用的研究资料还不多。这主要是因为在非哺乳类中还很难鉴别瘦素的特性。早期对欧洲海鲈和虹鳟的研究表明基因重组的瘦素直接作用于 FSH 和 LH 的释放（Peyon 等，2001；Weil 等，2003）。在对虹鳟的研究中，这种作用只在配子生成开始之后才出现，表明瘦素并不是引发促性腺激素轴起作用的唯一因素（Weil 等，2003）。尽管鱼类瘦素现在已经被鉴别并且能够生产出来，但是，对它调控生殖和摄食的神经内分泌线路的作用以及瘦素受体的表达等方面还有必要做深入的研究（Huising 等，2006；Nagasaka 等，2006；Murashita 等，2008；Yacobovitz 等，2008）。很可能和哺乳类一样，瘦素将会在鱼类的生长/营养轴和生殖轴之间的相互作用中扮演一个重要的角色。

2.4.2 胰岛素和胰岛素样生长因子（IGF）

目前已经知道胰岛素（insulin）是一个必需的合成代谢激素（anabolic hormone），对脊椎动物保持葡萄糖稳态起着关键作用。胰岛素在胰脏中产生，在激素原转化酶的作用下由胰岛素原前体分子（proinsulin precursor molecule）转化而成。但是，胰岛素原亦在发育的脑内表达，参与早期的形态发生（morphogenesis），起着存活因子的作用。这个时期，未加工的胰岛素原刺激增殖和细胞在发育的脑和视网膜内存活（de la Rosa 等，1994；Hernandez-Sanchez 等，2006；Papasani 等，2006）。在硬骨鱼类中，胰岛素主要从布洛克曼小体（Brockmann body，胰脏小块组织）分泌出来，但如同在罗非鱼（*Oreochromis niloticus*）中，亦在脑和脑垂体内少量产生（Hrytsenko 等，2007）。虽然鱼脑产生胰岛素已经讨论多时（Plisetskaya 等，1993），但最近的研究表明胰岛素 mRNA 和胰岛素在几个脑区中表达，主要是下丘脑、视顶盖、小脑和延脑（Hrytsenko 等，2007）。

IGF-I 是一种内分泌激素，主要在肝脏中产生，亦以旁分泌/自分泌方式在靶组织中产生。IGF-I 由 GH 的刺激而产生，它通过和特异性 IGF 受体结合后而介导其主要作用；IGF 受体存在于许多类型的细胞，实际上是所有的组织中，特别是肌肉、软骨、骨骼、肝脏、肾脏、脑、皮肤和肺。此外，IGF-I 亦在脑内表达，主要是嗅球、海马和小脑（Werther 等，1990）。

在鲤鱼和褐鳟（*Salmo trutta fario*）的脑内能检测到胰岛素和 IGF-I 受体结合以及受体内在的酪氨酸激酶活性。两者的配体都以剂量依存关系刺激外源物质的磷酸化作用（Leibush 等，1996）。学者们对褐鳟幼鱼和成鱼 IGF-I 受体的分布进行了深入的研究。和 [^{125}I] IGF-I 的结合主要分布在前脑的嗅球、下丘脑和丘脑，而在小脑和视顶盖分布得最多（Smith 等，2005）。有学者认为 IGF-I 受体这种分布是和早期发育与成年期鱼脑的生长密切联系的。在斑马鱼中，IGF-I 通过促进细胞存活和细胞周期进程（cell cycle progression）而调控胚胎发育和生长（Schlueter 等，2007）。

许多研究都认为雌激素受体和 IGF-I 受体在哺乳动物的脑内有密切的相互作用（Quesada 等，2007）。目前关于鱼类在这方面的研究资料还很少，有些研究表明 E_2 在许多组织主要是脑和脑垂体中调控 IGF-I 和 IGF-I 受体，其结果影响到许多生理功能，而不仅是生殖（Filby 等，2006）。

IGF-I 在脑垂体内产生的证据来自对尼罗罗非鱼的研究。在所有的个体中，IGF-I 和 IGF-I mRNA 都存在于大部分 ACTH 细胞中；而在数量变动的个体中，IGF-I 和 IGF-I mRNA 存在于 GH 细胞中（Eppler 等，2007）。在 α-MSH 细胞中只检测到 IGF-I mRNA，而检测不到 IGF-I，表明 IGF-I 在合成后就立即被释放出来。MSH 细胞释放出来的 IGF-I 可以加入到肝脏衍生的内分泌 IGF-I 当中，起着自分泌/旁分泌作用，介导一种负反馈的作用机理。局部的 IGF-I 能以自分泌或者旁分泌的方式调节脑垂体激素合成与释放，还能够预防细胞凋亡（apoptosis）和刺激内分泌细胞增殖（Eppler 等，2007）。在马苏大麻哈鱼（*Oncorhynchus masu*）的性成熟期间，IGF-I 能直接调控 GTH 亚单位基因的表达，特别是 IGF-I 能够在不同的生殖时期对 sGnRH 诱导 GTH 亚单位的基因表达产生不同的影响（Ando 等，2006；Furukuma 等，2008）。

2.4.3 甲状腺激素受体

和甲状腺激素受体（TRs）结合后，甲状腺激素（THs）起着对生长、分化、变态和生殖等方面的多效作用（pleiotropic effects）。THs 是酪氨酸的碘化衍生物，有两种主要类型：甲状腺素（thyroxin，T_4），含有四个碘残基；三碘甲腺原氨酸（triiodothyronine，T_3），只含有三个碘残基，是功能较强的起主要生物学作用的甲状腺激素。一些甲状腺激素高度复杂的作用机理是由于甲状腺激素反馈作用于脑与脑垂体复合体而产生的（Bernal，2007；Alkemade 等，2008）。和类固醇受体相似，甲状腺激素受体亦属于核受体超家族。哺乳类有两个 TR 基因，即 TRα 和 TRβ；TRβ 因剪接变体（splice variant）而产生 TRβ1 和 TRβ2。鱼类的 THs 亦参与发育、生长和代谢的调控。在鱼类和两栖类的变态以及鲑鳟鱼类由一龄降海幼鲑向二龄鲑转变（parr-smolt transformation）的过程中，THs 都起着至关重要的作用（Power 等，2001；Eales，2006）。在一些硬骨鱼类的相关研究中都曾证明 TRα1 和 TRβ 型亦产生剪接变体（Essner 等，1999；Marchand 等，2004；Kawakami 等，2007；Nelson 和 Habibi，2009）。

甲状腺激素显著影响斑马鱼脑的发育（Essner 等，1999），在胚胎发育期，TRα1 的超量表达干扰了后脑图式发育（hindbrain patterning）。这个作用和视黄酸受体（retinoic acid receptor）控制同源异型基因表达（hox gene expression）受到抑制有关（Essner 等，1999）。在星康吉鳗（*Conger myriaster*）中鉴别到两个 TRαs 和两个 TRβs；在成年鱼中，TRs 在许多组织中广泛表达，主要是在脑和脑垂体中强烈表达（Kawakami 等，2003，2007），但还没有它们精确表达部位的详细资料。最近对日本鳗（*Anguilla japonica*）的研究报道 TRβ mRNA 在整个腺脑垂体中强烈表达，大多是相当于 TSH、ACTH、GH 和/或 GTH 激素细胞分布的部位（Kawakami 等，2007）。显然，还有必要对 TH 在成年的和发育中的鱼脑的作用做进一步研究。

2.5 结 束 语

近 30 年来，我们对生理过程的了解取得极大的进展，特别是在性类固醇激素的反馈作用方面。但是在取得完全清晰的阐述之前，还需要做许多深入的研究。有关进一步掌握性类固醇激素在鱼脑中作用的未来研究领域，诸如，雌激素和雄激素各自真正作用的脑区是哪些？它们的靶标是什么？在何时和在何处起作用？脑芳化酶的作用是什么？如何调控芳化酶的作用？这些问题都有待回答。

关于类固醇激素膜受体的作用亦存在着许多"未知框"（black box，即不清楚之处），如它们的定位和它们表达的调控。可以期望在这些研究领域会有令人鼓舞的发现。希望创建新的研究手段使我们能够研究解决这些课题，以及其他诸如生长/代谢轴、应邀反应轴和生殖轴之间的相互关系。这些轴间的相互关系将整合起来而形成稳态，以防止在生理状态或者外界环境不良时出现有害的反应。

最后，近年来，分子生物学技术的普及，促进了分子生物学和免疫学等技术手段的应用而取得许多研究成果。没有这些技术手段就不可能在脑里面取得精确的研究资料。中枢神经系统组织的复杂性和它的异质性（heterogeneity），要求我们在细胞和亚细胞水平研究它们的形态学关系以及在信使和后蛋白质水平开展研究。PCR 技术的可行性和敏感性在很多情况下都提供了极为重要的帮助，但是亦不要忘记重要的具有功能的分子仍然是蛋白质。

O. 凯

参考文献

Abraham I M, Han S K, Todman M G, Korach K S, Herbison A E. 2003. Estrogen receptor beta mediates rapid estrogen actions on gonadotropin-releasing hormone neurons in vivo. *J. Neurosci*, 23: 5771-5777.

Abraham I M, Todman M G, Korach K S, Herbison A E. 2004. Critical *in vivo* roles for classical estrogen receptors in rapid estrogen actions on intracellular signaling in mouse brain. *Endocrinology*, 145: 3055-3061.

Adolf B, Chapouton P, Lam C S, Topp S, Tannhauser B, Strahle U, Gotz M, Bally-Cuif L. 2006. Conserved and acquired features of adult neurogenesis in the zebrafish telencephalon. *Dev. Biol*, 295: 278-293.

Alkemade A, Visser T J, Fliers E. 2008. Thyroid hormone signaling in the hypothalamus. *Curr. Opin. Endocrinol. Diabetes Obes*, 15: 453-458.

Ando H, Luo Q, Koide N, Okada H, Urano A. 2006. Effects of insulin-like growth factor I on GnRH-induced gonadotropin subunit gene expressions in masu salmon pituitary cells at different stages of sexual maturation. *Gen. Comp. Endocrinol*, 149: 21-29.

Andreassen T K, Skjoedt K, Anglade I, Kah O, Korsgaard B. 2003. Molecular cloning, characterisation, and tissue distribution of oestrogen receptor alpha in eelpout (*Zoarces viviparus*). *Gen. Comp. Endocrinol*, 132: 356-368.

Anglade I, Zandbergen A M, Kah O. 1993. Origin of the pituitary innervation in the goldfish. *Cell Tissue Res*, 273: 345-355.

Anglade I, Pakdel F, Bailhache T, Petit F, Salbert G, Jego P, Valotaire Y, Kah O. 1994. Distribution of estrogen receptor-immunoreactive cells in the brain of the rainbow trout (*Oncorhynchus mykiss*). *J. Neuroendocrinol*, 6: 573-583.

Anglade I, Mazurais D, Douard V, Le Jossic-Corcos C, Mananos E L, Michel D, Kah O. 1999. Distribution of glutamic acid decarboxylase mRNA in the forebrain of the rainbow trout as studied by *in situ* hybridization. *J. Comp. Neurol*, 410: 277-289.

Bass A H, Segil N, Kelley D B. 1986. Androgen binding in the brain and electric organ of a mormyrid fish. *J. Comp. Physiol [A]*. 159: 535-544.

Begay V, Valotaire Y, Ravault J P, Collin J P, Falcon J. 1994. Detection of estrogen receptor mRNA in trout pineal and retina: Estradiol-17 beta modulates melatonin production by cultured pineal photoreceptor cells. *Gen. Comp. Endocrinol*, 93: 61-69.

Bentivoglio M, Mazzarello P. 1999. The history of radial glia. *Brain Res. Bull*, 49: 305-315.

Bernal J. 2007. Thyroid hormone receptors in brain development and function. *Nat. Clin. Pract. Endocrinol. Metab*, 3: 249-259.

Blazquez M, Piferrer F. 2005. Sea bass (*Dicentrarchus labrax*) androgen receptor: cDNA cloning, tissue-specific expression, and mRNA levels during early development and sex differentiation. *Mol. Cell Endocrinol*, 237: 37-48.

Breton B, Sambroni E. 1996. Steroid activation of the brain-pituitary complex gonadotropic function in the triploid rainbow trout *Oncorhynchus mykiss*. *Gen. Comp. Endocrinol*, 101: 155-164.

Burmeister S S, Kailasanath V, Fernald R D. 2007. Social dominance regulates androgen and estrogen receptor gene expression. *Horm. Behav*, 51: 164-170.

Bury N R, Sturm A, Le Rouzic P, Lethimonier C, Ducouret B, Guiguen Y, Robinson-Rechavi M, Laudet V, Rafestin-Oblin M E, Prunet P. 2003. Evidence for two distinct functional glucocorticoid receptors in teleost fish. *J. Mol. Endocrinol*, 31: 141-156.

Callard G V, Petro Z, Ryan K J. 1978. Phylogenetic distribution of aromatase and other androgen-converting enzymes in the central nervous system. *Endocrinology*, 103: 2283-2290.

Callard G V, Petro Z, Ryan K J. 1981. Estrogen synthesis *in vitro* and *in vivo* in the brain of a marine teleost (Myoxocephalus). *Gen. Comp. Endocrinol*, 43: 243-255.

Carroll J S, Meyer C A, Song J, Li W, Geistlinger T R, Eeckhoute J, Brodsky A S, Keeton E K, Fertuck K C, Hall G F, Wang Q, Bekiranov S, et al. 2006. Genome-wide analysis of estrogen receptor binding sites. *Nat. Genet*, 38: 1289-1297.

Caviola E, Dalla Valle L, Belvedere P, Colombo L. 2007. Characterisation of three variants of estrogen receptor beta mRNA in the common sole, *Solea solea*L. (Teleostei). *Gen. Comp. Endocrinol*, 153: 31-39.

Cerda-Reverter J M, Canosa L F, Peter R E. 2006. Regulation of the hypothalamic melanin-concentrating hormone neurons by sex steroids in the goldfish: Possible role in the modulation of luteinizing hormone secretion. *Neuroendocrinology*, 84: 364-377.

Chang J P, Peter R E. 1983. Effects of dopamine on gonadotropin release in female goldfish, *Carassius auratus*. *Neuroendocrinology*, 36: 351-357.

Chang J P, Yu K L, Wong A O, Peter R E. 1990. Differential actions of dopamine receptor subtypes on gonadotropin and growth hormone release *in vitro* in goldfish. *Neuroendocrinology*, 51: 664-674.

Chaube R, Joy K P. 2003. *In vitro* effects of catecholamines and catecholestrogens on brain tyrosine hydroxylase activity and kinetics in the female catfish *Heteropneustes fossilis*. *J. Neuroendocrinol*, 15: 273-279.

Colombe L, Fostier A, Bury N, Pakdel F, Guiguen Y. 2000. A mineralocorticoid-like receptor in the rainbow trout, *Oncorhynchus mykiss*: Cloning and characterization of its steroid binding domain. *Steroids*, 65: 319-328.

Corpechot C, Robel P, Axelson M, Sjovall J, Baulieu E E. 1981. Characterization and measurement of dehydroepiandrosterone sulfate in rat brain. *Proc. Natl. Acad. Sci. USA*, 78: 4704-4707.

Crim L W, Evans D M. 1983. Influence of testosterone and/or luteinizing hormone releasing hormone analogue on precocious sexual development in the juvenile rainbow trout. *Biol. Reprod*, 29: 137-142.

Davis R E, Morrell J I, Pfaff D W. 1977. Autoradiographic localization of sex steroid-concentrating cells in the brain of the teleost *Macropodus opercularis* (Osteichthyes: Belontiidae). *Gen. Comp. Endocrinol*, 33: 496-505.

dela Rosa E J, Bondy C A, Hernandez-Sanchez C, Wu X, Zhou J, Lopez-Carranza A, Scavo L M, de Pablo F. 1994. Insulin and insulin-like growth factor system components gene expression in the chicken retina from early neurogenesis until late development and their effect on neuroepithelial cells. *Eur. J. Neurosci*, 6: 1801-1810.

de Leeuw R, Smit-van Dijk W, Zigterman J W, van der Loo J C, Lambert J G, Goos H J. 1985. Aromatase, estrogen 2-hydroxylase, and catechol-O-methyltransferase activity in isolated, cultured gonadotropic cells of mature African catfish, *Clarias gariepinus* (Burchell). *Gen. Comp. Endocrinol*, 60: 171-177.

Donaldson E M, McBride J R. 1967. The effects of hypophysectomy in the rainbow trout *Salmo gairdnerii* (Rich.) with special reference to the pituitary-interrenal axis. *Gen. Comp. Endocrinol*, 9: 93-101.

Dubois E A, Florijn M A, Zandbergen M A, Peute J, Goos H J. 1998. Testosterone accelerates the development of the catfish GnRH system in the brain of immature African catfish (*Clarias gariepinus*). *Gen. Comp. Endocrinol*, 112: 383-393.

Ducouret B, Tujague M, Ashraf J, Mouchel N, Servel N, Valotaire Y, Thompson E B. 1995. Cloning of a teleost fish glucocorticoid receptor shows that it contains a deoxyribonucleic acid-binding domain different from that of mammals. *Endocrinology*, 136: 3774-3783.

Eales J G. 2006. Modes of action and physiological effects of thyroid hormones in fish. In: Reinecke M, Ed. *Fish Endocrinology*. Science Publishers: Enfield, NH 767-808.

Eppler E, Shved N, Moret O, Reinecke M. 2007. IGF-I is distinctly located in the bony fish pituitary as revealed for *Oreochromis niloticus*, the Nile tilapia, using real-time RT-PCR, *in situ* hybridisation and immunohistochemistry. *Gen. Comp. Endocrinol*, 150: 87-95.

Essner J J, Johnson R G, HackettJr P B. 1999. Overexpression of thyroid hormone receptor alpha 1 during zebrafish embryogenesis disrupts hindbrain patterning and implicates retinoic acid receptors in the control of hox gene expression. *Differentiation*, 65: 1-11.

Felip A, Zanuy S, Pineda R, Pinilla L, Carrillo M, Tena-Sempere M, Gómez A. 2009. Evidence for two distinct KiSS genes in non-placental vertebrates that encode kisspeptins with different gonadotropin-releasing activities in fish and mammals. *Mol. Cell Endocrinol*, in press.

Filby A L, Tyler C R. 2005. Molecular characterization of estrogen receptors 1, 2a, and 2b and their tissue and ontogenic expression profiles in fathead minnow (*Pimephales promelas*). *Biol. Reprod*, 73: 648-662.

Filby A L, Thorpe K L, Tyler C R. 2006. Multiple molecular effect pathways of an environmental oestrogen in fish. *J. Mol. Endocrinol*, 37: 121-134.

Fine M L, Keefer D A, Leichnetz G R. 1982. Testosterone uptake in the brainstem of a sound-producing fish. *Science*, 215: 1265-1267.

Fine M L, Chen F A, Keefer D A. 1996. Autoradiographic localization of dihydrotestosterone and testosterone concentrating neurons in the brain of the oyster toadfish. *Brain Res*, 709: 65-80.

Finn P D, Cunningham M J, Pau K Y, Spies H G, Clifton D K, Steiner R A. 1998. The stimulatory effect of leptin on the neuroendocrine reproductive axis of the monkey. *Endocrinology*, 139: 4652-4662.

Forlano P M, Deitcher D L, Bass A H. 2005. Distribution of estrogen receptor alpha mRNA in the brain and inner ear of a vocal fish with comparisons to sites of aromatase expression. *J. Comp. Neurol*, 483: 91-113.

Fox H E, White S A, Kao M H, Fernald R D. 1997. Stress and dominance in a social fish. *J. Neurosci*, 17: 6463-6469.

Franceschini I, Lomet D, Cateau M, Delsol G, Tillet Y, Caraty A. 2006. Kisspeptin immunoreactive cells of the ovine preoptic area and arcuate nucleus co-express estrogen receptor alpha. *Neurosci. Lett*, 401: 225-230.

Fraser E J, Bosma P T, Trudeau V L, Docherty K. 2002. The effect of water temperature on the GABAergic and reproductive systems in female and male goldfish (*Carassius auratus*). *Gen. Comp. Endocrinol*, 125: 163-175.

Fu K Y, Chen C Y, Lin C T, Chang W M. 2008. Molecular cloning and tissue distribution of three estrogen receptors from the cyprinid fish *Varicorhinus barbatulus*. *J. Comp. Physiol* [*B*]. 178: 189-197.

Furukuma S, Onuma T, Swanson P, Luo Q, Koide N, Okada H, Urano A, Ando H. 2008. Stimulatory effects of insulin-like growth factor 1 on expression of gonadotropin subunit genes and release of follicle-stimulating hormone and luteinizing hormone in masu salmon pituitary cells early in gametogenesis. *Zoolog. Sci*, 25: 88-98.

Gamba M, Pralong F P. 2006. Control of GnRH neuronal activity by metabolic factors: The role of leptin and insulin. *Mol. Cell Endocrinol*, 254-255: 133-139.

Gao Q, Horvath T L. 2008. Cross-talk between estrogen and leptin signaling in the hypothalamus. *Am. J. Physiol. Endocrinol. Metab*, 294: E817-E826.

Germain P, Staels B, Dacquet C, Spedding M, Laudet V. 2006. Overview of nomenclature of nuclear receptors. *Pharmacol. Rev*, 58: 685-704.

Gomez J M. 2007. Serum leptin, insulin-like growth factor-I components and sex-hormone binding globulin. Relationship with sex, age and body composition in healthy population. *Protein Pept. Lett*, 14: 708-711.

Goos H J, van der Loo J C, Smit-van Dijk W, de Leeuw R. 1985. Steroid aromatase, 2-hydroxylase and COMT activity in gonadotropic cells of the African catfish, *Clarias gariepinus*. *Cell Biol. Int. Rep*, 9: 529

Gorelick D A, Watson W, Halpern M E. 2008. Androgen receptor gene expression in the developing and adult zebrafish brain. *Dev. Dyn*, 237: 2987-2995.

Goulis D G, Tarlatzis B C. 2008. Metabolic syndrome and reproduction: I. testicular function. *Gynecol. Endocrinol*, 24: 33-39.

Greenwood A K, Fernald R D. 2004. Social regulation of the electrical properties of gonadotropin-releasing hormone neurons in a cichlid fish (*Astatotilapia burtoni*). *Biol. Reprod*, 71: 909-918.

Greenwood A K, Butler P C, White R B, DeMarco U, Pearce D, Fernald R D. 2003. Multiple corticosteroid receptors in a teleost fish: Distinct sequences, expression patterns, and transcriptional activities. *Endocrinology*, 144: 4226-4236.

Greytak S R, Callard G V. 2007. Cloning of three estrogen receptors (ER) from killifish (*Fundulus heteroclitus*): Differences in populations from polluted and reference environments. *Gen. Comp. Endocrinol*, 150: 174-188.

Halm S, Kwon J Y, Rand-Weaver M, Sumpter J P, Pounds N, Hutchinson T H, Tyler C R. 2003. Cloning and gene expression of P450 17alpha-hydroxylase, 17, 20-lyase cDNA in the gonads and brain of the fathead minnow *Pimephales promelas*. *Gen. Comp. Endocrinol*, 130: 256-266.

Halm S, Martinez-Rodriguez G, Rodriguez L, Prat F, Mylonas C C, Carrillo M, Zanuy S. 2004. Cloning, characterisation, and expression of three oestrogen receptors (ERalpha, ERbeta1 and ERbeta2) in the European sea bass, *Dicentrarchus labrax*. *Mol. Cell Endocrinol*, 223: 63-75.

Hanna R, Pang Y, Thomas P, Zhu Y. 2006. Cell-surface expression, progestin binding, and rapid nongenomic signaling of zebrafish membrane progestin receptors alpha and beta in transfected cells. *J. Endocrinol*, 190: 247-260.

Harbott L K, Burmeister S S, White R B, Vagell M, Fernald R D. 2007. Androgen receptors in a cichlid fish, *Astatotilapia burtoni*: Structure, localization, and expression levels. *J. Comp. Neurol*, 504: 57-73.

Hawkins M B, Thornton J W, Crews D, Skipper J K, Dotte A, Thomas P. 2000. Identification of a third distinct estrogen receptor and reclassification of estrogen receptors in teleosts. *Proc. Natl. Acad. Sci. USA*, 97: 10751-10756.

Hernandez-Sanchez C, Mansilla A, de la Rosa E J, de Pablo F. 2006. Proinsulin in development: New roles for an ancient prohormone. *Diabetologia*, 49: 1142-1150.

Hojo Y, Hattori T A, Enami T, Furukawa A, Suzuki K, Ishii H T, Mukai H, Morrison J H, Janssen W G, Kominami S, Harada N, Kimoto T, et al. 2004. Adult male rat hippocampus synthesizes estradiol from pregnenolone by cytochromes P45017alpha and P450 aromatase localized in neurons. *Proc. Natl. Acad. Sci. USA*, 101: 865-870.

Hojo Y, Murakami G, Mukai H, Higo S, Hatanaka Y, Ogiue-Ikeda M, Ishii H, Kimoto T, Kawato S. 2008. Estrogen synthesis in the brain-role in synaptic plasticity and memory. *Mol. Cell Endocrinol*, 290: 31-43.

Hossain M S, Larsson A, Scherbak N, Olsson P E, Orban L. 2008. Zebrafish androgen receptor: Isolation, molecular, and biochemical characterization. *Biol. Reprod*, 78: 361-369.

Hrytsenko O, WrightJr J R, Morrison C M, Pohajdak B. 2007. Insulin expression in the brain and pituitary cells of tilapia (*Oreochromis niloticus*). *Brain Res*, 1135: 31-40.

Hsu H J, Hsiao P, Kuo M W, Chung B C. 2002. Expression of zebrafish cyp11a1 as a maternal transcript and in yolk syncytial layer. *Gene Expr. Patterns*, 2: 219-222.

Huising M O, Geven E J, Kruiswijk C P, Nabuurs S B, Stolte E H, Spanings F A, Verburg-van Kemenade B M, Flik G. 2006. Increased leptin expression in common Carp (*Cyprinus carpio*) after food intake but not after fasting or feeding to satiation. *Endocrinology*, 147: 5786-5797.

Jiang J Q, Young G, Kobayashi T, Nagahama Y. 1998. Eel (*Anguilla japonica*) testis 11beta-hydroxylase gene is expressed in interrenal tissue and its product lacks aldosterone synthesizing activity. *Mol. Cell Endocrinol*, 146: 207-211.

Joy K P, Senthilkumaran B. 1998. Annual and diurnal variations in, and effects of altered photoperiod and temperature, ovariectomy, and estradiol-17 beta replacement on catechol-O-methyltransferase level in brain regions of the catfish, *Heteropneustes fossilis*. *Comp. Biochem. Physiol. C Pharmacol. Toxicol. Endocrinol*, 119: 37-44.

Joy K P, Senthilkumaran B, Sudhakumari C C. 1998. Periovulatory changes in hypothalamic and pituitary monoamines following GnRH analogue treatment in the catfish *Heteropneustes fossilis*: A study correlating changes in plasma hormone profiles. *J. Endocrinol*, 156: 365-372.

Kah O, Dubourg P, Martinoli M G, Rabhi M, Gonnet F, Geffard M, Calas A. 1987. Central GABAergic innervation of the pituitary in goldfish: A radioautographic and immunocytochemical study at the electron microscope level. *Gen. Comp. Endocrinol*, 67: 324-332.

Kah O, Dulka J G, Dubourg P, Thibault J, Peter R E. 1987. Neuroanatomical substrate for the inhibition of gonadotrophin secretion in goldfish: Existence of a dopaminergic preoptico-hypophyseal pathway. *Neuroendocrinology*, 45: 451-458.

Kah O, Trudeau V L, Sloley B D, Chang J P, Dubourg P, Yu K L, Peter R E. 1992. Influence of GABA on gonadotrophin release in the goldfish. *Neuroendocrinology*, 55: 396-404.

Kallivretaki E, Eggen R I, Neuhauss S C, Kah O, Segner H. 2007. The zebrafish, brain-specific, aromatase cyp19a2 is neither expressed nor distributed in a sexually dimorphic manner during sexual differentiation. *Dev. Dyn*, 236: 3155-3166.

Kanda S, Akazome Y, Matsunaga T, Yamamoto N, Yamada S, Tsukamura H, Maeda K, Oka Y. 2008. Identification of KiSS-1 product kisspeptin and steroid-sensitive sexually dimorphic kisspeptin neurons in medaka (*Oryzias latipes*). *Endocrinology*, 149: 2467-2476.

Katzenellenbogen B S, Choi I, Delage-Mourroux R, Ediger T R, Martini P G, Montano M, Sun J, Weis K, Katzenellenbogen J A. 2000. Molecular mechanisms of estrogen action: Selective ligands and receptor pharmacology. *J. Steroid Biochem. Mol. Biol*, 74: 279-285.

Kawakami Y, Tanda M, Adachi S, Yamauchi K. 2003. cDNA cloning of thyroid hormone receptor betas from the conger eel, *Conger myriaster*. *Gen. Comp. Endocrinol*, 131: 232-240.

Kawakami Y, Adachi S, Yamauchi K, Ohta H. 2007. Thyroid hormone receptor beta is widely expressed in the brain and pituitary of the Japanese eel, *Anguilla japonica*. *Gen. Comp. Endocrinol*, 150: 386-394.

Khan I A, Thomas P. 1999. GABA exerts stimulatory and inhibitory influences on gonadotropin II secretion in the Atlantic croaker (*Micropogonias undulatus*). *Neuroendocrinology*, 69: 261-268.

Kim Y S, Stumpf W E, Sar M. 1978. Topography of estrogen target cells in the forebrain of goldfish, *Carassius auratus*. *J. Comp. Neurol*,. 182: 611-620.

Kim Y S, Sar M, Stumpf W E. 1979. Estrogen target cells in the pituitary of platyfish, *Xiphophorus maculatus*. *Cell Tissue Res*, 198: 435-440.

Kim Y S, Stumpf W E, Sar M. 1979. Topographical distribution of estrogen target cells in the forebrain of platyfish, *Xiphophorus maculatus*, studied by autoradiography. *Brain Res*, 170: 43-59.

Kitahashi T, Ogawa S, Parhar I S. 2009. Cloning and expression of kiss2 in the zebrafish and medaka. *Endocrinology*, 150: 821-831.

Klein-Hitpass L, Schorpp M, Wagner U, Ryffel G U. 1986. An estrogen-responsive element derived from the 5′flanking region of the Xenopus vitellogenin A2 gene functions in transfected human cells. *Cell*, 46: 1053-1061.

Knoebl I, Fitzpatrick M S, Schreck C B. 1996. Characterization of a glucocorticoid receptor in the brains of chinook salmon, *Oncorhynchus tshawytscha*. *Gen. Comp. Endocrinol*, 101: 195-204.

Kuiper G G, Shughrue P J, Merchenthaler I, Gustafsson J A. 1998. The estrogen receptor beta subtype: A novel mediator of estrogen action in neuroendocrine systems. *Neuroendocrinol*, 19: 253-286.

Kurokawa T, Uji S, Suzuki T. 2005. Identification of cDNA coding for a homologue to mammalian leptin from pufferfish, *Takifugu rubripes*. *Peptides*, 26: 745-750.

Lee P C, Goodrich M, Struve M, Yoon H I, Weber D. 1992. Liver and brain glucocorticoid receptor in rainbow trout, *Oncorhynchus mykiss*: Down-regulation by dexamethasone. *Gen. Comp. Endocrinol*, 87: 222-231.

Lee Y R, Tsunekawa K, Moon M J, Um H N, Hwang J I, Osugi T, Otaki N, Sunakawa Y, Kim K, Vaudry H, Kwon H B, Seong J Y, et al. 2009. Molecular evolution of multiple forms of kisspeptins and GPR54 receptors in vertebrates. *Endocrinology*, Jan 22. [Epub ahead of print].

Leibush B, Parrizas M, Navarro I, Lappova Y, Maestro M A, Encinas M, Plisetskaya E M, Gutierrez J. 1996. Insulin and insulin-like growth factor-I receptors in fish brain. *Regul. Pept*, 61: 155-161.

Lethimonier C, Flouriot G, Kah O, Ducouret B. 2002. The glucocorticoid receptor represses the positive autoregulation of the trout estrogen receptor gene by preventing the enhancer effect of a C/EBPbeta-like protein. *Endocrinology*, 143: 2961-2974.

Lethimonier C, Tujague M, Kern L, Ducouret B. 2002. Peptide insertion in the DNA-binding domain of fish glucocorticoid receptor is encoded by an additional exon and confers particular functional properties. *Mol. Cell Endocrinol*, 194: 107-116.

Linard B, Bennani S, Saligaut C. 1995. Involvement of estradiol in a catecholamine inhibitory tone of gonadotropin release in the rainbow trout (*Oncorhynchus mykiss*). *Gen. Comp. Endocrinol*, 99: 192-196.

Linard B, Anglade I, Corio M, Navas J M, Pakdel F, Saligaut C, Kah O. 1996. Estrogen receptors are expressed in a subset of tyrosine hydroxylase-positive neurons of the anterior preoptic region in the rainbow trout. *Neuroendocrinology*, 63: 156-165.

Liu X, Su H, Zhu P, Zhang Y, Huang J, Lin H. 2009. Molecular cloning, characterization and expression pattern of androgen receptor in *Spinibarbus denticulatus*. *Gen. Comp. Endocrinol*, 160: 93-101.

Lovejoy J C, Sainsbury A. 2009. Sex differences in obesity and the regulation of energy homeostasis. *Obes. Rev*, 10: 154-167.

Ma C H, Dong K W, Yu K L. 2000. cDNA cloning and expression of a novel estrogen receptor beta-subtype in goldfish (*Carassius auratus*). *Biochim. Biophys. Acta*, 1490: 145-152.

Mananos E L, Anglade I, Chyb J, Saligaut C, Breton B, Kah O. 1999. Involvement of gamma-aminobutyric acid in the control of GTH-1 and GTH-2 secretion in male and female rainbow trout. *Neuroendocrinology*, 69: 269-280.

Marchand O, Duffraisse M, Triqueneaux G, Safi R, Laudet V. 2004. Molecular cloning and developmental expression patterns of thyroid hormone receptors and T3 target genes in the turbot (*Scophtalmus maximus*) during post-embryonic development. *Gen. Comp. Endocrinol*, 135: 345-357.

Marsh K E, Creutz L M, Hawkins M B, Godwin J. 2006. Aromatase immunoreactivity in the bluehead wrasse brain, *Thalassoma bifasciatum*: Immunolocalization and co-regionalization with arginine vasotocin and tyrosine hydroxylase. *Brain Res*, 1126: 91-101.

Menuet A, Guigen Y, Teitsma C, Pakdel F, Mañanos E, Mazurais D, Kah O, Anglade I. 1999. Localization of nuclear androgen receptor messengers in the brain and pituitary of rainbow trout. In: Norberg B, Kjesby O S, Taranger G L, Andersson E, Stefansson S O, Ed. *Reproductive Physiology of Fish*. Bergen, John Grieg AS.

Menuet A, Anglade I, Flouriot G, Pakdel F, Kah O. 2001. Tissue-specific expression of two structurally different estrogen receptor alpha isoforms along the female reproductive axis of an oviparous species, the rainbow trout. *Biol. Reprod*, 65: 1548-1557.

Menuet A, Pellegrini E, Anglade I, Blaise O, Laudet V, Kah O, Pakdel F. 2002. Molecular characterization of three estrogen receptor forms in zebrafish: Binding characteristics, transactivation properties, and tissue distributions. *Biol. Reprod*, 66: 1881-1892.

Menuet A, Anglade I, Le Guevel R, Pellegrini E, Pakdel F, Kah O. 2003. Distribution of aromatase mRNA and protein in the brain and pituitary of female rainbow trout: Comparison with estrogen receptor alpha. *J. Comp. Neurol*, 462: 180-193.

Menuet A, Pellegrini E, Brion F, Gueguen M M, Anglade I, Pakdel F, Kah O. 2005. Expression and estrogen-dependent regulation of the zebrafish brain aromatase gene. *J. Comp. Neurol*, 485: 304-320.

Metivier R, Penot G, Hubner M R, Reid G, Brand H, Kos M, Gannon F. 2003. Estrogen receptor-alpha directs ordered, cyclical, and combinatorial recruitment of cofactors on a natural target promoter. *Cell*, 115: 751-763.

Metivier R, Gallais R, Tiffoche C, Le Peron C, Jurkowska R Z, Carmouche R P, Ibberson D, Barath P, Demay F, Reid G, Benes V, Jeltsch A, et al. 2008. Cyclical DNA methylation of a transcriptionally active promoter. *Nature*, 452: 45-50.

Mindnich R, Deluca D, Adamski J. 2004. Identification and characterization of 17 beta-hydroxysteroid dehydrogenases in the zebrafish, *Danio rerio*. *Mol. Cell Endocrinol*, 215: 19-30.

Montero M, Le Belle N, King J A, Millar R P, Dufour S. 1995. Differential regulation of the two forms of gonadotropin-releasing hormone (mGnRH and cGnRH-II) by sex steroids in the European female silver eel (*Anguilla anguilla*). *Neuroendocrinology*, 61: 525-535.

Mouriec K, Pellegrini E, Anglade I, Menuet A, Adrio F, Thieulant M L, Pakdel F, Kah O. 2008. Synthesis of estrogens in progenitor cells of adult fish brain: Evolutive novelty or exaggeration of a more general mechanism implicating estrogens in neurogenesis?. *Brain Res. Bull*, 75: 274-280.

Mouriec K, Gueguen M L, Manuel C, Percevault F, Thieulant M L, Pakdel F, Kah O. 2009. Androgens upregulate cyp19a1b (aromatase B) gene expression in the brain of zebrafish (*Danio rerio*) through estrogen receptors. *Biol. Reprod*, in press.

Mourot B, Nguyen T, Fostier A, Bobe J. 2006. Two unrelated putative membrane-bound progestin receptors, progesterone membrane receptor component 1 (PGMRC1) and membrane progestin receptor (mPR) beta, are expressed in the rainbow trout oocyte and exhibit similar ovarian expression patterns. *Reprod. Biol. Endocrinol*, 4: 6.

Munoz-Cueto J A, Burzawa-Gerard E, Kah O, Valotaire Y, Pakdel F. 1999. Cloning and sequencing of the gilthead sea bream estrogen receptor cDNA. *DNA Seq*, 10: 75-84.

Murashita K, Uji S, Yamamoto T, Ronnestad I, Kurokawa T. 2008. Production of recombinant leptin and its effects on food intake in rainbow trout (*Oncorhynchus mykiss*). *Comp. Biochem. Physiol. B Biochem. Mol. Biol*, 150: 377-384.

Muriach B, Carrillo M, Zanuy S, Cerda-Reverter J M. 2008. Distribution of estrogen receptor 2 mRNAs (Esr2a and Esr2b) in the brain and pituitary of the sea bass (*Dicentrarchus labrax*). *Brain Res*, 1210: 126-141.

Muriach B, Cerda-Reverter J M, Gomez A, Zanuy S, Carrillo M. 2008. Molecular characterization and central distribution of the estradiol receptor alpha (ERalpha) in the sea bass (*Dicentrarchus labrax*). *J. Chem. Neuroanat*, 35: 33-48.

Nagasaka R, Okamoto N, Ushio H. 2006. Increased leptin may be involved in the short life span of ayu (*Plecoglossus altivelis*). *J. Exp. Zoolog. A Comp. Exp. Biol*, 305: 507-512.

Nagler J J, Cavileer T, Sullivan J, Cyr D G, Rexroad C. 2007. The complete nuclear estrogen receptor family in the rainbow trout: Discovery of the novel ERalpha2 and both ERbeta isoforms. *Gene*, 392: 164-173.

Navarro V M, Castellano J M, Garcia-Galiano D, Tena-Sempere M. 2007. Neuroendocrine factors in the initiation of puberty: The emergent role of kisspeptin. *Rev. Endocr. Metab. Disord*, 8: 11-20.

Navas J M, Anglade I, Bailhache T, Pakdel F, Breton B, Jego P, Kah O. 1995. Do gonadotrophin-releasing hormone neurons express estrogen receptors in the rainbow trout? A double immunohistochemical study. *J. Comp. Neurol*, 363: 461-474.

Nelson E R, Habibi H R. 2009. Thyroid receptor subtypes: Structure and function in fish. *Gen. Comp. Endocrinol*, 161: 90-96.

Nilsson S, Makela S, Treuter E, Tujague M, Thomsen J, Andersson G, Enmark E, Pettersson K, Warner M, Gustafsson J A. 2001. Mechanisms of estrogen action. *Physiol. Rev*, 81: 1535-1565.

Olivereau M, Olivereau J. 1979. Effect of estradiol-17 beta on the cytology of the liver, gonads and pituitary, and on plasma electrolytes in the female freshwater eel. *Cell Tissue Res*, 199: 431-454.

Olsson P E, Berg A H, von Hofsten J, Grahn B, Hellqvist A, Larsson A, Karlsson J, Modig C, Borg B, Thomas P. 2005. Molecular cloning and characterization of a nuclear androgen receptor activated by 11-ketotestosterone. *Reprod. Biol. Endocrinol*, 3: 37.

OSullivan A J. 2009. Does oestrogen allow women to store fat more efficiently? A biological advantage for fertility and gestation. *Obes. Rev*, 10: 168-177.

Pakdel F, Le Gac F, Le Goff P, Valotaire Y. 1990. Full-length sequence and *in vitro* expression of rainbow trout estrogen receptor cDNA. *Mol. Cell Endocrinol*, 71: 195-204.

Pakdel F, Feon S, Le Gac F, Le Menn F, Valotaire Y. 1991. *In vivo* estrogen induction of hepatic estrogen receptor mRNA and correlation with vitellogenin mRNA in rainbow trout. *Mol. Cell Endocrinol*, 75: 205-212.

Pakdel F, Petit F, Anglade I, Kah O, Delaunay F, Bailhache T, Valotaire Y. 1994. Overexpression of rainbow trout estrogen receptor domains in *Escherichia coli*: Characterization and utilization in the production of antibodies for immunoblotting and immunocytochemistry. *Mol. Cell Endocrinol*, 104: 81-93.

Pakdel F, Metivier R, Flouriot G, Valotaire Y. 2000. Two estrogen receptor (ER) isoforms with different estrogen dependencies are generated from the trout ER gene. *Endocrinology*, 141: 571-580.

Pang Y, Thomas P. 2009. Involvement of estradiol-17beta and its membrane receptor, G protein coupled receptor 30 (GPR30) in regulation of oocyte maturation in zebrafish, Danio rario. *Gen. Comp. Endocrinol*, 161: 58-61.

Pang Y, Dong J, Thomas P. 2008. Estrogen signaling characteristics of Atlantic croaker G protein-coupled receptor 30 (GPR30) and evidence it is involved in maintenance of oocyte meiotic arrest. *Endocrinology*, 149: 3410-3426.

Papasani M R, Robison B D, Hardy R W, Hill R A. 2006. Early developmental expression of two insulins in zebrafish (*Danio rerio*). *Physiol. Genomics*, 27: 79-85.

Parhar I S, Tosaki H, Sakuma Y, Kobayashi M. 2001. Sex differences in the brain of goldfish: gonadotropin-releasing hormone and vasotocinergic neurons. *Neuroscience*, 104: 1099-1110.

Pasmanik M, Callard G V. 1985. Aromatase and 5 alpha-reductase in the teleost brain, spinal cord, and pituitary gland. *Gen. Comp. Endocrinol*, 60: 244-251.

Pasmanik M, Callard G V. 1988. A high abundance androgen receptor in goldfish brain: Characteristics and seasonal changes. *Endocrinology*, 123: 1162-1171.

Pasmanik M, Schlinger B A, Callard G V. 1988. *In vivo* steroid regulation of aromatase and 5 alpha-reductase in goldfish brain and pituitary. *Gen. Comp. Endocrinol*, 71: 175-182.

Pellegrini E, Menuet A, Lethimonier C, Adrio F, Gueguen M M, Tascon C, Anglade I, Pakdel F, Kah O. 2005. Relationships between aromatase and estrogen receptors in the brain of teleost fish. *Gen. Comp. Endocrinol*, 142: 60-66.

Pellegrini E, Mouriec K, Anglade I, Menuet A, Le Page Y, Gueguen M M, Marmignon M H, Brion F, Pakdel F, Kah O. 2007. Identification of aromatase-positive radial glial cells as progenitor cells in the ventricular layer of the forebrain in zebrafish. *J. Comp. Neurol*, 501: 150-167.

Pepels P P, Van Helvoort H, Wendelaar Bonga S E, Balm P H. 2004. Corticotropin-releasing hormone in the teleost stress response: Rapid appearance of the peptide in plasma of tilapia (*Oreochromis mossambicus*). *J. Endocrinol*, 180: 425-438.

Peter R E, Paulencu C R. 1980. Involvement of the preoptic region in gonadotropin release-inhibition in goldfish, *Carassius auratus*. *Neuroendocrinology*, 31: 133-141.

Peyon P, Zanuy S, Carrillo M. 2001. Action of leptin on *in vitro* luteinizing hormone release in the European sea bass (*Dicentrarchus labrax*). *Biol. Reprod*, 65: 1573-1578.

Plisetskaya E M, Bondareva V M, Duan C, Duguay S J. 1993. Does salmon brain produce insulin?. *Gen. Comp. Endocrinol*, 91: 74-80.

Popa S M, Clifton D K, Steiner R A. 2008. The role of kisspeptins and GPR54 in the neuroendocrine regulation of reproduction. *Annu. Rev. Physiol*, 70: 213-238.

Power D M, Llewellyn L, Faustino M, Nowell M A, Bjornsson B T, Einarsdottir I E, Canario A V, Sweeney G E. 2001. Thyroid hormones in growth and development of fish. *Comp. Biochem. Physiol. C Toxicol. Pharmacol*, 130: 447-459.

Prunet P, Sturm A, Milla S. 2006. Multiple corticosteroid receptors in fish: From old ideas to new concepts. *Gen. Comp. Endocrinol*, 147: 17-23.

Quesada A, Romeo H E, Micevych P. 2007. Distribution and localization patterns of estrogen receptor-beta and insulin-like growth factor-1 receptors in neurons and glial cells of the female rat substantia nigra: Localization of ERbeta and IGF-1R in substantia nigra. *J. Comp. Neurol*, 503: 198-208.

Rakic P. 1978. Neuronal migration and contact guidance in the primate telencephalon. *Postgrad. Med. J.* 54 (Suppl 1): 25-40.

Roa J, Aguilar E, Dieguez C, Pinilla L, Tena-Sempere M. 2008. New frontiers in kisspeptin/GPR54 physiology as fundamental gatekeepers of reproductive function. *Front. Neuroendocrinol*, 29: 48-69.

Robel P, Young J, Corpechot C, Mayo W, Perche F, Haug M, Simon H, Baulieu E E. 1995. Biosynthesis and assay of neurosteroids in rats and mice: Functional correlates. *J. Steroid Biochem. Mol. Biol*, 53: 355-360.

Robertson S A, Leinninger G M, MyersJr M G. 2008. Molecular and neural mediators of leptin action. *Physiol. Behav*, 94: 637-642.

Robinson-Rechavi M, Escriva Garcia H, Laudet V. 2003. The nuclear receptor superfamily. *J. Cell Sci*, 116: 585-586.

Safe S, Kim K. 2008. Non-classical genomic estrogen receptor (ER) /specificity protein and ER/activating protein-1 signaling pathways. *J. Mol. Endocrinol*, 41: 263-275.

Sakamoto H, Ukena K, Tsutsui K. 2001. Activity and localization of 3beta-hydroxysteroid dehydrogenase/Delta5-Delta4-isomerase in the zebrafish central nervous system. *J. Comp. Neurol*, 439: 291-305.

Salbert G, Bonnec G, Le Goff P, Boujard D, Valotaire Y, Jego P. 1991. Localization of the estradiol receptor mRNA in the forebrain of the rainbow trout. *Mol. Cell Endocrinol*, 76: 173-180.

Salbert G, Atteke C, Bonnec G, Jego P. 1993. Differential regulation of the estrogen receptor mRNA by estradiol in the trout hypothalamus and pituitary. *Mol. Cell Endocrinol*, 96: 177-182.

Saligaut C, Linard B, Mananos E L, Kah O, Breton B, Govoroun M. 1998. Release of pituitary gonadotrophins GTH I and GTH II in the rainbow trout (*Oncorhynchus mykiss*): Modulation by estradiol and catecholamines. *Gen. Comp. Endocrinol*, 109: 302-309.

Schlueter P J, Peng G, Westerfield M, Duan C. 2007. Insulin-like growth factor signaling regulates zebrafish embryonic growth and development by promoting cell survival and cell cycle progression. *Cell Death Differ*, 14: 1095-1105.

Schreibman M P, Pertschuk L P, Rainford E A, Margolis-Kazan H, Gelber S J. 1982. The histochemical localization of steroid binding sites in the pituitary gland of a teleost (the platyfish). *Cell Tissue Res*, 226: 523-530.

Siegel N, Hoegg S, Salzburger W, Braasch I, Meyer A. 2007. Comparative genomics of ParaHox clusters of teleost fishes: Gene cluster breakup and the retention of gene sets following whole genome duplications. *BMC Genomics*, 8: 312.

Singleton D W, Khan S A. 2003. Xenoestrogen exposure and mechanisms of endocrine disruption. *Front. Biosci*, 8: s110-s118.

Smith A, Chan S J, Gutierrez J. 2005. Autoradiographic and immunohistochemical localization of insulin-like growth factor-I receptor binding sites in brain of the brown trout, *Salmo trutta*. *Gen. Comp. Endocrinol*, 141: 203-213.

Smith J T, Acohido B V, Clifton D K, Steiner R A. 2006. KiSS-1 neurones are direct targets for leptin in the ob/ob mouse. *J. Neuroendocrinol*, 18: 298-303.

Socorro S, Power D M, Olsson P E, Canario A V. 2000. Two estrogen receptors expressed in the teleost fish, *Sparus aurata*: CDNA cloning, characterization and tissue distribution. *J. Endocrinol*, 166: 293-306.

Sperry T S, Thomas P. 1999. Characterization of two nuclear androgen receptors in Atlantic croaker: Comparison of their biochemical properties and binding specificities. *Endocrinology*, 140: 1602-1611.

Sperry T S, Thomas P. 1999. Identification of two nuclear androgen receptors in kelp bass (*Paralabrax clathratus*) and their binding affinities for xenobiotics: Comparison with Atlantic croaker (*Micropogonias undulatus*) androgen receptors. *Biol. Reprod*, 61: 1152-1161.

Sperry T S, Thomas P. 2000. Androgen binding profiles of two distinct nuclear androgen receptors in Atlantic croaker (*Micropogonias undulatus*). *J. Steroid Biochem. Mol. Biol*, 73: 93-103.

Steinke D, Hoegg S, Brinkmann H, Meyer A. 2006. Three rounds (1R/2R/3R) of genome duplications and the evolution of the glycolytic pathway in vertebrates. *BMC Biol*, 4: 16.

Stolte E H, de Mazon A F, xLeon-Koosterziel A F, Jesiak M, Bury N R, Sturm A, Savelkoul H F J, Verburg van Kemenade B M L, Flik G. 2008. Corticosteroid receptors involved in stress regulation in common carp, *Cyprinus carpio*. *J. Endocrinol*, 198: 403-417.

Strobl-Mazzulla P H, Moncaut N P, Lopez G C, Miranda L A, Canario A V, Somoza G M. 2005. Brain aromatase from pejerrey fish (*Odontesthes bonariensis*): CDNA cloning, tissue expression, and immunohistochemical localization. *Gen. Comp. En-*

docrinol, 143: 21-32.

Strobl-Mazzulla P H, Lethimonier C, Gueguen M M, Karube M, Fernandino J I, Yoshizaki G, Patino R, Strussmann C A, Kah O, Somoza G M. 2008. Brain aromatase (Cyp19A2) and estrogen receptors, in larvae and adult pejerrey fish *Odontesthes bonariensis*: Neuroanatomical and functional relations. *Gen. Comp. Endocrinol*, 158: 191-201.

Sturm A, Bury N, Dengreville L, Fagart J, Flouriot G, Rafestin-Oblin M E, Prunet P. 2005. 11-deoxycorticosterone is a potent agonist of the rainbow trout (*Oncorhynchus mykiss*) mineralocorticoid receptor. *Endocrinology*, 146: 47-55.

Takeo J, Yamashita S. 1999. Two distinct isoforms of cDNA encoding rainbow trout androgen receptors. *J. Biol. Chem*, 274: 5674-5680.

Takeo J, Yamashita S. 2000. Rainbow trout androgen receptor-alpha fails to distinguish between any of the natural androgens tested in transactivation assay, not just 11-ketotestosterone and testosterone. *Gen. Comp. Endocrinol*, 117: 200-206.

Teitsma C A, Bailhache T, Tujague M, Balment R J, Ducouret B, Kah O. 1997. Distribution and expression of glucocorticoid receptor mRNA in the forebrain of the rainbow trout. *Neuroendocrinology*, 66: 294-304.

Teitsma C A, Anglade I, Toutirais G, Munoz-Cueto J A, Saligaut D, Ducouret B, Kah O. 1998. Immunohistochemical localization of glucocorticoid receptors in the forebrain of the rainbow trout (*Oncorhynchus mykiss*). *J. Comp. Neurol*, 401: 395-410.

Teitsma C A, Anglade I, Lethimonier C, Le Drean G, Saligaut D, Ducouret B, Kah O. 1999. Glucocorticoid receptor immunoreactivity in neurons and pituitary cells implicated in reproductive functions in rainbow trout: A double immunohistochemical study. *Biol. Reprod*, 60: 642-650.

Tena-Sempere M. 2006. KiSS-1 and reproduction: Focus on its role in the metabolic regulation of fertility. *Neuroendocrinology*, 83: 275-281.

Teves A C, Granneman J C, van Dijk W, Bogerd J. 2003. Cloning and expression of a functional estrogen receptor-alpha from African catfish (*Clarias gariepinus*) pituitary. *J. Mol. Endocrinol*, 30: 173-185.

Thomas P. 2000. Chemical interference with genomic and nongenomic actions of steroids in fishes: Role of receptor binding. *Mar. Environ. Res*, 50: 127-134.

Thomas P. 2008. Characteristics of membrane progestin receptor alpha (mPRalpha) and progesterone membrane receptor component 1 (PGMRC1) and their roles in mediating rapid progestin actions. *Front. Neuroendocrinol*, 29: 292-312.

Thomas P, Pang Y, Zhu Y, Detweiler C, Doughty K. 2004. Multiple rapid progestin actions and progestin membrane receptor subtypes in fish. *Steroids*, 69: 567-573.

Thomas P, Dressing G, Pang Y, Berg H, Tubbs C, Benninghoff A, Doughty K. 2006. Progestin, estrogen and androgen G-protein coupled receptors in fish gonads. *Steroids*, 71: 310-316.

Thomas P, Pang Y, Dong J, Groenen P, Kelder J, de Vlieg J, Zhu Y, Tubbs C. 2007. Steroid and G protein binding characteristics of the seatrout and human progestin membrane receptor alpha subtypes and their evolutionary origins. *Endocrinology*, 148: 705-718.

Timmers R J, Granneman J C, Lambert J G, van Oordt P G. 1988. Estrogen-2-hydroxylase in the brain of the male African catfish, *Clarias gariepinus*. *Gen. Comp. Endocrinol*, 72: 190-203.

Todo T, Ikeuchi T, Kobayashi T, Nagahama Y. 1999. Fish androgen receptor: CDNA cloning, steroid activation of transcription in transfected mammalian cells, and tissue mRNA levels. *Biochem. Biophys. Res. Commun*, 254: 378-383.

Tomy S, Wu G C, Huang H R, Dufour S, Chang C F. 2007. Developmental expression of key steroidogenic enzymes in the brain of protandrous black porgy fish, *Acanthopagrus schlegeli*. *J. Neuroendocrinol*, 19: 643-655.

Touhata K, Kinoshita M, Tokuda Y, Toyohara H, Sakaguchi M, Yokoyama Y, Yamashita S. 1999. Sequence and expression of a cDNA encoding the red sea bream androgen receptor. *Biochim. Biophys. Acta*, 1449: 199-202.

Trudeau V L. 1997. Neuroendocrine regulation of gonadotrophin II release and gonadal growth in the goldfish, *Carassius auratus*. *Rev. Reprod*, 2: 55-68.

Trudeau V L, Somoza G M, Nahorniak C S, Peter R E. 1992. Interactions of estradiol with gonadotropin-releasing hormone and thyrotropin-releasing hormone in the control of growth hormone secretion in the goldfish. *Neuroendocrinology*, 56: 483-490.

Trudeau V L, Spanswick D, Fraser E J, Lariviere K, Crump D, Chiu S, MacMillan M, Schulz R W. 2000. The role of amino acid neurotransmitters in the regulation of pituitary gonadotropin release in fish. *Biochem. Cell Biol*, 78: 241-259.

Tujague M, Saligaut D, Teitsma C, Kah O, Valotaire Y, Ducouret B. 1998. Rainbow trout glucocorticoid receptor overexpression in *Escherichia coli*: Production of antibodies for western blotting and immunohistochemistry. *Gen. Comp. Endocrinol*, 110: 201-211.

Urushitani H, Nakai M, Inanaga H, Shimohigashi Y, Shimizu A, Katsu Y, Iguchi T. 2003. Cloning and characterization of estrogen receptor alpha in mummichog, *Fundulus heteroclitus*. *Mol. Cell Endocrinol*, 203: 41-50.

Vetillard A, Benanni S, Saligaut C, Jego P, Bailhache T. 2002. Localization of tyrosine hydroxylase and its messenger RNA in the brain of rainbow trout by immunocytochemistry and *in situ* hybridization. *J. Comp. Neurol*, 449: 374-389.

Vetillard A, Ferriere F, Jego P, Bailhache T. 2006. Regulation of salmon gonadotrophin-releasing hormone gene expression by sex steroids in rainbow trout brain. *J. Neuroendocrinol*, 18: 445-453.

Wang Y, Ge W. 2004. Cloning of zebrafish ovarian P450c17 (CYP17, 17alpha-hydroxylase/17, 20-lyase) and characterization of its expression in gonadal and extragonadal tissues, *Gen. Comp. Endocrinol*, 135: 241-249.

Weil C, Le Bail P Y, Sabin N, Le Gac F. 2003. *In vitro* action of leptin on FSH and LH production in rainbow trout (*Onchorynchus mykiss*) at different stages of the sexual cycle. *Gen. Comp. Endocrinol*, 130: 2-12.

Weltzien F A, Pasqualini C, Sebert M E, Vidal B, Le Belle N, Kah O, Vernier P, Dufour S. 2006. Androgen-dependent stimulation of brain dopaminergic systems in the female European eel (*Anguilla anguilla*). *Endocrinology*, 147: 2964-2973.

Werther G A, Abate M, Hogg A, Cheesman H, Oldfield B, Hards D, Hudson P, Power B, Freed K, Herington A C. 1990. Localization of insulin-like growth factor-I mRNA in rat brain by *in situ* hybridization-relationship to IGF-I receptors. *Mol. Endocrinol*, 4: 773-778.

Wintermantel T M, Campbell R E, Porteous R, Bock D, Grone H J, Todman M G, Korach K S, Greiner E, Perez C A, Schutz G, Herbison A E. 2006. Definition of estrogen receptor pathway critical for estrogen positive feedback to gonadotropin-releasing hormone neurons and fertility. *Neuron*, 52: 271-280.

Wuertz S, Nitsche A, Jastroch M, Gessner J, Klingenspor M, Kirschbaum F, Kloas W. 2007. The role of the IGF-I system for vitellogenesis in maturing female sterlet, *Acipenser ruthenus* Linnaeus, 1758. *Gen. Comp. Endocrinol*, 150: 140-150.

Xia Z, Patino R, Gale W L, Maule A G, Densmore L D. 1999. Cloning, *in vitro* expression, and novel phylogenetic classification of a channel catfish estrogen receptor. *Gen. Comp. Endocrinol*, 113: 360-368.

Yacobovitz M, Solomon G, Gusakovsky E E, Levavi-Sivan B, Gertler A. 2008. Purification and characterization of recombinant pufferfish (*Takifugu rubripes*) leptin. *Gen. Comp. Endocrinol*, 156: 83-90.

Zhu Y, Rice C D, Pang Y, Pace M, Thomas P. 2003. Cloning, expression, and characterization of a membrane progestin receptor and evidence it is an intermediary in meiotic maturation of fish oocytes. *Proc. Natl. Acad. Sci. USA*, 100: 2231-2236.

Zupanc G K, Hinsch K, Gage F H. 2005. Proliferation, migration, neuronal differentiation, and long-term survival of new cells in the adult zebrafish brain. *J. Comp. Neurol*, 488: 290-319.

第二部分 鱼类神经内分泌系统的功能

第 3 章 GnRH 系统和生殖的神经内分泌调控

本章综述现有的关于神经内分泌通路调控硬骨鱼类脑垂体促性腺激素——卵泡刺激素（FSH）和促黄体生成激素（LH）合成与释放的研究进展，着重介绍促性腺激素释放激素（GnRH）及其受体的多种类型以及它们在脑和脑垂体中的作用。最近的研究证明，一系列神经递质、神经肽和其他因子起着抑制性和/或刺激性调控 GnRH-FSH/LH 系统的作用。本章将论述这些因子在性腺不同发育时期的特征和作用。最后，本章还会探讨在青春期和性成熟与产卵期间的性类固醇激素和抑制性神经递质多巴胺如何协调有序地激活 GnRH-FSH/LH 系统的功能。

3.1 导　言

鱼类的生殖活动受到卵泡刺激素（FSH）和促黄体生成激素（LH）作用的调控。FSH 和 LH 属于糖蛋白激素家族，是异源二聚体（heterodimeric）的糖蛋白，由一个共同的 α-亚基（α-糖蛋白，αGP）和一个激素特异性的 β-亚基组成。LH 已经在许多硬骨鱼类当中被鉴别，而对 FSH 的研究还比较少，因为它是在近几年才被鉴别出来。虽然 LH 和 FSH 的精确作用尚未完全弄清楚，但可以肯定的是它们是由脑垂体不同的细胞产生，在鱼类生殖周期的不同时期表现不同的表达型式，作用于性腺产生性类固醇激素及其他性腺因子，对配子的发育和成熟起着重要的调控作用（Devlin 和 Nagahama，2002；Yaron 等，2003；Kusakabe 等，2006；Zmora 等，2007）。以大部分是对鲑鳟鱼类进行的研究结果表明 FSH 参与配子生成的调控，因为它刺激 17β-雌二醇的产生并促使卵黄原蛋白渗入到正在发育的卵母细胞中。在雄鱼，FSH 刺激塞托利细胞（Sertoli cell）增殖并保持数量正常的精子发生（spermatogenesis）。LH 在生殖周期的早期很少检测到甚至检测不到，但在性腺成熟之前刺激性腺类固醇激素生成，并参与卵母细胞发育成熟、排卵和排精。

关于鱼类 LH 与 FSH 合成与释放的神经内分泌调控研究已经取得许多进展（Peter 等，1986；Yaron，1995；Yaron 等，2003），其中的一些研究是由于期望调控养殖鱼类的生殖活动而开展的。由于下丘脑和脑垂体解剖学上的一些特征和生殖方式的多样性，硬骨鱼类成为值得重视的研究模式群体（model group）。例如，鱼类和哺乳类不同，下丘脑的神经纤维直接分布到腺脑垂体细胞，硬骨鱼类没有下丘脑-脑垂体门脉系统（见本书第 1 章）。鱼类促性腺激素分泌活动的神经内分泌调控和其他脊椎动物相似，都是由促性腺激素释放激素（GnRH）和类固醇激素的反馈作用所调控。但是，鱼类具有一些不同的特点。和哺乳类相比，硬骨鱼类 GnRH 肽和 GnRH 受体显示较多的亚型（Lethimonier 等，2004）。多巴胺（DA）在鱼类中起着促性腺激素释放的抑制因子作用，和哺乳类不同，它是调控青春期和确定性成熟与排卵时间的关键性因子（Peter 等，1986；Vidal 等，2004）。与哺乳类一样，在鱼类中亦鉴定出一系列影响 LH 和 FSH 分泌活动的神经递质、神经激素和激素。这样就增加了生殖调控的复杂程度，从而激发了学者们更多的兴趣去阐明这些调节剂是如何起作用的。

本章的目的是综述由下丘脑因子、外周激素以及脑垂体内产生的局部调节剂对 LH 和 FSH 进行神经内分泌调节方面取得的最新研究进展。包括介绍不同类型的 GnRH 和 GnRH 受体以及它们在脑和脑垂体中的作用，探讨各种神经递质、神经肽和其他因子对 GnRH-FSH/LH 系统的影响。最后，本章还将阐述性类固醇激素和抑制性神经递质 DA 在青春期以及性成熟和产卵时协调激活 GnRH-FSH/LH 系统所起的作用。最近的研究表明促性腺激素和 GnRHs 还在脑垂体外表达，但本综述将不包括这部分内容（Wong 和 Zohar，2004；Andreu-Vieyra 等，2005；So 等，2005）。

3.2 促性腺激素释放激素（GnRH）

3.2.1 GnRH 的多重性

GnRH 是紧密联系的十肽，由分离的但系统发生有关联的基因产生。至今在脊椎动物中已经鉴定出 GnRH 肽 14 个不同的变体（variant），其中 8 个是在硬骨鱼类中发现的（见表 3.1；Lethimonier 等，2004；Kah 等，2007）。系统分析前 GnRH 原（preproGnRH）认为 GnRHs 可以划分为四类：GnRH1、GnRH2、GnRH3 和 GnRH4（Lethimonier 等，2004；Silver 等，2004；Tello 等，2008）。硬骨鱼类的 GnRH 包含前三类的成员，而七鳃鳗的 GnRH 属于第四类。GnRH2 和 GnRH3 的氨基酸序列是保守的，而 GnRH1 的结构变化很大，在不同种的脊椎动物当中都不相同。所有的脊椎动物都有两个或三个不同的 GnRH 类型。不同类型的 GnRH 在鱼的脑和脑垂体中的分布有所不同（Lethimonier 等，2004；见本书第 2 章）。GnRH 在下丘脑的视前区（POA）产生，可以认为是下丘脑-脑垂体（hypophysiotropic）激素，因为硬骨鱼类所有的 POA 神经元实际上都延伸到脑垂体。GnRH 对于生殖来说是必不可少的，但越来越多的证据表明 GnRH 还有其他的神经调控作用（Soga 等，2005；Kah 等，2007）。

GnRH 1 类是种族特异性类型，在硬骨鱼类中包括哺乳动物 GnRH（mGnRH）、鲷鱼 GnRH（sbGnRH）、银汉鱼 GnRH（pjGnRH）、鲶鱼 GnRH（catGnRH）、鲱鱼 GnRH（hGnRH）、白鱼 GnRH（wfGnRH）（见表 3.1；Lethimonier 等，2004；Millar 等，2004；Kah 等，2007）。GnRH1 主要存在于 POA、尾下丘脑和分布到脑垂体的神经纤维中。由于相当大量存在于脑和脑垂体中以及其含量和 mRNA 的季节性变动，GnRH1 很明显是由视前区产生并起着诱导脑垂体释放促性腺激素和促进性腺发育的作用（Kah 等，2007；Nocillado 等，2007）。

表 3.1 在硬骨鱼类中鉴定的 8 种 GnRH 变体的汇总

位置	1	2	3	4	5	6	7	8	9	10
GnRH1 类										
哺乳类 GnRH	焦谷	组	色	丝	酪	甘	亮	精	脯	甘-H2
银汉鱼 GnRH	—	—	—	—	苯丙	—	—	丝	—	—
鲷鱼 GnRH	—	—	—	—	—	—	—	丝	—	—
鲶鱼 GnRH	—	—	—	—	组	—	—	天冬酰胺	—	—
鲱鱼 GnRH	—	—	—	—	组	—	—	丝	—	—
白鱼 GnRH	—	—	—	—	—	—	蛋	天冬酰胺	—	—
GnRH2 类										
鸡 GnRH-Ⅱ	—	—	—	—	组	—	色	酪	—	—
GnRH3 类										
鲑鱼 GnRH	—	—	—	—	—	—	色	亮	—	—

GnRH2 类是一个「组^{5}色^{7}酪^8」GnRH 或称为鸡Ⅱ GnRH（cGnRH-Ⅱ）。这个 GnRH 出现在所有有颌脊椎动物的中脑顶盖（midbrain tegmentum），起着调控生殖行为和/或食欲与代谢的作用（Volkoff 和 Peter，1999；Kah 等，2007）。已经证明 cGnRH-Ⅱ 纤维分布在一些鱼类的脑垂体中，包括金鱼（*Carassius auratus*）、欧洲鳗鱼（*Anguilla anguilla*）、非洲胡子鲶（*Clarias gariepinus*）、罗非鱼（*Oreochromis mossambicus*）、杂交的条纹狼鲈（*Morone saxatilus*）、舌齿鲈（*Dicentrarchus labrax*）和太平洋鲱鱼（*Clupea harengus*）（Peter 等，1986；Canosa 等，2008）。一些研究表明 cGnRH-Ⅱ 能诱导脑垂体释放 LH（Chang 等，2009）。

GnRH3 类是鱼类特有的，编码一个肽，即鲑鱼 GnRH（sGnRH）。这个 GnRH 分布在一些鱼类的端神经（terminal nerve，TN）和下丘脑。GnRH3 在 TN 的功能还不清楚，但推想可能对生殖行为起作用（Ogawa 等，2006）。GnRH1 和 GnRH3 的神经元成分在一些鱼类的前脑区，包括 POA 出现重叠现象（Gonzalez-Martínezz 等，2002；Kah 等，2007）。鲤科鱼类，包括金鱼、拟鲤（*Rutilus rutilus*）、斑马鱼（*Danio rerio*）和胖头鲹（*Pimephales promelas*），只有 GnRH2 和 GnRH3 两个亚型（Lin 和 Peter，1996；Penlington 等，1997；Steven 等，2003；Filby 等，2008）。在这些鱼类中，GnRH1 类可能在进化过程中丢失，它的许多功能已由 GnRH3 类来担当。在鲤科鱼类中，GnRH3 分布在 TN、端脑和 POA，起着下丘脑调控脑垂体类型的作用。

3.2.2 GnRH 受体

对硬骨鱼类 GnRH 受体的鉴定已经进行了大量研究（Millar 等，2004；Lethimonier 等，2004；Kah 等，2007；Tello 等，2008）。GnRH 受体是 G-蛋白偶联体家族的成员，根据其核苷酸序列可划分为三个类型（类型 1、类型 2 和类型 3）（Millar 等，2004；

Tello 等，2008）。类型 1 GnRH 受体（GnRHr1）存在于哺乳类和鱼类，类型 2 GnRH 受体（GnRHr2）主要存在于两栖类和人类，类型 3 GnRH 受体（GnRHr3）主要存在于鲈形目鱼类（Levavi-Sivan 和 Avitan，2005；Flanagan 等，2007；Tello 等，2008）。硬骨鱼类由于在进化过程中从四足类分离后发生基因组复制（genome duplication），使 GnRH 受体变得颇为复杂，迄今所研究的鱼类都在每一个 GnRH 受体亚型中鉴别出一个以上的受体（Lethimonier 等，2004；Flanagan 等，2007；Tello 等，2008），如东方红鳍鲀（*Fugu ruprides*）和黑青斑河鲀（*Tetraodon nigrovidis*）、马苏大麻哈鱼（*Oncorhynchus masou*）和舌齿鲈曾被报道有 5 个 GnRH 受体，而斑马鱼有 4 个不同的 GnRH 受体（Kah 等，2007；Tello 等，2008）。

由于 GnRH 受体的多重性，参与调控 LH 和 FSH 释放的那些受体亚型有时就不那么确定。如尼罗罗非鱼（*Oreochromis niloticus*），在一个促性腺激素细胞中就鉴定出三个类型的 GnRH 受体（Parhar 等，2005）。早期对金鱼的研究表明，有两个类型的 GnRH 受体在脑垂体中表达，一个受体类型较好地和 GnRH 促进 LH 释放相联系，而另一个受体类型和生长激素的释放相关（Illing 等，1999）。Levavi-Sivan 等（2006）在尼罗罗非鱼中鉴定出两个类型的 GnRH 受体，它们的分布情况有所不同。GnRHr3 型受体主要分布在和生殖功能相关的脑区，而且在含有 LH 和 FSH 细胞的脑垂体后部高度表达，表明这个受体类型对调节生殖活动起重要作用。相反，GnRHr1 型受体在脑内广泛分布，从嗅球到延脑以及脑垂体背前部和后部，表明这个受体类型可能参与感觉信息传入和生长调控（Levavi-Sivan 等，2006）。在妊丽鱼（*Astatotilapia burtoni*）中，GnRH 受体亚型 GnRH-R1SHS 和促性腺激素细胞共定位，而另一个 GnRH 受体亚型 GnRH-R2PEY 和生长激素细胞共定位，表明这些受体具有不同的功能（Flanagan 等，2007）。这种鱼类的第三个 GnRH 受体亚型 GnRH-R3PEY，和产生 GnRH 的神经元共定位，可能起着调控 GnRH 产生的反馈作用。所以，要阐明硬骨鱼类不同的 GnRH 受体亚型在脑和脑垂体介导的各种反应和功能，还需做进一步深入研究。

3.2.3 GnRH 的作用

采用在体注射、离体脑垂体或分离的促性腺细胞孵育已经证明 GnRH 对 LH 合成与释放的刺激作用（Yaron 等，2003；Ando 和 Urano，2005；Chang 等，2009）。我们已经充分了解介导这个作用的信号通道（Chang 等，2009），但对 GnRH 促进 FSH 合成与释放的作用还极少研究。这主要是因为应用于硬骨鱼类 FSH 定量的免疫分析方法还不多，FSHβ mRNA 表达测定的不断增加可以弥补这个缺陷。目前的研究已经表明 GnRH 亦能刺激银大麻哈鱼（*Oncorhynchus kisutch*）（Dickey 和 Swanson，2000）、虹鳟（*Oncorhynchus mykiss*）（Vacher 等，2000）、马苏大麻哈鱼（Ando 和 Urano，2005）和尼罗罗非鱼（Levavi-Sivan 等，2006；Aizen 等，2007）等 FSH 的释放。值得注意的是，GnRH 并不能影响舌齿鲈（Mateos 等，2002）和青春前期真鲷（*Pagrus major*）（Kumakura 等，2003）*fshb* 的表达。学者们曾推想这可能是有些鱼类在早期发育期间 *fshb* 的组成性表达（constitutive expression）所致（Swanson 等，2003）。

已经清楚的是，GnRH 和 GnRH 受体起着调控性腺生长的起始和整个生殖季节周期的整合作用（Schulz 和 Goos，1999）。例如，曾报道一些鱼类随着视前区 GnRH 含量增加和 GnRH 细胞体的增大而性腺发育成熟（Amano 等，1994；Van Der Kraak 等，1998；Yaron 等，2003）。最近的研究采用实时定量 RT-PCR 技术对一些处于未成熟的和性腺恢复发育的雌雄性鱼类在表达部位测定 GnRH 和 GnRH 受体 mRNAs，以了解它们在性腺发育过程中所起的作用（Amano 等，2006；Mohammed 和 Khan，2006；Moles 等，2007；Nocillado 等，2007；Filby 等，2008；Martínezz-Chavez 等，2008）。这项研究技术能够准确地表明在 POA 的 GnRH 增加是和性腺发育密切相关的。例如，鲻鱼（*Mugil cephalus*）在青春期发育开始时，GnRH1、GnRH2 和 GnRH3 的表达增强（Nocillado 等，2007）。GnRH1 表达的增强是和它所在的神经元分布神经纤维到脑垂体以及它促进 GTH 释放的作用一致的，但其他两种 GnRH 类型表达发生变化的重要意义还不清楚。同样，胖头鲹在性腺发育时 GnRH3 的表达增强，而这种 GnRH 类型是起着下丘脑-脑垂体类型的作用。GnRH2 的表达亦增强，但目前学界对它在性腺发育早期的作用了解得很少。

最近的研究表明，在金鱼排卵期间，GnRH2 和 GnRH3 出现差别性控制作用（differential regulation）（Canosa 等，2008）。根据它们的表达型式，可以认为 GnRH3 调控 LH 的分泌，而 GnRH2 和产卵行为有关。采用前列腺素 F2α 诱导发生产卵行为的未排卵鱼，GnRH2 在前脑和中脑的表达增强，而 GnRH3 在前脑的表达减弱，这进一步支持以上结论。

3.3　GnRH、FSH 和 LH 的类固醇反馈调控

性腺类固醇对硬骨鱼类 LH 和 FSH 的合成与释放起着重要的调控作用（Goos，1987；Trudeau，1997；Van Der Kraak 等，1998；Yaron 等，2003；Levavi-Sivan 等，2006）。已经证明睾酮和 17β-雌二醇的正反馈和负反馈的调节作用。性类固醇对 LH 和 FSH 释放的反馈调控包括在下丘脑和脑垂体水平的作用，是相当复杂的。除了影响 LH 和 FSH 在脑垂体内的合成之外，睾酮和 17β-雌二醇亦影响 GnRH 系统和其他控制 LH 合成与释放的神经内分泌因子。由于硬骨鱼类脑和脑垂体内高水平的芳化酶，使得难以区分睾酮的作用究竟是由雄激素受体介导还是睾酮在芳化酶作用下转化为雌激素后通过雌激素受体来介导。因此，经常都要用不可芳化的雄激素，如 11-酮雄烷二酮（11-ketoandrostenedione，11-KA）、11-酮睾酮（11-ketotestosterone，11-KT）和睾酮与 17β-雌二醇进行比较研究，以便区分通过这两种受体所起的作用如何。其他的类固醇包括诱导性腺成熟的类固醇 17，20β-双羟黄体酮（17，20βP）和糖皮质激素皮质醇都参与 LH 和 FSH 释放的调控（Levavi-Sivan 等，2006），但这些作用对于睾酮和 17β-雌二醇的主导作用来说是次要的。

3.3.1 负反馈作用

切除性腺和置换类固醇激素是常用的技术手段，用以确定硬骨鱼类性类固醇激素对 LH 和 FSH 分泌活动的影响。这些研究表明性类固醇激素的作用明显取决于性腺发育的阶段。在卵黄生成和精子生成的后期，将性腺切除能增强金鱼（Kobayashi 和 Stacey，1990）、非洲鲶鱼（de Leeuw 等，1986）、大西洋绒须石首鱼（*Micropgonias undulates*）（Khan 等，1999）、印度鲶鱼（囊鳃鲶）（*Heteropneustes fossilis*）（Senthilkumaran 和 Joy，1996）、杂交条纹狼鲈（Klenke 和 Zohar，2003）、舌齿鲈（Mateos 等，2002）以及一些鲑鳟鱼类（Bommelaer 等，1981；Larson 和 Swanson，1997）LH 的分泌活动，显著地证明类固醇的负反馈作用（negative feedback）。切除性腺对 LH 分泌的刺激作用能为睾酮或 17β-雌二醇的处理所逆转。其他的对虹鳟（Saligaut 等，1998；Chyb 等，1999；Vetillard 等，2003）、金鱼（Kobayashi 等，2000）、银大麻哈鱼（Larson 和 Swanson，1997）、大西洋绒须石首鱼（Banerjee 和 Khan，2008）和舌齿鲈（Mateos 等，2002）的研究表明，依赖于类固醇的负反馈作用环亦调控 FSH 的分泌活动。同样，采用睾酮或 17β-雌二醇可以还原性腺切除对 FSH 分泌活动的影响。

介导类固醇对 LH 和 FSH 分泌活动的负反馈作用有几个作用机理。一些研究表明，类固醇的负反馈作用主要不是在 FSH 与 LH 合成的脑垂体水平介导，因为离体的研究表明 LH 和 FSH 对睾酮或 17β-雌二醇的表达反应通常是没有变化或者是增强的（Dickey 和 Swanson，1998；Huggard-Nelson 等，2002；Levavi-Sivan 等，2006）。一系列研究证明，采用睾酮或 17β-雌二醇后能降低 GnRH mRNA 的水平，说明负反馈作用的主要目标是脑的 GnRH 神经元（Vacher 等，2002；Levavi-Sivan 等，2006）。

此外，类固醇的负反馈作用可能是通过影响 FSH 与 LH 释放的其他调控因子的作用而介导的。例如，类固醇负反馈作用可能由减弱氨基丁酸（GABA）的刺激影响而介导，或者通过增强多巴胺能对 GnRH 释放的抑制作用而介导。在多巴胺能的神经元和 GABA 能的神经元中存在着类固醇受体支持这些可能性（Linard 等，1995；Vetillard 等，2003）。LH 合成的抑制作用可能和类固醇激素对刺激性的性腺多肽如激活蛋白（activin）的影响有关联。

3.3.2 正反馈作用

许多研究表明睾酮或 17β-雌二醇对 LH 与 FSH 的合成起着正反馈作用（positive feedback）（Trudeau 等，1991；Mateos 等，2002；Banerjee 和 Khan，2008）。这种正反馈作用通常出现在未成熟的鱼体内，虽然有时亦会出现在成年的鱼体内（Schulz 等，1995）。类固醇的正反馈作用直接在脑垂体水平并通过对 GnRH 系统的影响而介导。

许多研究表明幼鱼脑垂体 LH 含量受到雌激素和可芳化雄激素的影响而增加，非芳化雄激素的影响很小（Crim 等，1981；Borg 等，1998；Dickey 和 Swanson，1998）。这说明这些影响是由睾酮芳化作用衍生的 17β-雌二醇在促性腺激素细胞中介导的。对金鱼和鳗鱼脑垂体的研究表明睾酮能增加 LH mRNA 水平（Huggard 等，1996；Vidal 等，

2004；Aroua 等，2007）。对于金鱼，17β-雌二醇增加所有促性腺激素亚单位 mRNA 的水平，但睾酮增加 *lhb*-亚单位而降低 *fshb*-亚单位的 mRNA 水平（Huggard 等，1996；Sohn 等，2001；Huggard-Nelson 等，2002）。对欧洲鳗鱼脑垂体的离体研究表明非芳化雄激素而不是 17β-雌二醇能使 LHβ-亚单位的 mRNA 增加（Huang 等，1997；Aroua 等，2007）。综合这些研究结果说明睾酮和 17β-雌二醇调控 LH 和 FSH 的正（和负）反馈作用具有充分的证据，但不同种鱼类的作用机理和作用部位亦存在着明显的多样性（Mateos 等，2002；Aroua 等，2007）。

性类固醇对 GnRH 系统亦有刺激作用并进而刺激 FSH 和 LH 的释放。睾酮和 11-酮睾酮能使许多鱼类脑的 GnRH 含量增加（Amano 等，1994；Breton 和 Sambroni，1996；Borg 等，1998；Dubois 等，1998）。17β-雌二醇亦作用于 GnRH 受体，例如，在尼罗罗非鱼中，17β-雌二醇增强 gnrhr1 和 gnrhr3 的表达（Levevi-Sivan 等，2006）。

3.4 单胺和氨基酸神经递质

3.4.1 多巴胺

对金鱼的研究表明，损伤前视前室周核（anterior nucleus preopticus periventricularis，NPP），破坏脑垂体柄，或者将脑垂体移植到一个受体鱼后都能使 LH 分泌活动明显而持续地增强（Peter 等，1986；Trudeau，1997）。这些反应支持了这个观点：金鱼的 LH 分泌活动受到一种促性腺激素释放的抑制性因子紧张作用（tonic action）的调控。接着的研究证明多巴胺（dopamine，DA）在金鱼体内起着促性腺激素释放的抑制因子的作用，它直接抑制基础的以及 GnRH-刺激的 LH 释放（Peter 等，1986；Trudeau，1997）。之后的研究表明 DA 的作用在不同的鱼类中存在着明显的种族差别。在鲤科鱼类的相关研究中已证明 DA 对 LH 释放起着有力的抑制作用（Peter 等，1986；Blazquez 等，1988a）；对于鲑鳟鱼类，DA 的抑制作用不明显（Van Der Kraak 等，1986；Breton 等，1998；Saligaut 等，1998）；而对于大西洋绒须石首鱼，DA 没有抑制作用（Copeland 和 Thomas，1989）。

现已清楚地确认 DA 通过几个作用机理抑制基础的和 GnRH-刺激的 LH 分泌活动（Van Der Kraak 等，1998；Popesku 等，2008；Chang 等，2009）。DA 的主要作用是干扰介导 LH 释放的细胞内 GnRH 的信号通路（Yaron 等，2003）。采用金鱼（Chang 等，1990）、虹鳟（Saligaut 等，1999；Vacher 等，2000）、尼罗罗非鱼（Levavi-Sivan 等，1995）、非洲鲶鱼（Van Asselt 等，1990）、鳗鱼（Vidal 等，2004）和鲻鱼（Aizen 等，2005）的脑垂体进行离体实验，结果证明是 DA 的 D2 受体而不是 D1 受体参与对 LH 分泌的抑制作用。一些研究未能检测到 DA 对 LH 合成的直接作用（Melamed 等，1998；Yaron 等，2003），而最近对欧洲鳗鱼的研究则证明 DA 抑制性地控制 GnRH 诱导 LH 的合成（Vidal 等，2004）。对金鱼的研究表明 DA 亦能够通过阻断脑垂体神经末梢肽类的合成或者抑制它的释放而对 GnRH 神经元起抑制作用（Yu 和

Peter，1990）。在 POA，DA 通过类 D1 受体作用机理而抑制 GnRH（Yu 和 Peter，1990，1992）。DA 亦作用于促性腺激素细胞水平，通过下调 GnRH 的受体合成（Kumakura 等，2003；Levavi-Sivan 等，2004）和 GnRH 的结合容量（de Leeuw 等，1989）而抑制 GnRH 的活性。这些多方面的作用说明 DA 能够对许多硬骨鱼类的 LH 释放起到明显的抑制作用。

有关 DA 参与调控 FSH 释放的研究很少。对虹鳟的研究表明 DA 能调控 FSH 的释放，这种作用在生殖周期后期当促性腺激素细胞对 GnRH 高度敏感时最为明显（Vacher 等，2000，2002）。儿茶酚胺合成抑制剂 α-甲基-p-酪氨酸（α-methyl-p-tyrosine）能使性成熟虹鳟的 FSH 浓度增加，但对卵黄生成的虹鳟没有影响（Saligaut 等，1998；Vacher 等，2000，2002）。DA D2 受体的同功物溴麦角环肽（bromocriptine）能抑制孵育的性成熟虹鳟促性腺激素细胞基础的 FSH 释放活动，但对卵黄发生的虹鳟没有影响。溴麦角环肽能阻抑 sGnRH-诱导卵黄发生的和性成熟的虹鳟促性腺激素细胞 FSH 释放活动（Vacher 等，2002）。

DA 亦会和其他的激素与神经递质相互作用以调控 GTH 的释放。在这些相互作用当中最重要的是 17β-雌二醇对 LH 分泌活动的负反馈调控作用。17β-雌二醇对于维持多巴胺能抑制虹鳟和金鱼 LH 释放的作用是必不可少的（Trudeau 等，1993a；Linard 等，1995；Saligaut 等，1999；Vacher 等，2002）。例如，在卵黄发生时，17β-雌二醇能激活儿茶酚胺周转并增强 DA 抑制 LH 释放的作用（Saligaut 等，1992）。对尼罗罗非鱼的研究表明，增加 17β-雌二醇的浓度，能使在体的和离体的 DA-2 受体 mRNA 水平升高，并且抑制 LH 和 FSH 释放以及在体的 lhb 单位 mRNA 水平（Levavi-Sivan 等，2006）。对于虹鳟，视前区的多巴胺能神经元能表达雌激素受体（Linard 等，1996），为 DA – 17β-雌二醇的相互作用提供神经解剖的基础。最近发现鲻鱼启动子 drd2 具有一个功能的 17β-雌二醇反应元件（Nocillado 等，2005），支持了 17β-雌二醇负反馈调控 LH 分泌是通过作用于多巴胺能神经元的假说。总之，17β-雌二醇对卵黄生成的和产卵前的鱼负反馈抑制 LH 和 FSH 的分泌活动显然是通过多巴胺能的抑制作用而介导的（Levavi-Sivan 等，2006）。

DA 亦影响 GABA 和 NPY 等 LH 释放的刺激性因子的作用。清除 DA 导致 GABA 合成和 GABA 合成酶谷氨酸脱羧酶 67（glutamic acid decarboxylase 67，GAD67）增加，从而使 GABA 促进 LH 释放的作用增强（Trudeau，1997；Trudeau 等，2000；Popesku 等，2008）。DA 亦能阻抑 NPY-诱导离体孵育的金鱼脑垂体释放 LH（Peng 等，1993a）。

已有充分的证据表明清除多巴胺能的抑制作用并同时增强 GnRH 的刺激作用，组成重要的神经内分泌作用机理，使得许多硬骨鱼类在排卵前大量分泌 LH 并导致排卵（Peter 等，1988）。这项理论成果已经在水产养殖中应用。许多鱼类由于强烈的内源性多巴胺抑制作用而抵制了应用外源 GnRH 促进 LH 释放的作用，结果是亲鱼不能排卵（Yaron，1995）。对于有强烈多巴胺能抑制作用的鱼类，广泛采用的诱导产卵的方法是同时使用强有力的多巴胺受体拮抗物和 GnRH 高活性的类似物（Peter 等，1988；

Yaron，1995）。对于雄性金鱼，减弱 DA 的抑制作用对于性外激素诱导 LH 分泌亦有密切关系（Dulka 等，1992）。

3.4.2 5-羟色胺

5-羟色胺（serotonin，5-hydroxytryptamine，5-HT）能增强一些硬骨鱼类的 LH 分泌活动（Kreke 和 Dietrich，2007；Popesku 等，2008）。在金鱼体内注射 5-HT 能使血清 LH 水平明显增加，但在第三脑室注射 5-HT 没有作用，表明 5-HT 对 LH 分泌活动的刺激作用是在脑垂体水平上（Somoza 等，1988）。对大西洋绒须石首鱼单独使用 5-HT 并不能提高 LH 水平，但能增强 GnRH 促进 LH 释放的作用（Khan 和 Thomas，1992）。对金鱼（Somoza 和 Peter，1991；Wong 等，1998）和大西洋绒须石首鱼（Khan 和 Thomas，1992）进行脑垂体碎片灌流的研究表明 5-HT 能刺激 LH 释放。对于金鱼和大西洋绒须石首鱼，5-HT 的作用能被 5-HT 的 T2 型受体拮抗物 ketanserin 所阻断。5-HT 亦能刺激性成熟雌金鱼视前区和脑垂体释放 GnRH（Yu 等，1991）。对于真鲷，5-HT 能诱导幼鱼和成鱼的 POA 分泌 sbGnRH，但对脑垂体的 GnRH 水平没有影响（Senthilkumaran 等，2001）。5-HT 的这个作用似乎是由 5-HT2 型受体所介导，因为用 ketanserin 能将它阻断。总之，这些研究证明 5-HT 能刺激 LH 释放，其作用可能发生在下丘脑和脑垂体水平上。

最近的研究表明鱼类在低氧情况下性腺发育明显退化。对大西洋绒须石首鱼的研究表明低氧和 GnRH mRNA 在 POA 的表达降低、LH 对 GnRH 刺激释放的反应性减弱、下丘脑 5-HT 含量和 5-HT 生物合成酶色氨酸羟化酶（tryptophan hydroxylase）活性降低都有密切关系（Thomas 等，2007）。下丘脑 5-HT 水平的药理性替换使神经内分泌功能恢复，表明刺激性的 5-羟色胺能神经内分泌通路是大西洋绒须石首鱼低氧诱发生殖活动受到抑制的主要作用部位。其他的研究表明，多氯联苯（polychlorinated biphenyls，PCB）对鱼类生殖的负面作用是通过 5-羟色胺能系统的影响（Thomas，2008），从而增强了这个调控鱼类 LH 分泌的神经内分泌通道的重要性。

3.4.3 去甲肾上腺素

一系列研究证明去甲肾上腺素（noradrenaline）通过 $\alpha 1$-类受体作用于脑垂体刺激金鱼、虹鳟和印度鲶鱼的 LH 释放活动（Chang 等，1991；Linard 等，1995；Senthilkumaran 和 Joy，1996；Wong 等，1998）。去甲肾上腺素亦刺激下丘脑产生 GnRH 并释放到脑垂体内（Yu 和 Peter，1992）。

3.4.4 褪黑激素

褪黑激素（melatonin）由松果体在夜间分泌，对调控哺乳动物生殖活动起着十分重要的作用。对于褪黑激素在鱼类神经内分泌调控 LH 或 FSH 分泌活动中的作用还很少研究。采用松果体切除和施用褪黑激素的研究证明松果体影响鱼类生殖活动，但具体的反应情况变化很大（Khan 和 Thomas，1996；Sebert 等，2008）。褪黑激素使印度鲶鱼脑垂体内促性腺激素细胞的数量减少，形状变小（Sundararaj 和 Keshavanath，1976）。对性腺充分发育的大西洋绒须石首鱼注射和脑垂体碎片离体孵育都证明褪黑激素能增强 LH 的分泌活动（Khan 和 Thomas，1996）。但是在印度鲶鱼腹腔注射褪黑激素使血浆中 LH 水平降低（Senthilkumaran 和 Joy，1995）。对银鳗鱼进行长期的褪黑激素处理能使脑的酪氨酸羟化酶（tyrosine hydroxylase，TH，DA 合成的限速酶）mRNA 表达增加（Sebert 等，2008）。褪黑激素刺激视前区的多巴胺能系统，从而增强 LH 和 FSH 合成与释放的抑制性调控作用。褪黑激素对鳗鱼两种 GnRH 类型（mGnRH 和 cGnRH Ⅱ）mRNA 的表达没有作用，但能使 LHβ 和 FSHβ mRNA 的表达降低。对于阐明褪黑激素调控鱼类生殖活动的作用，还需要做许多研究。

3.4.5 γ-氨基丁酸

氨基酸神经递质 γ-氨基丁酸（γ-aminobutyric acid，GABA）能刺激硬骨鱼类释放 LH（Trudeau 等，2000；Popesku 等，2008）。在体研究表明施用 GABA 或者 GABA 同功物能使金鱼（Kah 等，1992；Martyniuk 等，2007）、大西洋绒须石首鱼（Khan 和 Thomas，1999）和虹鳟（Mañanos 等，1999）血浆 LH 含量增加。亦有研究证明 GABA 能刺激虹鳟释放 FSH（Mañanos 等，1999）。

采用雄性虹鳟分散的脑垂体细胞进行的研究表明 GABA 能刺激 LH 和 FSH 的基础性分泌活动，并能增强 sGnRH 刺激这两种激素分泌的作用（Mañanos 等，1999）。但是，对雄性虹鳟的分散脑垂体细胞，GABA 并不影响基础性的或者 sGnRH 诱导的 LH 分泌活动。采用金鱼和大西洋绒须石首鱼分散的脑垂体细胞进行的研究却未能证明 GABA 直接影响 LH 的释放（Khan 和 Thomas，1999）。接着的研究表明在雌金鱼体内，GABA 对 LH 释放的刺激作用和脑垂体碎片与下丘脑的神经末梢释放的 GnRH 刺激作用密切相关（Kah 等，1992）。对真鲷的研究表明 GABA 刺激离体的 POA 薄片释放 sGnRH，但对脑垂体薄片没有作用（Senthilkumaran 等，2001）。GABA 和 GABA 的同功物蝇蕈醇（muscimol）能刺激处于生殖最后期的雌性七鳃鳗产生和释放 GnRH（Root 等，2004）。除了影响 GnRH 神经元，GABA 另一个可能的作用机理是参与多巴胺能神经元的调节进而影响鱼类 LH 的释放。已有证据表明 GABA 影响金鱼的多巴胺能神经元，但对非洲鲶鱼和大西洋绒须石首鱼没有作用（Trudeau 等，2000），说明这些影响对于 GnRH 神经元的作用似乎并不是重要的。

对金鱼的研究表明 GABA 的作用取决于类固醇所处环境和鱼类的生殖时期（Kah 等，1992；Trudeau 等，1993a、b）。GABA 在金鱼性腺发育早期刺激 LH 释放，而在性腺成熟期或性腺退化期没有作用。17β-雌二醇能减弱 GABA 刺激 LH 释放的作用（Trudeau 等，1993a）。睾酮能恢复 GABA 对这些鱼类刺激 LH 释放的作用（Kah 等，1992；Trudeau 等，1993a）。相反，GABA 能刺激性成熟的雌虹鳟释放 LH，而值得注意的是，这些鱼的睾酮水平高而 17β-雌二醇水平下降（Mañanos 等，1999）。再者，睾酮和 17β-雌二醇都能够调控性腺退化鱼的端脑和下丘脑内 GABA 的合成作用（Trudeau 等，1993a）。

最近对金鱼的研究表明 GABA 能诱导活化蛋白（activin）βa mRNA 快速而大量地增加，并伴随着 LH 释放的增加（Martyniuk 等，2007）。由于活化蛋白能刺激 LH 释放，这可能是 GABA 刺激鱼类 LH 释放的一种新的作用机理。

3.5 神 经 肽 类

3.5.1　Kisspeptins

Kisspeptins 是由 kiss1 基因编码的一个神经肽家族，它们是 kisspeptin 受体（kiss 1r）的内源性配体，而 kisspeptin 受体是 G 蛋白-偶联受体 GPR54。Kisspeptin 对哺乳类生殖轴的维持起着主导作用，特别是对启动青春期至关重要（Seminara 和 Kaiser，2005；Kauffman 等，2007）。许多研究正在证明 kisspeptin 对鱼类的重要作用。采用基因组数据库信息，kiss 直向同源序列已经分别在斑马鱼（Van Aerle 等，2008；Biran 等，2008；Kitahashi 等，2009）、河豚（Van Aerle 等，2008）和青鳉（*Oryzias latipes*；Kanda 等，2008；Kitahashi 等，2009）体内被鉴定出来。此外，单个类型的 gpr54 cDNA 已经从尼罗罗非鱼（Parhar 等，2004）、鲻鱼（Nocillado 等，2007）、军曹鱼（*Rachycentron canadum*）（Mohamed 等，2007）和胖头鲹（Filby 等，2008）中被克隆。两个 gpr54 类型已从斑马鱼中被鉴定出来（Biran 等，2008）。

一些研究指出 kisspeptin 对鱼类生殖功能的重要性。早期研究在尼罗罗非鱼的 GnRH1、GnRH2 和 GnRH3 神经元中鉴别出 gpr54 的 mRNA（Parhar 等，2004）。在斑马鱼和青鳉的脑内，采用原位杂交和激光捕捉显微切割技术（laser capture microdissection）结合实时 PCR 的研究表明 kiss1 在腹-中缰（ventro-medial habenula）和室周下丘脑核表达（Kitahashi 等，2009）。kiss2 在后结节核和室周下丘脑核表达。在青鳉的脑部鉴定出两个表达 kiss1 基因的神经元群体，它们都定位于下丘脑核，即后室周核（nucleus posterioris periventricularis，NPPv）和腹结节核（nucleus ventral tuberis，NVT），这些核都是 LH 和 FSH 释放的重要调控部位。

Elizur（2009）总结研究结果并评估 kiss 和 GPR54 的表达和青春期启动的关系。综合鱼类 GPR54 表达的研究结果表明它们参与了青春期的调控。例如，成熟的尼罗罗非鱼雄鱼脑的 GPR54 水平要明显高于不成熟的雄鱼（Parhar 等，2004）。对于雌性鲻鱼，性腺发育早期 GPR54 在脑的表达水平最高，在性期发育中期以及后期表达降低（Nocillado 等，2007）。GPR54 在鲻鱼脑的表达型式和三种 GnRHs 很相似。对胖头鲹、尼罗罗非鱼和斑马鱼的研究表明 GPR54 表达的高峰出现在性腺开始发育时，而在生殖成熟时下降（Biran 等，2008；Filby 等，2008；Martínezz-Chavez 等，2008）。对胖头鲹的研究表明 GPR54 表达和 GnRH 基因密切相关，特别是调控脑垂体的 gnrh3 型（Filby 等，2008）。定量实时 PCR 的分析表明在斑马鱼发育与青春期开始时，kiss1、kiss2、gnrh2 和 gnrh3 mRNA 水平明显增高，并在成年期保持高水平（Kitahashi 等，2009）。

对性成熟雌斑马鱼进行的研究提供了关于 kisspeptin/GPR54 系统生理功能重要性的

最有说服力的证据，即施用 kiss2 而不是 kiss1 使脑垂体的 *fshb* 和 *lhb* mRNA 显著增加（Kitahashi 等，2009）。是 kiss2 通过视前的 GnRH 神经元激活促性腺激素，或者是通过 GPR54 在促性腺激素细胞水平上的作用？这个问题还需要研究，因为外周的 kisspeptin 有可能通过 *gpr*54-b 在鱼脑垂体直接刺激促性腺激素细胞的分泌活动。给青春期胖头鲹腹腔注射哺乳类 kisspeptin-10 后，脑的 *gnrh*3 表达增强（Filby 等，2008）。总之，上述的研究都提供证据表明 kisspeptin 参与刺激鱼类青春期的启动和性腺发育成熟。

3.5.2 脑垂体腺苷酸环化酶激活多肽

在对金鱼的研究中发现，脑垂体腺苷酸环化酶激活多肽（pituitary adenylate cyclase-activating polypeptide，PACAP）的下丘脑免疫反应神经纤维直接分布到脑垂体，终止在促性腺激素细胞附近。采用 PACAP 和 PACAP 拮抗剂 PACAP6-38 进行离体和在体的研究表明 PACAP 能直接作用于促性腺激素细胞刺激 LH 释放（Wong 等，2000；Chang 等，2001；Sawisky 和 Chang，2005）。在对尼罗罗非鱼的研究中，PACAP 使 GPα、LHβ 和 FSHβ 的 mRNA 水平增加（Yaron 等，2003）。还没有研究 PACAP 参与 FSH 释放的调控。

3.5.3 神经肽 Y

神经肽 Y（neuropeptide Y，NPY）能刺激金鱼、鲤鱼、虹鳟和舌齿鲈 LH 的释放（Peng 等，1990；Breton 等，1991；Cerdá-Reverter 等，1999）。对金鱼的研究表明，NPY 通过 Y2 受体直接作用于脑垂体细胞刺激 LH 释放，又通过在脑垂体的末梢增加 GnRH 释放而间接刺激 LH 释放（Peng 等，1993a）。NPY 能刺激性未成熟真鲷离体的下丘脑视前区碎片释放 sbGnRH，但对脑垂体没有作用（Senthilkumaran 等，2001）。在这项研究中，NPYY1 和 NPYY2 特异性同等物的作用是相同的。免疫组织化学研究表明 NPY 是通过 Y1 受体影响鲶鱼（*Clarias batrachus*）的 GnRH-LH 系统（Mazumdar 等，2006）。在对尼罗罗非鱼的研究中，NPY 能增强 *lhb* 的表达，但对 *fshb* 没有作用（Yaron 等，2003）。对雌雄同体雌性先熟的双带锦鱼（*Thalassoma bifasciatum*），NPY 亦能够诱导性反转（Kramer 和 Imbriano，1997）。

很明显，NPY 对 LH 分泌活动的刺激作用受到鱼体生殖和能量状态的影响。在对舌齿鲈施行慢性禁食使胰岛素和果糖显著减少而造成负能量的状态下，血浆中 LH 水平以依赖于 NPY 的剂量而增加。相反，正的能量状态能压制 NPY 刺激 LH 分泌活动的能力（Cerdá-Reverter 等，1999）。同样，从禁食的鱼体内取得的分散脑垂体细胞在缺乏基本营养素而受限制的介质中孵育，要比在 L-15 介质中孵育的表现出对 NPY 更高的反应性。性类固醇似乎是影响 NPY 出现差别调节作用的候选物。但是，禁食鱼血浆中睾酮和 17β-雌二醇与摄食鱼相比较并没有明显差别。对金鱼和虹鳟的研究表明对 NPY 的反应性取决于鱼的生殖状态，而 NPY 对 LH 释放的刺激作用受到睾酮和 17β-雌二醇的影响（Peng 等，1993b，1994）。在影响 NPY-诱导 LH 分泌的营养状态和生殖季节之间可能存在着复杂的相互作用。

3.5.4 其他的神经肽

其他的一系列神经肽参与 LH 和 FSH 释放的调控。生长素释放肽（ghrelin）直接作用于金鱼脑垂体的细胞使 LH 释放（Unniappan 和 Peter，2004）。从中枢和外围注射生长素释放肽能刺激 LH 释放和 *lhb* 表达，表明生长素释放肽亦能参与 LH 的合成。幼鱼生长素释放肽能刺激鱼类的摄食活动（见本书第 9 章），这些研究结果表明生长素释放肽可能是联系食物摄取、生长和生殖生理的信号之一（Unniappan 和 Peter，2004，2005）。

包括一个 C-端-精-苯丙-NH_2 序列的神经肽（RF 酰胺-肽）已经在许多脊椎动物（包括鱼类）的脑内鉴别。这个肽类家族对四足类起着促性腺激素释放-抑制激素作用，但它们刺激红大麻哈鱼（*Oncorhynchus nerka*）（Amano 等，2006）脑垂体的细胞释放 LH 和 FSH。此外，RF 酰胺-肽使金鱼减弱 LH 的释放（Chang 等，2009）。

分泌粒蛋白-Ⅱ（secretogranin-Ⅱ，Sg-Ⅱ）及其蛋白水解产物 secretoneurin（SN）是金鱼 LH 释放和产生的潜在旁分泌调节剂。金鱼脑垂体细胞表达 Sg-Ⅱ mRNA，而 SN 能增强脑垂体碎片释放 LH 的作用（Blazquez 等，1998b；Zhao 等，2006）。

黑色素浓集激素（MCH）是一个环七十肽（cyclic heptadecapeptide），在下丘脑内产生，能调节金鱼的 LH 释放（Cerdá-Reverter 等，2006）。鲑鱼 MCH 能刺激分散的脑垂体细胞释放 LH，表明它直接作用于促性腺激素细胞。17β-雌二醇和睾酮能使 MCH 的表达增强，表明 MCH 可能参与类固醇对 LH 分泌活动的正反馈作用环（Cerdá-Reverter 等，2006）。

神经脑垂体激素中的加压催产素（vasotocin）和硬骨鱼催产素（isotocin）是血管升压素（vasopressin）和催产素（oxytocin）肽类家族的成员，对鱼类的生殖特别是生殖行为起着重要的调节作用（Balment 等，2006；Popesku 等，2008）。但这些激素是否影响 LH 和 FSH 的释放，还不是很清楚。

3.6 蛋白质激素

3.6.1 胰岛素样生长因子Ⅰ

胰岛素样生长因子Ⅰ（insulin-like growth factor Ⅰ，IGF-Ⅰ）是硬骨鱼类促性腺激素分泌的重要调节剂（见表 3.2）。IGF-Ⅰ 能使幼年欧洲鳗鱼脑垂体细胞的 LH 含量增加和促进 LH 释放（Huang 等，1998，1999）。对虹鳟的脑垂体细胞同时给予 IGF-Ⅰ 和 sGnRH，能提高 sGnRH 促进 LH 和 FSH 释放的敏感性（Weil 等，1999）。对银大麻哈鱼的脑垂体细胞，IGF-Ⅰ 能增强 GnRH 刺激 FSH 释放的作用，并使细胞的 LH 与 FSH 含量升高（Baker 等，2000）。这些研究结果都证明 IGF-Ⅰ 对 LH 和 FSH 的产生起着调控作用，特别是通过和 GnRH 的相互作用。

表 3.2　IGF-I 对 LH 和 FSH 分泌活动的影响

种类	试验条件	对 IGF-I 的反应	参考文献
幼年雌性鳗鱼（Anguilla anguilla）	脑垂体细胞的离体孵育	增加 LH 释放及脑垂体细胞的含量	Huang 等（1998，1999）
虹鳟（Oncorhynchus mykiss）	雌雄两性脑垂体细胞的离体孵育	对 LH 和 FSH 的释放没有作用，但增强 GnRH 的作用，特别是在配子生成的早期	Weil 等（1999）
银大麻哈鱼（Oncorhynchus kisutch）	脑垂体细胞的离体孵育	对 FSH 的释放没有作用，但增强 GnRH 的作用	Baker 等（2000）
马苏大麻哈鱼（Oncorhynchus masou）	初生脑垂体细胞的离体孵育	IGF-I 差别地调节 sGnRH 诱导 GTH 亚基基因的表达，在性成熟时取决于生殖的不同时期	Ando 等（2006）；Furukama 等（2008）

对鱼类在性成熟的不同时期进行的研究可以深入了解 IGF-I 的作用。对马苏大麻哈鱼初生的脑垂体细胞进行孵育的研究表明 IGF-I 在配子生成早期增强 GPα、FSHβ 和 LHβ mRNA 的表达，但从性成熟的或产卵的鱼体内取得的脑垂体细胞孵育则没有这些作用（Ando 等，2006；Furukuma 等，2008）。在雌鱼的配子生成早期，IGF-I 刺激 FSH 和 LH 释放，但对 FSH 与 LH 亚基 mRNA 水平没有刺激作用。IGF-I 加上 sGnRH 能刺激雌雄两性在各个生殖时期的 FSH 与 LH 释放，但对 LH 与 FSH 亚基 mRNA 水平则在不同的亚基和生殖时期有不同的影响。这些研究结果和早期对虹鳟脑垂体的研究结果一致，即 IGF-I 对 LH 和 FSH 释放的刺激作用在鱼的配子生成早期要比在鱼的性成熟期显著得多（Weil 等，1999）。总之，这些研究结果表明 IGF-I 可能起着一个信号分子的作用，在鱼类青春期启动时将生长与营养状态的信息传送到促性腺轴（gonadotropic axis）。

3.6.2　激活蛋白和促滤泡素抑制素

一些研究已经说明激活蛋白（activin）及其结合蛋白促滤泡素抑制素（follistatin）对金鱼 LH 和 FSH 分泌的调控作用。激活蛋白 βB 是激活蛋白亚基的主要型式，在金鱼的脑垂体中表达（Lau 和 Ge，2005）。研究表明基因重组的金鱼激活蛋白刺激 *fshb* 的表达而阻抑 *lhb* 表达（Yuen 和 Ge，2004；Yam 等，1999；Cheng 等，2007）。促滤泡素抑制素亦在金鱼的脑垂体表达，并且表现出能逆转激活蛋白的作用。特别是促滤泡素抑制素能降低基础的 *fshb* 表达而增强 *lhb* 表达，并且能抑制激活蛋白诱导的 LH 与 FSH 亚基 mRNA 水平的变化（Yuen 和 Ge，2004），最近采用斑马鱼初生脑垂体细胞孵育的研究证实了在金鱼的研究结果，即激活蛋白抑制斑马鱼 *fshb* 的表达，阻抑 *lhb* 的表达，而促滤泡素抑制素能使 *fshb* 和 *lhb* 的反应逆转过来（Lin 和 Ge，2009）。

由于激活蛋白和促滤泡素抑制素能在脑垂体中共表达，表明 FSH 和 LH 的相对表达水平会受到这两种蛋白质之间平衡状况的影响。离体和在体研究都表明激活蛋白 βB 在脑垂体中的表达水平是稳定的，因而可以推想促滤泡素抑制素在脑垂体的激活蛋白系统中起着重要的调控作用。为了测试这个推想，Cheng 等（2007）检测了睾酮和 17β-雌二醇对促滤泡素抑制素、激活蛋白 βB 和 FSHβ 表达的影响。对在生理范围浓度中孵育的脑垂体细胞，睾酮和 17β-雌二醇能显著增强促滤泡素抑制素和 LHβ 的表达。但是，睾酮和 17β-雌二醇对激活蛋白 βB 和 FSHβ 的表达没有任何影响。性类固醇能调节促滤泡素抑制素的表达而对激活蛋白没有作用，表明在内分泌或者神经内分泌调控的支配下，促滤泡素抑制素在这个激活蛋白系统中起着重要的作用（见图 3.1；Cheng 等，2007）。

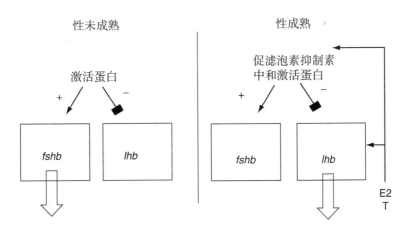

图 3.1　激活蛋白、激活蛋白-结合蛋白促滤泡素抑制素和性类固醇对性未成熟与性成熟金鱼脑垂体 *fshb* 和 *lhb* 表达的相互作用影响模式（Cheng 等，2007，有改动）

对于性未成熟鱼，激活蛋白刺激 *fshb* 表达而抑制 *lhb* 表达。对于性成熟鱼，17β-雌二醇、睾酮和激活蛋白诱导促滤泡素抑制素表达。促滤泡素抑制素中和激活蛋白的作用，使 *fshb* 的表达降低。17β-雌二醇和睾酮亦作用于 LH 细胞增强 *lhb* 的表达，而在促滤泡素抑制素存在的情况下，激活蛋白对 *lhb* 的抑制作用减弱。E_2，17β-雌二醇；T，睾酮。

3.7　性腺发育的综合神经内分泌调控

研究者对于全面了解调控硬骨鱼类关键的生殖过程如青春期、性成熟期和产卵等的神经内分泌通路是饶有兴趣的。很明显，这些生殖过程包含调控 LH 和 FSH 释放的多种神经内分泌通路的协调作用，其中性类固醇和多巴胺起着重要作用。硬骨鱼类生殖发育的调控过程中亦必须整合外界环境的信息，如温度、光周期等，这就进一步增加了复杂程度。有关硬骨鱼类生殖内分泌调控显著多样性的证据正在不断增加，而差不多占脊椎动物 50% 的鱼类展现出动物界最为丰富多样的生殖策略是不足为奇的。

一些研究提到 DA 在青春期的定时方面起调控作用。例如，清除 DA 的抑制作用对于幼年欧洲鳗鱼 GnRH 刺激 LH 基因表达的作用是重要的（Vidal 等，2004）。幼年的大西洋棘白鲳（*Chaetodipterus faber*）在青春期下丘脑的多巴胺能活性降低（Marcano 等，1995）。对鲻鱼的研究表明在青春期开始后，脑垂体的 DA-D2 受体丰度下降（Nocillado 等，2007）。DA 水平和能量状态亦有联系，饥饿使肺鱼（*Protopterus annectens*）端脑和间脑的 DA 水平增加，而这可能导致生殖发育退化。进一步的研究表明用氟哌啶醇（haloperidol）阻断 DA 受体后能启动锦鱼（*Thalassoma duperrey*）的性反转（Larson 等，2003）。相反，使用 DA 的同等物阿朴吗啡（apomorphine）能制止锦鱼在社会条件允许的情况下完成性反转。但是，DA 对鱼类青春期的重要性并不是普遍存在的。对于杂交条纹狼鲈幼鱼和真鲷雌性幼鱼，没有证据表明 DA 在青春期对 LH 的分泌活动起调控作用（Holland 等，1998；Kumakura 等，2003）。

尽管类固醇对幼年硬骨鱼类脑垂体 LH 和 FSH 含量起着正反馈的调控作用，但这些激素的释放通常是有限的。例如对鳗鱼用 17β-雌二醇预处理后，单独用 GnRH 并不能诱导 LH 释放（Dufour 等，1988）。在鳗鱼（Vidal 等，2004）和杂交狼鲈（Holland 等，1998）的相关研究中发现，GnRH 对 LH 释放的刺激作用必须有睾酮参与，对于增强脑垂体细胞对 GnRH 的敏感性，睾酮似乎起着促进作用（Vidal 等，2004）。虽然目前学界对介导睾酮刺激作用的机理还了解不多，但很可能是由于增强了 GnRH 受体的表达和/或调整了 GnRH-受体偶联与信号通路（Trudeau 等，1991；Lo 和 Chang，1998）。对非洲鲶鱼的研究表明来自睾丸的可芳化雄激素能使和青春期相联系的免疫反应促性腺激素细胞增加，以及使脑垂体内 LH mRNA 和蛋白质含量增加（Covaco 等，2001）。总之，这些研究成果一致说明雄激素对一些硬骨鱼类的幼鱼在青春期激活脑-脑垂体-性腺轴起着关键作用。

研究者们对类固醇调控 FSH 释放的反馈作用还没有深入了解。关于类固醇激素对 FSH 释放起负反馈作用的文献居多。提高 17β-雌二醇的水平似乎是和 FSHβ mRNA 水平下降与血浆中 FSH 含量降低同时发生的（Larson 和 Swanson，1997；Dickey 和 Swanson，1998；Sohn 等，1998；Mateos 等，2002；Banerjee 和 Khan，2008）。对于未成熟的鱼，这些作用则是变化不定的。对于未成熟的或者性腺发育早期的银大麻哈鱼，17β-雌二醇对血浆 FSH 或 FSHβ mRNA 水平没有影响（Dickey 和 Swanson，1995），而对于虹鳟（Breton 等，1997）和大西洋鲑鱼（*Salmo salar*）（Borg 等，1998），类固醇对脑垂体和血浆的 FSH 水平表现出正反馈作用。对雌银鳗的研究表明，类固醇的在体处理对 FSH mRNA 水平没有影响，尽管离体孵育的脑垂体细胞用 17β-雌二醇处理后 FSH mRNA 水平略有增加。对金鱼脑垂体细胞的研究得到同样的结果（Huggard-Nelson 等，2002）。综合这些研究结果说明 17β-雌二醇在脑垂体和下丘脑中水平的反应存在着明显的差别，可能在不同种类和不同性腺发育时期对 FSH 产生有差别的调控作用（differential regulation）。

性腺类固醇和类固醇反馈调节的变化对于介导季节的/性成熟的 LH 与 FSH 释放有效性起着重要作用。17β-雌二醇对 LH mRNA 表达的影响取决于性腺发育时期，17β-雌二醇对幼年金鱼和大西洋绒须石首鱼的 LH mRNA 起着正反馈作用，但对性腺发育成熟的鱼没有作用（Kobayashi 等，2000；Banerjee 和 Khan，2008）。对于性成熟的鱼，切除性腺并不影响 LH 分泌，而 17β-雌二醇对 LH mRNA 表达没有作用。在这两种鱼中，雌激素对性成熟鱼反馈的主要影响是阻断 GnRH 诱导 LH 释放。在这些鱼类性成熟的最后阶段，血浆雌激素含量降低，黄体酮含量升高，以促进卵母细胞成熟和排卵。雌激素水平降低亦就减少了对 GnRH 释放的负反馈作用影响，从而使 LH 大量释放与诱导排卵。

光周期能调节类固醇对脑垂体 LH 与 FSH 表达的反馈调控作用。在对三棘刺鱼（*Gaoterosteus aculeatus*）的研究中，在长日（long-day）和短光周期下，被阉割雄鱼的 LHβ mRNA 水平要比对照组（假手术）雄鱼的更低些，而用 11-KA 和 T 处理使 LHβ mRNA 表达水平增强，表现为正反馈的作用机理（Hellqvist 等，2008）。在长光周期下亦出现对 *fshb* 表达的正反馈作用，阉割后降低 *fshb* 表达，而用雄激素处理后恢复 *fshb* 表达。但在短光周期下出现不同的反应。在短光周期下阉割使 *fshb* 水平增加，而使用 11-KA 和 11-KT 处理后降低了 *fshb* 的表达，表明在这种情况下出现对 *fshb* 调控的负反馈作用机理（Hellqvist 等，2008）。在不同的光周期下，正反馈作用和负反馈作用的变化并不出现于哺乳类动物之中，在它们那里只是负反馈作用的有效性发生改变。

3.8 展　　望

采用分子生物学的技术手段极大地促进了鱼类神经内分泌学的发展。这些技术有助于证明 GnRH 及其受体的多重性（multiplicity），并在鉴定新出现的具有下丘脑与脑垂体功能的调节剂如 kisspeptin 方面迅速取得进展。对调控通道中各个成员的鉴定才刚开始（见图 3.2），现在需要阐明这些调节剂的作用。曾经注重基因产物的鉴定，接着有必要加强对这些蛋白质的存在度（presence）和功能的研究。例如，对硬骨鱼类 FSH（某种程度上包括 LH）的研究就曾经被用于生物测试的纯化蛋白质的有效性所牵制。现在，随着表达 LH 和 FSH（Aizen 等，2007；Zmora 等，2007；Kazeto 等，2008）和其他激素（Yuen 和 Ge，2004）方法的建立，将提供新的机会去测试同源性蛋白质的活性，亦就减轻了对相关异源性（哺乳类的）激素测试的担心。基因重组 FSH 的有效性亦提供了机会去建立新的免疫测定方法以测定这种激素含量的变动，这将填补有关这方面很大的知识空白。鱼类基因在哺乳类细胞中表达的技术已经应用于 GnRH 受体（Tello 等，2008），这为相关受体和信号通路的功能研究提供新的机会。

学者们越来越深刻地认识到 GnRH-FSH/LH 系统受到来自下丘脑和外周多种输入因子的调控。对所有这些输入因子进行鉴别以确定哪些是起主要作用的，哪些可能是多余而不那么重要的，这将是很有意义的挑战。分子生物学研究技术手段将再次为我们提供重要的有效帮助。至今还很少采用这些先进技术研究基因沉默或者过表达以及影响蛋白质功能的获得或丧失。要深刻地分析掌握调控基因转录的因子，亦可以从启动子区（promoter region）的序列分析和基因表达的转录调控剂的共有序列（consensus sequence）的鉴定中实现。最近，各种类固醇激素受体亚型的证明和非传统膜受体的鉴定使得学者们需要重新评定类固醇激素的作用。各种类固醇激素选择性的同等物与拮抗物的有效性可以显著地加深我们对类固醇如何影响 GnRH-FSH/LH 系统的了解。

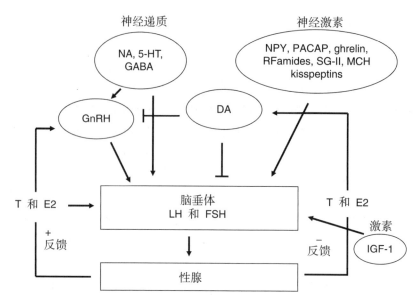

图 3.2 调控硬骨鱼类 LH 和 FSH 释放的主要神经内分泌因子总结示意图

大量的刺激性因子包括 GnRH、神经递质、神经激素和外周蛋白质激素。主要的抑制性因子是多巴胺。在不同发育时期，性类固醇激素睾酮和雌二醇对 LH 与 FSH 的释放可能起着正的和负的反馈作用。激活蛋白和它的结合蛋白对 LH 和 FSH 的释放起着差别性影响不包括在本图内。在整个性腺发育成熟过程中，各种因子的相对重要性发生变化。5-HT，5-羟色胺；DA，多巴胺；E_2，17β-雌二醇；GABA，γ-氨基丁酸；GnRH，促性腺激素释放激素；IGF-I，胰岛素生长因子-I；MCH，黑色素浓集激素；NA，去甲肾上腺素；NPY，神经肽 Y；PACAP，脑垂体腺苷酸环化酶激活多肽；RFamides，RF 酰胺肽；SG-II，分泌粒蛋白-II；T，睾酮。

硬骨鱼类生殖策略的多样性使它们成为神经内分泌调控生殖研究非常有意义的模式群体。迄今为止的研究还只局限于相当少数的种类，将来对采取各种各样生殖策略的鱼类进行研究会给我们带来很多好处。目前的许多研究将研究对象集中于季节性生殖的鱼类，而它们通常都是卵巢发育同步性的类群。今后对卵巢发育不同步的持续性生殖的鱼类进行研究，例如，研究基因组序列已经完成的斑马鱼和青鳉，将会带来新的理念。斑马鱼和青鳉已广泛用于发育生物学中的生物医学研究，它们在 GnRH-FSH/LH 系统的个体发生研究中亦会起很好的作用。最后，不要忽视环境因子的重要作用，例如温度和光周期对硬骨鱼类生殖功能所起的调控作用。在 GnRH-FSH/LH 系统的研究中，采用新的技术手段将会敞开研究环境因子如何影响鱼类生殖活动的大门（Hellqvist 等，2008）。同样，许多人造的化合物都能够破坏鱼类的内分泌稳态。这些化合物的功能通常都模拟类固醇，而类固醇参与 GnRH-FSH/LH 系统正的和负的反馈调控作用。因而，这些人造化合物影响鱼类生殖的神经内分泌调控是不足为奇的（Van Der Kraak 等，1992；Thomas，2008）。研究表明人造化合物和低氧能影响 5-羟色胺能系统（Thomas 等，2007），说明它们可能影响到多个靶标。很明显，令人振奋的事情在于有新的机会去研究阐明关于神经内分泌调控鱼类生殖活动的各种基础理论性和实际应用性的课题。

G. Van Der 克拉克

参考文献

Aizen J, Meiri I, Tzchori I, Levavi-Sivan B, Rosenfeld H. 2005. Enhancing spawning in the grey mullet (*Mugil cephalus*) by removal of dopaminergic inhibition. *Gen. Comp. Endocrinol*, 142: 212-222.

Aizen J, Kasuto H, Golan M, Zakay H, Levavi-Sivan B. 2007. Tilapia follicle-stimulating hormone (FSH): Immunochemistry, stimulation by gonadotropin-releasing hormone, and effect of biologically active recombinant FSH on steroid secretion. *Biol. Reprod*, 76: 692-700.

Amano M, Hyodo S, Urano A, Okumoto N, Kitamura S, Ikuta K, Suzuki Y, Aida K. 1994. Activation of salmon gonadotropin-releasing hormone synthesis by 17-methyl-testosterone administration in yearling masu salmon, *Oncorhynchus masou*. *Gen. Comp. Endocrinol*, 95: 374-380.

Amano M, Moriyama S, Iigo M, Kitamura S, Amiya N, Yamamori K, Ukena K, Tsutsui K. 2006. Novel fish hypothalamic neuropeptides stimulate the release of gonadotrophins and growth hormone from the pituitary of sockeye salmon. *J. Endocrinol*, 188: 417-423.

Ando H, Urano A. 2005. Molecular regulation of gonadotropin secretion by gonadotropin-releasing hormone in salmonid fishes. *Zool. Sci*, 22: 379-389.

Ando H, Luo Q, Koide N, Okada H, Urano A. 2006. Effects of insulin-like growth factor I on GnRH-induced gonadotropin subunit gene expressions in masu salmon pituitary cells at different stages of sexual maturation. *Gen. Comp. Endocrinol*, 149: 21-29.

Andreu-Vieyra C V, Buret A G, Habibi H R. 2005. Gonadotropin-releasing hormone induction of apoptosis in the testes of goldfish (*Carassius auratus*). *Endocrinology*, 146: 1588-1596.

Aroua S, Weltzien F A, Belle N L, Dufour S. 2007. Development of real-time RT-PCR assays for eel gonadotropins and their application to the comparison of *in vivo* and *in vitro* effects of sex steroids. *Gen. Comp. Endocrinol*, 153: 333-343.

Baker D M, Davies B, Dickhoff W W, Swanson P. 2000. Insulin-like growth factor I increases follicle-stimulating hormone (FSH) content and gonadotropin-releasing hormone-stimulated FSH release from coho salmon pituitary cells *in vitro*. *Biol. Reprod*, 63: 865-871.

Balment R J, Lu W, Weybourne E, Warne J K. 2006. Arginine vasotocin a key hormone in fish physiology and behaviour: A review with insights from mammalian models. *Gen. Comp. Endocrinol*, 147: 9-16.

Banerjee A, Khan I A. 2008. Molecular cloning of FSH and LH b subunits and their regulation by estrogen in Atlantic croaker. *Gen. Comp. Endocrinol*, 155: 827-837.

Biran J, Ben-Dor S, Levavi-Sivan B. 2008. Molecular identification and functional characterization of the kisspeptin/kisspeptin receptor system in lower vertebrates. *Biol. Reprod*, 79: 776-786.

Blazquez M, Bosma P T, Fraser E J, Van Look K J W, Trudeau V L. 1998. Fish as models for the neuroendocrine regulation of reproduction and growth. *Comp. Biochem. Physiol. Part C*, 119. 345-364.

Blazquez M, Bosma P T, Chang J P, Docherty K, Trudeau V L. 1998. Gamma aminobutyric acid up-regulates the expression of a novel secretogranin-II messenger ribonucleic acid in the goldfish pituitary. *Endocrinology*, 139: 4870-4880.

Bommelaer M C, Billard R, Breton B. 1981. Changes in plasma gonadotropin after ovariectomy and estradiol supplementation at different stages at the end of the reproductive cycle in the rainbow trout (*Salmo gairdneri* R.). *Reprod. Nutr. Dev*, 21: 989-997.

Borg B, Antonopoulu E, Mayer I, Andersson E, Berglund I, Swanson P. 1998. Effects of gonadectomy and androgen treatments on pituitary and plasma levels of gonadotropins in mature male Atlantic salmon, *Salmo salar*, parr-positive feedback control of both gonadotropins. *Biol. Reprod*, 58: 814-820.

Breton B, Sambroni E. 1996. Steroid activation of the brain-pituitary complex gonadotropic function in the triploid rainbow trout *Oncorhynchus mykiss*. *Gen. Comp Endocrinol*, 101: 155-164.

Breton B, Mikolajczyk T W, Popek, Bieniarz K, Epler P. 1991. Neuropeptide Y stimulates *in vivo* gonadotropin secretion in teleost fish. *Gen. Comp. Endocrinol*, 84: 277-283.

Breton B, Sambroni É, Govoroun M, Weil C. 1997. Effects of steroids on GTH I and GTH II secretion and pituitary concentration in the immature rainbow trout *Oncorhynchus mykiss*. *C. R. Acad. Sci. Paris, Life Sci*, 320: 783-789.

Breton B, Govoroun M, Mikolajczyk T. 1998. GTH I and GTH II secretion profiles during the reproductive cycle in female rainbow trout: Relationship with pituitary responsiveness to GnRH-A stimulation. *Gen. Comp. Endocrinol*, 111: 38-50.

Canosa L F, Stacey N, Peter R E. 2008. Changes in brain mRNA levels of gonadotropin-releasing hormone, pituitary adenylate cyclase activating polypeptide, and somatostatin during ovulatory luteinizing hormone and growth hormone surges in goldfish. *Am. J. Physiol. Regul. Integr. Comp. Physiol*, 295: R1815-R1821.

Cerdá-Reverter J M, Sorbera L A, Carrillo M, Zanuy S. 1999. Energetic dependence of NPY-induced LH secretion in a teleost fish (*Dicentrarchus labrax*). *Am. J. Physiol. Reg. Int. Comp. Physiol*, 277: R1627-R1634.

Cerdá-Reverter J M, Canosa L F, Peter R E. 2006. Regulation of the hypothalamic melanin-concentrating hormone neurons by sex steroids in the goldfish: Possible role in the modulation of luteinizing hormone secretion. *Neuroendocrinology*, 84: 364-377.

Chang J P, Yu K L, Wong A O L, Peter R E. 1990. Differential actions of dopamine receptor subtypes on gonadotropin and growth-hormone release *in vitro* in goldfish. *Neuroendocrinology*, 51: 664-674.

Chang J P, Van Goor F, Acharya S. 1991. Influences of norepinephrine, and adrenergic agonists and antagonists on gonadotropin secretion from dispersed pituitary cells of goldfish, *Carassius auratus*. *Neuroendocrinology*, 54: 202-210.

Chang J P, Wirachowsky N R, Kwong P, Johnson J D. 2001. PACAP stimulation of gonadotropin-II secretion in goldfish pituitary cells: Mechanism of actions and interaction with gonadotropin releasing hormone signaling. *J. Neuroendocrinol*, 13: 540-550.

Chang J P, Johnson J D, Sawisky G R, Grey C L, Mitchell G, Booth M, Volk M M, Parks S K, Thompson E, Goss G G, Klausen C, Habibi H R. 2009. Signal transduction in multifactorial neuroendocrine control of gonadotropin secretion and synthesis in teleosts - studies on the goldfish model. *Gen. Comp. Endocrinol*, 161: 42-52.

Cheng G F, Yuen C W, Ge W. 2007. Evidence for the existence of a local activin follistatin negative feedback loop in the goldfish pituitary and its regulation by activin and gonadal steroids. *J. Endocrinol*, 195: 373-384.

Chyb J, Mikolajczyk T, Breton B. 1999. Post-ovulatory secretion of pituitary gonadotropins GTH I and GTH II in the rainbow trout (*Oncorhynchus mykiss*): Regulation by steroids and possible role of non-steroidal gonadal factors. *J. Endocrinol*, 163: 87-97.

Copeland P A, Thomas P. 1989. Control of gonadotropin release in Atlantic croaker: Evidence for a lack of dopaminergic inhibition. *Gen. Comp. Endocrinol*, 74: 474-483.

Covaco J E B, van Baal J, van Dijk W, Hassing G A M, Goos H J Th, Schulz R W. 2001. Steroid hormones stimulate gonadotrophs in juvenile male African catfish (*Clarias gariepinus*). *Biol. Reprod*, 64: 1358-1365.

Crim L W, Peter R E, Billard R. 1981. Onset of gonadotropic hormone accumulation in the immature trout pituitary gland in response to estrogen or aromatizable androgen steroid hormones. *Gen. Comp. Endocrinol*, 44: 374-381.

de Leeuw R, Wurth Y A, Zandbergen M A, Peute J, Goos H J. 1986. The effects of aromatizable androgens, non-aromatizable androgens, and estrogens on gonadotropin release in castrated African catfish, *Clarias gariepinus* (Burchell): A physiological and ultrastructural study. *Cell Tissue Res*, 243: 587-594.

de Leeuw R, Habibi H R, Nahorniak C S, Peter R E. 1989. Dopaminergic regulation of pituitary gonadotrophin-releasing hormone receptor activity in the goldfish (*Carassius* auratus). *J. Endocrinol*, 121: 239-247.

Devlin R, Nagahama Y. 2002. Sex determination and sex differentiation on fish: An overview of genetic, physiological and environmental influences. *Aquaculture*, 208: 191-364.

Dickey J T, Swanson P. 1998. Effects of sex steroids on gonadotropin (FSH and LH) regulation in coho salmon (*Oncorhynchus kisutch*). *J. Mol. Endocrinol*, 21: 291-306.

Dickey J T, Swanson P. 2000. Effects of salmon gonadotropin-releasing hormone on follicle stimulating hormone secretion and subunit gene expression in coho salmon (*Oncorhynchus kisutch*). *Gen. Comp. Endocrinol*, 118: 436-449.

Dubois E A, Florijn M A, Zandbergen M A, Peute J, Goos H J. T. 1998. Testosterone accelerates the development of the catfish GnRH system in the brain of immature African catfish (*Clarias gariepinus*). *Gen. Comp. Endocrinol*, 112: 383-393.

Dufour S, Lopez E, Le Menn F, Le Belle N, Baloche S, Fontaine Y A. 1988. Stimulation of gonadotropin release and of ovarian development, by the administration of a gonadoliberin agonist and of dopamine antagonists, in female silver eel pretreated with estradiol. *Gen. Comp. Endocrinol*, 70: 20-30.

Dulka J G, Sloley B D, Stacey N E, Peter R E. 1992. A reduction in pituitary dopamine turnover is associated with sex pheromone-induced gonadotropin secretion in male goldfish. *Gen. Comp. Endocrinol*, 86: 496-505.

Elizur A. 2009. The KiSS1/GPR54 system in fish. *Peptides*, 30: 164-170.

Filby A L, Aerle R V, Duitman J, Tyler C R. 2008. The kisspeptin/gonadotropin-releasing hormone pathway and molecular signaling of puberty in fish. *Biol. Reprod*, 78: 278-289.

Flanagan C A, Chen C C, Coetsee M, Famputha S, Whitlock K E, Bredenkamp N, Grosenick L, Fernald R D, Illing N. 2007. Expression, structure, function, and evolution of gonadotropin-releasing hormone (GnRH) receptors GnRH-R1SHS and GnRH-R2PEY in the teleost, *Astatotilapia burtoni*. *Endocrinology*, 148: 5060-5071.

Furukuma S, Onuma T, Swanson P, Luo Q, Koide N, Okada H, Urano A, Ando H. 2008. Stimulatory effects of insulin-like growth factor 1 on expression of gonadotropin subunit genes and release of follicle-stimulating hormone and luteinizing hormone in masu salmon pituitary cells early in gametogenesis. *Zoolog. Sci*, 25: 88-98.

Gonzalez-Martinez D, Zmora N, Zanuy S, Sarasquete C, Elizur A, Kah O, Munoz-Cueto J A. 2002. Developmental expression of three different prepro-GnRH (gonadotrophin-releasing hormone) messengers in the brain of the European sea bass (*Dicentrarchus labrax*). *J. Chem. Neuroanat*, 23: 255-267.

Goos H J. 1987. Steroid feedback on pituitary gonadotropin secretion. In: Idler D R,

Crim L W, Walsh J, Ed. *Proceedings of the Third International Symposium on Reproductive Physiology of Fish* St. John's University: St. John, NS, Canada, 16-20.

Hellqvist A, Schmitz M, Borg B. 2008. Effects of castration and androgen treatment on the expression of FSH-beta and LH-beta in the three-spine stickleback, *Gasterosteus aculeatus*- Feedback differences mediating the photoperiodic maturation response?. *Gen. Comp. Endocrinol*, 158: 178-182.

Holland M C, Hassin S, Zohar Y. 1998. Effects of long-term testosterone, gonadotropin-releasing hormone agonist, and pimozide treatments on gonadotropin II levels and ovarian development in juvenile female striped bass (*Morone saxatilis*). *Biol. Reprod*, 59: 1153-1162.

Huang Y S, Schmitz M, Le Belle N, Chang C F, Quérat B, Dufour S. 1997. Androgens stimulate gonadotropin-II β subunit in eel pituitary cells *in vitro*. *Mol. Cell. Endocrinol*, 131: 157-166.

Huang Y S, Rousseau K, Le Belle N, Vidal B, Burzawa-Gérard E, Marchelidon J, Dufour S. 1998. Insulin-like growth factor-I stimulates gonadotrophin production from eel pituitary cells: A possible metabolic signal for induction of puberty. *J. Endocrinol*, 159: 43-52.

Huang Y S, Rousseau K, Le Belle N, Vidal B, Burzawa-Gérard E, Marchelidon J, Dufour S. 1999. Opposite effects of insulin-like growth factors (IGF-Is) on gonadotropin (GTH-II) and growth hormone (GH) production by primary culture of European eel (*Anguilla anguilla*) pituitary cells. *Aquaculture*, 177: 73-83.

Huggard D, Khakoo Z, Kassam G, Mahmoud S S, Habibi H R. 1996. Effect of testosterone on maturational gonadotropin subunit messenger ribonucleic acid levels in the goldfish pituitary. *Biol. Reprod*, 54: 1184-1191.

Huggard-Nelson D L, Nathwani P S, Kermouni A, Habibi H R. 2002. Molecular characterization of LH-beta and FSH-beta subunits and their regulation by estrogen in the goldfish pituitary. *Mol. Cell. Endocrinol*, 188: 171-193.

Illing N, Troskie B E, Nahorniak C S, Hapgood J P, Peter R E, Millar R P. 1999. Two gonadotropin-releasing hormone receptor subtypes with distinct ligand selectivity and differential distribution in brain and pituitary in the goldfish (*Carassius auratus*). *Proc. Natl. Acad. Sci. USA*, 96: 2526-2531.

Kah O, Trudeau V L, Sloley B D, Chang J P, Dubourg P, Yu K L, Peter R E. 1992. Influence of GABA on gonadotrophin release in the goldfish. *Neuroendocrinology*, 55: 396-404.

Kah O, Lethimonier C, Somoza G, Guilgur L G, Vaillant C, Lareyre J J. 2007. GnRH and GnRH receptors in metazoa: A historical, comparative, and evolutive perspective. *Gen. Comp. Endocrinol*, 153: 346-364.

Kanda S, Akazome Y, Matsunaga T, Yamamoto N, Yamada S, Tsukamura H, Maeda K, Oka Y. 2008. Identification of KiSS-1 product kisspeptin and steroid-sensitive sexually dimorphic kisspeptin neurons in medaka (*Oryzias latipes*). *Endocrinology*, 149: 2467-2476.

Kauffman A S, Clifton D K, Steiner R A. 2007. Emerging ideas about kisspeptin-GPR54 signaling in the neuroendocrine regulation of reproduction. *Trends Neurosci*, 30: 504-511.

Kazeto Y, Kohara M, Miura T, Miura C, Yamaguchi S, Trant J M, Adachi S, K. Yamauchi K. 2008. Japanese eel follicle-stimulating hormone (FSH) and luteinizing hormone (LH): Production of biologically active recombinant FSH and LH by Drosophila S2 cells and their differential actions on the reproductive biology. *Biol. Reprod*, 79: 938-946.

Khan I A, Thomas P. 1992. Stimulatory effects of serotonin on maturational gonadotropin release in the Atlantic croaker, *Micropogonias undulatus*. *Gen. Comp. Endocrinol*, 88: 388-396.

Khan I A, Thomas P. 1996. Melatonin influences gonadotropin II secretion in the Atlantic croaker (*Micropogonias undulatus*). *Gen. Comp. Endocrinol*, 104: 231-242.

Khan I A, Thomas P. 1999. GABA exerts stimulatory and inhibitory influences on gonadotropin II secretion in the Atlantic croaker (*Micropogonias undulatus*). *Neuroendocrinology*, 69: 261-268.

Khan I A, Hawkins M B, Thomas P. 1999. Gonadal stage-dependent effects of gonadal steroids on gonadotropin II secretion in the Atlantic croaker (*Micropogonias undulatus*). *Biol. Reprod*, 61: 834-841.

Kitahashi T, Ogawa S, Parhar I S. 2009. Cloning and expression of kiss2 in the zebrafish and medaka. *Endocrinology*, 150: 821-823.

Klenke U, Zohar Y. 2003. Gonadal regulation of gonadotropin subunit expression and pituitary LH protein content in female hybrid striped bass. *Fish Physiol. Biochem*, 28: 25-27.

Kobayashi M, Stacey N E. 1990. Effects of ovariectomy and steroid hormone implantation on serum gonadotropin levels in female goldfish. *Zool. Sci*, 7: 715-721.

Kobayashi M A, Sohn Y C, Yoshiura Y A, Aida K A. 2000. Effects of sex steroids on the mRNA levels of gonadotropin subunits in juvenile and ovariectomized goldfish *Carassius auratus*. *Fisher. Sci*, 66: 223-231.

Kramer C R, Imbriano M A. 1997. Neuropeptide Y (NPY) induced gonad reversal in the protogynous bluehead wrasses, *Thalassoma bifasciatum* (Teleostei: Labridea). *J. Exp. Zool*, 279: 133-144.

Kreke N, Dietrich D R. 2007. Physiological endpoints for potential SSRI interactions in fish. *Crit. Rev. Toxicol*, 38: 215-247.

Kumakura N, Okuzawa K, Gen K, Kagawa H. 2003. Effects of gonadotropin-releasing hormone agonist and dopamine antagonist on hypothalamus-pituitary-gonadal axis of pre-pubertal female red seabream (*Pagrus major*). *Gen. Comp. Endocrinol*, 131: 264-273.

Kusakabe M, Nakamura I, Evans J, Swanson P, Young G. 2006. Changes in mRNAs encoding steroidogenic acute regulatory protein, steroidogenic enzymes and receptors for gonadotropins during spermatogenesis in rainbow trout testes. *J. Endocrinol*, 189: 541-554.

Larsen D A, Swanson P. 1997. Effects of gonadectomy on plasma gonadotropins I and II in coho salmon, *Oncorhynchus kisutch*. *Gen. Comp. Endocrinol*, 108: 152-160.

Larson E T, Norris D O, Grau E G, Summers C H. 2003. Monoamines stimulate sex reversal in the saddleback wrasse. *Gen. Comp. Endocrinol*, 130: 289-298.

Lau M T, Ge W. 2005. Cloning of Smad2, Smad3, Smad4, and Smad7 from the goldfish pituitary and evidence for their involvement in activin regulation of goldfish FSHβ promoter activity. *Gen. Comp. Endocrinol*, 141: 22-38.

Lethimonier C, Madigou T, Munoz-Cueto J A, Lareyre J J, Kah O. 2004. Evolutionary aspects of GnRHs, GnRH neuronal systems and GnRH receptors in teleost fish. *Gen. Comp. Endocrinol*, 135: 1-16.

Levavi-Sivan B, Avitan A. 2005. Sequence analysis, endocrine regulation, and signal transduction of GnRH receptors in teleost fish. *Gen. Comp. Endocrinol*, 142: 67-73.

Levavi-Sivan B, Biran J, Fireman E. 2006. Sex steroids are involved in the regulation of gonadotropin-releasing hormone and dopamine D2 receptors in female tilapia pituitary. *Biol. Reprod*, 75: 642-650.

Levavi-Sivan B, Ofir M, Yaron Z. 1995. Possible sites of dopaminergic inhibition of gonadotropin release from the pituitary of a teleost fish, tilapia. *Mol. Cell Endocrinol*, 109: 87-95.

Levavi-Sivan B, Safarian H, Rosenfeld H, Elizur A, Avitan A. 2004. Regulation of gonadotropin-releasing hormone (GnRH) -receptor gene expression in tilapia: Effect of GnRH and dopamine. *Biol. Reprod*, 70: 1545-1551.

Lin S W, Ge W. 2009. Differential regulation of gonadotropins (FSH and LH) and growth hormone (GH) by neuroendocrine, endocrine, and paracrine factors in the zebrafish - An *in vitro* approach. *Gen. Comp. Endocrinol*, 160: 183-193.

Lin X W, Peter R E. 1996. Expression of salmon gonadotropin-releasing hormone (GnRH) and chicken GnRH-II precursor messenger ribonucleic acids in the brain and ovary of goldfish. *Gen. Comp. Endocrinol*, 101: 282-296.

Linard B, Bennani S, Saligaut C. 1995. Involvement of estradiol in a catecholamine inhibitory tone of gonadotropin release in rainbow trout (*Oncorhyncus mykiss*). *Gen. Comp. Endocrinol*, 99: 192-196.

Linard B, Anglade I, Corio M, Navas J M, Pakdel F, Saligaut C, Kah O. 1996. Estrogen receptors are expressed in a subset of tyrosine hydroxylase-positive neurons of the anterior preoptic region in the rainbow trout. *Neuroendocrinology*, 63: 156-165.

Lo A, Chang J P. 1998. *In vitro* application of testosterone potentiates gonadotropin-releasing hormone-stimulated gonadotropin-II secretion from cultured goldfish pituitary cells. *Gen. Comp. Endocrinol*, 111: 334-346.

Mañanos E L, Anglade I, Chyb J, Saligaut C, Breton B, Kah O. 1999. Involvement of gamma-aminobutyric acid in the control of GTH-1 and GTH-2 secretion in male and female rainbow trout. *Neuroendocrinology*, 69: 269-280.

Marcano D, Guerrero H Y, Gago N, Cardillo E, Requena M, Ruiz L. 1995. Monoamine metabolism in the hypothalamus of the juvenile teleost fish, *Chaetodipterus faber*. In: Goetz F W, Thomas P, Ed. *Proceedings of the Fifth International Symposium on Reproductive Physiology of Fish*. University of Texas: Austin, TX 64-66.

Martinez-Chavez C C, Minghetti M, Migaud H. 2008. GPR54 and rGnRH I gene expression during the onset of puberty in Nile tilapia. *Gen. Comp. Endocrinol*, 156: 224-233.

Martyniuk C J, Chang J P, Trudeau V L. 2007. The effects of GABA agonists on glutamic acid decarboxylase, GABA-transaminase, activin, salmon gonadotrophin-releasing hormone and tyrosine hydroxylase mRNA in the goldfish (*Carassius auratus*) neuroendocrine brain. *J. Neuroendocrinol*, 19: 390-396.

Mateos J, Mananos E, Carrillo M, Zanuy S. 2002. Regulation of follicle stimulating hormone (FSH) and luteinizing hormone (LH) gene expression by gonadotropin-releasing hormone (GnRH) and sexual steroids in the Mediterranean Sea bass. *Comp. Biochem. Physiol. B Biochem. Mol. Biol*, 132: 75-86.

Mazumdar M, Lal B, Sakharkar A J, Deshmukh M, Singru P S, Subhedar N. 2006. Involvement of neuropeptide Y Y1 receptors in the regulation of LH and GH cells in the pituitary of the catfish, *Clarias batrachus*: An immunocytochemical study. *Gen. Comp. Endocrinol*, 149: 190-196.

Melamed P, Rosenfeld H, Elizur A, Yaron Z. 1998. Endocrine regulation of gonadotropin and growth hormone gene transcription in fish. *Comp. Biochem. Physiol. C Pharmacol. Toxicol. Endocrinol*, 119: 325-338.

Millar R P, Lu Z L, Pawson A J, Flanagan C A, Morgan K, Maudsley S R. 2004. Gonadotropin-releasing hormone receptors. *Endocr. Rev*, 25: 235-275.

Mohamed J S, Khan I A. 2006. Molecular cloning and differential expression of three

GnRH mRNAs in discrete brain areas and lymphocytes in red drum. *J. Endocrinol*, 188: 407-416.

Mohamed J S, Benninghoff A D, Holt G J, Khan I A. 2007. Developmental expression of the G protein-coupled receptor 54 and three GnRH mRNAs in the teleost fish cobia. *J. Mol. Endocrinol*, 38: 235-244.

Moles G, Carrillo M, Mañanósa E, Mylonas C C, Zanuy S. 2007. Temporal profile of brain and pituitary GnRHs, GnRH-R and gonadotropin mRNA expression and content during early development in European sea bass (*Dicentrarchus labrax* L.). *Gen. Comp. Endocrinol*, 150: 75-86.

Nocillado J N, Levavi-Sivan B, Avitan A, Carrick F, Elizur A. 2005. Isolation of dopamine D2 receptor (D2R) promoters in *Mugil cephalus*. *Fish Physiol. Biochem*, 31: 149-152.

Nocillado J N, Levavi-Sivan B, Carrick F, Elizur A. 2007. Temporal expression of G protein-coupled receptor 54 (GPR54), gonadotropin-releasing hormones (GnRH), and dopamine receptor D2 (drd2) in pubertal female grey mullet, *Mugil cephalus*. *Gen. Comp. Endocrinol*, 150: 278-287.

Ogawa S, Akiyama G, Kato S, Soga T, Sakuma Y, Parhar I S. 2006. Immunoneutralization of gonadotropin-releasing hormone type-Ⅲ suppresses male reproductive behavior of cichlids. *Neurosci. Lett*, 403: 201-205.

Parhar I S, Ogawa S, Sakuma Y. 2004. Laser-captured single digoxigenin-labeled neurons of gonadotropin-releasing hormone types reveal a novel G protein-coupled receptor (Gpr54) during maturation in cichlid fish. *Endocrinology*, 145: 3613-3618.

Parhar I S, Ogawa S, Sakuma Y. 2005. Three GnRH receptor types in laser-captured single cells of the cichlid pituitary display cellular and functional heterogeneity. *Proc. Natl. Acad. Sci. USA*, 102: 2204-2209.

Peng C, Huang Y P, Peter R E. 1990. Neuropeptide Y stimulates growth hormone and gonadotropin release from the goldfish pituitary *in vitro*. *Neuroendocrinology*, 52: 28-34.

Peng C, Humphries S, Peter R E, Rivier J E, Blomqvist A G, Larhammar D. 1993. Actions of goldfish neuropeptide Y on secretion of growth hormone and gonadotropin-II in female goldfish. *Gen. Comp. Endocrinol*, 90: 306-317.

Peng C, Trudeau V, Peter R E. 1993. Seasonal variation of neuropeptide Y actions on growth hormone and gonadotropin-II secretion in the goldfish: Effects of sex steroids. *J. Neuroendocrinol*, 5: 273-280.

Peng C, Gallin W, Peter R E, Blomqvist A G, Larhammar D. 1994. Neuropeptide-Y gene expression in the goldfish brain: Distribution and regulation by ovarian steroids. *Endocrinology*, 134: 1095-1103.

Penlington M C, Williams M A, Sumpter J P, Rand-Weaver M, Hoole D, Arme C. 1997. Isolation and characterisation of mRNA encoding the salmon and chicken-II type gonadotrophin-releasing hormones in the teleost fish *Rutilus rutilus* (Cyprinidae). *J. Mol. Endocrinol*, 19: 337-346.

Peter R E, Chang J P, Nahorniak C S, Omeljaniuk R J, Sokolowska M, Shih S H, Billard R. 1986. Interactions of catecholamines and GnRH in regulation of gonadotropin secretion in teleost fish. *Recent Progress Hormone Res*, 42: 513-548.

Peter R E, Lin H R, Van Der Kraak G. 1988. Induced ovulation and spawning of cultured freshwater fish in China: Advances in application of GnRH analogues and dopamine antagonists. *Aquaculture*, 74: 1-10.

Popesku J T, Martyniuk C J, Mennigen J, Xiong H, Zhang D, Xia X, Cossins A R, Trudeau V L. 2008. The goldfish (*Carassius auratus*) as a model for neuroendocrine signalling. *Mol. Cell. Endocrinol*, 293: 43-56.

Root A R, Sanford J D, Kavanaugh S I, Sower S A. 2004. *In vitro* and *in vivo* effects of GABA, muscimol, and bicuculline on lamprey GnRH concentration in the brain of the sea lamprey (*Petromyzon marinus*). *Comp. Biochem. Physiol. A Mol. Integr. Physiol*, 138: 493-501.

Saligaut C, Garnier D H, Bennani S, Salbert G, Bailhache T, Jego P. 1992. Effects of estradiol on brain aminergic turnover of the female rainbow trout (*Oncorhynchus mykiss*) at the beginning of the vitellogenesis. *Gen. Comp. Endocrinol*, 88: 209-216.

Saligaut C, Linard B, Mañanós E L, Kah O, Breton B, Govoroun M. 1998. Release of pituitary gonadotrophins GTH I and GTH II in the rainbow trout (*Oncorhynchus mykiss*): Modulation by estradiol and catecholamines. *Gen. Comp. Endocrinol*, 109: 302-309.

Saligaut C, Linard B, Breton B, Anglade I, Bailhache T, Kah O, Jego P. 1999. Brain aminergic systems in salmonids and other teleosts in relation to steroid feedback and gonadotropin release. *Aquaculture*, 177: 13-20.

Sawisky G R, Chang J P. 2005. Intracellular calcium involvement in pituitary adenylate cyclase-activating polypeptide stimulation of growth hormone and gonadotrophin secretion in goldfish pituitary cells. *J. Neuroendocrinol*, 17: 353-371.

Schulz R, Goos H J Th. 1999. Puberty in male fish: Concepts and recent developments with special reference to the African catfish (*Clarias gariepinus*). *Aquaculture*, 177: 5-12.

Schulz R W, Bogerd J, Bosma P T, Peute J, Rebers F E M, Zandbergen M A, Goos H J Th. 1995. In: Goetz F W, Thomas P, Ed. *Proceedings of the Fifth International Symposium on Reproductive Physiology of Fish*. University of Texas: Austin, TX 2-6.

Sebert M E, Legros C, Weltzien F A, Malpaux B, Chemineau P, Dufour S. 2008. Melatonin activates brain dopaminergic systems in the eel with an inhibitory impact on reproductive function. *J. Neuroendocrinol*, 20: 917-929.

Seminara S B, Kaiser U B. 2005. New gatekeepers of reproduction: GPR54 and its cognate ligand, KiSS-1. *Endocrinology*, 146: 1686-1688.

Senthilkumaran B, Joy K P. 1995. Effects of melatonin, p-chlorophenylalanine, and α-methyltyrosine on plasma gonadotropin level and ovarian activity in the catfish, *Heteropneustes fossilis*: A study correlating changes in hypothalamic monoamines. *Fish Physiol. Biochem*, 14: 471-480.

Senthilkumaran B, Joy K P. 1996. Effects of administration of some monoamine-synthesis blockers and precursors on ovariectomy-induced rise in plasma gonadotropin II in the catfish, *Heteropneustes fossilis*. *Gen. Comp. Endocrinol*, 101: 220-226.

Senthilkumaran B, Okuzawa K, Gen K, Kagawa H. 2001. Effects of serotonin, GABA and neuropeptide Y on seabream gonadotropin releasing hormone release *in vitro* from preoptic-anterior hypothalamus and pituitary of red seabream, *Pagrus major*. *J. Neuroendocrinol*, 13: 395-400.

Silver M R, Kawauchi H, Nozaki M, Sower S A. 2004. Cloning and analysis of the lamprey GnRH-III cDNA from eight species of lamprey representing the three families of Petromyzoniformes. *Gen. Comp. Endocrinol*, 139: 85-94.

So W K, Kwok H F, Ge Y. 2005. Zebrafish gonadotropins and their receptors: II. Cloning and characterization of zebrafish follicle-stimulating hormone and luteinizing hormone subunits - their spatial-temporal expression patterns and receptor specificity. *Biol. Reprod*, 72: 1382-1396.

Soga T, Ogawa S, Millar R P, Sakuma Y, Parhar I S. 2005. Localization of the three GnRH types and GnRH receptors in the brain of a cichlid fish: Insights into their neuroendocrine and neuromodulator functions. *J. Comp. Neurology*, 487: 28-41.

Sohn Y C, Yoshiura Y, Kobayashi M, Aida K. 1998. Effect of sex steroids on the mRNA levels of gonadotropin I and II subunits in the goldfish *Carassius auratus*. *Fish. Sci*, 64: 715-721.

Sohn Y C, Kobayashi M, Aida K. 2001. Regulation of gonadotropin beta subunit gene expression by testosterone and gonadotropin-releasing hormones in the goldfish, *Carassius auratus*. *Comp. Biochem. Physiol. B Biochem. Mol. Biol*, 129: 419-426.

Somoza G M, Peter R E. 1991. Effects of serotonin on gonadotropin and growth hormone release from *in vitro* perfused goldfish pituitary fragments. *Gen. Comp. Endocrinol*, 82: 103-110.

Somoza G M, Yu K L, Peter R E. 1988. Serotonin stimulates gonadotropin release in female and male goldfish, *Carassius auratus*, L. *Gen. Comp. Endocrinol*, 72:

364-382.

Steven C, Lehnen N, Kight K, Ijiri S, Klenke U, Harris W A, Zohar Y. 2003. Molecular characterization of the GnRH system in zebrafish (*Danio rerio*): Cloning of chicken GnRH-II, adult brain expression patterns and pituitary content of salmon GnRH and chicken GnRH-II. *Gen. Comp. Endocrinol*, 133: 27-37.

Sundararaj B I, Keshavanath P. 1976. Effects of melatonin and prolactin treatment in hypophyseal-ovarian system in *Heteropneustes fossilis* (Bl.). *Gen. Comp. Endocrinol*, 29: 84-96.

Swanson P, Dickey J T, Campbell B. 2003. Biochemistry and physiology of fish gonadotropins. *Fish Physiol. Biochem*, 28: 53-59.

Tello J A, Wu S, Rivier J E, Sherwood N M. 2008. Four functional GnRH receptors in zebrafish: Analysis of structure, signalling, synteny and phylogeny. *Integ. Comp. Biology*, 48: 570-587.

Thomas P. 2008. The endocrine system. In: Di Giulio R T, Hinton D E, Ed. *The Toxicology of Fishes*. CRC Press: Boca Raton, FL 457-488.

Thomas P, Rahman M S, Khan I A, Kummer J A. 2007. Widespread endocrine disruption and reproductive impairment in an estuarine fish population exposed to seasonal hypoxia. *Proc Biol. Sci*, 274: 2693-2701.

Trudeau V L. 1997. Neuroendocrine regulation of gonadotrophin II release and gonadal growth in the goldfish, *Carassius auratus*. *Rev. Reprod*, 2: 55-68.

Trudeau V L, Peter R E, Sloley B D. 1991. Testosterone and estradiol potentiate the serum gonadotropin response to gonadotropin-releasing hormone in goldfish. *Biol. Reprod*, 44: 951-960.

Trudeau V L, Sloley B D, Peter R E. 1993. GABA stimulation of gonadotropin secretion in the goldfish: Involvement of GABA receptors, dopamine, and sex steroids. *Am. J. Physiol*, 265: R348-R355.

Trudeau V L, Sloley B D, Peter R E. 1993. Testosterone enhances GABA and taurine but not N-methyl-D-L-aspartate stimulation of gonadotropin secretion in the goldfish: Possible sex steroid feedback mechanisms. *J. Neuroendocrinol*, 5: 129-136.

Trudeau V L, Spanswick D, Fraser E J, Lariviere K, Crump D, Chiu S, MacMillan M, Schulz R W. 2000. The role of amino acid neurotransmitters in the regulation of pituitary gonadotropin release in fish. *Biochem. Cell Biol*, 78: 241-259.

Unniappan S, Peter R E. 2004. *In vitro* and *in vivo* effects of ghrelin on luteinizing hormone and growth hormone release in goldfish. *Am. J. Physiol. Regul. Integr. Comp. Physiol*, 286: R1093-R1101.

Unniappan S, Peter R E. 2005. Structure, distribution and physiological functions of ghrelin in fish. *Comp. Biochem. Physiol. A Mol. Integr. Physiol*, 140: 396-408.

Vacher C, Mananos E L, Breton B, Marmignon M H, Saligaut C. 2000. Modulation of pituitary dopamine D1 or D2 receptor and secretion of follicle stimulating hormone and luteinizing hormone during the annual reproductive cycle of female rainbow trout. *J. Neuroendocrinol*, 12: 1219-1226.

Vacher C, Ferrière F, Marmignon M H, Pellegrini E, Saligaut C. 2002. Dopamine D2 receptors and secretion of FSH and LH: Role of sexual steroids on the pituitary of the female rainbow trout. *Gen. Comp. Endocrinol*, 127: 198-206.

van Aerle R, Kille P, Lange A, Tyler C R. 2008. Evidence for the existence of a functional Kiss1/Kiss1 receptor pathway in fish. *Peptides*, 29: 57-64.

Van Asselt L A, Goos H J, de Leeuw R, Peter R E, Hol E M, Wassenberg F P, Van Oordt P G. 1990. Characterization of dopamine D2 receptors in the pituitary of the African catfish, *Clarias gariepinus*. *Gen. Comp. Endocrinol*, 80: 107-115.

Van Der Kraak G, Donaldson E M, Chang J P. 1986. Dopamine involvement in the regulation of gonadotropin secretion in coho salmon. *Can. J. Zool*, 64: 1245-1248.

Van Der Kraak G, Munkittrick K R, McMaster M E, Portt C B, Chang J P. 1992. Exposure to bleached kraft pulp mill effluent disrupts the pituitary-gonadal axis of white sucker at multiple sites. *Tox. Appl. Pharmacol*, 115: 224-233.

Van Der Kraak G J, Chang J P, Janz D M. 1998. Reproduction. In: Evans D H, Ed. *The Physiology of Fishes*, *Second Edition.* CRC Press: Boca Raton, FL 465-488.

Vetillard A, Atteke C, Saligaut C, Jego P, Bailhache T. 2003. Differential regulation of tyrosine hydroxylase and estradiol receptor expression in the rainbow trout brain. *Mol. Cell. Endocrinol*, 199: 37-47.

Vidal B, Pasqualini C, Le Belle N, Holland M C, Sbaihi M, Vernier P, Zohar Y, Dufour S. 2004. Dopamine inhibits luteinizing hormone synthesis and release in the juvenile European eel: A neuroendocrine lock for the onset of puberty. *Biol. Reprod*, 71: 1491-1500.

Volkoff H, Peter R E. 1999. Actions of two forms of gonadotropin releasing hormone and GnRH antagonists on spawning behavior of the goldfish *Carassius auratus*. *Gen. Comp. Endocrinol*, 116: 347-355.

Weil C, Carré F, Blaise O, Breton B, Le Bail P Y. 1999. Differential effect of insulin-like growth factor I on *in vitro* gonadotropin (I and II) and growth hormone secretions in rainbow trout (*Oncorhynchus mykiss*) at different stages of the reproductive cycle. *Endocrinology*, 140: 2054-2062.

Wong A O, Li W S, Lee E K, Leung M Y, Tse L Y, Chow B K, Lin H R, Chang J P. 2000. Pituitary adenylate cyclase activating polypeptide as a novel hypophysiotropic factor in fish. *Biochem. Cell Biol*, 78: 329-343.

Wong A O L, Murphy C K, Chang J P, Neumann C M, Lo A, Peter R E. 1998. Direct actions of serotonin on gonadotropin-II and growth hormone release from goldfish pituitary cells: Interactions with gonadotropin-releasing hormone and dopamine and further evaluation of serotonin receptor specificity. *Fish Physiol. Biochem*, 19: 23-34.

Wong T T, Zohar Y. 2004. Novel expression of gonadotropin subunit genes in oocytes of the gilthead seabream (*Sparus aurata*). *Endocrinology*, 145: 5210-5220.

Yam K M, Yoshiura Y, Kobayashi M, Ge W. 1999. Recombinant goldfish activin B stimulates gonadotropin-Ib but inhibits gonadotropin-IIb expression in the goldfish, *Carassius auratus*. *Gen. Comp. Endocrinol*, 116: 81-89.

Yaron Z. 1995. Endocrine control of gametogenesis and spawning induction in the carp. *Aquaculture*, 129: 49-73.

Yaron Z, Gur G, Melamed P, Rosenfeld H, Elizur A, Levavi-Sivan B. 2003. Regulation of fish gonadotropins. *Intl. Rev. Cytol*, 225: 131-185.

Yu K L, Peter R E. 1990. Dopaminergic regulation of brain gonadotropin-releasing hormone in male goldfish during spawning behaviour. *Neuroendocrinology*, 52: 276-283.

Yu K L, Peter R E. 1992. Adrenergic and dopaminergic regulation of gonadotropin-releasing hormone release from goldfish preoptic-anterior hypothalamus and pituitary *in vitro*. *Gen. Comp. Endocrinol*, 85: 138-146.

Yu K L, Rosenblum P M, Peter R E. 1991. *In vitro* release of gonadotropin releasing hormone from the brain preoptic-anterior hypothalamic region and pituitary of female goldfish. *Gen. Comp. Endocrinol*, 81: 256-267.

Yuen C W, Ge W. 2004. Follistatin suppresses FSHb but increases LHb expression in the goldfish-evidence for an activin-mediated autocrine/paracrine system in fish pituitary. *Gen. Comp. Endocrinol*, 135: 108-115.

Zhao E, Basak A, Trudeau V L. 2006. Secretoneurin stimulates goldfish pituitary luteinizing hormone production. *Neuropeptides*, 40: 275-282.

Zmora N, Kazeto Y, Kumar R S, Schulz R W, Trant J M. 2007. Production of recombinant channel catfish (*Ictalurus punctatus*) FSH and LH in S2 Drosophila cell line and an indication of their different actions. *J. Endocrinol*, 194: 407-416.

第4章 鱼类生长激素的调控：一个包括下丘脑、外周和局部自分泌/旁分泌信号的多因子模式

对于鱼类，生长激素（GH）影响许多生理功能，包括身体生长、能量代谢、生殖、摄食、渗透压调节和免疫。GH的合成与释放受到脑和外周组织的神经内分泌因子调控。下丘脑的调控剂影响互相之间的表达，形成一个调控GH的互相作用的网络。GH的释放受到生长抑素（somatostatin）的紧张性（tonically）抑制，而类胰岛素生长因子（insulinlike growth factor）起着主要反馈作用的调节剂作用，但GH的实际释放量反映总的抑制性与刺激性影响的平衡。性类固醇和营养状况亦调节下丘脑因子的表达及其对脑垂体的作用。脑垂体内的调节剂，包括GH、GTH抑制素（inhibin）/激活素（activin），提供自分泌/旁分泌控制GH合成与分泌的作用。在生长激素细胞水平，对多种神经内分泌因子的受体表达能整合来自不同调节剂的调控信号。由不同调节剂使用的不同而又会重叠的信号级联放大（signaling cascades）能进行配体的和功能的特异性GH调节。

4.1 导　　言

生长激素（GH）最早是根据它刺激长骨（long bone）延长的能力来鉴别的。如在本章和本书其他章节所论述的，鱼类GH是一种多效性（pleiotropic）激素，具有多种不同的生理功能。GH的作用通常被认为通过GH刺激生长调节素（somadomedin）（即胰岛素样生长因子，IGFs）的释放而实现，但是，GH亦能不依赖于IGF而直接起作用（Nordgarden等，2006；Wong等，2006）。腺脑垂体是GH的主要来源，但亦存在着脑垂体外的GH（如在免疫细胞、性腺和脑内）。本章论述下丘脑因子、外周激素和局部脑垂体内的信号对鱼类脑垂体GH合成与分泌的神经内分泌调控。由于鱼类脑垂体外的GH合成与释放的调控作用还不太清楚，本章将不做论述。

4.2　生长激素和生长激素受体

GH、催乳激素（PRL）和生长乳素（SL）是细胞因子（cytokine）家族的成员。通常认为SL和PRL是在有颌类进化的早期由祖先GH衍生而来。SL是特有的鱼类模式，因为在陆地侵入（land invasion）时期，四足类动物丢失了这个基因（Fukamachi和Meyer，2007）。硬骨鱼类的GH是一个21-23KDa的单链多肽蛋白质，和其他四足类的GH相似，包含两个高度保守的对生物学功能起重要作用的分子内二硫键，以及一个N-连接的糖基化（glycosylation）位点。但是，骨鳔鱼类含有一个附加的不成对的半胱

氨酸残基，其功能尚不清楚。鱼类的 GH 基因结构要比四足类有更多的变化。较高等脊椎动物的典型 5 个外显子/4 个内含子结构出现在七鳃鳗、鲤形目和鲶形目鱼类中，而 6 个外显子/5 个内含子结构出现在鲑形目、鲈形目和鲀形目鱼类中。后一种构型的形成是由于插入了第 5 个内含子和第 5 个外显子分为两个（5 和 6）。已报道鱼类 GH 基因的 5′侧翼区（5′-flanking region）含有多个转录调控元件（transcription regulatory elements），包括生长激素因子-1（GHF-1/pit-1）、激活蛋白 1（AP-1）和 cAMP 反应元件（CRE）。值得注意的是，cDNA 克隆显示一些鱼类（如斜带石斑鱼，*Epinephelus coioides*）存在一个 GH 转录体，而其他研究证明一些鱼类存在两个 GH 转录体（如虹鳟 *Orcorhynchus mykiss* 和金鱼 *Carassius auratus*）。用模式鱼类（如斑马鱼 *Danio rerio*，青鳉 *Oryzias melastigma* 和东方红鳍鲀 *Takifugu rubripes*）的基因数据库和其他脊椎动物的做比较，通常认为在现今硬骨鱼类的进化期间发生过两次基因组复制（genome duplication）。在这个过程中，鱼类基因组的四倍体作用（tetraploidization）可能导致一些硬骨鱼类（如鲑鳟鱼类）出现 GH 转录体的多个复制品（综述见 Kawauchi 和 Sower，2006）。

与哺乳类一样，鱼类 GH 受体（GHRs）属于 I 型细胞因子受体家族，包括单个细胞外区、跨膜区和细胞内区。细胞外区有保守的半胱氨酸残基和包含 GH 结合位点的两个反平行 β-折叠（antiparallel β-sheets）组成的三明治形结构，而细胞内区具有两个保守的富含脯氨酸序列，Box1 和 Box2。Box1 是一个詹纳斯激酶 2（Janus kinase 2，JAK2）结合位点，对于 GHR 信号是必需的，而 Box2 参与受体的内在化作用（internalization）（综述见 Lichanska 和 Waters，2008）。已报道硬骨鱼的 GHR 有两个亚型，GHR1 和 GHR2，是基因复制的结果。GHR1 和四足类 GHRs 的结构比较相近，包含 6～7 个细胞外的半胱氨基残基；而 GHR2 是硬骨鱼类特有的，只含有 4～5 个细胞外的半胱氨基残基。GHR1 和 GHR2 都是真正的 GH 受体（Fukamachi 和 Meyer，2007）。对金头鲷（*Sparus aurata*）的研究表明，GHR1 而不是 GHR2 能激发表达系统中的 c-fos 启动子活性，表明这两个 GHR 亚型可能具有不同的信号转导系统。GHR1 和 GHR2 的组织分布不同，同样，GHR1 和 GHR2 mRNA 在同一个组织中表达的强度亦有所不同（Jiao 等，2006）。此外，性腺类固醇和糖皮质激素以及渗透压和盐度变化都能不同程度地影响 GHR1 和 GHR2 mRNA 水平（Pierce 等，2007）。鱼类存在着 GHR 两种亚型可能是介导 GH 多种组织特异性作用和其他调控信号进行差异性调整的作用机理的一部分。

4.3 鱼类生长激素的生物学功能

如同在对其他脊椎动物的研究中所报道的，GH 从前脑垂体释放，起着鱼类身体生长和代谢的主要调节剂作用。将哺乳类 GH 施用于鱼类能有效地刺激鱼类体重和体长增长，包括虹鳟、鲑鱼和鲤鱼（Peng 和 Peter，1997）。使用哺乳类动物的 GH 进行基因转移（transgenesis）研究（如鲤鱼，*Cyprinus carpio*）和鱼类 GH 转基因（transgene）研究（如鲑鱼和罗非鱼）（Zbikowska，2003）进一步证实了 GH 的促生长作用。在这些早期研究中，GH 诱导生长增长期间经常会观察到肥满度（Condition factor，以体重×100/体长³表示）降低，表明鱼类在体重增长的同时体形会变得"瘦长"些。代表种类如虹鳟，用 GH 处理后，其身体各种组织的蛋白质合成增加。在这个过程中，脂解作用（lipolysis）亦被激活，血液循环中的游离脂肪酸和甘油水平迅速升高，而这种分解代谢作用可能是由于 GH 对肝脏三酰甘油（triacylglycerol）脂肪酶和乙酰-辅酶 A 羧化酶（carboxylase）活性产生的不同作用所引起（Bjornsson，1997）。在模式鱼类的寻食期间，GH 亦起着调节行为型式的作用。对鲑鳟鱼类（如鳟鱼）用 GH 处理后能增强其食欲和主要的摄食行为，同时使其对捕食者的回避反应减弱。类似的行为亦出现在 GH 过表达的转基因鱼中，表明生长增强对代谢的需求可能会导致其在寻食期间增加冒险行为（Sundstrom 等，2004）。这种行为的变化可能是 GH 调节中枢神经系统内多巴胺能活性/神经元回路的结果（Bjornsson 等，2002）。

和哺乳类相似，GH 被认为是一个"辅助促性腺激素"（co-GTH），GH 亦和鱼类促性腺激素轴相互作用以促进性成熟、配子生成和性腺类固醇生成。GH 的生殖功能主要是由性腺内 GH 受体表达所介导，如虹鳟和罗非鱼（*Oreochromis mossambicus*）。在性腺水平，GH 亦局部产生并通过直接作用于卵巢组织而诱导类固醇生成。有些鱼类的 GH 亦能通过增强 GTH 的刺激作用而间接影响类固醇产生（如金鱼）。这种刺激作用可能是 GH 激活依赖于 cAMP 的串联而诱导卵巢芳化酶活性的结果（Kajimura 等，2004；Li 等，2005）。除生殖功能外，GH 对海水的适应性亦是必不可少的，溯河洄游鲑鱼由一龄降海幼鲑向二龄鲑转变期间经常都能观察到其血清 GH 水平升高。GH 能增强鱼类对高渗性应激反应的耐受，主要是增强鳃部氯细胞增殖，刺激鳃 Na^+/K^+-ATP 酶活性，激活 Na^+、K^+、$2Cl^-$-协同转运蛋白（cotransporter）和参与渗透压调节的离子通道蛋白［如囊性纤维化跨膜传导调节通道蛋白（cystic fibrosis transmembrane conductance regulator channels，CFTR）］（Sakamoto 和 McCormick，2006；Makino 等，2007）。GH 的这些刺激作用能为皮质醇进一步增强；在鱼类处于渗透压应激反应期间，皮质醇是来自下丘脑-脑垂体-肾上腺（HPA）轴的主要信号；这种协同作用部分是由于 GH 诱导了皮质醇受体在鳃部表达而引起（Pelis 和 McCormick，2001）。由于 IGF-I 能模拟 GH 诱导氯细胞增殖与 Na^+/K^+-ATP 酶活性，在高渗性应激反应期间，鱼血液循环中 IGF-I 水平升高，在鳃上皮能检测到 IGF-I mRNA 和 IGF-I 的结合位点，因此，通常都认为 IGF-I 作为内分泌和自分泌/旁分泌的成员参与了鱼类 GH 调节渗透压的功能（Sakamoto 和 McCormick，2006）。

广盐性鱼类在高渗性应激反应期间，GH 释放作用增强的同时亦激活免疫功能，这和哺乳动物 GH 的免疫调节作用一致。脑垂体切除的鱼类模型（如斑点鮰 *Ictalurus punctatus* 和虹鳟）能够阻抑免疫反应（如由减少 Ig-分泌的白细胞所引起），而用 GH 置换（replacement）后能部分地逆转这种情况（Yada，2007）。此外，施用 GH 能增强鱼类抵抗细菌感染的能力和人为弧菌病的存活力（Sakai 等，1997）。GH 的免疫保护作用可以归因于：①它刺激抗体产生和免疫细胞增殖；②激发白细胞的吞噬活性和非特异性细胞毒素的活性；③诱导超氧化物的产生和溶菌酶（lysozyme）的活性；④通过血浆铜蓝蛋白（ceruloplasmin）的产生而发挥抗炎作用（anti-inflammatory action）（Yada，2007，见本书第 7 章的综述）。

4.4 生长激素分泌与合成的调控

4.4.1 总的思考和模式

在对哺乳类的研究中，破坏下丘脑和脑垂体的联系后，GH 基础性释放减少，表明在 GH 分泌的神经内分泌调控中刺激性影响的支配地位。生长激素释放激素（GHRH）和生长抑素（SS）分别是主要的刺激因子和抑制因子。GHRH 和 SS 从正中隆起释放进入门脉血液中形成一个"双重调节剂"的调控系统。在低谷时，SS 抑制作用使 GH 水平降低，而在每次 GH 阵发式释放（episodic release）期间，GHRH 增加释放以诱导 GH 大量释放。GHRH 和 SS 神经元亦形成一个互相控制的网络。SS 神经元阻抑 GHRH 释放，而 GHRH 增强 SS 分泌活动。这些相互作用再加上 GH 的反馈作用，形成 GH 脉冲式分泌（pulsatile secretion）的基础。这个"双重调节剂"的调控系统可能还是过于简单了，因为其他一些神经内分泌调节剂，如脑垂体腺苷酸环化酶激活多肽（PACAP）和生长素释放肽（ghrelin）都已经显示出对调控 GH 分泌活动的重要作用（Lengyel，2006；Goldenberg 和 Barkan，2007）。

和哺乳类不同，鱼类 GH 释放的神经内分泌调控，特别是刺激性下丘脑因子的参与，必然是多因子的（multifactorial）（Rousseau 和 Dufour，2004）。在鱼类模型中，GHRH 不是主要的 GH 释放因子（Montero 等，2000），在脑垂体水平，GH 分泌活动处于紧张的抑制当中，而不是刺激性的控制（Rousseau 和 Dufour，2004）。对草鱼（*Ctenopharyngodon idellus*）的研究已证明 GH 的脉冲式分泌（Zhang 等，1994），但鱼类 GH 阵发式分泌活动的普遍性还没有确定。主要原因是难以在足够的紧密时间间隔中重复地从鱼体中获得血样，以便恰当地评定 GH 在鱼体内释放的动态。另一方面，已经在一些关于鲑鳟鱼类（Bjornsson 等，2002）和鲤科鱼类（Marchant 和 Peter，1986；Zhang 等，1994）的研究中清楚地证明 GH 释放的昼夜型式。这种昼夜型式的一般特点是在光照阶段血液 GH 水平出现一个或多个不规则的波峰，而血液的 GH 水平通常在黑暗阶段比较高。血液 GH 水平在 24 h 出现的波峰型式亦受到温度、摄食、光周期和/或生长发育的影响。例如，驯养在短光周期和低温度下的金鱼在黑暗阶段开始时血清 GH 出现一个波峰，而驯养在长光周期和较高温度下的金鱼则没有这样的 GH 波峰（Marchant 和 Peter，1986）。GH 分泌活动经常受到食物不足的影响，并且能为摄食所驱动（见本书第 9 章）。最近报道了点蓝子鱼（*Siganus guttatus*）幼鱼 GH mRNA 在脑垂体表达的昼夜型式，其 GH mRNA 水平在光照阶段要较低于黑暗阶段（Ayson 和 Takamura，2006）。基于点蓝子鱼 GH mRNA 表达的昼夜型式和其他一些鱼类血清 GH 水平昼夜型式的相似性，人们推想它们可能是由一个共同的因子所调节。对鳟鱼的研究已经报道褪黑激素能刺激离体的脑垂体组织释放 GH（Falcon 等，

2003）。对于褪黑激素是否确实参与鱼类 GH mRNA 水平昼夜节律的产生和 GH 释放，还没有进行检验。此外，对模式鱼类如金鱼（Johnson 等，2002）和金头鲷（Chan 等，2004a）的研究都曾报道 GH 释放和 GH 基因转录之间的分离（dissociation），这表明参与 GH 调控的作用机理是相当复杂的。

鱼类血液 GH 水平亦存在年/季节的变动，但是，鱼类最大的体质生长年周期通常并不和血液循环中 GH 水平最高的时期密切联系，部分原因是 IGF 参与介导 GH 对鱼体生长的影响以及 GH 对生殖活动的作用。例如，金头鲷的生长率和血液 IGF-I 水平互相联系，但滞后于血液的 GH 水平（Mingarro 等，2002）。同样，金鱼最大的生长率在 7 月（夏天），血清的高 GH 水平一直保持到秋末，而最高的血清 GH 水平出现在春季产卵期（Marchant 和 Peter，1986）。血清 GH 水平的升高通常出现在性腺生长发育、卵黄生成和排卵前 LH 大量释放的鱼类当中，例如白亚口鱼（*Catostomus commersoni*）、鲤鱼、金头鲷、大西洋庸鲽（*Hippoglossus hippoglossus*）、大西洋鲑鱼（*Salmo salar*）、大麻哈鱼（*Oncorhynchus keta*）和虹鳟（综述见 Canosa 等，2008）。GH 血清水平的变化是和 GH 作为共-GTH（co-GTH）在鱼类季节性生殖周期中所起的作用密切联系的。鲤鱼脑垂体的 GH mRNA 和蛋白质水平在适应于夏季的鱼中要比适应于冬季的鱼更高些（Figueroa 等，2005），表明鱼类脑垂体的 GH 合成作用亦存在着季节变动。

4.4.2　CNS 的下丘脑信号

4.4.2.1　抑制物

1. 生长抑素（SS）

在硬骨鱼类的相关研究中，脑垂体的异位自我移植（ectopic autotransplant）后血液的 GH 水平通常会升高，而 GH 在脑垂体组织或细胞孵育中的基础性分泌都是高的，表明脑垂体 GH 释放经常处于紧张性抑制作用的调控之中。对金鱼的研究表明 SS 起着这种紧张性抑制的调控作用，用电射频（electroradiofrequency）损伤有 SS 神经元细胞体伸进脑垂体的室周区，明显地使血清 GH 水平升高。在同样的研究中，下丘脑和脑垂体 SS 含量的季节型式亦大致是血清 GH 水平的反映（Marchant 等，1989；见图 4.1）。金鱼脑垂体细胞原代培养物（primary culture）的基础性 GH 释放通常都在秋末冬初（10 月到 12 月）高于春天（3 月到 5 月），这或许是从内源性 SS 的抑制状态中释放的反映；相反，对于 SS 降低离体 GH 基础性分泌的相对能力方面，没有检测到明显的季节性差别（Yunker W. K. 和 Chang J. P.，未发表结果；见图 4.1）。半胱胺是一种 SS 的抑制剂，能提高草鱼血清的 GH 水平，亦证明 SS 起着紧张性抑制草鱼 GH 释放的调控作用（Xiao 和 Lin，2003）。瘤棘鲆（*Psetta maxima*）在体的 GH 释放紧张性抑制作用十分强烈，使离体的脑垂体基础性 GH 释放能力特别强，达到最大的释放容量（Rousseau 等，2001）。

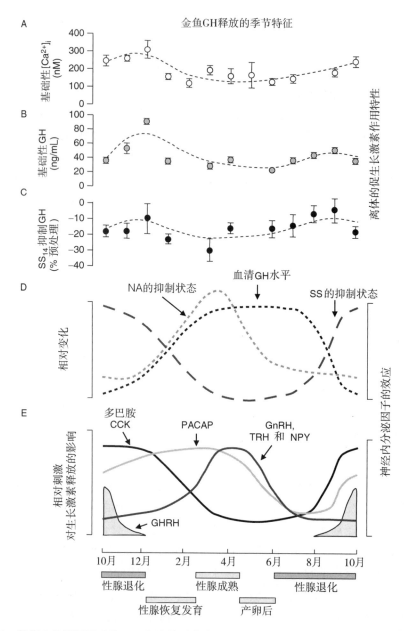

图 4.1　神经内分泌调控在体 GH 释放的季节变化对离体的 GH 在生长激素细胞水平的反应

（A）细胞内 Ca^{2+}（$[Ca^{2+}]_i$）的季节变化；（B）离体的基础性 GH 释放；由生长抑素-14（SS-14）处理金鱼脑垂体细胞原代培养物引起的 GH 抑制作用幅度（C）并和金鱼在生殖周期 GH 在体释放情况一同表示（D）。血浆 GH 水平的季节性能和生长抑素（SS）、去甲肾上腺素（NA）（D）以及（E）多巴胺、缩胆囊肽（CCK）、脑垂体腺苷酸环化酶激活多肽（PACAP）、神经肽 Y（NPY）、甲状腺素释放激素（TRH）、生长激素释放激素（GHRH）与促性腺激素释放激素（GnRH）等神经内分泌输入的季节变化相互联系。金鱼脑垂体孵育物中的基础性 GH 释放的季节变化和 $[Ca^{2+}]_i$ 紧密地相互联系，但没有和血清 GH 水平相互联系，表明神经内分泌的调节剂对血液循环中 GH 水平的季节变化起着重要作用。在体的 GH 释放在任何一个时间点上的幅度都是作用于脑垂体细胞水平的多个抑制性与刺激性作用取得平衡的结果。（参照 Chang 和 Habibi，2002，有改动）（见书后彩图）

SS 是一个多功能和多成员的肽类家族。它的组织分布、系统发生和脑垂体的功能已经广泛地在别的地方做了综述，而外周来源的 SS 对脑垂体 GH 分泌活动的调控并不起生理作用（Nelson 和 Sheridan，2005；Klein 和 Sheridan，2008）。对于下丘脑调控脑垂体的 SS，传统的分子是 SS-14，它是一个最早在哺乳类当中被发现的由 14 个氨基酸组成的长肽。SS-14 在所有的脊椎动物包括鱼类当中是高度保守的。在鱼类中已鉴定多达三个编码前 SS 肽原（PSSs）的分离的基因，最近有学者对它们的分子进化做了综述（Tostivint 等，2008）。SS-14 由前生长抑制素肽-1 原（PSS-I）编码。哺乳类 PSS-I 加工形成 SS-14 而其 N-端延伸成 SS-28。但是，PSS-I 在鱼类中只产生 SS-14。PSS-II 在鱼类中是独特的，编码不同长度的 SS 肽（通常是 22～28 个氨基酸），它们在 C-端都具有 [酪^7，甘^{10}] SS-14。金鱼 PSS-II 在金鱼中产生金鱼脑（gb）SS-28（Lin 等，2000）。但虹鳟有两个 PSS-II 肽，即 PSS-II′和 PSS-II″，分别产生鲑鱼 SS-28 和 SS-25（Holloway 等，2000）。PSS-III 编码 [脯^2] SS-14 和它的变体，并被认为是哺乳类皮质素抑素（cortistatin）的直向同原物。鱼类编码所有三个 PSS 类型的 cDNA 都已经在调控脑垂体的下丘脑中发现，其中 PSS-I 和 PSS-II 亦曾定位在脑垂体内（Lin 和 Peter，2001；Cerdà-Reverter 和 Canosa，本书第 1 章）。同样，三种 PSS 类型的信息都存在于斜带石斑鱼的下丘脑和脑垂体内（Xing 等，2005）。这些研究结果表明所有三个 PSS 的产物对一些鱼类脑垂体功能的神经内分泌调控起着重要作用。

PSS-I、PSS-II 和 PSS-III 的产物在脑垂体水平都是有活性的。SS-14 是许多鱼类基础性 GH 分泌活动的有力抑制剂，包括金鱼、虹鳟、罗非鱼、鲑鱼、鳗鱼和大菱鲆（Rousseau 等，1998，2001；Lin 等，2000）。同样，SS-14 能有效地减弱/阻抑刺激性的 GH 分泌活动，包括一些鱼类对多巴胺（DA）、GnRH、PACAP、NPY、生长素释放肽和/或 CRF 等的反应（综述见 Lin 和 Peter，2001）。对于金鱼，SS-14 和 [脯^2] SS-14 对阻抑 GH 的基础性分泌活动具有同等效能，而 gbSS-28 是三种 SS 类型中效能最强的。另一方面，对于抑制 DA 和 PACAP 刺激 GH 释放的反应，gbSS-28 的效能则较低于 SS-14 和 [脯^2] SS-14。此外，鲑鱼 SS-25 和鲶鱼 SS-22 对金鱼 GH 的分泌活动没有影响（Peter 和 Chang，1999；Lin 等，2000）。这些研究结果表明，金鱼的生长激素细胞能够选择性地对本种类不同的 PSS 产物产生反应，亦能够辨别其他鱼类不同的 SS 类型。综合这些研究结果表明，不同的 SS 同等型（isoform）可能会有差别地调节 GH 对各种不同的神经内分泌刺激物的反应，以及在整合神经内分泌调控 GH 释放过程中选择性地发挥神经内分泌作用（见第 4.5 节）。

哺乳类有 5 个 SS 受体（SST）的亚型（SST$_{1-5}$），每个受体亚型都和不同组合的细胞内第二信使系统相偶联。到目前为止，鱼类 SSTs 克隆的是 SST$_{1-3}$ 和 SST$_5$ 群，并且在一种鱼的单个受体亚型中有多种信息（Lin 等，2000；Nelson 和 Sheridan，2005）。例如，虹鳟的 SST$_1$ 有两个变体（1A 和 1B）（Nelson 和 Sheridan，2006）。在金鱼中曾经鉴别出 8 个 SST（gfSST$_{1A}$、gfSST$_{1B}$、gfSST$_2$、gfSST$_{3A}$、gfSST$_{3B}$、gfSST$_{5A}$、gfSST$_{5B}$、gfSST$_{5C}$），并有多个 SSTs 在脑垂体表达，其中 gfSST$_2$ 和 gfSST$_5$ mRNA 最为丰富。在 COS-7 细胞中表达时，gfSST$_2$ 能被 SS-14 和 [脯^2] SS-14 激活，但 gbSS-28 没有这种作用；然而 gfSST$_5$ 对 gbSS-28 的亲和力要强于 SS-14 和 [脯^2] SS-14（Lin 等，2000；Lin 和 Peter，2001）。这样，金鱼

脑垂体 SST 亚型的补充就能够介导三个内源 SS 类型的有差别的作用。

有关鱼类不同的 SS 类型如何作用于细胞内信号水平以抑制 GH 释放的资料主要是由金鱼脑垂体细胞的相关研究提供的（综述见 Chang 等，2000；见表 4.1）。基础性 GH 分泌活动对细胞内 Ca^{2+} 水平（$[Ca^{2+}]_i$）和 cAMP 水平是敏感的。尽管 SS-14、[脯^2] SS-14 和 gbSS-28 都会使 cAMP 降低，但是它们影响 cAMP 产生的相对效能（relative potency）和基础性 GH 释放的阻抑并没有关系（Yunker 等，2003）。此外，SS-14 并不影响金鱼生长激素细胞的基础 $[Ca^{2+}]_i$。这些结果意味着在其他信号成员中的作用亦可能参与介导基础性 GH 释放（Yunker 和 Chang，2004）。由 cAMP、Ca^{2+}、蛋白激酶 C（PKC）、一氧化氮（NO）和花生四烯酸（AA）等信号级联放大的活化作用而引发 SS-14 有效地抑制 GH 反应，表明 SS-14 能通过细胞内下游各种途径以抑制刺激性的 GH 释放（Kwong 和 Chang，1997；Yunker 和 Chang，2001，2004）。值得注意的是 [脯^2] SS-14 的活动范围和 SS-14 十分相似，而 gdSS-28 对 cAMP/PKC 激活的抑制 GH 释放的作用要小得多（Yunker 等，2003）。依赖于 cAMP 的 GH 释放的差别作用可以解释 gbSS-28 和 SS-14 在 DA 和 PACAP 影响下阻抑 GH 释放的不同效能。三种 SS 同等物亦可以作用于第二信使的产生水平以影响刺激性的 GH 释放。在鲑鱼 GnRH、DA 和 PACAP 影响下，SS-14 抑制 GH 的反应是和 $[Ca^{2+}]_i$ 对这些配体反应最大变幅（meximal amplitude）的降低相联系的。相反，由鸡（c）GnRH-II 诱导 Ca^{2+} 信号不受 SS-14 影响，而 SS-14 直接受 PKC 激活的阻抑 GH 反应甚至和 $[Ca^{2+}]_i$ 的增加相关（Yunker 和 Chang，2004）。因而，不管 SS 对第二信使产生水平的影响如何，SS 在这些信使下游部位的作用都是相当重要的。

表 4.1　GH-释放抑制物在鱼类生长激素细胞内的信号转导引起 GH 分泌和 GH 基因表达的作用

信号成员	IGF	SS-14	[脯^2] SS-14	gbSS-28	NA	5-HT
PI3K	↑					
MAPK	↑					
CAM	↑*					
途径的作用下游						
Ca^{2+}		↓	↓	↓/↔	↓*	↔
NO		↓	↓	↔		
AA		↓			↓	
cAMP		↓	↓	↔	↓*	
PKC		↓	↓	↓/↔	↓	↓?
诱导信号产生						
cAMP		↓			↓	
Ca^{2+}		↓/↑			↓	

以金鱼和草鱼脑垂体细胞的研究报告为基础。"*"表示有 GH 基因表达的各个信号途径参与。（符号：↑，增加/活化；↓，降低/抑制；↔，小或没有作用；?，以原始数据为基础的发现）（缩写：IGF，胰岛素样生长因子；SS-14，生长抑素-14；[脯^2] SS-14，[脯^2] 生长抑素-14 的变体；gbSS-28，金鱼脑生长抑素-28；NA，去甲肾上腺素；5-HT，5-羟色胺；PI3K，磷脂酰肌醇-3-激酶；MAPK，促分裂原活化蛋白激酶；CAM，钙调蛋白；NO，一氧化氮；AA，花生四烯酸；cAMP，环 AMP；PKC，蛋白质激酶 C）

尽管对 GH 释放起着有力的抑制作用，但 SS-14 并不影响罗非鱼和虹鳟 GH mRNA 水平的稳态（Melamed 等，1998）。是否对其他鱼类亦是这样以及是否所有 SS 同等物都是这样，目前还不清楚。

2. 5-羟色胺（5-HT）

对金鱼的研究表明，5-HT 降低离体脑垂体碎片和脑垂体细胞的基础性 GH 释放，此外，5-HT 抑制 sGnRH 和 DA 诱导的 GH 释放（Somoza 和 Peter，1991；Wong 等，1998）。5-HT 的这些作用是由 5-HT2 类受体介导的。5-HT 是否同样影响其他鱼类的 GH 分泌活动，目前还不清楚。在金鱼、虹鳟、纹电鳐（*Torpedo marmorata*）、非洲鲶鱼（*Clarias gariepinus*）、大西洋绒须石首鱼（*Micropogonias undulatus*）等的腺脑垂体中都能检测到 5-HT 的免疫反应神经纤维（Kah 和 Chambolle，1983；Frankenhuis-van den Heuvel 和 Nieuwenhuys，1984；Bonn 和 König，1990；Corio 等，1991；Khan 和 Thomas，1993）。此外，在印度鲶鱼（*Heteropneustes fossilis*）（Joy 等，1998）和金鱼（Sloley 等，1991）的脑垂体中曾测量到 5-HT 含量，表明脑垂体细胞直接接触到 5-HT。显然，由下丘脑产生的 5-HT 能直接作用于鱼类脑垂体水平抑制 GH 分泌活动。但是，在剑尾鱼的促性腺激素细胞和脑垂体中间部细胞中亦曾检测到 5-HT 的免疫活性（Margolis-Nunno 等，1986），表明有些鱼类可能产生局部脑垂体来源的 5-HT。目前对介导 5-HT 抑制 GH 分泌活动的细胞内信号系统还了解得不多，但初步的研究结果表明 5-HT 作用于 PKC 下游，但不会影响 $[Ca^{2+}]_i$ 增加（Yu 等，2008；见表 4.1）。

3. γ 氨基丁酸（GABA）

对金鱼的研究表明，GABA 抑制在体的 GH 释放，但不影响离体的脑垂体释放 GH。由于 GABA 亦抑制鱼类脑的 DA 周转，因而 GABA 有可能通过阻抑下丘脑的多巴胺能输入到金鱼的生长激素细胞而间接起作用（Trudeau 等，2006），进而减弱 D1 受体介导 DA 刺激 GH 释放的作用（Wong 等，1993）。

4.4.2.2 刺激物

1. 生长激素释放激素（GHRH）

GHRH 是胰高血糖素（glucagon）肽类超家族的成员之一，亦包括 PACAP 在内。免疫组织化学研究表明，GHRH 神经纤维由视前区发出，分布到腺脑垂体（见本书第 1 章）。合成的鲤鱼 GHRH 能刺激虹鳟、罗非鱼和金鱼在体和离体的 GH 释放作用（Peng 和 Peter，1997）。GHRH 刺激脑垂体碎片和细胞制品增强 GH 释放的能力表明 GHRH 直接作用于脑垂体生长激素细胞。但是，GHRH 刺激 GH 释放的效能相当低，其增强 GH 释放的能力并非一致地得到证实。对于金鱼，GHRH 只在非常有限的季节性生殖周期才是有作用的（Chang 和 Habibi，2002）。因此，和哺乳类不同，GHRH 可能不是鱼类主要的 GH 释放的神经内分泌刺激物（综述见 Montero 等，2006）。

有学者曾经推想鱼类的 GHRH 和 PACAP 是由相同的基因编码的，但最近的研究表明 GHRH 和 PACAP 由各自不同的基因编码，尽管在 PACAP 基因中亦有一个类 GHRH

肽（称为 PACAP 相关肽）（Tam 等，2007）。金鱼 GHRH 能激活表达金鱼 GHRH 受体的中国仓鼠卵巢细胞产生 cAMP，并且能以剂量依存关系刺激金鱼脑垂体细胞释放 GH（Lee 等，2007；见表 4.2）。这些结果和对哺乳类的研究结果一致，GHRH 通过依赖于 cAMP 的途径刺激 GH 释放。这些研究引发了这样的思考：先前观察研究的 GHRH 实际上可能是类 GHRH 肽的作用而不是真正的 GHRH 作用。

表4.2　GH-释放因子在鱼类生长激素细胞中的细胞内信号转导引起 GH 释放和 GH 基因表达的作用概要

信号成员	sGnRH	cGnRH-Ⅱ	DA	PACAP	GHRH	生长素释放肽	GH	LH
主要通路								
AC/cAMP/PKA	×	×	↑	↑*	↑?		↑*	
PLC				↑		↑?		
VSCC	↑	↑	↑	↑*		↑?		
PKC	↑, ×*	↑, ×*	×	↑/×				
CAM	↑	↑		↑*				
AA	×	×	↑					
NO/cGMP	↑	↑	↑	×				
PI3K							↑*	
JAK2							↑*	
ERK/MAPK	↑*	↑*				↑?	↑*	
Na^+/K^+ 逆向转运	↑	↑	×					
细胞内 Ca^{2+} 成分								
IP3-灵敏的	↑	×		↲				
TMB 8-灵敏的	↑	↑	↑	↑				
毒胡萝卜素灵敏的	×	×	↲	↲				
BHQ-灵敏的	×	×		↑				
CPA-灵敏的	×	×						
咖啡因-灵敏的	↑	↑	×	↑				
利阿诺定-灵敏的	×	↑	×	×				
硝苯呋海因-灵敏的	×			×				
cADP 核糖				×				
线粒体	↲	×		×				

以金鱼和草鱼脑垂体细胞的研究报告为基础。"＊"表示参与 GH 基因表达的各个信号通路。符号：↑，增加/活化；↓，降低/抑制；×，没有参加；↲，间接调节；?，基于初步的发现。[缩写词：sGnRH，鲑鱼促性腺激素释放激素；cGnRH-Ⅱ，鸡Ⅱ型促性腺激素释放激素；DA，多巴胺；PACAP，脑垂体腺苷酸环化酶激活多肽；GHRH，生长激素释放激素；GH，生长激素；LH，促黄体激素；E_2，雌二醇；AC，腺苷酸环化酶；cAMP，环 AMP；PKA，蛋白激酶 A；PLC，磷脂酶 C；VSCC，电压灵敏 Ca^{2+} 通道；PKC，蛋白激酶 C；CAM，钙调蛋白；AA，花生四烯酸；NO，一氧化氮；cGMP，环 GMP；PI3K，磷脂酰肌醇-3-激酶；JAK2，詹纳斯激酶2；MAPK，促分裂原活化蛋白激酶；IP3，肌醇1,4,5-三磷酸；TMB8，8-(N,N-二乙氨基)-辛基-3,4,5-三甲氧苯甲酸；CPA，cycloplanzonic acid；cADP Ribose，环 ADP 核糖]

2. 促性腺激素释放激素（GnRH）

在鱼类中已经鉴别出多个 GnRHs，而且鱼类的 GnRH 类型代表了由 Silver 等（2004）提出的分类标准中的所有 4 个类别。通常每种鱼类都有 2 种或 3 种 GnRH 类型，而视前区的 1 种类型（GnRH1 或 GnRH3）释放到脑垂体（Kah 等，2007；亦见 Cardà-Reverter 和 Canosa，本书第 1 章）。GnRH2（cGnRH-Ⅱ）是中脑特有的 GnRH 类型，由神经板（neural plate）后面的室周室管膜细胞（circumventricular ependymal cell）衍生，但是，一个前嗅区衍生 GnRH2 的神经元群体却出现在斑马鱼的视前区（Palevitch 等，2007）。这表明对于有些鱼类，GnRH2 亦可能具有下丘脑调节脑垂体的功能。对金鱼、鲤鱼、罗非鱼、虹鳟和草鱼的研究都已证明 GnRH 刺激 GH 释放，而非洲鲶鱼和欧洲鳗鱼 GnRH 没有这方面作用（Peng 和 Peter，1997；Bosma 等，1997；Rousseau 等，1999）。GnRH 从脑垂体细胞制品中增强 GH 释放的能力以及 GnRH 受体存在于金鱼和银汉鱼（*Odontesthes bonariensis*）的生长激素细胞之中，证明 GnRH 直接作用于生长激素细胞（Peter 和 Chang，1999；Stefano 等，1999）。GnRH 增强 GH 释放的能力存在着季节差异。在金鱼和虹鳟的性成熟期，GH 对 GnRH 的反应最强，而在性退化期最弱（Chang 和 Habibi，2002）。GH 对 GnRH 反应性依赖于季节/生殖周期的变动，部分原因可能是 GnRH 受体（GnRH）表达的变化（Jodo 等，2005）。对金鱼的研究表明，GH 大量释放和产卵时 GTH 的迅速增加是同时发生的（Peter 和 Chang，1999）。如上所述，GH 起着增强性腺类固醇生成的作用。这些研究结果以及 GnRH 拮抗物降低性成熟金鱼血清 GH 水平的能力（Peter 和 Chang，1999），不仅证明 GnRH 是 GH 分泌活动的生理调节剂，而且表明 GnRH 为 GH 和 GTH 分泌活动之间在两种激素必须同时增强释放时（如在卵母细胞成熟期间）提供了神经内分泌调控的联系。金鱼的两种 GnRH 类型，即 sGnRH（GnRH3 型）和 cGnRH-Ⅱ（GnRH2 型）都释放到脑垂体中，虽然目前学界对 cGnRH-Ⅱ 的来源尚未完全弄清楚。对于金鱼，sGnRH 和 cGnRH-Ⅱ 都能同样有效地刺激 GH 分泌活动。

许多属于 A 级家族的 G-蛋白偶联受体 GnRHRs 都已在鱼类中克隆。有些鱼类曾被发现多达 5 种 GnRHRs（Guilgur 等，2006），有 4 种 GnRHR 类型出现在鱼类脑垂体，如马苏大麻哈鱼（*Oncorhynchus masu*）（Jodo 等，2003）、舌齿鲈（*Dicentrarchus labrax*）（Moncaut 等，2005）。在金鱼中曾经克隆属于同一个 GnRHR 亚型（1 型）的 2 个 GnRHR 类型。一个类型的受体对基于原位杂交和药理性质的 GH 调控来说是最好的（Illing 等，1999），但是，鱼类生长激素细胞具有多于一个 GnRHR 类型的可能性不应忽视。有关介导 GnRH 刺激 GH 释放的细胞内信号级联放大的资料都来自对金鱼、鲤鱼和罗非鱼的研究，其中以金鱼的资料最为详细。调动细胞内的 Ca^{2+} 贮存以升高 $[Ca^{2+}]_i$，细胞外钙通过电压灵敏 Ca^{2+} 通道（voltage-sensitive Ca^{2+} channels，VSCC）进入并激活钙调蛋白（CAM）激酶以及 PKC 的活化等是通常参与的信号通路（见图 4.2）。此外，GnRH 诱导金鱼 GH 释放还包括一氧化氮合酶/NO/cGMP 和 PKC 下游的 Na^+/H^+ 逆向转运（antiport）。这些反应元件是否参与其他鱼类的 GnRH 信号通路，目前还不清楚。此外,sGnRH 和 cGnRH-Ⅱ采用不同的药理作用特有的细胞内 Ca^{2+} 贮存。尽管共用一些相同的 Ca^{2+} 贮存,但 sGnRH 单独利用一个 IP3-灵敏库和一个线粒体的 Ca^{2+} 库,而只有 cGnRH-Ⅱ使用一个对利阿诺定灵敏(ryanodine-sensitive)但对咖啡因不灵敏(caffeine-insensitive)的 Ca^{2+} 贮存(综述见 Chang 等,2000;见表 4.2)。

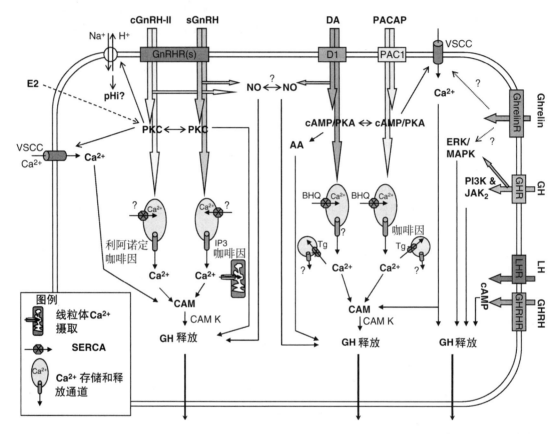

图 4.2 介导 sGnRH、cGnRH-Ⅱ、DA、PACAP、GH、LH、GHRH 等直接作用于鱼类生长激素细胞的受体后信号级联放大的模式示意

本模式以对金鱼与草鱼的研究报告为基础。在生长激素细胞水平,sGnRH 和 cGnRH-Ⅱ刺激 GH 释放是通过 PKC 级联偶联 Ca^{2+} 经过电压灵敏 Ca^{2+} 通道(VSCC)进入和利阿诺定-与 IP3-灵敏的细胞内 Ca^{2+}($[Ca^{2+}]_i$)贮存的转移。随着 $[Ca^{2+}]_i$ 升高,通过钙调蛋白(CAM)和钙调蛋白激酶(CAMK)的活化而引发 GH 胞吐作用(exocytosis)。两种 GnRH 通过 PKC 的刺激作用是在雌二醇(E_2)的调控下,并可能还有在膜水平的 Na^+/H^+ 交换以及依赖于一氧化氮作用机理的参与。和 GnRH 不同,DA 和 PACAP 的刺激作用是由 cAMP/PKA 级联介导的,它们偶联 Ca^{2+} 通过 VSCC 进入和对丁基羟基苯醚(BHQ)灵敏的 $[Ca^{2+}]_i$ 从 Ca^{2+} 贮存转移以及咖啡因的作用。BHQ 是一种肌质内质网 Ca^{2+} ATP 酶(SERCA)的抑制剂。此外,DA 和 PACAP 的作用亦为毒胡罗卜素(Tg)-灵敏的 SERCA 所调节。虽然 cAMP 依赖的作用机理参与由 LH 和 GnRH 刺激而引发的 GH 释放,ERK/MAPK 和 JAK2/PI3K 级联通过激活 GH 受体(GHR)仍是介导 GH 自调节(autoregulation)的主要通道。目前,对生长素释放肽诱导 GH 释放的信号作用机理还不清楚,但可能包含 Ca^{2+} 通过 VSCC 进入和 ERK/MAPK 通路的平行激活。(根据 Canosa 等,2007,有改动)(见书后彩图)

GnRH 亦能提高石斑鱼、马苏大麻哈鱼、金鱼、草鱼和红大麻哈鱼（Oncorhynchus nerka）等的脑垂体 GH 水平，但对罗非鱼没有作用（Melamed 等，1998；Taniyama 等，2000；Li 等，2002；Bhandari 等，2003；Ran 等，2004；Klausen 等，2005）。马苏大麻哈鱼 GnRH 增加脑垂体 GH mRNA 水平的能力是季节性的，只在 3 月性腺成熟开始时才有这种作用，GnRH 刺激 Pit-1 基因表达进而导致 GH 基因表达的增强（Onuma 等，2005）。对金鱼的研究表明，是细胞外信号调节激酶（signal-regulated-kinase，ERK）而不是 PKC 的激活介导 GnRH 对 GH 基因表达的影响（Klausen 等，2005）。在促使 $[Ca^{2+}]_i$ 增加的反应中都曾观察到 GH mRNA 水平的增加与降低，Ca^{2+} 的来源和 Ca^{2+} 的转移作用机理可能是关键性的（Johnson 等，2002）。

3. 多巴胺（DA）

鱼类包括金鱼和虹鳟的腺脑垂体亦有多巴胺能的神经纤维分布（见本书第 1 章）。DA 能刺激金鱼、罗非鱼、草鱼、虹鳟、非洲鲶鱼和石斑鱼在体和离体的 GH 释放（综述见 Canosa 等，2007）。药理学的、放射性配体结合的（radioligand binding）和荧光配体结合成像（fluorescence ligand binding imaging）的研究表明，DA 通过生长激素细胞的 DA-D1 受体刺激 GH 释放（Chang 等，2000）。注射 DA-D1 受体的拮抗物能降低金鱼血清 GH 水平（Wong 等，1993），支持了 DA 具有调节 GH 释放的生理作用的研究结果；此外，用电化学技术在金鱼和印度鲶鱼的脑垂体提取物中曾检测到 DA（Sloley 等，1991；Joy 等，1998）。值得注意的是，虹鳟接受 GH 植入物 7 天，下丘脑内 DA 的代谢物二羟苯乙酸（dihydroxy phenylacetic acid，DOPAC）水平提高（Johansson 等，2004）。由于 DOPAC 升高通常都被认为是多巴胺能神经元活化的一个指标，因此，这些研究结果表明多巴胺能神经元系统可能是 GH 反馈作用调控的一个靶标。

DA 刺激金鱼 GH 分泌活动的能力在全年生殖周期的各个时间段都可以检测到，但是，在每年性腺退化时期效能最高，而在产卵前最低（Chang 和 Habibi，2002）。对鲤科鱼类的研究表明 DA 刺激 GH 释放主要由腺苷酸环化酶（AC）/cAMP/PKA 信号通路介导，随后刺激细胞外 Ca^{2+} 经过 VSCC 进入和 CAM 激酶活化（见图 4.2）。AA 转移的活化和通过脂肪加氧酶途径（lipoxygenase pathway）的代谢作用是 cAMP 的 DA 信号下游另一个组成。最近对金鱼的研究结果表明，DA 亦利用 NOS/NO 信号和细胞内钙贮存去刺激 GH 释放（Wong 等，2001；Chang 等，2003；Mitchell 等，2008）；但是，DA-和 GnRH-灵敏的 Ca^{2+} 贮存具有不同的药理学特性，表明这两类神经内分泌调节剂使用不同的细胞内 Ca^{2+} 贮存（Chang 等，2003；Mitchell 等，2008；见表 4.2）。

DA 亦能使石斑鱼、虹鳟和罗非鱼脑垂体 GH mRNA 水平升高（Melamed 等，1998；Ran 等，2004）。在对罗非鱼的研究中直接观察到 cAMP/PKA 参与介导 DA 对 GH 合成作用的影响（Melamed 等，1998）。对其他鱼类的情况如何，目前还不清楚，但 cAMP-反应元件参与了大麻哈鱼（Oncorhynchus tshawytscha）GH 启动子活性的调控（Wong 等，1996）。

4. 脑垂体腺苷酸环化酶激活多肽（PACAP）

PACAP 神经纤维分布于金鱼、草鱼、欧洲鳗鱼和日本䲢（Uranoscopus japonicus）

的脑垂体远侧部（综述见 Wong 等, 2000, 2005; Cerdà-Reverter 和 Canosa, 本书第 1 章）。PACAP 曾在许多鱼类中被分离和克隆, 包括草鱼、斑马鱼、日本鳀、石斑鱼、鲱形白鲑（*Coregonus clupeaformis*）、北极茴鱼（*Thymallus arcticus*）、大西洋鲽（*Pleuronectes ferrugineus*）、大西洋庸鲽、斑点鮰、赤魟（*Dasyatis akajei*）和非洲鲶鱼以及一种软骨硬鳞鱼类高首鲟鱼（*Ascipenser transmontanus*）（McRory 等, 1995; Matsuda 等, 1998; Wei 等, 1998; Small 和 Nonneman, 2001; Jiang 等, 2003; Wang 等, 2003; Matsuda 等, 2005; Sze 等, 2007）。最近采用核磁共振（nuclear magnetic resonance, NMR）技术建立了草鱼 PACAP 的三维分子结构, 发现它和哺乳类的高度相似（Sze 等, 2007）。有些鱼类（如草鱼）是单个拷贝基因编码 PACAP, 另一些鱼类（如斑马鱼）是两个拷贝。通常, 两个 PACAP（氨基酸为～38 和 27）和四足类相应的 PACAP 高度同源。PACAP 能刺激金鱼、鲑鱼、欧洲鳗鱼、瘤棘鲆、鲤鱼和草鱼释放 GH, 而鱼类 PACAP 和哺乳类 PACAP38 通常都是有效的。和 GHRH 相比较, PACAP 刺激 GH 释放的效能要更稳定和较强些, 这表明对调控鱼类 GH 的分泌活动, PACAP 要比 GHRH 更为重要。对于金鱼, PACAP 在生殖周期的所有时间段都能刺激 GH 释放, 但存在刺激效能的差异。PACAP 作用的最大效能是在金鱼产卵前性腺成熟时（Chang J. P., 未发表结果; 见图 4.1）。上述研究结果表明, PACAP 是鱼类 GH 释放的一种生理调节剂, 但功能试验的缺失将有助于强化这个结论。

关于介导 PACAP 刺激 GH 释放的信号转导通路的资料主要来自对草鱼（Wong 等, 2005）和金鱼（Wong 等, 2000; Mitchell 等, 2008; Sawisky 和 Chang, 2005; 见表 4.2）的研究。PACAP 和 I 型 PACAP（PAC-I）受体结合后激活 AC/cAMP/PKA, 通过增强 Ca^{2+} 经 VSCC 进入和 Ca^{2+} 从细胞内贮存释放而使 $[Ca^{2+}]_i$ 增加, 并且激活 Ca^{2+}/CAM 激酶（见图 4.2）。DA 和 GnRH 利用的 Ca^{2+} 贮存具有不同的药理学特性。和 DA 不同, PACAP 不使用 NOS/NO 信号。尽管有学者曾证明磷脂酶 C（PLC）参与 PACAP 的作用, 但这是否引发传统的 IP_3/Ca^{2+} 和 PKC 信号的活化, 或者通过新的通路, 是有争议的。在对金鱼的一项研究中, PKC 参与 GH mRNA 表达的说法是被驳倒的（Klausen 等, 2005）, 而 PKC 表明参与另外一项研究中（Wong 等, 2007）。PAC-I 受体的剪接变体（splice variant）存在于斑马鱼和金鱼中（Fradinger 等, 2005; Kwok 等, 2006）, 这可以解释 PACAP 信号出现的差别。通过激活 cAMP/PKA、VSCC 和 CAM 激酶, PACAP 亦可以使草鱼脑垂体的 GH mRNA 水平升高（Wong 等, 2005）。

5. 神经肽 Y（NPY）

在金鱼的相关研究中最早证明 NPY 是一个 GH 的释放因子。NPY 存在于下丘脑和 NPY 免疫反应神经纤维分布于生长激素细胞附近, 表明 NPY 具有调控 GH 释放的作用。用人和金鱼的 NPY 对金鱼脑垂体细胞和脑垂体碎片进行研究的结果表明, NPY 通过 Y2 受体直接作用于脑垂体细胞刺激 GH 释放, 和/或间接通过脑垂体内的神经末梢释放 GnRH 而刺激 GH 释放。NPY 刺激 GH 释放的能力是季节性的, 最强的作用出现在金鱼性腺成熟/产卵前的脑垂体（Chang 和 Habibi, 2002; 见图 4.1）。在印度鲶鱼（*Clarias batrachus*）颅内注射 NPY，会使其通过激活 Y1 受体而减弱脑垂体 GH 免疫反应，这是 GH 胞吐作用增强而使脑垂体 GH 贮存细胞耗尽的结果（Mazumdar 等，2006）。Y1 受体是否直接在脑垂体细胞水平起作用还不清楚。介导 NPY 使 GH 释放的直接与间接作用的受体信号作用机理以及 NPY 是否影响 GH 的合成，还需进一步研究。

6. 甲状腺素-释放激素（TRH）

TRH 转录体和/或免疫反应存在于鱼类下丘脑，表示这种肽类亦具有调控脑垂体细胞的功能（见本书第 1 章）。TRH 能刺激金鱼在体 GH 释放和金鱼离体的脑垂体释放 GH（Trudeau 等，1992）。产卵前的金鱼离体脑垂体对 TRH 刺激 GH 释放的反应程度要大于性腺退化的金鱼离体脑垂体，表明存在着对 TRH 作用的季节性生殖影响（Chang 和 Habibi，2002）。但是，在对罗非鱼的研究中观察到，TRH 能提高血清 GH 水平，而对 GH 从脑垂体释放没有影响，表明 TRH 对 GH 释放的作用是间接的（Melamed 等，1998）。TRH 的 GH-释放作用并不直接作用于生长激素细胞可以解释在欧洲鳗鱼和瘤棘鲆脑垂体细胞中得到的负面发现（negative finding）（Rousseau 等，2001）。另一方面，使用 TRH 能刺激鲤鱼脑垂体细胞从头（*de novo*）合成 GH 蛋白质，表明 TRH 作用于脑垂体细胞水平刺激 GH 的产生（Kagabu 等，1998）。

7. 促肾上腺皮质素-释放因子（CRF）

CRF 是另一个鱼类调控脑垂体的重要因子（见本书第 1 章）。CRF 能刺激欧洲鳗鱼脑垂体细胞培养物释放 GH，表明 CRF 直接作用于脑垂体细胞水平（Rousseau 等，1999）。目前还不清楚 CRF 是否对其他鱼类亦有类似促进 GH 释放的作用。

8. 缩胆囊肽（CCK）

CCK 是一个厌食性因子（anorexigenic factor）（见本书第 9 章），CCK 纤维出现在鱼类的前脑和腺脑垂体远侧部。腹腔注射或颅内注射 CCK−8 硫酸盐能促进 GH 释放，它还能刺激金鱼脑垂体释放 GH，表明它直接作用于脑垂体水平（Himick 等，1993）。CCK 是否直接作用于生长激素细胞还不清楚，但 CCK 能使金鱼前脑的 PSS-I mRNA 减少（Canosa 和 Peter，2004），表明 CCK 能通过改变 SS 神经元活性而间接地增加 GH 释放。鱼类的生殖周期能影响 CCK 的 GH 释放效能，从性腺退化鱼中取得的脑垂体，对 CCK 的反应性最强（见图 4.1）。由于在消化道内的高表达水平，外周的 CCK 可能是神经内分泌调控 GH 释放的肽类重要来源。

9. 胃泌素释放肽（gastrin-releasing peptide，GRP）

GRP 是铃蟾肽（BBS）肽类家族的一个成员，已经从许多鱼类中分离出来（Volkoff 等，2005）。GRP 是一个有效能的厌食肽（见本书第 9 章）。在金鱼的相关研究中，BBS 刺激离体的脑垂体碎片释放 GH（Himick 等，1993）。此外，在腺脑垂体远侧部检测到 BBS/GRP 的结合位点，GRP-免疫反应纤维亦在腺脑垂体中间部和远侧部的分界面以及前脑被检测到。在脑垂体中亦发现 GRP mRNA（Himick 等，1995；Himick 和 Peter，1995）。这些研究结果表明 GRP 通过直接作用于金鱼脑垂体水平起着一个调控 GH 释放的下丘脑因子作用。此外，用 BBS 处理使前脑的 PSS-Ⅰ 和 PSS-Ⅱ mRNA 水平降低，表明 SS 表达和神经元活性的下调是 BBS 刺激 GH 释放的部分作用机理（Canosa 和 Peter，2004）。但是，GRP mRNA 亦在外周组织中广泛表达，包括皮肤和消化道。因此，GRP 通过血液循环携带，亦可能起着外周信号的作用。至于 GRP 是否对其他鱼类亦起着 GH-释放作用，目前还不清楚。

10. 甘丙肽（Galanin）

甘丙肽已经在圆口类（如七鳃鳗，*Lampetra fluviatilis*）、软骨鱼类（如小点猫鲨，*Scyliorhinus canicula*）和硬骨鱼类［如虹鳟、弓鳍鱼（*Amia calva*）和金鱼］中被鉴别，此外，甘丙肽的纤维和/或 mRNA 存在于鲑鱼和金鱼等鱼类的下丘脑和脑垂体内（见本书第 1 章和第 9 章）。在体使用甘丙肽能刺激银大麻哈鱼（*Oncorhynchus kisutch*）释放 GH（Diez 等，1992）。这些研究结果表明，鱼类下丘脑的甘丙肽通过直接刺激脑垂体释放 GH 而起着 GH 释放因子的作用。但甘丙肽在鱼类生长激素细胞水平的直接作用还有待于证实。

4.4.2.3 其他的因子

其他一些"非传统的"GH 释放的神经内分泌调节剂已经鉴定，但对于它们的作用还有争论，了解得亦不多。

1. 谷氨酸盐（Glutamate）

注射谷氨酸盐的同等物 N-甲基-D-天冬氨酸（N-methyl-D-aspartate，NMDA）使金鱼血清 GH 含量降低，用雌二醇处理能使这个反应增强（Trudeau 等，1996）。相反，NMDA 能刺激预先经过类固醇处理（steroid-primed）的未成熟虹鳟分泌 GH（Holloway 和 Leatherland，1997）。学界对于谷氨酸盐调节鱼类 GH 释放的作用一直是有争议的。

2. RF 酰胺（RFamides）

具有 LPXRF 酰胺模体（motif）的 RF 酰胺肽已在四足类和鱼类的脑内被鉴定，包括圆口类（如海七鳃鳗，*Petromyzon marinus*）和硬骨鱼类（如金鱼和弹涂鱼，*Periophthalmus modestus*）。免疫组织化学和 mRNA 的研究表明，下丘脑的 RF 酰胺肽神经分布于腺脑垂体的远侧部和中间部，说明这种肽类具有调控硬骨鱼类脑垂体的功能（Osugi 等，2006；Cerdà-Reverter 和 Canosa，本书第 1 章）。三种金鱼 RF 酰胺刺激红大麻哈鱼脑垂体细胞培养物释放 GH（Amano 等，2006），表明这些肽起着 GH 释放肽的作用。另一方面，两种合成的海七鳃鳗 RF 酰胺抑制七鳃鳗离体的 GH mRNA 表达（Moriyama 等，2007）。这些研究结果表明 RF 酰胺能直接作用于生长激素细胞，但它们调控生长激素细胞的功能还需进一步阐明。

3. 利尿钠肽（natriuretic peptides）

C-型利尿钠肽（CNP）的基因能在罗非鱼的脑垂体中表达。在对罗非鱼脑垂体细胞培育物的研究中，罗非鱼的 CNP、心房钠尿肽（atrial natriuretic peptide，ANP）和 B-型利尿钠肽（BNP）能刺激 GH mRNA 表达，而 ANP 和 BNP 能增加 GH 释放。值得注意的是 ANP 和 BNP 能增加细胞内 cGMP 水平，而 CNP 没有作用。这些研究结果表明 ANP 和 BNP 能够通过 cGMP 调节硬骨鱼类生长激素细胞的功能。由于罗非鱼脑的 ANP 和 BNP 水平非常低，脑和外周组织是不是影响生长激素细胞功能的利尿钠肽来源，目前还不清楚（Fox 等，2007a）。

4.4.3　外周器官/组织的信号

从外周器官/组织释放的控制鱼类生长激素分泌的调节剂，有些是符合传统的"反馈"调控模式的，另外的则可能不是。

4.4.3.1　抑制剂

1. 胰岛素样生长因子（IGF）

IGF 参与传统的"长环负反馈"调控 GH 的分泌活动（Wong 等，2006）。鱼类 IGF 和 IGF-I 受体的分子结构是高度保守的。IGF-I 和 IGF-II 在硬骨鱼类和板鳃鱼类中表达，而只有 IGF-I 在无颌类中表达。GH 能增强一些鱼类肝脏 IGF-I 基因表达，包括鲤鱼、金头鲷、鲑鱼、金鱼、虹鳟和罗非鱼（Wong 等，2006；Chen 等，2007）。IGF-I 受体存在于鱼类的脑垂体中（Fruchtman 等，2002；Filby 和 Tyler，2007），IGF-I 和/或 IGF-II 能使虹鳟（Weil 等，1999）和草鱼（Huo 等，2005）脑垂体细胞 GH 释放减少和 GH mRNA 水平降低。这些结果表明血液循环中的 IGFs 与哺乳类一样，对鱼类生长激素细胞的功能起着负反馈调控作用。IGF-I 影响 GH 释放是和 PI3K 与 MAPK 的活化相联系的（Fruchtman 等，2001）；而 IGF-I 和 IGF-II 抑制 GH 合成是由 CAM/钙依赖磷酸酶（calcineurin）作用机理与 CAM 基因表达的上调介导的（Huo 等，2005；见表 4.1）。IGF-I 受体亦在鱼的脑中表达，包括褐鳟（*Salmo trutta*）、胖头鲦（*Pimephales promelas*）和石斑鱼（Smith 等，2005；Kuang 等，2005；Filby 和 Tyler，2007）；此外，IGF-I 表达神经元出现在幼年罗非鱼的脑内（Shved 等，2007）。IGF 是否亦作用于下丘脑调节物释放的水平以影响鱼类生长激素细胞的功能，目前还不清楚。值得注意的是，IGF-I mRNA 在胖头鲦的脑垂体中表达（Filby 和 Tyler，2007），这使得 IGF 可能参与局部的自分泌/旁分泌以调节鱼类生长激素细胞的功能（见第 4.4.4 节）。

2. 去甲肾上腺素（NA）

NA 使金鱼血清 GH 水平降低，并且直接使金鱼和草鱼脑垂体细胞原代培养物的基础 GH 分泌活动减弱。NA 通过 $\alpha2$-肾上腺素能（adrenergic）受体亦能在脑垂体水平抑制 GH 对 GnRH、DA 和 PACAP 的反应。NA 的抑制作用是在 PKC 活化、Ca^{2+} 和 AA 转移以及 cAMP 产生的下游水平完成的。此外，NA 阻抑 PACAP 的作用以升高 cAMP 和 $[Ca^{2+}]_i$，表明 NA 亦能作用于第二信使产生的水平（见表 4.1）。在对草鱼的研究中，$NA\alpha2$ 的抑制作用能降低 GH mRNA、GH 初级转录体和 GH 启动子的活性，表明 NA 通过下调 GH 基因转录亦能阻抑 GH 合成。有趣的是，NA 抑制基础的 GH 释放虽然和 $[Ca^{2+}]_i$ 的释放没有联系，但清除 NA 之后引起 GH 释放的反弹增加却和 $[Ca^{2+}]_i$ 的增加相关。NA 阻抑 GnRH 和 PACAP 刺激 GH 分泌之后都能观察到一个较强的 GH 释放增加的反弹，而这是和相对于基础状态较高的［Ca^{2+}］$_i$相联系的（Wang 等，2007；Wong 等，2007）。这些研究结果引起了这种可能性，即 NA 阻抑 Ca^{2+} 转移细胞内信号分子的能力以增加［Ca^{2+}］$_i$，因而当 NA 的抑制作用清除时，就发生［Ca^{2+}］$_i$ 突然迅速的"过度回弹"，其幅度取决于神经内分泌调节剂所产生的刺激强度。由于在鱼类（如金鱼和虹鳟）脑垂体内检测到的 NA 量相当低，NA 作为对脑垂体生长激素细胞起作用的下丘脑因子，其重要性尚存在疑问。另外，硬骨鱼类在应激状态下，血液循环中 NA 含量能显著增加，在这种情况下，体循环的 NA 能到达脑垂体而影响 GH 的分泌活动。脑和外周的 NA 对调控 GH 分泌活动的相对重要性还需继续研究。

4.4.3.2 刺激剂

1. 生长素释放肽（ghrelin）

学者们曾对鱼类和其他脊椎动物生长素释放肽的结构和功能做了详细的综述（Unniappan 和 Peter，2005；Kaiya 等，2008）。简而言之，生长素释放肽已经从几种辐鳍鱼类和两种鲨鱼中克隆出来。鱼类的生长素释放肽和哺乳类的相似，是 n-酰基化的（n-acylated）。尽管 n-octanylation（C-8）被认为是盛行的修饰，但 C-8 和 C-10（decanylated）型都能在罗非鱼体内纯化得到（Fox 等，2007b）。此外，两个生长素释放肽受体［GH 促分泌受体，（GH secretagogue receptor，GHS-R）］的剪接变体存在于一些鱼类中（如黑棘鲷，*Acanthopagrus schlegeli*）（Chan 和 Cheng，2004）。生长素释放肽是一种食欲性因子，在消化道产生，释放到血液中，在食物摄取和 GH 释放之间形成链节。虹鳟在食物长期缺少期间血液循环中的生长素释放肽和 GH 都增加，这是支持上述见解的依据之一（Jönsson 等，2007；Volkoff 等，本书第 9 章）。胃是生长素释放肽的主要来源，而在鱼类血液中可以测定生长素释放肽的含量。由于生长素释放肽 mRNA 存在于虹鳟（Kaiya 等，2008）和金鱼（Unniappan 和 Peter，2005）的下丘脑中，而不存在于其他鱼类（如金头鲷）（Yeung 等，2006）的下丘脑中，所以，对于下丘脑是不是影响脑垂体的生长素释放肽的另一个来源，是有争议的。然而，生长素释放肽和其他下丘脑因子如 GRP/BBS 等的相互作用仍是下丘脑网络调控鱼类 GH 水平在餐前和餐后变化的一部分。

生长素释放肽能刺激一些鱼类脑垂体细胞培养物释放 GH，而和 GHS-R 的拮抗物共同使用会阻抑这种作用（Unniappan 和 Peter，2005；Kaiya 等，2008）。这些研究结果表明生长素释放肽通过激活 GHS-Rs 而直接作用于生长激素细胞刺激 GH 释放。生长素释放肽亦能提高大多数（不是全部）曾研究过的鱼类的 GH mRNA 水平（综述见 Unniappan 和 Peter，2005）。用 GHS-R 的人造同等物 GRP6 和 L163、L540 刺激金头鲷 GHS-R1a 在 HEK293 细胞中表达，证明 PLC 的刺激、［Ca^{2+}］$_i$ 增加以及 VSCC 与 ERK1/2 的活化可能是和生长素释放肽作用于鱼类相关的作用机理（Chan 等，2004b）。和 Ca^{2+} 信号通道参与生长素释放肽作用于鱼类 GH 释放相一致的是，金鱼的生长素释放肽能使金鱼生长激素细胞的［Ca^{2+}］$_i$ 增加（Grey 和 Chang，2009；见表 4.2）。

2. 甲状腺素

如同在第4.4.2.2节中对促甲状腺素释放激素所分析的，TRH诱导的在体GH释放不能都归因于TRH直接作用脑垂体水平。甲状腺素受体存在于鱼类的脑垂体中，如鲤鱼和胖头鲦，表明甲状腺素直接影响生长激素细胞的功能。用T_3/T_4处理能诱导虹鳟离体释放GH以及GH基因在罗非鱼和虹鳟脑垂体内的表达支持了这个见解。此外，GH通过激活外周组织的5′-单脱碘酶（5′-monodeiodinase）而增加甲状腺外的T_3产生。相反，用T_4和T_3处理能抑制鳗鱼在体与离体脑垂体GH的释放和合成，表明在甲状腺素如何影响生长激素细胞方面存在着种类差异。总而言之，这些研究结果表明甲状腺素参与鱼类脑垂体GH分泌与产生的反馈调控作用环（负反馈或正反馈取决于不同的种类）。TSH能直接刺激罗非鱼释放GH，T_3能增强GHRH诱导虹鳟的GH释放（Melamed等，1998；Rousseau等，2002；Wong等，2006；Filby和Tyler，2007），这些研究结果进一步证明存在着复杂的TRH/GH/甲状腺轴相互作用。

3. 皮质醇（cortisol）

和GH一样，皮质醇参加能量代谢、鳃离子转运活动、渗透压调节、免疫功能和应激反应等的调控。能反映这两种激素之间紧密联系的是，皮质醇影响GH的分泌活动，反过来也一样，GH影响皮质醇的释放。皮质醇能直接作用于脑垂体细胞水平，而糖皮质素受体亦在鱼类如胖头鲦的脑垂体中表达（Filby和Tyler，2007）。皮质醇能增强罗非鱼脑垂体培养物释放GH（Uchida，2004）。在虹鳟的脑垂体细胞中，地塞米松（dexamethasone）能增强基础的以及GHRH诱导的GH分泌活动（Luo和McKeown，1991）。此外，皮质醇亦能通过限制IGF-I对生长激素细胞的反馈作用活性而间接影响GH释放。用地塞米松处理能降低鲑鱼肝细胞培养物中由GH诱导的IGF-I mRNA增加程度（Pierce等，2005），而注射皮质醇使罗非鱼肝的IGF-I mRNA水平降低，但提高血液循环中IGF-结合蛋白水平（Kajimura等，2003）。投喂添加皮质醇的饲料亦能使斑点叉尾鮰脑垂体的GH mRNA水平升高（Peterson和Small，2005），虽然皮质醇对GH基因表达的作用部位还不清楚。相反，用GH处理能使虹鳟血液循环中皮质醇水平升高（Biga等，2004），使大西洋鲑鱼鳃的糖皮质素受体亲和力和容量增强（Shrimpton和McCormick，1998）。这些研究结果表明GH增强皮质醇的释放和活性，而皮质醇通过直接作用于脑垂体水平，或者通过降低IGF-I负反馈作用强度的间接作用，起着对GH释放与合成的正反馈调控作用。

4. 性类固醇

在性腺发育成熟期间和季节生殖周期，血清GH水平发生变化，表明性类固醇调节GH分泌活动（Makino等，2007）。支持这个见解的是，在体的雌二醇（E_2）处理能提高罗非鱼、虹鳟和金鱼血清的GH水平。在体用睾酮（T）处理亦能增加金鱼血清的GH水平，而这种作用需要T芳化为E_2（综述见Canosa等，2007）。值得注意的是，脑垂体芳化酶活性在杜父鱼（*Myoxocephalus*）中和生长激素细胞有关系（Olivereau和Callard，1985），而在罗非鱼中和促性腺激素细胞有关联（Melamed等，1999），表明T

向 E_2 的局部转变受到不同因素的影响。然而，用 T 和 11-KT 做同样的处理对银大麻哈鱼的血清 GH 水平没有影响，表明对雄激素的反应存在着种族的差别（Larsen 等，2004）。虽然用 E_2 做离体的处理并不影响金鱼脑垂体细胞释放 GH，但用 E_2 和 T 处理后能提高鲤鱼脑垂体细胞基础的 GH 分泌活动（Melamed 等，1998），这表明性类固醇对血清 GH 水平的影响可以通过直接作用于基础的 GH 释放而实现（至少在一些鱼类中是这样的）。性类固醇亦可以通过调节其他外周因子的产生和/或它们对 GH 分泌活动的影响而间接起作用。E_2 的离体处理能使金头鲷肝脏的 IGF-I 和 IGF-II mRNA 水平降低（Carnevali 等，2005），而用乙炔基雌二醇（ethinylestradiol，EE）处理能使胖头鲅肝脏 GHR 和 IGF-I 表达降低（Filby 和 Tyler，2007）。另外，T 和 11-KT 使银大麻哈鱼血清 IGF-结合蛋白和 IGF-I 水平增加（Larsen 等，2004）。还有，E_2 能降低虹鳟的血浆 SS-14 水平和脑垂体对 SS-14 的反应性（Holloway 等，1997），并下调金鱼脑垂体的 SST_2 mRNA 水平（Cardenas 等，2003）。由于 GH 能增强许多鱼类性腺类固醇生成和血清中 E 和/或 T 的水平，可以设想性类固醇参与正反馈的调节作用环。在这个模型中，GH 引起血液循环中性类固醇水平升高，在一些鱼类中降低了 GH 诱导的肝脏 IGF 产生，而在其他一些鱼类中提高 IGF-结合蛋白水平。这就减少了 IGF 反馈作用抑制 GH 的释放。此外，性类固醇降低脑垂体的 SST 表达，减少 SS 对 GH 分泌的抑制作用。这种累积作用的结果是使 GH 分泌活动增强。

性类固醇能影响一些鱼类的 GH 合成，但这种作用甚至在同一个种类中都有争议。性类固醇（E_2，雄激素或者合成的雌激素同等物）对欧洲鳗鱼、成年罗非鱼、虹鳟、鲤鱼、幼年大西洋鲑、马苏大麻哈鱼等的 GH mRNA 水平没有影响（Yadetie 和 Male，2002；Onuma 等，2005；Wong 等，2006）。相反，E_2 的在体处理使金鱼脑垂体 GH 含量增加（Zou 等，1997），而用 E_2 和异源雌激素（xenoestrogen）做离体处理能使虹鳟脑垂体培养物的 GH mRNA 水平升高（Elango 等，2006）。这些研究结果表明性类固醇激素能刺激 GH 合成。另一方面，用 E_2 和 T 进行在体处理使罗非鱼脑垂体的 GH mRNA 水平降低（Melamed 等，1998；Shved 等，2007），表明性类固醇对 GH 的合成在鱼类中会有不同的影响。生命时期、性别、种类差异、神经内分泌的内环境以及直接与间接的作用等等可能都有助于解释各项研究结果的多样性。除了这些外周作用之外，性类固醇亦能作用于中枢以影响鱼类 GH 的分泌活动（见第 4.5 节）。

5. 瘦素（瘦蛋白，Leptin）

瘦素已经从鱼类包括河鲀（Kurokawa 等，2005；Yacobovitz 等，2008）和鲤鱼（Huising 等，2006）中克隆出来。和哺乳类相似，鱼类瘦素通过阻抑 NPY 在下丘脑表达而抑制食物摄取，如金鱼（Volkoff 等，2003）和鳟鱼（Murashita 等，2008）。因为脑腔内注射瘦素能提高缩胆囊肽（CCK）在金鱼脑内的表达（Volkoff 等，2003），CCK 和 NPY 都能够刺激鱼类脑垂体释放 GH（Peter 和 Chang，1999），因此，可以认为瘦素能通过在 CNS 内的间接作用而起到调控 GH 释放的作用。在对模式鱼类的研究中还未曾报道瘦素直接作用于脑垂体水平调控 GH 释放和合成。但是，瘦素能引发鱼类脑垂体细胞催乳激素（如在罗非鱼）和生长乳素（如在海鲈）的分泌活动（Peyon 等，2003；Tipsmark 等，2008），表明除生长激素细胞外，瘦素亦能作用于其他靶细胞（综述见 Volkoff，2006；见本书第 9 章）。

4.4.4　脑垂体内的自分泌/旁分泌信号

4.4.4.1　LH 和 GH 局部作用引起的脑垂体内反馈作用

在硬骨鱼类脑垂体远侧部通常都能观察到内分泌细胞的区带分布（zonal distribution），而促性腺激素细胞和生长激素细胞的紧密靠近，表明它们能通过自分泌/旁分泌作用机理而互相作用。最近对鲤鱼脑垂体的研究证明存在着一个 GH 调控的脑垂体内反馈作用环（Wong 等，2006）。在这个模型中，促性腺激素细胞释放 LH，通过依赖 cAMP 的作用机理而引起生长激素细胞的 GH 分泌活动。LH 的这种旁分泌作用能为 GH-诱导 GH 分泌以及通过 JAK_2/MAPK 和 JAK_2/PI3K 级联直接作用于生长激素细胞的 GH 基因表达而进一步增强（见图 4.2、表 4.2）。同时，GH 释放的局部增加起着负反馈作用以阻抑邻近的促性腺激素细胞分泌 LH。鲤鱼脑垂体由 GH-诱导 GH 释放的自身调节作用（autoregulation）和哺乳类的"超短反馈"（ultra-short feedback）不同，后者是在脑垂体水平，GH 阻抑 GH 分泌活动。值得注意的是曾报道在虹鳟整个脑垂体培养物中出现类似的用 GH 处理抑制 GH 释放的事例（Agustsson 和 Bjornsson，2000）。这些不一致的结果表明在脑垂体水平的 GH 局部作用方面可能存在着种族特异性的差别。在鲤鱼的脑垂体细胞中，GH-释放因子包括 GnRH、PACAP 和 DA 能诱导 GH 基因表达，而这些刺激作用能为清除分泌到培养介质中的内源 GH 所起的免疫中和作用（immunoneutralization）所阻断（Wong 等，2006）。很明显，GH 的局部释放不仅能在脑垂体水平调控 GH 的分泌和释放，而且还有助于使生长激素细胞保持对下丘脑因子刺激的敏感性。

4.4.4.2　脑垂体的激活蛋白和促滤泡素抑制素

和哺乳类相似，激活蛋白（activin）/促滤泡素抑制素（follistatin）系统参与鱼类生殖功能的调控。除了参与类固醇生成和卵母细胞成熟之外，激活蛋白亦在金鱼的脑垂体水平起作用，调节 GTH 分泌和基因表达（Ge，2000）。激活蛋白亚单位、激活蛋白受体和激活蛋白-结合蛋白促滤泡素抑制素都在鱼类（如金鱼）的脑垂体中表达，而这种脑垂体内的激活蛋白/促滤泡素抑制素系统是和 Smad2 与 Smad3 激活在功能上偶联的（Yuen 和 Ge，2004；Lau 和 Ge，2005）。对于哺乳类，激活蛋白主要在促性腺激素细胞中表达，而激活蛋白一直是抑制 GH 释放的；但对于鱼类，激活蛋白在生长激素细胞而不在促性腺激素细胞中表达，激活蛋白直接作用于脑垂体水平刺激 GH 分泌活动（Ge 和 Peter，1994）。这些研究结果表明激活蛋白通过自分泌/旁分泌作用起着鱼类生长激素细胞局部调节剂的作用。在金鱼的相关研究中，激活蛋白调节 LH 分泌和基因表达，而这些调节作用为促滤泡素抑制素所阻断（Yuen 和 Ge，2004）。由于 LH 在鲤科鱼类中起着 GH 分泌与合成的局部调节剂作用（见第 4.4.4.1 节），激活蛋白可能通过调节 LH 的局部产生而间接影响 GH 的释放。

4.4.4.3 IGF-I 在脑垂体内局部产生

哺乳类 IGF-I 的自分泌/旁分泌作用在靶组织内的局部产生已经得到充分阐明，并已成为生长调节素假说（somatomedin hypothesis）修正模型的基础（Kaplan 和 Cohen，2007）。鱼类模型的 IGF-I 转录体和免疫反应能在脑垂体内鉴别，其信号与在促皮质激素细胞内检测到的是一致的。有些鱼类（如罗非鱼）的 IGF-I 的转录体亦在生长激素细胞和促性腺激素细胞中被检测到（Melamed 等，1999；Eppler 等，2007）。此外，IGF-I 结合位点普遍在各种鱼类的脑垂体内表达（Fruchtman 等，2002），而 IGF-I 在脑垂体的局部产生能起到抗凋亡的（antiapoptotic）作用，以保持生长激素细胞种群的稳定（Melamed 等，1999）。这些研究结果证明了脑垂体 IGF-I 除了包括由肝脏产生的 IGF-I 长环反馈作用之外的对 GH 的调节作用。哺乳类 IGF-I 和胰岛素之间通过它们各自受体的功能联系已经充分阐明。由于胰岛素能刺激 GH 基因在鲤鱼脑垂体细胞内表达（Huo 等，2005），且最近已证明胰岛素能在罗非鱼脑垂体内表达（Hrytsenko 等，2007），因此，胰岛素亦可能起着 GH 表达的局部调节剂作用。IGF-I 和胰岛素之间的功能相互作用可能揭开通过自分泌/旁分泌的 GH 调控作用新的一页，值得进一步研究。

4.5 在下丘脑和脑垂体水平 GH 调节剂的功能相互作用

上述许多 GH 释放的神经内分泌调节剂亦在脑垂体水平发生相互作用。这些相互作用是复杂的，因为 GH 参与多个生理作用系统。例如，整合摄食和饱食的代谢调控需要复杂的 GH 释放的神经内分泌系统参与，而整合生长、生殖、能量代谢和分配的是包括 GnRH 和类固醇在内的另外一个系统。在第 1 章和第 2 章已经综述了许多这类神经内分泌系统之间实质相互作用的资料。本节将着重论述以下课题：①生长激素细胞区分和综合多种神经内分泌信号的能力；②性类固醇作用于调控生长激素细胞功能的中枢部位（central sites）；③在介导 GH 对食物缺乏和摄食的反应方面生长素释放肽和其他 GH 调节剂之间的关系。

4.5.1 在信号转导水平神经内分泌信号的整合

由于多种神经内分泌因子直接影响鱼类生长激素细胞的激素释放与合成，一个值得提出的问题是：这些作用在生长激素细胞中是如何整合和区分的？一个可能性是各种不同调节因子的受体在生长激素细胞当中有差别地表达。但是，对金鱼生长激素细胞进行 Ca^{2+} 图像和荧光-受体配体的研究表明，DA、PACAP 和 GnRH 的受体都存在于相同的细胞中，这些多种调节剂的受体在单个细胞中是同时表达的，至少在有些生长激素细胞中是这样的。在金鱼和鲤鱼中，GH 对 GnRH 和 DA，以及 PACAP 和 GnRH 的最大反应是叠加的（additive），而对 GnRH 和 cGnRH-II，以及 PACAP 和 DA 则不是叠加的。所以，除受体的特异性之外，在后受体细胞内信号水平必须重视配体和功能的选择性（Chang 等，2000；Wong 等，2006）。

如第 4.4.1 节所述，介导 sGnRH、cGnRH-Ⅱ、DA、PACAP 和生长素释放肽的刺激作用的通路是不同的，虽然有部分重叠（见表 4.2）。由药理学处理直接激活 PKC 和 PKA 的最大 GH 反应在金鱼和鲤鱼中是叠加的（Chang 等，2000），表明两个主要的级联导致 GH 释放：①PKC 和它的下游成员，Na^+/H^+ 交换；②PKA 和随后的 AA 活化。这个假说可以解释为什么 GH 对 DA 或 PACAP 和 GnRH 的反应是叠加的，而在两 GnRH 之间或者 DA 和 PACAP 之间是不重叠的。虽然 PKC 和 PKA 对 $[Ca^{2+}]_i$ 在生长激素细胞内的增加是相联系的，但存在着多种药理学性质不同的 Ca^{2+} 贮存引起另一个水平的复杂性，使 Ca^{2+} 的转移和 Ca^{2+} 的影响出现差别。对于 GnRH、DA 和 PACAP 增加 GH 合成相互作用的能力如何，目前还没有研究。此外，NOS/NO/cGMP、PI3K、ERK 和 MAPK 信号如何和 PKC 与 cAMP/PKA 通路相互作用以调控 GH 释放与合成，还有待于进一步研究。

关于在生长激素细胞水平对 GH 释放的抑制性神经内分泌调节剂相互作用的研究要比关于刺激性因子的研究少得多。与刺激性因子一样，由抑制性调控剂如 SS-14、[脯^2] SS-14、gbSS-28、NE 和 5-HT 影响的细胞内信号目标是部分不相同的（见第 4.4.2 节，表 4.1）。细胞内信号的不同成为说明三种 SSs、NA 和 5-HT 调控 GH 对神经内分泌刺激剂释放反应能力的基础。此外，这些差别亦可以解释在金鱼脑垂体细胞的初步实验中，在有 SS-14 的情况下，NA 进一步降低基础 GH 释放的能力（Yunker W. K. 和 Chang J. P.，未发表研究结果）。最近已经报道一些关于 NA 如何抑制 PACAP 诱导 GH 合成的研究结果（见第 4.4.3.1 节关于去甲肾上腺素的叙述），但是，要了解细胞内信号水平的作用是如何抑制 GH 合成的，还需要做许多研究工作。

4.5.2　性类固醇对下丘脑 GH 调节剂的影响

性类固醇影响 GH 释放和 GH 基因表达的部分功能存在于下丘脑水平，即性类固醇能够影响参与神经内分泌调控生长激素细胞的神经元活性。在金鱼的相关研究中，E_2 使 NPY 和 PACAP 在前脑的表达增强（Canosa 等，2007），并提高下丘脑视前区的 DA 周转率（Trudeau 等，1993a）。此外，E_2 增加黑棘鲷（*Acanthopagrus schlegeli*）（Lee 等，2004）脑细胞培养物 GnRH 的含量和 GnRH 释放。因此，E_2 对这种鱼类 GH 释放的刺激作用是和增强这四种 GH 释放的神经内分泌因子的神经元活性相关的。同样，E_2 亦增强 DA 在虹鳟脑内的周转（Saligaut 等，1992），这和 E_2 与 DA 增强虹鳟 GH mRNA 水平的作用是一致的（见第 4.4.2.2 节和第 4.4.3.2 节）。

除影响 DA 周转之外，E_2 亦影响其他神经递质系统。E_2 能降低罗非鱼脑的 5-HT 含量（Tsai 等，2000）和减弱虹鳟脑垂体 5-HT 的周转（Hernandez-Rauda 和 Aldegunde，2002）。同样，E_2 也能增加 NA 在金鱼端脑/视前区和下丘脑的周转（Trudeau 等，1993b）。这些研究结果表明 E_2 对 GH 释放的刺激性影响是通过减弱 GH 释放的抑制性神经内分泌系统的活性而介导的。相反，用 E_2 处理后增加 PPS-Ⅰ 和 PPS-Ⅲ 在金鱼前脑的表达（Canosa 等，2002）。这项研究和其他的发现相一致，表明是由反馈调控而引起次级性的增强 GH 分泌活动。

除了影响 GABA、SS 和谷氨酸盐的活性之外（见第 4.4.2.2 节、第 4.4.2.1 节和第4.4.2.3节)，性类固醇亦调节其他神经内分泌调节剂的能力以影响GH分泌。T增强金鱼GH对NPY的反应。同样，E_2增强GnRH和TRH促进GH释放的能力，但减弱对DA在体的反应。目前尚未研究性类固醇如何影响生长激素细胞对NPY和TRH的反应，但对GnRH已经有一些研究进展。用E_2处理离体的金鱼脑垂体细胞，对总的细胞GH含量没有影响，但增强sGnRH、cGnRH-Ⅱ和PKC激活因子的GH释放能力，而不影响Ca^{2+}离子载体（ionophore）。这些研究结果表明，E_2直接作用于脑垂体水平选择性地调节对GH释放（即PKC），而不取决于GH合成的特异性信号级联效应，以增强GH对GnRH的反应（综述见Canosa等，2007；见表4.2）。因此，性类固醇调控生长激素细胞的功能可能归因于多重水平，包括：①GH合成；②脑垂体对神经内分泌调节剂的反应性；③在CNS内下丘脑网络的活性；④GH-IGF反馈环的不同部位。

4.5.3 在摄食与食物不足的情况下GH调节剂在介导变化与生长激素细胞活性之间的相互作用

在摄食和饥饿期间，促进GH释放的神经内分泌因子之间的相互作用是复杂的，因为有些下丘脑的调节剂亦影响摄食行为和饱食。长期缺乏食物使许多鱼类血清GH和脑垂体mRNA水平增加，如罗非鱼（Weber和Grau，1999）、金鱼（Unniappan和Peter，2005）、石斑鱼（Pedroso等，2006）、蓝子鱼（Ayson等，2007）、大西洋鲑（Wilkinson等，2006）。对金鱼禁食会使其下丘脑的生长素释放肽和NPY mRNA增加，而使CCK mRNA降低。生长素释放肽和NPY mRNA增加可能和饥饿时GH分泌与合成的增加有关。一种厌食性因子CCK mRNA的下调，连同一种食欲性因子生长素释放肽的增加，可能是反映"饥饿期"将会得到的食物期待（见本书第9章）。在金鱼的相关研究中曾报道一个过渡性的餐后血清GH水平增加，在30 min时达到高峰，而在1.5 h后回复到餐前水平（Himick和Peter，1994）。从对金鱼的实验中可以深入了解介导这些激素水平变化的调控GH的神经内分泌因子之间的相互作用。在金鱼的相关实验中，前脑CCK mRNA水平有短时间的餐后增加，而注射CCK使前脑PSS-Ⅰ水平降低。这些研究结果表明餐后GH增加释放可能部分是由CCK诱导SS神经元活性降低所介导的。在摄食后的3 h，脑的生长素释放肽和PPS-Ⅱ mRNA水平下降。施用BBS使脑的PSS-Ⅰ和PSS-Ⅱ mRNA减少。注射生长素释放肽阻抑脑的PSS-Ⅱ mRNA水平，增加PSS-Ⅰ mRNA并阻抑BBS诱导的PSS-Ⅰ降低。结果是，BBS和生长素释放肽的结合处理抵消了互相间对PSS-Ⅰ的作用，而生长素释放肽和BBS只使脑PSS-Ⅱ mRN降低（Canosa等，2005）。这些研究结果表明增强生长素释放肽和BBS的神经元活性介导了和摄食相关的血清GH水平短暂升高。正是由于生长素释放肽和BBS的结合处理对GH释放的直接影响，以及通过阻抑PSS-Ⅱ神经元活性的间接作用而刺激生长激素细胞的活性。

4.5.4 其他的相互作用

PACAP和SS的神经元系统亦互相作用并且相互交叉调节。对金鱼注射gbSRIF和[脯^2] SS-14能增加前脑PACAP mRNA水平，而注射SS-14使前脑PACAP mRNA水平降低。相反，PACAP 使前脑 SS-14 和［脯^2］SS-14 mRNA 水平降低（Canosa 等，2007）。虽然和这些变化相关的生理状态还不清楚，但这些研究结果证明 PACAP 和 SS 的神经元活性不仅直接作用于脑垂体水平以调控生长激素细胞的活性，而且亦能通过影响其他神经内分泌调控神经元的活性而间接起作用。

最近，cGnRH-Ⅱ 在金鱼中亦被鉴别为一个厌食性因子（Matsuda 等，2008）。sGnRH 和 cGnRH-Ⅱ 亦能诱导雌金鱼的产卵行为（Volkoff 和 Peter，1999）。在自然条件下，只有排卵的雌金鱼会出现产卵行为，而在产卵时通常是停止摄食的。所以，GnRH 神经元活性在脑不同部位的变化不仅整合与驱动 GH 和 GTH 在排卵时的大量释放，而且诱导产卵行为和阻抑摄食活动（见第 3 章关于神经内分泌调控生殖活动的综述）。如上所述，E_2 能增加黑棘鲷脑神经元培养物的 GnRH 释放和 GnRH 含量，表明 E_2 对 GnRH 网络的正反馈作用亦是整合 GH 与 GTH 分泌活动的神经内分泌线路的一部分。在这些作用过程中，脑 GnRH 线路、性类固醇和其他神经内分泌因子如何相互作用调控生长激素细胞的功能，对于今后深入了解复杂的神经内分泌调控 GH 释放的作用机理，将是一个富有成效的模型。

4.6 结 束 语

GH 的释放由来自下丘脑和外周组织的多个神经内分泌因子所调控。由于 GH 生物学作用的特性，GH 释放和合成的综合性神经内分泌调控必然是复杂的，其相互作用发生在脑和脑垂体水平。在脑垂体水平发生的刺激性和抑制性神经内分泌调节剂作用的季节性变化，至少部分是由于性类固醇的作用所引起，导致血液循环中 GH 水平的季节性变动，而这又受到一定时期内摄食和食物可获得性的影响。在脑垂体水平释放的局部因子提供了 GH 释放与合成的进一步旁分泌/副分泌调控。在生长激素细胞的水平，由各种不同的神经内分泌因子使用不同组合的信号转导作用机理成为整合配体和功理特性的基础。掌握生长激素细胞复杂的神经内分泌调控作用亦具有重要的环境保护和水产养殖意义。许多环境的污染物作用于核受体和/或影响类固醇的代谢作用。因此，内分泌损害物（endocrine disruptor）能在脑-脑垂体轴的多个层次通过作用于性类固醇和甲状腺激素的受体而影响神经内分泌对 GH 释放与合成的调控作用，进而影响到 GH 直接或间接介导的许多生理功能，包括生长、能量代谢和生殖活动。同样，加强对于鱼类还缺乏深入了解的一些领域的研究，例如，免疫细胞因子和应激反应激素等对神经内分泌调控 GH 的影响，对于水产养殖生产中最大限度地提高养殖鱼类生长效率和健康水平是十分必要的。

J. P. 张

A. O. L. 王

参考文献

Agustsson T, Bjornsson B T. 2000. Growth hormone inhibits growth hormone secretion from the rainbow trout pituitary *in vitro*. *Comp. Biochem. Physiol. C Toxicol. Pharmacol*, 126: 299-303.

Amano M, Moriyama S, Iigo M, Kitamura S, Amiya N, Yamamori K, Ukena K, Tsutsui K. 2006. Novel fish hypothalamic neuropeptides stimulate the release of gonadotrophins and growth hormone from the pituitary of sockeye salmon. *J. Endocrinol*, 188: 417-423.

Ayson F G, Takemura A. 2006. Daily expression patterns for mRNAs of GH, PRL, SL, IGF-I and IGF-II in juvenile rabbitfish, *Siganus guttatus*, during 24-h light and dark cycles. *Gen. Comp. Endocrinol*, 149: 261-268.

Ayson F G, de Jesus-Ayson E G, Takemura A. 2007. mRNA expression patterns for GH, PRL, SL, IGF-I and IGF-II during altered feeding status in rabbitfish, *Siganus guttatus*. *Gen. Comp. Endocrinol*, 150: 196-204.

Bhandari R K, Taniyama S, Kitahashi T, Ando H, Yamauchi K, Zohar Y, Ueda H, Urano A. 2003. Seasonal changes of responses to gonadotropin-releasing hormone analog in expression of growth hormone/prolactin/somatolactin genes in the pituitary of masu salmon. *Gen. Comp. Endocrinol*, 130: 55-63.

Biga P R, Cain K D, Hardy R W, Schelling G T, Overturf K, Roberts S B, Goetz F W, Ott T L. 2004. Growth hormone differentially regulates muscle myostatin1 and -2 and increases circulating cortisol in rainbow trout (*Oncorhynchus mykiss*). *Gen. Comp. Endocrinol*, 138: 32-41.

Bjornsson B. 1997. The biology of salmon growth hormone: From day light to dominance. *Fish Physiol. Biochem*, 17: 9-24.

Bjornsson B, Johnsson J, Benedet S, Einarsdottir I, Hildahl J, Agustsson T, Johnsson E. 2002. Growth hormone endocrinology of salmonids: Regulatory mechanisms and mode of action. *Fish Physiol. Biochem*, 27: 227-242.

Bonn U, König B. 1990. Serotonin-immunoreactive neurons in the brain of *Eigenmannia lineata* (Gymnotiformes, Teleostei). *J. Hirnforsch*, 31: 297-306.

Bosma P T, Kolk S M, Rebers F E, Lescroart O, Roelants I, Willems P H, Schulz R W. 1997. Gonadotrophs but not somatotrophs carry gonadotrophin-releasing hormone receptors: Receptor localisation, intracellular calcium, and gonadotrophin and GH release. *J. Endocrinol*, 152: 437-446.

Canosa L F, Peter R E. 2004. Effects of cholecystokinin and bombesin on the expression of preprosomatostatin-encoding genes in goldfish forebrain. *Regul. Pept*, 121: 99-105.

Canosa L F, Lin X, Peter R E. 2002. Regulation of expression of somatostatin genes by sex steroid hormones in goldfish forebrain. *Neuroendocrinology*, 76: 8-17.

Canosa L F, Unniappan S, Peter R E. 2005. Periprandial changes in growth hormone release in goldfish: Role of somatostatin, ghrelin, and gastrin-releasing peptide. *Am. J. Physiol. Regul. Integr. Comp. Physiol*, 289: R125-133.

Canosa L F, Chang J P, Peter R E. 2007. Neuroendocrine control of growth hormone in fish. *Gen. Comp. Endocrinol*, 151: 1-26.

Canosa L F, Stacey N, Peter R E. 2008. Changes in brain mRNA levels of gonadotropin-releasing hormone, pituitary adenylate cyclase activating polypeptide, and somatostatin during ovulatory luteinizing hormone and growth hormone surges in goldfish. *Am. J. Physiol. Regul. Integr. Comp. Physiol*, 295: R1815-1821.

Cardenas R, Lin X, Canosa L F, Luna M, Aramburo C, Peter R E. 2003. Estradiol reduces pituitary responsiveness to somatostatin (SRIF-14) and down-regulates the expression of somatostatin SST2 receptors in goldfish pituitary. *Gen. Comp. Endocrinol*, 132: 119-124.

Carnevali O, Cardinali M, Maradonna F, Parisi M, Olivotto I, Polzonetti-Magni A M, Mosconi G, Funkenstein B. 2005. Hormonal regulation of hepatic IGF-I and IGF-II gene expression in the marine teleost *Sparus aurata*. *Mol. Reprod. Dev*, 71: 12-18.

Chan C B, Cheng C H. 2004. Identification and functional characterization of two alternatively spliced growth hormone secretagogue receptor transcripts from the pituitary of black seabream *Acanthopagrus schlegeli*. *Mol. Cell. Endocrinol*, 214: 81-95.

Chan C B, Fung C K, Fung W, Tse M C, Cheng C H. 2004. Stimulation of growth hormone secretion from seabream pituitary cells in primary culture by growth hormone secretagogues is independent of growth hormone transcription. *Comp. Biochem. Physiol. C Toxicol. Pharmacol*, 139: 77-85.

Chan C B, Leung P K, Wise H, Cheng C H. 2004. Signal transduction mechanism of the seabream growth hormone secretagogue receptor. *FEBS Lett*, 577: 147-153.

Chang J P, Habibi H R. 2002. Intracellular integration of multifactorial neuroendocrine regulation of goldfish somatotrophe functions. In: Small B, MacKinlay D, Ed. *Developments in understanding fish growth*. University of British Columbia: Vancouver 5-14. Symposium Proceedings of the International Congress on the Biology of Fish.

Chang J P, Johnson J D, Van Goor F, Wong C J H, Yunker W K, Uretsky A D, Taylor D, Jobin R M, Wong A O L, Goldberg J I. 2000. Signal transduction mechanisms mediating secretion in goldfish gonadotropes and somatotropes. *Biochem. Cell Biol*, 78: 139-153.

Chang J P, Wong C J, Davis P J, Soetaert B, Fedorow C, Sawisky G. 2003. Role of Ca^{2+} stores in dopamine- and PACAP-evoked growth hormone release in goldfish.

Mol. Cell. Endocrinol, 206: 63-74.

Chen M H, Li Y H, Chang Y, Hu S Y, Gong H Y, Lin G H, Chen T T, Wu J L. 2007. Co-induction of hepatic IGF-I and progranulin mRNA by growth hormone in tilapia, *Oreochromis mossambiccus*. *Gen. Comp. Endocrinol*, 150: 212-218.

Corio M, Peute J, Steinbusch H W. 1991. Distribution of serotonin- and dopamine-immunoreactivity in the brain of the teleost *Clarias gariepinus*. *J. Chem. Neuroanat*, 4: 79-95.

Diez J M, Giannico G, McLean E, Donaldson E M. 1992. The effect of somatostatin (SRIF-14, 25, 28), galanin and anti-SRIF on plasma growth hormone levels in coho salmon (*Oncorhynchus kituch*, Walbaum). *J. Fish Biol*, 40: 877-893.

Elango A, Shepherd B, Chen T T. 2006. Effects of endocrine disrupters on the expression of growth hormone and prolactin mRNA in the rainbow trout pituitary. *Gen. Comp. Endocrinol*, 145: 116-127.

Eppler E, Shved N, Moret O, Reinecke M. 2007. IGF-I is distinctly located in the bony fish pituitary as revealed for *Oreochromis niloticus*, the Nile tilapia, using real-time RT-PCR, *in situ* hybridisation and immunohistochemistry. *Gen. Comp. Endocrinol*, 150: 87-95.

Falcon J, Besseau L, Fazzari D, Attia J, Gaildrat P, Beauchaud M, Boeuf G. 2003. Melatonin modulates secretion of growth hormone and prolactin by trout pituitary glands and cells in culture. *Endocrinology*, 144: 4648-4658.

Figueroa J, Martín R S, Flores C, Grothusen H, Kausel G. 2005. Seasonal modulation of growth hormone mRNA and protein levels in carp pituitary: Evidence for two expressed genes. *J. Comp. Physiol* [*B*]. 175: 185-192.

Filby A L, Tyler C R. 2007. Cloning and characterization of cDNAs for hormones and/or receptors of growth hormone, insulin-like growth factor-I, thyroid hormone, and corticosteroid and the gender-, tissue-, and developmental-specific expression of their mRNA transcripts in fathead minnow (*Pimephales promelas*). *Gen. Comp. Endocrinol*, 150: 151-163.

Fox B K, Naka T, Inoue K, Takei Y, Hirano T, Grau E G. 2007. *In vitro* effects of homologous natriuretic peptides on growth hormone and prolactin release in the tilapia, *Oreochromis mossambicus*. *Gen. Comp. Endocrinol*, 150: 270-277.

Fox B K, Riley L G, Dorough C, Kaiya H, Hirano T, Grau E G. 2007. Effects of homologous ghrelins on the growth hormone/insulin-like growth factor-I axis in the tilapia, *Oreochromis mossambicus*. *Zoolog. Sci*, 24: 391-400.

Fradinger E A, Tello J A, Rivier J E, Sherwood N M. 2005. Characterization of four receptor cDNAs: PAC1, VPAC1, a novel PAC1 and a partial GHRH in zebrafish. *Mol. Cell. Endocrinol*, 231: 49-63.

Frankenhuis-van den Heuvel T H, Nieuwenhuys R. 1984. Distribution of serotonin-immunoreactivity in the diencephalon and mesencephalon of the trout, *Salmo gairdneri*. Cellbodies, fibres and terminals. *Anat. Embryol.* (*Berl.*), 169: 193-204.

Fruchtman S, McVey D C, Borski R J. 2002. Characterization of pituitary IGF-I receptors: Modulation of prolactin and growth hormone. *Am. J. Physiol. Regul. Integr. Comp. Physiol*, 283: R468-476.

Fruchtman S, Gift B, Howes B, Borski R. 2001. Insulin-like growth factor-I augments prolactin and inhibits growth hormone release through distinct as well as overlapping cellular signaling pathways. *Comp. Biochem. Physiol. B Biochem. Mol. Biol*, 129: 237-42.

Fukamachi S, Meyer A. 2007. Evolution of receptors for growth hormone and somatolactin in fish and land vertebrates: Lessons from the lungfish and sturgeon orthologues. *J. Mol. Evol.* 65: 359-372.

Ge W. 2000. Roles of the activin regulatory system in fish reproduction. *Can. J. Physiol. Pharmacol*, 78: 1077-1085.

Ge W, Peter R E. 1994. Activin-like peptides in somatotrophs and activin stimulation of growth hormone release in goldfish. *Gen. Comp. Endocrinol*, 95: 213-221.

Goldenberg N, Barkan A. 2007. Factors regulating growth hormone secretion in humans. *Endocrinol. Metab. Clin. North. Am*, 36: 37-55.

Grey C L, Chang J P. 2009. Ghrelin-induced growth hormone release from goldfish pituitary cells involves voltage-gated calcium channels. *Gen. Comp. Endocrinol*, 160: 148-157.

Guilgur L G, Moncaut N P, Canário A V, Somoza G M. 2006. Evolution of GnRH ligands and receptors in gnathostomata. *Comp. Biochem. Physiol. A Mol. Integr. Physiol*, 144: 272-283.

Hernandez-Rauda R, Aldegunde M. 2002. Effects of acute 17alpha-methyltestosterone, acute 17beta-estradiol, and chronic 17alpha-methyltestosterone on dopamine, norepinephrine and serotonin levels in the pituitary, hypothalamus and telencephalon of rainbow trout (*Oncorhynchus mykiss*). *J. Comp. Physiol* [*B*]. 172: 659-667.

Himick B A, Peter R E. 1994. Bombesin acts to suppress feeding behavior and alter serum growth hormone in goldfish. *Physiol. Behav*, 55: 65-72.

Himick B A, Peter R E. 1995. Bombesin-like immunoreactivity in the forebrain and pituitary and regulation of anterior pituitary hormone release by bombesin in goldfish. *Neuroendocrinology*, 61: 365-376.

Himick B A, Golosinski A A, Jonsson A C, Peter R E. 1993. CCK/gastrin-like immunoreactivity in the goldfish pituitary: Regulation of pituitary hormone secretion by CCK-like peptides *in vitro*. *Gen. Comp. Endocrinol*, 92: 88-103.

Himick B A, Vigna S R, Peter R E. 1995. Characterization and distribution of bombesin binding sites in the goldfish hypothalamic feeding center and pituitary. *Regul. Pept*, 60: 167-176.

Holloway A C, Leatherland J F. 1997. The effects of N-methyl-D, L-aspartate and gonadotropin-releasing hormone on *in vitro* growth hormone release in steroid-primed immature rainbow trout, Oncorhynchus mykiss. *Gen. Comp. Endocrinol*, 107: 32-43.

Holloway A C, Sheridan M A, Leatherland J F. 1997. Estradiol inhibits plasma somatostatin 14 (SRIF-14) levels and inhibits the response of somatotrophic cells to SRIF-14 challenge *in vitro* in rainbow trout, Oncorhynchus mykiss. *Gen. Comp. Endocrinol*, 106: 407-414.

Holloway A C, Melroe G T, Ehrman M M, Reddy P K, Leatherland J F, Sheridan M A. 2000. Effect of 17beta-estradiol on the expression of somatostatin genes in rainbow trout (*Oncorhynchus mykiss*). *Am. J. Physiol. Regul. Integr. Comp. Physiol*, 279: R389-393.

Hrytsenko O, WrightJr J R, Morrison C M, Pohajdak B. 2007. Insulin expression in the brain and pituitary cells of tilapia (*Oreochromis niloticus*). *Brain Res*, 1135: 31-40.

Huising M O, Geven E J, Kruiswijk C P, Nabuurs S B, Stolte E H, Spanings F A, Verburg-van Kemenade B M, Flik G. 2006. Increased leptin expression in common Carp (*Cyprinus carpio*) after food intake but not after fasting or feeding to satiation. *Endocrinology*, 147: 5786-5797.

Huo L, Fu G, Wang X, Ko W K, Wong A O. 2005. Modulation of calmodulin gene expression as a novel mechanism for growth hormone feedback control by insulin-like growth factor in grass carp pituitary cells. *Endocrinology*, 146: 3821-3835.

Illing N, Troskie B E, Nahorniak C S, Hapgood J P, Peter R E, Millar R P. 1999. Two gonadotropin-releasing hormone receptor subtypes with distinct ligand selectivity and differential distribution in brain and pituitary in the goldfish (*Carassius auratus*). *Proc. Natl. Acad. Sci. USA*, 96: 2526-2531.

Jiang Y, Li W S, Xie J, Lin H R. 2003. Sequence and expression of a cDNA encoding both pituitary adenylate cyclase activating polypeptide and growth hormone-releasing hormone in grouper (*Epinephelus coioides*). *Sheng Wu Hua Xue Yu Sheng Wu Wu Li Xue Bao (Shanghai)*, 35: 864-872.

Jiao B, Huang X, Chan C B, Zhang L, Wang D, Cheng C H. 2006. The co-existence of two growth hormone receptors in teleost fish and their differential signal transduction, tissue distribution and hormonal regulation of expression in seabream. *J. Mol. Endocrinol*, 36: 23-40.

Jodo A, Ando H, Urano A. 2003. Five different types of putative GnRH receptor gene are expressed in the brain of masu salmon (*Oncorhynchus masou*). *Zoolog. Sci*, 20: 1117-1125.

Jodo A, Kitahashi T, Taniyama S, Bhandari R K, Ueda H, Urano A, Ando H. 2005. Seasonal variation in the expression of five subtypes of gonadotropin-releasing hormone receptor genes in the brain of masu salmon from immaturity to spawning. *Zoolog. Sci*, 22: 1331-1338.

Johansson V, Winberg S, Jönsson E, Hall D, Björnsson B T. 2004. Peripherally administered growth hormone increases brain dopaminergic activity and swimming in rainbow trout. *Horm. Behav*, 46: 436-443.

Johnson J D, Klausen C, Habibi H R, Chang J P. 2002. Function-specific calcium stores selectively regulate growth hormone secretion, storage, and mRNA level. *Am. J. Physiol. Endocrinol. Metab*, 282: E810-819.

Jönsson E, Forsman A, Einarsdottir I E, Kaiya H, Ruohonen K, Björnsson B T. 2007. Plasma ghrelin levels in rainbow trout in response to fasting, feeding and food composition, and effects of ghrelin on voluntary food intake. *Comp. Biochem. Physiol. A Mol. Integr. Physiol*, 147: 1116-1124.

Joy K P, Senthilkumaran B, Sudhakumari C C. 1998. Periovulatory changes in hypothalamic and pituitary monoamines following GnRH analogue treatment in the catfish *Heteropneustes fossilis*: A study correlating changes in plasma hormone profiles. *J. Endocrinol*, 156: 365-372.

Kagabu Y, Mishiba T, Okino T, Yanagisawa T. 1998. Effects of thyrotropin-releasing hormone and its metabolites, Cyclo (His-Pro) and TRH-OH, on growth hormone and prolactin synthesis in primary cultured pituitary cells of the common carp, *Cyprinus carpio*. *Gen. Comp. Endocrinol*, 111: 395-403.

Kah O, Chambolle P. 1983. Serotonin in the brain of the goldfish, *Carassius auratus*. An immunocytochemical study. *Cell Tissue Res*, 234: 319-333.

Kah O, Lethimonier C, Somoza G, Guilgur L G, Vaillant C, Lareyre J J. 2007. GnRH and GnRH receptors in metazoa: A historical, comparative, and evolutive perspective. *Gen. Comp. Endocrinol*, 153: 346-364.

Kaiya H, Miyazato M, Kangawa K, Peter R E, Unniappan S. 2008. Ghrelin: A multifunctional hormone in non-mammalian vertebrates. *Comp. Biochem. Physiol. A Mol. Integr. Physiol*, 149: 109-128.

Kajimura S, Hirano T, Visitacion N, Moriyama S, Aida K, Grau E G. 2003. Dual mode of cortisol action on GH/IGF-I/IGF binding proteins in the tilapia, *Oreochromis mossambicus*. *J. Endocrinol*, 178: 91-99.

Kajimura S, Kawaguchi N, Kaneko T, Kawazoe I, Hirano T, Visitacion N, Grau E G, Aida K. 2004. Identification of the growth hormone receptor in an advanced teleost, the tilapia (*Oreochromis mossambicus*) with special reference to its distinct expression pattern in the ovary. *J. Endocrinol*, 181: 65-76.

Kaplan S A, Cohen P. 2007. The somatomedin hypothesis 2007: 50 years later. *J. Clin. Endocrinol. Metab*, 92: 4529-4535.

Kawauchi H, Sower S A. 2006. The dawn and evolution of hormones in the adenohypophysis. *Gen. Comp. Endocrinol*, 148: 3-14.

Khan I A, Thomas P. 1993. Immunocytochemical localization of serotonin and gonadotropin-releasing hormone in the brain and pituitary gland of the Atlantic croaker *Micropogonias undulatus*. *Gen. Comp. Endocrinol*, 91: 167-180.

Klausen C, Tsuchiya T, Chang J P, Habibi H R. 2005. PKC and ERK are differentially involved in gonadotropin-releasing hormone-induced growth hormone gene expression in the goldfish pituitary. *Am. J. Physiol. Regul. Integr. Comp. Physiol*, 289: R1625-1633.

Klein S E, Sheridan M A. 2008 Somatostatin signaling and the regulation of growth and metabolism in fish. *Mol. Cell. Endocrinol*, 286: 148-154.

Kuang Y M, Li W S, Lin H R. 2005. Molecular cloning and mRNA profile of insulin-like growth factor type 1 receptor in orange-spotted grouper, *Epinephelus coioides*. *Acta Biochim. Biophys. Sin. (Shanghai)*, 37: 327-334.

Kurokawa T, Uji S, Suzuki T. 2005. Identification of cDNA coding for a homologue to mammalian leptin from pufferfish, *Takifugu rubripes*. *Peptides*, 26: 745-750.

Kwok Y Y, Chu J Y, Vaudry H, Yon L, Anouar Y, Chow B K. 2006. Cloning and characterization of a PAC1 receptor hop-1 splice variant in goldfish (*Carassius auratus*). *Gen. Comp. Endocrinol*, 145: 188-196.

Kwong P, Chang J P. 1997. Somatostatin inhibition of growth hormone release in goldfish: Possible targets of intracellular mechanisms of action. *Gen. Comp. Endocrinol*, 108: 446-456.

Larsen D A, Shimizu M, Cooper K A, Swanson P, Dickhoff W W. 2004. Androgen effects on plasma GH, IGF-I, and 41-kDa IGFBP in coho salmon (*Oncorhynchus kisutch*). *Gen. Comp. Endocrinol*, 139: 29-37.

Lau M T, Ge W. 2005. Cloning of Smad2, Smad3, Smad4, and Smad7 from the goldfish pituitary and evidence for their involvement in activin regulation of goldfish FSHbeta promoter activity. *Gen. Comp. Endocrinol*, 141: 22-38.

Lee L T, Siu F K, Tam J K, Lau I T, Wong A O, Lin M C, Vaudry H, Chow B K. 2007. Discovery of growth hormone-releasing hormones and receptors in nonmammalian vertebrates. *Proc. Natl. Acad. Sci. USA*, 104: 2133-2138.

Lee Y H, Du J L, Shih Y S, Jeng S R, Sun L T, Chang C F. 2004. *In vivo* and *in vitro* sex steroids stimulate seabream gonadotropin-releasing hormone content and release in the protandrous black porgy, *Acanthopagrus schlegeli*. *Gen. Comp. Endocrinol*, 139: 12-19.

Lengyel A M. 2006. Novel mechanisms of growth hormone regulation: Growth hormone-releasing peptides and ghrelin. *Braz. J. Med. Biol. Res*, 39: 1003-1011.

Li W S, Lin H R, Wong A O. 2002. Effects of gonadotropin-releasing hormone on growth hormone secretion and gene expression in common carp pituitary. *Comp. Biochem. Physiol. B Biochem. Mol. Biol*, 132: 335-341.

Li W S, Chen D, Wong A O, Lin H R. 2005. Molecular cloning, tissue distribution, and ontogeny of mRNA expression of growth hormone in orange-spotted grouper (*Epinephelus coioides*). *Gen. Comp. Endocrinol*, 144: 78-89.

Lichanska A M, Waters M J. 2008. New insights into growth hormone receptor function and clinical implications. *Horm. Res*, 69: 138-145.

Lin X, Peter R E. 2001. Somatostatins and their receptors in fish. *Comp. Biochem. Physiol. B Biochem. Mol. Biol*, 129: 543-550.

Lin X, Otto C J, Cardenas R, Peter R E. 2000. Somatostatin family of peptides and its receptors in fish. *Can. J. Physiol. Pharmacol*, 78: 1053-1066.

Luo D, McKeown B A. 1991. The effect of thyroid hormone and glucocorticoids on carp growth hormone-releasing factor (GRF)-induced growth hormone (GH) release in rainbow trout (*Oncorhynchus mykiss*). *Comp. Biochem. Physiol. A*, 99: 621-626.

Makino K, Onuma T A, Kitahashi T, Ando H, Ban M, Urano A. 2007. Expression of hormone genes and osmoregulation in homing chum salmon: A minireview. *Gen. Comp. Endocrinol*, 152: 304-309.

Marchant T A, Peter R E. 1986. Seasonal variations in body growth rates and circulating levels of growth hormone in the goldfish, *Carassius auratus*. *J. Exp. Zool*, 237: 231-239.

Marchant T A, Dulka J G, Peter R E. 1989. Relationship between serum growth hormone levels and the brain and pituitary content of immunoreactive somatostatin in the goldfish, *Carassius auratus L*. *Gen. Comp. Endocrinol*, 73: 458-468.

Margolis-Nunno H, Halpern-Sebold L, Schreibman M P. 1986. Immunocytochemical changes in serotonin in the forebrain and pituitary of aging fish. *Neurobiol. Aging*, 7: 17-21.

Matsuda K, Yoshida T, Nagano Y, Kashimoto K, Yatohgo T, Shimomura H, Shioda S, Arimura A, Uchiyama M. 1998. Purification and primary structure of pituitary adenylate cyclase activating polypeptide (PACAP) from the brain of an elasmobranch, stingray, *Dasyatis akajei*. *Peptides*, 19: 1489-1495.

Matsuda K, Nagano Y, Uchiyama M, Onoue S, Takahashi A, Kawauchi H, Shioda S. 2005. Pituitary adenylate cyclase-activating polypeptide (PACAP) -like immunoreactivity in the brain of a teleost, *Uranoscopus japonicus*: Immunohistochemical relationship between PACAP and adenohypophysial hormones. *Regul. Pept*, 126: 129-136.

Matsuda K, Nakamura K, Shimakura S I, Miura T, Kageyama H, Uchiyama M, Shioda S, Ando H. 2008. Inhibitory effect of chicken gonadotropin-releasing hormone II on food intake in the goldfish, *Carassius auratus. Horm. Behav*, 54: 83-89.

Mazumdar M, Lal B, Sakharkar A J, Deshmukh M, Singru P S, Subhedar N. 2006. Involvement of neuropeptide Y Y1 receptors in the regulation of LH and GH cells in the pituitary of the catfish, *Clarias batrachus*: An immunocytochemical study. *Gen. Comp. Endocrinol*, 149: 190-196.

McRory J E, Parker D B, Ngamvongchon S, Sherwood N M. 1995. Sequence and expression of cDNA for pituitary adenylate cyclase activating polypeptide (PACAP) and growth hormone-releasing hormone (GHRH) -like peptide in catfish. *Mol. Cell. Endocrinol*, 108: 169-177.

Melamed P, Rosenfeld H, Elizur A, Yaron Z. 1998. Endocrine regulation of gonadotropin and growth hormone gene transcription in fish. *Comp. Biochem. Physiol. C Pharmacol. Toxicol. Endocrinol*, 119: 325-338.

Melamed P, Gur G, Rosenfeld H, Elizur A, Yaron Z. 1999. Possible interactions between gonadotrophs and somatotrophs in the pituitary of tilapia: Apparent roles for insulin-like growth factor I and estradiol. *Endocrinology*, 140: 1183-1191.

Mingarro M, Vega-Rubin de Celis S S, Astola A, Pendon C, Perez-Sanchez J. 2002. Endocrine mediators of seasonal growth in gilthead sea bream (*Sparus aurata*): The growth hormone and somatolactin paradigm. *Gen. Comp. Endocrinol*, 128: 102-111.

Mitchell G, Sawisky G R, Grey C L, Wong C J, Uretsky A D, Chang J P. 2008. Differential involvement of nitric oxide signaling in dopamine and PACAP stimulation of growth hormone release in goldfish. *Gen. Comp. Endocrinol*, 155: 318-327.

Moncaut N, Somoza G, Power D M, Canário A V. 2005. Five gonadotrophin-releasing hormone receptors in a teleost fish: Isolation, tissue distribution and phylogenetic relationships. *J. Mol. Endocrinol*, 34: 767-779.

Montero M, Yon L, Kikuyama S, Dufour S, Vaudry H. 2000. Molecular evolution of the growth hormone-releasing hormone/pituitary adenylate cyclase-activating polypeptide gene family. Functional implication in the regulation of growth hormone secretion. *J. Mol. Endocrinol*, 25: 157-168.

Moriyama S, Kasahara M, Amiya N, Takahashi A, Amano M, Sower S A, Yamamori K, Kawauchi H. 2007. RFamide peptides inhibit the expression of melanotropin and growth hormone genes in the pituitary of an Agnathan, the sea lamprey, *Petromyzon marinus*. *Endocrinology*, 148: 3740-3749.

Murashita K, Uji S, Yamamoto T, Ronnestad I, Kurokawa T. 2008. Production of recombinant leptin and its effects of food intake in rainbow trout (*Oncorhynchus mykiss*). *Comp. Biochem. Physiol. B Biochem. Mol. Biol*, 150: 377-384.

Nelson L E, Sheridan M A. 2005. Regulation of somatostatins and their receptors in fish. *Gen. Comp. Endocrinol*, 142: 117-133.

Nelson L E, Sheridan M A. 2006. Insulin and growth hormone stimulate somatostatin receptor (SSTR) expression by inducing transcription of SSTR mRNAs and by upregulating cell surface SSTRs. *Am. J. Physiol. Regul. Integr. Comp. Physiol*, 291: R163-169.

Nordgarden U, Fjelldal P G, Hansen T, Björnsson B T, Wargelius A. 2006. Growth hormone and insulin-like growth factor-I act together and independently when regulating growth in vertebral and muscle tissue of Atlantic salmon postsmolts. *Gen. Comp. Endocrinol*, 149: 253-260.

Olivereau M, Callard G. 1985. Distribution of cell types and aromatase activity in the sculpin (*Myoxocephalus*) pituitary. *Gen. Comp. Endocrinol*, 58: 280-290.

Onuma T, Ando H, Koide N, Okada H, Urano A. 2005. Effects of salmon GnRH and sex steroid hormones on expression of genes encoding growth hormone/prolactin/somatolactin family hormones and a pituitary-specific transcription factor in masu salmon pituitary cells *in vitro*. *Gen. Comp. Endocrinol*, 143: 129-141.

Osugi T, Ukena K, Sower S A, Kawauchi H, Tsutsui K. 2006. Evolutionary origin and divergence of PQRFamide peptides and LPXRFamide peptides in the RFamide peptide family. Insights from novel lamprey RFamide peptides. *FEBS J*, 273: 1731-1743.

Palevitch O, Kight K, Abraham E, Wray S, Zohar Y, Gothilf Y. 2007. Ontogeny of the GnRH systems in zebrafish brain: *In situ* hybridization and promoter-reporter expression analyses in intact animals. *Cell Tissue Res*, 327: 313-322.

Pedroso F L, de Jesus-Ayson E G, Cortado H H, Hyodo S, Ayson F G. 2006. Changes in mRNA expression of grouper (*Epinephelus coioides*) growth hormone and insulin-like growth factor I in response to nutritional status. *Gen. Comp. Endocrinol*, 145: 237-246.

Pelis R M, McCormick S D. 2001. Effects of growth hormone and cortisol on Na^+-K^+-$2Cl^-$ cotransporter localization and abundance in the gills of Atlantic salmon. *Gen. Comp. Endocrinol*, 124: 134-143.

Peng C, Peter R E. 1997. Neuroendocrine regulation of growth hormone secretion and growth in fish. *Zool. Stud*, 36: 79-89.

Peter R E, Chang J P. 1999. Brain regulation of growth hormone secretion and food intake in fish. In: Rao P D P, Peter R E, Ed. *Regulation of the Vertebrate Endocrine System.* Plenum: New York, 55-67.

Peterson B C, Small B C. 2005. Effects of exogenous cortisol on the GH/IGF-I/IGFBP network in channel catfish. *Domest. Anim. Endocrinol*, 28: 391-404.

Peyon P, Vega-Rubín de Celis S, Gómez-Requeni P, Zanuy S, Pérez-Sánchez J, Carrillo M. 2003. *In vitro* effect of leptin on somatolactin release in the European sea bass (*Dicentrarchus labrax*): Dependence on the reproductive status and interaction with NPY and GnRH. *Gen. Comp. Endocrinol*, 132: 284-292.

Pierce A L, Fukada H, Dickhoff W W. 2005. Metabolic hormones modulate the effect of growth hormone (GH) on insulin-like growth factor-I (IGF-I) mRNA level in primary culture of salmon hepatocytes. *J. Endocrinol*, 184: 341-349.

Pierce A L, Fox B K, Davis L K, Visitacion N, Kitahashi T, Hirano T, Grau E G. 2007. Prolactin receptor, growth hormone receptor, and putative somatolactin receptor in Mozambique tilapia: Tissue specific expression and differential regulation by salinity and fasting. *Gen. Comp. Endocrinol*, 154: 31-40.

Ran X Q, Li W S, Lin H R. 2004. Stimulatory effects of gonadotropin-releasing hormone and dopamine on growth hormone release and growth hormone mRNA expression in *Epinephelus coioides*. *Sheng Li Xue Bao*, 56: 644-650.

Rousseau K, Dufour S. 2004. Phylogenetic evolution of the neuroendocrine control of growth hormone: Contribution from teleosts. *Cybium*, 28: 181-198.

Rousseau K, Huang Y S, Le Belle N, Vidal B, Marchelidon J, Epelbaum J, Dufour S. 1998. Long-term inhibitory effects of somatostatin and insulin-like growth factor 1 on growth hormone release by serum-free primary culture of pituitary cells from European eel (*Anguilla anguilla*). *Neuroendocrinology*, 67: 301-309.

Rousseau K, Le Belle N, Marchelidon J, Dufour S. 1999. Evidence that corticotropin-releasing hormone acts as a growth hormone-releasing factor in a primitive teleost, the European eel (*Anguilla anguilla*). *J. Neuroendocrinol*, 11: 385-392.

Rousseau K, Le Belle N, Pichavant K, Marchelidon J, Chow B K, Boeuf G, Dufour S. 2001. Pituitary growth hormone secretion in the turbot, a phylogenetically recent teleost, is regulated by a species-specific pattern of neuropeptides. *Neuroendocrinology*, 74: 375-385.

Rousseau K, Le Belle N, Sbaihi M, Marchelidon J, Schmitz M, Dufour S. 2002. Evidence for a negative feedback in the control of eel growth hormone by thyroid hormones. *J. Endocrinol*, 175: 605-613.

Sakai M, Kajita Y, Kobayashi M, Kawauchi H. 1997. Immunostimulating effect of growth hormone: *In-vivo* administration of growth hormone in rainbow trout enhances resistance to *Vibrio anguillarum* infection. *Vet. Immunol. Immunopathol*, 57: 147-152.

Sakamoto T, McCormick S D. 2006. Prolactin and growth hormone in fish osmoregulation. *Gen. Comp. Endocrinol*, 147: 24-30.

Saligaut C, Garnier D H, Bennani S, Salbert G, Bailhache T, Jego P. 1992. Effects of estradiol on brain aminergic turnover of the female rainbow trout (*Oncorhynchus mykiss*) at the beginning of vitellogenesis. *Gen. Comp. Endocrinol*, 88: 209-216.

Sawisky G R, Chang J P. 2005. Intracellular calcium involvement in pituitary adenylate cyclase-activating polypeptide stimulation of growth hormone and gonadotropin secretion in goldfish pituitary cells. *J. Neuroendocrinol*, 17: 353-371.

Shrimpton J M, McCormick S D. 1998. Regulation of gill cytosolic corticosteroid receptors in juvenile Atlantic salmon: Interaction effects of growth hormone with prolactin and triiodothyronine. *Gen. Comp. Endocrinol*, 112: 262-274.

Shved N, Berishvili G, DCotta H, Baroiller J F, Segner H, Eppler E, Reinecke M. 2007. Ethinylestradiol differentially interferes with IGF-I in liver and extrahepatic sites during development of male and female bony fish. *J. Endocrinol*, 195: 513-523.

Silver M R, Kawauchi H, Nozaki M, Sower S A. 2004. Cloning and analysis of the lamprey GnRH-III cDNA from eight species of lamprey representing the three families of Petromyzoniformes. *Gen. Comp. Endocrinol*, 139: 85-94.

Sloley B D, Trudeau V L, Dulka J G, Peter R E. 1991. Selective depletion of dopamine in the goldfish pituitary caused by domperidone. *Can. J. Physiol. Pharmacol*, 69: 776-781.

Small B C, Nonneman D. 2001. Sequence and expression of a cDNA encoding both pituitary adenylate cyclase activating polypeptide and growth hormone-releasing hormone-like peptide in channel catfish (*Ictalurus punctatus*). *Gen. Comp. Endocrinol*, 122: 354-363.

Smith A, Chan S J, Gutiérrez J. 2005. Autoradiographic and immunohistochemical localization of insulin-like growth factor-I receptor binding sites in brain of the brown trout, *Salmo trutta*. *Gen. Comp. Endocrinol*, 141: 203-213.

Somoza G M, Peter R E. 1991. Effects of serotonin on gonadotropin and growth hormone release from *in vitro* perifused goldfish pituitary fragments. *Gen. Comp. Endocrinol*, 82: 103-110.

Stefano A V, Vissio P G, Paz D A, Somoza G M, Maggese M C, Barrantes G E. 1999. Colocalization of GnRH binding sites with gonadotropin-, somatotropin-, somatolactin-, and prolactin-expressing pituitary cells of the pejerrey, *Odontesthes bonariensis*, in vitro. *Gen. Comp. Endocrinol*, 116: 133-139.

Sundstrom L F, Lohmus M, Johnsson J I, Devlin R H. 2004. Growth hormone transgenic salmon pay for growth potential with increased predation mortality. *Proc. Biol.*

Sci,. 271 (Suppl 5): S350-352.

Sze K H, Zhou H, Yang Y, He M, Jiang Y, Wong A O. 2007. Pituitary adenylate cyclase-activating polypeptide (PACAP) as a growth hormone (GH) -releasing factor in grass carp: II. Solution structure of a brain-specific PACAP by nuclear magnetic resonance spectroscopy and functional studies on GH release and gene expression. *Endocrinology*, 148: 5042-5059.

Tam J K, Lee L T, Chow B K. 2007. PACAP-related peptide (PRP) - molecular evolution and potential functions. *Peptides*, 28: 1920-1929.

Taniyama S, Kitahashi T, Ando H, Kaeriyama M, Zohar Y, Ueda H, Urano A. 2000. Effects of gonadotropin-releasing hormone analog on expression of genes encoding the growth hormone/prolactin/somatolactin family and a pituitary-specific transcription factor in the pituitaries of prespawning sockeye salmon. *Gen. Comp. Endocrinol*, 118: 418-424.

Tipsmark C K, Strom C N, Bailey S T, Borski R J. 2008. Leptin stimulates pituitary prolactin release through an extracellular signal-regulated kinase-dependent pathway. *J. Endocrinol*, 196: 275-281.

Tostivint H, Lihrmann I, Vaudry H. 2008. New insight into the molecular evolution of the somatostatin family. *Mol. Cell. Endocrinol*, 286: 5-17.

Trudeau V L, Somoza G M, Nahorniak C, Peter R E. 1992. Interactions of estradiol with gonadotropin-releasing hormone and thyrotropin-releasing hormone in the control of growth hormone secretion in the goldfish. *Neuroendocrinology*, 56: 483-490.

Trudeau V L, Sloley B D, Wong A O, Peter R E. 1993. Interactions of gonadal steroids with brain dopamine and gonadotropin-releasing hormone in the control of gonadotropin-II secretion in the goldfish. *Gen. Comp. Endocrinol*, 89: 39-50.

Trudeau V L, Sloley B D, Peter R E. 1993. Norepinephrine turnover in the goldfish brain is modulated by sex steroids and GABA. *Brain Res*, 624: 29-34.

Trudeau V L, Sloley B D, Kah O, Mons N, Dulka J G, Peter R E. 1996. Regulation of growth hormone secretion by amino acid neurotransmitters in the goldfish (I): Inhibition by N-methyl-D, L-aspartic acid. *Gen. Comp. Endocrinol*, 103: 129-137.

Trudeau V L, Kah O, Chang J P, Sloley B D, Dubourg P, Fraser E J, Peter R E. 2000. The inhibitory effects of γ-aminobutyric acid (GABA) on growth hormone secretion in the goldfish are modulated by sex steroids. *J. Exp. Biol*, 203: 1477-1485.

Trudeau V L, Spanswick D, Fraser E J, Lariviere K, Crump D, Chiu S, MacMillan M, Schulz R W. 2000. The role of amino acid neurotransmitters in the regulation of pituitary gonadotropin release in fish. *Biochem. Cell Biol*, 78: 241-259.

Tsai C L, Wang L H, Chang C F, Kao C C. 2000. Effects of gonadal steroids on brain serotonergic and aromatase activity during the critical period of sexual differentiation in tilapia, *Oreochromis mossambicus*. *J. Neuroendocrinol*, 12: 894-898.

Uchida K, Yoshikawa-Ebesu J S, Kajimura S, Yada T, Hirano T, Gordon Grau E. 2004. *In vitro* effects of cortisol on the release and gene expression of prolactin and growth hormone in the tilapia, *Oreochromis mossambicus*. *Gen. Comp. Endocrinol*, 135: 116-125.

Unniappan S, Peter R E. 2005. Structure, distribution and physiological functions of ghrelin in fish. *Comp. Biochem. Physiol. A Mol. Integr. Physiol*, 140: 396-408.

Volkoff H. 2006. The role of neuropeptide Y, orexins, cocaine and amphetamine-related transcript, cholecystokinin, amylin and leptin in the regulation of feeding in fish. *Comp. Biochem. Physiol. A Mol. Integr. Physiol*, 144: 325-331.

Volkoff H, Peter R E. 1999. Actions of two forms of gonadotropin releasing hormone and a GnRH antagonist on spawning behavior of the goldfish *Carassius auratus*. *Gen. Comp. Endocrinol*, 116: 347-355.

Volkoff H, Eykelbosh A J, Peter R E. 2003. Role of leptin in the control of feeding of goldfish: Interactions with cholecystokinin, neuropeptide Y and orexin-A, and modulation by fasting. *Brain Res*, 972: 90-109.

Volkoff H, Canosa L F, Unniappan S, Cerdá-Reverter J M, Bernier N J, Kelly S P, Peter R E. 2005. Neuropeptides and the control of food intake in fish. *Gen. Comp. Endocrinol*, 142: 3-19.

Wang X, Chu M M, Wong A O. 2007. Signaling mechanisms for alpha2-adrenergic inhibition of PACAP-induced growth hormone secretion and gene expression grass carp pituitary cells. *Am. J. Physiol. Endocrinol. Metab*, 292: E1750-1762.

Wang Y, Wong A O, Ge W. 2003. Cloning, regulation of messenger ribonucleic acid expression, and function of a new isoform of pituitary adenylate cyclase-activating polypeptide in the zebrafish ovary. *Endocrinology*, 144: 4799-4810.

Weber G M, Grau E G. 1999. Changes in serum concentrations and pituitary content of the two prolactins and growth hormone during the reproductive cycle in female tilapia, *Oreochromis mossambicus*, compared with changes during fasting. *Comp. Biochem. Physiol. C Pharmacol. Toxicol. Endocrinol*, 124: 323-335.

Wei Y, Martin S C, Heinrich G, Mojsov S. 1998. Cloning and functional characterization of PACAP-specific receptors in zebrafish. *Ann. N. Y. Acad. Sci*, 865: 45-48.

Weil C, Carre F, Blaise O, Breton B, Le Bail P Y. 1999. Differential effect of insulin-like growth factor I on *in vitro* gonadotropin and growth hormone secretions in rainbow trout (*Oncorhynchus mykiss*) at different stages of the reproductive cycle. *Endo-

crinology, 140: 2054-2062.

Wilkinson R J, Porter M, Woolcott H, Longland R, Carragher J F. 2006. Effects of aquaculture related stressors and nutritional restriction on circulating growth factors (GH, IGF-I and IGF-II) in Atlantic salmon and rainbow trout. *Comp. Biochem. Physiol. A Mol. Integr. Physiol*, 145: 214-224.

Wong A O, Le Drean Y, Liu D, Hu Z Z, Du S J, Hew C L. 1996. Induction of chinook salmon growth hormone promoter activity by the adenosine 3′, 5′-monophosphate (cAMP)-dependent pathway involves two cAMP-response elements with the CGTCA motif and the pituitary-specific transcription factor Pit-1. *Endocrinology*, 137: 1775-1784.

Wong A O, Li W, Leung C Y, Huo L, Zhou H. 2005. Pituitary adenylate cyclase-activating polypeptide (PACAP) as a growth hormone (GH) -releasing factor in grass carp. I. Functional coupling of cyclic adenosine 3′, 5′-monophosphate and Ca^{2+}/calmodulin-dependent signaling pathways in PACAP-induced GH secretion and GH gene expression in grass carp pituitary cells. *Endocrinology*, 146: 5407-5424.

Wong A O, Zhou H, Jiang Y, Ko W K. 2006. Feedback regulation of growth hormone synthesis and secretion in fish and the emerging concept of intrapituitary feedback loop. *Comp. Biochem. Physiol. A Mol. Integr. Physiol*, 144: 284-305.

Wong A O, Chuk M C, Chan H C, Lee E K. 2007. Mechanisms for gonadotropin-releasing hormone potentiation of growth hormone rebound following norepinephrine inhibition in goldfish pituitary cells. *Am. J. Physiol. Endocrinol Metab*, 292: E203-214.

Wong A O L, Chang J P, Peter R E. 1993. *In vitro* and *in vivo* evidence that dopamine exerts growth hormone releasing activity in the goldfish, *Carassius auratus*. *Am. J. Physiol*, 264: E925-932.

Wong A O L, Murthy C K, Chang J P, Neumann C M, Lo A, Peter R E. 1998. Direct actions of serotonin on gonadotropin-II and growth hormone release from goldfish pituitary cells: Interactions with gonadotropin-releasing hormone and dopamine and further evaluation of serotonin receptor specificity. *Fish Physiol. Biochem*, 19: 22-34.

Wong A O L, Li W S, Lee E K Y, Leung M Y, Tse L Y, Chow B K C, Lin H R, Chang J P. 2000. Pituitary adenylate cyclase-activating polypeptide as a novel hypophysiotropic factor in fish. *Biochem. Cell Biol*, 78: 329-343.

Wong C J, Johnson J D, Yunker W K, Chang J P. 2001. Caffeine stores and dopamine differentially require Ca^{2+} channels in goldfish somatotropes. *Am. J. Physiol. Regul. Integr. Comp. Physiol*, 280: R494-R503.

Xiao D, Lin H R. 2003. Effects of cysteamine - a somatostatin-inhibiting agent - on serum growth hormone levels and growth in juvenile grass carp (*Ctenopharyngodon idellus*). *Comp. Biochem. Physiol. A Mol. Integr. Physiol*, 134: 93-99.

Xing Y, Wensheng L, Haoran L. 2005. Polygenic expression of somatostatin in orange-spotted grouper (*Epinephelus coioides*): Molecular cloning and distribution of the mRNAs encoding three somatostatin precursors. *Mol. Cell. Endocrinol*, 241: 62-72.

Yacobovitz M, Solomon G, Gusakovsky E E, Levavi-Sivan B, Gertler A. 2008. Purification and characterization of recombinant pufferfish (*Takifugu rubripes*) leptin. *Gen. Comp. Endocrinol*, 156: 83-90.

Yada T. 2007. Growth hormone and fish immune system. *Gen. Comp. Endocrinol*. 152: 353-358.

Yadetie F, Male R, 2002: Effects of 4-nonylphenol on gene expression of pituitary hormones in juvenile Atlantic salmon (*Salmo salar*). *Aquat. Toxicol*, 58: 113-129.

Yeung C M, Chan C B, Woo N Y, Cheng C H. 2006. Seabream ghrelin: CDNA cloning, genomic organization and promoter studies. *J. Endocrinol*, 189: 365-379.

Yu Y, Wong A O, Chang J P. 2008. Serotonin interferes with Ca^{2+} and PKC signaling to reduce gonadotropin-releasing hormone-stimulated GH secretion in goldfish pituitary cells. *Gen. Comp. Endocrinol*, 159: 58-66.

Yuen C W, Ge W. 2004. Follistatin suppresses FSHbeta but increases LHbeta expression in the goldfish - evidence for an activin-mediated autocrine/paracrine system in fish pituitary. *Gen. Comp. Endocrinol*, 135: 108-115.

Yunker W K, Chang J P. 2001. Somatostatin actions on a protein kinase C-dependent growth hormone secretagogue cascade. *Mol. Cell. Endocrinol*, 175: 193-204.

Yunker W K, Chang J P. 2004. Somatostatin-14 actions on dopamine- and pituitary adenylate cyclase-activating polypeptide-evoked Ca^{2+} signals and growth hormone secretion. *J. Neuroendocrinol*, 16: 684-694.

Yunker W K, Smith S, Graves C, Unniappan S, Rivier J E, Peter R E, Chang J P. 2003. Endogenous hypothalamic somatostatins differentially regulate growth hormone secretion from goldfish pituitary somatotropes *in vitro*. *Endocrinology*, 144: 4031-4041.

Zbikowska H M. 2003. Fish can be first - advances in fish transgenesis for commercial applications. *Transgenic Res*, 12: 379-389.

Zhang W M, Lin H R, Peter R E. 1994. Episodic growth hormone secretion in the grass carp, *Ctenopharyngodon idellus*. *Gen. Comp. Endocrinol*, 95: 337-341.

Zou J J, Trudeau V L, Cui Z, Brechin J, Mackenzie K, Zhu Z, Houlihan D F, Peter R E. 1997. Estradiol stimulates growth hormone production in female goldfish. *Gen. Comp. Endocrinol*, 106: 102-112.

第5章 神经内分泌调控鱼类催乳激素和生长乳素的分泌活动

本章介绍神经内分泌调控硬骨鱼类两种同源的脑垂体激素——催乳激素（prolactin，PRL）和生长乳素（somatolactin，SL）的合成与分泌活动。PRL 和 SL 分别在硬骨鱼类腺脑垂体的吻端远侧部（RPD）和中间部（PI）产生。PRL 和 SL 合成与分泌活动受到两个拮抗性的下丘脑神经激素的调控。每个下丘脑神经激素的特异性作用由特异性的 G-蛋白偶联受体所决定。PRL 和 SL 通过特异性詹纳斯激酶（Janus Kinase，JAK）联系的受体而参与许多生理功能。本章简要归纳了这些激素和它们受体的分子特征和生物学功能，并对下丘脑刺激性和抑制性因子的作用模式，包括我们最近发现的 PRL-释放肽（PrRP）的作用做了论述。

5.1 导 言

催乳激素（PRL）与生长乳素（SL）和生长激素一起组成属于 I 级螺旋细胞因子超家族（helical cytokines superfamily）中的一个脑垂体激素家族。这个超家族中的所有成员都有一个相似的三维折叠进入一束四个 α-螺旋中，信号通过相联系的受体并激活一个相似的细胞内信号级联放大（signaling cascade）（Huising 等，2006）。PRLs 和 GHs 已经在脊椎动物所有各纲的代表种类中鉴别出来，除了 PRL 不存在于软骨鱼类和无颚类之外；而 SL 只存在于硬骨鱼类（Kawauchi 等，2002）。在硬骨鱼类中，PRL、GH 和 SL 细胞分布于脑垂体的三个不同的区，PRL 在吻端远侧部（RPD），GH 在近端远侧部（PPD），SL 在中间部（PI）。

最早是在雌兔的乳腺中确认 PRL 的功能是生乳作用，并因此而命名。此后，在各种不同的脊椎动物相关研究中曾报道 PRL 数量不断增多的生物学功能。按照 Nicoll（1993）、Bole-Feysot 等（1998）对 PRL 功能所做的综述，可把多达 300 种功能归纳为七类：①水和电解质平衡；②生长和发育；③内分泌和代谢；④脑和行为；⑤生殖；⑥免疫调节和保护；⑦和疾病状态的病理学相联系的作用。PRL 的渗透压调节作用对于栖息在不同盐度水域中的鱼类特别重要。GHs 的主要作用是通过胰岛素样生长因子（IGF）的介导而刺激脊椎动物的生长，并促进性成熟和生殖功能，还能调节一些硬骨鱼类对海水的适应性。SL 是鱼类特有的脑垂体激素，具有许多和 PRL 相似的生理学作用，包括对背景的适应性、能量稳态、脂肪代谢、渗透压调节、应激反应、生殖活动的某些方面以及酸-碱调节等。PRL、SL 和 GH 通过和它们特异性单个跨膜-区段受体的相互作用和二聚作用（dimerizing）而发挥它们的生物学作用。这些受体属于细胞因子 I 级受体超家族，没有内在的酪蛋白激酶，使用 JAK（詹纳斯激酶）-STAT（信号转导和转录激活蛋白，signal transducers and activators of transcription）的级联作为一个主要的信号通路（Bole-Feysot 等，1998）。因此，这些激素和它们的受体可以认为是协同进化过程中在基因复制及其后趋异（divergence）的结果。

脑垂体激素的合成与分泌活动受到下丘脑神经激素的调控，这些激素通常分为两类：释放激素和抑制激素。每一种下丘脑神经激素的特异性作用是由对脑垂体某种靶细胞的高度特异性受体所决定的。这些受体都属于 G-蛋白偶联受体（GPCR），它们具有七个跨膜区并结合 G-蛋白以起激活或抑制作用。

本章将概括介绍 PRL 和 SL 的分子特征和生物学功能，并综述下丘脑神经激素对硬骨鱼类 PRL 和 SL 分泌活动的调控作用。

5.2 催乳激素分泌活动的神经内分泌调控

5.2.1 催乳激素及其受体

5.2.1.1 催乳激素

催乳激素（prolactin，PRL）是一个单链的多肽，在有颌类脑垂体的远侧部产生；无颌类没有 PRL（Kawauchi 等，2002）。大多数硬骨鱼类的 PRL 细胞都聚集成一个几乎完全同源的团块，位于脑垂体 RPD 的最前部，只有少量其他类群的内分泌细胞，即肾上腺皮质激素细胞（ACTH）沿着神经垂体的边缘分布。所以，这一分布区可以分离出来而得到一群完全同源的 PRL 细胞用于孵育或灌流，以便进行离体的孵育实验，研究神经激素和其他因子的作用。

硬骨鱼类 PRL 基因已在鲤鱼（*Cyprinus carpio*，X52881）、大麻哈鱼（*Oncorhynchus tshawytscha*，S66606）、莫桑比克罗非鱼（*Oreochromis mossambicus*，X92380）和金头鲷（*Sparus aurata*，AJ509807）中鉴别。所有已知的 PRL 基因都由 5 个外显子和 4 个内含子组成，与 SLs 和主要的 GHs 一样。只有高等硬骨鱼类的 GHs 由 6 个外显子组成，在硬骨鱼类多样化（diversification）过程中，其中第 5 个内含子插入到其他 GH 基因的第 5 个外显子中。硬骨鱼类 PRL 基因的 TATA 框（TATA box）大约 20 bp，上游从转录起始位点（transcriptional initiation site）开始。在虹鳟（*Oncorhynchus mykiss*，X95907）、罗非鱼（X92380）和金头鲷（AJ509807）PRL 基因的 5′-侧翼区（flanking region）已经检测到几个 Pit-1/GHF-1 结合的契合序列（consensus sequence）。

已经从超过 30 种硬骨鱼类中获得了鱼类 PRL 的蛋白质和 cDNA 序列。有些鱼类曾被确定 PRL 有两个结构变体，它们只在鲑鳟鱼类和鲤科鱼类中有少数氨基酸置换而有所不同，表明在这两个同种型（isoform）之间没有功能的不同。相反，罗非鱼的两个 PRL，即 PRL_{177}（P09318）和 PRL_{188}（P09319）的键长度不同，且只有 69% 的同一性。PRL_{188} 要比 PRL_{177} 和其他鱼类的 PRL 更相似些。所以，一些功能的差别就显得比较明显。在氨基酸序列的比较方面，硬骨鱼类 PRL 最明显的特征是缺少一个四足类、肺鱼（*Protopterus aethiopicus*）和鲟鱼（*Acipenser gueldenstaedtii*）都具有的 N-端二硫键，表明辐鳍鱼类在趋异过程中丢失了这个区段。

采用逆转录酶-多聚酶链式反应（reverse transcriptase-polymerase chain reaction，RT-PCR）在脑垂体外组织中，例如金头鲷的肝脏、肠道和性腺而不是脑（Santos等，1999），金鱼（Imaoka等，2000）和虹鳟（Sakamoto等，2003a）的肝脏、肾脏、脾脏、鳃、肌肉、性腺和脑而不是肠，都曾证明PRL基因的表达。PRL可能在这些脑垂体以外的组织中起着自分泌或旁分泌的作用，这是值得今后研究的领域。

5.2.1.2 PRL 受体

PRL和细胞表面的受体结合，这类受体是细胞因子/血细胞生成素受体（cytokine/hematopoietin receptor）超家族的成员。哺乳类有长的、中间的和短的PRLR类型，它们由单个PRLR基因通过可变剪接（alternative splicing）和启动子的使用而产生（Freeman等，2000）。这些同种型在长度、细胞内区域的组成以及信号转导作用机理方面有所不同。和配体结合后引发由一个PRL分子和两个受体分子组成同型二聚体（homodimer）。在PRLR的细胞外区域，对于合适的折叠和受体的运送（trafficking），需要两对二硫键和一个Trp-Ser-Xaa-Trp-Ser模体（WS模体），虽然它们不是结合本身所需求的（Bole-Feysot等，1998；Freeman等，2000）。在细胞内区域，有两个相对保守的区，称为box1和box2。Box1是近膜的（membrane-proximal）、富含脯氨酸的模体，它是转导分子所识别的分子契合折叠（consensus folding）所需要的。Box2比较不保守，在短型的催乳激素受体中是缺失的。PRL和一个受体分子结合是位点1，而第二个受体分子和PRL结合是位点2。在这些位点上，PRLR分子和细胞质的酪氨酸激酶、JAK2以及另一个激酶如STAT相联系。受体的二聚化（dimerization）诱导Jak激酶活化，接着是受体的磷酸化作用（phosphorylation）和STAT。磷酸化的STAT二聚化并传导到核而和PRL-反应基因上的特异性启动子元件（promoter element）结合。此外，PRL亦激活Ras/Ref/MAP激酶通路，它可能参与PRL的增生效应（proliferative effect）。（见图5.1）

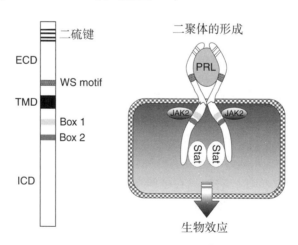

图5.1 PRLR（左侧）和PRL与PRLR作用机理（右侧）的代表性图解

PRL诱导两个PRL受体二聚化。ECD，细胞外区域；TMD，跨膜区域；ICD，细胞内区域。PRL和PRL二聚体的受体结合通过JAKs（詹纳斯激酶）的活化而激活转录因子的STAT（信号转导及转录激活蛋白）级别。配体引起受体的二聚化从而诱导和JAKs相关联的交互酪氨酸磷酸化作用，使磷酸化的酪氨酸残基处在受体的胞质尾区（cytoplasmic tail）。

在硬骨鱼类中，PRLRs 已经在罗非鱼（Sandra 等，1995；Prunet 等，2000）、金鱼（Tse 等，2000）、虹鳟（Prunet 等，2000）、金头鲷（Santos 等，2001）、牙鲆（*Paralichthys olivaceus*）（Higashimoto 等，2001）、鲤鱼（San Martin 等，2004）和东方红鳍鲀（*Takifugu rubripes*；Lee 等，2006）中克隆出来。多个 PRLR mRNA 已在金鱼（Tse 等，2000）和金头鲷（Santos 等，2001）中检测到。San Martin 等（2007）首次鉴定鲤鱼催乳激素受体基因，它表达转录体同种型，编码短型的 PRLR，但和人类与啮齿类的不同，并不和交替的启动子（alternative promoter）相联系。最近 Huang 等（2007）在河鲀和斑马鱼（*Danio rerio*）的基因组中而不是在其他脊椎动物基因组中鉴别出两个 PRLR 基因，并从黑棘鲷（*Acanthopagrus schlegeli*）和尼罗罗非鱼（*Oreochromis niloticus*）中分离出两个 PRLR cDNA。传统的一个命名为 PRLR1，而新鉴定的命名为 PRLR2。这两个 PRLRs 都和哺乳类长型的 PRLRs 相似，但只有约 30% 的相似性。这些受体的差别主要是在细胞内区域的 box2 以及在 PRLR2 的细胞内区缺乏酪氨酸残基。硬骨鱼类的 PRLRs 和哺乳类的一样都具有重要的信号转导的结构特征，而鱼类如金鱼（Tse 等，2000）和金头鲷（Santos 等，2001）的 PRLRs 的 WS 模体在第 5 位有一个氨基酸置换（丝氨酸换苏氨酸）。

在鱼类渗透压调节的组织中检测到 PRLR 较高水平的表达，如罗非鱼（Sandra 等，1995，2000；Prunet 等，2000；Pierce 等，2007）、金鱼（Tse 等，2000）、虹鳟（Prunet 等，2000）、金头鲷（Santos 等，2001；Huang 等，2007）、牙鲆（Higashimoto 等，2001）、河鲀（Lee 等，2006）和大西洋鲑鱼（*Salmo salar*）（Kiilerich 等，2007）等的鳃、肾脏和肠。在有些鱼类的脑、性腺、肝脏、肌肉、皮肤、脾脏、头肾、淋巴细胞和骨骼中亦检测到较低水平的表达（Sandra 等，1995，2000；Tse 等，2000；Higashimoto 等，2001；Santos 等，2001；Huang 等，2007）。在黑棘鲷的大多数组织中，PRLR1 的表达水平都较高于 PRLR2，只有鳃是例外。在金头鲷的肾脏中，PRLR1 的表达为雌二醇和皮质醇上调，但不受睾酮的影响；而 PRLR2 的表达为雌二醇和睾酮下调，但不受皮质醇的影响（Huang 等，2007）。鱼类 PRLRs 在身体组织中的广泛分布以及和类固醇激素不同的基因表达型式是和 PRL 的多种不同功能紧密联系的。

5.2.2 催乳激素的功能

5.2.2.1 水和电解质平衡

Pickford 和 Phillips（1959）首先证明施用羊的 PRL 能使切除脑垂体的青鳉（*Fundulus heteroclitus*）在淡水中存活。此后，有学者曾经证明 PRL 通过刺激离子保留和防止水分流入渗透压调节器官而对许多硬骨鱼类适应淡水生活起了关键作用。主要的渗透压调节器官是鳃、鳃盖膜、皮肤、胃肠道、肾脏和膀胱。驯养在淡水中的鱼的 PRL 细胞活性要较高于在海水中的鱼，而当水介质中渗透压降低时，PRL 从离体脑垂体的释放增加（Nagahama 等，1975；Grau 等，1994）。由于在这个领域已发表许多出色的综述文章，我们在此不予重述，但仍向读者们推荐下列的综述：Hirano，1986；Bern 和

Madsen, 1992; McCormick, 2001; Manzon, 2002; Sakamoto 和 McCormick, 2006。最近的研究表明 PRLR 在主要渗透压调节器官的表达水平比较高（Sandra 等, 2000; Prunet 等, 2000; Tse 等, 2000; Higashimoto 等, 2001; Santos 等, 2001; Pierce 等, 2007）。驯养在海水中的罗非鱼, PRLR 在鳃中的表达要低于驯养在淡水中的罗非鱼（Shiraishi 等, 1999; Prunet 等, 2000; Pierce 等, 2007）。广盐性的硬骨鱼类可能都具有 PRL 和 PRLR 系统, 使它们能够适应环境中渗透压的忽然变化, 随时准备在渗透压调节器官中表达 PRLRs。

5.2.2.2 生长和发育

PRL 的促生长作用在哺乳类的相关研究中已有很多报道, 但在鱼类的研究中还比较少涉及。对于罗非鱼（Shepherd 等, 1997）, 是 PRL_{177} 而不是 PRL_{188} 具有促进生长的作用, 其证据是增加 [H^3] 胸腺嘧啶核苷（thymidine）和 [S^{35}]-硫酸盐结合到鳃软骨中以及刺激肝脏 IGF-I mRNA 的表达。此外, 放射受体测定（radio-receptor assay）表明 PRL_{177} 能够在罗非鱼 GH 的受体中置换 GH, 但 PRL_{188} 没有这个作用。Sandra 等（1995）证明基因重组的 PRLR 对 PRL_{188} 比对 PRL_{177} 有较强的亲和力。Leona 等（2001）证明 PRL 能抑制攀鲈（*Anabas testudineys*）肝脏参与脂肪酸生物合成的几种酶类活性, 而用 PRL 处理银大麻哈幼鱼能导致明显的脂类耗尽（Sheridan, 1986）。

PRL 对哺乳类早期发育的作用已充分阐明, PRL 对硬骨鱼类的早期发育亦可能有影响。在刚孵化的罗非鱼幼鱼中可检测到 PRL_{188} 的基因转录体（gene transcript）, 而在孵化前一天胚胎的脑垂体中能检测到 PRL_{177} 的基因转录体（Ayson 等, 1994）。在虹鳟胚胎发育过程中（脑垂体器官生成前和之后）和孵化后都能检测到 PRL 基因（Yang 等, 1999）。在金头鲷的胚胎和孵化后幼鱼中可检测到 PRL mRNA; 而在胚胎的囊胚期首次检测到 PRLR 基因, 它在孵化后 2 天还一直保持较高的水平（Santos 等, 2003）。这些研究结果表明 PRL 存在于鱼类胚胎和幼鱼的正在发育的脑垂体内, 而 PRLR mRNA 和其蛋白质亦存在于鱼类胚胎中, 并且在幼鱼组织中广泛分布（Power, 2005）。这些研究结果证明了 PRL 对硬骨鱼类的促生长作用。

催乳激素对两栖类的蝌蚪有抗变态的（antimetamorphic）作用。在牙鲆的相关研究中, 羊 PRL 能够抵消三碘甲腺原氨酸（T_3）刺激变态前幼鱼背鳍鳍条的吸附作用, 并且亦会延缓背鳍鳍条的解吸附作用（desorption）而不影响眼睛迁移和下沉栖息的效率。但是, 羊 GH 没有这些作用。在持续的变态期间, PRL 和 GH 基因都增加表达。这些研究结果表明 PRL 和甲状腺素都参与牙鲆发育的调控作用（De Jesus 等, 1994）。

5.2.2.3 免疫调控作用

内分泌-免疫相互作用已经成为硬骨鱼类重要的研究领域之一。皮质醇、性类固醇和脑垂体激素包括 PRL 都已证明能影响一些硬骨鱼类的免疫功能（Harris 和 Bird, 2000）。PRL 能刺激白细胞的有丝分裂发生（mitogenesis）（Sakai 等, 1996a; Yada 等, 2002a）、呼吸爆发活动（Sakai 等, 1996b）和吞噬作用（phagocytosis）（Kajita 等, 1992; Sakai 等, 1995; Narnaware 等, 1998）, 并能增加血浆 IgM 的滴度（Yada 等,

2002a）。在罗非鱼的脾脏、头肾和血液循环中的淋巴细胞中都能检测到 PRLR mRNA （Sandra 等，2000）。在罗非鱼的淋巴组织和细胞中能检测到两种 PRLs 和 PRL 受体的表达，而且驯养在海水中的鱼的表达水平要高于驯养在淡水中的鱼。这些研究结果表明 PRL 的免疫调节作用似乎并不依赖于其渗透压调节作用。

5.2.2.4 **行为**

目前，学者们对 PRL 对两栖类行为的影响研究得比较清楚，但对鱼类的相关研究还很少。在展现对亲本的扇动行为（fanning behavior）期间，性成熟的雄性三棘刺鱼（*Gasterosteus aculeatus*）脑垂体 PRL 细胞的合成与释放速率，经定量电镜分析是明显增高的（Slijkhuis 等，1984），而且 PRL 能刺激性成熟三棘刺鱼雄鱼的亲本扇动行为（de Ruiter 等，1986）。在对罗非鱼进行的实验中，PRL 能诱导皮肤腺体转化变形（transformation）以产生黏液向幼鱼提供营养（Ogawa 等，1970）。给海水鳗鱼第四脑室注射 PRL 能影响其饮水行为，抑制水分摄入（Kozaka 等，2003）。

5.2.2.5 **生殖**

目前已报道 PRL 刺激卵巢和精巢的类固醇生成（Rubin 和 Specker，1992）。在鲑鳟鱼类的性成熟期间，血浆 PRL 和脑垂体 mRNA 水平升高（Ogasawara 等，1996；Onuma 等，2003）。在罗非鱼性成熟时亦观察到 PRL 水平升高（Tacon 等，2000）。已经证明 GnRH 和 E_2 能刺激鱼类在体和离体的 PRL 释放作用（Weber 等，1997；Kagabu 等，1998）。此外，在有些硬骨鱼类的性腺中能检测到 PRLR mRNA（Sandra 等，2000；Tse 等，2000；Higashimoto 等，2001；Santos 等，2001）。这些研究结果表明 PRL 在鱼类生殖过程的不同时期起着一定的作用。

5.2.3 **下丘脑肽影响催乳激素的分泌活动**

对于硬骨鱼类，下丘脑神经激素、血浆因子和血浆重量摩尔渗透压浓度（osmolality）调控 PRL 从脑垂体分泌的活动。血浆重量摩尔渗透压浓度对 PRL 的分泌起着直接而重要的调控作用，在低渗的情况下刺激 PRL 释放（Kaneko 和 Hirano，1993；Shepherd 等，1999；Weber 等，2004）。从其他组织分泌的几种血浆因子，可以划分为刺激性和抑制性两类。刺激性因子包括 E_2（Barry 和 Grau，1986；Williams 和 Wigham，1994a；Weber 等，1997；Kagabu 等，1998）、血管紧张素 II（angiotensin II）（Eckert 等，2003）、利尿钠肽（natriuretic peptide，Fox 等，2007）和 IGFs（Fruchtman 等，2000）。抑制性因子包括皮质醇（Borski 等，1991；Williams 和 Wigham，1994a；Uchida 等，2004）、乌本苷（ouabain）（Kajimura 等，2005）、硬骨鱼紧张肽 II（urotensin II）（Grau 等，1982；Leedom 等，2003）和血管活性肠多肽（vasoactive intestinal polypeptide，VIP）（Brinca 等，2003）。已经充分阐明多巴胺（DA）和生长抑素（SS）对鱼类 PRL 释放起着负的下丘脑调控作用。此外，一些常见的下丘脑神经肽，如 GnRH、神经肽 Y（NPY）、copeptine 和 PACAP 都曾被报道能增强 PRL 的释放能力。最近，一种新的神经肽称为 PRL-释放肽（PrRP），在哺乳类中发现后我们接着鉴定了硬骨鱼类的

PrRP，并证明这种新的下丘脑神经肽具有特异性的 PRL-释放作用（Moriyama 等，2002，2007）。这里综述下丘脑神经肽包括我们新发现的 PrRP 对 PRL 分泌活动的调控作用（见图 5.2）。

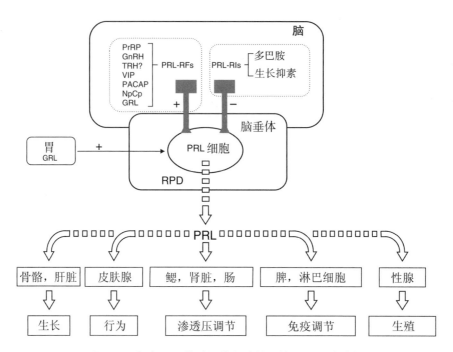

图 5.2　鱼类 PRL 从脑垂体释放的调控和 PRL 的功能

在各种神经肽和神经递质的多功能调控下，PRL 从 PRL-产生细胞中释放出来。这些刺激性/抑制性的调节剂通过从脑/下丘脑的直接神经分布而分送到脑垂体 RPD 的 PRL-产生细胞。PRL-RFs，催乳激素释放因子；PRL-RIs，催乳激素-释放抑制剂；PrRP，催乳激素释放肽；GnRH，促性腺激素释放激素；TRH，促甲状腺激素释放激素；VIP，血管活性肠肽；PACAP，脑垂体腺苷酸环化酶激活多肽；NpCp，Copeptine；GRL，生长素释放肽；RPD，吻端远侧部。

5.2.3.1　抑制性因子

1. 多巴胺（DA）

DA 属于中枢神经系统神经递质中的儿茶酚胺（catecholamine）一类。自 1970 年了解 DA 起着 PRL 释放的抑制剂作用之后，DA 已经被确认为哺乳类和鸟类 PRL 基因表达和 PRL 释放的主要调节剂（Al Kahtane 等，2003）。DA 的两个受体亚型 D_1 和 D_2 受体是两个不同的膜蛋白，属于 GPCR 超家族（Oliveira 等，1994）。D_1 受体和腺苷酸环化酶刺激作用联系，介导 GH 从金鱼脑垂体释放（Wong 等，1993）；而 D_2 受体抑制腺苷酸环化酶活性，从而抑制哺乳类 PRL 释放（Jose 等，1999）。

给金鱼注射 DA 的同功物使 PRL 细胞肥大，细胞核增大。细胞质内的颗粒数量在金鱼的对照组和处理组都一样。罗非鱼经 DA 处理后，PRL 细胞的粗面内质网数量增加而分泌颗粒数量减少（Hazineh 等，1997）。对虹鳟用 DA 和脑垂体一起孵育减少了 PRL

释放到孵育介质中。DA 的前体，L-多巴能降低在体脑垂体的 PRL 含量，而多巴胺的抑制剂能增加脑垂体 PRL 的总含量以及随后进行脑垂体孵育的 PRL 释放量（James 和 Wigham，1984）。这些研究结果表明 DA 抑制硬骨鱼类 PRL 细胞的活性。

Johnston 和 Wigham（1988）曾设想 G_i 蛋白依存的同功物、Ca^{2+}、钙调蛋白和 cAMP 参与虹鳟调控 PRL 的作用机理。GTP 增强 DA 对离体 PRL 释放的抑制作用，而 DA 亦降低脑垂体的 cAMP 含量。毛喉素（forskolin）增加离体的 PRL 释放和 cAMP 含量，但这个作用被 DA 阻止，并且不发生在无 Ca^{2+} 的介质中。cAMP 类似物在低 Ca^{2+} 介质中增加 PRL 合成，但不明显影响释放。钙离子载体（calcium ionophore）增加 PRL 释放，但用电压依赖 Ca^{2+} 通道（voltage-dependent Ca^{2+} channel）的抑制剂并不出现这种反应。钙调蛋白抑制剂能增加在体的 PRL 合成和脑垂体 PRL 含量，并且提高脑垂体 cAMP 水平（Johnston 和 Wigham，1988）。这样，DA 可能激活 PRL 细胞的 D_2 受体，使腺苷酸环化酶活性降低，进而抑制 cAMP 的产生，cAMP 量的减少就会引起 PRL 分泌的抑制作用。

给金鱼注射 DA（0.5～50 μg/g 体重）和用 1～100 μM 浓度进行脑垂体细胞孵育能剂量依存地降低 PRL 基因表达。DA 显著抑制金鱼 PRL 启动子活性。DA 抑制启动子活性的相关反应区可能出现在-188 bp 内，这正是位于金鱼 PRL 基因上的两个推定的 pit-1 结合位点（Tse 等，2008）。这些研究结果证明 DA 起着硬骨鱼类 PRL 基因转录和 PRL 分泌活动的强有力的负调节剂作用。

2. 生长抑素（SS）

许多鱼类已被证明存在多个 SS 基因。SS-14 是保守的，在脊椎动物各类群的代表种类中都具有同一的初级结构。SS 受体（SST）亦属于 GPCR 超家族。鱼类共有 4 个不同的 SST 亚型（Zupanc 等，1999；Slagter 等，2004）。

SS 免疫反应定位于硬骨鱼类的脑和脑垂体内（Dubois 等，1979；Olivereau 等，1984a、b；Grau 等，1985；Marchant 等，1989）。在脑垂体中，SS-免疫反应纤维延伸到鲤鱼、金鱼、罗非鱼、青鳉和弹涂鱼（*Periophthalmus modestus*）PPD 的 GH 细胞（Kah 等，1982；Olivereau 等，1984a、b；Grau 等，1985），亦延伸到舌齿鲈（*Dicentrarchus labrax*）RPD 的 ACTH 细胞（Power 等，1996）。编码鱼类 SSTs 的 mRNA 广泛分布在脑垂体中（Sheridan 等，2000；Slagter 等，2004）。

SS 能抑制罗非鱼（Grau 等，1982，1985；Helms 等，1991）、花鳉（*Poecilia latipinna*）（Wigham 和 Batten，1984）和虹鳟（Williams 和 Wigham，1994b）孵育的脑垂体组织或者 RPD 释放 PRL。将这些硬骨鱼类孵育的脑垂体或者 RPD 置于低渗的介质中会引起新合成的 PRL 急剧释放。把 SS 加入介质中，SS 能以剂量依存的方式迅速阻止或减少 PRL 释放。SS 的拮抗物能阻断 SS 对虹鳟的抑制作用（Williams 和 Wigham，1994b）。

SS 能减弱毛喉素刺激的 cAMP 水平增高，就与它对 PRL 释放的迅速抑制作用一样。SS 可以减弱腺苷酸环化酶对毛喉素直接刺激引起的反应（Helms 等，1991）。在对罗非鱼进行的实验中，SS 对 PRL 释放的抑制作用能为钙离子载体 A23187 完全阻止。PRL

释放亦可以被 Ca^{2+} 从孵育介质中清除而受到抑制，甚至有离子载体时亦一样（Grau 等，1982）。

5.2.3.2 刺激性肽素

1. 催乳激素释放肽（prolactin-relasing peptide，PrRP）

Hinuma 等（1998）从牛的下丘脑提取物中鉴定出一个新的具有促进 PRL 释放活性的下丘脑神经肽，它是孤独 GPCR 的配体。这个神经肽被命名为 PrRP，具有两个分子类型，一个由 31 个氨基酸组成的肽（PrRP31）和 C-端 20 个氨基酸残基（PrRP20），它们属于所谓的"精氨酰-苯丙胺酰-酰胺肽"（arginyl-phenylalanyl-amide peptides，RFa）。至今已在脊椎动物的 6 个纲中鉴定到 PrRP 的同源物（见图 5.3）。

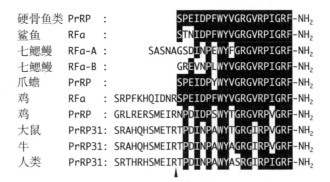

图 5.3　PrRP 及其同源物的排比

最重要的氨基酸和硬骨鱼类 PrRP20 的相同。硬骨鱼类：鲫鱼（AB020024）、大麻哈鱼（Moriyama 等，2002）、大西洋鲑鱼（NM_001123641）、罗非鱼（Seale 等，2002）、斑马鱼（EU117421）。鲨鱼 RFa（AB433893），七鳃鳗（*Petromyzon marinus*）RFa-A 和 RFa-B（Moriyama 等，2007），爪蟾 PrRP（AB251344），鸡 PrRP（EF418015）和 RFa（NM_001114503），大鼠（NM_001101647），牛（NM_174790），人（NM_015893）。

在硬骨鱼类中，哺乳类 PrRP20 的同源物首次从日本鲫鱼（*Carassius auratus langsdorfi*）（Fujimoto 等，1998）的脑部被分离出来；接着从大麻哈鱼（*Oncorhynchus keta*；Moriyama 等，2002）和莫桑比克罗非鱼（Seale 等，2002）的脑部被分离出来。前原 PrRP cDNA 曾经从鲫鱼（Satake 等，1999）、大麻哈鱼（Moriyama 等，2002）、罗非鱼（Seale 等，2002）和大西洋鲑鱼（Montefusco-Siegmund 等，2006）等的脑中被克隆出来。它们编码一个单独的由 20 个氨基酸残基组成同一序列的 PrRP 片段，在本章中就把它称为硬骨鱼类 PrRP（tPrRP）（见图 5.3）

PrRP-免疫反应（tPrRP-ir）细胞体出现在下丘脑的后部，而 PrRP-ir 纤维从下丘脑广泛延伸到金鱼（Wang 等，2000a）、虹鳟（Moriyama 等，2002）和网纹花鳉（*Poecilia reticulata*）（Amano 等，2007）的脑中。采用 RT-PCR 在鲫鱼（Satake 等，1999）、弹

涂鱼（Sakamoto 等，2005）、大西洋鲑鱼（Montefusco-Siegmund 等，2006）和金鱼（Kelly 和 Peter，2006）的下丘脑检测到 PrRP mRNA 的存在。在虹鳟（Moriyama 等，2002）和网纹花鳉的脑部（Amano 等，2007），少量 PrRP-ir 纤维伸入到脑垂体，在靠近 RPD 的 PRL 细胞和 PI 的 SL 细胞的位置中止。在大西洋鲑鱼（Montefusco-Siegmund 等，2006）的相关研究中，虽然在脑垂体没有观察到 PrRP-ir 纤维，但在合成 PLR 的脑垂体 RPD 亦发现 PrRP-ir 细胞体和 PrRP mRNA。此外，在弹涂鱼（Sakamoto 等，2005）的外周组织如肝脏、消化道和卵巢中检测到 PrRP mRNA 的表达，而较低程度的表达还出现在弹涂鱼的皮肤、肾脏以及鲤科鱼类的视网膜（Wang 等，2000b）。值得注意的是，PRL mRNA 和 PrRP mRNA 的分布相一致，它亦在弹涂鱼的这些器官中被检测到。tPrRP 和 PRL 的广泛分布表明它们参与硬骨鱼类的一系列生理功能（见表 5.1）。

表 5.1　硬骨鱼类 PrRP 的功能

作用	鱼类	参考文献
PRL 释放	虹鳟	Moriyama 等（2002）；Sakamoto 等（2003 b）
	罗非鱼	Seale 等（2002）
	弹涂鱼	Sakamoto 等（2005）
	金鱼	Kelly 和 Peter（2006）
SL 释放	虹鳟	Moriyama 等（2002）
GH 抑制	虹鳟	Moriyama 等（2002）
渗透压调节	弹涂鱼	Sakamoto 等（2005）
	金鱼	Fujimoto 等（2006）；Kelly 和 Peter（2006）
陆地适应	弹涂鱼	Sakamato 等（2005）
内脏肌肉收缩	鲫鱼	Fujimoto 等（1998）
心血管调节	虹鳟	Sakamoto 等（2003）
食物摄取	金鱼	Kelly 和 Peter（2006）
对光的反应	鲫鱼	Wang 等（2000 b）

PRL，催乳激素；SL，生长乳素；GH，生长激素。

如上所述，PrRP 原先被鉴定为孤独 GPCR 的配体（Hinuma 等，1998）。有关 PrRP 受体基因同源物的结构信息现在已经在大鼠、小鼠、豚鼠、鸡、河豚、斑马鱼和鲟鱼的相关研究中得到（Lagerström 等，2005）。它们分为两个亚型，和 NPY 受体有一个共同的祖先。对于硬骨鱼类 PrRP 受体的分布还没有相关报道。

对罗非鱼，用 tPrRP（100 nM）孵育脑垂体能明显地刺激罗非鱼两种 PRL（PRL_{188} 和 PRL_{177}）类型的释放，但不能刺激 GH 释放。然而，tPrRP 对 PRL 释放的作用还比不上低渗性介质刺激 PRL 的大量释放。相反，哺乳类 PrRPs 对 PRL 释放没有作用。对预先用 E_2 孵育的脑垂体，tPrRP 刺激 PRL 释放的效能和鸡 GnRH 相当。给淡水中的雌罗非鱼腹腔注射 0.1 mg/g 体重的 tPrRP，1 h 后血液循环中的 PRL 明显增加，但其对雄鱼没有这个作用（Seale 等，2002）。

在对虹鳟进行的实验中，用浓度为 10 pM 到 100 nM 的 tPrRP 对脑垂体进行灌流试验，证明刺激 PRL 的最大释放是浓度 100 pM，而刺激 SL 的最大释放是浓度 10 nM 和 100 nM，但 GH 的释放不受 tPrRP 影响。给虹鳟腹腔注射 tPrRP，剂量为 50 ng/g 体重和 500 ng/g 体重，血浆中的 PRL 水平和 SL 水平分别在注射 3 h 和 9 h 后升高。相反，注射剂量为 500 ng/g 体重，1 h 后血浆 GH 水平降低（Moriyama 等，2002）。对虹鳟通过背大动脉导管一次动脉内注射 tPrRP（40 nmol/kg 体重），注射后 2 min 血浆的 PRL 水平迅速增长，注射后 8 h 脑垂体 PRL mRNA 水平上升。相反，血浆的 SL 水平降低，GH 和 SL mRNA 水平没有明显影响（Sakamoto 等 2003b）。在金鱼腹腔注射 tPrRP 25 ng/g 体重 8 h 后脑垂体 PRL mRNA 水平明显升高。腹腔注射 tPrRP 250 ng/g 体重导致脑垂体 PRL mRNA 的表达明显升高，但和注射 25 ng/g 体重的结果没有明显差别（Kelly 和 Peter，2006）。Fujimoto 等（2006）证明 tPrRP 能使脑垂体 PRL mRNA 水平增长，而抗-tPrRP 使它降低。此外，tPrRP 使鳃的水分流入减少而抗-tPrRP 使之增长；tPrRP 还能使鳞片的黏膜细胞层伸展。

这些研究结果表明 tPrRP 参与硬骨鱼类特别是淡水硬骨鱼类 PRL 细胞活性或 PRL 分泌和渗透压平衡的生理调控作用。但是，哺乳类下丘脑的 PrRP 对脑垂体 PRL 细胞的作用如何还不清楚（Taylor 和 Samson，2001）。不过，PrRP mRNA、PrRP 肽以及受体在中枢神经系统的广泛分布表明它们参与脑的许多功能（Sun 等，2005）。Kelly 和 Peter（2006）发现 tPrRP 参与调控金鱼的食欲和水矿物平衡（hydromineral balance）。给金鱼腹腔和脑室内注射 PrRP 能引起剂量依存地减少食物摄取。要阐明 PrRP 对鱼类脑的功能还需要做进一步的研究。

2. 促性腺激素释放激素（gonadotropin-releasing hormone，GnRH）

GnRH 是十肽，通过下丘脑-脑垂体轴来调控生殖活动。对于哺乳类，GnRH 亦是 GH 和 PRL 分泌强有力的刺激剂（Blackwell 等，1986；Robberecht 等，1992）。GnRH 及其基因已在整个脊索动物中鉴定（见本书第 3 章）。

许多硬骨鱼类的 sbGnRH 细胞体主要分布在视前区，而 sbGnRH 免疫反应纤维主要延伸到金头鲷（Gothilf 等，1996）、罗非鱼（Parhar，1997）、海鲈和条斑星鲽（Amano 等，2002）脑垂体的 PPD。银汉鱼（Vissio 等，1999）和尼罗尖吻鱼（*Lates niloticus*）（Mousa 和 Mousa，2003）的 GnRH-ir 纤维延伸到脑垂体的三个区，即 RPD、PPD 和 PI，并且紧密联系着。

GnRH 通过和细胞膜上的 GPCRs 结合而发挥作用（见本书第 3 章）。它们通过激活一个或多个 G-蛋白而介导它们在细胞内的功能（Sealfon 等，1997；Millar 等，2004）。对罗非鱼的研究发现，除 LH 和 FSH 细胞外，三种 GnRH-Rs 亦出现在 PRL、GH、TSH、黑色素细胞刺激激素、ACTH 和 SL 的细胞中（Parhar 等，2005）。成熟的雄罗非鱼的 GnRH-R1 和 GnRH-R2 转录体在 PRL 细胞的亦明显比较高（Parhar 等，2005）。在银汉鱼的 PRL 细胞中能检测到 GnRH 结合位点（Stefano 等，1999）。有关 GnRH-免疫反应纤维和 GnRH-Rs 的这些定位研究结果表明 GnRH 可能是一个调控 PRL 的下丘脑神经肽。

对于罗非鱼，三种天然的 GnRH 类型都能刺激离体的脑垂体 RPD 释放 PRL，其作用能力的顺序是 cGnRH Ⅱ > sGnRH > sbGnRH。此外，一种哺乳类 GnRH 类似物能刺激在等渗性或高渗性介质中孵育的脑垂体 RPD 释放 PRL，而高渗性介质在正常情况下是抑制 PRL 释放的。用睾酮或 E_2 进行共孵育能增强脑垂体 RPD 对 GnRH 的反应作用（Weber 等，1997）。对罗非鱼亦曾研究 GnRH 调控 PRL 细胞分泌活动的作用机理（Tipsmark 等，2005）。GnRH 通过增强罗非鱼磷脂酸 C（Phospholipase C，PLC）、肌醇三磷酸（inositol triphosphate，IP3）和细胞内钙（Ca_i^{2+}）的信号刺激 PRL 释放，cGnRH-Ⅱ 诱导分散的罗非鱼 PRL 细胞迅速而剂量依存地 Ca_i^{2+} 增加。Ca_i^{2+} 信号能被 U73122——一种 PLC-依赖的磷酸肌醇（phosphoinositide）水解作用抑制剂抵消。相应地，这个抑制剂能阻抑 cGnRH-Ⅱ 诱 $tPRL_{188}$ 的分泌活动，表明 PLC 的激活介导了 cGnRH-Ⅱ 刺激 PRL 分泌的作用。用 Ca^{2+} 拮抗剂，8-(N, N-二乙氨基)-辛基-3, 4, 5-三甲氧苯甲酸盐酸盐 [8-(N, N-diethylamino)-octyl-3, 4, 5-trimethoxybenzoate hydrochloride]，一种 Ca^{2+} 从细胞内贮存释放的抑制剂做预先处理，会阻抑 cGnRH-Ⅱ 对 Ca^{2+} 的影响。Ca^{2+} 拮抗物会阻抑 $tPRL_{188}$ 对 cGnRH-Ⅱ 刺激分泌的反应性。这些研究结果表明 GnRH 通过增加 Ca_i^{2+} 而诱导它的促进 PRL 释放的作用。此外，Ca_i^{2+} 的增加可能是由 PLC/IP3 诱导 Ca^{2+} 从细胞内贮存的转移而衍生，并一起通过 L-型电压控制的 Ca^{2+} 通道（L-type voltage-gated Ca^{2+} channel）而流入。

对于正在生长和性成熟的马苏大麻哈鱼（*Oncorhynchus masou*），GnRH 能增强编码脑垂体 GH、PRL 和 SL 的基因在特别的季节里的表达，这可能和这些激素的生理作用有关（Bhandari 等，2003）。Onuma 等（2005）进一步表明，sGnRH 直接调节 Pit-1 的合成和 PRL 与 SL 基因的表达；但是，GnRH 调控 GH/PRL/SL 家族激素的基因可能是间接的，特别是在配子发生的后期。

3. 促甲状腺素释放激素（thyrotropin-releasing hormone，TRH）

TRH 除了促甲状腺素的释放作用之外，在许多脊椎动物中还具有一系列激素的和神经递质/神经调节剂的功能，包括促进 GH 和 PRL 释放的作用（Nillni 和 Sevarino，1999）。鲤鱼的 TRH-免疫反应纤维存在于神经元突起，从视前核延伸到下丘脑的外侧隐窝核（nucleus recessus lateralis）和脑垂体内。在脑垂体内的纤维主要局限于神经叶（neural lobe）区，有些分布在神经叶的纤维紧密靠近前叶（Hamano 等，1996）。

TRH 通过细胞表面的属于 GPCR 超家族的受体相互作用而启动它的功能（Sun 等，2003）。

已经充分阐明 TRH 刺激哺乳类在体和离体的 PRL 释放（Grosvenor 和 Mena，1980）。但是，TRH 对硬骨鱼类 PRL 细胞的生理作用还有些模糊之处，即种类的特异性或者条件的差异性。在一定的条件下，例如渗透压和 E_2 预先孵育，TRH 能够刺激

PRL 释放。在花鳉的相关研究中，TRH 刺激 PRL 释放到高渗性的介质中，但在低渗性介质中没有作用（Wigham 和 Batten，1984）；对罗非鱼，如果预先用 E_2 处理，TRH 能够起刺激作用（Barry 和 Grau，1986）。对虹鳟的研究曾报道有三种不同的结果，TRH 没有作用（James 和 Wigham，1984）或者有刺激作用（Prunet 和 Gonnet，1986），或者有抑制作用（Williams 和 Wigham，1994a）。对鲤鱼，用 TRH 1～100 nM 的浓度处理原代培养脑垂体细胞，TRH 能剂量依存地增加释放新合成的 PRL（Kagabu 等，1998）。相反，Tse 等（2008）证明 TRH 能通过下调 PRL 基因启动子的转录而降低金鱼原代培养脑垂体细胞的 PRL mRNA 水平。这些有争议的研究结果表明，TSH 可能不是硬骨鱼类调控 PRL 释放的关键性因子。

4. 其他的下丘脑肽

（1）血管活性肠肽（vasoactive intestinal polypeptide，VIP）是一个在大脑皮质（cerebral cortex）、下丘脑和前脑垂体（anterior pituitary）以及其他组织中合成的神经肽。已经证明下丘脑的 VIP 能激活腺苷酸环化酶-cAMP 通道并和 GPCR 结合（Robbercht 等，1979）而刺激哺乳类和鸟类在体与离体的 PRL 释放（Mezety 和 Kiss，1985）。VIP 能在高渗性和低渗性介质中明显地抑制罗非鱼两种 PRLs——PRL_{177} 和 PRL_{188} 的分泌。在高渗性介质中，300 nM VIP 抑制两种 PRLs 分泌的 47%，而在低渗性介质中，300 nM VIP 抑制它们分泌的 27%（Kelly 等，1988）。VIP 对罗非鱼 PRL 分泌的抑制作用和已知的 VIP 对四足类动物 PRL 分泌的刺激作用形成鲜明的对照。

（2）脑垂体腺苷酸环化酶激活多肽（PACAP）最初是从羊下丘脑中分离出来的，是一种新的下丘脑调控脑垂体的神经肽，能激活大鼠脑垂体细胞培养物的腺苷酸环化酶（Miyata 等，1989），参与许多生理过程，包括下丘脑调控哺乳类 GH、GTH 和 PRL 细胞的作用。PACAP 的一级结构在脊椎动物当中高度保守（Adams 等，2002），PACAP 受体属于 GPCR 超家族，已经从金鱼脑和脑垂体中鉴定出来（Chow 等，1997；Wong 等，1998）。PACAP-免疫反应细胞体主要分布在间脑，它们的纤维延伸到金鱼（Wong 等，1998），日本䲢（*Uranoscopus japonicus*）（Matsuda 等，1997）和欧洲鳗鱼（*Anguilla anguilla*；Montero 等，1998）的 PPD 内，并且发现它能刺激金鱼、欧洲鳗鱼和大麻哈鱼（Parker 等，1997；Wirachowsky 等，2000；Sawisky 和 Chang 2005；Wong 等，2005）等脑垂体细胞培养物释放 GH 和 GTH。此外，Matsuda 等（2008）确定 PACAP-免疫反应纤维分布在神经垂体紧密靠近含有 PRL 和 SL 的细胞内以及 PACAP 受体免疫反应的位置。用 PACAP 处理能以剂量依存方式（10^{-9}～10^{-7} M）增加 PRL-和 SL-免疫反应细胞的免疫印迹区（immunoblot area）；用 10^{-7} M PACAP 能明显增强 SL mRNA 的表达，但对 PRL mRNA 没有影响；用 10^{-8} M PACAP 灌流分离的金鱼脑垂体细胞能增加细胞内钙的移动并使这些脑垂体细胞对 PRL 和 SL 产生免疫反应。这些研究结果表明，PACAP 作为下丘脑调控脑垂体的因子，其不仅调控 GH 和 GTH，亦调控金鱼脑垂体的 PRL 和 SL。

（3）生长素释放肽（ghrelin，GRL）是最近鉴别的 GH-释放肽，从胃提取物中分离出来，成为 GPCK 超家族的 GH 促分泌受体的内源配体（Kojima 等；1999）。GRL 主要在肠管内表达，而在鱼类脑和下丘脑的表达程度要稍小一些（Kojima 等，1999；Kaiya 等，2003a、b、c；Unniappan 等，2002）。和胃产生的 GRL 一起，下丘脑 GRL 似乎亦参与 GH 从脑垂体分泌以及摄食行为的调节（Unniappan 等，2002）。在牛蛙的相关研究中（Kaiya 等，2001）亦曾报道同源的 GRL 刺激 PRL 释放，但其对大鼠没有这个作用（Kojima 等，1999）。罗非鱼和鳗鱼 GRL 能以剂量依存方式刺激离体的罗非鱼脑垂体培养物释放 GH 和 PRL（Kaiya 等，2003b、c）。但是，虹鳟的 GRL 只刺激在体和离体的 GH 释放而不刺激 PRL 释放（Kaiya 等，2003a）。这些研究结果表明 GRL 刺激 PRL 释放的作用可能是种类特异性的（species-specific）。

（4）Copeptine（NpCp）。在大鼠中，39 个氨基酸的糖肽（glycopeptide）组成神经脑垂体血管加压素-后叶激素运载蛋白（vasopressin neurophysin）前体的羧基端（carboxyterminus）可能是一个 PRL 的释放因子（Nagy 等，1998）。最近，Flores 等（2007）分离出了硬骨鱼催产素（isotocin）前体 NpCp 的 C-端肽，它是在鲤鱼后叶激素运载蛋白（NP）和 copeptin（CP）之间未分裂的区域。鲤鱼 NpCp-免疫反应纤维大量出现在下丘脑，并和脑垂体 RPD 的 PRL 细胞直接接触。用鲤鱼 NpCp（0.02～2μg）孵育鲤鱼脑垂体的 RPD 30 min，介质中的 PRL 以剂量依存方式增加。但是，这个肽对 PRL 释放的生理学意义还需做进一步研究。

（5）利尿钠肽（natriuretic peptide，NPs）由一个家族的激素组成，和脊椎动物心血管和体液稳态的保持有密切关系（Takei，2000；Toop 和 Donald，2004；Potter 等，2006）。在一些硬骨鱼类中已经鉴别心房钠尿肽（atrial natriuretic peptide，ANP）、B-型 NP（BNP）、心室 NP（VNP）和 4 种 C 型 NPs（CNP）（Inoue 等，2003，2005）。CNP 基因在罗非鱼脑内强烈表达，在脑垂体内轻微表达，而 ANP 和 BNP 基因几乎不在脑和脑垂体中表达。NP 受体（NPRs）已经在哺乳类的下丘脑和前脑垂体中鉴别（Potter 等，2006）。在脑垂体内存在着 NPRs 表明在血液循环中的或者局部产生的 NPs 对脑垂体激素的分泌活动起调控作用。但是，没有一种鳗鱼的 NPs 对日本鳗鱼（Anguilla japonica）的 PRL 释放起作用（Eckert 等，2003）。罗非鱼的 ANP 和 BNP 能刺激驯养在淡水中罗非鱼的分散脑垂体细胞释放 PRL 和 GH，而罗非鱼 CNP 对 PRL 释放没有影响。ANP 和 BNP 都能有效地提高细胞的 cGMP 的积累，而 CNP 没有这个作用。脑的 CNP 可能不参与罗非鱼 PRL 的分泌活动（Fox 等，2007）。

5.3　生长乳素分泌活动的神经内分泌调控

5.3.1　生长乳素和它的受体

5.3.1.1　生长乳素和基因

生长乳素（somatolactin，SL）原先是在大西洋鳕鱼（*Gadus morhua*）（Rand-Weaver 等，1991a）和鲽鱼（Ono 等，1990）的脑垂体中发现的。SL 细胞定位于所有曾经研究过的硬骨鱼类脑垂体的 PI 和过碘酸-希夫(PAS)-阳性反应细胞内，只有鲑鳟鱼类是例外（Rand-Weaver 等，1991b）。SL 的 cDNA 和基因已经在许多硬骨鱼类包括鲟鱼类和肺鱼类中分离出来，但是脊椎动物的其他各纲都没有 SL。所有 SL 基因都由 5 个外显子组成，和所有已知的 PRL 和 GH 基因一样，但不包括高等硬骨鱼类的 GH 基因。它们有 1 个 TATA 框和 4 个 Pit-1/GHF-1 结合位点。对虹鳟的免疫组织化学研究表明 Pit-1 蛋白定位在脑垂体中间叶的产生 SL 细胞的核内（Ono 等，1994）。

Zhu 等（2004）发现斑马鱼 SL 的两个共生同源基因（paralogous gene）（分别命名为 SLα 和 SLβ），它们在 PI 的不同细胞中表达。以序列的相似性为基础，SLβs 包括金鱼 SL（CAU72940）、斑点鮰（*Ictalurus punctatus*）SL（AF062744）、鳗鱼 SL（AAU63884）和虹鳟 SLP（Yang 和 Chen，2003）。其他的 SLs，包括金头鲷两个非常相似的 SLs（Y11144，L49205），属于 SLα。两个 SL 亚型有 35%～48% 的序列相同。SLα 存在于所有鱼类之中，但 SLβ 只存在于一些低等的硬骨鱼类如鲶鱼、金鱼、鲑鳟鱼类和鳗鱼之中。SLβ 基因可能在硬骨鱼类的多样化（diversification）过程中丢失。

SLs 亦在鲟鱼（AB017766）和肺鱼（AB017200）中被鉴定出来（Amemiya 等，1999）。皱唇鲨（*Triakis scyllium*）脑垂体 PI 的一些细胞能和抗-鲑鱼 SL 血清染色并曾克隆它们的 cDNA 片段（未发表结果）。但是，没有证据表明 SL 存在于四足类动物和无颌类之中，表明 SL 基因出现于进化到有颌类的过程中，而在进化到四足类动物期间丢失。

所有的 SLs 至少有 6 个半胱氨酸（Cys）残基，能在和四足类 PRLs 相似的位置形成 3 个二硫键。SLs 是糖蛋白，有 1～3 个 N-糖基化作用（N-glycosylation）位点；而鲑鳟鱼类和金鱼是例外，它们没有 N-糖基化作用位点（Takayama 等，1991；Yang 等，1997）。缺少糖基化作用位点可能是鲑鳟鱼类脑垂体的 PI 没有 PAS-阳性反应细胞的原因。

5.3.1.2　生长激素受体（SLR）

已经在马苏大麻哈鱼（*Oncorhynchus masou*）（Fukada 等，2005）、青鳉、河鲀（Fukamachi 等，2005）、金头鲷（AY573601）和鳗鱼（AB180476）中鉴定 SLR，证明 SLR 是细胞因子受体 I 型的同型二聚体群（homodimeric group）的成员，和硬骨鱼类的两个 GHRs（GHR1 和 GHR2）与 PRLR 同属一类。SLRs 和脊椎动物的 GHRs 与 PRLRs

分别有38%～58%和28%～33%的序列具有同一性，并和脊椎动物GHR包括FGEFS模体有共同的特征，在细胞外区有6个半胱氨酸残基，1个单个跨膜区，细胞内区的框1和框2区。但是，系统发生的分析表明鲑鱼和青鳉（*Oryzias latipes*）的SLR属于GHR1谱系。此外，GHR2和GH结合，GHR1主要和SL结合（Fukada等，2004，2005）。这些研究结果表明金头鲷和鳗鱼的GHR1命名是不恰当的，应该是SLR（Fukamachi和Meyer，2007）。

在银大麻哈鱼的脑、脑垂体、鳃、心脏、头肾、后肾、脾脏、肝脏、肌肉、脂肪和性腺中都检测到SLR mRNA，而最高的水平出现在肝脏和脂肪组织中（Fukada等，2005）。在青鲷的肝脏中亦观察到SLR mRNA的较高水平（Fukamachi等，2005）。罗非鱼GHR1在脂肪、肝脏和肌肉中高度表达，表明了它们的代谢功能。GHR1在皮肤的表达亦较高，这和SL对载色素细胞（chromatophore）的调节作用一致。这些研究结果支持GHR1是SL受体的假说。

5.3.2 生长乳素的功能

曾经通过免疫组织化学方法观察SL细胞的形态学，用放射免疫分析测定血浆和脑垂体SL水平的变化以及SL mRNA在不同生理和环境条件下的表达水平等确定SL的生理作用。研究结果表明，SL参与对环境变化的适应（Ono和Kawauchi，1994）、鲑鱼降海洄游的适应性转变（smoltification）（Rand-Weaver和Swanson，1993）、对背景和降低照明度的适应（Zhu和Thomas，1995，1998；Zhu等，1999；Canepa等，2006）、应激反应（Kakizawa等，1995a；Johnson等，1997）、生殖生理某些方向的调控（Planas等，1992；Rand-Weaver等，1992；Rand-Weaver和Swanson，1993；Olivereau和Rand-Weaver等，1994a、b；Vissio等，2002）、酸碱平衡（Kakizawa等，1995a，1996，1997a、b）、钙的调节（Kakizawa等，1993，1995a、b）和磷酸盐代谢（Kakizawa等，1995a）。SL还参与生长（Duan等，1993；Company等，2001）和能量代谢活动（Rand-Weaver等，1992；Kaneko等，1993；Kakizawa等，1995a；Company等，2001；Mingarro等，2002；Vega-Rubin de Celis等，2004）。

但是，对这些公认的作用，亦出现一些有争论的研究结果。虹鳟SLR（Fukada等，2005）和罗非鱼GHR1（Fukamachi等，2006；Fukamachi和Meyer，2007；Pierce等，2007）在皮肤的表达高，这和SL对载色素细胞（chromophore）的调节功能一致。根据分析和受体结合的研究，在前述研究中罗非鱼GHR1被认为是一个SLR，但是，亦有研究报道认为背景颜色并不影响虹鳟血浆中的SL水平（Kakizawa等，1995a），而SL分泌活动的季节变化和水温的联系要比和光周期的联系更为紧密（Rand-Weaver等，1995）。应激反应引起血液循环中SL水平增高在虹鳟的两个品系中亦明显不同（Rand-Weaver等，1993），并且，红拟石首鱼（*Sciaenops ocellatus*）、大西洋绒须石首鱼（*Micropogonias undulatus*）（Zhu和Thomas，1995）、大西洋庸鲽（*Hippoglossus hippoglossus*）和侧枝鲽（*Parophrys vetulus*）（Johnson等，1997）的血浆SL浓度和不同的盐度、外界的钙浓度或生殖状态之间并没有联系。亦没有证据表明SL参与虹鳟对禁食反应的积极作用（Pottinger 等，2003）。但是，这些有争议的研究结果应该根据有两种不同的 SLs，即 SL α 和 SL β，以及有低等的硬骨鱼类如鲤科、鲑科和鳗鱼等来进行综述。随着被鉴定的激素及其受体的数量不断增加，阐明这些配体/受体的结构和功能是鱼类内分泌学者们面临的重大挑战。

如前所述，SL 参与鱼类生殖的一些活动。但是，除了对银大麻哈鱼性腺类固醇生成的刺激作用（Planas 等，1992）之外，没有进一步实验证明的报道。

Lu 等（1995）检测鲑鱼 SL 对美洲拟鲽（*Pseudopleuronectes americanus*）肾近侧小管（renal proximal tubule）细胞的肾跨上皮运输（transepithelial transport）P_i 和 Ca^{2+} 的影响。SL 能在其生理的作用水平（12.5 ng/mL）以剂量依存方式刺激 P_i 重吸收。SL 增加到 200 ng/mL 时，Ca^{2+} 的流出没有变化。SL 诱导的 P_i 重吸收被一个高度特异性的蛋白激酶 A 抑制剂所抵消。此外，用 SL 处理 1 h 和 2 h 后，cAMP 的产生和释放明显增加。这些研究数据表明 SL 通过 cAMP-依赖的通路直接刺激肾的 P_i 重吸收。

采用纯化的 SL 在红拟石首鱼证明 SL 对离体的黑色素体（melanosome）聚集的影响（Zhu 和 Thomas，1997）。给处在黑色背景水族箱中的红拟石首鱼肌肉注射 SL（1 nmol/g 体重），2 min 内鱼的皮肤变灰白色，而在随后的 30 min 逐渐恢复它的黑体色。用 1 μM SL 离体孵育，10 min 内鳞片黑色素细胞的黑色素体完全聚集在一起。而且，对基因重组的斑马鱼 SL β 以 1ng/mL 的浓度进行处理，能以浓度相关的方式诱导斑马鱼皮肤的黑色素体聚集（Nguyen 等，2006）。此外，对青鳉"体色干扰"（color interfere，ci）相关的一个基因突变体的鉴定可以确认 SL 对硬骨鱼类体色的调控起着必不可少的作用（Fukamachi 等，2004）。这些变突体除了在戴色素细胞增殖和形态发生方面有所欠缺之外，没有任何明显的形态上和生理上的缺点。这个突变确认为 SL 基因缺失 11-碱基，造成 SL 位于 C-端上游 230 氨基酸的 91-氨基酸的平截（truncation）。这个有缺陷的 SL 突变体的体色干扰表现为明显增加白色素细胞（leukophore）的数量而减少可见的黄色素细胞（xanthophore）的数量。在形态的体色适应于不同背景期间，SL 的转录发生明显的变化。这些研究结果证明 SL 影响戴色素细胞的发育。体色干扰突变体的鱼能组成性地增加白色素细胞的数量而相应减少可见的黄色素细胞的数量。Fukamachi 等（2006）认为 SL 调节在一些鱼类的黄色素细胞是保守的功能，而青鳉特异性的和白色素细胞依赖作用并不是 SL 在体色调控方面的进化表现。支持这个观点的是虹鳟的"钴"变种（"cobalt" variant）非常明显的较灰白体色，该变种在 SL-生产细胞分布的脑垂体中间部（PI）有缺陷（Kaneko 等，1993），还有 SLs 在青鳉、红拟石首鱼和大西洋绒须石首鱼表达的相似性，即黑暗照明（dark illumination）或者黑色背景都能明显增强 SL 的表达（Zhu 和 Thomas，1995；Fukamachi 等，2005）。

虹鳟的 SLR 和罗非鱼的 GHR1 在脂肪、肝脏和肌肉中高度表达，表明了它们的代谢功能。Vega-Rubin de Celis 等（2003）给金头鲷幼鱼一次腹腔注射海鲈 SL（0.1 μg/g 体重），并不影响血液循环中 IGF-I 的含量和氮-氨的排泄，但增加二氧化碳的排出和氧的摄入，从而降低呼吸商（respiratory quotient，RQ，CO_2 排出/O_2 摄入）。SL 亦能够抑制肝脏乙酰辅酶 A 羧化酶（acetyl-coenzyme A carboxylase）的活性。这些研究结果表明

SL 参与能量稳态和增强脂类代谢活动。对虹鳟"钴"变种的研究亦表明 SL 参与脂类的代谢活动，该变种的脑垂体缺少 SL-生产细胞，体内积存大量脂肪组织（Kaneko 等，1993），在肝脏和肌肉中含有许多甘油三酯和胆固醇（Yada 等，2002b）。

5.3.3 下丘脑调控生长乳素的分泌活动

和 SL 的生理功能一样，对神经内分泌调控 SL 从脑垂体释放的作用机理还研究得很少（见图 5.4）。通过几种下丘脑激素神经元在脑垂体的组织学观察，表明下丘脑可能调控 SL 的分泌活动。已经鉴定下列肽类的免疫反应神经纤维和末梢紧密靠近在腺脑垂体 PI 的 SL 细胞：硬头鳟（Parhar 和 Iwata，1994）、尼罗尖吻鲈（Mousa 和 Mousa，2003）、银汉鱼（Vissio 等，1999；Stefano 等，1999）和金头鲷（González-Martínezz 等，2002）的 GnRH，金鱼和虹鳟（Wang 等，2000b；Moriyama 等，2002）的 PrRP，日本䲢（Matsuda 等，2005a、b）和金鱼（Matsuda 等，2008）的 PACAP。此外，在 SL 细胞中还检测到 GnRHR 和 PACAPR。

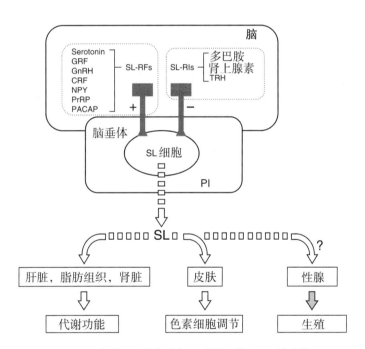

图 5.4　鱼类 SL 从脑垂体释放的调控和 SL 的功能

在各种神经肽和神经递质的多功能调节下，SL 从 SL-产生细胞中释放。通过从下丘脑的直接神经分布，这些刺激性/抑制性的调节剂从脑运送到在腺脑垂体 PI 的 SL-产生细胞。SL-RFs，生长乳素释放因子；SL-RIs，生长乳素释放抑制剂；GRF，生长激素释放因子；GnRH，促性腺激素释放激素；CRF，肾上腺皮质激素释放因子；NPY，神经肽 Y；PrRP，催乳激素释放肽；PACAP，脑垂体腺苷酸环化酶激活多肽；TRH，促甲状腺素释放激素；PI，中间部。

1. 抑制因子

Kakizawa 等（1997a）首次研究虹鳟下丘脑因子对 SL 从脑垂体器官培养物分泌活动的影响。浓度为 30 μM 和 300 μM 的多巴胺、30 μM 和 300 μM 的肾上腺素、100nM 的 TRH 都能抑制 SL 释放。

2. 刺激肽类

Kakizawa 等（1997a）研究下丘脑因子对 SL 从离体的虹鳟脑垂体培养物在有或没有 DA 的情况下分泌活动的影响。GnRH、5-羟色胺和肾上腺皮质激素释放因子能刺激被 DA-抑制的脑垂体释放 SL。Taniyama 等（2000）证明 GnRH 类似物能增强产卵前或性成熟的红大麻哈鱼 SL 基因的表达。此外，sbGnRH 对离体孵育 3 h 的 SL 基础性释放没有影响（Peyon 等，2003）。

脊椎动物包括鱼类的 NPY 能参与调节各种神经内分泌轴。猪 NPY 单独对舌齿鲈（*Dicentrarchus labrax*）离体脑垂体的 SL 释放没有影响；但 NPY 能剂量依存地增强被瘦蛋白（瘦素）诱导的舌齿鲈青春前期稍后的 SL 释放，而对青春后期舌齿鲈 SL 释放没有影响（Peyon 等，2001，2003）。

tPrRP 能有效地刺激虹鳟脑垂体释放 SL（Moriyama 等，2002）。在腹腔注射 tPrRP 后血浆的 SL 水平升高，但其水平和注射剂量没有关系。SL 水平的增加要比 PRL 水平的增加迟些出现。PrRP 对 SL 释放的作用可能是间接的或者由其他的一些作用机理介导。的确，tPrRP 对离体 SL 释放的刺激作用要比对 PRL 释放的刺激作用小一点。Sakamoto 等（2003b）亦曾报道在动脉内注射 tPrRP 后脑垂体的 SL mRNA 和血浆的 SL 水平并不增加。

PACAP 亦能有效地刺激金鱼脑垂体释放 SL（Matsuda 等，2008）。用 PACAP 处理离体的金鱼分散脑垂体细胞，能以剂量依存方式增加 SL 免疫反应区域。PACAP 亦刺激 SL mRNA 的表达，但对 PRL mRNA 没有作用。这表明 PACAP 不仅对 GH 和 GTH，亦对金鱼的 SL 起着强有力的下丘脑调控脑垂体因子的作用。

哺乳动物的 GnRHa 能明显增加雄性产卵前的红大麻哈鱼（*Oncorhynchus nerka*）脑垂体 SL mRNA 水平，而对雌鱼没有作用，而且，GnRHa 亦不能诱导雄鱼与雌鱼 GH 和 PRL mRNA 水平的明显增加（Taniyama 等，2000）。

5.4 结　束　语

本章介绍了神经内分泌调控两种同源的脑垂体激素 PRL 和 SL 合成与分泌的最前沿的研究进展。和作为哺乳类模式种类的啮齿动物在这研究领域的迅速发展相对比，对鱼类的研究要少得多，尽管我们现在已经获得许多脑垂体激素以及受体的分子信息。关于硬骨鱼类 PRL 和 SL 的功能研究，已经进行了激素产生细胞的形态学观察，如脑垂体和血浆中激素水平的变化以及在一定条件下配体与其相应受体表达水平的变化。然而，还很少采用同源的激素和/或神经激素进行研究。所以，在做出结论之前，对有争论的结果必须重新进行严密的分析。激素、受体以及突变体的基因分析必将为硬骨鱼类的激素调控生理学带来新的见识，如同对 SL 功能的阐述那样，尽管这样的好机遇不可能出现在每一个事例中。必须强调的是，要得到结论性回答的普遍方案是在蛋白质水平进行研究，即在体和离体检测同源性神经激素和脑垂体激素的作用。

目前已经证明许多下丘脑神经激素参与 PRL 和 SL 的调控。再者，这些神经激素和脑垂体激素亦在硬骨鱼类中展现多方面的功能（见表 5.1）。单个的一种神经激素表现出不同的功能，或者有时候正好是相对立的功能，或者是没有功能，这不仅取决于种类或者个体发育（ontogeny），而且关系到许多其他的因素，包括处理的时间、生殖状态、环境状况和施用的途径等。所有这些对于确定神经激素对 PRL 和 SL 合成和/或释放的影响都起着重要的作用。对于鱼类，必须考虑把温度当作一个主要的环境因素，它能影响激素及其相应受体的表达和/或功能。所以，这些因素能够激活或者抑制和特异性受体结合后的细胞功能。许多化学信使在系统发生中保持着相同的或者相似的功能，但亦并非都是这样。在适应新的环境之后，生物有机体内的一些化学信使可能发生变化、修饰或者接收其他的功能。确实有许多例子表示配体及其相应的受体展现其明显的保守性而不顾及强烈的突变和选择的压力。此外，为了适应性的需求，它们的功能亦会展现明显的多样性。通过组织表达特异性受体的不同的定位、选择性的受体结合后细胞内传导通路的分化，以及由各种不同的因子调控激素的作用等，都有可能引起作用的多样化。

川内浩司

S. A. 索沃

森山俊介

参考文献

Adams B A, Lescheid D W, Vickers E D, Crim L W, Sherwood N M. 2002. Pituitary adenylate cyclase-activating polypeptide in sturgeon, whitefish, grayling, flounder and halibut: cDNA sequence, exon skipping and evolution. *Regul. Pept*, 109: 27-37.

Al Kahtane A, Chaiseha Y, El Halawani M. 2003. Dopaminergic regulation of avian prolactin gene transcription. *J. Mol. Endocrinol*, 31: 185-196.

Amano M, Oka Y, Yamanome T, Okuzawa K, Yamamori K. 2002. Three GnRH systems in the brain and pituitary of a pleuronectiform fish, the barfin flounder *Verasper moseri*. *Cell Tissue Res*, 309: 323-329.

Amano M, Oka Y, Amiya N, Yamamori K. 2007. Immunohistochemical localization and ontogenic development of prolactin-releasing peptide in the brain of the ovoviviparous fish species *Poecilia reticulata* (guppy). *Neurosci. Lett*, 413: 206-209.

Amemiya Y, Sogabe Y, Nozaki M, Takahashi A, Kawauchi H. 1999. Somatolactin in the white sturgeon and African lungfish and its evolutionary significance. *Gen. Comp. Endocrinol*, 114: 181-190.

Ayson F G, Kaneko T, Hasegawa S, Hirano T. 1994. Differential expression of two prolactin and growth hormone genes during early development of tilapia (*Oreochromis mossambicus*) in fresh water and seawater: Implications for possible involvement in osmoregulation during early life stages. *Gen. Comp. Endocrinol*, 95: 143-152.

Barry T P, Grau E G. 1986. Estradiol-17beta and thyrotropin-releasing hormone stimulate prolactin release from the pituitary gland of a teleost fish *in vitro*. *Gen. Comp. Endocrinol*, 62: 306-314.

Bern H A, Madsen S S. 1992. A selective survey of the endocrine system of the rainbow trout (*Oncorhynchus mykiss*) with emphasis on the hormonal regulation of ion balance. *Aquaculture*, 100: 237-262.

Bhandari R K, Taniyama S, Kitahashi T, Ando H, Yamauchi K, Zohar Y, Ueda H, Uranob A. 2003. Seasonal changes of responses to gonadotropin-releasing hormone analog in expression of growth hormone/prolactin/somatolactin genes in the pituitary of masu salmon. *Gen. Comp. Endocrinol*, 130: 55-63.

Blackwell R E, Rodgers-Neame N T, BradleyJr E L, Asch R H. 1986. Regulation of human prolactin secretion by gonadotropin-releasing hormone *in vitro*. *Fertil. Steril*, 46: 26-31.

Bole-Feysot C, Goffin V, Edery M, Binart N, Kelly P A. 1998. Prolactin (PRL) and its receptor: Actions, signal transduction pathways and phenotypes observed in PRL receptor knockout mice. *Endocr. Rev*, 19: 225-268.

Borski R J, Helms L M, Richman N H, Grau E G. 1991. Cortisol rapidly reduces prolactin release and cAMP and $^{45}Ca^{2+}$ accumulation in the cichlid fish pituitary *in vitro*. *Proc. Natl. Acad. Sci. USA*, 88: 2758-2762.

Brinca L, Fuentes J, Power D M. 2003. The regulatory action of estrogen and vasoactive intestinal peptide on prolactin secretion in sea bream (*Sparus aurata*, L.). *Gen. Comp. Endocrinol*, 131: 117-125.

Canepa M M, Pandolfi M, Maggese M C, Vissio P G. 2006. Involvement of somatolactin in background adaptation of the cichlid fish *Cichlasoma dimerus*. *J. Exp. Zool*, 305: 410-419.

Chow B K C, Yuen T T H, Chan K W. 1997. Molecular evolution of vertebrate VIP receptors and functional characterization of a VIP receptor from goldfish *Carassius auratus*. *Gen. Comp. Endocrinol*, 105: 176-185.

Company R, Astola A, Pendon C, Valdivia M M, Perez-Sanchez J. 2001. Somatotropic regulation of fish growth and adiposity: Growth hormone (GH) and somatolactin

(SL) relationship. *Comp. Biochem. Physiol*, 130C: 435-445.

De Jesus E G, Hirano T, Inui Y. 1994. The antimetamorphic effect of prolactin in the Japanese flounder. *Gen. Comp. Endocrinol*, 93: 44-50.

de Ruiter A J, Wendelaar Bonga S E, Slijkhuis H, Baggerman B. 1986. The effect of prolactin on fanning behavior in the male three-spined stickleback, *Gasterosteus aculeatus*L. *Gen. Comp. Endocrinol*, 64: 273-283.

Duan C, Dugua S J, Plisetskaya E M. 1993. Insulin-like growth factor I (IGF-I) mRNA expression in coho salmon, *Oncorhynchus kisutch*: Tissue distribution and effects of growth hormone/ prolactin family proteins. *Fish Physiol. Biochem*, 11: 371-379.

Dubois M P, Billard R, Breton B, Peter R E. 1979. Comparative distribution of somatostatin, LH-RH, neurophysin, and alpha-endorphin in the rainbow trout: An immunocytological study. *Gen. Comp. Endocrinol*, 37: 220-232.

Eckert S M, Hirano T, Leedom T A, Takei Y, Grau G E. 2003. Effects of angiotensin II and natriuretic peptides of the eel on prolactin and growth hormone release in the tilapia. *Oreochromis mossambicus. Gen. Comp. Endocrinol*, 130: 333-339.

Flores C M, Munoz D, Soto M, Kausel G, Romero A, Figueroa J. 2007. Copeptin, derived from isotocin precursor, is a probable prolactin releasing factor in carp. *Gen. Comp. Endocrinol*, 150: 343-354.

Fox B K, Naka T, Inoue K, Takei Y, Hirano T, Grau E G. 2007. *In vitro* effects of homologous natriuretic peptides on growth hormone and prolactin release in the tilapia, *Oreochromis mossambicus. Gen. Comp. Endocrinol*, 150: 270-277.

Freeman M E, Kanyicska B, Lerant A, Nagy G. 2000. Prolactin: Structure, function, and regulation of secretion. *Physiol. Rev*, 80: 1523-1631.

Fruchtman S, Jackson L, Borski R. 2000. Insulin-like growth factor I disparately regulates prolactin and growth hormone synthesis and secretion: Studies using the teleost pituitary model. *Endocrinology*, 141: 2886-2894.

Fujimoto M, Takeshita K, Wang X, Takabatake I, Fujisawa Y, Teranishi H, Ohtani M, Muneoka Y, Ohta S. 1998. Isolation and characterization of a novel bioactive peptide, Carassius RF-amide (C-RFa), from the brain of the Japanese crucian carp. *Biochem. Biophys. Res. Commun*, 242: 436-440.

Fujimoto M, Sakamoto T, Kanetoh T, Osaka M, Moriyama S. 2006. Prolactin-releasing peptide is essential to maintain the prolactin levels and osmotic balance in freshwater teleost fish. *Peptides*, 27: 1104-1109.

Fukada H, Ozaki Y, Pierce A L, Adachi S, Yamauchi K, Hara A, Swanson P, Dickhoff W W. 2004. Salmon growth hormone receptor: Molecular cloning, ligand specificity, and response to fasting. *Gen. Comp. Endocrinol*, 139: 61-71.

Fukada H, Ozaki Y, Pierce A L, Adachi S, Yamauchi K, Hara A, Swanson P,

Dickhoff W W. 2005. Identification of the salmon somatolactin receptor, a new member of the cytokine receptor family. *Endocrinology*, 146: 2354-2361.

Fukamachi S, Meyer A. 2007. Evolution of receptors for growth hormone and somatolactin in fish and land vertebrates: Lessons from the lungfish and sturgeon orthologues. *J. Mol. Evol*, 65: 359-372.

Fukamachi S, Sugimoto M, Mitani H, Shima A. 2004. Somatolactin selectively regulates proliferation and morphogenesis of neural-crest derived pigment cells in medaka. *Proc. Natl. Acad. Sci. USA*, 101: 10661-10666.

Fukamachi S, Yada T, Mitani H. 2005. Medaka receptors for somatolactin and growth hormone: Phylogenetic paradox among fish growth hormone receptors. *Genetics*, 171: 1875-1883.

Fukamachi S, Wakamatsu Y, Mitani H. 2006. Medaka double mutants for color interfere and leucophore free: Characterization of the xanthophore-somatolactin relationship using the leucophore free gene. *Dev. Genes Evol*, 216: 152-157.

González-Martínez D, Zmora N, Mañanos E, Saligaut D, Zanuy S, Zohar Y, Elizur A, Kah O, Muñoz-Cueto J A. 2002. Immunohistochemical localization of three different prepro-GnRHs in the brain and pituitary of the European sea bass (*Dicentrarchus labrax*) using antibodies to the corresponding GnRH-associated peptides. *J. Comp. Neurol*, 446: 95-113.

Gothilf Y, Muñoz-Cueto J A, Sagrillo C A, Selmanoff M, Chen T T, Kah O, Elizur A, Zohar Y. 1996. Three forms of gonadotropin-releasing hormone in a perciform fish (*Sparus aurata*): Complementary deoxyribonucleic acid characterization and brain localization. *Biol. Reprod*, 55: 636-645.

Grau E G, Nishioka R S, Bern H A. 1982. Effects of somatostatin and urotensin II on tilapia pituitary prolactin release and interactions between somatostatin, osmotic pressure, Ca^{2+}, and adenosine 3, 5-monophosphate in prolactin release *in vitro*. *Endocrinology*, 110: 910-914.

Grau E G, Nishioka R S, Young G, Bern H A. 1985. Somatostatin-like immunoreactivity in the pituitary and brain of three teleost fish species: Somatostatin as a potential regulator of prolactin cell function. *Gen. Comp. Endocrinol*, 59: 350-357.

Grau E G, RichmannIII N H, Borski R J. 1994. Osmoreception and a simple endocrine reflex of the prolactin cell of the tilapia *Oreochromis mossambicus*. In: Davey K G, Peter R E, Tobe S S, Ed. *Perspectives in Comparative Endocrinology*. National Research Council of Canada: Ottawa, 251-256.

Grosvenor C E, Mena F. 1980. Evidence that TRH and a hypothalamic prolactin-releasing factor may function in the release of prolactin in the lactating rat. *Endocrinology*, 107: 863-868.

Hamano K, Yoshida K, Suzuki M, Asahida K. 1996. Changes in thyrotropin-releasing hormone concentration in the brain and levels of prolactin and thyroxin in the serum during spawning migration of the chum salmon, *Oncorhynchus keta*. *Gen. Comp. Endocrinol*, 101: 275-281.

Harris J, Bird D J. 2000. Modulation of the fish immune system by hormones. *Vet. Immunol. Immunopathol*, 77: 163-176.

Hazineh A, Shin S H, Reifel G, Pang S C, Van der Kraak G J. 1997. Dopamine causes ultrastructural changes in prolactin cells of tilapia (*Oreochromis niloticus*). *Cell Mol. Life. Sci*, 53: 452-458.

Helms L M, Grau E G, Borski R J. 1991. Effects of osmotic pressure and somatostatin on the cAMP messenger system of the osmosensitive prolactin cell of a teleost fish, the tilapia (*Oreochromis mossambicus*). *Gen. Comp. Endocrinol*, 83: 111-117.

Higashimoto Y, Nakao N, Ohkubo T, Tanaka M, Nakashima K. 2001. Structure and tissue distribution of prolactin receptor mRNA in Japanese flounder (*Paralichthys olivaceus*): Conserved and preferential expression in osmoregulatory organs. *Gen. Comp. Endocrinol*, 123: 170-179.

Hinuma S, Habata Y, Fujii R, Kawamata Y, Hosoya M, Fukusumi S, Kitada C, Masuo Y, Asano T, Matsumoto H, Sekiguchi M, Kurokawa T, Nishimura O, Onda H, Fujino M. 1998. A prolactin-releasing peptide in the brain. *Nature*, 393: 272-276.

Hirano T. 1986. The spectrum of prolactin action in teleosts. In: Ralph C L, Ed. *Comparative Endocrinology: Developments and Directions*. A. R. Liss: New York, 53-74.

Huang X, Jiao B, Fung C K, Zhang Y, Ho W K, Chan C B, Lin H, Wang D, Cheng C H. 2007. The presence of two distinct prolactin receptors in sea bream with different tissue distribution patterns, signal transduction pathways and regulation of gene expression by steroid hormones. *J. Endocrinol*, 194: 373-392.

Huising M O, Kruiswijk C P, Flik G. 2006. Phylogeny and evolution of class-I helical cytokines. *J. Endocrinol*, 189: 1-25.

Imaoka T, Matsuda M, Mori T. 2000. Extrapituitary expression of the prolactin gene in the goldfish, African clawed frog and mouse. *Zool. Sci*, 17: 791-796.

Inoue K, Naruse K, Yamagami S, Mitani H, Suzuki N, Takei Y. 2003. Four functionally distinct C-type natriuretic peptides found in fish reveal evolutionary history of the natriuretic peptide system. *Proc. Natl. Acad. Sci. USA*, 100: 10079-10084.

Inoue K, Sakamoto T, Yuge S, Iwatani H, Yamagami S, Tsutsumi M, Hori H, Cerra M C, Tota B, Suzuki N, Okamoto N, Takei Y. 2005. Structural and functional evolution of three cardiac natriuretic peptides. *Mol. Biol. Evol*, 2005: 2428-2434.

James V A, Wigham Y. 1984. Evidence for dopaminergic and serotonergic regulation of prolactin cell activity in the trout *Salmo gairdneri*. *Gen. Comp. Endocrinol*, 56: 231-239.

Johnson L L, Norberg B, Willis M L, Zebroski H, Swanson P. 1997. Isolation, characterization, and radioimmunoassay of Atlantic halibut somatolactin and plasma levels during stress and reproduction in flatfish. *Gen. Comp. Endocrinol*, 105: 194-209.

Johnston L R, Wigham T. 1988. The intracellular regulation of prolactin cell function in the rainbow trout, *Salmo gairdneri*. *Gen. Comp. Endocrinol*, 71: 284-289.

Jose P A, Yu P Y, Yamaguchi I, Eisner G M, Mouradian M M, Felder C C, Felder R A. 1999. Dopamine D1 receptor regulation of phospholipase C. *Hypertens. Res*, 18: 39-42.

Kagabu Y, Mishiba T, Okino T, Yanagisawa T. 1998. Effects of thyrotropin-releasing hormone and its metabolites, Cyclo (His-Pro) and TRH-OH, on growth hormone and prolactin synthesis in primary cultured pituitary cells of the common carp, *Cyprinus carpio*. *Gen. Comp. Endocrinol*, 111: 395-403.

Kah O, Chambolle P, Dubourg P, Dubois M P. 1982. Immunocytochemical distribution of somatostatin in the forebrain of two teleosts, the goldfish (*Carassius auratus*) and Gambusia sp. *CR Seances Acad Sci III*, 294: 519-524.

Kaiya H, Kojima M, Hosoda H, Koda A, Yamamoto K, Kitajima Y, Matsumoto M, Minamitake Y, Kikuyama S, Kangawa K. 2001. Bullfrog ghrelin is modified by n-octanoic acid at its third threonine residue. *J. Biol. Chem*, 276: 40441-40448.

Kaiya H, Kojima M, Hosoda H, Moriyama S, Takahashi A, Kawauchi H, Kangawa K. 2003. Peptide purification, complementary deoxyribonucleic acid (DNA) and genomic DNA cloning, and functional characterization of ghrelin in rainbow trout. *Endocrinology*, 144: 5215-5226.

Kaiya H, Kojima M, Hosoda H, Riley L G, Hirano T, Grau E G, Kangawa K. 2003. Identification of tilapia ghrelin and its effects on growth hormone and prolactin release in the tilapia, *Oreochromis mossambicus*. *Comp. Biochem. Physiol. B & Biochem. Mol. Biol*, 135: 421-429.

Kaiya H, Kojima M, Hosoda H, Riley L G, Hirano T, Grau E G, Kangawa K. 2003. Amidated fish ghrelin: Purification, cDNA cloning in the Japanese eel and its biological activity. *J. Endocrinol*, 176: 415-423.

Kajimura S, Seale A P, Hirano T, Cooke I M, Grau E G. 2005. Physiological concentrations of ouabain rapidly inhibit prolactin release from the tilapia pituitary. *Gen. Comp. Endocrinol*, 143: 240-250.

Kajita Y, Sakai M, Kobayashi M, Kawauchi H. 1992. Enhancement of non-specific cytotoxic activity of leucocytes in rainbow trout *Oncorhynchus mykiss* injected with growth hormone. *Fish Shellfish Immunol*, 2: 155-157.

Kakizawa S, Kaneko T, Hasegawa S, Hirano T. 1993. Activation of somalactin cells in the pituitary of the rainbow trout *Oncorhynchus mykiss* by low environmental calcium. *Gen. Comp. Endocrinol*, 91: 298-306.

Kakizawa S, Kaneko T, Hasegawa S, Hirano T. 1995. Effects of feeding, fasting, background adaptation, acute stress, and exhaustive exercise on the plasma somatolactin concentrations in rainbow trout. *Gen. Comp. Endocrinol*, 98: 137-146.

Kakizawa S, Kaneko T, Ogasawara T, Hirano T. 1995. Change in plasma somatolactin levels during spawning migration of chum salmon (*Oncorhynchus keta*). *Fish Physiol. Biochem*, 14: 93-101.

Kakizawa S, Kaneko T, Hirano T. 1996. Elevation of plasma somatolactin concentrations during acidosis in rainbow trout (*Oncorhynchus mykiss*). *J. Exp. Biol*, 199: 1043-1051.

Kakizawa S, Kaneko T, Hirano T. 1997. Effects of hypothalamic factors on somatolactin secretion from the organ-cultured pituitary of rainbow trout. *Gen. Comp. Endocrinol*, 105: 71-78.

Kaneko T, Hirano T. 1993. Role of prolactin and somatolactin in calcium regulation in fish. *J. Exp. Biol*, 184: 31-45.

Kaneko T, Kakizawa S, Yada T. 1993. Pituitary of "cobalt" variant of the rainbow trout separated from the hypothalamus lacks most pars intermedial and neurohypophysial tissue. *Gen. Comp. Endocrinol*, 92: 31-40.

Kawauchi H, Suzuki K, Yamazaki T, Moriyama S, Nozaki M, Yamaguchi K, Takahashi A, Youson J, Sower S A. 2002. Identification of growth hormone in the sea lamprey, an extant representative of a group of the most ancient vertebrates. *Endocrinology*, 143: 4916-4921.

Kelly K M, Nishioka R S, Bern H A. 1988. Novel effect of vasoactive intestinal polypeptide and peptide histidine isoleucine: Inhibition of *in vitro* secretion of prolactin in the tilapia, *Oreochromis mossambicus*. *Gen. Comp. Endocrinol*, 72: 98-106.

Kelly S P, Peter R E. 2006. Prolactin-releasing peptide, food intake, and hydromineral balance in goldfish. *Am. J. Physiol. Regul. Integr. Comp. Physiol*, 291: R1474-R1481.

Kiilerich P, Kristiansen K, Madsen S S. 2007. Hormone receptors in gills of smolting Atlantic salmon, *Salmo salar*: expression of growth hormone, prolactin, mineralocorticoid and glucocorticoid receptors and 11beta-hydroxysteroid dehydrogenase type 2. *Gen. Comp. Endocrinol*, 152: 295-303.

Kojima M, Hosoda H, Date Y, Nakazato M, Matsuo H, Kangawa K. 1999. Ghrelin is a growth-hormone-releasing acylated peptide from stomach. *Nature*, 402: 656-660.

Kozaka T, Fujii Y, Ando M. 2003. Central effects of various ligands on drinking behavior in eels acclimated to seawater. *J. Exp. Biol*, 206: 687-692.

Lagerström M C, Fredriksson R, Bjarnadóttir T K, Fridmanis D, Holmquist T, Andersson J, Yan Y L, Raudsepp T, Zoorob R, Kukkonen J P, Lundin L G, Klovins J, et al. 2005. Origin of the prolactin-releasing hormone (PRLH) receptors: Evidence of coevolution between PRLH and a redundant neuropeptide Y receptor during vertebrate evolution. *Genomics*, 85: 688-703.

Lee K M, Kaneko T, Aida K. 2006. Prolactin and prolactin receptor expressions in a marine teleost, pufferfish *Takifugu rubripes*. *Gen. Comp. Endocrinol*, 146: 318-328.

Leedom T A, Hirano T, Grau E G. 2003. Effect of blood withdrawal and angiotensin II on prolactin release in the tilapia, *Oreochromis mossambicus*. *Comp. Biochem. Physiol. A & Mol. Integr. Physiol*, 135: 155-163.

Leena S, Shameena B, Oommen O V. 2001. *In vivo* and *in vitro* effects of prolactin and growth hormone on lipid metabolism in a teleost, *Anabas testudineus* (Bloch). *Comp. Biochem. Physiol. B*, 128: 761-766.

Lu M, Swanson P, Renfro J L. 1995. Effect of somatolactin and related hormones on phosphate transport by flounder renal tubule primary cultures. *Am. J. Physiol*, 268: R577-R582.

Lynn S G, Shepherd B S. 2007. Molecular characterization and sex-specific tissue expression of prolactin, somatolactin and insulin-like growth factor-I in yellow perch (*Perca flavescens*). *Comp. Biochem. PhysiolB*, 147: 412-427.

Manzon L A. 2002. The role of prolactin in fish osmoregulation. *Gen. Comp. Endocrinol*, 125: 291-310.

Marchant T A, Dulka J G, Peter R E. 1989. Relationship between serum growth hormone levels and the brain and pituitary content of immunoreactive somatostatin in the goldfish. *Carassius auratus*L. *Gen. Comp. Endocrinol*, 73: 458-468.

Matsuda K, Takei Y, Katoh J, Shioda S, Arimura A, Uchiyama M. 1997. Isolation and structural characterization of pituitary adenylate cyclase activating polypeptide (PACAP) -like peptide from the brain of a teleost, stargazer, *Uranoscopus japonicus*. *Peptides*, 18: 723-727.

Matsuda K, Nagano Y, Uchiyama M, Onoue S, Takahashi A, Kawauchi H, Shioda S. 2005. Pituitary adenylate cyclase-activating polypeptide (PACAP) -like immunoreactivity in the brain of a teleost, *Uranoscopus japonicus*: Immunohistochemical relationship between PACAP and adenohypophysial hormones. *Regul. Pept*, 126:

129-136.

Matsuda K, Nagano Y, Uchiyama M, Takahashi A, Kawauchi H. 2005. Immunohistochemical observation of pituitary adenylate cyclase-activating polypeptide (PACAP) and adenohypophysial hormones in the pituitary of a teleost, *Uranoscopus japonicus*. *Zool. Sci*, 22: 71-76.

Matsuda K, Nejigaki Y, Satoh M, Shimaura C, Tanaka M, Kawamoto K, Uchiyama M, Kawauchi H, Shioda S, Takahashi A. 2008. Effect of pituitary adenylate cyclase-activating polypeptide (PACAP) on prolactin and somatolactin release from the goldfish pituitary *in vitro*. *Regul. Pept*, 145: 72-79.

McCormick S D. 2001. Endocrine control of osmoregulation in teleost fish. *Am. Zool*, 41: 781-794.

Mezey E, Kiss J Z. 1985. Vasoactive intestinal polypeptide-containing neurons in the paraventricular nucleus may participate in regulating prolactin secretion. *Proc. Natl. Acad. Sci. USA*, 82: 245-247.

Millar R P, Lu Z L, Pawson A J, Flanagan C A, Morgan K, Maudsley S R. 2004. Gonadotropin-releasing hormone receptors. *Endocrinol. Rev*, 25: 235-275.

Mingarro M, Vega-Rubín de Celis S, Astola A, Pendón C, Valdivia M M, Pérez-Sánchez J. 2002. Endocrine mediators of seasonal growth in gilthead sea bream (*Sparus aurata*): The growth hormone and somatolactin paradigm. *Gen. Comp. Endocrinol*, 128: 102-111.

Miyata A, Arimura A, Dahl R R, Minamino N, Uehara A, Jiang L, Culler M D, Coy D H. 1989. Isolation of a novel 38 residue-hypothalamic polypeptide which stimulates adenylate cyclase in pituitary cells. *Biochem. Biophys. Res. Commun*, 164: 567-574.

Montefusco-Siegmund R A, Romero A, Kausel G, Muller M, Fujimoto M, Figueroa J. 2006. Cloning of the prepro C-RFa gene and brain localization of the active peptide in Salmo salar. *Cell Tissue Res*, 325: 277-285.

Montero M, Yon L, Rousseau K, Arimura A, Fournier A, Dufour S, Vaudry H. 1998. Distribution, characterization, and growth hormone-releasing activity of pituitary adenylate cyclase-activating polypeptide in the European eel, *Anguilla anguilla*. *Endocrinology*, 139: 4300-4310.

Moriyama S, Ito T, Takahashi A, Amano M, Sower S A, Hirano T, Yamamori K, Kawauch H. 2002. A homolog of mammalian PRL-releasing peptide (fish arginyl-phenylalanyl-amide peptide) is a major hypothalamic peptide of PRL release in teleost fish. *Endocrinology*, 143: 2071-2079.

Moriyama S, Kasahara M, Amiya N, Takahashi A, Amano M, Sower S A, Yamamori K, Kawauchi H. 2007. RFamide peptides inhibit the expression of melanotropin and growth hormone genes in the pituitary of an Agnathan, the sea lamprey, *Petromyzon marinus*. *Endocrinology*, 148: 3740-3749.

Mousa A, Mousa S A. 2003. Immunohistochemical localization of gonadotropin releasing hormones in the brain and pituitary gland of the Nile perch, *Lates niloticus* (Teleostei, Centropomidae). *Gen. Comp. Endocrinol*, 130: 245-255.

Nagahama Y, Nishioka R, Bern H A, Gunther R L. 1975. Control of prolactin secretion in teleosts, with special reference to *Gillichthys mirabilis* and *Tilapia mossambica*. *Gen. Comp. Endocrinol*, 25: 166-188.

Nagy G, Mulchahey J J, Smyth D J, Neull J D. 1988. The glycopeptide moiety of vasopressin-neurophysin precursor is neurohypophysial prolactin releasing factor. *Biochem. Biophys. Res. Commun*, 151: 524-529.

Narnaware Y K, Kelly S P, Woo N Y. 1998. Stimulation of macrophage phagocytosis and lymphocyte count by exogenous prolactin administration in silver sea bream (*Sparus sarba*) adapted to hyper- and hypo-osmotic salinities. *Vet. Immunol. Immunopathol*, 61: 387-391.

Nguyen N, Sugimoto M, Zhu Y. 2006. Production and purification of recombinant somatolactin beta and its effects on melanosome aggregation in zebrafish. *Gen. Comp. Endocrinol*, 145: 182-187.

Nicoll C S. 1993. Role of prolactin and placental lactogens in vertebrate growth and development. In: Schreibman M P, Scanes C G, Pang P K T, Ed. *The Endocrinology of Growth, Development, and Metabolism in Vertebrates*. Academic Press: New York, 183-219.

Nillni E A, Sevarino K A. 1999. The biology of pro-thyrotropin-releasing hormone-derived peptides. *Endo. Rev*, 20: 599-648.

Ogasawara T, Sakamoto T, Hirano T. 1996. Prolactin kinetics during freshwater adaptation of maturing chum salmon. *Zool. Sci*, 13: 443-447.

Ogawa M. 1970. Effects of prolactin on epidermal mucous cells of goldfish, *Carassius auratus* L. *Can. J. Zool*, 48: 501

Oliveira L, Paiva A C, Sander C, Vriend G. 1994. A common step for signal transduction in G protein-coupled receptors. *Trends Pharmacol. Sci*, 15: 170-172.

Olivereau M, Ollevier F, Vandesande F, Olivereau J. 1984. Somatostatin in the brain and the pituitary of some teleosts. Immunocytochemical identification and the effect of starvation. *Cell Tissue Res*, 238: 289-296.

Olivereau M, Ollevier F, Vandesande F, Verdonck W. 1984. Immunocytochemical identification of CRF-like and SRIF-like peptides in the brain and the pituitary of cyprinid fish. *Cell Tissue Res*, 237: 379-382.

Olivereau M, Rand-Weaver M. 1994. Immunoreactive somatolactin cells in the pituitary of young, migrating, spawning and spent chinook salmon, *Oncorhynchus tshawytscha*. *Fish Physiol. Biochem*, 13: 141-151.

Olivereau M, Rand-Weaver M. 1994. Immunocytochemical study of the somatolactin cells in the pituitary of Pacific salmons, *Oncorhynchus nerka* and *O. keta*, at some stages of the reproductive cycle. *Gen. Comp. Endocrinol*, 93: 28-35.

Ono M, Kawauchi H. 1994. The somatolactin gene. In: Farrell A P, Randall D J, Sherwood N M, Hew C L, Ed. *Fish Physiology*. Vol. 13. Academic Press: San Diego, 159-177.

Ono M, Takayama Y, Rand-Weaver M, Sakata S, Yasunaga T, Noso T, Kawauchi H. 1990. cDNA cloning of somatolactin, a pituitary protein related to growth hormone and prolactin. *Proc. Natl. Acad. Sci. USA*, 87: 4330-4334.

Ono M, Harigai T, Kaneko T, Sato Y, Ihara S, Kawauchi H. 1994. Pit-1/GH factor-1 involvement in the gene expression of somatolactin. *Mol. Endocrinol*, 8: 109-115.

Onuma T, Kitahashi T, Taniyama S, Saito D, Ando H, Urano A. 2003. Changes in expression of genes encoding gonadotropin subunits and growth hormone/prolactin/somatolactin family hormones during final maturation and freshwater adaptation in prespawning chum salmon. *Endocrine*, 20: 23-34.

Onuma T, Ando H, Koide N, Okada H, Urano A. 2005. Effects of salmon GnRH and sex steroid hormones on expression of genes encoding growth hormone/prolactin/somatolactin family hormones and a pituitary-specific transcription factor in masu salmon pituitary cells *in vitro*. *Gen. Comp. Endocrinol*, 143: 129-141.

Parhar I S. 1997. GnRH in tilapia: Three genes, three origins and their roles. In: Parhar I S, Sakuma Y, Ed. *GnRH Neurons: Gene to Behavior*. Brain Shuppan: Tokyo, 99-122.

Parhar I S, Iwata M. 1994. Gonadotropin releasing hormone (GnRH) neurons project to growth hormone and somatolactin cells in the Steelhead trout. *Histochemistry*, 102: 195-203.

Parhar I S, Ogawa S, Sakuma Y. 2005. Three GnRH receptor types in laser-captured single cells of the cichlid pituitary display cellular and functional heterogeneity. *Proc. Natl. Acad. Sci. USA*, 102: 2204-2209.

Parker D B, Power M E, Swanson P, River J, Sherwood N M. 1997. Exon skipping in the gene encoding pituitary adenylate cyclase-activating polypeptide in salmon alters the expression of two hormones that stimulate growth hormone release. *Endocrinology*, 138: 414-423.

Peyon P, Zanuy S, Carrillo M. 2001. Action of leptin on *in vitro* luteinizing hormone release in the European sea bass (*Dicentrarchus labrax*). *Biol. Reprod*, 65:

1573-1578.

Peyon P, Vega-Rubin de Celis S, Gomez-Requeni P, Zanuy S, Perez-Sanchez J, Carrillo M. 2003. In vitro effect of leptin on somatolactin release in the European sea bass (*Dicentrarchus labrax*): Dependence on the reproductive status and interaction with NPY and GnRH. *Gen. Comp. Endocrinol*, 132: 284-292.

Pickford G E, Phillips J G. 1959. Prolactin, a factor in promoting survival of hypophysectomized killifish in fresh water. *Science*, 130: 454-455.

Pierce A L, Fox B K, Davis L K, Visitacion N, Kitahashi T, Hirano T, Grau E G. 2007. Prolactin receptor, growth hormone receptor, and putative somatolactin receptor in Mozambique tilapia: Tissue specific expression and differential regulation by salinity and fasting. *Gen Comp. Endocrinol*, 154: 31-40.

Planas J V, Swanson P, Rand-Weaver M, Dickhoff W W. 1992. Somatolactin stimulates *in vitro* gonadal steroidogenesis in coho salmon, *Oncorhynchus kisutch*. *Gen. Comp. Endocrinol*, 87: 1-5.

Potter L R, Abbey-Hosch S, Dickey D M. 2006. Natriuretic peptides, their receptors, and cyclic guanosine monophosphate-dependent signaling functions. *Endocrinol. Rev*, 27: 47-72.

Pottinger T G, Rand-Weaver M, Sumpter J P. 2003. Overwinter fasting and re-feeding in rainbow trout: Plasma growth hormone and cortisol levels in relation to energy mobilisation. *Comp. Biochem. Physiol. B & Biochem. Mol. Biol*, 136: 403-417.

Power D M. 2005. Developmental ontogeny of prolactin and its receptor in fish. *Gen. Comp. Endocrinol*, 142: 25-33.

Power D M, Canario A V, Ingleton P M. 1996. Somatotropin release-inhibiting factor and galanin innervation in the hypothalamus and pituitary of sea bream (*Sparus aurata*). *Gen. Comp. Endocrinol*, 101: 264-274.

Prunet P, Gonnet F. 1986. *Prolactin secretion in the rainbow trout: Effect of osmotic pressure and characterization of a hypothalamic PRL releasing activity In*: Neuroendocrinology. 114, Abstract of the 1st International Congress of Neuroendocrinology. Karger, Basel.

Prunet P, Sandra O, Le Rouzic P, Marchand O, Laudet V. 2000. Molecular characterization of the prolactin receptor in two fish species, tilapia *Oreochromis niloticus* and rainbow trout, *Oncorhynchus mykiss*: A comparative approach. *Can. J. Physiol. Pharmacol*, 78: 1086-1096.

Rand-Weaver M, Swanson P. 1993. Plasma somatolactin levels in coho salmon (*Oncorhynchus kisutch*) during smoltification and sexual maturation. *Fish Physiol. Biochem*, 11: 175-182.

Rand-Weaver M, Noso T, Muramoto K, Kawauchi H. 1991. Isolation and characterization of somatolactin, a new protein related to growth hormone and prolactin from Atlantic cod (*Gadus morhua*) pituitary glands. *Biochemistry*, 30: 1509-1515.

Rand-Weaver M, Baker B I, Kawauchi H. 1991. Cellular localization of somatolactin in the pars intermedia of some teleost fish. *Cell Tissue Res*, 263: 207-215.

Rand-Weaver M, Swanson P, Kawauchi H, Dickhoff W W. 1992. Somatolactin, a novel pituitary protein: Purification and plasma levels during reproductive maturation of coho salmon. *J. Endocrinol*, 133: 393-403.

Rand-Weaver M, Kawauchi H, Ono M. 1993. Evolution of the structure of the growth hormone and prolactin family. In: Schreibman M P, Scanes C G, Pang P K T, Ed. *The Endocrinology of Growth, Development, and Metabolism in Vertebrates*. Academic Press: New York, 13-42.

Rand-Weaver M, Pottinger T G, Sumpter J P. 1995. Pronounced seasonal rhythms in plasma somatolactin levels in rainbow trout. *J. Endocrinol*, 146: 113-119.

Robbercht P, Deschodt-Lanckman M, Camus J C, de-Neef P, Lambert M, Christophe. 1979. VIP activation of rat anterior pituitary adenylate cyclase. *FEBS Lett*, 103: 229-233.

Robberecht W, Andries M, Denef C. 1992. Stimulation of prolactin secretion from rat pituitary by luteinizing hormone-releasing hormone: Evidence against mediation by angiotensin II acting through a (Sar1-Ala8)-angiotensin II-sensitive receptor. *Neuroendocrinology*, 56: 185-194.

Rubin D A, Specker J L. 1992. *In vitro* effects of homologous prolactins on testosterone production by testes of tilapia (*Oreochromis mossambicus*). *Gen. Comp. Endocrinol*, 87: 189-196.

Sakai M, Kobayashi M, Kawauchi H. 1995. Enhancement of chemiluminescent responses of phagocytic cells from rainbow trout, *Oncorhynchus mykiss*, by injection of growth hormone. *Fish Shellfish Immunol*, 5: 375-379.

Sakai M, Kobayashi M, Kawauchi H. 1996. *In vitro* activation of fish phagocytic cells by GH, prolactin and somatolactin. *J. Endocrinol*, 151: 113-118.

Sakai M, Kobayashi M, Kawauchi H. 1996. Mitogenic effect of growth hormone and prolactin on chum salmon *Oncorhynchus keta*leukocytes *in vitro*. *Ver. Immunol. Immunopathol*, 53: 185-189.

Sakamoto T, McCormick S D. 2006. Prolactin and growth hormone in fish osmoregulation. *Gen. Comp. Endocrinol*, 147: 24-30.

Sakamoto T, Fujimoto M, Ando M. 2003. Fishy tales of prolactin-releasing peptide. *Int. Rev. Cytol*, 225: 91-130.

Sakamoto T, Agustsson T, Moriyama S, Itoh T, Takahashi A, Kawauchi H, Bjoarnsson B Th, Ando M. 2003. Intra-arterial injection of prolactin-releasing peptide elevates prolactin gene expression and plasma prolactin levels in rainbow trout. *J. Comp. Physiol. B*, 173: 333-337.

Sakamoto T, Amano M, Hyodo S, Moriyama S, Takahashi A, Kawauchi H, Ando M. 2005. Expression of prolactin-releasing peptide and prolactin in the euryhaline mudskippers (*Periophthalmus modestus*): prolactin-releasing peptide as a primary regulator of prolactin. *J. Mol. Endocrinol*, 34: 825-834.

Sandra O, Sohm F, De Luze A, Prunet P, Edery M, Kelly P A. 1995. Expression cloning of a cDNA encoding a fish prolactin receptor. *Proc. Natl. Acad. Sci. USA*, 92: 6037-6041.

Sandra O, Le Rouzic P, Cauty C, Edery M, Prunet P. 2000. Expression of the prolactin receptor (tiPRL-R) gene in tilapia *Oreochromis niloticus*: Tissue distribution and cellular localization in osmoregulatory organs. *J. Mol. Endocrinol*, 24: 215-224.

San Martin R, Caceres P, Azocar R, Alvarez M, Molina A, Vera M I, Ktauskopf M. 2004. Seasonal environmental changes modulate the prolactin receptor expression in an eurythermal fish. *J. Cell. Biochem*, 92: 42-52.

San Martin R, Hurtado W, Quezada C, Reyes A E, Vera M I, Krauskopf M. 2007. Gene structure and seasonal expression of carp fish prolactin short receptor isoforms. *J. Cell. Biochem*, 100: 970-980.

Santos C R, Brinca L, Ingleton P M, Power D M. 1999. Cloning, expression, and tissue localisation of prolactin in adult sea bream (*Sparus aurata*). *Gen. Comp. Endocrinol*, 114: 57-66.

Santos C R A, Ingleton P M, Cavaco J E B, Kelly P A, Edery M, Power D M. 2001. Cloning, characterization, and tissue distribution of prolactin receptor in the sea bream (*Sparus aurata*). *Gen. Comp. Endocrinol*, 121: 32-47.

Santos C R, Cavaco J E, Ingleton P M, Power D M. 2003. Developmental ontogeny of prolactin and prolactin receptor in the sea bream (*Sparus aurata*). *Gen. Comp. Endocrinol*, 132: 304-314.

Satake H, Minakata H, Wang X, Fujimoto M. 1999. Characterization of a cDNA encoding a precursor of Carassius RFamide, structurally related to a mammalian prolactin-releasing peptide. *FEBS Lett*, 446: 247-250.

Sawisky G R, Chang J P. 2005. Intracellular calcium involvement in pituitary adenylate cyclase-activating polypeptide stimulation of growth hormone and gonadotropin secretion in goldfish pituitary cells. *J. Neuroendocrinol*, 17: 353-371.

Seale A P, Itoh T, Moriyama S, Takahashi A, Kawauchi H, Sakamoto T, Fujimoto M, Riley L G, Hirano T, Grau E G. 2002. Isolation and characterization of a homologue of mammalian prolactin-releasing peptide from the tilapia brain and its effect on prolactin release from the tilapia pituitary. *Gen. Comp. Endocrinol*, 125: 328-339.

Sealfon S C, Weinstein H, Millar R P. 1997. Molecular mechanisms of ligand interaction with the gonadotropin-releasing hormone receptor. *Endocrinol. Rev*, 18: 180-205.

Shepherd B S, Sakamoto T, Nishioka R S, Richman N H, Mori I, Madsen S S, Chen T T, Hirano T, Bern H A, Grau E G. 1997. Somatotropic actions of the homologous growth hormone and prolactins in the euryhaline teleost, tilapia, *Oreochromis mossambicus*. *Proc. Natl. Acad. Sci. USA*, 94: 2068-2072.

Shepherd B S, Sakamoto T, Hyodo S, Nishioka R S, Ball C, Bern H A, Grau E G. 1999. Is the primitive regulation of pituitary prolactin ($tPRL_{177}$ and $tPRL_{188}$) secretion and gene expression in the euryhaline tilapia (*Oreochromis mossambicus*) hypothalamic or environmental?. *J. Endocrinol*, 161: 121-129.

Sheridan M A. 1986. Effects of thyroxin, cortisol, growth hormone, and prolactin on lipid metabolism of coho salmon, *Oncorhynchus kisutch*, during smoltification. *Gen. Comp. Endocrinol*, 64: 220-238.

Sheridan M A, Kittilson J D, Slagter B J. 2000. Structure-function relationships of the signaling system for the somatostatin peptide hormone family. *Am. Zool*, 40: 269-286.

Shiraishi K, Matsuda M, Mori T, Hirano T. 1999. Changes in expression of prolactin and cortisol receptor genes during early life-stages of euryhaline tilapia (*Oreochromis mossambicus*) in fresh water and seawater. *Zool. Sci*, 16: 139-146.

Slagter B J, Kittilson J D, Sheridan M A. 2004. Somatostatin receptor subtype 1 and subtype 2 mRNA expression is regulated by nutritional state in rainbow trout (*Oncorhynchus mykiss*). *Gen. Comp. Endocrinol*, 139: 236-244.

Slijkhuis H, de Ruiter A J, Baggerman B, Wendelaar Bonga S E. 1984. Parental fanning behavior and prolactin cell activity in the male three-spined stickleback *Gasterosteus aculeatus* L. *Gen. Comp. Endocrinol*, 54: 297-307.

Stefano A V, Vissio P G, Paz D A, Somoza G M, Maggese M C, Barrantes G E. 1999. Colocalization of GnRH binding sites with gonadotropin-, somatotropin-, somatolactin-, and prolactin-expressing pituitary cells of the pejerrey, *Odontesthes bonariensis*, in vitro. *Gen. Comp. Endocrinol*, 116: 133-139.

Sun B, Fujiwara K, Adachi S, Inoue K. 2005. Physiological roles of prolactin-releasing peptide. *Regul. Pept*, 126: 27-33.

Sun Y, Lu X, Gershengorn M C. 2003. Thyrotropin-releasing hormone receptors - similarities and differences. *J. Mol. Endocrinol*, 30: 87-97.

Tacon P, Baroiller J F, Le Bail P Y, Prunet P, Jalabert B. 2000. Effect of egg deprivation on sex steroids, gonadotropin, prolactin, and growth hormone profiles during the reproductive cycle of the mouthbrooding cichlid fish *Oreochromis niloticus*. *Gen. Comp. Endocrinol*, 117: 54-65.

Takayama Y, Ono M, Rand-Weaver M, Kawauchi H. 1991. Greater conservation of somatolactin, a presumed pituitary hormone of the growth hormone/prolactin family, than of growth hormone in teleost fish. *Gen. Comp. Endocrinol*, 83: 366-374.

Takei Y. 2000. Structural and functional evolution of the natriuretic peptide system in vertebrates. *Int. Rev. Cytol*, 194: 1-66.

Taniyama S, Kitahashi T, Ando H, Kaeriyama M, Zohar Y, Ueda H, Urano A. 2000. Effects of gonadotropin-releasing hormone analog on expression of genes encoding the growth hormone/prolactin/somatolactin family and a pituitary-specific transcription factor in the pituitaries of prespawning sockeye salmon. *Gen. Comp. Endocrinol*, 118: 418-424.

Taylor M M, Samson W K. 2001. The prolactin releasing peptides: RFamide peptides. *Cell. Mol. Life Sci*, 58: 1206-1215.

Tipsmark C K, Weber G M, Strom C N, Grau E G, Hirano T, Borski R J. 2005. Involvement of phospholipase C and intracellular calcium signaling in the gonadotropin-releasing hormone regulation of prolactin release from lactotrophs of tilapia (*Oreochromis mossambicus*). *Gen. Comp. Endocrinol*, 142: 227-233.

Toop T, Donald J A. 2004. Comparative aspects of natriuretic peptide physiology in non-mammalian vertebrates: A review. *J. Comp. Physiol. B*, 174: 189-204.

Tse D L Y, Chow B K C, Chan C B, Lee L T O, Cheng C H K. 2000. Molecular cloning and expression studies of a prolactin receptor in goldfish (*Carassius auratus*). *Life Sci*, 66: 593-605.

Tse M C L, Wong G K P, Xiao P, Cheng C H K, Chan K M. 2008. Down-regulation of goldfish (*Carassius auratus*) prolactin gene expression by dopamine and thyrotropin releasing hormone. *Gen. Comp. Endocrinol*, 155: 729-741.

Uchida K, Yoshikawa-Ebesu J S, Kajimura S, Yada T, Hirano T, Grau G E. 2004. *In vitro* effects of cortisol on the release and gene expression of prolactin and growth hormone in the tilapia, *Oreochromis mossambicus*. *Gen. Comp. Endocrinol*, 135: 116-125.

Unniappan S, Lin X, Cervini L, Rivier J, Kaiya H, Kangawa K, Peter R E. 2002. Goldfish ghrelin: Molecular characterization of the complementary deoxyribonucleic acid, partial gene structure and evidence for its stimulatory role in food intake. *Endocrinology*, 143: 4143-4146.

Vega-Rubín de Celis S, Gómez P, Calduch-Giner J A, Médale F, Pérez-Sánchez J. 2003. Expression and characterization of European sea bass (*Dicentrarchus labrax*) somatolactin: Assessment of *in vivo* metabolic effects. *Mar. Biotechnol*, 5: 92-101.

Vega-Rubin de Celis S, Rojas P, Gomez-Requeni P, Albalat A, Gutierrez J, Medale F, Kaushik S J, Navarro I, Perez-Sanchez J. 2004. Nutritional assessment of somatolactin function in gilthead sea bream (*Sparus aurata*): Concurrent changes in somatotropic axis and pancreatic hormones. *Comp. Biochem. Biophys.* A 138: 533-542.

Vissio P G, Stefano A V, Somoza G M, Maggese M C, Paz D A. 1999. Close association among GnRH (Gonadotropin-releasing hormone) fibers and GTH, GH, SL and PRL expressing cells in pejerrey, *Odontesthes bonariensis* (Teleostei, Atheriniformes). *Fish Physiol. Biochem*, 21: 121-127.

Vissio P G, Andreone L, Paz M D, Aaggese M C, Somoza G M, Strussmann C A. 2002. Relation between the reproductive status and somatolactin cell activity in the pituitary of pejerrey, *Odontesthes bonariensis* (Atheriniformes). *J. Exp. Zool*, 293: 492-499.

Wang X, Morishita F, Matsushima O, Fujimoto M. 2000. Immunohistochemical localization of C-RF-amide, a FMRF-related peptide, in the brain of the goldfish, *Carassius auratus*. *Zool. Sci*, 17: 1067-1074.

Wang X, Morishita F, Matsushima O, Fujimoto M. 2000. Carassius RFamide, a novel FMRFa-related peptide, is produced within the retina and involved in retinal information processing in cyprinid fish. *Neurosci. Lett*, 289: 115-118.

Weber G M, Powell J F F, Park M, Fischer W H, Craig A G, Rivier J E, Nanakorn U, Parhar I S, Ngamvongchon S, Grau E G, Sherwood N M. 1997. Evidence that gonadotropinreleasing hormone (GnRH) functions as a prolactin-releasing factor in a teleost fish (*Oreochromis mossambicus*) and primary structures for three native GnRH molecules. *J. Endocrinol*, 155: 121-132.

Weber G M, Seale A P, RichmanIII N H, Stetson M H, Hirano T, Grau E G. 2004. Hormone release is tied to changes in cell size in the osmoreceptive prolactin cell of a euryhaline teleost fish, the tilapia, *Oreochromis mossambicus*. *Gen. Comp Endocrinol*, 138: 8-13.

Wigham T, Batten T F. 1984. *In vitro* effects of thyrotropin-releasing hormone and somatostatin on prolactin and growth hormone release by the pituitary of *Poecilia latipinna*. I. An electrophoretic study. *Gen. Comp. Endocrinol*, 55: 444-449.

Williams A J, Wigham T. 1994. The regulation of prolactin cells in the rainbow trout (*Oncorhynchus mykiss*). I. Possible roles for thyrotropin-releasing hormone (TRH) and oestradiol. *Gen. Comp. Endocrinol*, 93: 388-397.

Williams A J, Wigham T. 1994. The regulation of prolactin cells in the rainbow trout (*Oncorhynchus mykiss*). 2. Somatostatin. *Gen Comp. Endocrinol*, 93: 398-405.

Wirachowsky N R, Kwong P, Yunker W K, Johnson J D, Chang J D. 2000. Mechanisms of action of pituitary adenylate cyclase-activating polypeptide (PACAP) on growth hormone release from dispersed goldfish pituitary cells. *Fish Physiol. Biochem*, 23: 201-214.

Wong A O, Chang J P, Peter R E. 1993. Characterization of D1 receptors mediating dopamine-stimulated growth hormone release from pituitary cells of the goldfish, *Carassius auratus*. *Endocrinology*, 133: 577-584.

Wong A O L, Leung M Y, Shea W L C, Tse L Y, Chang J P, Chow B K C. 1998. Hypophysiotropic action of pituitary adenylate cyclase-activating polypeptide (PACAP) in the goldfish: Immunohistochemical demonstration of PACAP in the pituitary, PACAP stimulation of growth hormone release from pituitary cells, and molecular cloning of pituitary type I PACAP receptor. *Endocrinology*, 139: 3465-3479.

Wong A O L, Li W, Leung C Y, Huo L, Zhou H. 2005. Pituitary adenylate cyclase activating polypeptide (PACAP) and growth hormone (GH) -releasing factor in grass carp. I. functional coupling of cyclic adenosine 35'-monophosphate and Ca^{2+}/calmodulin-dependent signaling pathways in PACAP-induced GH secretion and GH gene expression in grass carp pituitary cells. *Endocrinology*, 146: 5407-5424.

Yada T, Uchida K, Kajimura S, Azuma T, Hirano T, Grau E G. 2002. Immunomodulatory effects of prolactin and growth hormone in the tilapia, *Oreochromis mossambicus*. *J. Endocrinol*, 173: 483-492.

Yada T, Moriyama S, Suzuki Y, Azuma T, Takahashi A, Hirose S, Naito N. 2002. Relationships between obesity and metabolic hormones in the "cobalt" variant of rainbow trout. *Gen. Comp. Endocrinol*, 128: 36-43.

Yang B Y, Chen T T. 2003. Identification of a new growth hormone family protein, somatolactin-like protein, in the rainbow trout (*Oncorhyncus mykiss*) pituitary gland. *Endocrinology*, 144: 850-857.

Yang B Y, Arab M, Chen T T. 1997. Cloning and characterization of rainbow trout (*Oncorhynchus mykiss*) somatolactin cDNA and its expression in pituitary and nonpituitary tissues. *Gen. Comp. Endocrinol*, 106: 271-280.

Yang B Y, Greene M, Chen T T. 1999. Early embryonic expression of the growth hormone family protein genes in the developing rainbow trout, *Oncorhynchus mykiss*. *Mol. Reprod. Dev*, 53: 127-134.

Zhu Y, Thomas P. 1995. Red drum somatolactin: Development of a homologous radioimmunoassay and plasma levels after exposure to stressors or various backgrounds. *Gen. Comp. Endocrinol*, 99: 275-288.

Zhu Y, Thomas P. 1997. Studies on the physiology of somatolactin secretion in red drum and Atlantic croaker. *Fish Physiol. Biochem*, 17: 271-278.

Zhu Y, Thomas P. 1998. Effects of light on plasma somatolactin levels in red drum (*Sciaenops ocellatus*). *Gen. Comp. Endocrinol*, 111: 76-82.

Zhu Y, Yoshiura Y, Kikuchi K, Aida K, Thomas P. 1999. Cloning and phylogenetic relationship of red drum somatolactin cDNA and effects of light on pituitary somatolactin mRNA expression. *Gen. Comp. Endocrinol*, 113: 69-79.

Zhu Y, Stiller J W, Shaner M P, Baldini A, Scemama J L, Capehart A A. 2004. Cloning of somatolactin alpha and beta cDNAs in zebrafish and phylogenetic analysis of two distinct somatolactin subtypes in fish. *J. Endocrinol*, 182: 509-518.

Zupanc G K, Siehler S, Jones E M, Seuwen K, Furuta H, Hoyer D, Yano H. 1999. Molecular cloning and pharmacological characterization of a somatostatin receptor subtype in the gymnotiform fish *Apteronotus albifrons*. *Gen. Comp. Endocrinol*, 115: 333-345.

第6章 促肾上腺皮质激素轴、促黑色素激素轴和促甲状腺激素轴对鱼类应激反应的调控和作用

许多下丘脑因子参与鱼类脑垂体促肾上腺皮质激素细胞、促黑色素激素细胞和促甲状腺激素细胞分泌活动的调控作用。在这些因子当中，肾上腺皮质激素、释放因子（CRF）和促甲状腺激素、释放因子（TRH）刺激，而多巴胺通常抑制所有三个下丘脑-脑垂体轴。CRF系统还在内分泌对应激物（stressor）反应的调控中起主要作用。一般来说，促肾上腺皮质激素轴、促黑色素激素轴和促甲状腺激素轴对应激的反应是种类特异性的，并且取决于给该系统造成的冲击、它的持续时间以及鱼类稳态的复原。在这些内分泌轴之间已经确认存在着相互的作用和反馈的影响。我们设想广泛而多方向的沟通以及在肾上腺皮质激素、促黑色素激素和促甲状腺激素轴之间的相互作用已形成一个"应激网"（"stress web"），作为它们的主要功能，在能量代谢方面起着良好的协调作用。本章综述促肾上腺皮质激素、促黑色素激素和促甲状腺激素轴的调控通道、靶标、作用、它们之间的相互作用以及每个轴参与应激反应的证据。

6.1 导　言

一个"轴"是指以概念的方式来描述与规范一个内分泌系统是怎样被控制的。本章涉及下丘脑-脑垂体轴。下丘脑-脑垂体-肾间腺（HPI）轴和下丘脑-脑垂体-甲状腺（HPT）轴包括下丘脑的释放因子，它们通过脑垂体的远侧部，调控外周皮质醇和甲状腺素生产腺体的分泌量。由于没有鉴定到一个真实的（*Bona fide*）外周内分泌靶标，脑垂体中叶的促黑色素激素细胞就成为促黑色素激素轴的最终靶器官。

正如Ball等（1980）所认为的，脊椎动物下丘脑-脑垂体系统可能是一个由脊索动物的祖先构造进化而来，其脑的腹壁是全身的神经内分泌活动的位置。现存硬骨鱼类的典型组织结构是缺少正中隆起，下丘脑的神经元，特别是视前区（NPO）和外侧结节核（NLT）的神经元直接延伸到脑垂体远侧部（Peter和Fryer, 1984）。许多下丘脑因子同时起着神经递质的作用，参与调控脑垂体促肾上腺皮质激素细胞、促黑色素激素细胞和促甲状腺激素细胞的分泌活动。有些下丘脑因子能调控所有三个轴的作用，表明存在着协同的调控作用。

事实上，现已确认在促肾上腺皮质激素、促黑色素激素和促甲状腺激素轴之间存在着多种相互作用和反馈作用。这些相互作用主要发生在下丘脑和外周的靶腺体。因此，坚持由三个分开的平行活动的轴组成一个模式的调控系统是不容易的。当我们正确评估这个复杂的网络在一个应激物开始产生影响后是如何发挥作用以维持体内稳态

或者重新建立稳定状态时，我们要想进一步了解这种交流沟通和相互作用的生理意义，就必须对三个轴有更为全面的理解。所以，本章的目的是综述现今所掌握的鱼类促肾上腺皮质激素、促黑色素激素和促甲状腺激素轴在基础条件下的以及对应激反应的调控通路和功能。本章亦会着重介绍促肾上腺皮质激素轴、促黑色素激素轴和促甲状腺激素轴相互作用的生理过程。

6.2 脑垂体细胞的下丘脑调控

6.2.1 促肾上腺皮质激素细胞的下丘脑调控

在脑垂体远侧部的促肾上腺皮质激素细胞分泌产生促肾上腺皮质激素（ACTH）和由前体阿黑皮素原（proopiomelanocortin，POMC）衍生的非-乙酰化阿片样 β-内啡肽（non-acetylated opioid β-endorphin，βEND）。已有一些研究确认鱼类的下丘脑因子调控脑垂体 ACTH 的分泌，而对参与 βEND 从脑垂体远侧部特异性释放的调控作用机理还了解得很少。

6.2.1.1 刺激性因子

在影响 ACTH 分泌的多个因子当中（见图6.1），CRF 是硬骨鱼类 ACTH 分泌的主要调节剂（Lederis 等，1994；Flik 等，2006）。对一些鱼类用 CRF 离体灌流（superfusion）远侧部细胞或者整个脑垂体能剂量依存地刺激 ACTH 分泌（见表6.1）。CRF 能在至今曾研究过的硬骨鱼类视前核（NPO）内表达，并在鱼类的一些下丘脑区延伸到脑垂体（Okawara 等，1992；Pepels 等，2002；Aldermen 和 Bernier，2007）。对电鳗（*Apteronatus leptorhynchus*）进行免疫组织化学结合神经元神经束追踪（neuronal tract tracing）的研究提供直接证据表明 NPO 的 CRF-免疫反应细胞亚群（subpopulation）神经分布在脑垂体（Zupane 等，1999）。此外，对十多种硬骨鱼类的免疫组织化学研究表明 CRF-免疫反应神经纤维分布在脑垂体的吻端远侧部（RPD），特别是邻近 ACTH 细胞的位置（Pepels 等，2002；见本书第1章综述）。

有少量的证据亦表明促肾上腺皮质激素细胞内的非-乙酰化 βEND 受到 CRF 的刺激性调控。用羊的 CRF 刺激鲤鱼离体的脑垂体远侧部能显著增加 βEND 释放（van den Burg 等，2001）。尽管非-乙酰化 βEND 是血液循环中 βEND 丰富的类型，而且这种阿片样物质可能起着各种生理作用（见第6.3.1节），但据我们所知，还没有其他的研究去查证参与这种肽类分泌活动的调控作用机理。

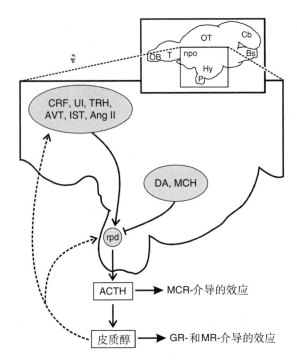

图6.1　影响硬骨鱼类促肾上腺皮质激素轴活性的主要因子的汇总示意

在上方插图框内的中矢切面表示脑的主要分区。放大的是视前区和下丘脑，是下丘脑调控脑垂体促肾上腺皮质激素轴的主要脑区。ACTH是皮质醇分泌的主要促分泌激素，但一些次级因子亦可能参与调控皮质醇的合成与分泌。ACTH和皮质醇的作用由不同的受体亚型介导，而皮质醇能对脑垂体的促肾上腺皮质激素细胞和一些下丘脑调控脑垂体的因子起负反馈作用。要注意这个汇总示意图不是表示普遍化（generalization），而是一个根据不同鱼类得到的实验结果构建的模式（详细内容见正文）。实线箭头表示刺激作用，T线表示抑制作用，虚线箭头表示负反馈作用。AngⅡ，血管紧张素Ⅱ；AVT，精氨酸加压催产素；Bs，脑干；Cb，小脑；CRF，促肾上腺皮质激素释放因子；DA，多巴胺；GR，糖皮质素受体；Hy，下丘脑；IST，硬骨鱼催产素；MCH，黑色素浓集激素；MCR，黑皮质素受体；MR，盐皮质素受体；NPO，视前核；OB，嗅球；OT，视顶盖；P，脑垂体；rpd，吻端远侧部；T，端脑；TRH，促甲状腺激素释放激素；UI，硬骨鱼紧张肽Ⅰ。

表6.1　影响硬骨鱼类ACTH分泌的因子

因子	种类	报道的反应	有效的或试验的剂量	实验观察/评论	参考文献
CRF	金鱼（Carassius auratus）	刺激	2～100 nM	羊CRF：刺激离体灌注的远侧部细胞分泌ACTH	Fryer等（1983）
	金鱼（Carassius auratus）	刺激	10～500 nM	白亚口鱼和人/鼠CRF：刺激离体灌注的前脑垂体细胞分泌ACTH	Lederis等（1994）
	虹鳟（Oncorhynchus mykiss）	刺激	0.1～1000 nM	人/鼠CRF：刺激离体孵育30 min，远侧部碎片分泌ACTH	Baker等（1996）

续表 6.1

因子	种类	报道的反应	有效的或试验的剂量	实验观察/评论	参考文献
CRF	虹鳟 (*Oncorhynchus mykiss*)	刺激	0.1 pM～0.1 μM ED_{50} 0.8 pM	人/鼠 CRF：刺激离体灌注脑垂体分泌 ACTH	Pierson 等（1996）
	金头鲷 (*Sparus auratus*)	刺激	0.01～1000 nM ED_{50} 1.5 nM	人/鼠 CRF：刺激离体灌注脑垂体分泌 ACTH	Rotlant 等（2000b，2001）
	罗非鱼 (*Oreochromis mossambicus*)	刺激	100 nM	罗非鱼和人/鼠 CRF：同等效应的刺激离体灌注脑垂体分泌 ACTH	van Enchevort 等（2000）
	鲤鱼 (*Cyprinus carpio*)	刺激	100 nM	羊 CRF：只在有 DA 的情况下刺激离体灌注的远侧部分泌 ACTH	Metz 等（2004）
UI	金鱼 (*Carassius auratus*)	刺激	0.5～10 nM	白亚口鱼 UI：刺激离体灌注远侧部细胞分泌 ACTH。UI 刺激 ACTH 分泌的效能是羊 CRF 或蛙皮降压肽（sauvagine）的 2～3 倍	Fryer 等（1983）
SVG	金鱼 (*Carassius auratus*)	刺激	2～100 nM	刺激离体灌注远侧部细胞分泌 ACTH	Fryer 等（1983）
AVT	金鱼 (*Carassius auratus*)	刺激	0.063～4 nM	刺激离体灌注远侧部细胞分泌 ACTH	Fryer 等（1983）Lederis 等（1994）
	虹鳟 (*Oncorhynchus mykiss*)	刺激	0.1 pM～1 μM ED_{50} 0.2 nM	刺激离体灌注脑垂体分泌 ACTH	Pierson 等（1996）
	虹鳟 (*Oncorhynchus mykiss*)	刺激	0.1～100 nM	刺激离体孵育 30 min，远侧部碎片分泌 ACTH	Baker 等（1996）
IST	金鱼 (*Carassius auratus*)	刺激	0.063～4 nM	刺激离体灌注远侧部细胞分泌 ACTH	Fryer 等（1985）Lederis 等（1994）
	虹鳟 (*Oncorhynchus mykiss*)	刺激	0.1 pM～1 μM ED_{50} 100 nM	刺激离体灌注脑垂体分泌 ACTH	Pierson 等（1996）

续表 6.1

因子	种类	报道的反应	有效的或试验的剂量	实验观察/评论	参考文献
AVP	金鱼（Carassius auratus）	刺激	0.025～10 nM	刺激离体灌注远侧部细胞分泌 ACTH	Fryer 等（1985）Lederis 等（1994）
血管紧张素	金鱼（Carassius auratus）	刺激	0.5～200 nM	鲑鱼 Ang I、人 Ang I 和 Ang II 刺激离体灌注远侧部细胞分泌 ACTH。Ang I 的效能约为 Ang II 的 1/10	Weld 和 Fryer（1987，1988）
TRH	金鱼（Carassius auratus）	没有作用	2.76 μM（1μg/mL）	TRH 试验对离体灌注远侧部细胞分泌 ACTH 的作用	Fryer 等（1983）
	金头鲷（Sparus auratus）	刺激	50 nM	刺激离体灌注远侧部细胞分泌 ACTH	Rotllant 等（2000 b）
MCH	虹鳟（Oncorhynchus mykiss）	抑制	0.01～100 nM	鲑鱼 MCH：抑制离体孵育 30 min 远侧部基础的和 CRF-诱导的 ACTH 分泌活动	Baker 等（1985）
DA	金鱼（Carassius auratus）	抑制	没有使用	细胞计量分析：腹腔注射 DA 拮抗物的在体处理，使促肾上腺皮质激素细胞和细胞核肥大	Olivereau 等（1988）
	鲤鱼（Cyprinus carpio）	抑制	10 μM	抑制离体灌注的远侧部碎片分泌 ACTH	Metz 等（2004）

ACTH，促肾上腺皮质激素；Ang，血管紧张素；AVP，精氨酸血管加压素；AVT，精氨酸加压催产素；CRF，促肾上腺皮质激素释放因子；DA，多巴胺；ip，腹腔内；IST，硬骨鱼催产素；MCH，黑色素浓集激素；TRH，促甲状腺激素释放激素；SVG，鲑皮降压肽；UI，硬骨鱼紧张肽 I。

硬骨鱼紧张肽（UI）是 CRF 家族肽中的另外一个成员，亦刺激鱼类 ACTH 的分泌活动。事实上，异源肽的效能是明显不同的。例如，白亚口鱼（Catostomus commersoni）的 UI 对刺激金鱼脑垂体远侧细胞分泌 ACTH 的效能要比羊 CRF 强得多（Fryer 等，1983）。UI 的促肾上腺皮质激素作用和 CRF 一样，被非选择性的 CRF 受体拮抗剂 α-螺旋 CRF$_{(9-41)}$ [α-helical CRF$_{(9-41)}$] 所逆转（Weld 等，1987）。为数不多的描述 UI 在脑分布的研究表明它在斑马鱼的 NPO 表达（Aldeman 和 Bernier，2007），亦在白亚口鱼（Yulis 等，1986）、金鱼（Fryer，1989）和斑马鱼（Aldeman 和 Bernier，2007）有脑垂体延伸的下丘脑核中表达。但是，至今还没有明确的证据表明

UI-免疫反应神经纤维分布到鱼类脑垂体的吻端远侧部（RPD）。在哺乳类的相关研究中发现，尾皮质素（urocortin）（UI 的直向同源物）能在离体刺激 ACTH 分泌，但并不参与在体的 ACTH 分泌活动的调控（Oki 和 Sasano，2004）。总之，对于确定鱼类内源 UI 作为一个促肾上腺皮质激素因子的相对重要性，还需做进一步的研究。

如同在哺乳类的相关研究中所了解的，已有充分的证据表明神经垂体激素亦参与鱼类 ACTH 分泌活动的调控。精氨酸加压催产素（AVT）和硬骨鱼催产素（IST）分别和哺乳类的精氨酸血管加压素（arginine vasopressin）和催产素（oxytocin）同源，能刺激离体的金鱼（Fryer 等，1985）和虹鳟（Baker 等，1996；Pierson 等，1996）脑垂体释放 ACTH。在对金鱼的研究中，同源的 AVT 和 IST 刺激 ACTH 释放的最大活性大约是异质的 CRF 和 UI 的一半（Lederis 等，1994），而且和哺乳类的情况不同（Aguilera 等，2008），AVT 和 IST 都不会增强 CRF-相关肽的促肾上腺皮质激素的活性（Fryer 等，1995）。在虹鳟的相关研究中，AVT 和 IST 亦具有比人/鼠 CRF 更低的促进 ACTH-释放的活性，但 AVT 能明显地增强 CRF 促进 ACTH 释放的作用（Baker 等，1996；Pierson 等，1996）。Pierson 等（1996）采用一种药理学的研究方法亦提供证据表明一种 V1-型的血管加压素受体参与介导 AVT-刺激虹鳟的脑垂体释放 ACTH。

在硬骨鱼类的脑内，AVT 和 IST 几乎全部在视前区（NPO）的小细胞和大细胞的神经元中表达，这些神经元中的一部分还延伸到神经垂体（见本书第 1 章）。但有证据表明有些鱼类 AVT 和 IST 的免疫反应纤维和腺脑垂体的促肾上腺素皮质激素细胞直接接触（Batten，1986；Batten 等，1999）。有一些硬骨鱼类的 AVT-免疫反应神经元亦和 CRF 神经元在 NPO 中共定位（Ando 等，1999；Pepels 等，2002；Huising 等，2004）。对于哺乳类，大细胞 AVP 的分泌活动进入外周血液循环取决于渗透的刺激，但小细胞 AVP 的分泌活动进入脑垂体的动脉血液循环并不依赖于渗透状态，并对特异类型急性应激物引起反应而增加分泌，从而增强 CRF 对 ACTH 分泌的刺激作用（Aguilera 等，2008）。同样，血浆的重量摩尔渗透压浓度（osmolality）是鱼类神经脑垂体 AVT 分泌强有力的刺激剂（见第 8 章），而 AVT mRNA 受到各种急性应激物作用而增加在视前区的表达，但这些急性应激物并不需要和渗透的干扰相联系（见第 6.5 节）。有些鱼类 IST 的脑垂体含量和血浆水平受到特异性应激物的作用亦会增加（Mancera 等，2008），但有关 IST 可能对鱼类应激反应起作用的证据还很少。

血管紧张素（Ang）和促甲状腺释放激素（TRH）是影响鱼类 ACTH 释放活动的另外一些因子。血管紧张素Ⅰ（AngⅠ）和血管紧张素Ⅱ（AngⅡ）都能刺激灌注的金鱼脑垂体远侧部细胞分泌 ACTH（Weld 和 Fryer，1987），但它们并不增强 CRF 或 UI 的促肾上腺皮质激素的作用（Weld 和 Fryer，1988）。有趣的是，金鱼视前区许多 AVT-免疫反应神经分泌神经元亦对 AngⅡ产生免疫反应，而且这些神经纤维有些还分布到腺脑垂体的远侧部（Yamada 等，1990）。对哺乳类的研究结果已经充分阐明 AngⅡ能刺激 ACTH 的分泌活动以及肾素血管紧张素系统（renin-angiotensin system）对应激反应的调节作用（Saavedra 和 Benicky，2007），而血管紧张素对鱼类促肾上腺皮质激素细胞的调控是否存在着生理作用还有待研究。虽然 TRH 能够刺激灌注的金头鲷（Sparus auratus）脑垂体分泌 ACTH（Rotllant 等，2000b），但对灌注的金鱼脑垂体远侧部细胞的 ACTH 分泌活动没有作用（Fryer 等，1983）。此外，至今还没有证据表明 TRH-免疫反应纤维分布到硬骨鱼类的脑垂体吻端远侧部（RPD）（见 Cerda-Reverter 和 Canosa 的综述）。最后，单胺类的 5-羟色胺（Winberg 等，1997；Hoglund 等，2002b）和细胞因子白细胞介素 1β（Holland 等，2002；Metz 等，2006a；见本书第 7 章）都能刺激鱼类的 HPI 轴，这些因子是否直接作用于促肾上腺皮质激素细胞，还有待研究确定。采用原位杂交对受体 mRNAs 进行研究将能够回答这个问题。

6.2.1.2 抑制性因子

有些下丘脑因子能抑制鱼类促肾上腺皮质激素细胞释放 ACTH。黑色素浓集激素（MCH）对离体的虹鳟脑垂体远侧部基础的和 CRF-诱导的 ACTH 分泌活动是有力的抑制剂（Baker 等，1985，1986）。大多数 MCH-阳性的纤维延伸到神经脑垂体，它们的产物对背景颜色的适应起着重要作用（Kawauchi，2006），而免疫细胞化学研究的证据表明一些 MCH 纤维分布到 RPD（Naito 等，1986；Batten 等，1999）。此外，适应于光亮背景的虹鳟要比适应于黑暗背景的在血液循环中有较高的 MCH 水平以及较低的血浆 ACTH 和皮质醇水平（Baker 和 Rance，1981；Gilham 和 Baker，1985）。

细胞学和生理学的证据表明多巴胺（DA）能抑制鱼类 ACTH 基础的分泌活动。在金鱼腹腔注射 DA 拮抗剂诱导促肾上腺皮质激素细胞的细胞学变化和多巴胺对 ACTH 合成与分泌的抑制作用相一致（Olivereau 等，1988）。对另一些鲤科鱼类如鲤鱼进行离体的灌注研究表明 ACTH 的在体释放是在多巴胺能的抑制作用控制下进行的，而 CRF 只能在多巴胺抑制的轻微强度时才能刺激 ACTH 分泌（Metz 等，2004）。对几种硬骨鱼类的研究已经证明脑垂体存在着多巴胺能的神经分布（见本书第 1 章综述）。DA 对鲤鱼 ACTH 分泌活动的抑制作用是否是硬骨鱼类促肾上腺皮质激素细胞分泌活动的普遍特点，还有待证实。

6.2.2 下丘脑对促黑色素激素细胞的调控

在脑垂体中间部的促黑色素激素细胞产生 α-促黑激素（α-melanocyte-stimulating hormone，αMSH）和 POMC 衍生的 βEND。这两种激素都受到后-翻译乙酰化作用（post-translational acetylation）而深刻影响它们的功能（见第 6.3.1.2 节和第 6.3.2.1 节）。表 6.2 和表 6.3 表示 TRH 和 CRF 神经肽家族的成员（CRF、UI、蛙皮降压肽）对 αMSH 和 βEND 释放的刺激作用以及 DA 与 MCH 的抑制作用（见图 6.2）前后一致的研究结果。将这些结果和两栖类大批因子参与促黑色素激素细胞分泌活动的调控进行对比，就会得到更深刻的理解。

对于南非爪蟾（Xenopus laevis），来自交叉上核（suprachiasmatic nucleus，NSC）和更多部位的下丘脑因子（CRF、urocortin1、TRH、DA、GABA、NPY），来自下丘脑外的肾上腺素和 5-羟色胺（5-HT）以及自分泌的信息［嘌呤、Ca^{2+}、脑衍生的神经营养因子（brain-derived neurotrophin factor, BDNF）] 都参与 αMSH 释放的刺激性或抑制性调控作用（综述见 Jenks 等, 2007）。对于爪蟾广泛的调节作用以释放一种内分泌产物也许是难以认同的，或者认为这种使黑色素体（melanosome）在皮肤促黑色素激素细胞内扩散是一种简单的外周作用。确实，多种调控通路能表示激素的多种功能。尽管还不能肯定，但硬骨鱼类 αMSH 和 βEND 的多重作用是显而易见的（见第 6.3.2 节综述）。我们可以断定对硬骨鱼类促黑色素激素细胞的研究要比对两栖类的研究欠缺得多，而发现新的调控因子则是可以期待的。

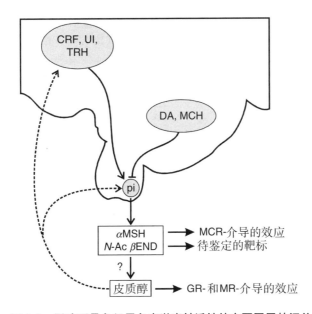

图 6.2　影响硬骨鱼促黑色素激素轴活性的主要因子的汇总示意

表示放大的视前区和下丘脑是参与下丘脑调控脑垂体促黑色素激素轴的主要脑区。促肾上腺皮质激素（ACTH）是分泌皮质醇的主要促分泌激素，但 αMSH 和 N-AcβEND 在有些鱼类中亦参与调控皮质醇的释放（种族特异性用"?"表示）。αMSH 的作用由黑皮质素（melanocortin）受体（MCR）介导，而皮质醇由不同的糖皮质激素受体亚型介导。N-AcβEND 的靶标还有待鉴定。皮质醇能对脑垂体的促黑色素激素细胞以及一些下丘脑调控脑垂体因子起负反馈作用。要注意这个汇总示意图不是表示普遍化，而是一个根据不同鱼类得到的实验结果构建的模式（详见正文）。实线箭头表示刺激作用，T 线表示抑制作用，虚线箭头表示负反馈作用。CRF, 促肾上腺皮质激素释放因子；DA, 多巴胺；GR, 糖皮质激素受体；MCH, 黑色素浓集激素；MR, 盐皮质激素受体；pi, 中间部；TRH, 促甲状腺素释放激素；UI, 硬骨鱼紧张肽 I。

表 6.2 影响硬骨鱼类 αMSH 分泌活动的因子

因子	种类	报道的效应	有效的或试验的剂量	试验的观察/评论	参考文献
CRF	金鱼 (Carassius auratus)	刺激	1～64 nM	羊CRF；刺激离体灌注的 PI 细胞分泌 αMSH	Tran 等（1990）
	罗非鱼 (Oreochromis mossambicus)	刺激	1～1000 nM EC$_{50}$ 10 nM	羊CRF；刺激离体灌注的 PI 细胞分泌 αMSH	Lamers 等（1991）
	金头鲷 (Sparus auratus)	刺激	0.01～1000 nM EC$_{50}$ 12 nM	人/鼠 CRF；刺激离体灌注的脑垂体分泌 αMSH	Rotllant 等（2000b, 2001）
	罗非鱼 (Oreochromis mossambicus)	刺激	100 nM	罗非鱼和鼠 CRF；同等的刺激离体灌注的脑垂体分泌 αMSH	van Enckevort 等（2000）
	赤鲷 (Pagrus pagrus)	刺激	0.01～100 nM	刺激离体灌注的脑垂体分泌 αMSH。只有 100 nM 剂量有效	van der Salm 等（2004）
	鲤鱼 (Cyprinus carpio)	刺激	0.01～100 nM	羊CRF；刺激离体的 PI 碎片分泌 αMSH。对 CRF 离体的反应状态有明显的个体间差别	van den Burg 等（2005）
UI	金鱼 (Carassius auratus)	刺激	1～64 nM	白亚口鱼 UI；刺激离体灌注的 PI 细胞分泌 αMSH。UI 刺激 αMSH 分泌的效能是 CRF 或蛙皮降压肽的 2～3 倍	Tran 等（1990）
SVG	金鱼 (Carassius auratus)	刺激	1～64 nM	刺激离体灌注的 PI 细胞分泌 αMSH	Tran 等（1990）
TRH	金鱼 (Carassius auratus)	刺激	0.5～250 nM	刺激离体灌注的 PI 细胞分泌 αMSH	Tran 等（1989）
	金鱼 (Carassius auratus)	刺激	1～10000 nM EC$_{50}$ 6.9 nM	刺激离体灌注的 PI 碎片分泌 αMSH	Omeljaniuk 等（1989）
	罗非鱼 (Oreochromis mossambicus)	刺激	1～1000 nM EC$_{50}$ 158 nM	刺激离体灌注的 PI 碎片分泌 αMSH	Lamers 等（1991, 1994）
	金头鲷 (Sparus auratus)	刺激	50 nM	刺激离体灌注的脑垂体分泌 αMSH	Rotllant 等（2000b）
	鲤鱼 (Cyprinus carpio)	刺激	0.01～10 nM	刺激离体灌注的部分脑垂体分泌 αMSH	van den Burg 等（2003）
	赤鲷 (Pagrus pagrus)	刺激	0.01～100 nM EC$_{50}$ 0.7 nM	刺激离体灌注的脑垂体分泌 αMSH	van der Salm 等（2004）
AVT	金鱼 (Carassius auratus)	没有作用	19.0 nM (20 ng/mL)	AVT 试验对离体灌注 PI 细胞刺激 αMSH 分泌的作用	Tran 等（1989, 1990）

续表 6.2

因子	种类	报道的反应	有效的或试验的剂量	实验观察/评论	参考文献
IST	金鱼 (Carassius auratus)	没有作用	20.7 nM (20 ng/mL)	IST 试验对离体灌注 PI 细胞对 αMSH 分泌的作用	Tran 等（1989，1990）
DA	欧洲鳗鲡 (Anguilla anguilla)	抑制	没有数据	细胞计量分析：腹腔注射 DA 拮抗剂 pimozide 增强促黑色素激素细胞的活性	Olivereau（1978）
	虹鳟 (Oncorhynchus mykiss)	抑制	10 μM	离体孵育 PI 18 h 阻抑 αMSH 的分泌	Barber 等（1987）
	金鱼 (Carassius auratus)	抑制	没有数据	细胞计量分析：腹腔注射各种 DA 拮抗剂增强促黑色素激素细胞的活性	Olivereau 等（1987）
	金鱼 (Carassius auratus)	抑制	0.1 nM～100 μM	阻抑离体灌注 PI 碎片基础的和 TRH-刺激 αMSH 分泌活动	Omeljaniuk 等（1989）
DA	罗非鱼 (Oreochromis mossambicus)	抑制	3 nM～10 μM	阻抑离体灌注的 PI 碎片分泌 αMSH	Lamers 等（1991）
	罗非鱼 (Oreochromis mossambicus)	双向反应	0.01 pM～10 μM	在低-pH 驯养的鱼，低 DA 水平刺激而高水平阻抑离体灌注 PI 碎片分泌 αMSH	Lamers 等（1997）
	赤鲷 (Pagrus pagrus)	抑制	0.01～100 nM	阻抑离体灌注脑垂体分泌 αMSH	van der Salm 等（2004）
MCH	虹鳟 (Oncorhynchus mykiss)	抑制	没有数据	细胞计量和血浆分析：在体腹腔注射鲑鱼 MCH 抑制促黑色素激素细胞活性及降低血浆 αMSH 水平	Barber 等（1986）
	欧洲鳗鲡 (Anguilla anguilla)	抑制	没有数据	半个脑垂体离体孵育 18 h，内源 MCH 被吸收而增强 αMSH 分泌活动	Barber 等（1987）
	虹鳟 (Oncorhynchus mykiss)	抑制	没有数据	PI 离体孵育 18 h 内源 MCH 被吸收而增强 αMSH 分泌活动	Barber 等（1987）
	罗非鱼 (Oreochromis mossambicus)	双向反应	0.01～10 μM	鲑鱼 MCH；0.01～1 μM 抑制而 10 μM 刺激离体灌注 PI 碎片分泌 αMSH	Gröneveld 等（1995b）
	赤鲷 (Pagrus pagrus)	抑制	0.01～100 nM	阻抑离体灌注脑垂体分泌 αMSH	van der Salm 等（2004）

αMSH，α-促黑激素；AVT，精氨酸加压催产素；CRF，促肾上腺皮质释放因子；DA，多巴胺；ip，腹腔注射；IST，硬骨鱼催产素；MCH，促黑色素浓集激素；TRH，促甲状腺激素释放激素；SVG，蛙皮降压肽；UI，硬骨鱼紧张肽 I。

表 6.3　影响硬骨鱼类 β-内啡肽分泌活动的因子

因子	种类	报道的效应	有效的或试验的剂量	试验的观察/评论	参考文献
CRF	鲤鱼 (Cyprinus carpio)	刺激	1 nM	羊CRF；离体灌注 PI 刺激非-乙酰化 βEND 分泌	van den Burg 等（2001）
	鲤鱼 (Cyprinus carpio)	没有作用	1 nM	羊CRF；试验离体灌注 PI 分泌 βEND 的影响	van den Burg 等（2001）
	鲤鱼 (Cyprinus carpio)	刺激	0.01～10 nM	羊CRF；离体灌注 PI 刺激 N-乙酰化 βEND 分泌	van den Burg 等（2005）
	金头鲷 (Sparus auratus)	抑制	0.01～1000 nM ED_{50} 16 nM	人/鼠 CRF；离体灌注脑垂体刺激 N-乙酰化 βEND 分泌	Rotllant 等（2005）
TRH	鲤鱼 (Cyprinus carpio)	刺激	0.01～10 nM	离体灌注部分脑垂体刺激 N-乙酰化 βEND 分泌	van den Burg 等（2003）
阿片样物质 βEND	鲤鱼 (Cyprinus carpio)	抑制	2 nM	预先用鲤 βEND [1-29] 孵育 2 h，羊 CRF 抑制离体灌注 PI 碎片分泌 N-乙酰化 βEND	van den Burg 等（2005）

CRF，促肾上腺皮质激素释放因子；βEND，β-内啡肽；TRH，促甲状腺激素释放激素。

6.2.2.1　TRH 和 CRF 系统

下丘脑因子 TRH 和 CRF 不仅影响促肾上腺皮质激素细胞轴（见第 6.2.1 节），亦影响促黑色素激素细胞的分泌活动。促黑色素激素细胞对 TRH 和 CRF 刺激的敏感性在和应激物作用时发生变化（Lamers 等，1994；Rotllant 等，2000b；van den Burg 等，2003），表明这些释放因子和它们的靶标参与应激的反应。CRF 的生物学活性为 CRF 结合蛋白（CRF-BP）所调节，而 CRF-BP 是一个下丘脑来源的、在系统发生上高度保守的蛋白质（Huising 和 Flik，2005；Westphal 和 Seasholtz，2006）。在鲤鱼脑内，包含 CRF-BP 和 CRF 的神经纤维（存在于不同神经元群体分开的轴突中）从 NPO 发出而延伸到脑垂体的中间部，表明促黑色素激素细胞至少受到 CRF 系统两种成分的影响（Huising 等，2004）。有趣的是，用甲状腺处理能使鲤鱼下丘脑 CRF-BP mRNA 的表达增强，而这是预料中的和血浆皮质醇基础的水平下降相联系（Geven 等，2006）。离体的试验亦得到类似的结果（Geven 等，2009）。这些研究证实下丘脑 CRF-BP 起着一个信号分子（或者当它起着蛋白伴侣作用时成为信号调控分子）的作用，从而使甲状腺系统能够和促黑色素激素轴与促肾上腺皮质激素轴互相沟通。

Van den Burg 等（2005）的研究为 CRF 影响 αMSH 和 N-乙酰化 βEND（N-AcβEND）离体释放提供个体间发生变异性的唯一数据（见表 6.2）。对驯养在三个不同环境温度（15℃、22℃和29℃）中的鲤鱼进行 CRF-刺激垂体释放 αMSH 和 N-AcβEND 的评定。除最大的反应量度（magnitude）外，CRF 剂量反应的型式也依温度而有很大差别，有 2/3 的试验鱼对 CRF 完全没有反应。由于 TRH 能连贯地刺激 αMSH 和 N-AcβEND 释放，而 CRF 能刺激阿片样物质 βEND 从脑垂体远侧部释放（van den Burg 等，2005），因而可以排除假象的可能性。值得注意的是，TRH 和 CRF 明显是从两个不同的亚细胞群体刺激 N-AcβEND 释放。急性应激反应的鲤鱼有两个不同的内分泌型式：表现血浆的皮质醇、αMSH、N-AcβEND 水平都升高的鱼和表现只有血浆的皮质醇水平升高的鱼（van den Burg 等，2005）。显然，这些研究结果表明促黑色素激素细胞对 CRF 的敏感性决定于发生的应激反应。然而，还要决定 CRF 的反应在第一个位置为什么是不同的。通常，我们应当注意到群体效应（group effect），反映有反应的和没有反应的个体；鱼类和其他脊椎动物一样，面对新生事物和挑战时会具有一种持续的从胆小受惊到大胆冒险的"性格特征"（Wilson 等，1994）

6.2.2.2 MCH 和 DA

MCH 可以看作一种 αMSH 的内源拮抗物，是外周皮肤的载黑色素细胞（melanophore）和脑垂体促黑色素激素细胞分泌 αMSH 的抑制剂。除了抑制虹鳟由 CRH 诱导 ACTH 释放的作用之外（Baker 等，1985），MCH 主要参与促黑色素激素细胞轴的作用。鱼类 MCH 主要定位于下丘脑的两个核——外侧结节核（NLT）和外侧隐窝核（NRL），而只有 NLT 延伸到脑垂体（Kawauchi 和 Baker，2004）。驯养在白色背景中的莫桑比克罗非鱼 NLT 内的而不是 NRL 内的前 MCH 原（ppMCH）mRNA 水平要比驯养在黑色背景中的鱼增加许多。驯养在酸性（pH = 3.5）的水中，NLT 神经元的 ppMCH 表达亦有所增加，和血浆的 ACTH 与皮质醇水平增加有正的联系（Gröneveld 等，1995b）。较轻微的应激作用（pH = 4.0）和高渗性状态对 NLT 和 NRL 的 ppMCH 表达没有影响，血浆中 ACTH 和皮质醇水平亦不会出现明显变化。相反，对鱼进行重复干扰诱导的应激反应使 NRL 内的 ppMCH mRNA 的表达增加 45%，而 NLT 内表达的不受影响（Gröneveld 等，1995b）。这些研究结果反映下丘脑应激物-特异性的 MCH-神经元亚群在调控皮肤色素形成和应激反应中的活性。

在 20 世纪，对哺乳类的研究已证明下丘脑的 MCH 对食物摄入和能量平衡方面起着食欲性因子的作用（Qu 等，1996）。MCH 亦影响鱼类的摄食行为，但在神经中枢注射条斑星鲽（*Verasper moseri*）或人的 MCH 对金鱼起厌食性作用（Matsuda 等，2006），而免疫中和（immunoneutralization）脑的 MCH 能增强食物的摄取（Matsuda 等，2009）。要注意的是，研究 MCH 介导金鱼厌食性作用的通道表明 MCH 通过 MC4R 信号通道增强 αMSH 的厌食作用并阻抑 NPY 和生长素释放肽在间脑的合成（Shimakura 等，2008）。但是，MCH 并不是鱼类唯一的厌食性因子。CRF、UI 和 TRH 都已确认是调控食物摄取和能量平衡的肽类（Jensen，2001；Joseph-Bravo，2004；Bernier，2006）。总而言之，由 MCH、TRH 和 CRF-相关肽类对促黑色素激素轴的下丘脑调控应该以摄食和能量稳态作为主要内容加以阐述，而不是生理学的皮肤颜色变化。

DA 对本章所讨论的三个轴都起着抑制作用。但是，通过高亲和力的 DI 受体在罗非鱼促黑色素激素细胞上的表达，DA 能对应激反应的鱼起刺激作用，由此可以反映多巴胺系统的可塑性（Lamers 等，1997）。有趣的是，低亲和力的 D2 受体在这些鱼类中并不下调，因此，多巴胺受体的净作用取决于 DA 的局部浓度，而这再次为较高层次脑中心的传入输入（affesent input）所决定。

最后，在离体灌注期间，αMSH 基础的与非刺激的释放是超时的降低（Lamers 等 1991；van den Burg 等，2005）。这和 ACTH 从脑垂体远侧部的基础释放、促甲状腺素细胞的活性变为非抑制的以及自体移植的脑垂体移植物的去神经（denervation）增加形成对比（Metz 等，2004）（见第 6.2.3.1 节）。这表明促黑色素激素细胞是处在一种原位净刺激的调控中，而促肾上腺皮质激素细胞和促甲状腺素细胞处在一种净抑制的调控中。这或许和脑垂体中间部与远侧部神经分布之间的差别有关。目前对这些研究的生理学联系还不清楚。

6.2.3　下丘脑对促甲状腺素细胞的调控

在整个哺乳纲，下丘脑的酰胺三肽（amidated tripeptide）TRH（促甲状腺素-释放激素）激活前脑垂体的促甲状腺素细胞释放促甲状腺素（或甲状腺刺激激素，TSH）。TSH 刺激参与甲状腺激素生成的多种通路（Dunn 和 Dunn，2001；Klaren 等，2007a）而增加甲状腺素（T_4）激素原（prohormone）的分泌。T_4 衍生具有生物活性的 3,5,3′-三碘甲腺原氨酸（T_3），它以负反馈作用环抑制脑垂体 TSH 和下丘脑 TRH 的表达和分泌活动。在传统的哺乳类反馈作用系统，负的甲状腺激素反应元件存在或者靠近启动子区，使得 T_3 能够阻抑 TSH α-和 β-亚单位以及 TRH 的基因表达。尽管对鱼类很少有深入的相关研究，但鱼类的这种调控过程肯定复杂得多。甲状腺激素和 TSH 负反馈调控 TRH 的释放（Gorbman 等，1983），但 TRH 不能被认定为是一个普通的和唯一的促甲状腺素细胞因子。实际上，正如下面阐述的，有许多因子影响脑垂体甲状腺素细胞的活性（见图 6.3）。

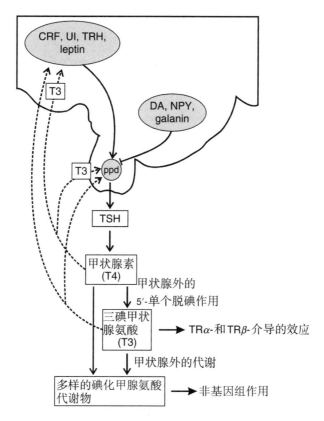

图 6.3 影响甲状腺素从甲状腺分泌的主要因子汇总示范图

要注意的是,这个汇总示意图不是表示普遍化,而是一个根据不同鱼类得到的实验结果构建的模式。甲状腺素(T_4)为脑垂体促甲状腺素细胞吸收,但它在细胞内脱碘而成为T_3,最后起负反馈作用。对下丘脑的促甲状腺素神经元设定一个相似的作用机理。图6.4详细总结了甲状腺外的代谢通道。实线箭头表示刺激作用,T线表示抑制作用,虚线箭头表示负反馈作用。CRF,促肾上腺皮质激素释放因子;DA,多巴胺;NPY,神经肽Y;ppd,近端远侧部;TR,甲状腺激素受体;TRH,促甲状腺素释放激素;TSH,促甲状腺激素;UI,硬骨鱼紧张肽I。

6.2.3.1 促甲状腺素释放激素

TRH作为促甲状腺素细胞的因子,其作用在哺乳类的相关研究中已经得到充分阐明(Nillni和Savarino,1999),但在鱼类的研究领域还一直是模糊不清的(见表6.4)。在一些硬骨鱼类的研究中已经观察到TRH"传统的"刺激促甲状腺素细胞的作用,但亦存在着抑制作用和负的结果,这就要转而关注另一些强有力的下丘脑促甲状腺素细胞的因子,例如CRF,已经报道它具有促甲状腺素细胞的作用。在早期的脊椎动物相关研究中,CRF要比TRH更多地被认为是主要的下丘脑促甲状腺素细胞的因子(Lovejoy和Balment,1999;Seasholtz等,2002;De Groef等,2006)。但是,在综述有关TRH作为鱼类TSH-释放激素的证据时或者持反对意见时都必须小心谨慎;而且,学者们一直在寻找证明TRH对鱼类起促甲状腺激素细胞因子的作用的可靠的事例。

表6.4 TRH对硬骨鱼类促甲状腺激素（TSH）分泌活动的影响

上标的数字表示在报道鱼的种类或科时使用体长（L, cm）-体重（W, g）关系（$W=aL^b$）的中间参数值来评定有效的剂量（http://www.fishbase.org/search.php.version04/2008）。

种类	报道的反应	有效的或试验的剂量	实验观察/评论	参考文献
北极红点鲑（Salvelinus alpinus）	刺激	0.001～1 μg/g体重	一次腹腔注射剂量；对饥饿的红点鲑注射后1～5 h，血浆T_4水平增加	Eales 和 Himick（1988）
虹鳟（Oncorhynchus mykiss）	刺激	0.7 μg/g体重	一次腹腔注射剂量；注射后2 h，血浆T_4水平增加	同上
金鱼（Carassius auratus）	刺激	2×2 μg/g体重[a]	间隔1 h 2次腹腔注射；注射后1 h 脑垂体远侧部嗜碱性细胞脱颗粒（$n=1$，硫脲产生同样结果）	Kaul 和 Vollrath（1974）
长颌大口鰕虎鱼（Chasmichthys dolichognathus）（鰕虎鱼科）	刺激	0.005～0.7 μg/g体重[b]	一次腹腔注射剂量；注射后4～10 h 脑垂体尾远侧部嗜碱性细胞（暂定为TSH细胞）肥大	Tsuneki 和 Feinholm（1975）
翠鳢（Channa punctata）	刺激	2～6 μg/g体重	一次腹腔注射剂量；注射后6 h 甲状腺过氧化物酶活性增加（较高剂量为抑制作用）	Bhattacharya 等（1979）
日本鳗鱼（Anguilla japonica）	刺激	10 nM	脑垂体细胞离体孵育后6 h，增加TSHβ-亚基mRNA的表达	Han 等（2004）
鳙鱼（Aristichthys nobilis）	刺激	0.01～10 nM	脑垂体细胞离体孵育后6～36 h，TRH剂量依存地增加TSHβ-亚基mRNA的表达	Chatterjee 等（2001）；Chowdhury 等（2004）
肺鱼（Protopterus ethiopicus）	没有作用	4×0.005～0.45 μg/g体重	24 h 间隔4次腹腔注射；注射后24 h 下咽部（"甲状腺"）24 h 和^{131}I结合没有作用	Gorbman 和 Hyder（1973）
罗非鱼（Oreochromis niloticus × O. aureus）	没有作用	5 ng/g体重	一次腹腔注射；注射后24 h 在体血浆T_4水平没有影响（用βTSH处理起刺激作用）	Melamed 等（1995）

续表 6.4

种类	报道的反应	有效的或试验的剂量	实验观察/评论	参考文献
金鱼 (Carassius auratus)	没有作用	6~250 μg/g 体重	一次腹腔注射剂量；注射后 4 h 在体没有诱导突眼性甲状腺肿（TSH 诱导突眼性甲状腺肿）	Wildmeister 和 Horster (1971)
	没有作用	没有报道	腹腔注射一个"大"剂量 TRH，注射后 1~24 h 对血浆 T_4 水平没有影响	Peter 和 McKeown (1975)
鲤鱼 (Cyprinus carpio)	没有作用	100 nM	离体孵育 24 h，脑垂体细胞新合成的 TSHα 和 β-亚单位的释放没有作用	Kagabu 等 (1998)
	没有作用	100 nM	完整的脑垂体离体孵育 6~36 h 对 TSHβ-亚单位 mRNA 表达没有作用	Geven 等 (2009)
银大麻哈鱼 (Oncorhynchus kisutch)	没有作用	0.1~100 nM	脑垂体细胞孵育 6 h，对 TSH 分泌活动没有作用	Larsen 等 (1998)
虹鳟 (Oncorhynchus mykiss)	抑制	3~7 μg/g 体重	没有报道 TRH 使用方法，"明显降低"（$p<0.001$）在体血清 T_4 水平（βTSH 处理后增加）	Bromage 等 (1996)
金鱼 (Carassius auratus)	抑制	2×0.3~3 μg/g 体重	间隔 12 h 2 次腹腔注射；在咽部与头肾在体和 ^{131}I 结合 24 h 内减少 60%	Peter 和 McKeown (1975)
	抑制	没有报道	间隔 12 h 2 次腹腔注射"大"剂量 TRH，注射后 12 h，引起处理鱼血浆 T_4 水平"明显受抑制"	同上
翠鳢 (Channa punctata)	抑制	8~12 μg/g 体重	一次腹腔注射；注射后 6 h 甲状腺过氧化物酶活性下降（较低剂量起刺激作用）	Bhattacharya 等 (1979)

续表6.4

种类	报道的反应	有效的或试验的剂量	实验观察/评论	参考文献
网纹花鳉 (*Poecilia reticulata*)	抑制	0.2 μg/g 体重[d]	间隔24 h 3次腹腔注射；注射后24 h，甲状腺胶体每个单位面积的干重增加23%（用干涉量度学方法测定），甲状腺滤泡的胶体不具液泡，上皮细胞的高度低	Bromage (1973)
	抑制	没有说明	下颌离体孵育24 h，介质中含有剑尾鱼处在 28 μM TRH 36 h 的脑垂体的"射气（emanation）"，结果是甲状腺胶体每个单位面积的干重增加40%（用干涉量度学方法测定）	同上

a 根据使用的剂量（2 × 300 μg 每条鱼）估算正常剂量，报道体长（18～20 cm），参数值 $a = 0.0295, b = 2.900$。

b 根据使用剂量（0.1～2 μg 每条鱼）估算，报道体长（6.6～13 cm），参数值 $a = 0.0094, b = 3.0160$。

c 根据使用剂量（1～2 μg 每条鱼）估算，设定成年雄鱼体长为30 cm，参数值 $a = 0.0118, b = 3.006$。

d 根据使用剂量（40 μg 每条鱼）估算，设定体长为3 cm，参数 $a = 0.0081, b = 3.1492$。

BW, 体重；ip, 腹腔内；TRH, 促甲状腺素释放激素；TSH, 促甲状腺激素。

在临床实践中，脑垂体对TRH反应的试验是一次静脉内注射剂量为3～7 μg TRH 每公斤病人体重。15～30 min 后测定血清TSH含量，亦可以完美地在注射后间隔3 h 测定。健康的有甲状腺的病人使用TRH后15～40 min 内血清TSH水平会升高5倍。几小时后，血清T_3和T_4达到它们的高峰浓度，而TSH回复到注射前的水平（Faglia, 1998; van Tijn 等, 2007）。由于对硬骨鱼类还没有同源的血清-TSH测定技术，间接的试验示值读数（readouts），例如TSH β-亚基 mRNA 表达、甲状腺放射性碘化合物的掺入以及血浆中T_3和T_4的浓度都经常用来评估TRH刺激鱼类TSH释放的效价（见表6.4）。但是，由于甲状腺的活性以及使用TRH的效价，测定的结果可能会有争议。

重要的是在利用这些代表参数时要考虑到以下几方面：首先，适当地在体使用TRH处理后，使血浆的甲状腺激素水平升高，将会反馈影响脑垂体，使TSH的表达和分泌恢复到正常水平。特别是使用多种剂量，间隔不同时间和天数时，不仅甲状腺激素对脑垂体TSH的负反馈作用将会成为一个混杂因子，促甲状腺素细胞TSH贮存的衰竭也会成为一个混杂的因子。这就会导致难以从混合的刺激性和抑制性的作用中解析出单纯的 TRH 诱导作用,即 TRH 和甲状腺激素对单纯的 TSH 分泌活动的影响 (Rabello 等,1974;Staub 等,1987;Faglia,1998)。其次,并非所有的试验示值读数,如放射性碘化合物的摄取和突眼性甲状腺肿 (exophthalmos)① 都可以设想用作 TRH 对促甲状腺素细胞作用的相同的指标作用。的确,曾报道人在被 TRH 处理后对于甲状腺摄取放射性碘是颇不敏感的 (Haigler 等,1972)。另外要考虑的是甲状腺对放射性的摄取是多样而分开的作用过程的净结果 (Dunn 和 Dunn,2001)。甲状腺能有效地从血液中清除碘化合物,因此,在一次大剂量处理后,在早的时间点取样测定掺入的碘化合物将主要反映腺体的净运送容量。但在迟些时间取样测定,得到的是参与甲状腺合成的酶与细胞活动过程的结果。确实,给莫桑比克罗非鱼腹腔注射一个剂量的 ^{125}I,咽下部的甲状腺和放射性的碘化物结合主要是在注射后 24 h,而血浆中放射性标志的甲状腺激素的显现要滞后一些,出现在注射后 48 h (Geven 等,2007)。同样,用催乳激素处理底鳉 (*Fundulus heteroclitus*),血浆 T_4 水平明显降低,并没有一个相伴的甲状腺摄取放射性碘化物的变化 (Grau 和 Stetson,1977a)。没有关于碘化物和甲状腺激素动态更多的信息。甲状腺放射性碘含量的降低可以解释为是一个没有活性的腺体,或者同样的意思,是一个有活性而碘化物高周转的腺体。

表 6.4 列出了腹腔注射 TRH 对促甲状腺素细胞影响的各项研究结果。很明显的是只有两种硬骨鱼类——金鱼和鲤鱼是由两个独立的研究人员分开进行研究的。在 13 项不同鱼类的研究中有 7 项报道 TRH 是刺激性地影响促甲状腺激素细胞分泌活动。这些研究亦实现了至少两种 TRH 临床试验的标准,即单独一次剂量(或者在一个短时间间隔的两个剂量)的给药和处理后几小时内测定试验的示值读数。此外,在体使用的有效剂量范围是 0.001~6 μg/g 体重,以及在莫桑比克罗非鱼测定的血浆 TRH 浓度的范围 (300 pM,或大约 100 ng/L) (Lamers 等,1994)。

有 5 种鱼类用 TRH 处理后没有反应。对金鱼腹腔注射高剂量的 TRH 既不会出现突眼性甲状腺肿 (Wildmeister 和 Horster,1971),亦不会使血浆 T_4 水平升高 (Peter 和 McKeown,1975)。和这些结果不同的是,Kaul 和 Vollrath (1974) 观察到低剂量 TRH 能使金鱼脑垂体 RPD 的类似促甲状腺素细胞的嗜碱性细胞脱粒。在他们调控的实验中,TSH 能引起突眼性甲状腺肿,而用甲状腺拮抗剂硫脲处理能模拟假定的 TSH 细胞出现脱粒。要调和这些有争议的研究结果,也许可以认为 TRH 引起了嗜碱性细胞脱粒,尽管这只是在金鱼的一个样品中得到的结果 (Kaul 和 Vollrath,1974)。比起血浆的 T_4 浓度和突眼性甲状腺肿,一个细胞学的反应似乎是 TSH 作用的更为直接的指标。然而可

---

① 突眼性甲状腺肿是患 Graves 病的甲状腺功能亢进病人的典型症状。这是一种自身免疫的疾病,TSH 受体为刺激的免疫球蛋白激活。症状的表现是,经过几周到几个月时间,组织液 (interstitial fluid) 累积于眼眶外肌内,使眼睛突出。鱼类突眼性甲状腺肿是在实验处理后几小时出现(见表 6.4),表明这是一个非常不同的病因学。而更令人惊奇的是硬骨鱼类头颅的眼窝解剖构造,要比哺乳类的骨骼结构更少受到局限。Eales (1979) 亦阐述过鱼类的突眼性甲状腺肿。

以争论的是 TRH 对金鱼是起着促甲状腺素细胞的作用，但导致 TSH 的释放并不适宜于诱导甲状腺轴下游的变化。我们亦要评论和赞赏 Melamed 等（1995）对罗非鱼杂交种使用的 TRH 剂量要低于其他大多数研究者使用的剂量，从而导致不适宜地提高在体血浆的 T_4 水平。Gorbman 和 Hyder（1973）报道肺鱼甲状腺对放射性碘摄取没有变化可能是受到上述的 TRH 长期处理而产生回缩的影响。

只有鲤鱼和银大麻哈鱼缺少 TRH 诱导的促甲状腺素细胞的作用。Larsen 等（1998）设立了一个"金本位"（gold standard），他采用同源的免疫测量技术测定银大麻哈鱼脑垂体离体的 TSH 分泌活动。他们通过观察认为 TRH 并不刺激 TSH 释放是从其他内分泌因子得到的正面结果而证实的。此外，TRH 对 TSH 亚单位的表达没有促甲状腺素细胞的作用是在两个对鲤鱼独立的研究中检测到的（Kajabu 等，1998；Geven 等，2009）。

在金鱼的脑腔内注射剂量为 0.2～1 μg TRH 每条鱼的研究结果没有包括在表 6.4 内，注射后 5 min 到 4 h，血浆中 T_4 水平没有受到影响（Crim 等，1978）。可以提出这样的疑问：在脑腔内使用 TRH 是否能使 TRH 在脑垂体内产生充分的活性？还必须指出的是，在这些研究的试验设计中，TRH 对促甲状腺素细胞起抑制作用的测定并不符合临床测试 TRH 作用的标准，并且受到一个或多个上述的回缩影响。总而言之，还没有令人信服的证据说明 TRH 对鱼类起着抑制促甲状腺素细胞的作用。

6.2.3.2　其他的下丘脑因子

表 6.5 展示了除 TRH 之外的其他许多（下丘脑的）促甲状腺素细胞的因子。最早进行下丘脑调控硬骨鱼类甲状腺轴的研究是鱼的自体移植（autotransplant），即去神经的异位的脑垂体。这些鱼类具有激活的甲状腺，并为组织学指标、放射性碘化合物摄取增加和血浆 T_4 水平增加等所证实。此外，脑垂体移植物中的促甲状腺素细胞表现得比它们完整的没有移植的配对体（counterparts）更为有活力（Ball 等，1963；Olivereau 和 Ball，1966；Higgins 和 Ball，1970；Peter，1972；Grau 和 Stetson，1977b）。TSH 的分泌活动是在脑因子的抑制性调控之下的概念，是通过电解损毁（electrolytic lesioning）一定的下丘脑区或者脑垂体柄而建立的。这个过程和甲状腺的激活一致（Ball 等，1963；Peter，1970，1971；Ball 等，1972；Peter 和 McKeown，1975；Pickford 等，1981），以致有学者得出结论认为脑垂体的促甲状腺素细胞受到下丘脑的甲状腺抑制因子的紧张性抑制（Grau 和 Stetson，1978；Olivereau 等，1988），这个抑制因子最可能是 DA。一个抑制性的 DA D2 型受体在虹鳟脑垂体远侧部的表达支持这个观点（Vacher 等，2003）。生长抑素在脊椎动物当中亦是很明确地抑制 TRH 诱导 TSH 释放的抑制剂（Hedge 等，1981；De Groef 等，2005）。然而，还缺少生长抑素在硬骨鱼类中起类似作用的实验证据（Byamungu 等，1991）。

表6.5　TRH以外的其他因子影响硬骨类促甲状腺素的分泌活动

因子	种类	报道的效应	有效的或试验的剂量	试验的观察/评论	参考文献
CRF	银大麻哈鱼 (*Oncorhynchus kisutch*)	刺激	0.01～100 nM	羊CRF；刺激离体孵育的脑垂体细胞6 h分泌TSH	Larsen等（1998）
	欧洲鳗鱼 (*Anguilla anguilla*)	刺激	没有报道	TSHβ-亚基mRNA在孵育的脑垂体细胞中表达	Rousseau等（1999）
	鲤鱼 (*Cyprinus carpio*)	没有作用	100 nM	羊CRF；对离体孵育完整的脑垂体6～36 h，TSHβ-单位mRNA的表达没有影响	Geven等（2000）
UI	银大麻哈鱼 (*Oncorhynchus kisutch*)	刺激	0.01～100 nM	鲤鱼UI；刺激离体孵育脑垂体细胞6 h，分泌TSH	Lasren等（1998）
SVG	银大麻哈鱼 (*Oncorhynchus kisutch*)	刺激	1～100 nM	青蛙SVG；刺激离体孵育脑垂体细胞6 h，分泌TSH	Lasren等（1998）
DA	金鱼 (*Carassius auratus*)	抑制	没有说明	细胞计量技术分析：在体腹腔注射DA拮抗物，促甲状腺素细胞的细胞体和核肥大，TSH略微增加	Olivereau等（1998）
	底鳉 (*Fundulus heteroclitus*)	抑制	没有说明	颅内注射6-OH DA，特异性破坏多巴胺能纤维，导致血清T_4水平升高	Grau和Stetson等（1978）
βEND	鳙鱼 (*Aristichthys nobilis*)	刺激		TSHβ-亚基单位mRNA在培养的脑垂体细胞中表达	Chowdhury等（2004）
PGEI, PGF2α	囊鳃鲶 (*Heteropneustes fossilis*)	刺激	100 μg/(鱼·天)	增加甲状腺[131]I掺入和血浆TSH水平，降低脑垂体TSH含量	Singh和Singh（1977）
甘丙肽	鳙鱼 (*Aristichthys nobilis*)	抑制		TSHβ-亚基mRNA在培养的脑垂体细胞中表达	Chowdhury等（2004）

续表 6.5

因子	种类	报道的反应	有效的或试验的剂量	实验观察/评论	参考文献
NPY	鳙鱼 (*Aristichthys nobilis*)	抑制		TSHβ-亚基 mRNA 在培养的脑垂体细胞中表达	Chowdhury 等（2004）
瘦素	鳙鱼 (*Aristichthys nobilis*)	刺激		TSHβ-亚基 mRNA 在培养的脑垂体细胞中表达	Chowdhury 等（2004）
GHRH	银大麻哈鱼 (*Oncorhynchus kisutch*)	没有作用		鲑鱼 GHRH, TSH 从培养的脑垂体细胞分泌	Larsen 等（1998）
GnRH	底鳉 (*Fundulus heteroclitus*)	没有作用		"对甲状腺功能没有可测定的效应"	Brown 等（1982）
	银大麻哈鱼 (*Oncorhynchus kisutch*)	没有作用		鲑鱼 GnRH, TSH 从培养的脑垂体细胞分泌	Larsen 等（1998）
	银大麻哈鱼 (*Oncorhynchus kisutch*)	没有作用		鲑鱼 GnRH，在体血浆 TSH 水平	Moriyama 等（1997）

CRF, 促肾上腺皮质激素释放因子；DA, 多巴胺；βEND, β-内啡肽；GHRH, 生长激素释放素；GnRH, 促性腺激素释放激素；ip, 腹腔注射；NPY, 神经肽 Y；PGE1, 前列腺素 E1；PGF2α, 前列腺素 F2α；SVG, 蛙皮降压肽；TSH, 促甲状腺素；UI, 硬骨鱼紧张肽 I。

  CRF 家族的几个成员，即 CRF、UI、蛙皮降压肽，对鲑鳟鱼类和鳗鱼具有促甲状腺素细胞因子的作用（Larsen 等，1998；Rousseau 等，1999）。CRF 的促甲状腺素细胞因子的作用已经在其他非哺乳类脊椎动物中观察到（De Groef 等，2006）。促甲状腺素细胞在非哺乳类脊椎动物中表达 CRF-2 型受体（De Groef 等，2003，2006），但最近对小鼠进行的实验所得结果表明一个小亚型的促甲状腺素细胞表达 CRF-1 型受体（Westphal 等，2009）。CRF 的促甲状腺素细胞作用只在对大麻哈鱼（*Oncorhynchus tshawytscha*）进行脑腔内注射 CRF 及其拮抗物 α-螺旋 CRF$_{9-41}$ 时得到部分证实，其研究结果分别是对血浆 T$_4$ 水平没有作用和刺激作用（Clements 等，2002）。鲤鱼 CRF 参与 TSH 分泌活动的调控作用可能是间接地从 T$_4$ 诱导下丘脑 CRF-结合蛋白（CRF-BP）在体与离体的表达而得到证明。然而，在培育的鲤鱼脑垂体内，TSHβ mRNA 的表达对于 CRF 和 TRH 都是没有反应的（Kagabu 等，1998；Geven 等，2006，2009），而鲤鱼下丘脑促甲状腺素细胞因子的鉴定还有待完成。

  除 CRF 和 TRH 外，其他假定的促甲状腺素细胞的因子只对一种鱼类进行过一次研究。Chowdhury 等（2004）曾观察到 NPY、甘丙肽、瘦素和 βEND 的促甲状腺素细胞因子的刺激性作用。βEND 的作用值得注意，因为它亦是促黑色素激素细胞轴的一个信号分子（见第 6.3.2.2 节）。要准确地确定这些因子的促甲状腺素细胞的功能，已经进行的研究还是太少了些。

6.3 促肾上腺皮质激素细胞、促黑色素激素细胞和促甲状腺素细胞分泌的靶标和功能

在所有的有颌脊椎动物中,促肾上腺皮质素细胞和促黑色素激素细胞的分泌物都是由一个共同的前体蛋白质 POMC 衍生(Kawauchi 和 Sower, 2006)。大的 POMC 分子由激素原转变酶(prohormone convertase, PCs)作用而进行蛋白质剪切(proteolytic cleavage),并且是组织特异性的(tissue-specific)(Zhou 等,1993; Tanaka, 2003)。在肾上腺皮质激素细胞中,POMC 由 PC1 加工的主要产物是 ACTH 和 β-促脂解素(β-lipotrophic hormone,β-LPH)。β-LPH 由 PC2 进一步加工而产生阿片样物质 βEND。在促黑色素激素细胞中,由 PC1 和 PC2 引起较广泛的分解而将 ACTH 转变为 αMSH 和促肾上腺皮质激素样中间叶肽(corticotropin-like intermediate lobe peptide, CLIP)。β-LPH 亦在促黑色素激素细胞内由 PC2 加工而形成不同的 βENDs,它们主要是/大部分是在 N-端乙酰化。不同的是,促甲状腺素细胞只产生一种产物 TSH。TSH 和促卵泡激素(FSH)与促黄体激素(LH)一起属于脑垂体异源二聚体糖蛋白家族。这些激素由两个化学结构不同的亚基组成,其中的 α-亚基完全是一样的,而 β-亚基是独特的,决定蛋白质的特征和其特异性功能(Kawauchi 和 Sower, 2006)。

6.3.1 促肾上腺皮质激素细胞的分泌物:ACTH 和非酰化 βEND 的靶标和功能

6.3.1.1 ACTH

ACTH 和其他 POMC 衍生的黑皮质素(melanocortin)的第一个靶标是黑皮质素受体(MCR)。MCRs 属于 7-跨膜区 G-蛋白偶联受体超家族,刺激腺苷酸环化酶和 cAMP 信号级联(Mountjoy 等,1992)。在一些鱼类及其他的脊椎动物中已经鉴定 5 个 MCRs(MC1R-MC5R)(Logan 等,2003; Klovins 等,2004a; Flik 等,2006)。ACTH 是唯一的天然配体在鱼类和哺乳类结合与激活 MC2R(Klovins 等,2004a),而 ACTH 与 MSHs 和其他 4 个 MCRs 结合(Schioth 等,2005)。然而,和哺乳类 MC1R、MC3R、MC4R 和 MC5R 亚型都分别不同地识别 MSHs 的情况不一样,在鱼类中,这 4 个 MCR 亚型对 ACTH$_{1-24}$ 比对不同的 MSHs 有更大的亲和力(Haitina 等,2004; Klovins 等,2004a)。所以,在鱼类以及鸡类(Barimo 等,2004)中,不同 MCRs 的结合特性表明 ACTH 是一个在系统进化上比 MSH 肽更早些的配体,而 MCRs 对 MSH 肽的选择性识别是在较高等脊椎动物的进化中逐渐形成的(Cerda-Reverter 等,2005; Schioth 等,2005)。

ACTH 对于鱼类的主要功能是调控头肾的肾间腺细胞糖皮质激素的类固醇生成(Donaldson, 1981; Wendelaar Bonga 等,1997)。尽管交感神经纤维(Arends 等,1999)、αMSH(Lamers 等,1992)、N-AcβEND(Balm 等,1995)、血管紧张肽Ⅱ(Perrott 和 Balment, 1990)、心房利尿钠肽(Arnold-Reed 和 Balment, 1991)、UI(Kelsall 和 Balment, 1998)和其他一些因子刺激一些鱼类皮质醇的分泌活动(Schreck 等,1989; Mommsen 等,1999),但 ACTH 还是被认为在应激反应的急性期间皮质醇释

放的主要刺激剂（Flik 等，2006）。作为回应，非-ACTH 的促皮质激素信号很可能增强 ACTH 在对特异性应激物的反应中以及在应激反应的慢性期间的类固醇生成作用（见第 6.3.2 节）。在细胞水平，ACTH 和在头肾内的 MC2R 结合后激活酶的通路，将胆固醇转变为糖皮质激素（Huising 等，2005；Hagen 等，2006；Aluru 和 Vijayan，2008）。特别是，ACTH 通过刺激性类固醇生成的急性调节蛋白（StAR）（Hagen 等，2006；Aluru 和 Vijayan，2008）、细胞色素 P450 侧链裂解酶（P450scc）（Aluru 和 Vijayan，2008）和 11β-羟化酶（P450cll）（Hagen 等，2006）而增加肾间腺细胞的类固醇生成容量。尽管 StAR 促进胆固醇跨过线粒体膜的运输，P450scc 将胆固醇转变为孕烯醇酮（糖皮质激素合成的第一个限速酶步骤），以及 P450c$_{11}$ 将 11-脱氧皮质醇变为皮质醇（Mommsen 等，1999）。同样，ACTH 对 MC2R 在肾间腺组织中表达的刺激作用很可能有助于在急性应激物起作用后增强肾间腺组织的类固醇生成容量（Aluru 和 Vijayan，2008）。

除了 MC2R-介导 ACTH 的作用之外，ACTH 比 MSHs 对鱼类所有的 MCR 亚型都有更强的亲和力（Schioth 等，2005），表明 ACTH 在鱼类中的生理功能可能比哺乳类具有更明显的多样性。例如，尽管 αMSH 参与黑色素体在鱼类皮肤的扩散（见第 6.3.2 节），MC1R 是 αMSH 在皮肤的促黑色素激素细胞中起促黑色素激素因子作用的主要受体，它在河鲀（*Takifugu rubripes*）中对 ACTH 的亲和力要比 αMSH 高 10 倍（Klovins 等，2004）。ACTH 在一些鱼类中的黑色素体扩散的特点亦表明它包含 αMSH 的序列，可能对这些鱼类色素扩散的调控起作用（Fujii，2000；van der Salm 等，2005）。在外周，ACTH 亦可能参与儿茶酚胺从头肾嗜铬细胞释放的调控（Reid 等，1996）、应激物介导性腺类固醇产生的调节（Aluru 和 Vijayan，2008）以及 T$_4$ 从肾甲状腺滤泡释放的调节（Geven 等，2009）。在脑内，MC2R、MC4R 和 MC5R 的广泛表达（Ringholm 等，2002；Cerda-Reverter 等，2003a、b，2005；Haitina 等，2004；Klovins 等，2004a）以及相当大的 ACTH 免疫反应活动（Vallarino 等，1989；Olivereau 和 Olivereau，1990b；Metz 等，2004）表明 ACTH 具有各种不同的中枢作用。在较高等脊椎动物的脑内，POMC 加工的最终产物似乎是 αMSH 和 βEND，而至少在虹鳟的脑内，ACTH 的浓度和 αMSH 与 βEND 的浓度处在相同的范围内（Vallarino 等，1989）。因此，尽管 ACTH 在鱼脑中的特异性功能还不清楚，以及似乎存在着种族特异性的差别，但 ACTH 免疫反应活性以及 MC4R 与 MC5R 表达神经元在中枢的定位可以表明，ACTH 既参与下丘脑调控脑垂体的作用，而且和哺乳类一样，亦参与一系列自分泌和行为功能的调控作用（Bertolini 等，2009）。

6.3.1.2 非乙酰化的 β 内啡肽（βEND）

已经阐明内啡肽具有明显的多样生理功能。虽然目前对 βEND 在鱼类中的功能还了解得很少，但已有的证据表明这个肽参与脑垂体激素分泌的调控以及免疫和应激反应的调节。βEND 具有多个化学类型以及不同的生物活性。例如，新生的 β-内啡肽是一种强有力的麻醉剂，但在肽的 N-端乙酰化后（N-Ac-βEND）会失去阿片样物质的活性（Akil 等，1981）。鲤鱼脑垂体中间部分泌全长的和一系列截断的（trucated）N-乙酰化 βEND（见第 6.3.2.2 节），而阿片样物质（非乙酰化的）βEND 从脑垂体远侧部释放

(van den Burg 等，2001）。对于其他鱼类，有关 βEND 在脑垂体的活动和 βEND 释放的资料还很少（Mosconi 等，1998）。

阿片样物质肽，如非乙酰化 βEND 能和 4 个阿片样物质受体类型结合［delta、kappa、mu 和伤害感受（nociceptin）受体（NOP）］，它们是在脊椎动物进化早期作为基因复制的结果而出现的（Dreborg 等，2008）。在哺乳类中，βEND 优先和 delta 与 Mu 受体结合（Przewlocki 和 Przewlocki，2001）。在斑马鱼中，βEND 亦和 delta 受体结合，但要准确鉴别鱼类和阿片样物质结合特性还需进行许多研究（Rodríguez 等，2000；Gonzalez-Nunez 等，2006）。至今确定硬骨鱼类阿片样物质受体表达型式的研究还很少。在斑马鱼中，delta 和 mu 受体广泛分布在胚胎发育全过程的中枢神经系统和外周器官（Sanchez-Simon 和 Rodríguez，2008）。在鲤鱼中，delta、mu 和 kappa 受体组成性地在下丘脑、脑垂体的远侧部和中间部、头肾、肾脏、胸腺、脾脏和白细胞中表达（Chadzinska 等，2009；亦见本书第 7 章）。

非乙酰化 βEND 对脑垂体激素分泌活动的调控作用是由于这种阿片样肽沿着下丘脑调控脑垂体的神经纤维穿行于 NPO 和脑垂体之间的分布状况而得以证明。对鲤鱼的研究表明，视前区表现为非乙酰化 βEND 的主要染色部位，而神经部（pars nervosa）进入到脑垂体内的神经纤维是非乙酰化 βEND 阳性的（Metz 等，2004；Flik 等，2006）。此外，将鲤鱼的促黑色素激素细胞用同源的 βEND（1-29）进行预温育，会逆转 CRF 对 N-ACβEND 分泌活动的刺激作用（van den Burg 等，2005）。

在免疫系统，非乙酰化 βEND 调控虹鳟、鲤鱼（Watanuki 等，2000）和翠鳢（*Channa punctatus*）（Singh 和 Rai，2008）吞噬细胞的多种功能，而用选择性的阿片样受体拮抗物能逆转这些作用。在鲤鱼中亦能够主动调节阿片样物质受体基因在对免疫和应激刺激起反应的腹腔白细胞和头肾吞噬细胞中的表达（Chadzinska 等，2009）。

6.3.2 促黑色素激素细胞分泌物：αMSH 和 N-乙酰化 βENDs 的靶标和功能

6.3.2.1 αMSH

在脑垂体中间部促黑色素激素细胞内的前体 POMC 的一个重要产物是 αMSH。脊椎动物 αMSH 的后翻译乙酰化（post-translational acetylation）产生 3 个 αMSH 类型：des-（去）、mono-（单）和 di(二)-乙酰化 αMSH。鱼类所有 3 个类型 αMSH 都出现在血浆和神经脑垂体中间叶的提取物中，而莫桑比克罗非鱼三种类型 αMSH 的释放可能是分别各自调节的（Lamers 等，1991）。三种不同类型 αMSH 的功能意义源自它们不同的潜能和半衰期（Rudman 等，1983）。

鱼类和哺乳类的黑皮质素肽 αMSH 能和 4 个不同的黑皮质素受体 MC1R、MC3R、MC4R 和 MC5R 结合并激活（Metz 等，2006b）。在它们当中，MC1R 对 αMSH 具有最大的亲和力，它虽然表达广泛，但主要是在富含载黑色素细胞的器官中表达（van der Salm 等，2005；Selz 等，2007）。MC3R 在哺乳类许多脑区和外周器官中表达，但在鱼类中的分布还不清楚，除了在板鳃鱼类白斑角鲨（*Squalus acanthias*）的许多脑区有微弱的表达型式之外（Klovins 等，2004b）。MC4R 在鱼脑内大量表达，并且偶尔被报道在外周组织中表达（Metz 等，2006b）。对金鱼的详细作图表明 MC4R 在脑的神经内分泌和食物摄取调控的脑区有高水平的表达（Cerda-Reverter 等，2003c）。最后，一些研究证明 MC5R 在鱼类许多外周组织中以及许多不同的脑区中表达（Cerda-Reverter 等，2003b；Haitina 等，2004；Metz 等，2005）。

顾名思义，αMSH 参与皮肤的颜色形成。鱼类对适应黑色背景的反应是 αMSH 刺激皮肤黑色素细胞内的黑色素体扩散，它覆盖大的表面积而使鱼的体表呈现较黑的体色（van der Salm 等，2005；Amiya 等，2007）。和其他脊椎动物一样，αMSH 对鱼类颜色适应性的调节作用是由 MC1R 介导的。在斑马鱼胚胎敲除 MC1R 的表达会导致黑色素体聚集，在黑暗中也不会扩散（Richardson 等，2008）。野灰蛋白信号传递肽（agouti-signling pepitide, ASP）是一种内源的 MC1R 拮抗物，它几乎独特地在腹部皮肤表达，亦表现出参与决定鱼类背-腹的色素型式。

虽然有争议，但 αMSH 可能参与硬骨鱼类另一个生理作用是调控皮质醇从肾间腺细胞分泌出来。在莫桑比克罗非鱼纯化的脑垂体 PI 提取物中已知含有 αMSH，表现出适度的离体促肾上腺皮质激素的活性（Lamers 等，1992）。此外，尽管任何一种促肾上腺皮质激素因子的作用都是独自发挥的，而 βEND 能明显增强来自罗非鱼脑垂体一个 αMSH 分馏物的促肾上腺皮质激素的活性（Balm 等，1995）。相反，虽然鲤鱼脑垂体的 PI 含有至少一种未经鉴定而具有促肾上腺皮质激素功能的因子，αMSH 和 βEND 单独或两者结合，都不会刺激鲤鱼的皮质醇释放（Metz 等，2005）。同样，尽管在鳟鱼的肾间腺存在着 MC4R（Haitina 等，2004），αMSH 或者 NDP-MSH（一种 MC4R 的同等物）都不会刺激皮质醇从虹鳟的肾间腺分泌出来，而 SHU9119（一种 MC4R/MC3R 拮抗物）并不会影响 ACTH 刺激离体的皮质醇生成量（Aluru 和 Vijayan，2008）。

最近的证据亦表明 αMSH 调控硬骨鱼类的食物摄取和脂类代谢。金鱼脑腔内注射 MC4R 同等物 NDP-MSH 和非特异性的同等物 melanotanⅡ（MTⅡ）能剂量依存地抑制食物摄取，而特异性的 MC4R 拮抗物 HS024 能刺激食欲（Cerda-Reverter 等，2003a，c）。同样，在虹鳟的相关研究中发现，虽然中枢给予 MTⅡ降低了食物摄取，但 HS024 和 MC3/4R 拮抗物 SHU9119 具有相反的作用（Schjolden 等，2009）。有趣的是，MTⅡ对金鱼的食欲性作用能被 CRF 受体拮抗物 α-螺旋 CRF（9–14）抵消，而免疫组织化学证据表明含有 αMSH 的神经纤维或末梢和下丘脑内含有 CRF 的神经元紧密并列在一起（Matsuda 等，2008）。总之，这些研究结果表明，αMSH 在鱼类中通过中枢的 MC4R 信号起着一种紧张性抑制食物摄取的作用，而黑皮质素的厌食性作用是由 CRF-信号通路介导的。设想 αMSH 能跨过鱼类的血-脑屏障，与哺乳类一样（Strand，1995），那么作为一些应激物反应特性的血浆 αMSH 水平的升高（见第 6.5 节）就会对食物摄取的中枢调控起一定作用。在虹鳟的相关研究中观察到的 αMSH 刺激肝脏三酰甘油脂肪酶（triacylglycerol lipase）活性和增加血液循环中的脂肪酸（Yada 等，2000），表明血液循环中的 αMSH 参与脂类的转移。和将潜在的代谢活动与厌食性作用都归因于鱼类 αMSH 相一致的是虹鳟的"钴变种"，它们缺少脑垂体 PI 的大部分和促黑色素激素细胞，其特征是大量积累脂肪，肝脏三酰甘油脂肪酶降低，变成过度饱食（hyperphagic）（Yada 等，2002）。

近期的研究表明了 αMSH 对鲤鱼的促甲状腺素作用。αMSH 以亚微摩尔（submicromolar）的浓度能刺激离体而有功能的肾脏和头脏的甲状腺滤泡释放甲状腺素（T_4）（Geven 等，2009）。用 T_4 处理的鲤鱼，其血浆的 αMSH 水平增长 30%（Geven 等，2006）。这些研究表明 αMSH 的正反馈作用机理能诱导 T_4 释放。处在慢性应激反应之中的鲤鱼，其血浆 αMSH 水平的增加表明它们参与食物摄取和能量平衡的调控（Metz 等，2005；Flik 等，2006），这亦可能是通过甲状腺素和基础代谢率的调控而得以实现的。

6.3.2.2 乙酰化 β-内啡肽

尽管一些鱼类在应激反应期间脑垂体的促黑色素激素细胞能产生和释放几种 N-Ac-βENDs，但目前对这些肽类的靶标和功能还了解得很少。至今，乙酰化内啡肽特异性的刺激性受体还没有鉴定出来（Flik 等，2006）。相反，一些证据表明乙酰化的阿片样物质肽类能导致阿片样物质受体拮抗物的产生（Rene 等，1998；Bennett 等，2005）。这样的现象是否实际适用于 N-Ac-βENDs 的促黑色素激素细胞来源还不清楚，但这对未来的研究是非常有用的，可以观察鱼类的阿片样物质受体在中枢和外周的主要表达形式（Sanchez-Simon 和 Rodríguez，2008；Chadzinska 等，2009）。此外，人们可能会说一个为"MSH-目的"而加工 POMC 的细胞还需要钝化在 POMC 加工过程中固有的潜在副产品（阿片样物质）。

6.3.3 促甲状腺素的靶标和功能

6.3.3.1 促甲状腺素结构和功能

促甲状腺素（TSH）是甲状腺功能的关键性调控因子。TSH 的 β-亚基已经在许多种硬骨鱼类中鉴定，其 mRNA 几乎是独特地在脑垂体中表达（Ito 等，1993；Salmon 等，1993；Martin 等，1999；Yashiura 等，1999；Chatterjee 等，2001；Han 等，2004；Lema 等，2008）。例外的是，在大西洋鲑鱼（Salmo salar）肝脏和肾脏中 TSHβ mRNA 的 T_4 敏感表达（Mortensen 和 Arukwe，2006）。许多学者采用异体的抗-人 TSHβ 抗体研究和阐述硬骨鱼类脑垂体远侧部 TSHβ 的分布和局部解剖学（Kasper 等，2006）。

在真骨鱼类的总目内，成熟的 TSHβ-亚基蛋白质的序列同一性变动很大，从 53% 到 97%（Marchelidon 等，1991；Lema 等，2008），而有些例子是两种鱼类 TSHβ-亚基的序列同一性就如同于鱼类和哺乳类的亚基同一性。这种多样性可以解释牛的和大麻哈鱼的 TSH 刺激夏威夷鹦嘴鱼（Scarus dubius）离体甲状腺组织释放 T_4 的能力为何相差 1000 倍（Swanson 等，1988），以及异源的促性腺激素对尼罗罗非鱼（Oreochromis niloticus）和底鳉（Brown 等，1985；Byamungu 等，1991）刺激甲状腺的活性。

用异源的 TSH 处理能够一致性地提高鱼类在体血浆的 T_4 水平（Grau 和 Stetson，1977；Brown 和 Stetson，1983；Specker 和 Richman，1984；Brown 等，1985；Nishioka 等，1985；Leatherland，1987；Inui 等，1989；Byamungu 等，1990；Bandyopadhyay 和 Bhattacharga，1993）以及增加鱼类离体孵育的甲状腺组织释放 T_4（Bonnin，1971；

Jackson 和 Sage，1973；Grau 等，1986；Swanson 等，1988；Okimoto 等，1991；Geven 等，2007）。目前对 TSH 在鱼类甲状腺滤泡中的细胞靶标还研究得不多。可以设想，鱼类甲状腺激素生成的大部分过程是和哺乳类生物化学的和细胞的途径相似的（Klaren 等，2007a）。银大麻哈鱼 TSH 对甲状腺细胞（thyrocyte）明显地起着营养作用，用异源的 TSH 处理能增加甲状腺细胞滤泡的高度（一种传统的测定甲状腺活性增强的方法）。在亚细胞水平，甲状腺过氧化物酶（thyroid peroxidase，TPO）是对甲状腺激素合成起关键性作用的酶类，用牛的 TSH 离体处理翠鳢能刺激它的活性（Chakraborti 和 Bhattacharya，1978；Bhattacharya 等，1979）。

不同于下丘脑因子和甲状腺激素的负反馈作用，有关调控促甲状腺素细胞的 TSH 表达和分泌的资料不多。鲑鳟鱼类脑垂体 TSHβ mRNA 的表达在性未成熟的鱼要比性成熟的鱼更高（Ito 等，1993；Martin 等，1999），表明甲状腺激素参与性成熟。但是，这种关联在欧洲鳗鱼中如果不是没有的话，亦是很不明显的（Han 等，2004；Aroua 等，2005）。还有，甲状腺激素参与性成熟的见解可以为银大麻哈鱼的一龄降海幼鲑比降海二龄鲑鱼对甲状腺组织的反应性更强（Specker 和 Kobuke，1995）以及罗非鱼脑垂体的促甲状腺素细胞内存在着雌激素和雄激素受体（Arai 等，2001）所证实。

6.3.3.2 促甲状腺的受体

TSH 受体（TSHR）在甲状腺细胞的基底外侧膜表达，但定位于甲状腺外。确实，在阐明其 cDNA 的初级序列之后，曾报道丰富的 TSHR mRNA 特别地在许多硬骨鱼类雄性与雌性的性腺中表达（Kumar 等，2000；Goto-Kazeto 等，2003；Vischer 和 Bogerd，2003；Rocha 等，2007）。只有玫瑰大麻哈鱼（*Oncorhynchus rhodurus*）TSHR 的表达严格地局限于下咽区的甲状腺组织（Oba 等，2000）。这些研究结果都是采用反转录 PCR（RT-PCR）技术获得的，而要指出的是这种技术可能过于敏感，它可能会处理 TSHR 基因的"非常规转录（illegitimate trancription）"（sic），而这并不是蛋白表达的恰当的生理学水平（Rapoport 等，1998）。然而，TSHR 存在于哺乳类甲状腺外的组织中已得到充分的阐明，因为 mRNA 的表达已经为免疫组织化学和配体-结合分析测定所得到的正面研究结果证实（综述见 Davies 等，2002）。

TSHR 在性腺中的表达随着发育阶段和季节周期而变化，这可以解释为甲状腺激素参与一些鱼类的性腺发育成熟（Ito 等，1993；Martin 等，1999；Rocha 等，2007）。另外，必须仔细说明 TSHR mRNA 表达的器官和组织，而不只是鱼类甲状腺典型位置的下咽区。正如前面指出的，有功能的异位（heterotopic）甲状腺滤泡存在于许多器官中，而 TSHR mRNA 的表达能很好地反映 TSHR 在甲状腺细胞而不是其他类型细胞中的正常细胞位置。

6.4 皮质醇和甲状腺激素的靶标、功能和反馈作用

6.4.1 皮质醇

硬骨鱼类肾间腺分泌的主要糖皮质激素是皮质醇（Mommsen 等，1999）。尽管对于不同应激物的反应使血液循环中的可的松（cortisone）和皮质醇一样增加，但可的松是没有活性的，它原本是皮质醇在靶组织中 $11\text{-}\beta$ 氧化的产物（Patino 等，1985，1987）。此外，和四足类不同，硬骨鱼类看来缺乏产生盐皮质激素醛固酮的合成能力（Prunet 等，2006）。这样，皮质醇就被认为是硬骨鱼类主要的糖皮质激素和盐皮质激素。但对于虹鳟，11-脱氧皮质酮（11-deoxycorticosterone，DOC）是一个盐皮质激素受体的强有力的同类物（Sturm 等，2005），在雄鱼生殖周期结束时，其血浆浓度增加 $10\sim50$ 倍，达到高峰浓度时和基础的血浆皮质醇水平相当（Milla 等，2008）。值得注意的是，DOC 和它的外源同类物醛固酮对激活虹鳟盐皮质激素受体具有同等的效能，而且有学者曾经认为 DOC 是盐皮质激素受体的一个祖先内源性配体（Bury 和 Sturm，2007）。

6.4.1.1 靶标：糖皮质激素和盐皮质激素受体

硬骨鱼类糖皮质激素的靶标是配体-激活的转录因子，包括糖皮质激素受体（GR）和盐皮质激素受体（MR）。GRs 和 MRs 都由一个祖先的皮质类固醇受体（corticosteroid receptor，CR）经过有颌类出现早期的基因组复制过程进化而来（Thornton，2001）。在大约 3.35 亿年前的第二次全基因组复制过程中，大多数硬骨鱼类具有两个不同的 GR，但只有一个复制的 MR 得以保留（Colombe 等，2000；Bury 等，2003；Greenwood 等，2003；Stolte 等，2006，2008）。在一些硬骨鱼类中还曾鉴别到一个复制 GR 基因的剪接变体（splice variant）（Bury 和 Sturm，2007）。所以，在大多数硬骨鱼类中，皮质类固醇信号是通过三个 GRs 和一个 MR 而得到。相反，斑马鱼有一个 GR 基因和 MR 基因以及一个 $GR\beta$ 剪接变体。这个剪接变体缺少反式激活作用（transactivational）的活性，但表现为一个 $GR\alpha$ 反式激活作用的显性负调控（dominant-negative）的抑制物（Alsop 和 Vijayan，2008；Schaaf 等，2008）。虽然有关皮质类固醇的非基因组信号的资料在鱼类中是缺乏的，但对其他脊椎动物的研究证明存在着膜联系的糖皮素受体（Tasker 等，2006）。

反式激活作用试验证明硬骨鱼类的 CRs 对皮质类固醇激素是有选择性的，表明鱼类的多个 CRs 可能介导不同的生理功能。对于虹鳟（Sturm 等，2005）和伯氏朴丽鱼（*Haplochromis bustoni*；Greenwood 等，2003），MRs 对皮质醇的灵敏性是 GRs 的 $10\sim100$ 倍。相反，对于鲤鱼，MR 的灵敏性是在两个 GRs 的中间位置（Stolte 等，2008）。此外，虹鳟（Bury 等，2003）和鲤鱼（Stolte 等，2008）的，但不是伯氏朴丽鱼（Greenwood 等，2003）的 GR2 同等型对皮质醇的灵敏性要明显强于 GR1 同等型。总之，CR 反式激活作用能力的这些差别可以使基础的和应激引起的皮质醇水平得到差别性调节，CRs 对硬骨鱼类类固醇的选择性显然是种类特异性的。

GRs 和 MRs 在硬骨鱼类中的普遍分布表明由皮质醇介导或者调节多种功能。GRs 已经确认在肝脏、鳃、肠、肾脏、脾脏、心脏、骨骼肌、性腺、白细胞和红细胞中表达（Mommsen 等，1999；Shrimpton 和 McCormick，1999；Vijayan 等，2003；Takahashi 等，2001；Vazzana 等，2008）。GR-免疫反应和 GR mRNA 表达细胞分布在 CNS 的许多区，主要聚集在背端脑和具有神经内分泌功能的核，包括 NPO、NLT、下丘脑下叶和尾神经内分泌系统（CNSS；Teitsma 等，1997，1998；Repele 等，2004；Marley 等，2008；Stolte 等，2008；见本书第 2 章）。GRs 和 MRs 都在脑垂体的远侧部和中间部表达（Pepels 等，2004；Stolte 等，2008）。特别是，GRs 曾定位于 RPD 的促肾上腺皮质激素细胞、PPD 的生长激素细胞和促性腺激素细胞（Teitsma 等，1998；Pepels 等，2004；Kitahashi 等，2007；Stolte 等，2008）。虽然不像 GRs 那样表现出明显的特征，但 MRs 在虹鳟和伯氏朴丽鱼的脑和外周组织中广泛表达（Greenwood 等，2003；Sturm 等，2005）。在整个胚胎发育过程中都能观察到 GR 和 MR 的转录体，它们表现出有差别的表达型式（Tagawa 等，1997；Alsop 和 Vijayan，2008）。总而言之，虽然每个 CR 亚型的特异性生理功能还不清楚，但多种受体结合的差异性表达并表现出不同的反式激活能力将会使硬骨鱼类 CR 信号出现高度的灵活性。

6.4.1.2 功能

皮质醇在鱼类中参与调控许多生理功能，影响能量代谢、生长、生殖、水盐平衡和免疫系统（详细综述见 Wendelaar Bonga，1997；Mommsen 等，1999；Norris 和 Hobbs，2006）。尤其是皮质醇通过增加肝脏葡萄糖生成，刺激蛋白水解和脂解能力而影响鱼类的能量代谢（De Boeck 等，2001；Aluru 和 Vijayan，2007）。介导皮质醇在鱼类代谢作用中的关键靶基因包括肝糖异生酶烯醇丙酮酸磷酸羧激酶（phosphoenolpyruvate carboxykinase，PEPCK）、参与蛋白质代谢的基因（谷氨酰胺合成酶、精氨酸酶、氨基转移酶）、蛋白质分解代谢（组织蛋白酶 D、谷氨酸脱羧酶）和脂解的酶类三酰甘油脂肪酶（Sheridan，1986；Mommsen 等，1999；Aluru 和 Vijayan，2007）。除分解代谢作用外，皮质醇通过其作用于生长激素-胰岛素样生长因子轴而阻抑鱼体生长（Kajimura 等，2003；Peterson 和 Small，2005；Pierce 等，2005），而且，它通过对神经内分泌通路的影响而调节食物摄取（Bernier，2006；见本书第 9 章）。皮质醇通过多种作用阻抑性发育成熟和生殖活动，包括阻抑血浆促性腺激素水平、抑制性腺类固醇激素生成和降低肝脏卵黄蛋白原合成等（Contreras-Sanchez 等，1998；Lethimonier 等，2000；Pankhurst 和 Van Der Kraak，2000；Consten 等，2001）。有趣的是，作为皮质醇在应激物作用之后恢复体内液体平衡中起重要作用的一部分，它能促进硬骨鱼类进行离子摄取和离子分解（McCormick，2001）。皮质醇促进离子摄取部分地是通过刺激淡水 Na^+/K^+-ATP 酶同等型 NKAα1a 的表达来实现的，而促进离子分泌是通过刺激海水 Na^+/K^+-ATP 酶同等型 NKAα1b 的表达，以及囊性纤维化跨膜传导调节蛋白（cystic fibrosis transmembrane conductance regulator，CFTR）的阴离子通道来实现的（Kiilerich 等，2007；McCormick 等，2008）。很明显，皮质醇诱导转运蛋白酶类的亚型，但必须了解它是上皮电化学的组成（即一个负的跨上皮电位在一个紧密的淡水鳃上皮上，而一个正的跨上皮电位在一个渗漏的海水鳃上皮上），决定净的离子流动。皮质醇对免疫系统的影响亦是复杂的。皮质醇通常压抑免疫系统，这种作用取决于参与的免疫细胞类型和涉及的特异性参数（见本书第7章）。

对于鳟鱼和鲤鱼，MR 对盐皮质激素 DOC 要比对皮质醇表现更高的敏感性，因而认为 DOC 可能是鱼类 MRs 的一个生理学配体（Sturm 等，2005；Stolte 等，2008）。虽然在基础的情况下，DOC 在血液循环中的浓度要比皮质醇低得多，有一些因子可能参与降低皮质醇的生物利用率（bioavailability）（Prunet 等，2006）。例如，将皮质醇转变为没有活性的可的松的酶类，11β-羟类固醇脱氢酶（11-β-hydroxysteroid dehydrogenase）在鱼类组织中广泛表达（Jiang 等，2003；Kusakabe 等，2003）。尽管 DOC 在鱼类中的功能还不清楚，但它在雄性与雌性虹鳟生殖周期结束时的血浆浓度明显升高（Campbell 等，1980；Milla 等，2008），而且有证据表明 DOC 在精子排放期间起作用（Milla 等，2008）。相反，DOC 并不参与卵母细胞的水合作用与成熟（Milla 等，2006）和渗透压调节（McCormick 等，2008）。

6.4.1.3 负反馈调节

通过对多个信号因子的一些反馈作用以及在 HPI 轴的不同水平上，血浆皮质醇亦调节应激反应。腹腔内的皮质醇植入物能抑制 CRF 在金鱼 NPO 的表达（Bernier 等，1999，2004）。同样，使用甲吡酮（metyrapone）进行药理性的肾上腺切除术清除皮质醇的负反馈作用后，白亚口鱼（Morley 等，1991）和金鱼（Bernier 和 Peter，2001）CRF 基因在 NPO 的表达增加，以及 CRF-免疫反应在鳗鱼（Olivereau 和 Olivereau，1990a）脑和脑垂体中增强。相反，尽管明显的证据表明 CRF-免疫反应神经元和 GRs 在虹鳟 NPO 中共表达（Teitsma 等，1998；见本书第2章），但皮质醇在调控虹鳟 CRF 基因表达中的作用还是不明确的。对虹鳟埋植皮质醇后，在隔离期间能使应激反应引起 CRF mRNA 在 NPO 升高的水平降低下来，但并不妨碍和禁闭相关的 CRF 基因表达的增强（Doyon 等，2006）。此外，用 GR 拮抗物钌-486（Ru-486）处理隔离的和禁闭的鳟鱼，其结果都是 NPO 的 CRF mRNA 水平降低（Doyon 等，2006）。最后，慢性提高虹鳟血浆的皮质醇水平能和 NPO 的 CRF mRNA 水平的升高联系起来（Madison B. N.，Tavakoli S. 和 Bernier N. T.，未发表结果）。

再者，尽管已经知道 CRF 在哺乳类（Shepard 等，2000）和爪蟾（*Xenopus*）（Yao 和 Denver，2007）中的表达是由脑不同部位的皮质类固醇激素进行差异性调控的，但至今还不清楚鱼类是否由 GCs 类似的细胞类型特异性地影响 CRF 表达。皮质醇亦抑制来自金鱼（Fryer 和 Lederis，1998）和鳗鱼（Olivereau 和 Olivereau，1990a）NPO 的 AVT 合成与在脑垂体内的释放，以及金鱼下丘脑 UI 的 mRNA 水平（Bernier 和 Peter）。在脑垂体水平，皮质醇抑制灌注的金鱼促肾上腺皮质激素细胞的基础 ACTH-释放以及 CRF 和 UI 的 ACTH-释放活性（Fryer 等，1984）。外源的皮质类固醇亦能压制金头鲷（Rotllant 等，2000a）和褐鳟（Pickering 等，1987）的 ACTH 水平，并且能剂量依存地降低虹鳟脑垂体 POMC 表达和血浆 ACTH 水平（Madison B. N., Tavakoli S. 和 Bernier N. J., 未发表结果）。皮质醇能抑制莫桑比克罗非鱼离体促黑色素激素细胞由 CRF-刺激的 αMSH 释放（Balm, 1993），但对金头鲷血浆的 αMSH 水平没有影响（Rotllant 等, 2000a）。在肾间腺细胞，皮质醇能通过银大麻哈鱼的副分泌反馈作用环而抑制其自身的分泌活动（Bradford 等, 1992），但对金头鲷没有这个作用（Rotllant 等, 2000a）。最后，对鱼类的一些研究表明皮质醇能介导 GR 蛋白质的降解（Shrimpton 和 Randall, 1994；Aluru 和 Vijayan, 2007）。

总之，虽然有充分的证据表明皮质醇通过一些负反馈作用的通路能够限制内分泌应激反应的幅度和持续时间，但皮质醇及其受体调控 HPI 轴内各种靶标的转录机（transriptional machinery）的分子作用机理还有待于阐明。

6.4.2 甲状腺激素

6.4.2.1 甲状腺素和甲状腺外的脱碘作用

甲状腺，或者更准确地说是甲状腺的功能单位，即甲状腺滤泡，是脊椎动物甲状腺素激素原的唯一来源。鱼类和其他脊椎动物不同，通常没有一个密集的甲状腺，而是一个松散的，没有包囊的甲状腺滤泡组织。甲状腺滤泡典型的是散布在咽下区腹大动脉附近，但也有许多硬骨鱼类有功能的和内分泌活性的滤泡分布于其他的身体部位（Baker-Cohen, 1959；Geven 等, 2007）。特别是肾的组织，是异位甲状腺滤泡的常见部位，对有些鱼类是有生物活性甲状腺的独一无二的位置。鲤鱼咽下区、肾脏和头肾的甲状腺滤泡都对 T_4 产生免疫反应，只有分布在肾脏和头肾的异位甲状腺滤泡能够表现其积累碘的功能以及在 TSH 与其他内分泌因子刺激下离体地分泌 T_4（Geven 等, 2007, 2008）。

甲状腺素没有或者很少生物活性，它是在正常或者非禁食情况下甲状腺滤泡分泌的甲状腺素的主要类型（Eales 和 Brown, 1993）。T_4 分泌出来后化学转变为强有力生物活性的 3,5,3′-三碘甲腺原氨酸（3,5,3′-triiodothyronine, T_3）是甲状腺激素代谢的关键性步骤（综述见 Kohrle, 1999, 2002），而甲状腺素受体对 T_3 的亲和力要比对 T_4 高 10 倍左右。化学转变包括从激素原分子的 5′-位置去掉一个碘原子，并被两个碘化甲状腺氨酸脱碘酶（iodothyronine deiodinase）的同等型，即 D_1 和 D_2 催化。所以，任何细胞或组织表达一个 5′-脱碘活性就能产生有生物活性的甲状腺素，并且在适宜的细胞挤压通路（cellular extrusion pathway）的情况下，可以认为是血液循环中 T_3 的一个来源。这样就对甲状腺外组织 T_3 的全身供给形成关键的作用。

5′脱碘酶 D_1 和 D_2 广泛表达，特别是在鱼的肝脏和肾脏显示高的 T_4 脱碘作用活性。这些器官亦成为甲状腺状态重要的决定因素（Mol 等, 1997；Van der Geyton 等, 1998），而脱碘酶的活性是有关试验的示值读数。对整个器官匀浆或亚细胞分馏物采用放射性标志的底物，能在离体可靠地测定脱碘作用，但必须验证测定的状况，因为硬骨鱼类脱碘酶类的生物化学特征和它们的哺乳类直向同源物会有明显的不同（Klaren

等，2005；Arjona 等，2008）。严格说来，下丘脑-脑垂体-甲状腺（HPT）轴的最终产物只是一个激素原，按照 HPT 轴的理论模型是不完整的，它既不包括有生物活性的甲状腺激素，亦不认可在甲状腺外 T_4 转化为 T_3。

和后翻译修饰（post-translational modification）的一些肽类激素调控的功能相似，甲状腺激素接受一系列生物化学反应（见图 6.4）。所产生的脱碘的、结合的、脱羧的和/或脱氨的碘化甲状腺氨酸代谢产物都具有不同的生物化学和生理学特性（Visser, 1994b；Wu 等，2005）。一定的代谢通路，例如 T_3 的脱碘作用和接合作用（conjugation），都分别清楚地参与激素信号的终止和身体的清除。然而，另外方面的作用，主要是非基因组的，可以归因于多种甲状腺素代谢产物（这将在第 6.4.2.4 节中讨论）。

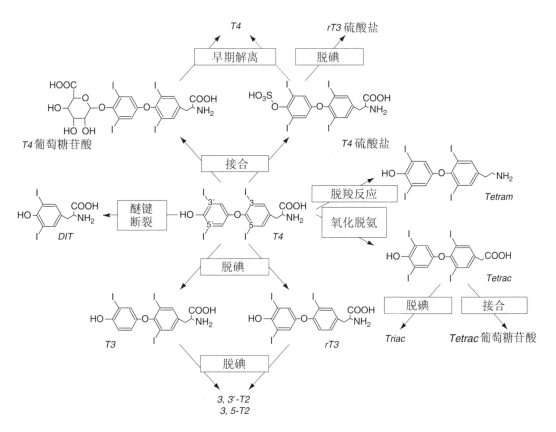

图 6.4 甲状腺激素的代谢通路（修改自 Kohrle 等，1987）

在图中，甲状腺素（T_4）选做中间代谢物，但大多数通路适用于其他的碘化甲状腺氨酸种类。注意：T_4 硫酸盐对硫酸酯酶并不容易受到影响而早期解离（deconjugation）（如虚线箭头所示），或者 D_1 对 5'-脱碘作用，但 T_3 硫酸盐容易受到硫酸酯酶的影响。缩写：DIT，二碘酪氨酸；Tetrac，3,3',5,5'-四碘甲状腺乙酸；Tetram，3,3',5,5'-四碘甲状腺原氨酸胺；Triac，3,5,3'-三碘甲状腺乙酸。

6.4.2.2 甲状腺激素信号取决于受体和细胞的摄取

甲状腺素受体（TRs）是配体依赖的转录子因子（ligand-dependent trancription factor），属于2型核受体家族，和9-cir-视黄酸受体，RXR形成二聚体（Harvey 和 Williams，2002；Tata，2002）[①]。大多数脊椎动物有两个分开的基因编码 TRα 和 TRβ 同等型，但已经确认鱼类有三个基因，分别编码两个 TRα 和一个 TRβ（Yamano 等，1994；Yamano 和 Inui，1995；Liu 等，2000；Tang 等，2008）。多个有功能的受体是由于选择性的 mRNA 剪接和交替的启动子所造成（Lazar，1993；Marchand 等，2001；Nelson 和 Habibi，2009）。由于特殊的基因组结构和转录调节（transcriptional regulation），TRs 的表达能够短暂地和季节性地进行精确的细微调整。通常，TR mRNA 实际上能在鱼体的每一个组织中表达（Yamano 和 Miwa，1998；Marchand 等，2004；Nelson 和 Habibi，2006；Filby 和 Tyler，2007），这就是甲状腺激素多效而具有普遍作用的原因。

T_3 的大多数分子靶标定位在基因组内。对 T_3 起反应的基因携带一个由一对靠近启动子区转录起始位点（transcription initiation site）的 DNA 六聚体半位点（hexamer half-site）组成的甲状腺应答元件（thyroid response element，TRE）。TRE 能和 TRs 结合成单体（monomer）或同型二聚体（homodimer），但最活跃的 TR 构型是一个和 RXR 形成的异源二聚体（heterodimer）。取决于存在一个正的或负的 TRE，T_3 能激活或抑制基因表达。T_3 的结合使一个由协阻抑物蛋白质（corepressor protein）从 TR-RXR-TRE 复合物中解离，并且结合一个辅激活物（coactivator）。许多辅激活物具有一个内在的组蛋白乙酰转移酶（histone acetyltransferase）活性，能修饰染色质结构以便进入 RNA 聚合酶，然后进行转录（Yen，2001）。有关 T_3-依赖的转录阻抑（transcriptional suppression）的分子作用机理尚未得到充分阐明。

有关在鱼类基因中（正的）TREs 的试验数据还很少。它们包括生长激素和脂解的酶，脂蛋白脂肪酶（Sternbery 和 Moav，1999；Almuly 等，2000；Oku 等，2002）。除了有一项研究试图确定 T_3 在比目鱼变态过程中的靶基因之外（Marchand 等，2004），目前学者们对鱼类 T_3-应答基因还了解得很少。河鲀和斑马鱼基因组的汇集和 TRE 半位点 AGGT（C/A）A 契合序列（consensus sequence）所提供的大量资料，必将有助于在计算机芯片上（in silico）确定 TREs 和 T_3-靶基因的位置和作用。

除甲状腺受体外，同样重要的是在靶细胞中的甲状腺激素转运蛋白（transporter）质膜的所有组成成分（repertoire）。由于碘化甲状腺氨酸分子的亲脂性（lipophilic），传统上都认为它是以扩散方式进行跨膜转运。但是，甚至早期的研究都已经表明甲状腺激素在细胞内的聚集是能量-依赖的（Hennemann，2005）。此外，核磁共振研究表明碘化甲状腺氨酸在脂双层内优先分隔（preferential partition），而被动的跨膜扩散，如果有的话，主要是依赖于膜的磷脂成分和胆固醇含量（Lai 等，1985；Chehin 等，1999）。

[①] 通过网址 www.NURSA.org 可以得到核受体信号图集（Nuclear Receptor Signaling Atlas），包含甲状腺激素受体广泛和详细的资料（Margolis，2008）。

扩散通路的通透性不容易调节，通常是以一种低效的方式调节激素的进入。在过去的几年里鉴定了系列甲状腺激素的转运蛋白（Jansen 等，2005），它们在甲状腺激素信号中的关键作用已经在严重病态的病人中得到明确的证实，原因可以归咎于功能异常的突变型转运蛋白（Friesema 等，2004；Jansen 等，2008）。所以，未受干扰的甲状腺激素信号不仅需要有生物活性的配体和受体，还需要完整的跨膜转运通路（Henemann 等，1998）。

有机的阴离子转运蛋白多肽（organic anion transporter polypeptide，OATP）家族的成员和单羧化转运蛋白（monocarboxylate transporter，MCT）具有特异性的甲状腺激素转运能力（Jansen 等，2005；Hagenbuch，2007）。我们有一些不那么贴切的关于这些转运蛋白参与鱼类甲状腺激素信号的资料。几种 MCT 蛋白质，如特异性地在人体内转运甲状腺激素的 MCT8 和 MCT10（Friesema 等，2008）已经在硬骨鱼类基因组中鉴别出来（Liu 等，2008），但我们并不了解它们在甲状腺生理学中的在体作用。值得注意的是，从美洲拟鲽（*Pseudopleuronectes americanus*）肾脏中克隆到一个有机的阴离子转运蛋白 fOat，和鼠与人的 Oat1 和 Oat3 同源（Wolff 等，1997；Aslamkhan 等，2006），即 OATP 家族的两个成员在哺乳类中介导甲状腺激素的转运（Hagenbuch 和 Meier，2004；Hagenbuch，2007）。至于 fOat 蛋白质是否有能力通过质膜转运甲状腺激素，还有待研究确定。

6.4.2.3 靶标和功能

哺乳类的基因为 T_3 正调控的是 1 型脱碘酶、α-肌球蛋白重链、几种 ATP 酶（Na^+/K^+-ATP 酶 α-亚单位、肌质网 Ca^{2+}-ATP 酶）、参与神经元发育的蛋白质（髓鞘碱性蛋白、浦肯野细胞蛋白质 pcp2）、脂肪生成的酶类（苹果酸酶、葡萄糖-6-磷酸盐脱氧酶）、糖异生酶-烯醇丙酮酸磷酸羧激酶（PERCK），以及大鼠生长激素。负的调控基因是 TSH α 和 β-亚基、TRH、2 型脱碘酶、催乳激素以及人的生长激素（Harvey 和 Williams，2002；Konig 和 Neto，2002）。只有一项研究试图确定硬骨鱼类 T_3 的靶基因（Marchand 等，2004），但目前对鱼类的 T_3-应答基因还了解得很少。传统上，甲状腺激素的作用被认为是 T_3 单独的基因组作用，但这个观点忽略了有关 T_3（Davis 和 Davis，1996；Bassett 等，2003；Goglia，2005）和其他碘化甲状腺氨酸代谢产物（见图 6.4）非基因组作用的报道，这将在下一节中讨论。

6.4.2.4 甲状腺激素代谢产物的生物学活性

对哺乳类和一些鱼类的研究曾报道多种甲状腺激素代谢产物的生物学作用。从鳟鱼和鲑鱼肝脏中分离出来的核和一系列甲状腺激素类似物结合，尽管它们的亲和力有三个幅度等级的差别（Darling 等，1982；Bres 和 Eales，1986；Leeson 等，1998）。不通过核受体和基因组介导的作用实例包括由 T_4 和 T_3-异构体 3, 3′, 5′-三碘甲腺原氨酸（rT_3）在培养细胞系中诱导促分裂原活化蛋白激酶（mitogen-activated protein kinase，MAPK）通路（Lin 等，1999），由二碘甲腺原氨酸的 3,5-T_2 和 3, 3′-T_2 对哺乳类线粒体的呼吸作用和底鳉 2 型脱碘酶的刺激（Lanni 等，1994；Goglia，2005；Garcia-G 等，

2007），以及一碘甲腺原氨酸胺（monoiodothyronamine）T_1AM 的收缩能效应（inotropic effect）（Scanlan 等，2004）。硫酸盐的和葡萄糖醛酸的甲状腺激素结合物在代谢和转运的通路中具有不同的反应（Visser，1994a；Wu 等，2005；van der Heide 等，2007）。在金头鲷的相关研究中发现参与甲状腺激素结合与早期解离的酶类特异性活性能对渗透性变化起反应（Klaren 等，2007b），表明这是结合的碘化甲腺氨酸的一种生理作用。

除了鱼类天然的甲状腺激素外，有必要去推测碘化甲腺氨酸代谢产物新的功能。的确，可以设想进化的压力能够促使碘化甲腺氨酸分子的多种代谢产物都具有生物功能。在淡水和海水的环境中，碘和硒（一种脱碘酶含硒蛋白质的必要成分）以痕量存在，这使得甲状腺激素的作用依赖于两种痕量元素。T_4 酚环的单一脱碘作用产生生物活性很强的 T_3，这就已经说明一个碘原子对一个分子的生物学作用的影响。T_3 的碘原子能够继续被除去而产生二碘化甲腺原氨酸或一碘化甲腺原氨酸以及它们的衍生物，所有这些都具有不同的立体结构特征和不同的生物活性。对于碘化甲腺氨酸的代谢产物都具有一种生物学活性，可以说是充分开发利用痕量元素的一种"经济"途径。

6.4.2.5 负反馈作用

脑垂体促甲状腺素细胞的 TSH 分泌活动和血浆甲状腺激素之间存在着传统的负反馈作用。早期的研究已经表明试验性地引起甲状腺功能减退（hypothyroidism）后能导致脑垂体促甲状腺素细胞过度生长和脱粒（Barrington 和 Matty，1995；Haider，1975）。最近对几种硬骨鱼类的研究表明 T_3 和 T_4 都能下调在体和离体的脑垂体 TSHB-亚单位 mRNA 的含量（Larsen 等，1997；Pradel-Balade 等，1997，1999；Sohn 等，1999；Yoshiura 等，1999；Chatterjee 等，2001；Manchado 等，2008）。因此，硬骨鱼类似乎并不符合哺乳类的反馈作用模型，因为对于哺乳类只有从局部脑垂体的 T_4 脱碘而衍生的 T_3 才是脑垂体 TSH 分泌活动的主要抑制剂（Silva 等，1978；Christoffolete 等，2006）。这很可能反映了在脑垂体促甲状腺素细胞质膜中甲状腺激素转运通路的不同特性。

6.5 促肾上腺皮质激素轴、促黑色素激素轴和促甲状腺素激素轴对应激反应的作用

促肾上腺皮质激素、促黑色素激素和促甲状腺素激素轴都参与对应激物的内分泌反应。尽管在这一领域的鱼类文献都集中于促肾上腺皮质激素轴对调控皮质醇分泌活动的直接作用，亦有一些证据表明促黑色素激素轴和促甲状腺素激素轴在中枢与外周和促肾上腺皮质激素轴相互作用以协调应激反应。本节将概括介绍每个轴参与应激反应的证据，以及它们的相互作用和协调作用。

6.5.1 促肾上腺皮质激素轴

脊椎动物促肾上腺皮质激素轴的基本功能是刺激肾上腺髓质或其相等物的类固醇生成组织生物合成与释放皮质类固醇。作为回应的是它们在应激反应中起重要作用的

结果（见第6.4.1.2节），提高血液循环中的皮质类固醇水平是应激反应的关键指标。因此，对许多鱼类进行研究的报道都认为对各种应激物的反应是血浆的皮质醇水平升高（Barton和Iwama，1991；Mommsen等，1999；Norris和Hobbs，2006）。一些研究表明促肾上腺皮质激素轴在皮质醇的应激反应中是血浆ACTH水平升高而后表现为血液循环中皮质醇增加。例如，手工操作和禁闭（Sumpter等，1986；Rotllant等，2001；Doyon等，2006）、拥挤（Rotllant等，2000b）、热休克（Pickering等，1986）、渗透压休克（Craig等，2005）和从属地位（subordination）（Hoglund等，2000）等都以血浆ACTH和皮质醇水平的增加为特征。同样，对手工操作、禁闭和拥挤的反应是脑垂体ACTH含量下降（Rotllant等，2000b，2001），而和禁闭（Gilchriest等，2000）或监禁（Metz等，2004）相联系的是脑垂体远侧部POMC基因表达增加，这些都表明对应激物的反应是POMC衍生的ACTH合成与释放的增加。一般来说，尽管少数急性应激物可能并不依赖于ACTH而刺激皮质醇释放（Balm等，1994；Arends等，1999），但对大多数应激物的常见反应是促肾上腺皮质激素轴和ACTH所起的作用。

6.5.2 促黑色素激素轴

一些研究表明促黑色素激素轴对鱼类的内分泌应激反应起作用。例如，北极红点鲑（*Salvelinus alpinus*）处于社会性从属地位时，血浆的 αMSH 水平升高，并和皮肤变黑呈现正的联系（Hoglund等，2000，2002a）。同样，处于从属地位的虹鳟，主要来源于促黑色素激素细胞轴的脑垂体 POMC mRNA 水平增加（Winberg和Lepage，1998）。虽然对手工操作、禁闭和慢性监禁（Sumpter等，1985；Arends等，1999；Rotllant等，2000b；Metz等，2005；van den Burg等，2005）的反应亦曾观察到血浆 αMSH 和 N-AcβEND 水平增加，但在另一些研究中对同样的应激物亦曾表明这些肽类在血液循环中的水平并不受到影响或者下降（Balm等，1995；Balm和Pottinger，1995；Ruame等，1999；Rotllant等，2001）。在对鲤鱼进行的研究中，αMSH 和 N-AcβEND 对应激反应作用的变化取决于促黑色素激素轴对 CRF 的反应状态（van den Burg等，2005）（见第6.2.2.1节）。总之，促黑色素激素轴对鱼类应激反应的作用依赖于应激物的性质和强度，参与的种类以及促黑色素激素对下丘脑调控脑垂体因子反应的状态。

目前学者们对鱼类促肾上腺皮质激素轴的血液循环中 αMSH 和 βEND 的靶标与作用尚未很好地阐明，但初步证据表明很可能是多重的相互作用。虽然在莫桑比克罗非鱼的相关研究中观察到 αMSH 的促肾上腺皮质激素活性在其他的硬骨鱼类中并不普遍存在（见第6.3.2节），但 MC4R 和 MC5R 在金鱼的视前区和结节下丘脑（Cerda-Reverter等，2003b、c）广泛的基因表达表明，如同 Matsuda 等（2008）所证明的，在 αMSH 和 CRF 系统之间存在着相互作用。外周的 αMSH 能通过血脑屏障（Banks和Kastin，1995），并且能够达到这些中枢的靶标。还有，需要进一步研究以确定血液循环中的 αMSH 对促肾上腺皮质激素轴的下丘脑调控脑垂体的调节剂是否起着反馈作用（Metz等，2005）。同样，除了对罗非鱼皮质醇分泌活动的增强作用之外，N-ACβEND 对应激反应的作用还是难以捉摸的（Flik等，2006）。相反，非乙酰化 βEND 对促肾上腺皮质激素轴的调节作用已经为鲤鱼下丘脑、脑垂体远侧部与头肾的 mu 和 delta 阿片样物质受体的表达和应激反应引起的调控作用所证明（Chadzinska 等，2009）。

6.5.3 促甲状腺激素轴

甲状腺素具有普遍性的功能，实际上鱼体的任何细胞都能表达甲状腺激素受体。按照这个看法，可以很容易地想到促甲状腺激素轴参与鱼类的内分泌应激反应。HPT 轴参与的证据主要是血浆甲状腺激素水平的改变。不过，要指出的是，不能把事情都说成一个模样。有些例子足以说明：对虹鳟进行手工操作或者抽取血样，使血浆的 T_3 水平增加而 T_4 水平下降（Todd 和 Eales，2002）；限制喂食使斑点鮰（*Ictalurus punctatus*）、虹鳟和金头鲷血浆的 T_3 和 T_4 水平降低（Farbridge 等，1992；Power 等，2000；Gaylord 等，2001）；虹鳟在保持高密度时血浆 T_3 水平下降（Leatherland，1993）；褐鳟（*Salmo trutta*）处在酸性水中时，T_4 水平升高，但 T_3 水平不受影响（Brown 等，1989）；渗透压变化能短暂地使虹鳟血浆总的 T_4 水平和总的 T_3 水平增加（Orozco 等，2002），以及使塞内加尔鳎（*Solea senegalensis*）游离的 T_4 水平增加（Arjona 等，2008）。T_4 和 T_3 典型的血浆水平并不平行地发生变化，亦不依赖于所测定的总的或游离的激素浓度，这表明在应激反应期间，甲状腺外的脱碘通道参与调控血浆的甲状腺激素浓度（Waring 和 Brown，1997；Orozco 等，2002；Lopez-Bojorquez 等，2007；Arjona 等，2008）。脊椎动物的甲状腺是激素原 T_4 唯一的内在来源，但外周的组织和器官是血液循环的或局部有生物活性的 T_4 的来源。这使得 HPT 轴和促肾上腺皮质激素轴与促黑色素激素轴有所不同，因为鱼类甲状腺状态的调控能很好地扩展到狭小而有限的 HPT 轴之外。

至少两个内分泌轴的终产物，即皮质醇和甲状腺素具有交互的影响（reciprocal effect）。在用这两种激素处理鱼类的实验中得到明显的表现（Young 和 Lin，1988；Vijayan 和 Leatherland，1992；Mustafa 和 Mackinnon，1999；Walpita 等，2007）。正如在鲤鱼中所表现的（Geven 等，2006，2009），通过共用脑的信号分子，可能产生双向的沟通，但外周的脱碘活性亦是一种靶标（Vijayan 和 Leatherlard，1992；Walpita 等，2007）。

有学者曾致力于整合鲤鱼甲状腺系统的中枢与外周调控作用（Geven 等，2006，2007，2009）。得到的结果是甲状腺素通过 CRF-BP 和 CRF 降低在体促肾上腺皮质激素轴的活性，并激活 αMSH 的分泌。多种外周因子，即 TSH、ACTH、αMSH 和皮质醇，能刺激鲤鱼离体的肾甲状腺（renal thyroid gland）释放 T_4。很可能有 GR 和 MC2R 参与（Metz 等，2005a）。有意思的是，鲤鱼的肾脏和头肾含有功能的甲状腺，并且形成一个中心，将来源于3个内分泌轴的外周信号整合起来。

6.5.4 在有机体水平上的整合：下丘脑调控脑垂体的因子对应激反应的在体作用

离体的研究结果表明，多种下丘脑因子能够刺激和抑制鱼类促肾上腺皮质激素轴、促黑色素激素轴和促甲状腺素激素轴的分泌活动（见第6.2节），而在体的研究提供的证据则说明这些潜在的下丘脑调控脑垂体的信号实际上是在各种不同的生理状态下起着调控脑垂体细胞分泌活动的作用。这种离体与在体之间的差别并非不重要，而离体数据的翻译对于一个有机体可能是不恰当的。在前一节中我们曾讨论脑垂体一旦失去神经支配与缺少神经内分泌输入后脑垂体细胞基础活性的变化，并且展示了内分泌综合体（combinations of endocrines）协同作用与许可作用的资料。此外，不同类型的应激物接受不同的应激敏感线路，汇聚于间脑的下丘脑释放的和抑制的神经元（Herman 等，2003），而下丘脑调节脑垂体各种信号的相对作用很可能是应激物特异性的。本节将综述在体研究下丘脑调控脑垂体的因子在调控促肾上腺皮质激素轴、促黑色素激素轴与促甲状腺激素轴的作用中所取得的研究成果。

除了已经确定 CRF 和 UI 作为重要的下丘脑调控脑垂体因子调节 ACTH、TSH 和 αMSH 的分泌以外，有充分的证据表明 CRF 系统调节鱼类的促肾上腺皮质激素轴。在虹鳟的相关研究中，隔离和禁闭（Ando 等，1999；Doyon 等，2003）、社会性从属地位（Doyon 等，2003）、血氨过多（hyperammonemia）（Ortega 等，2005）、低氧（Bernier 和 Craig，2005）和移入海水（Craig 等，2005）都以应激物特异性地和时间依赖性地增加血浆皮质醇和视前区 CRF mRNA 水平为特征（Bernier 等，2008）。虹鳟对所有上述的应激物，除社会性从属地位自相矛盾地表现出最明显的皮质醇反应之外，都和前脑 UI mRNA 水平的增加相关（Craig 等，2005；Bernier 等，2008）。在鲤鱼的相关研究中，24 h 监禁的应激使血浆皮质醇水平与下丘脑 CRF 基因表达增加，余下 UI mRNA 水平没有变化（Huising 等，2004）。相反，在川鲽（*Platichthys flesus*）的相关研究中，30 min 禁闭的应激，3 h 后下丘脑 UI mRNA 水平增加，而 CRF 基因表达没有变化（Lu 等，2004）。在马苏大麻哈鱼（*Oncorhynchus masou*）的相关研究中，血浆皮质醇和前脑 CRF 与 UI mRNA 水平的季节变化表明在产卵期间，UI 调节促肾上腺皮质激素轴的重要性大于 CRF（Westring 等，2008）。令人遗憾的是，上述研究中没有一项用定量分析来研究这些应激物是否影响到血液循环中的 αMSH、N-AcβEND、TSH 或甲状腺激素的水平，使我们留下这样的疑问：CRF 或 UI 对这些特异性应激物的反应是否会对促黑色素激素轴或促甲状腺素激素轴的调节起作用。不同的是，金头鲷在拥挤的应激反应后对促黑色素激素轴离体的 CRF 敏感性增加，表明 CRF 系统有可能参与 αMSH 释放的调控。

除 CRF-相关肽对调控促肾上腺皮质激素轴的刺激作用之外，CRF 系统的其他组分可能参与阻抑对应激物的内分泌反应和调节它们的持续时间。例如，对鲤鱼 24 h 禁闭的应激作用，使脑垂体 CRF1 型受体（CRFR1）mRNA 水平下调和下丘脑 CRF-BP 基因表达增加（Huising 等，2004）。对虹鳟反复地驱赶后，其视前区 CRF mRNA 增加，脑垂体 CRF-BP 基因表达增强（Doyon 等，2005）；而低氧和从属地位的应激，使脑 CRF-BP mRNA 水平区域特异性增加（Alderman 等，2008）。同样，对伯氏朴丽鱼给予一个月的社会性应激反应，脑 CRF 和脑垂体 CRFR1 水平降低，脑垂体 CRF-BP 增加（Chen 和 Fernald，2008）。甲状腺素对鲤鱼下丘脑 CRF-BP mRNA 水平的刺激作用亦表明两个轴之间在体的相互作用（Geven 等，2006，2008）（见第 6.2.2.1 节）。有意思的是，斑马鱼的 CRF-BP 表现明显地和 CRF 区域共表达，而比较少和 UI 共表达（Alderman 和 Bernier，2007）。总之，尽管还需要进行一些针对性的实验以直接证明 CRF-BP 参与鱼类 CRF 相关肽的信号活动和确定介导下丘脑调控脑垂体 CRF 和 UI 作用的特异性 CRF 受体亚型，但根据前述的研究结果就可以考虑确定 CRF-BP 和 CRF 受体在 3 个下丘脑-脑垂体轴的综合协调中起着重要的作用。

一些确定应激物影响视前区 AVT mRNA 表达特征的研究亦提供证据表明参与调节鱼类促肾上腺皮质激素轴的下丘脑调控脑垂体因子的多重性（multiplicity）。在虹鳟的相关研究中，急性监禁的应激反应使血浆皮质醇增加，但慢性应激反应并不和 NPO 小细胞神经元的 AVT mRNA 水平增加相关（Gilchriest 等，2000）。禁闭的应激反应特性亦使牙鲆血浆皮质醇和 NPO AVT mRNA 表达增加，但应激物是使牙鲆 NPO 大细胞而不是小细胞的 AVT 转录体发生变化（Bond 等，2007）。斑马鱼对社会性从属地位的应激反应使视前区小细胞 AVT 免疫反应显著增强（Larson 等，2006），这亦表明 AVT 对促肾上腺皮质激素轴的激活产生作用。

至今，只有离体的研究证明 TRH 能刺激鱼类促肾上腺皮质激素轴、促黑色素激素轴和促甲状腺素激素轴的分泌活动，而在体的研究证明 TRH 调控这 3 个轴的作用还很少。然而，罗非鱼在慢性酸化作用应激反应（Lamers 等，1994）后和金头鲷在拥挤应激反应（Rotllant 等，2000b）后促黑色素激素细胞离体对 TRH 的敏感性增加，表明 TRH 参与对应激物反应的 αMSH 释放的刺激作用。此外，鲤鱼在禁闭 24 h 的应激反应后，下丘脑 TRH 表达增加 2 倍多（Metz J. 和 Flik G.，未发表的观察结果），而我们已经知道这种应激物是和血浆皮质醇增加与下丘脑 CRF mRNA 表达水平增加相联系的（Huising 等，2004）。

对和下丘脑 MCH 核周体含量、整个脑垂体 MCH 的含量、血浆 MCH 的水平（Green 和 Baker，1991；Green 等，1991）以及下丘脑 MCH mRNA 水平（Gröneveld 等，1995a）等变化相联系的各种应激物进行的观察研究表明，MCH 在应激反应期间具有调控 ACTH 和 αMSH 分泌活动的潜在作用。然而，在应激反应的情况下，是否可以以及在多大程度上能够把抑制作用归因于 MCH 离体（见第 6.2.1 节和第 6.2.2 节）对促肾上腺皮质激素轴和促黑色素激素轴的调控，还有待于证明。同样，虽然各种应激物在鱼类中都和下丘脑（Overli 等，1999，2001；Gesto 等，2008）与脑垂体（Gesto 等，2008）的多巴胺能活性增加相联系，而 DA 能抑制 ACTH、αMSH 和 TSH 离体的分泌（见第 6.2 节），但还需做进一步的研究以确定多巴胺作为一个起抑制作用的因子在多大程度上影响脑垂体对内分泌应激反应的应激物-特异性作用。

除了它们对脑垂体分泌活动的作用之外，促肾上腺皮质激素轴、促黑色素激素轴和促甲状腺素激素轴的重要下丘脑调控脑垂体因子还参与多种调控功能。例如，在斑马鱼的脑内，CRF、UI、CRF-BP（Alderman 和 Bernier，2007）和 TRH（Diaz 等，2002）的广泛表达型式表明其对感觉的、自主的、行为的和神经内分泌功能的多种作用。由于 CRF 和 UI 在调节鱼类心血管系统（Le Mevel 等，2006）、通风（ventilation）（Le Mevel 等，2009）、行动（Clements 等，2002；Carpenter 等，2007）和食欲（Bernier，2006）方面的作用亦证明 CRF 系统在协调应激反应方面起着更为广泛的作用。除前脑外，鱼类还能独特地具有 CRF-相关肽的第二个主要来源——CNSS（caudal neurosecretory system）（McCrohan 等，2007），它能释放到血液循环中，并对应激物起反应（Bernier 等，2008）。鲤鱼的 CRF 和 CRF-BP 在头肾的嗜铬组织中大量表达，表明它和哺乳类相似，一个局部的 CRF 系统可能对皮质醇的释放起着调节作用（Huising 等，2007）。整合 CRF 和 TRH 的下丘脑调控脑垂体的功能作为更广阔的肽类生理作用将是未来研究的一个重要方向。

6.6 展 望

如果在我们的阐述和资料讨论中出现问题，那是同一个下丘脑信号分子，即 TRH 和 CRF-家族的肽类，对促肾上腺皮质激素轴、促黑色素激素轴和促甲状腺素激素轴的中枢调控起显著的作用，尽管其他的下丘脑因子亦如此。这三个轴的最终产物在外周以及在中枢都是交叉沟通的。要想持续保持一种应激反应调控的概念式模型作为平行-运转的内分泌轴是很困难的。我们提供充分的文献资料阐明 CRF 和 TRH 能刺激脑垂体主要激素 ACTH、αMSH 和 TSH 的释放，而它们都能够刺激外周释放的激素，如皮质醇和 T_4/T_3 的释放。αMSH 作为一个厌食性因子与脂解因子的作用，皮质醇作为一种糖皮质激素，以及甲状腺激素对基础代谢率的影响，所有这些都汇聚到能量稳态这个共同的名称中。要想调整那种认为只有两种下丘脑因子通过多种中间的信号因子调控如此基础性生理过程的错误见解，我们必须更加重视这些因子的特异性下丘脑定位，特别是从较高等的/其他的脑中心的传入输入（afferent input）它们所要连接的和已经连接的。把下丘脑看作一个动物生理状态相关中枢信号的整合中心的观点是正确的，但必须对这些信号的状态做详细的阐述。我们未来的研究方向应该朝着下丘脑的上游去确定新的作用因子——不是去扩展内分泌轴，而是更深刻地了解在一个基础水平上构建的应激网络（stress net）。通过这个网络适当调控能量的稳态/异态，对于脊椎动物，依据它们的关键信号分子及其多层次与高度整合特性的极端保守性，提供给生物有机体很大的柔性（flexibility），以对付不断变化的环境而获得进化成功，是至关重要且不可缺少的。

N. J. 伯尼尔

G. 弗利克

P. H. M. 克拉伦

参考文献

Aguilera G, Subburaju S, Young S, Chen J. 2008. The parvocellular vasopressinergic system and responsiveness of the hypothalamic pituitary adrenal axis during chronic stress. *Prog. Brain Res*, 170: 29-39.

Akil H, Young E, Watson S J, Coy D H. 1981. Opiate binding properties of naturally occurring N- and C-terminal modified β-endorphins. *Peptides*, 2: 289-292.

Alderman S L, Bernier N J. 2007. Localization of corticotropin-releasing factor, urotensin I, and CRF-binding protein gene expression in the brain of the zebrafish, *Danio rerio*. *J. Comp. Neurobiol*, 502: 783-793.

Alderman S L, Raine J C, Bernier N J. 2008. Distribution and regional stressor-induced regulation of corticotrophin-releasing factor binding protein in rainbow trout (*Oncorhynchus mykiss*). *J. Neuroendocrinol*, 20: 347-358.

Almuly R, Cavari B, Ferstman H, Kolodny O, Funkenstein B. 2000. Genomic structure and sequence of the gilthead seabream (*Sparus aurata*) growth hormone-encoding gene: Identification of minisatellite polymorphism in intron I. *Genome*, 43: 836-845.

Alsop D, Vijayan M M. 2008. Development of the corticosteroid stress axis and receptor expression in zebrafish. *Am. J. Physiol*, 294: R711-719.

Aluru N, Vijayan M M. 2007. Hepatic transcriptome response to glucocorticoid receptor activation in rainbow trout. *Physiol. Genomics*, 31: 483-491.

Aluru N, Vijayan M M. 2008. Molecular characterization, tissue-specific expression, and regulation of melanocortin 2 receptor in rainbow trout. *Endocrinology*, 149: 4577-4588.

Amiya N, Amano M, Takahashi A, Yamanome T, Yamamori K. 2007. Profiles of α-melanocyte-stimulating hormone in the Japanese flounder as revealed by a newly developed time-resolved fluoroimmunoassay and immunohistochemistry. *Gen. Comp. Endocrinol*, 151: 135-141.

Ando H, Hasegawa M, Ando J, Urano A. 1999. Expression of salmon corticotropin-releasing hormone precursor gene in the preoptic nucleus in stressed rainbow trout. *Gen. Comp. Endocrinol*, 113: 87-95.

Arai M, Assil I Q, Abou-Samra A B. 2001. Characterization of three corticotropin-releasing factor receptors in catfish: A novel third receptor is predominantly expressed in pituitary and urophysis. *Endocrinology*, 142: 446-454.

Arends R J, Mancera J M, Muñoz J L, Wendelaar Bonga S E, Flik G. 1999. The stress response of the gilthead sea bream (*Sparus aurata* L.) to air exposure and confinement. *J. Endocrinol*, 163: 149-157.

Arjona F J, Vargas-Chacoff L, Martín del Río M P, Flik G, Mancera J M, Klaren P H M. 2008. The involvement of thyroid hormones and cortisol in the osmotic acclimation of *Solea senegalensis*. *Gen. Comp. Endocrinol*, 155: 796-803.

Arnold-Reed D E, Balment R J. 1991. Atrial natriuretic factor stimulates in-vivo and in-vitro secretion of cortisol in teleosts. *J. Endocrinol*, 128: R17-R20.

Aroua S, Schmitz M, Baloche S, Vidal B, Rousseau K, Dufour S. 2005. Endocrine evidence that silvering, a secondary metamorphosis in the eel, is a pubertal rather than a metamorphic event. *Neuroendocrinology*, 82: 221-232.

Aslamkhan A G, Thompson D M, Perry J L, Bleasby K, Wolff N A, Barros S, Miller D S, Pritchard J B. 2006. The flounder organic anion transporter fOat has sequence, function, and substrate specificity similarity to both mammalian Oat1 and Oat3. *Am. J. Physiol. Regul. Integr. Comp. Physiol*, 291: R1773-1780.

Baker-Cohen K F. 1959. Renal and other heterotopic thyroid tissue in fishes. In: Gorbman A, Ed. *Proceedings of the Columbia University Symposium on Comparative Endocrinology*. Wiley, Cold Spring Harbor: 283-301.

Baker B I, Rance T A. 1981. Differences in concentrations of plasma-cortisol in the trout and the eel following adaptation to black or white backgrounds. *J. Endocrinol*, 89: 135-140.

Baker B I, Bird D J, Buckingham J C. 1985. Salmonid melanin-concentrating hormone inhibits corticotropin release. *J. Endocrinol*, 106: R5-R8.

Baker B I, Bird D J, Buckingham J C. 1986. Effects of chronic administration of melanin-concentrating hormone on corticotrophin, melanotrophin, and pigmentation in the trout. *Gen. Comp. Endocrinol*, 63: 62-69.

Baker B I, Bird D J, Buckingham J C. 1996. In the trout, CRH and AVT synergize to stimulate ACTH release. *Regul. Peptides*, 67: 207-210.

Ball J N, Olivereau M, Kallman K D. 1963. Secretion of thyrotrophic hormone by pituitary transplants in a teleost fish. *Nature*, 199: 618-620.

Ball J N, Olivereau M, Slicher A M, Kallman K D. 1965. Functional capacity of ectopic pituitary transplants in the teleost *Poecilia formosa*, with a comparative discussion on the transplanted pituitary. *Phil. Trans. R. Soc. London B*, 249: 69-99.

Ball J N, Baker B I, Olivereau M, Peter R E. 1972. Investigations on hypothalamic control of adenohypophysial functions in teleost fishes. *Gen. Comp. Endocrinol. Suppl*, 3: 11-21.

Ball J N, Batten T F C, Young G. 1980. Evolution of hypothalamo-adenohypophysial systems in lower vertebrates. In: Ishii S, Hirano T, Wada M, Ed. *Hormones, Adaptation and Evolution*. Japan Scientific Societies Press/Springer-Verlag: Tokyo/Berlin.

Balm P H M, Pottinger T G. 1995. Corticotrope and melanotrope POMC-derived peptides in relation to interrenal function during stress in rainbow trout (*Oncorhynchus mykiss*). *Gen. Comp. Endocrinol*, 98: 279-288.

Balm P H M, Gröneveld D, Lamers A E, Wendelaar Bonga S E. 1993. Multiple actions of melanotropic peptides in the teleost *Oreochromis mossambicus* (Tilapia). *Annals N. Y. Acad. Sci*, 680: 448-450.

Balm P H M, Pepels P, Helfrich S, Hovens M L M, Wendelaar Bonga S E. 1994. Adrenocorticotropic hormone in relation to interrenal function during stress in tilapia (*Oreochromis mossambicus*). *Gen. Comp. Endocrinol*, 96: 347-360.

Balm P H M, Hovens M L, Wendelaar Bonga S E. 1995. Endorphin and MSH in concert form the corticotropic principle released by tilapia (*Oreochromis mossambicus*; Teleostei) melanotropes. *Peptides*, 16: 463-469.

Bandyopadhyay S, Bhattacharya S. 1993. Purification and properties of an Indian major carp (*Cirrhinus mrigala*, Ham.) pituitary thyrotropin. *Gen. Comp. Endocrinol*, 90: 192-204.

Banks W A, Kastin A J. 1995. Permeability of the blood-brain-barrier to melanocortins. *Peptides*, 16: 1157-1161.

Barber L D, Baker B I, Penny J C, Eberle A N. 1987. Melanin concentrating hormone inhibits the release of αMSH from teleost pituitary glands. *Gen. Comp. Endocrinol*, 65: 79-86.

Barimo J F, Steele S L, Wright P A, Walsh P J. 2004. Dogmas and controversies in the handling of nitrogenous wastes: Ureotely and ammonia tolerance in early life stages of the gulf toadfish, *Opsanus beta*. *J. Exp. Biol*, 207: 2011-2020.

Barrington E J W, Matty A J. 1955. The identification of thyrotrophin-secreting cells in the pituitary gland of the minnow (*Phoxinus phoxinus*). *Q. J. Microsc. Sci*, 96: 193-201.

Barton B A, Iwama G K. 1991. Physiological changes in fish from stress in aquaculture with emphasis on the response and effects of corticosteroids. *Ann. Rev. Fish Dis*, 1: 3-26.

Bassett J H D, Harvey C B, Williams G R. 2003. Mechanisms of thyroid hormone receptor-specific nuclear and extra nuclear actions. *Mol. Cell. Endocrinol*, 213: 1-11.

Batten T F. 1986. Ultrastructural characterization of neurosecretory fibres immunoreactive for vasotocin, isotocin, somatostatin, LHRH, and CRF in the pituitary of a teleost fish, *Poecilia latipinna*. *Cell Tissue Res*, 244: 661-672.

Batten T F C, Moons L, Vandesande F. 1999. Innervation and control of the adenohypophysis by hypothalamic peptidergic neurons in teleost fishes: EM immunohistochemical evidence. *Microsc. Res. Techn*, 44: 19-35.

Bennett M A, Murray T F, Aldrich J V. 2005. Structure-activity relationships of arodyn, a novel acetylated kappa opioid receptor antagonist. *J. Pept. Res*, 65: 322-332.

Bernier N J. 2006. The corticotropin-releasing factor system as a mediator of the appetite-suppressing effects of stress in fish. *Gen. Comp. Endocrinol*, 146: 45-55.

Bernier N J, Craig P M. 2005. CRF-related peptides contribute to stress response and regulation of appetite in hypoxic rainbow trout. *Am. J. Physiol*, 289: R982-R990.

Bernier N J, Peter R E. 2001. Appetite-suppressing effects of urotensin I and corticotropin-releasing hormone in goldfish (*Carassius auratus*). *Neuroendocrinology*, 73: 248-260.

Bernier N J, Lin X, Peter R E. 1999. Differential expression of corticotropin-releasing factor (CRF) and urotensin I precursor genes, and evidence of CRF gene expression regulated by cortisol in goldfish brain. *Gen. Comp. Endocrinol*, 116: 461-477.

Bernier N J, Bedard N, Peter R E. 2004. Effects of cortisol on food intake, growth, and forebrain neuropeptide Y and corticotropin-releasing factor gene expression in goldfish. *Gen. Comp. Endocrinol*, 135: 230-240.

Bernier N J, Alderman S L, Bristow E N. 2008. Heads or tails? Stressor-specific expression of corticotropin-releasing factor and urotensin I in the preoptic area and caudal neurosecretory system of rainbow trout. *J. Endocrinol*, 196: 637-648.

Bertolini A, Tacchi R, Vergoni A V. 2009. Brain effects of melanocortins. *Pharmacol. Res*, 59: 13-47.

Bhattacharya S, Mukherjee D, Sen S. 1979. Role of synthetic mammalian thyrotropin releasing hormone on fish thyroid peroxidase activity. *Indian J. Exp. Biol*, 17: 1041-1043.

Bond H, Warne J M, Balment R J. 2007. Effect of acute restraint on hypothalamic pro-vasotocin mRNA expression in flounder, *Platichthys flesus*. *Gen. Comp. Endocrinol*, 153: 221-227.

Bonnin J P. 1971. Cultures organotypiques de thyroïdes d'un poisson Téléostéen marin: *Gobius niger* L. Effets de la TSH et de la prolactine. *Compt. Rend. Seanc. Soc. Biol. Fil*, 165: 1284-1291.

Bradford C S, Fitzpatrick M S, Schreck C B. 1992. Evidence for ultra-short-loop feedback in ACTH-induced interrenal steroidogenesis in coho salmon: Acute self-suppression of cortisol secretion *in vitro*. *Gen. Comp. Endocrinol*, 87: 292-299.

Brent G A, Harney J W, Chen Y, Warne R L, Moore D D, Larsen P R. 1989. Mutations of the rat growth hormone promoter which increase and decrease response to thyroid hormone define a consensus thyroid hormone response element. *Mol. Endocrinol*,

3: 1996-2004.

Bres O, Eales J G. 1986. Thyroid hormone binding to isolated trout (*Salmo gairdneri*) liver nuclei *in vitro*: Binding affinity, capacity, and chemical specificity. *Gen. Comp. Endocrinol*, 61: 29-39.

Bromage N R. 1975. The effects of mammalian thyrotropin-releasing hormone on the pituitary-thyroid axis of teleost fish. *Gen. Comp. Endocrinol*, 25: 292-297.

Bromage N R, Whitehead C, Brown T J. 1976. Thyroxine secretion in teleosts and effects of TSH, TRH and other peptides. *Gen. Comp. Endocrinol*, 29: 246 (conference abstract).

Brown C L, Stetson M H. 1983. Prolactin - thyroid interaction in *Fundulus heteroclitus*. *Gen. Comp. Endocrinol*, 50: 167-171.

Brown C L, Grau E G, Stetson M H. 1982. Endogenous gonadotropin does not have heterothyrotropic activity in *Fundulus heteroclitus*. *Am. Zool*, 22: 854 (conference abstract).

Brown C L, Grau E G, Stetson M H. 1985. Functional specificity of gonadotropin and thyrotropin in *Fundulus heteroclitus*. *Gen. Comp. Endocrinol*, 58: 252-258.

Brown J A, Edwards D, Whitehead C. 1989. Cortisol and thyroid hormone responses to acid stress in the brown trout, *Salmo trutta* L. *J. Fish Biol*, 35: 73-84.

Bury N R, Sturm A. 2007. Evolution of the corticosteroid receptor signalling pathway in fish. *Gen. Comp. Endocrinol*, 153: 47-56.

Bury N R, Sturm A, LeRouzic P, Lethimonier C, Ducouret B, Guiguen Y, Robinson-Rechavi M, Laudet V, Rafestin-Oblin M E, Prunet P P. 2003. Evidence for two distinct functional glucocorticoid receptors in teleost fish. *J. Mol. Endocrinol*, 31: 141-156.

Byamungu N, Corneillie S, Mol K, Darras V, Kühn E R. 1990. Stimulation of thyroid function by several pituitary hormones results in an increase in plasma thyroxine and reverse triiodothyronine in tilapia (*Tilapia nilotica*). *Gen. Comp. Endocrinol*, 80: 33-40.

Byamungu N, Mol K, Kühn E R. 1991. Somatostatin increases plasma T_3 concentrations in *Tilapia nilotica* in the presence of increased plasma T_4 levels. *Gen. Comp. Endocrinol*, 82: 401-406.

Campbell C M, Fostier A, Jalabert B, Truscott B. 1980. Identification and quantification of steroids in the serum of rainbow trout during spermiation and oocyte maturation. *J. Endocrinol*, 85: 371-378.

Carpenter R E, Watt M J, Forster G L, Øverli Ø, Bockholt C, Renner K J, Summers C H. 2007. Corticotropin releasing factor induces anxiogenic locomotion in trout and alters serotonergic and dopaminergic activity. *Horm. Behav*, 52: 600-611.

Cerdá-Reverter J M, Schiöth H B, Peter R E. 2003. The central melanocortin system regulates food intake in goldfish. *Regul. Pept*, 115: 101-113.

Cerdá-Reverter J M, Ling M K, Schiöth H B, Peter R E. 2003. Molecular cloning, characterization and brain mapping of the melanocortin 5 receptor in the goldfish. *J. Neurochem*, 87: 1354-1367.

Cerdá-Reverter J M, Ringholm A, Schiöth H B, Peter R E. 2003. Molecular cloning, pharmacological characterization and brain mapping of the melanocortin 4 receptor in the goldfish: Involvement in the control of food intake. *Endocrinology*, 144: 2336-2349.

Cerdá-Reverter J M, Haitina T, Schiöth H B, Peter R E. 2005. Gene structure of the goldfish agouti-signaling protein: A putative role in the dorsal-ventral pigment pattern of fish. *Endocrinology*, 146: 1597-1610.

Chadzinska M, Hermsen T, Savelkoul H F J, Verburg-van Kemenade B M L. 2009. Cloning of opioid receptors in common carp (*Cyprinus carpio* L.) and their involvement in regulation of stress and immune response. *Brain Behav. Immun*, 23: 257-266.

Chakraborti P, Bhattacharya S. 1978. Bovine TSH-stimulation of fish thyroid peroxidase activity and role of thyroxine thereon. *Experientia*, 34: 136-137.

Chatterjee A, Hsieh Y L, Yu J Y L. 2001. Molecular cloning of cDNA encoding thyroid stimulating hormone β subunit of bighead carp *Aristichthys nobilis* and regulation of its gene expression. *Mol. Cell. Endocrinol*, 174: 1-9.

Chehín R N, Issé B G, Rintoul M R, Farías R N. 1999. Differential transmembrane diffusion of triiodothyronine and thyroxine in liposomes: Regulation by lipid composition. *J. Membrane Biol*, 167: 251-256.

Chen C C, Fernald R D. 2008. Sequences, expression patterns and regulation of the corticotropin-releasing factor system in a teleost. *Gen. Comp. Endocrinol*, 157: 148-155.

Chowdhury I, Chien J T, Chatterjee A, Yu J Y L. 2004. *In vitro* effects of mammalian leptin, neuropeptide-Y, β-endorphin and galanin on transcript levels of thyrotropin β and common α subunit mRNAs in the pituitary of bighead carp (*Aristichthys nobilis*). *Comp. Biochem. Physiol. B*, 139: 87-98.

Christoffolete M A, Ribeiro R, Singru P, Fekete C, da Silva W S, Gordon D F, Huang S A, Crescenzi A, Harney J W, Ridgway E C, Larsen P R, Lechan R M, et al. 2006. Atypical expression of type 2 iodothyronine deiodinase in thyrotrophs explains the thyroxine-mediated pituitary TSH feedback mechanism. *Endocrinology*, 147: 1735-1743.

Clements S, Schreck C B, Larsen D A, Dickhoff W W. 2002. Central administration of corticotropin-releasing hormone stimulates locomotor activity in juvenile chinook salmon (*Oncorhynchus tshawytscha*). *Gen. Comp. Endocrinol*, 125: 319-327.

Colombe L, Fostier A, Bury N, Pakdel F, Guiguen Y. 2000. A mineralocorticoid-like receptor in the rainbow trout, *Oncorhynchus mykiss*: Cloning and characterization of its steroid binding domain. *Steroids*, 65: 319-328.

Consten D, Bogerd J, Komen J, Lambert J G D, Goos H J T. 2001, Long-term cortisol treatment inhibits pubertal development in male common carp, *Cyprinus carpio* L. *Biol. Reprod*, 64: 1063-1071.

Contreras-Sanchez W M, Schreck C B, Fitzpatrick M S, Pereira C B. 1998. Effects of stress on the reproductive performance of rainbow trout (*Oncorhynchus mykiss*). *Biol. Reprod*, 58: 439-447.

Craig P M, Al-Timimi H, Bernier N J. 2005. Differential increase in forebrain and caudal neurosecretory system corticotropin-releasing factor and urotensin I gene expression associated with seawater transfer in rainbow trout. *Endocrinology*, 146: 3851-3860.

Crim J W, Dickhoff W W, Gorbman A. 1978. Comparative endocrinology of piscine hypothalamic hypophysiotropic peptides: Distribution and activity. *Am. Zool*, 18: 411-424.

Darling D S, Dickhoff W W, Gorbman A. 1982. Comparison of thyroid hormone binding to hepatic nuclei of the rat and a teleost (*Oncorhynchus kisutch*). *Endocrinology*, 111: 1936-1943.

Davies T, Marians R, Latif R. 2002. The TSH receptor reveals itself. *J. Clin. Invest*, 110: 161-164.

Davis P J, Davis F B. 1996. Nongenomic actions of thyroid hormone. *Thyroid*, 6: 497-504.

De Boeck G, Alsop D, Wood C. 2001. Cortisol effects on aerobic and anaerobic metabolism, nitrogen excretion, and whole-body composition in juvenile rainbow trout. *Physiol. Biochem. Zool*, 74: 858-868.

De Groef B, Goris N, Arckens L, Kühn E R, Darras V M. 2003. Corticotropin-releasing hormone (CRH)-induced thyrotropin release is directly mediated through CRH receptor type 2 on thyrotropes. *Endocrinology*, 144: 5537-5544.

De Groef B, Vandenborne K, Van As P, Darras V M, Kühn E R, Decuypere E, Geris K L. 2005. Hypothalamic control of the thyroidal axis in the chicken: Over the boundaries of the classical hormonal axes. *Domest. Anim. Endocrinol*, 29: 104-110.

De Groef B, Van der Geyten S, Darras V M, Kühn E R. 2006. Role of corticotropin-releasing hormone as a thyrotropin-releasing factor in non-mammalian vertebrates. *Gen. Comp. Endocrinol*, 146: 62-68.

Díaz M L, Becerra M, Manso M J, Anadón R. 2002. Distribution of thyrotropin-releasing hormone (TRH) immunoreactivity in the brain of the zebrafish (*Danio rerio*). *J. Comp. Neurol*, 450: 45-60.

Donaldson E M. 1981. The pituitary-interrenal axis as an indicator of stress in fish. In: Pickering A D, Ed. *Stress and Fish*. Academic Press: London, 11-47.

Doyon C, Gilmour K M, Trudeau V L, Moon T W. 2003. Corticotropin-releasing factor and neuropeptide Y mRNA levels are elevated in the preoptic area of socially subordinate rainbow trout. *Gen. Comp. Endocrinol*, 133: 260-271.

Doyon C, Trudeau V L, Moon T W. 2005. Stress elevates corticotropin-releasing factor (CRF) and CRF-binding protein mRNA levels in rainbow trout (*Oncorhynchus mykiss*). *J. Endocrinol*, 186: 123-130.

Doyon C, Leclair J, Trudeau V L, Moon T W. 2006. Corticotropin-releasing factor and neuropeptide Y mRNA levels are modified by glucocorticoids in rainbow trout, *Oncorhynchus mykiss. Gen. Comp. Endocrinol*, 146: 126-135.

Dreborg S, Sundström G, Larsson T A, Larhammar D. 2008. Evolution of vertebrate opioid receptors. *Proc. Natl. Acad. Sci. USA*, 105: 15487-15492.

Dunn J T, Dunn A D. 2001. Update on intrathyroidal iodine metabolism. *Thyroid*, 11: 407-414.

Eales J G. 1979. Thyroid functions in cyclostomes and fishes. In: Barrington E J W, Ed. *Hormones and Evolution* Vol. 1. Academic Press: New York, 341-436.

Eales J G, Brown S B. 1993. Measurement and regulation of thyroidal status in teleost fish. *Rev. Fish Biol. Fish*, 3: 299-347.

Eales J G, Himick B A. 1988. The effects of TRH on plasma thyroid hormone levels of rainbow trout (*Salmo gairdneri*) and arctic charr (*Salvelinus alpinus*). *Gen. Comp. Endocrinol*, 72: 333-339.

Faglia G. 1998. The clinical impact of the thyrotropin-releasing hormone test. *Thyroid*, 8: 903-908.

Farbridge K J, Flett P A, Leatherland J F. 1992. Temporal effects of restricted diet and compensatory increased dietary intake on thyroid function, plasma growth hormone levels and tissue lipid reserves of rainbow trout *Oncorhynchus mykiss. Aquaculture*, 104: 157-174.

Filby A L, Tyler C R. 2007. Cloning and characterization of cDNAs for hormones and/or receptors of growth hormone, insulin-like growth factor-I, thyroid hormone, and corticosteroid and the gender-, tissue-, and developmental-specific expression of their mRNA transcripts in fathead minnow (*Pimephales promelas*). *Gen. Comp. Endocrinol*, 150: 151-163.

Flik G, Klaren P H M, van den Burg E H, Metz J R, Huising M O. 2006. CRF and stress in fish. *Gen. Comp. Endocrinol*, 146: 36-44.

Friesema E C H, Grueters A, Biebermann H, Krude H, von Moers A, Reeser M, Barrett T G, Mancilla E E, Svensson J, Kester M H A, Kuiper G G J M, Balkassmi S, et al. 2004. Association between mutations in a thyroid hormone transporter and severe X-linked psychomotor retardation. *Lancet*, 364: 1435-1437.

Friesema E C H, Jansen J, Jachtenberg J W, Visser W E, Kester M H A, Visser T J. 2008. Effective cellular uptake and efflux of thyroid hormone by human monocarboxylate transporter 10. *Mol. Endocrinol*, 22: 1357-1369.

Fryer J, Lederis K, Rivier J. 1983. Urotensin I, a CRF-like neuropeptide, stimulates ACTH release from the teleost pituitary. *Endocrinology*, 113: 2308-2310.

Fryer J, Lederis K, Rivier J. 1984. Cortisol inhibits the ACTH-releasing activity of urotensin I, CRF and sauvagine observed with superfused goldfish pituitary cells. *Peptides*, 5: 925-930.

Fryer J, Lederis K, Rivier J. 1985. ACTH-releasing activity of urotensin I and ovine CRF: Interactions with arginine vasotocin, isotocin and arginine vasopressin. *Regul. Peptides*, 11: 11-15.

Fryer J N. 1989. Neuropeptides regulating the activity of goldfish corticotropes and melanotropes. *Fish Physio. Biochem*, 7: 21-27.

Fryer J N, Lederis K. 1988. Comparison of actions of posterior pituitary hormones in corticotropin secretion in mammals and fishes. In: Yoshida S, Share L, Ed. *Recent Progress in Posterior Pituitary Hormones*. Elsevier: Amsterdam, 337-344.

Fujii R. 2000. The regulation of motile activity in fish chromatophores. *Pigment Cell Res*, 13: 300-319.

García-G C, López-Bojorquez L, Nunez J, Valverde-R C, Orozco A. 2007. 3, 5-Diiodothyronine *in vivo* maintains euthyroidal expression of type 2 iodothyronine deiodinase, growth hormone, and thyroid hormone receptor β1 in the killifish. *Am. J. Physiol. Regul. Integr. Comp. Physiol*, 293: R877-R883.

Gaylord T G, MacKenzie D S, GatlinIII D M. 2001. Growth performance, body composition and plasma thyroid hormone status of channel catfish (*Ictalurus punctatus*) in response to short-term feed deprivation and refeeding. *Fish Physiol. Biochem*, 24: 73-79.

Gesto M, Soengas J L, Miguez J M. 2008. Acute and prolonged stress responses of brain monoaminergic activity and plasma cortisol levels in rainbow trout are modified by PAHs [naphthalene, beta-naphthoflavone and benzo(a)pyrene] treatment. *Aquat. Toxicol*, 86: 341-351.

Geven E J W, Verkaar F, Flik G, Klaren P H M. 2006. Experimental hyperthyroidism and central mediators of stress axis and thyroid axis activity in common carp (*Cyp-*

rinus carpio L.). *J. Mol. Endocrinol*, 37: 443-452.

Geven E J W, Nguyen N K, van den Boogaart M, Spanings F A T, Flik G, Klaren P H M. 2007. Comparative thyroidology: Thyroid gland location and iodothyronine dynamics in Mozambique tilapia (*Oreochromis mossambicus* Peters) and common carp (*Cyprinus carpio* L.). *J. Exp. Biol*, 210: 4005-4015.

Geven E J W, Flik G, Klaren P H M. 2009. Central and peripheral integration of interrenal and thyroid axes signals in common carp (*Cyprinus carpio* L.). *J. Endocrinol*, 200: 117-123.

Gilchriest B J, Tipping D R, Hake L, Levy A, Baker B I. 2000. The effects of acute and chronic stresses on vasotocin gene transcripts in the brain of the rainbow trout (*Oncorhynchus mykiss*). *J. Neuroendocrinol*, 12: 795-801.

Gilham I D, Baker B I. 1985. A black background facilitates the response to stress in teleosts. *J. Endocrinol*, 105: 99-105.

Goglia F. 2005. Biological effects of 3,5-diiodothyronine (T_2). *Biochemistry (Mosc.)*, 70: 164-172.

Gonzalez-Nuñez V, Barrallo A, Traynor J R, Rodriguez R E. 2006. Characterization of opioid-binding sites in zebrafish brain. *J. Pharmacol. Exp. Ther*, 316: 900-904.

Gorbman A, Hyder M. 1973. Failure of mammalian TRH to stimulate thyroid function in the lungfish. *Gen. Comp. Endocrinol*, 20: 588-589.

Gorbman A, Dickhoff W W, Vigna S R, Clark N B, Ralph C L. 1983. *Comparative Endocrinology*. John Wiley & Sons, Inc. : New York.

Goto-Kazeto R, Kazeto Y, Trant J M. 2003. Cloning and seasonal changes in ovarian expression of a TSH receptor in the channel catfish, *Ictalurus punctatus*. *Fish Physiol. Biochem*, 28: 339-340.

Grau E G, Stetson M H. 1977. The effects of prolactin and TSH on thyroid function in *Fundulus heteroclitus*. *Gen. Comp. Endocrinol*, 33: 329-335.

Grau E G, Stetson M H. 1977. Pituitary autotransplants in *Fundulus heteroclitus*: Effect on thyroid function. *Gen. Comp. Endocrinol*, 32: 427-431.

Grau II E G, Stetson M H. 1978. Dopaminergic neurons and TSH release in *Fundulus heteroclitus*. *Am. Zool*, 18: 651 (conference abstract).

Grau E G, Helms L M H, Shimoda S K, Ford C A, LeGrand J, Yamauchi K. 1986. The thyroid gland of the Hawaiian parrotfish and its use as an *in vitro* model system. *Gen. Comp. Endocrinol*, 61: 100-108.

Green J A, Baker B I. 1991. The influence of repeated stress on the release of melanin-concentrating hormone in the rainbow trout. *J. Endocrinol*, 428: 261-266.

Green J A, Baker B I, Kawauchi H. 1991. The effect of rearing rainbow trout on black or white backgrounds on their secretion of melanin-concentrating hormone and their sensitivity to stress. *J. Endocrinol*, 128: 267-274.

Greenwood A K, Butler P C, White R B, DeMarco U, Pearce D, Fernald R D. 2003. Multiple corticosteroid receptors in a teleost fish: Distinct sequences, expression patterns, and transcriptional activities. *Endocrinology*, 144: 4226-4236.

Gröneveld D, Balm P H M, Martens G J M, Wendelaar Bonga S E. 1995. Differential melanin-concentrating hormone gene expression in two hypothalamic nuclei of the teleost tilapia in response to environmental changes. *J. Neuroendocrinol*, 7: 527-533.

Gröneveld D, Balm P H M, Wendelaar Bonga S E. 1995. Biphasic effect of MCH on α-MSH release from the tilapia (*Oreochromis mossambicus*) pituitary. *Peptides*, 16: 945-949.

Hagen I J, Kusakabe M, Young G. 2006. Effects of ACTH and cAMP on steroidogenic acute regulatory protein and P450 11β-hydroxylase messenger RNAs in rainbow trout interrenal cells: Relationship with *in vitro* cortisol production. *Gen. Comp. Endocrinol*, 145: 254-262.

Hagenbuch B. 2007. Cellular entry of thyroid hormones by organic anion transporting polypeptides. *Best Pract. Res. Clin. Endocrinol. Metab*, 21: 209-221.

Hagenbuch B, Meier P J. 2004. Organic anion transporting polypeptides of the OATP/ *SLC*21family: Phylogenetic classification as OATP/*SLCO*superfamily, new nomenclature and molecular/functional properties. *Pflügers Arch*, 447: 653-665.

Haider S. 1975. Pituitary cytology of radiothyroidectomised teleost *Heteropneustes fossilis* (Bloch.). *Endokrinologie*, 65: 300-307.

HaiglerJr E D, Hershman J M, PittmanJr J A. 1972. Response to orally administered synthetic thyrotropin-releasing hormone in man. *J. Clin. Endocrinol. Metab*, 35: 631-635.

Haitina T, Klovins J, Andersson J, Fredriksson R, Lagerström M C, Larhammar D, Larson E T, Schiöth H B. 2004. Cloning, tissue distribution, pharmacology and three-dimensional modelling of melanocortin receptors 4 and 5 in rainbow trout suggest close evolutionary relationship of these subtypes. *Biochem. J*, 380: 475-486.

Han Y S, Liao I C, Tzeng W N, Yu J Y L. 2004. Cloning of the cDNA for thyroid stimulating hormone β subunit and changes in activity of pituitary-thyroid axis during silvering of the Japanese eel, *Anguilla japonica*. *J. Mol. Endocrinol*, 32: 179-194.

Harvey C B, Williams G R. 2002. Mechanism of thyroid hormone action. *Thyroid*, 12: 441-446.

Hedge G A, Wright K C, Judd A. 1981. Factors modulating the secretion of thyrotropin and other hormones of the thyroid axis. *Environ. Health Perspect*, 38: 57-63.

Hennemann G. 2005. Notes on the history of cellular uptake and deiodination of thyroid hormone. *Thyroid*, 15: 753-756.

Hennemann G, Everts M E, de Jong M, Lim C F, Krenning E P, Docter R. 1998. The significance of plasma membrane transport in the bioavailability of thyroid hormone. *Clin. Endocrinol*, 48: 1-8.

Herman J P, Figueiredo H, Mueller N K, Ulrich-Lai Y, Ostrander M M, Choi D C, Cullinan W E. 2003. Central mechanisms of stress integration: Hierarchical circuitry controlling hypothalamo-pituitary-adrenocortical responsiveness. *Front. Neuroendocrinol*, 24: 151-180.

Higgins K M, Ball J N. 1970. Investigations on the hypothalamic control of thyroid-stimulating hormone secretion in the teleost *Poecilia latipinna*. *J. Endocrinol*, 48: xxix.

Höglund E, Balm P H M, Winberg S. 2000. Skin darkening, a potential social signal in subordinate arctic charr (*Salvelinus alpinus*): The regulatory role of brain monoamines and pro-opiomelanocortin-derived peptides. *J. Exp. Biol*, 203: 1711-1721.

Höglund E, Balm P H M, Winberg S. 2002. Behavioural and neuroendocrine effects of environmental background colour and social interaction in Arctic charr (*Salvelinus alpinus*). *J. Exp. Biol*, 205: 2535-2543.

Höglund E, Balm P H M, Winberg S. 2002. Stimulatory and inhibitory effects of 5-HT_{1A} receptors on adrenocorticotropic hormone and cortisol secretion in a teleost fish, the Arctic charr (*Salvelinus alpinus*). *Neurosci. Lett*, 324: 193-196.

Holland J W, Pottinger T G, Secombes C J. 2002. Recombinant interleukin-1β activates the hypothalamic-pituitary-interrenal axis in rainbow trout, *Oncorhynchus mykiss*. *J. Endocrinol*, 175: 261-267.

Huising M O, Flik G. 2005. The remarkable conservation of corticotropin-releasing hormone-binding protein (CRH-BP) in the honeybee (*Apis mellifera*) dates the CRH system to a common ancestor of insects and vertebrates. *Endocrinology*, 146: 2165-2170.

Huising M O, Metz J R, van Schooten C, Taverne-Thiele A J, Hermsen T, Verburg-van Kemenade B M L, Flik G. 2004. Structural characterisation of a cyprinid (*Cyprinus carpio* L.) CRH, CRH-BP and CRH-R1, and the role of these proteins in the acute stress response. *J. Mol. Endocrinol*, 32: 627-648.

Huising M O, Metz J R, de Mazon A F, Verburg-van Kemenade B M L, Flik G. 2005. Regulation of the stress response in early vertebrates. *Ann. N. Y. Acad. Sci*, 1041: 345-347.

Huising M O, van der Aa L M, Metz J R, de Mazon A F, Verburg-van Kemenade B M L, Flik G. 2007. Corticotropin-releasing factor (CRF) and CRF-binding protein expression in and release from the head kidney of common carp: Evolutionary conservation of the adrenal CRF system. *J. Endocrinol*, 193: 349-357.

Inui Y, Tagawa M, Miwa S, Hirano T. 1989. Effects of bovine TSH on the tissue thyroxine level and metamorphosis in prometamorphic flounder larvae. *Gen. Comp. Endocrinol*, 74: 406-410.

Ito M, Koide Y, Takamatsu N, Kawauchi H, Shiba T. 1993. cDNA cloning of the β subunit of teleost thyrotropin. *Proc. Natl. Acad. Sci. USA*, 90: 6052-6055.

Jackson R G, Sage M. 1973. A comparison of the effects of mammalian TSH on the thyroid glands of the teleost *Galeichthys felis* and the elasmobranch *Dasyatis sabina*. *Comp. Biochem. Physiol. A*, 44: 867-870.

Jansen J, Friesema E C H, Milici C, Visser T J. 2005. Thyroid hormone transporters in health and disease. *Thyroid*, 15: 757-768.

Jansen J, Friesema E C H, Kester M H A, Schwartz C E, Visser T J. 2008. Genotype-phenotype relationship in patients with mutations in thyroid hormone transporter MCT8. *Endocrinology*, 2184-2190.

Jenks B G, Kidane A H, Scheenen W J J M, Roubos E W. 2007. Plasticity in the melanotrope neuroendocrine interface of *Xenopus laevis*. *Neuroendocrinology*, 85: 177-185.

Jensen J. 2001. Regulatory peptides and control of food intake in non-mammalian vertebrates. *Comp. Biochem. Physiol.* A128: 471-479.

Jiang J Q, Wang D S, Senthilkumaran B, Kobayashi T, Kobayashi H K, Yamaguchi A, Ge W, Young G, Nagahama Y. 2003. Isolation, characterization and expression of 11 β-hydroxysteroid dehydrogenase type 2 cDNAs from the testes of Japanese eel (*Anguilla japonica*) and Nile tilapia (*Oreochromis niloticus*). *J. Mol. Endocrinol*, 31: 305-315.

Joseph-Bravo P. 2004. Hypophysiotropic thyrotropin-releasing hormone neurons as transducers of energy homeostasis. *Endocrinology*, 145: 4813-4815.

Kagabu Y, Mishiba T, Okino T, Yanagisawa T. 1998. Effects of thyrotropin-releasing hormone and its metabolites, cyclo (His-Pro) and TRH-OH, on growth hormone and prolactin synthesis in primary cultured pituitary cells of the common carp, *Cyprinus carpio*. *Gen. Comp. Endocrinol*, 111: 395-403.

Kajimura S, Hirano T, Visitacion N, Moriyama S, Aida K, Grau E G. 2003. Dual mode of cortisol action on GH/IGF-I/IFG binding proteins in the tilapia, *Oreochromis mossambicus*. *J. Endocrinol*, 178: 91-99.

Kasper R S, Shved N, Takahashi A, Reinecke M, Eppler E. 2006. A systematic immunohistochemical survey of the distribution patterns of GH, prolactin, somatolactin, β-TSH, β-FSH, β-LH, ACTH, and α-MSH in the adenohypophysis of *Oreochromis* niloticus, the Nile tilapia. *Cell Tissue Res*, 325: 303-313.

Kaul S, Vollrath L. 1974. The goldfish pituitary. I. Cytology. *Cell Tissue Res*, 154: 211-230.

Kawauchi H. 2006. Functions of melanin-concentrating hormone in fish. *J. Exp. Zool.* A305: 751-760.

Kawauchi H, Baker B I. 2004. Melanin-concentrating hormone signaling systems in fish. *Peptides*, 25: 1577-1584.

Kawauchi H, Sower S A. 2006. The dawn and evolution of hormones in the adenohypophysis. *Gen. Comp. Endocrinol*, 148: 3-14.

Kelsall C J, Balment R J. 1998. Native urotensins influence cortisol secretion and plasma cortisol concentration in the euryhaline flounder, *Platichthys flesus*. *Gen. Comp. Endocrinol*, 112: 210-219.

Kiilerich P, Kristiansen K, Madsen S S. 2007. Cortisol regulation of ion transporter mRNA in Atlantic salmon gill and the effect of salinity on the signaling pathway. *J. Endocrinol*, 194: 417-427.

Kitahashi T, Ogawa S, Soga T, Sakuma Y, Parhar I. 2007. Sexual maturation modulates expression of nuclear receptor types in laser-captured single cells of the cichlid (*Oreochromis niloticus*) pituitary. *Endocrinology*, 148: 5822-5830.

Klaren P H M, Haasdijk R, Metz J R, Nitsch L M C, Darras V M, Van der Geyten S, Flik G. 2005. Characterization of an iodothyronine 5′-deiodinase in gilthead seabream (*Sparus auratus*) that is inhibited by dithiothreitol. *Endocrinology*, 146: 5621-5630.

Klaren P H M, Geven E J W, Flik G. 2007. The involvement of the thyroid gland in teleost osmoregulation. In: Baldisserotto B, Mancera J M, Kapoor B G, Ed. *Fish Osmoregulation*. Science Publishers, Inc.: Enfield, USA, 35-65.

Klaren P H M, Guzmán J M, Reutelingsperger S J, Mancera J M, Flik G. 2007. Low salinity acclimation and thyroid hormone metabolizing enzymes in gilthead seabream (*Sparus auratus*). *Gen. Comp. Endocrinol*, 152: 215-222.

Klovins J, Haitina T, Fridmanis D, Kilianova Z, Kapa I, Fredriksson R, Gallo-Payet N, Schiöth H B. 2004. The melanocortin system in fugu: Determination of POMC/AGRP/MCR gene repertoire and synteny, as well as pharmacology and anatomical distribution of the MCRs. *Mol. Biol. Evol*, 21: 563-579.

Klovins J, Haitina T, Ringholm A, Löwgren M, Fridmanis D, Slaidina M, Stier S, Schiöth H B. 2004. Cloning of two melanocortin (MC) receptors in spiny dogfish: MC3 receptor in cartilaginous fish shows high affinity to ACTH-derived peptides while it has lower preference to α-MSH. *Eur. J. Biochem*, 271: 4320-4331.

Köhrle J. 1999. Local activation and inactivation of thyroid hormones: The deiodinase family. *Mol. Cell. Endocrinol*, 151: 103-119.

Köhrle J. 2002. Iodothyronine deiodinases. *Methods Enzymol*, 347: 125-167.

Köhrle J, Brabant G, Hesch R D. 1987. Metabolism of the thyroid hormones. *Horm. Res*, 26: 58-78.

König S, Neto V M. 2002. Thyroid hormone actions on neural cells. *Cell. Mol. Neurobiol*, 22: 517-544.

Kumar R S, Ijiri S, Kight K, Swanson P, Dittman A, Alok D, Zohar Y, Trant J M. 2000. Cloning and functional expression of a thyrotropin receptor from the gonads of a vertebrate (bony fish): Potential thyroid-independent role for thyrotropin in reproduction. *Mol. Cell. Endocrinol*, 167: 1-9.

Kusakabe M, Nakamura I, Young G. 2003. 11β-Hydroxysteroid dehydrogenase complementary deoxyribonucleic acid in rainbow trout: Cloning, sites of expression, and seasonal changes in gonads. *Endocrinology*, 144: 2534-2545.

Lai C S, Korytowski W, Niu C H, Cheng S Y. 1985. Transverse motion of spin-labeled 3,3',5-triiodo-L-thyronine in phospholipid bilayers. *Biochem. Biophys. Res. Commun*, 131: 408-412.

Lamers A E, Balm P H M, Haenen H E M G, Jenks B G, Wendelaar Bonga S E. 1991. Regulation of differential release of α-melanocyte-stimulating hormone forms from the pituitary of a teleost fish, Oreochromis mossambicus. *J. Endocrinol*, 129: 179-187.

Lamers A E, Flik G, Atsma W, Wendelaar Bonga S E. 1992. A role for di-acetyl α-melanocyte-stimulating hormone in the control of cortisol release in the teleost Oreochromis mossambicus. *J. Endocrinol*, 135: 285-292.

Lamers A E, Flik G, Wendelaar Bonga S E. 1994. A specific role for TRH in release of diacetyl α-MSH in tilapia stressed by acid water. *Am. J. Physiol*, 267: R1302-R1308.

Lamers A E, ter Brugge P J, Flik G, Wendelaar Bonga S E. 1997. Acid stress induces a D1-like dopamine receptor in pituitary MSH cells of Oreochromis mossambicus. *Am. J. Physiol*, 273: R387-R392.

Lanni A, Moreno M, Lombardi A, Goglia F. 1994. Rapid stimulation *in vitro* of rat liver cytochrome oxidase activity by 3,5-diiodo-L-thyronine and by 3,3'-diiodo-L-thyronine. *Mol. Cell. Endocrinol*, 99: 89-94.

Larsen D A, Dickey J T, Dickhoff W W. 1997. Quantification of salmon α- and thyrotropin (TSH) β-subunit messenger RNA by an RNase protection assay: Regulation by thyroid hormones. *Gen. Comp. Endocrinol*, 107: 98-108.

Larsen D A, Swanson P, Dickey J T, Rivier J, Dickhoff W W. 1998. *In vitro* thyrotropin-releasing activity of corticotropin-releasing hormone-family peptides in coho salmon, *Oncorhynchus kisutch*. *Gen. Comp. Endocrinol*, 109: 276-285.

Larson E T, O'Malley D M, MelloniJr R H. 2006. Aggression and vasotocin are associated with dominant-subordinate relationships in zebrafish. *Behav. Brain Res*, 167: 94-102.

Lazar M A. 1993. Thyroid hormone receptors: multiple forms, multiple possibilities. *Endocrine Rev*, 14: 184-193.

Le Mével J C, Mimassi N, Lancien F, Mabin D, Conlon J M. 2006. Cardiovascular actions of the stress-related neurohormonal peptides, corticotropin-releasing factor and urotensin-I in the trout *Oncorhynchus mykiss*. *Gen. Comp. Endocrinol*, 146: 56-61.

Le Mével J C, Lancien F, Mimassi N, Conlon J M. 2009. Central hyperventilatory action of the stress-related neurohormonal peptides, corticotropin-releasing factor and urotensin-I in the trout *Oncorhynchus mykiss*. *Gen Comp Endocrinol* doi: 10. 1016/ j. ygen, 2009.

Leatherland J F. 1987. Thyroid response to ovine thyrotropin challenge in cortisol- and dexamethasone-treated rainbow trout, *Salmo gairdneri*. *Comp. Biochem. Physiol. A*86: 383-387.

Leatherland J F. 1993. Stocking density and cohort sampling effects on endocrine interactions in rainbow trout. *Aquacult. Int*, 1: 137-156.

Lederis K, Fryer J N, Okawara Y, Schönrock C, Richter D. 1994. Corticotropin-releasing factors acting on the fish pituitary: Experimental and molecular analysis. In: Sherwood N M, Hew C L, Ed. *Fish Physiology*. Vol. XIII. Academic Press: San Diego, 67-100.

Leeson P D, Ellis D, Emmett J C, Shah V P, Showell G A, Underwood A H. 1988. Thyroid hormone analogs. Synthesis of 3'-substituted 3, 5-diiodo-L-thyronines and quantitative structure-activity studies of *in vitro* and *in vivo* thyromimetic activities in rat liver and heart. *J. Med. Chem*, 31: 37-54.

Lema S C, Dickey J T, Swanson P. 2008. Molecular cloning and sequence analysis of multiple cDNA variants for thyroid-stimulating hormone β subunit (TSHβ) in the fathead minnow (*Pimephales promelas*). *Gen. Comp. Endocrinol*, 155: 472-480.

Lethimonier C, Flouriot G, Valotaire Y, Kah O, Ducouret B. 2000. Transcriptional interference between glucocorticoid receptor and estradiol receptor mediates the inhibitory effect of cortisol on fish vitellogenesis. *Biol. Reprod*, 62: 1763-1771.

Lin H Y, Davis F B, Gordinier J K, Martino L J, Davis P J. 1999. Thyroid hormone induces activation of mitogen-activated protein kinase in cultured cells. *Am. J. Physiol*, 276: C1014-C1024.

Liu Q P, Dou S J, Wang G E, Li Z M, Feng Y. 2008. Evolution and functional divergence of monocarboxylate transporter genes in vertebrates. *Gene*, 423: 14-22.

Liu Y W, Lo L J, Chan W K. 2000. Temporal expression and T3 induction of thyroid hormone receptors α1 and β1 during early embryonic and larval development in zebrafish, *Danio rerio*. *Mol. Cell. Endocrinol*, 159: 187-195.

Logan D W, Bryson-Richardson R J, Pagan K E, Taylor M S, Currie P D, Jackson I J. 2003. The structure and evolution of the melanocortin and MCH receptors in fish and mammals. *Genomics*, 81: 184-191.

López-Bojorquez L, Villalobos P, García-G C, Orozco A, Valverde-R C. 2007. Functional identification of an osmotic response element (ORE) in the promoter region of the killifish deiodinase 2 gene (*FhDio2*). *J. Exp. Biol*, 210: 3126-3132.

Lovejoy D A, Balment R J. 1999. Evolution and physiology of the corticotropin-releasing factor (CRF) family of neuropeptides in vertebrates. *Gen. Comp. Endocrinol*, 115: 1-22.

Lu W, Dow L, Gumusgoz S, Brierly M J, Warne J M, McCrohan C R, Balment R J, Riccardi D. 2004. Coexpression of corticotropin-releasing hormone and urotensin I precursor genes in the caudal neurosecretory system of the Euryhaline flounder (*Platichthys flesus*): A possible shared role in peripheral regulation. *Endocrinology*, 145: 5786-5797.

Mancera J M, Vargas-Chacoff L, Garcia-Lopez A, Kleszczynska A, Kalamarz H, Martinez-Rodriguez G, Kulczykowska E. 2008. High density and food deprivation affect arginine vasotocin, isotocin and melatonin in gilthead sea bream (*Sparus auratus*). *Comp. Biochem. Physiol.* A149: 92-97.

Manchado M, Infante C, Asensio E, Planas J V, Cañavate J P. 2008 Thyroid hormones down-regulate thyrotropin β subunit and thyroglobulin during metamorphosis in the flatfish Senegalese sole (*Solea senegalensis* Kaup). *Gen. Comp. Endocrinol*, 155: 447-455.

Marchand O, Safi R, Escriva H, Van Rompaey E, Prunet P, Laudet V. 2001. Molecular cloning and characterization of thyroid hormone receptors in teleost fish. *J. Mol. Endocrinol*, 26: 51-65.

Marchand O, Duffraisse M, Triqueneaux G, Safi R, Laudet V. 2004. Molecular cloning and developmental expression patterns of thyroid hormone receptors and T3 target genes in the turbot (*Scophtalmus maximus*) during post-embryonic development. *Gen. Comp. Endocrinol*, 135: 345-357.

Marchelidon J, Huet J C, Salmon C, Pernollet J C, Fontaine Y A. 1991. Purification and characterization of the putative thyrotropic hormone subunits of a teleost fish, the eel (*Anguilla anguilla*). *C. R. Acad. Sci. Ser. III*, 313: 253-258.

Margolis R N. 2008. The nuclear receptor signaling atlas: Catalyzing understanding of thyroid hormone signaling and metabolic control. *Thyroid*, 18: 113-122.

Marley R, Lu W, Balment R J, McCrohan C R. 2008. Cortisol and prolactin modulation of caudal neurosecretory system activity in the euryhaline flounder *Platichthys flesus*. *Comp. Biochem. Physiol. A Mol. Integr. Physiol*, 151: 71-77.

Martin S A M, Wallner W, Youngson A F, Smith T. 1999. Differential expression of Atlantic salmon thyrotropin β subunit mRNA and its cDNA sequence. *J. Fish Biol*, 54: 757-766.

Matsuda K, Shimakura S, Maruyama K, Miura T, Uchiyama M, Kawauchi H, Shioda S, Takahashi A. 2006. Central administration of melanin-concentrating hormone (MCH) suppresses food intake, but not locomotor activity, in the goldfish, *Carassius auratus*. *Neurosci. Lett*, 399: 259-263.

Matsuda K, Kojima K, Shimakura S, Wada K, Maruyama K, Uchiyama M, Kikuyama S, Shioda S. 2008. Corticotropin-releasing hormone mediates α-melanocyte-stimulating hormone-induced anorexigenic action in goldfish. *Peptides*, 29: 1930-1936.

Matsuda K, Kojima K, Shimakura S, Takahashi A. 2009. Regulation of food intake by melanin-concentrating hormone in goldfish. *Peptides*. doi: 10. 1016/j. peptides. 2009. 02. 015.

McCormick S D. 2001. Endocrine control of osmoregulation in teleost fish. *Am. Zool*, 41: 781-794.

McCormick S D, Regish A, ODea M F, Shrimpton J M. 2008. Are we missing a mineralocorticoid in teleost fish? Effect of cortisol, deoxycorticosterone and aldosterone on osmoregulation, gill Na^+, K^+-ATPase activity and isoform mRNA levels in Atlantic salmon. *Gen. Comp. Endocrinol*, 157: 35-40.

McCrohan C R, Lu W, Brierley M J, Dow L, Balment R J. 2007. Fish caudal neurosecretory system: A model for the study of neuroendocrine secretion. *Gen. Comp. Endocrinol*, 153: 243-250.

Melamed P, Eliahu N, Levavi-Sivan B, Ofir M, Farchi-Pisanty O, Rentier-Delrue F, Smal J, Yaron Z, Naor Z. 1995. Hypothalamic and thyroidal regulation of growth hormone in tilapia. *Gen. Comp. Endocrinol*, 97: 13-30.

Metz J R, Huising M O, Meek J, Taverne-Thiele A J, Wendelaar Bonga S E, Flik G. 2004. Localisation, expression and control of adrenocorticotropic hormone in the nucleus preopticus and pituitary gland of common carp (*Cyprinus carpio* L.). *J. Endocrinol*, 182: 23-31.

Metz J R, Geven E J W, van den Burg E H, Flik G. 2005. ACTH, α-MSH and control of cortisol release: Cloning, sequencing and functional expression of the melanocortin-2 and melanocortin-5 receptor in *Cyprinus carpio*. *Am. J. Physiol*, 289: R814-R826.

Metz J R, Huising M O, Leon K, Verburg-van Kemenade B M L, Flik G. 2006.

Central and peripheral interleukin-1β and interleukin-1 receptor I expression and their role in the acute stress response of common carp, *Cyprinus carpio* L. *J. Endocrinol*, 191: 25-35.

Metz J R, Peters J J M, Flik G. 2006. Molecular biology and physiology of the melanocortin system in fish: A review. *Gen. Comp. Endocrinol*, 148: 150-162.

Milla S, Jalabert B, Rime H, Prunet P, Bobe J. 2006. Hydration of rainbow trout oocyte during meiotic maturation and *in vitro* regulation by 17, 20β-hydroxy-4-pregnen-3-one and cortisol. *J. Exp. Biol*, 209: 1147-1156.

Milla S, Terrien X, Sturm A, Ibrahim F, Giton F, Fiet J, Prunet P, Le Gac F. 2008. Plasma 11-deoxycorticosterone (DOC) and mineralocorticoid receptor testicular expression during rainbow trout *Oncorhynchus mykiss* spermiation: Implication with 17α, 20β-dihydroxyprogesterone on the milt fluidity?. *Reprod. Biol. Endocrinol*, 6: 19.

Mol K A, Van der Geyten S, Darras V M, Visser T J, Kühn E R. 1997. Characterization of iodothyronine outer ring and inner ring deiodinase activities in the blue tilapia, *Oreochromis aureus*. *Endocrinology*, 138: 1787-1793.

Mommsen T P, Vijayan M M, Moon T W. 1999. Cortisol in teleosts: Dynamics, mechanisms of action, and metabolic regulation. *Rev. Fish Biol. Fish*, 9: 211-268.

Moriyama S, Swanson P, Larsen D A, Miwa S, Kawauchi H, Dickhoff W W. 1997. Salmon thyroid-stimulating hormone: Isolation, characterization, and development of a radioimmunoassay. *Gen. Comp. Endocrinol*, 108: 457-471.

Morley S D, Schonrock C, Richter D, Okawara Y, Lederis K. 1991. Corticotropin-releasing factor (CRF) gene family in the brain of the teleost fish *Catostomus commersoni* (white sucker): Molecular analysis predicts distinct precursors for two CRFs and one urotensin I peptide. *Mol. Mar. Biol. Biotechnol*, 1: 48-57.

Mortensen A S, Arukwe A. 2006. The persistent DDT metabolite, 1, 1-dichloro-2, 2-bis (*p*-chlorophenyl) ethylene, alters thyroid hormone-dependent genes, hepatic cytochrome P4503A, and pregnane X receptor gene expressions in Atlantic salmon (*Salmo salar*) parr. *Environ. Toxicol. Chem*, 25: 1607-1615.

Mosconi G, Gallinelli A, Polzonetti-Magni A M, Facchinetti F. 1998. Acetyl salmon endorphin-like and interrenal stress response in male gilthead sea bream, *Sparus aurata*. *Neuroendocrinology*, 68: 129-134.

Mountjoy K G, Robbins L S, Mortrud M T, Cone R D. 1992. The cloning of a family of genes that encode the melanocortin receptors. *Science*, 257: 1248-1251.

Mustafa A, MacKinnon B M. 1999. Atlantic salmon, *Salmo salar* L., and Arctic char, *Salvelinus alpinus* (L.): Comparative correlation between iodine-iodide supplementation, thyroid hormone levels, plasma cortisol levels, and infection intensity with the sea louse *Caligus elongatus*. *Can. J. Zool*, 77: 1092-1101.

Naito N, Kawazoe I, Nakai Y, Kawauchi H, Hirano T. 1986. Coexistence of immunoreactivity for melanin-concentrating hormone and α-melanocyte-stimulating hormone in the hypothalamus of the rat. *Neurosci. Lett*, 70: 81-85.

Nelson E R, Habibi H R. 2006. Molecular characterization and sex-related seasonal expression of thyroid receptor subtypes in goldfish. *Mol. Cell. Endocrinol*, 253: 83-95.

Nelson E R, Habibi H R. 2009. Thyroid receptor subtypes: Structure and function in fish. *Gen. Comp. Endocrinol*, 161: 90-96.

Nillni E A, Sevarino K A. 1999. The biology of pro-thyrotropin-releasing hormone-derived peptides. *Endocrine Rev*, 20: 599-648.

Nishioka R S, Grau E G, Lai K V, Bern H A. 1985. Normal and induced development of the thyroid gland of coho salmon. *Aquaculture*, 45: 384-385.

Norris D O, Hobbs S L. 2006. The HPA axis and functions of corticosteroids in fishes. In: Reinecke M, Zaccone G, Kapoor B G, Ed. *Fish Endocrinology*. Vol. 2. Science Publishers, Enfield: USA, 721-765.

Oba Y, Hirai T, Yoshiura Y, Kobayashi T, Nagahama Y. 2000. Cloning, functional characterization, and expression of thyrotropin receptors in the thyroid of amago salmon (*Oncorhynchus rhodurus*). *Biochem. Biophys. Res. Commun*, 276: 258-263.

Okawara Y, Ko D, Morley S D, Richter D, Lederis K P. 1992. *In situ* hybridization of corticotropin-releasing factor-encoding messenger RNA in the hypothalamus of the white sucker, *Catostomus commersoni*. *Cell Tissue Res*, 267: 545-549.

Oki Y, Sasano H. 2004. Localization and physiological roles of urocortin. *Peptides*, 25: 1745-1749..

Okimoto D K, Tagawa M, Koide Y, Grau E G, Hirano T. 1991. Effects of various adenohypophyseal hormones of chum salmon on thyroxine release *in vitro* in the medaka, *Oryzias latipes*. *Zool. Sci*, 8: 567-573.

Oku H, Ogata H Y, Liang X F. 2002. Organization of the lipoprotein lipase gene of red sea bream *Pagrus major*. *Comp. Biochem. Physiol*. B131: 775-785.

Olivereau M. 1978. Effect of pimozide on the cytology of the eel pituitary. *Cell Tiss. Res*: 189: 231-239.

Olivereau M, Ball J N. 1966. Histological study of functional ectopic pituitary transplants in a teleost fish (*Poecilia formosa*). *Proc. R. Soc. Lond*. B164: 106-129.

Olivereau M, Olivereau J. 1990. Effect of pharmacological adrenalectomy on corticotropin-releasing factor-like and arginine vasotocin immunoreactivities in the brain and pituitary of the eel: immunocytochemical study. *Gen. Comp. Endocrinol*, 80:

199-215.

Olivereau M, Olivereau J M. 1990. Corticotropin-like immunoreactivity in the brain and pituitary of three teleost species (goldfish, trout and eel). *Cell Tissue Res*, 262: 115-123.

Olivereau M, Olivereau J, Lambert J. 1987. *In vivo* effect of dopamine antagonists on melanocyte-stimulating hormone cells of the goldfish (*Carassius auratus* L.) pituitary. *Gen. Comp. Endocrinol*, 68: 12-18.

Olivereau M, Olivereau J M, Lambert J F. 1988. Cytological responses of the pituitary (rostral pars distalis) and immunoreactive corticotropin-releasing factor (CRF) in the goldfish treated with dopamine antagonists. *Gen. Comp. Endocrinol*, 71: 506-515.

Omeljaniuk R J, Tonon M, Peter R E. 1989. Dopamine inhibition of gonadotropin and α-melanocyte stimulating hormone release *in vitro* from the pituitary of the goldfish (*Carassius auratus*). *Gen. Comp. Endocrinol*, 74: 451-467.

Orozco A, Villalobos P, Valverde-R C. 2002. Environmental salinity selectively modifies the outer-ring deiodinating activity of liver, kidney and gill in the rainbow trout. *Comp. Biochem. Physiol. A*131: 387-395.

Ortega V A, Renner K J, Bernier N J. 2005. Appetite-suppressing effects of ammonia exposure in rainbow trout associated with regional and temporal activation of brain monoaminergic and CRF systems. *J. Exp. Biol*, 208: 1855-1866.

ØverliØ, Harris C A, Winberg S. 1999. Short-term effects of fights for social dominance and the establishment of dominant-subordinate relationships on brain monoamines and cortisol in rainbow trout. *Brain Behav. Evol*, 54: 263-275.

ØverliØ, Pottinger T G, Carrick T R, Øverli E, Winberg S. 2001. Brain monoaminergic activity in rainbow trout selected for high and low stress responsiveness. *Brain Behav. Evol*, 57: 214-224.

Pankhurst N W, Van Der Kraak G. 2000. Evidence that acute stress inhibits ovarian steroidogenesis in rainbow trout *in vivo*, through the action of cortisol. *Gen. Comp. Endocrinol*, 117: 225-237.

Patiño R, Schreck C B, Redding J M. 1985. Clearance of plasma corticosteroids during smoltification of coho salmon, *Oncorhynchus kisutch*. *Comp. Biochem. Physiol. A*82: 531-535.

Patiño R, Redding J M, Schreck C B. 1987. Interrenal secretion of corticosteroids and plasma cortisol and cortisone concentrations after acute stress and during seawater acclimation in juvenile coho salmon (*Oncorhynchus kisutch*). *Gen. Comp. Endocrinol*, 68: 431-439.

Pepels P P L M, Meek J, Wendelaar Bonga S E, Balm P H M. 2002. Distribution and quantification of corticotropin-releasing hormone (CRH) in the brain of the teleost fish

Oreochromis mossambicus (Tilapia). *J. Comp. Neurol*, 453: 247-268.

Pepels P P L M, van Helvoort H, Wendelaar Bonga S E, Balm P H M. 2004. Corticotropin-releasing hormone in the teleost stress response: Rapid appearance of the peptide in plasma of tilapia (*Oreochromis mossambicus*). *J. Endocrinol*, 180: 425-438.

Perrott M N, Balment R J. 1990. The renin-angiotensin system and the regulation of plasma cortisol in the flounder, *Platichthys flesus*. *Gen. Comp. Endocrinol*, 78: 414-420.

Peter R E. 1970. Hypothalamic control of thyroid gland activity and gonadal activity in the goldfish, *Carassius auratus*. *Gen. Comp. Endocrinol*, 14: 334-356.

Peter R E. 1971. Feedback effects of thyroxine on the hypothalamus and pituitary of goldfish, *Carassius auratus*. *J. Endocrinol*, 51: 31-39.

Peter R E. 1972. Feedback effects of thyroxine in goldfish *Carassius auratus* with an autotransplanted pituitary. *Neuroendocrinology*, 10: 273-281.

Peter R E, Fryer J N. 1984. Endocrine function of the hypothalamus of actinopterygians. In: Davis R E, Northcutt R G, Ed. *Fish neurobiology*. University of Michigan Press: Ann Arbor, 165-201.

Peter R E, McKeown B A. 1975, Hypothalamic control of prolactin and thyrotropin secretion in teleosts, with special reference to recent studies on the goldfish. *Gen. Comp. Endocrinol*, 25: 153-165.

Peterson B C, Small B C. 2005. Effects of exogenous cortisol on the GH/IGF/IGFBP network in channel catfish. *Dom. Anim. Endocrinol*, 28: 391-404.

Pickering A D, Pottinger T G, Sumpter J P. 1986. Independence of the pituitary-interrenal axis and melanotroph activity in the brown trout, *Salmo trutta* L., under conditions of environmental stress. *Gen. Comp. Endocrinol*, 64: 206-211.

Pickering A D, Pottinger T G, Sumpter J P. 1987. On the use of dexamethasone to block the pituitary-interrenal axis in the brown trout, *Salmo trutta* L. *Gen. Comp. Endocrinol*, 65: 346-353.

Pickford G E, Knight W R, Knight J N, Gallardo R, Baker B I. 1981. Long-term effects of hypothalamic lesions on the pituitary and its target organs in the killifish *Fundulus heteroclitus*. 1. Effects on the gonads, thyroid, and growth. *J. Exp. Zool*, 217: 341-351.

Pierce A L, Fukada H, Dickhoff W W. 2005. Metabolic hormones modulate the effect of growth hormone (GH) on insulin-like growth factor-I (IGF-I) mRNA level in primary culture of salmon hepatocytes. *J. Endocrinol*, 184: 341-349.

Pierson P M, Guibbolini M E, Lahlou B. 1996. A V1-type receptor for mediating the neurohypophysial hormone-induced ACTH release in trout pituitary. *J. Endocrinol*, 149: 109-115.

Power D M, Melo J, Santos C R A. 2000. The effect of food deprivation and refeeding on the liver, thyroid hormones and transthyretin in sea bream. *J. Fish Biol*, 56: 374-387.

Pradet-Balade B, Schmitz M, Salmon C, Dufour S, Querat B. 1997. Down-regulation of TSH subunit mRNA levels by thyroid hormones in the European eel. *Gen. Comp. Endocrinol*, 108: 191-198.

Pradet-Balade B, Burel C, Dufour S, Boujard T, Kaushik S J, Querat B, Boeuf G. 1999. Thyroid hormones down-regulate thyrotropin beta mRNA level *in vivo* in the turbot (*Psetta maxima*). *Fish Physiol. Biochem*, 20: 193-199.

Prunet P, Sturm A, Milla S. 2006. Multiple corticosteroid receptors in fish: From old ideas to new concepts. *Gen. Comp. Endocrinol*, 147: 17-23.

Przewlocki R, Przewlocka B. 2001. Opioids in chronic pain. *Eur. J. Pharmacol*, 429: 79-91.

Qu D, Ludwig D S, Gammeltoft S, Piper M, Pelleymounter M A, Cullen M J, Mathes W F, Przypek J, Kanarek R, Maratos-Flier E. 1996. A role for melanin-concentrating hormone in the central regulation of feeding behaviour. *Nature*, 380: 243-247.

Rabello M M, Snyder P J, Utiger R D. 1974. Effects on pituitary-thyroid axis and prolactin secretion of single and repetitive oral doses of thyrotropin-releasing hormone (TRH). *J. Clin. Endocrinol. Metab*, 39: 571-578.

Rapoport B, Chazenbalk G D, Jaume J C, McLachlan S M. 1998. The thyrotropin (TSH) receptor: interaction with TSH and autoantibodies. *Endocrine Rev*, 19: 673-716.

Reid S G, Vijayan M M, Perry S F. 1996. Modulation of catecholamine storage and release by the pituitary-interrenal axis in the rainbow trout (*Oncorhynchus mykiss*). *J. Comp. Physiol. B*165: 665-676.

RenéF, Muller A, Jover E, Kieffer B, Koch B, Loeffler J P. 1998. Melanocortin receptors and δ-opioid receptor mediate opposite signalling actions of POMC-derived peptides in CATH. a cells. *Eur. J. Neurosci*, 10: 1885-1894.

Richardson J, Lundegaard P R, Reynolds N L, Dorin J R, Porteous D J, Jackson I J, Patton E E. 2008. mc1r Pathway regulation of zebrafish melanosome dispersion. *Zebrafish*, 5: 289-295.

Ringholm A, Fredriksson R, Poliakova N, Yan Y, J. H. Postlethwait J H, Larhammar D, Schiöth H B. 2002. One melanocortin 4 and two melanocortin 5 receptors from zebrafish show remarkable conservation in structure and pharmacology. *J. Neurochem*, 82: 6-18.

Rocha A, Gómez A, Galay-Burgos M, Zanuy S, Sweeney G E, Carrillo M. 2007. Molecular characterization and seasonal changes in gonadal expression of a thyrotropin receptor in the European sea bass. *Gen. Comp. Endocrinol*, 152: 89-101.

Rodriguez R E, Barrallo A, Garcia-Malvar F, McFadyen I J, Gonzalez-Sarmiento R, Traynor J R. 2000. Characterization of ZFOR1, a putative delta-opioid receptor from the teleost zebrafish (*Danio rerio*). *Neurosci. Lett*, 288: 207-210.

Rotllant J, Arends R J, Mancera J M, Flik G, Wendelaar Bonga S E, Tort L. 2000. Inhibition of HPI axis response to stress in gilthead sea bream (*Sparus aurata*) with physiological plasma levels of cortisol. *Fish Physiol. Biochem*, 23: 13-22.

Rotllant J, Balm P H M, Ruane N M, Perez-Sanchez J, Wendelaar Bonga S E, Tort L. 2000. Pituitary proopiomelanocortin-derived peptides and hypothalamus-pituitary-interrenal axis activity in gilthead sea bream (*Sparus aurata*) during prolonged crowding stress: Differential regulation of adrenocorticotropin hormone and α-melanocyte-stimulating hormone release by corticotropin-releasing hormone and thyrotropin-releasing hormone. *Gen. Comp. Endocrinol*, 119: 152-163.

Rotllant J, Balm P H M, Perez-Sanchez J, Wendelaar Bonga S E, Tort L. 2001. Pituitary and interrenal function in gilthead sea bream (*Sparus aurata* L., Teleostei) after handling and confinement stress. *Gen. Comp. Endocrinol*, 121: 333-342.

Rousseau K, Le Belle N, Marchelidon J, Dufour S. 1999. Evidence that corticotropin-releasing hormone acts as a growth hormone-releasing factor in a primitive teleost, the European eel (*Anguilla anguilla*). *J. Neuroendocrinol*, 11: 385-392.

Ruane N M, Wendelaar Bonga S E, Balm P H M. 1999. Differences between rainbow trout and brown trout in the regulation of the pituitary-interrenal axis and physiological performance during confinement. *Gen. Comp. Endocrinol*, 115: 210-219.

Rudman D, Hollins B M, Kutner M H, Moffitt S D, Lynn M J. 1983. Three types of α-melanocyte-stimulating hormone: Bioactivities and half-lives. *Am. J. Physiol*, 245: E47-54.

Saavedra J M, Benicky J. 2007. Brain and peripheral angiotensin II play a major role in stress. *Stress*, 10: 185-193.

Salmon C, Marchelidon J, Fontaine Y A, Huet J C, Querat B. 1993. Cloning and sequence of thyrotropin beta subunit of a teleost fish: The eel (*Anguilla anguilla*L.). *C. R. Acad. Sci. Ser. III*, 316: 749-753.

Sanchez-Simon F M, Rodriguez R E. 2008. Developmental expression and distribution of opioid receptors in zebrafish. *Neuroscience*, 151: 129-137.

Scanlan T S, Suchland K L, Hart M E, Chiellini G, Huang Y, Kruzich P J, Frascarelli S, Crossley D A, Bunzow J R, Ronca-Testoni S, Lin E T, Hatton D, et al. 2004. 3-Iodothyronamine is an endogenous and rapid-acting derivative of thyroid hormone. *Nature Med*, 10: 638-642.

Schaaf M J M, Champagne D, van Laanen I H C, van Wijk D C W A, Meijer A H,

Meijer O C, Spaink H P, Richardson M K. 2008. Discovery of a functional glucocorticoid receptor β-isoform in zebrafish. *Endocrinology*, 149: 1591-1599.

Schiöth H B, Haitina T, Ling M K, Ringholm A, Fredriksson R, Cerdá-Reverter J M, Klovins J. 2005. Evolutionary conservation of the structural, pharmacological, and genomic characteristics of the melanocortin receptor subtypes. *Peptides*, 26: 1886-1900.

Schjolden J, Schiöth H B, Larhammar D, Winberg S, Larson E T. 2009. Melanocortin peptides affect the motivation to feed in rainbow trout (*Oncorhynchus mykiss*). *Gen. Comp. Endocrinol*, 160: 134-138.

Schreck C B, Bradford C S, Fitzpatrick M S, Patiño R. 1989. Regulation of the interrenal of fishes: Non-classical control mechanisms. *Fish Physiol. Biochem*, 7: 259-265.

Seasholtz A F, Valverde R A, Denver R J. 2002. Corticotropin-releasing hormone-binding protein: Biochemistry and function from fishes to mammals. *J. Endocrinol*, 175: 89-97.

Selz Y, Braasch I, Hoffmann C, Schmidt C, Schultheis C, Schartl M, Volff J N. 2007. Evolution of melanocortin receptors in teleost fish: The melanocortin type 1 receptor. *Gene*, 401: 114-122.

Shepard J D, Barron K W, Myers D A. 2000. Corticosterone delivery to the amygdala increases corticotropin-releasing factor mRNA in the central amygdaloid nucleus and anxiety-like behavior. *Brain Res*, 861: 288-295.

Sheridan M A. 1986. Effects of thyroxin, cortisol, growth hormone, and prolactin on lipid metabolism of coho salmon, *Oncorhynchus kisutch*, during smoltification. *Gen. Comp. Endocrinol*, 64: 220-238.

Shimakura S, Miura T, Maruyama K, Nakamachi T, Uchiyama M, Kageyama H, Shioda S, Takahashi A, Matsuda K. 2008. Alpha-melanocyte-stimulating hormone mediates melanin-concentrating hormone-induced anorexigenic action in goldfish. *Horm. Behav*, 53: 323-328.

Shrimpton J M, McCormick S D. 1999. Responsiveness of gill Na^+/K^+-ATPase to cortisol is related to gill corticosteroid receptor concentration in juvenile rainbow trout. *J. Exp. Biol*, 202: 987-995.

Shrimpton J M, Randall D J. 1994. Downregulation of corticosteroid receptors in gills of coho salmon due to stress and cortisol treatment. *Am. J. Physiol*,. 267: R432-R438.

Silva J E, Dick T E, Larsen P R. 1978. The contribution of local tissue thyroxine monodeiodination to the nuclear 3,5,3'-triiodothyronine in pituitary, liver, and kidney of euthyroid rats. *Endocrinology*, 103: 1196-1207.

Singh A K, Singh T P. 1977. Thyroid activity and TSH levels in pituitary gland and blood serum in response to clomid, sexovid and prostaglandins treatment in *Heteropneustes fossilis* (Bloch). *Endokrinologie*, 70: 69-76.

Singh R, Rai U. 2008. β-Endorphin regulates diverse functions of splenic phagocytes through different opioid receptors in freshwater fish *Channa punctatus* (Bloch): An *in vitro* study. *Dev. Comp. Immunol*, 32: 330-338.

Sohn Y C, Yoshiura Y, Suetake H, Kobayashi M, Aida K. 1999. Isolation and characterization of the goldfish thyrotropin β subunit gene including the 5′-flanking region. *Gen. Comp. Endocrinol*, 115: 463-473.

Specker J L, Kobuke L. 1985. Thyroid physiology of juvenile coho salmon. *Aquaculture*, 45: 389-390.

Specker J L, Richman3rd N H. 1984. Environmental salinity and the thyroidal response to thyrotropin in juvenile coho salmon (*Oncorhynchus kisutch*). *J. Exp. Zool*, 230: 329-333.

Staub J J, Girard J, Mueller-Brand J, Noelpp B, Werner-Zodrow I, Baur U, Heitz P, Gemsenjaeger E. 1978. Blunting of TSH response after repeated oral administration of TRH in normal and hypothyroid subjects. *J. Clin. Endocrinol. Metab*, 46: 260-266.

Sternberg H, Moav B. 1999. Regulation of the growth hormone gene by fish thyroid retinoid receptors. *Fish Physiol. Biochem*, 20: 331-339.

Stolte E H, Verburg-van Kemenade B M L, Savelkoul H F J, Flik G. 2006. Evolution of glucocorticoid receptors with different glucocorticoid sensitivity. *J. Endocrinol*, 190: 17-28.

Stolte E H, de Mazon A F, Leon-Koosterziel K M, Jesiak M, Bury N R, Sturm A, Savelkoul H F J, van Kemenade B M L, Flik G. 2008. Corticosteroid receptors involved in stress regulation in common carp, *Cyprinus carpio*. *J. Endocrinol*, 198: 403-417.

Strand F L. 1999. New vistas for melanocortins. Finally, an explanation for their pleiotropic functions. *Ann. N. Y. Acad. Sci*, 897: 1-16.

Sturm A, Bury N, Dengreville L, Fagart J, Flouriot G, Rafestin-Oblin M E, Prunet P. 2005. 11-Deoxycorticosterone is a potent agonist of the rainbow trout (*Oncorhynchus mykiss*) mineralocorticoid receptor. *Endocrinology*, 146: 47-55.

Sumpter J P, Pickering A D, Pottinger T G. 1985. Stress-induced elevation of plasma α-MSH and endorphin in brown trout, *Salmo trutta* L. *Gen. Comp. Endocrinol*, 59: 257-265.

Sumpter J P, Dye H M, Benfey T J. 1986. The effects of stress on plasma ACTH, α-MSH, and cortisol levels in salmonid fishes. *Gen. Comp. Endocrinol*, 62: 377-385.

Swanson P, Grau E G, Helms L M, Dickhoff W W. 1988. Thyrotropic activity of salmon pituitary glycoprotein hormones in the Hawaiian parrotfish thyroid in vitro. J. Exp. Zool, 245: 194-199.

Tagawa M, Hagiwara H, Takemura A, Hirose S, Hirano T. 1997. Partial cloning of the hormone-binding domain of the cortisol receptor in Tilapia, Oreochromis mossambicus, and changes in the mRNA levels during embryonic development. Gen. Comp. Endocrinol, 108: 132-140.

Takahashi H, Sakamoto T, Hyodo S, Shepherd B S, Kaneko T, Grau E G. 2006. Expression of glucocorticoid receptor in the intestine of a euryhaline teleost, the Mozambique tilapia (Oreochromis mossambicus): Effect of seawater exposure and cortisol treatment. Life Sci, 78: 2329-2335.

Tanaka S. 2003. Comparative aspects of intracellular proteolytic processing of peptide hormone precursors: Studies of proopiomelanocortin processing. Zool. Sci, 20: 1183-1198.

Tang X, Liu X, Zhang Y, Zhu P, Lin H. 2008. Molecular cloning, tissue distribution and expression profiles of thyroid hormone receptors during embryogenesis in orange-spotted grouper (Epinephelus coioides). Gen. Comp. Endocrinol, 159: 117-124.

Tasker J G, Di S, Malcher-Lopes R. 2006. Minireview: Rapid glucocorticoid signaling via membrane-associated receptors. Endocrinology, 147: 5549-5556.

Tata J R. 2002. Signalling through nuclear receptors. Nature Rev. Mol. Cell Biol, 3: 702-710.

Teitsma C A, Bailhache T, Tujague M, Balment R J, Ducouret B, Kah O. 1997. Distribution and expression of glucocorticoid receptor mRNA in the forebrain of the rainbow trout. Neuroendocrinology, 66: 294-304.

Teitsma C A, Anglade I, Toutirais G, Muñoz-Cueto J A, Saligaut D, Ducouret B, Kah O. 1998. Immunohistochemical localization of glucocorticoid receptors in the forebrain of the rainbow trout (Oncorhynchus mykiss). J. Comp. Neurol, 401: 395-410.

Thornton J W. 2001. Evolution of vertebrate steroid receptors from an ancestral estrogen receptor by ligand exploitation and serial genome expansions. Proc. Natl. Acad. Sci. USA, 98: 5671-5676.

Todd K J, Eales J G. 2002. The effect of handling and blood removal on plasma levels and hepatic deiodination of thyroid hormones in adult male and female rainbow trout, Oncorhynchus mykiss. Can. J. Zool, 80: 372-375.

Tran T N, Fryer J N, Bennett H P, Tonon M C, Vaudry H. 1989. TRH stimulates the release of POMC-derived peptides from goldfish melanotropes. Peptides, 10: 835-

841.

Tran T N, Fryer J N, Lederis K, Vaudry H. 1990. CRF, urotensin I, and sauvagine stimulate the release of POMC-derived peptides from goldfish neurointermediate lobe cells. *Gen. Comp. Endocrinol*, 78: 351-360.

Tsuneki K, Fernholm B. 1975. Effect of thyrotropin-releasing hormone on the thyroid of a teleost, *Chasmichthys dolichognathus*, and a hagfish, *Eptatretus burgeri*. *Acta Zool*, 56: 61-65.

Vacher C, Pellegrini E, Anglade I, Ferriére F, Saligaut C, Kah O. 2003. Distribution of dopamine D_2 receptor mRNAs in the brain and the pituitary of female rainbow trout: An *in situ* hybridization study. *J. Comp. Neurol*, 458: 32-45.

Vallarino M, Delbende C, Bunel D T, Ottonello I, Vaudry H. 1989. Proopiomelanocortin (POMC) -related peptides in the brain of the rainbow trout, *Salmo gairdneri*. *Peptides*, 10: 1223-1230.

van den Burg E H, Metz J R, Arends R J, Devreese B, Vandenberghe I, Van Beeumen J, Wendelaar Bonga S E, Flik G. 2001. Identification of β-endorphins in the pituitary gland and blood plasma of the common carp (*Cyprinus carpio*). *J. Endocrinol*, 169: 271-280.

van den Burg E H, Metz J R, Ross H A, Darras V M, Wendelaar Bonga S E, Flik G. 2003. Temperature-induced changes in thyrotropin-releasing hormone sensitivity in carp melanotropes. *Neuroendocrinology*, 77: 15-23.

van den Burg E H, Metz J R, Spanings F A T, Wendelaar Bonga S E, Flik G. 2005. Plasma α-MSH and acetylated β-endorphin levels following stress vary according to CRH sensitivity of the pituitary melanotropes in common carp, *Cyprinus carpio*. *Gen. Comp. Endocrinol*, 140: 210-221.

Van der Geyten S, Mol K A, Pluymers W, Kühn E R, Darras V M. 1998. Changes in plasma T_3 during fasting/refeeding in tilapia (*Oreochromis niloticus*) are mainly regulated through changes in hepatic type II iodothyronine deiodinase. *Fish Physiol. Biochem*, 19: 135-143.

van der Heide S M, Joosten B J L J, Dragt B S, Everts M E, Klaren P H M. 2007. A physiological role for glucuronidated thyroid hormones: Preferential uptake by H9c2 (2-1) myotubes. *Mol. Cell. Endocrinol*, 264: 109-117.

van der Salm A L, Pavlidis M, Flik G, Wendelaar Bonga S E. 2004. Differential release of α-melanophore stimulating hormone isoforms by the pituitary gland of red porgy, *Pagrus pagrus*. *Gen. Comp. Endocrinol*, 135: 126-133.

van der Salm A L, Metz J R, Wendelaar Bonga S E, Flik G. 2005. Alpha-MSH, the melanocortin-1 receptor and background adaptation in the Mozambique tilapia, *Oreochromis mossambicus*. *Gen. Comp. Endocrinol*, 144: 140-149.

van Enckevort F H C, Pepels P P L M, Leunissen J A M, Martens G J M, Wendelaar Bonga S E, Balm P H M. 2000. *Oreochromis mossambicus* (tilapia) corticotropin-releasing hormone: cDNA sequence and bioactivity. *J. Neuroendocrinol*, 12: 177-186.

van Tijn D A, de Vijlder J J M, Vulsma T. 2007. Role of the thyrotropin-releasing hormone stimulation test in diagnosis of congenital central hypothyroidism in infants. *J. Clin. Endocrinol. Metab*, 93: 410-419.

Vazzana M, Vizzini A, Salerno G, Di Bella M L, Celi M, Parrinello N. 2008. Expression of a glucocorticoid receptor (DlGR1) in several tissues of the teleost fish *Dicentrarchus labrax*. *Tissue Cell*, 40: 89-94.

Vijayan M M, Leatherland J F. 1992. In vivo effects of the steroid analogue RU486 on some aspects of intermediary and thyroid metabolism of brook charr, *Salvelinus fontinalis*. *J. Exp. Zool*, 263: 265-271.

Vijayan M M, Raptis S, Sathiyaa R. 2003. Cortisol treatment affects glucocorticoid receptor and glucocorticoid-responsive genes in the liver of rainbow trout. *Gen. Comp. Endocrinol*, 132: 256-263.

Vischer H F, Bogerd J. 2003. Cloning and functional characterization of a testicular TSH receptor cDNA from the African catfish (*Clarias gariepinus*). *J. Mol. Endocrinol*, 30: 227-238.

Visser T J. 1994. Role of sulfation in thyroid hormone metabolism. *Chem. -Biol. Interact*, 92: 293-303.

Visser T J. 1994. Sulfation and glucuronidation pathways of thyroid hormone metabolism. In: Wu S Y, Visser T J, Ed. *Thyroid Hormone Metabolism: Molecular Biology and Alternate Pathways*. CRC Press: Boca Raton, 85-117.

Walpita C N, Grommen S V H, Darras V M, Van der Geyten S. 2007. The influence of stress on thyroid hormone production and peripheral deiodination in the Nile tilapia (*Oreochromis niloticus*). *Gen. Comp. Endocrinol*, 150: 18-25.

Waring C P, Brown J A. 1997. Plasma and tissue thyroxine and triiodothyronine contents in sublethally stressed, aluminum-exposed brown trout (*Salmo trutta*). *Gen. Comp. Endocrinol*, 106: 120-126.

Watanuki H, Gushiken Y, Takahashi A, Yasuda A, Sakai M. 2000. In vitro modulation of fish phagocytic cells by β-endorphin. *Fish Shellfish Immunol.*, 10: 203-212.

Weld M M, Fryer J N. 1987. Stimulation by angiotensins I and II of ACTH release from goldfish pituitary cell columns. *Gen. Comp. Endocrinol*, 68: 19-27.

Weld M M, Fryer J N. 1988. Angiotensin II stimulation of teleost adrenocorticotropic hormone release: Interactions with urotensin I and corticotropin-releasing factor. *Gen. Comp. Endocrinol*, 69: 335-340.

Weld M M, Fryer J N, Rivier J, Lederis K. 1987. Inhibition of CRF- and urotensin I-stimulated ACTH release from goldfish pituitary cell columns by the CRF analogue α-helical CRF (9-41). *Regul. Peptides*, 19: 273-280.

Wendelaar Bonga S E. 1997. The stress response in fish. *Physiol. Rev*, 77: 591-625.

Westphal N J, Seasholtz A F. 2006. CRH-BP: The regulation and function of a phylogenetically conserved binding protein. *Front. Biosci*, 11: 1878-1891.

Westphal N J, Evans R T, Seasholtz A F. 2009. Novel expression of type 1 corticotropin-releasing hormone receptor in multiple endocrine cell types in the murine anterior pituitary. *Endocrinology*, 150: 260-267.

Westring C G, Ando H, Kitahashi T, Bhandari R K, Ueda H, Urano A, Dores R M, Sher A A, Danielson P B. 2008. Seasonal changes in CRF-I and urotensin I transcript levels in masu salmon: Correlation with cortisol secretion during spawning. *Gen. Comp. Endocrinol*, 155: 126-140.

Wildmeister W, Horster F A. 1971. Die Wirkung von synthetischem Thyrotropin Releasing Hormone auf die Entwicklung eines experimentellen Exophthalmos beim Goldfish. *Acta Endocrinol*, 68: 363-366.

Wilson D S, Clark A B, Coleman K, Dearstyne T. 1994. Shyness and boldness in humans and other animals. *Trends Ecol. Evol*, 9: 442-446.

Winberg S, Lepage O. 1998. Elevation of brain 5-HT activity, POMC expression, and plasma cortisol in socially subordinate rainbow trout. *Am. J. Physiol. Regul*, 274: R645-R654.

Winberg S, Nilsson A, Hylland P, Soderstöm V, Nilsson G E. 1997. Serotonin as a regulator of hypothalamic-pituitary-interrenal activity in teleost fish. *Neurosci. Lett*, 230: 113-116.

Wolff N A, Werner A, Burkhardt S, Burkhardt G. 1997. Expression cloning and characterization of a renal organic anion transporter from winter flounder. *FEBS Lett*, 417: 287-291.

Wu S Y, Green W L, Huang W S, Hays M T, Chopra I J. 2005. Alternate pathways of thyroid hormone metabolism. *Thyroid*, 15: 943-958.

Yada T, Azuma T, Takahashi A, Suzuki Y, Hirose S. 2000. Effects of desacetyl-α-MSH on lipid mobilization in the rainbow trout, Oncorhynchus mykiss. *Zool. Sci*, 17: 1123-1127.

Yada T, Moriyama S, Suzuki Y, Azuma T, Takahashi A, Hirose S, Naito N. 2002. Relationships between obesity and metabolic hormones in the "cobalt" variant of rainbow trout. *Gen. Comp. Endocrinol*, 128: 36-43.

Yamada C, Noji S, Shioda S, Nakai Y, Kobayashi H. 1990. Intragranular colocalization of arginine vasopressin- and angiotensin II-like immunoreactivity in the hypothalamo-neurohypophysial system of the goldfish, *Carassius auratus*. *Zool. Sci*, 7: 257-263.

Yamano K, Inui Y. 1995. cDNA cloning of thyroid hormone receptor β for the Japanese flounder. *Gen. Comp. Endocrinol*, 99: 197-203.

Yamano K, Miwa S. 1998. Differential gene expression of thyroid hormone receptor α and β in fish development. *Gen. Comp. Endocrinol*, 109: 75-85.

Yamano K, Araki K, Sekikawa K, Inui Y. 1994. Cloning of thyroid hormone receptor genes expressed in metamorphosing flounder. *Dev. Genet*, 15: 378-382.

Yao M, Denver R J. 2007. Regulation of vertebrate corticotropin-releasing factor genes. *Gen. Comp. Endocrinol*,. 153: 200-216.

Yen P M. 2001. Physiological and molecular basis of thyroid hormone action. *Physiol. Rev*, 81: 1097-1142.

Yoshiura Y, Sohn Y C, Munakata A, Kobayashi M, Aida K. 1999. Molecular cloning of the cDNA encoding the β subunit of thyrotropin and regulation of its gene expression by thyroid hormones in the goldfish, *Carassius auratus*. *Fish Physiol. Biochem*, 21: 201-210.

Young G, Lin R J. 1988. Response of the interrenal to adrenocorticotropic hormone after short-term thyroxine treatment of coho salmon (*Oncorhynchus kisutch*). *J. Exp. Zool*, 245: 53-58.

Yulis C R, Lederis K, Wong K L, Fisher A W F. 1986. Localization of urotensin I- and corticotropin-releasing factor-like immunoreactivity in the central nervous system of *Catostomus commersoni*. *Peptides*, 7: 79-86.

Zhou A, Bloomquist B T, Mains R E. 1993. The prohormone convertases PC1 and PC2 mediate distinct endoproteolytic cleavages in a strict temporal order during proopiomelanocortin biosynthetic processing. *J. Biol. Chem*, 268: 1763-1769.

Zupanc G K H, Horschke I, Lovejoy D A. 1999. Corticotropin-releasing factor in the brain of the gymnotiform fish, *Apteronotus leptorhynchus*: Immunohistochemical studies combined with neuronal tract tracing. *Gen. Comp. Endocrinol*, 114: 349-364.

第7章 硬骨鱼类神经内分泌-免疫相互作用

7.1 神经内分泌-免疫相互作用

物理的、化学的和生物学的干扰会引起不可思议的生理学、内分泌学和免疫学的所有组成成分的反应。目前已经认识到神经内分泌系统和免疫系统以双向的方式相互作用。在这方面,病原体识别的状态传递到脑部,而免疫反应受到生理变化的影响。这种明晰的交流沟通必然需要一个信号分子和受体的共同体。这个网络包括皮质类固醇、传统的脑垂体激素、细胞因子、神经肽以及神经通路。了解这种交流表达的基本生物学意义有助于病理学的处理与治疗。本章着重综述参与组成硬骨鱼类一个完整模型的通路、受体和作用机理,以展示应激生理学和免疫学在系统发生过程中古老而独特的作用机理。

7.1.1 神经内分泌-免疫相互作用的导言

保持内部环境平衡的基础是双向生理过程的动态平衡。这种平衡经常受到内在的和外在的刺激作用或者物理的、化学的和生物学应激物如病原体的威胁。

大约在20年前,Blabock提到免疫系统具有感觉的功能,凭借"第六感觉"去检测病原体和肿瘤,而这是身体不能听到、感觉到、嗅到或尝到的(Blabock,1984)。要引发一个合适的神经内分泌反应,这种病原体检测需要信号到达神经系统。互相起作用的应激反应(Fast等,2008)和从属的等级地位(subordinate hierarchy)(Faisal等,1989)使鱼类的免疫功能降低,这意味着从内分泌系统发出的信号引起免疫系统的作用(Ader等,1995)。事实上,一些信号看起来都是一种进化上保守的交流沟通系统的变体。

鱼类和蕴藏着大量病原体的水环境密切接触。现存的硬骨鱼类是进化上最古老的脊椎动物的一部分,具备发达而有效的免疫系统。一个精确的细胞-细胞接触和体液因子(细胞因子)的交流沟通系统能够沟通病原体的识别和协调各种不同类型白细胞的适宜反应程度。细胞因子是多肽或者糖蛋白,通常具有非常低的组成性表达,它们的产生是短暂的,其作用一般是局部的(自分泌和副分泌而不是内分泌)。激素通常由一些数量有限的细胞类型产生并且朝向限定的靶细胞以诱导特异性的反应;而细胞因子由不同的细胞类型产生,在不同的靶细胞引发多种作用(多效性的),还能表现功能的丰余性(redundancy)(Vilcek,2003)。有意思的是,许多涉及细胞因子或激素的信号因子都属于结构上相关蛋白质的同一个分子家族,其成员都由一个共同的祖先分子衍生而来。此外,肽类激素表现出比原来设想的更为明显的多效性。由激素和细胞因子共同使用的细胞内通路，如詹纳斯激酶（JAK）、信号诱导物与转录激活剂（STAT）通路，可以说明它们作用的潜在丰余性。

虽然对于克服细胞感染，一个强有力的炎性细胞因子反应是不可缺少的，但我们还是要注重阐述平衡的交流沟通对于妥善处理应激反应与伤病治疗的必要，这可能像双刃剑那样，因为太强的炎症反应可能使寄主的组织受到伤害。此外，应激反应（病原体侵袭是一种严重的外在应激物）引起能量再分配，使皮质醇活性提高并能抑制免疫功能；而在生存受到威胁时，它们处于次级的重要性（Wendelarr Bonga，1997）。事实亦表明免疫反应和一龄降海幼鲑向二龄鲑转变（Maule等，1987）、温度或季节（Nakanishi，1986）、性别或昼夜节律（Nevid 和 Meier，1995）等相联系，突出显示相互作用的生理学重要性。

世界范围对蛋白质日益增长的要求，使集约水产养殖业在近数十年迅速发展。集约化鱼类养殖业带来的不良状况包括高养殖密度、非天然的养殖场地与社会状况、频繁的手工操作（如接种疫苗过程）、运输、过度投饵和不良的水质（如变动的温度、废弃产物）。这种状况会干扰鱼类正常的免疫机能（Engelsma 等，2003；Terova 等，2005），有利于病原体迅速传播。预防和/或有效地接种疫苗与传染性疾病的处理都要求对神经内分泌系统与免疫系统之间的精细双向交流沟通有深刻的了解。

7.1.2 鱼类的免疫性

硬骨鱼类已经发育形成一个有效保护的免疫系统。首先，它们产生了生物化学的屏障以防止病原体入侵，这就是皮肤的黏液层，它包含许多抗菌肽类、溶菌酶、凝集素和蛋白水解酶。其次，在病原体通过这个屏障后，一系列可溶性的细胞防御作用机理会被激活。鱼类在脊椎动物当中最早发展形成反映先天性和适应性免疫反应的免疫系统两大支臂。先天性免疫系统的功能是"一般性地（种属地）"识别病原体并防止病原体扩散，而适应性免疫系统是在特定的表面抗原（surface antigen）的基础上特异性地识别病原体，并通过产生特定的抗体和细胞毒性淋巴细胞（cytotoxic lymphocyte）以清除病原体侵染。重要的是，适应性免疫系统能够产生记忆。最后，综合的免疫反应在病原体清除后能够调控使之迅速终止，以防止宿主的并发性（collateral）损伤。

7.1.2.1 免疫器官和沟通

鱼类缺乏骨髓作为初级免疫器官，但它们分布在头肾的造血细胞能够产生髓样的（myeloid）和淋巴样的（lymphoid）免疫细胞（van Muiswinkel，1995）。鱼类缺乏淋巴结作为次级免疫器官。脾脏和头肾是免疫系统和抗原互相作用的部位，并含有产生抗体的淋巴细胞。鱼类的胸腺和哺乳类一样是T-淋巴细胞成熟的中心。和消化道、鳃与皮肤的上皮表面相联系的黏膜免疫系统形成防御的第一线，而和结缔组织联系的是稠密聚集的免疫细胞，可以攻击任何深入侵袭的感染物。

如上所述，免疫细胞通过细胞因子和相关的受体沟通，这对于迅速而有效地攻击病原体是至关重要的。微量的细胞因子能够产生强烈的炎症反应，而防止细胞损伤需要严密调控这种反应。免疫细胞产生不同类型的细胞因子，细胞因子的类型取决于感染的时期和特殊细胞类型的功能。我们把这些细胞因子分为白细胞介素（interleukins，ILs）和趋化因子（chemokines），前者是沟通白细胞之间的物质，后者是化学引诱物和细胞因子两者的缩略词（acronym），它监管细胞的迁移（趋化性）。细胞因子对细胞生长和分化能引发一系列广泛的作用。在检测病原体之后，细胞因子表达的动态能确保有效地清除病原体，同时使宿主受到的损伤最小。

7.1.2.2 先天性免疫反应：细胞类型和病原体识别

硬骨鱼类先天性免疫系统的主要细胞类型有髓样细胞、嗜中性粒细胞和巨噬细胞（来自单核细胞前体），它们都以吞噬作用为主要功能。这些细胞迁移到感染/炎症的部位，能杀死细菌和感染的细胞（Gómez 和 Balcazar，2008）。非特异性的细胞毒素细胞（non-specific cytotoxic cell，NCC）亦能杀死感染的细胞，鱼类的这种细胞相当于哺乳类的天然杀伤细胞（natural killer cell，NK）（Fischer 等，2006）。产生杀菌的或溶菌的蛋白质和酶类亦能杀死感染的细胞（Magnadottir，2006）。

病原体对致病力特别重要的特征通常并不在宿主中表达。这些特征统称为病原体相关分子型式（pathogen-associated molecular patterns，PAMPs），例如真菌 $\beta1,3$-葡聚糖（fungal $\beta1,3$-glucan），病毒双链核糖核酸（viral double stranded RNA），或者细菌细胞壁产物如脂多糖类（lipopolysaccharide，LPS）。它们为病原体的识别受体（PRR）所识别。PRRs 以可溶性体液的变体形式存在，和补体蛋白 C3 相似，或者表达为在免疫系统细胞上的膜受体，例如许多 Toll 样受体（TLRs）（Magnadottir，2006）。在哺乳类中，特殊的 TLRs 识别一个独立的传染物类群（例如 TLR1 和 TLR2 主要识别来自革兰氏阳性细菌的脂蛋白和肽聚糖，TLRs 识别病毒的双链 RNA，TLR4 识别来自革兰氏阴性细菌的 LPS，等等）（Purcell 等，2006）。对斑马鱼基因组数据库的分析能展示 TLRs 在哺乳类中表达的大部分直向同源物。

在激活时，这些识别的分子诱导一系列反应，直接朝向杀死病原体和感染的细胞，防止病原体扩散，并通过细胞因子和促炎因子的释放沟通感染的类型和严重性。通过调理作用（opsonization，病原体被吞噬细胞消化与降解的过程）能增加杀死病原体的数量，接着是对病原体的吞噬作用。此外，吞噬细胞还产生杀微生物的活性氧类别（reactive oxygen species，ROS）。还有，天然的细胞毒素细胞受到刺激能帮助杀死感染的细胞，而补体系统被激活后会攻击病原体的膜（Magnadottir，2006）。最后，急性期反应被激活，它由一系列血浆蛋白质组成［如血清淀粉状蛋白（SAA）、运铁蛋白、溶菌酶和 α-2-巨球蛋白（$\alpha2M$）］，限制病原体扩散，并启动组织的修复（Magnodottir，2006）。

通过细胞内通路，TLR 激活将会诱导促炎细胞因子的表达，这可能和在哺乳类系统中的阐述相似。有意思的是，由于 TLR 的特异性，其产生的细胞因子分布型将和感染的类型沟通（Purcell 等，2006）。例如，一次病毒攻击将诱导 1 型干扰素-α 和-β 产生，而这将"警告"其他细胞加强它们抗病毒的防御，刺激细胞毒素细胞活性以溶解感染的细胞，并且防止病毒的进一步增殖（Robertsen，2006）。

7.1.2.3 先天性免疫反应：细胞因子信号

如同 TLR 基因，在哺乳类中产生的许多细胞因子，已经被证实存在着鱼类的直向同源基因（见图 7.1）。虽然序列的同一性很低，但基本的模体和三维结构是保守的。在脊椎动物的免疫系统，促炎细胞因子（pro-inflammatory cytokines）包括肿瘤坏死因子 α（tumor necrosis factor alpha，TNFα）、白细胞介素 1β（IL-1β）和白细胞介素 6（IL-6）诱导急性期反应和趋化因子释放（Roitt 等，2006）。接着是白细胞介素 12（IL-12）释放，它转而刺激 2 型干扰素——干扰素 γ（IFN-γ）释放。和哺乳类相比较，鱼类促炎反应的特点是 TNF-α 和 IL-1β 表达是第一波峰，接着是趋化因子表达，而 IL-12 表达是高峰（Chadzinska 等，2008）。TNF-α 组成性地在头肾和鳃内产生，能被 LPS 诱导而刺激头肾的巨噬细胞。在一次免疫刺激之后，TNF-α 通过控制细胞内病原体的复制而抗感染，并诱导细胞的增殖。TNF-α 亦参与杀微生物的一氧化氮（NO）的产生，可能是通过调控可诱导的一氧化氮合成酶（iNOS）的表达来实现（Saeij 等，2003a）。巨噬细胞分泌 IL-1β 刺激胸腺细胞（T-淋巴细胞）增殖，启动急性期反应，并激活巨噬细胞和 T-淋巴细胞（Buonocore 等，2004）。在被细菌和病毒感染后，头肾巨噬细胞能诱导 IL-12 表达，表明 NCC 和 T-淋巴细胞的细胞溶解特性增强，以保证能清除被病毒感染的细胞（Huising 等，2006c；Nascimento 等，2007；Forlenza 等，2008）。IFN-γ 由天然细胞毒素细胞和 T-淋巴细胞表达，激活巨噬细胞以增强杀微生物活性，并刺激抗原的呈递（Zou 等，2005）。趋化因子被所有的免疫器官表达，如头肾、脾脏、胸腺和肾脏等，并且参与指示白细胞到达炎症部位（Huising 等，2003c）。

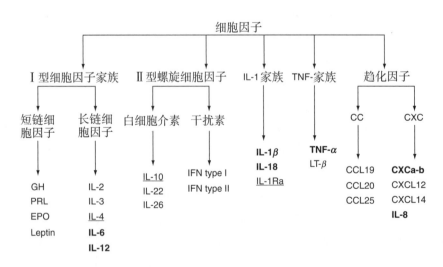

图 7.1 在几种硬骨鱼类中确定的细胞因子家族成员（见文中参考文献）

促炎细胞因子以黑体字表示，抗炎反应以下划线表示。GH，生长激素；PRL，催乳激素；EPO，红细胞生成素；1L，白细胞介素；IFN，干扰素；IL-1Ra，白细胞介素 1 受体拮抗物；TNF，肿瘤坏死因子；LT，淋巴毒素。

7.1.2.4 适应性免疫反应：细胞类型、反应和记忆

适应性免疫反应向脊椎动物免疫系统提供能识别和记忆特异性病原体的能力以及在随后遇到同一病原体时能激发较强和较快的反应。要激发这样的反应，必须识别一个特异性的病原体或者抗原。主要组织相容性复合体（major histocompatibility complex，MHC）蛋白质起着"路标"的作用，并在抗原-呈递细胞的表面展示一个抗原的断裂片段（Stet 等，2003）。T-细胞受体（TCR）是在 T-淋巴细胞（T-细胞）表面的一个分子，它通常和 MHC 分子结合起着识别抗原的作用。CD4 是一个共同受体，帮助 TCR 激活其 T-淋巴细胞，然后和抗原呈递细胞互相作用（Roitt 等，2006）。抗原呈递由巨噬细胞介导（Martin 等，2007），而 B-淋巴细胞产生抗体，其特点是免疫球蛋白在它们的细胞表面表达（Magnadottir 等，2005）。哺乳类展现 T-淋巴细胞（Th）功能的区分，分别由两种不同类型的 T-淋巴细胞实行：辅助 T-淋巴细胞（Th）支持 B-淋巴细胞的功能，而细胞毒性 T-淋巴细胞（Tc）攻击和杀死病毒感染细胞。Th-细胞的特点是以共同受体 CD4 和 MHC II 结合，而 Tc-细胞的特点是以 CD8 和 MHC I 结合。这种功能的区分在鱼类中并不很明确，虽然鱼类的白细胞亦表现出细胞毒的活性（Fischer 等，2006）。此外，硬骨鱼类的 T-淋巴细胞表达 CD4（Suetake 等，2004）、CD8（Hansen 和 Strassburger，2000）和 MHC I（Stet 等，2003）。

在 MHC 上的抗原-呈递细胞识别一个特别的抗原（如小的病原体碎片）将会引起特异性 B-淋巴细胞和 T-淋巴细胞的克隆扩充（Clonal expansion）。B-淋巴细胞将分泌特异性抗体（体液免疫）。这些抗体将攻击病原体并引起调理作用和病原体吞噬作用（Sanmartin 等，2008）。第一次的抗体遭遇将会诱发长命的记忆细胞保持受抗原刺激而产生较快与较激烈的第二次反应的能力（Kaattari 等，2002）。必须注意的是，和哺乳类相比，硬骨鱼类具有不同的免疫球蛋白（Ig）同种型（isotype）。在一次免疫冲击后，主要的 Ig 是一个四聚体的 IgM 分子。除了 IgD 分子和 Ig 重链变异如 IgZ（斑马鱼球蛋白）之外，还有 IgT（T，硬骨鱼类的）和嵌合体 IgM-IgZ（Hansen 等，2005；Savan 等，2005）。鱼类这些不同 Ig 分子的功能目前正在研究中。

抗原特异性 T-淋巴细胞能杀死感染的宿主细胞并帮助抗原特异性 B-淋巴细胞产生抗体（释放细胞因子，如 IFN-γ）（Cain 等，2002）。这种细胞免疫性在第二次刺激时亦表现出较强的效应，就如同第二次移植要比第一次移植较快地产生排异作用那样（Grether 等，2004）。

虽然在第二次遇到同一个病原体时适应性免疫反应迅速而强烈，但变温动物如鱼类在第一次遭遇病原体后需要数天到数周才能充分起作用，这取决于外界温度（Rijkers 等，1980；van Muiswinkel，1995）。所以，先天性免疫是极其重要的。如果没有一种恰当的从一开始就杀死病原体和防止病原体扩散的免疫反应，宿主就可能在适应性免疫反应激发之前就死于长时间的病原体感染（van Muiswinkel，1995；Raida 和 Buchmann，2007）。

7.1.2.5 免疫反应的终止

杀微生物蛋白质的产生，活性氧类别（ROS）以及血液循环中大量细胞毒性细胞

对于宿主是非常有害的，甚至可能致死。所以，要产生抗炎症细胞因子 IL-10 和 TGFβ（TGF-β）以阻止免疫反应的过度激活，并启动创伤愈合、组织重建与复原（见图 7.1）。IL-10 表达的高峰出现在炎症反应的后期（Pinto 等，2007），并和抑制促炎细胞因子 IL-1β 与趋化因子的表达相联系（Chadzinska 等，2008；Seppola 等，2008）。此外，TGF-β 在感染的后期亦增加（Tafalla 等，2005），下调 TNF-α-激活巨噬细胞的 NO 反应（Haddad 等，2008）。细胞凋亡是不可少的，以保持细胞群体的动态平衡，并能在炎症反应开始时受抑制以增加生命期限和吞噬细胞的有效作用，但它必须快速增强以终止开始时的作用（Elenkov 和 Chrousos，2006）。

7.1.2.6 和神经内分泌系统相互作用的前提

激素和受体在免疫细胞中广泛存在是在免疫的刺激与阻抑和内分泌功能的作用之间建立精确平衡的前提。神经内分泌系统要从免疫系统交互接收信号，内分泌细胞需要特异性的细胞因子受体，而细胞因子必须产生在内分泌细胞附近，或者必须释放到血液循环中。

硬骨鱼类免疫系统由直接的（副的）交感神经分布是可能的，正如交感的肾上腺髓质系统儿茶酚胺受体存在于免疫细胞（Roy 和 Rai，2008）。在较高等和较低等的脊椎动物中都发现激素受体存在于免疫细胞内，或者它们的存在为激素刺激后的功能反应所证实（见第 7.2 节）。影响神经内分泌功能的细胞因子主要由脑内的免疫相关细胞产生，如星形胶质细胞（astrocyte）或神经胶质细胞（glia cell）（Kelly 等，2003；Churchill 等，2006；Garden 和 Moller，2006）。如果从外周免疫器官产生，细胞因子要能够通过室周器（circumventricular organ，CVO）（见第 7.3 节）。

除内分泌作用之外，如果免疫细胞能够产生激素，则在免疫系统内出现副分泌和自分泌作用是可能的。传统的下丘脑和脑垂体激素在哺乳类免疫细胞中有节制但广泛地表达（Elenkow 和 Chrousos，2006）。这种现象在进化上是古老的。硬骨鱼类促肾上腺皮质素-释放因子（CRF）的免疫反应出现在鳃和皮肤的类吞噬细胞中（Mazon 等，2006）。POMC-衍生的 ACTH、β-内啡肽和 αMSH 都已证明存在于金鱼的胸腺内（Ottaviani 等，1995）。生长激素和催乳激素已经证实在白细胞中表达（Yada，2007）。其他的传统激素如瘦蛋白 I 和瘦蛋白 II 发现存在于胸腺和脾脏中（Huising 等，2006a）。

硬骨鱼类的胸腺是一个免疫和内分泌器官，造血组织位于内分泌组织附近，构建成典型的神经内分泌-免疫相互作用的部位。头肾嗜铬组织的细胞产生儿茶酚胺，而肾间细胞产生皮质醇，使头肾成为哺乳类肾上腺的同功能类似物。同时，头肾的造血组织产生淋巴样细胞（B-淋巴细胞和 T-淋巴细胞）和髓样细胞（吞噬细胞），使它成为哺乳类骨髓的同功能类似物（van Muiswinkel，1995）。

7.1.3 细胞因子家族和细胞因子与激素之间的系统进化关系

虽然界定一个特定的研究领域并不困难，但将激素和细胞因子分隔开似乎是不太容易的，因为细胞因子亦释放到血液循环中，其信号具有自分泌、旁分泌甚至内分泌的式样。最近完成的一些硬骨鱼类及其他脊椎动物的全基因组数据库使我们有可能去分析基因家族的系统进化关系。这些分析表明涉及细胞因子、激素或者生长因子的许多信号分子都属于结构上相关蛋白质相同的家族。这个家族的所有成员可能都从一个共同的祖先经过一系列连续的复制过程演化而来，尽管它们的序列保守水平变化很大（Huising 等，2006b）。Ⅰ型螺旋状细胞因子家族（helical cytokine family）和趋化因子是这种协同进化的有意思的例子，因而将在本章中进行较为详细的讨论。

7.1.3.1 细胞因子的分类和它们的受体

许多细胞因子和细胞因子受体都成群地分属不同的家族（见图7.1）。细胞因子的家族具有高度的序列同源性，但三级结构和存在着不同的家族模体对于完成这种分类亦是必要的。最重要的细胞因子家族有：①Ⅰ型细胞因子家族；②Ⅱ型螺旋状细胞因子；③IL-1家族。

在细胞因子的家族中，Ⅰ型细胞因子家族的配体和受体是最普遍的。Ⅰ型细胞因子受体家族在细胞外区具有一个保守的WSXWS模体，由许多重要的造血细胞因子（如IL-2、IL-3、IL-4、IL-5、IL-6、IL-7、IL-12和IL-13）、造血的生长因子［如集落刺激因子（colony-stimulating factor，CSFs）和红细胞生成素（EPO）］、生长因子（如生长激素、催乳激素）和饱食因子瘦蛋白等的受体组成。这个家族的配体（具有4个紧密包装的α-螺旋）和受体组成一个有意义的"内分泌"和"免疫"信号调色板（palette），可能是从相同的祖先前体分子协同进化而来（Boulay 等，2003；Huising 等，2006b）。它们在神经内分泌-免疫相互作用中所起的作用将在第7.2.4节中阐述。

Ⅱ型螺旋状细胞因子（HCⅡ），如IL-10、IL-26、IL-22和干扰素（Ⅰ型和Ⅱ型）都是同源二聚体，由两个细胞因子组成，每个细胞因子包含6个α-螺旋。HCⅡ及其受体的进化表明，对于四足类和硬骨鱼类宿主可能通过不同通路抵抗病原体而言，祖先作用机理的多样化是必然的（Lutfalla 等，2003）。

IL-1家族由促炎细胞因子如IL-1β和IL-18等组成，但亦有受体拮抗物，其特征是三维结构中有一个富含β链的折叠。少数IL-1家族的成员及其受体在鱼类中具有明确的直向同源物。但是，系统发育的分析支持在神经内分泌和免疫系统中共享信号作用机理的观点（Huising 等，2004b）。

7.1.3.2 趋化因子和趋化因子受体：免疫和神经内分泌信号协同进化的例子

趋化因子是通过趋化作用调控白细胞运输的重要介体。不同的白细胞由于不同的原因对不同类型的趋化剂表现出趋化性（chemotaxis）。趋化因子形成一个在结构与功能方面都和细胞因子相关的小超家族。根据它们在靠近NH_2-端保守半胱氨酸残基的型式，可以把趋化因子分为四类：CXC、CC、CX3C和C趋化因子（Laing 和 Secombes，2004）。值得注意的是，硬骨鱼类CC趋化因子的特点是有大量丰余序列（redundancy）。30个CC趋化因子是哺乳类特有的，而在硬骨鱼类中可以确定超过100个具有不同表达型式的CC趋化因子（De Vries 等，2006；Peatman 和 Liu，2007）。因此，不可能讨论所有的趋化因子以及它们在硬骨鱼类免疫中已知的作用。由于趋化因子的趋化性是一个非常古老的现象，而 CXC 趋化因子成为一个神经内分泌和免疫信号分子协同进化的极好例子，下面将对其做详述。

CXC 趋化因子的特点是头两个半胱氨酸残基被一个插入的氨基酸分隔开。根据存在着三肽模体 ELR 的情况，可以将 CXC 趋化因子进一步分类。这个模体直接出现在 CXC 模体之前。和硬骨鱼相比，哺乳类有一个复杂的趋化因子家族。至今已确定哺乳类有 16 个 CXC 趋化因子，系统的定名为 CXCL1—CXCL16。ELR 模体是哺乳类特有的进化上近期的现象，出现在 CXCL1—CXCL8。总而言之，这些趋化因子参与多形核嗜中性白细胞（polymorphonuclear neutrophils）的化学引诱作用过程。ELRCXC 趋化因子参与许多不同的作用过程，主要涉及淋巴细胞的化学引诱作用。

基因组数据库的发掘和分子分析展示了 5 个硬骨鱼类的 CXC 趋化因子。其中的 3 个（CXCa、CXCb、CXCc）和哺乳类 CXC 趋化因子群中的任何一个都不相同。CXCa 和 CXCb 的半胱氨酸残基间隔和基因结构是典型的 CXC 趋化因子，而且它们是哺乳类一些 CXC 趋化因子在功能上的类似物（Huising 等，2003c）。不同的是，鱼类另外 2 个 CXC 趋化因子在进化上很古老，是哺乳类 CXCL12 和 CXCL14 的直向同源物。有意思的是，它们主要在脑中表达（Huising 等，2004b、c）。CXCL12 及其受体 CXCR4（以及 CXCL14）的明显保守性和表达型式表明它们是原始 CXC 趋化因子/受体的现代后裔，亦表明这个配体/受体的重要功能。小鼠在敲除 CXCL12 或者 CXCR4 后，出现一个胚胎致死的表现型可以确认这一点（Ma 等，1998）。然而，破坏 CXCL12/CXCR4 信号是使一个小脑神经元亚型而不是白细胞的趋化作用受到损害。在进化树上重新安排 CXC 趋化因子家族的发生，并从结构和功能的特点推断，CXC 趋化因子是来源于脑而不是免疫系统，而且 CXC 趋化因子家族得以实行趋化的免疫功能只在脊椎动物免疫系统开始形成一个特化的有机结构的系统之后（Huising 等，2003b）。

CXC 趋化因子受体属于 7-螺旋 G-蛋白偶联受体的大家族。在哺乳类中已确定 6 个趋化因子受体（CXCR1—CXCR6）。CXC 趋化因子和它们受体的一个重要特征是它们之间存在着杂乱的相互作用，也就是说几个趋化因子可以共用一个受体，而大多数受体都能够识别好多个趋化因子。此外，正如下面要讨论的，G-蛋白偶联的趋化因子受体可以和神经肽的相似受体，如阿片样物质受体发生异聚化作用（heterodomerize）（Suzuki 等，2002）。硬骨鱼类含有 CXCR4 和 CXCR1/2 的同源物，但没有哺乳类其他的 CXCRs（Huising 等，2003b）。根据进化的保守性和系统进化树上整个类群相对短的分支长度来判断，CXCR4 是最古老的趋化因子受体。

7.2 神经内分泌因子对免疫的调节

数十年来，关于应激反应对免疫系统的影响已经在人体和模式动物中进行了广泛的研究（Elenkov 和 Chrousos，2005）。急性应激反应能起一些有益的作用，而慢性应激反应在哺乳类和硬骨鱼类中都会抑制一个最适的免疫反应（Weyts 等，1998b；Viveros-Paredes 等，2006；Edwards 等，2007；Fast 等，2008）。

对于研究免疫系统和神经内分泌系统之间的合作来说，硬骨鱼类是特别令人感兴趣的动物，因为头肾结合了免疫和内分泌的功能。这个组织构造表明脑-交感神经-嗜铬细胞轴的终产物肾上腺素（AD）和去甲肾上腺素（NA），和下丘脑-脑垂体-肾间腺轴的终产物皮质醇能够通过副分泌直接进入免疫系统的细胞，而且反过来亦是这样。所以，来自免疫系统的化学信号亦能够对嗜铬组织和肾间腺细胞起直接的副分泌作用。

7.2.1 交感神经的神经分布

哺乳类的淋巴器官是由副交感和交感神经的神经纤维分布的，推想它们能分别刺激和抑制一个免疫反应。此外，哺乳类的白细胞能表达大多数在神经系统中的肾上腺素能和胆碱能的成员，包括毒蕈碱性受体和烟碱性乙酰胆碱受体（Ach）（Kawashima 和 Fujii，2003）。由于鱼类免疫器官和细胞是否具有 Ach 受体或直接由胆碱能纤维分布的相关资料还非常少，我们将着重讨论免疫和肾上腺素能系统之间的交流沟通。

7.2.1.1 在免疫系统中的儿茶酚胺能神经分布和儿茶酚胺受体

起初阐述的肾上腺素能神经分布于小鼠的胸腺和脾脏，接着就延伸到其他的种类以及器官，如淋巴结、骨髓和消化道（Felten 等，1985）。例如，银大麻哈鱼（*Oncorhynchus kisutch*）的脾脏富于肾上腺素能神经元的神经分布。虽然这些神经分布进入脾脏并大量地和脾脏维管结构（vasculature）相联系，但实质上在脾脏中亦能观察到神经纤维（Flory，1989）。有两种类型的受体和儿茶酚胺结合：α-AR 和 β-AR 肾上腺素能受体。按照结合亲和力的不同，它们还可以进一步分为几个 α 和 β 亚型。在哺乳类中，先天性免疫细胞表达 α-ARs 和 β-ARs，而 β2-AR 是在 T-淋巴细胞和 B-淋巴细胞中表达的主要 AR。例外的是鼠的 TH2 细胞，它们和任何一个亚型受体都不表达（Nance 和 Sanders，2007）。特异的放射性配体试验证明 β-AR 在金鱼的头肾、脾脏和腹膜白细胞中表达，但配体效价的级系（rank order）并不和哺乳类任何已知的亚型等级相当（Jozefowski 和 Plytycz，1998）。同样，β-AR 已被证实存在于斑点鲖（*Ictalurus punctatus*）从头肾分离的膜和脾脏白细胞中（Finkenbine 等，2002）。最近阐明的硬骨鱼类不同组织中 β-ARs 的序列和特征（Dugan 等，2003；Fabbri 等，2008）将有助于它们在免疫细胞或组织中的鉴定，并将能够开展有关它们和其他 G-蛋白偶联受体的调节与相互作用的研究。

7.2.1.2 儿茶酚胺对先天性和适应性免疫的作用

用特异性 β-AR 的同等物和拮抗物进行的离体研究表明它们对几种硬骨鱼类先天性和适应性免疫具有明显的抑制作用，而 α-AR 同等物主要表现为刺激呼吸爆发和抗体产生（见表 7.1）。以 AD 和 NA 进行的检测通过不同的受体特点似乎产生了有差别的作用。

这些离体的作用为在体的研究所证实。在罗非鱼（*Oreochromis aureus*）的相关研究中，Chen 等（2002）发现冷的应激反应会引起儿茶酚胺的变化，而皮质醇抑制白细胞的吞噬活性和血浆的免疫球蛋白 M（IgM）水平；而在离体，皮质醇和异丙基肾上腺素

（isoproterenol）结合在一起，对降低吞噬作用具有叠加效果。最后，普萘洛尔（propranolol，一种 β-AR 拮抗物）延长了大底鳉（*Fundulus grendis*）鳞片异嫁接（allograft）的成活时间（Nevid 和 Meier，1995）。

表 7.1 离体的肾上腺素能受体同等物对鱼类免疫参数的影响

同等物	作用	种类	参考文献
NA	降低巨噬细胞的吞噬作用	翠鳢	Roy 和 Rai，2008
	刺激巨噬细胞的呼吸爆发		
AD	降低巨噬细胞的吞噬作用	翠鳢	Roy 和 Rai，2008
	刺激巨噬细胞的呼吸爆发		
	降低吞噬细胞的呼吸爆发	虹鳟	Flory 和 Bayne，1991
异丙基肾上腺素	降低吞噬细胞的呼吸爆发	虹鳟	Flory 和 Bayne，1991
	用 LPS、Con A、PHA 刺激后降低白细胞的增殖		
	降低白细胞的吞噬作用		Nanaware 等，1994
	降低抗体的反应		Flory，1990
脱羟肾上腺素	降低白细胞的吞噬作用	虹鳟	Nanaware 等，1994
	刺激吞噬细胞的呼吸爆发		Flory 和 Bayne，1991
	刺激抗体的反应		Flory，1990

NA，去甲肾上腺素；AD，肾上腺素；LPS，脂多糖；Con A，伴刀豆球蛋白 A；PHA，植物凝集素。

7.2.2 下丘脑-脑垂体-肾间腺（HPI）轴

应激反应激素皮质醇的释放是在 HPI 或者应激反应轴的控制下进行的，这和哺乳类的下丘脑-脑垂体-肾上腺（HPA）轴是功能上的类似物（见本书第 6 章）。当感受到应激反应信号时，在脑部起反应的下丘脑视前区将 CRF 释放到脑垂体内，CRF 信号为脑垂体远侧部促肾上腺皮质激素细胞的 CRF 受体亚型 1（CRF1）所接受。CRF 结合蛋白（CRF-BP）和 CRF 共同调节将 ACTH 释放到血液循环中（Huising 等，2004a；Metz 等，2004）。ACTH 接着诱导头肾的肾间细胞将皮质醇释放到血液循环中。皮质类固醇对哺乳类免疫平衡的重要性已经充分阐明（Elenkov 和 Chrousos，2006）。例如，主要包括皮质类固醇的抗炎作用，以及肾上腺切除和用糖皮质激素拮抗物 RU468 处理后能够防止在锥虫（*Trypanosoma*）感染期间出现的胸腺萎缩（Perez 等，2007）。

7.2.2.1 皮质醇受体类型及它们存在于免疫细胞中

皮质醇在通过细胞膜后就和细胞内的受体，即糖皮质激素受体（GR）结合。结合发生时，两个热激蛋白（HsP）HsP70 和 HsP90 是必要的蛋白伴侣（chaperone）（Pratt 和 Toft，1997）。这个激素-受体复合体易位到核内和糖皮质激素应答元件（GRE）结合以激活或者压制效应基因的转录（Stolte 等，2006）。有意思的是，GR 并不是唯一能和皮质醇结合的受体，亦不是一个单独类型的受体。在哺乳类中，GR 和盐皮质激素受体（MR）都能和皮质醇结合（Bridgham 等，2006）。此外，和哺乳类不同，鱼类具有复制的 GR 基因（GR1 和 GR2），它们都可以转录为有功能的蛋白质（Stolte 等，2006）。这似乎是在硬骨鱼类出现之前，但在四足类从鱼类谱系趋异之后的早期基因复制的结果。GR1 亦有两个选择性剪接变体（GR1a 和 GR1b），导致一个额外的外显子翻译，因而在 GR1a 的 DNA 结合区具有额外的 9 个氨基酸（Ducouret 等，1995；Stolte 等，2008c）。于是，在鱼类中有 4 个受体能和皮质醇结合，即 GR1a、GR1b、GR2 和 MR。它们诱导下游基因激活的能力取决于皮质醇的浓度。在虹鳟和鲤鱼中，GR2 能在皮质醇低浓度时（处于非应激状态的鱼）激活，而在应激状态下的鱼 GR1 受体只对高水平的皮质醇才是敏感的。

哺乳类的皮质醇受体存在于免疫系统的细胞中，对决定最后的免疫反应起作用（Heijnen，2007）。同样，鱼类的皮质醇受体存在于免疫细胞中，影响免疫反应（Maule 和 Schreck，1991）。鲤鱼所有 4 个皮质醇受体都在身体各部广泛表达，但 MR 在免疫系统的细胞和组织中几乎不表达。鲤鱼淋巴细胞对"敏感的" GR2 有相对较高的表达（Stolte 等，2008c）。如同在其他的器官，这些受体的表达和功能特征能为皮质醇的刺激所调节。对虹鳟的研究表明，脾脏白细胞在急性应激反应之后和 GR 的结合亲和力降低，虽然 GR 的数量有所增加（Maule 和 Schreck，1991）。相反，在海水中培育的虹鳟，皮质醇水平升高，而外周血液与头肾白细胞的 GR1、GR2 和 MR 表达都降低（Yada 等，2008）。值得注意的是，GR 的表达亦受到免疫刺激的影响。注射内毒素 LPS（革兰氏阴性细菌的细胞壁成分）使 GR 在金头鲷脾脏内的表达增加（Acerete 等，2007）。同样，对鲤鱼用酵母聚糖（酵母细胞壁的成分）诱导腹膜炎后 1 天和 2 天，GR1 在腹膜白细胞的表达增加，但 GR2 的表达不受影响（Stolte 等，2008a）。用 LPS 离体刺激头肾的吞噬细胞能迅速而短时间地诱导 GR1 表达，而 GR2 的表达减弱，意味着这是一种免疫-依赖性的 GR 反应（Stolte 等，2008c）。这可能是"应激反应"短暂高潮的反映，GR1 受体的表达被反馈性调控而增强敏感性，以便防止一个太强烈的促炎反应。

7.2.2.2　皮质醇对免疫反应的影响

糖皮质激素能调节哺乳类免疫防御的多方面，并且影响促炎与抗炎细胞因子平衡的逐次分泌以及平衡的细胞凋亡（Elenkov 和 Chrousos，2006），所以它被广泛地用做抗炎药物。同样，皮质醇对硬骨鱼类的免疫系统具有深刻而有差别的影响，由一个或几个存在于免疫细胞的 GRs 介导。

在免疫系统的先天性分支中，皮质醇离体刺激鲤鱼、罗非鱼和平鲷（*Sparus sarba*）头肾的白细胞，能明显压制其吞噬作用（Law 等，2001）。同样，皮质醇剂量依存地抑制金鱼巨噬细胞系的趋化性和吞噬作用，并强烈地抑制呼吸爆发活性（Wang 和 Belosevic，1995）。金头鲷头肾白细胞的呼吸爆发活性亦降低（Esteban 等，2004），而这种反应是由 GR 介导的（Vizzini 等，2007）。此外，皮质醇能影响免疫细胞产生细胞因子。的确，在一些离体的研究中证明了皮质醇抑制 LPS-诱导的急性期蛋白（acute phase protein），血清淀粉状蛋白 S（SAA）和促炎细胞因子 IL-1β、TNF-α、IL-12p35、IL-11 和 iNOS 等的表达（Saeij 等，2003b；Huising 等，2005；Fast 等，2008；Stolte 等，2008c）。

除调节细胞因子反应外，皮质醇还能影响免疫细胞的凋亡和增殖，以便有效地激活与钝化硬骨鱼类的免疫反应。皮质醇抑制虹鳟单核细胞/巨噬细胞系的增殖，而这个作用是 GR 依赖性的（Pagniello 等，2002）。同样，皮质醇能诱导平鲷巨噬细胞的凋亡，而大西洋鲑（*Salmo salar*）的巨噬细胞从应激反应的鱼体内分离出来之后用杀鲑气单胞菌（*Aeromonas salmonicida*）处理，其成活率降低（Fast 等，2008）。有意思的是嗜中性粒细胞为应激反应所保护。皮质醇的刺激能抑制（GR 依赖性的）鲤鱼嗜中性粒细胞的凋亡（Weyts 等，1998a）。在体的研究结果证实了这些离体的试验。应激反应使血液循环的 B-淋巴细胞减少，而粒细胞在血液循环中相对的百分比增加将近 1 倍（Engelsma 等，2003）。由于嗜中性粒细胞在防御的第一线十分重要，在急性应激反应（以及受损伤）的情况下，延长它们的生命期限以及在血液循环中保持它们较高的数量可能是有利的。

虽然研究得比较少，但在免疫系统的适应性免疫分支中亦发现应激反应诱导免疫的调节作用。对鲤鱼的研究表明，皮质醇的离体刺激能使 IgM 从脾脏、头肾和血管中的淋巴细胞分泌减少（Saha 等，2004）。在免疫接种（immunization）后，在体的温度应激反应能使抗体反应降低（Verburg-van Kemenade 等，1999）。皮质醇诱导的细胞凋亡能使 IgM 分泌和抗体产生减少。有研究曾报道血液、头肾、脾脏和胸腺的淋巴细胞的增殖受到抑制以及诱发细胞凋亡。这是一个 GR-依赖的过程，因为加入特异性的 GR 阻抑剂 RU486 就能够拯救 B-淋巴细胞（Weyts 等，1997）。

7.2.2.3 阿黑皮素原(POMC)-相关肽

POMC 衍生的肽类 ACTH、MSH、内啡肽和 POMC 的 N-端肽（NPP 或前-γ-MSH）主要分别由脑垂体远侧部和中间部的细胞分泌产生，并能分布到达血流中的免疫细胞或免疫组织中。亦曾有研究报道在免疫细胞中局部产生 POMC-衍生肽。虽然 POMC-衍生肽为免疫细胞表达的量小，但由于参与的细胞数量大，POMC-衍生肽的局部作用可能是重要的。

对应激反应诱导免疫调节作用的研究主要集中于皮质醇，关于 HPI 轴其他因子的资料很少。值得注意的是，ACTH 和类 ACTH 受体分子出现于肾间腺组织、胸腺和脾脏等免疫组织发育的早期（Mola 等，2005）。用 LPS 处理能诱导类 ACTH 分子在肾间腺和肝脏中产生，表明这种肽能在淋巴细胞完全成熟之前起作用。用 LPS 处理罗非鱼能抑制离体脑垂体释放 ACTH 与 α-MSH 以及头肾对 ACTH 的反应性（Balm 等，1995）。相应的，ACTH 参与增强吞噬细胞的氧化爆发活性（Bayne 和 Levy，1991）。

除了刺激 ACTH 从远侧部的细胞释放之外，GRF 亦能诱导 α-MSH 从脑垂体中间部细胞释放，而应激反应能使得 α-MSH 和内啡肽一起释放。ACTH 和 α-MSH 都具有免疫的调节作用。在胸腺区，POMC-衍生肽与细胞因子和胸腺内的凋亡细胞共定位，这表明神经内分泌细胞对胸腺、淋巴细胞的选择和凋亡起作用（Ottaviani 等，1995）。α-MSH 能诱导巨噬细胞超氧阴离子的产生与增强吞噬作用。用 α-MSH 处理鲤鱼的淋巴细胞还能增强对植物凝集素（phytohaemagglutinin，PHA）的促有丝分裂反应（Watanuki 等，2003）。此外，还有学者发现 α-MSH 对分离的吞噬细胞是没有影响的，但用 α-MSH 处理从混合的白细胞衍生的上清液能增强吞噬作用。这表明要使这种类型的免疫调节起作用，需要多于一种类型的免疫细胞（Harris 和 Bird，2000）。NPP 能增强虹鳟和鲤鱼离体巨噬细胞的吞噬作用。这些研究结果均被在体研究证实，施用 NPP 后巨噬细胞超氧化阴离子的产生增加，吞噬作用增强，表明 NPP 在低等脊椎动物中能激活吞噬细胞的作用（Sakai 等，2001）。

7.2.3 阿片样物质

对阿片样物质系统在较原始动物的系统进化研究中可以得到一个颇有兴趣的设想：这个系统是在无脊椎动物中作为免疫调节系统的一部分产生的，而当疼痛变成一种警诫的过程时，它们的镇痛特性才发展起来（Stefano 等，1998）。无脊椎动物和脊椎动物的抗微生物肽（如 ekelytin）都源自阿片样物质激素原（Salzet 等，2006）的这一事实支持了上述的观点。此外，白细胞能合成阿片样物质肽类，并具有阿片样物质受体（Sharp，2006）。所以，阿片样物质能直接通过激活淋巴细胞上的阿片样物质受体来影响免疫过程。然而，阿片样物质亦可以间接调节免疫反应，例如阿片样物质激活应激反应轴使皮质类固醇的产生增加，或者激活交感神经系统并随之促使儿茶酚胺释放（Mellon 和 Bayer，1998）。

7.2.3.1 阿片样物质肽和它们在免疫细胞内合成

阿片样物质肽的主要类群是由阿片样物质激素原如脑啡肽原（PENK）、前强啡肽（PDYN）和 POMC 的翻译后加工而产生的（见图 7.2）。POMC 是 α-内啡肽和 β-内啡肽（END）和几种非-阿片样物质肽的前体（Laurent 等，2004）。PENK 是亮氨酸-脑啡肽（LE）、蛋氨酸-脑啡肽（ME），蛋氨酸-脑啡肽-精^6-甘^7-亮^8，蛋氨酸-脑啡肽-精^6-苯丙^7，肽 E、B 和 F，以及抗菌的 enkelytin 的来源（Laurent 等，2004；Brogden 等，2005）。PDYN 产生强啡肽 A、B（rimorphin，DYN），α-和 β-新内啡肽（NEO）（Laurent 等，2004）。最近发现新的阿片样物质肽类群：内啡素（1 和 2，前体不清楚）和伤害感受肽（nociceptin）/孤肽（orphanin）FQ（从伤害感受肽原裂解而来）（Przewlocki 和 Przewlocka，2005）。所有阿片样物质激素原基因可能都来自一个共同的祖先，而在脊索动物进化过程中，PENK 基因进行了一系列复制以产生 POMC、伤害感受肽原和 PDYN（Danielson 和 Dores，1999）。有学者已经在几种鱼类中克隆到阿片样物质激素原并分析了它们的特征（Gonzalez-Nunez 等，2003a、b；Gonzalez-Nunez 等，2003）。

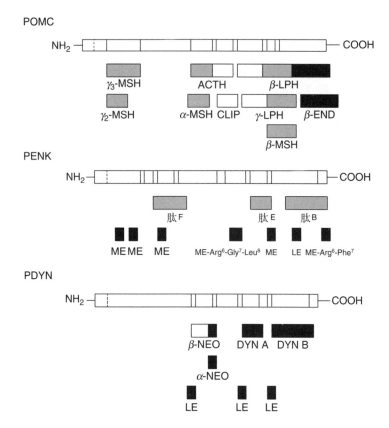

图7.2 产生阿片样物质肽与激素的激素原

ACTH，促肾上腺皮质激素；CLIP，促皮质素样中间叶肽；DYN，强啡肽；END，内啡肽；LE，亮氨酸-脑啡肽；LPH，促脂解素；ME，蛋氨酸-脑啡肽；MSH，促黑色素细胞激素；NEO，新内啡肽；POMC，阿黑皮素原；PENK，脑啡肽原；PDYN，前强啡肽。

7.2.3.2 阿片样物质受体和它们在免疫系统中的功能

在哺乳类中已经克隆到三种主要传统类型的阿片样物质受体并阐明了它们的药理学特征：类吗啡受体mu（MOR，μ，oprm），一个δ受体（DOR，δ，oprd）和一种kappa受体（KOR，K，oprk）。此外还有一个非传统的阿片样物质样受体——ORL-1（NOR，oprn）。以放射性配体的结合为基础，可以确定几个阿片样物质受体的亚型（Kieffer和Gaveriaux-Ruff，2002；Corbett等，2006）。各种阿片样物质受体的内源性配体列于表7.2中。

表7.2 哺乳类的阿片样物质受体类型和配体

受体	亚型	内源性配体
Mu	1–3	内啡素，β-END
Delta	1–2	脑啡肽，β-END
Kapper	1–3	强啡肽
ORL-1		伤害感受体

所有已知的阿片样物质受体都是 G-蛋白偶联受体，和同等物结合通过百日咳毒素敏感 G1 蛋白（pertrusis toxin-sensitive G1 protein）的活化而使腺苷酸环化酶活性和细胞内 cAMP 受到抑制（Kieffer 和 Gaveriaux-Ruff，2002）。阿片样物质受体可以和任何其他阿片样物质受体以及非阿片样物质受体如 β2-肾上腺素能受体甚至趋化因子受体形成同型和异型二聚体的受体复合体。同型二聚体化或异型二聚体化的不同类型的阿片样物质受体表明它们的药理学和信号转导已经发生变化（Jordan 等，2001；Gomes 等，2003；Corbett 等，2006）。

预测的阿片样物质蛋白质结构是高度同源的，以及具有几乎相同的基因组结构，表明它们源自一个共同的祖先基因（Stevens，2004）。对少数鱼类的研究（Darlison 等，1997；Barrallo 等，1998a、b，2000；Pinal-Seoane 等，2006）以及最近对鲤鱼（Chadzinska 等，2009a）的克隆而推定的阿片样物质受体和哺乳类的对应物是高度同源的。

在哺乳类的白细胞中发现 mu、δ 和 kappa，以及非传统的阿片样物质受体的结合部位（Sharp，2006）。在金鱼头肾白细胞对阿片样物质受体的鉴定表明，至少存在着两个不同的阿片剂-结合部位：一个 [^{3}H] 纳洛酮（naloxone）结合部和一个 [^{3}H] nal-trindole 结合部位（Chadzinska 等，1997；Jozefowski 和 Plytyez，1997）。纳洛酮结合部位和吗啡-敏感的以及阿片样物质肽-不敏感的 mu3 受体相似。选择性配体的结合亲和力低，在金鱼白细胞中存在的神经元-mu 类型和 δ 阿片样物质受体除外（Chadzinska 等，1997；Jozefowski 和 Plytyez，1997）。在鲤鱼的免疫器官中能观察到 mu、δ 和 kappa 基因的低组成性表达。用酵母聚糖在体诱导腹膜炎期间和离体 LPS-诱导的刺激之后，阿片样物质受体基因在白细胞中的表达是上调的（Chadzinska 等，2009a）。

在哺乳类中，阿片样物质调节先天性和获得性免疫反应，改变对许多传染物的抗性（McCarthy 等，2001；Eisenstein 等，2006；Singh 和 Rai，2008）。事实上，细胞因子的所有主要特性都为阿片样物质所共享，即免疫细胞以副分泌、自分泌和内分泌作用部位的产物，功能的丰余性、多效性以及剂量与时间依赖性的作用（Peterson 等，1998）。此外，阿片样物质受体同等物影响离体的吞噬作用和呼吸爆发活性（Eisenstein 等，2006）。同样，在硬骨鱼类中观察到对离体重要先天性免疫参数的明显影响（见表 7.3）。引人注目的是，β-END 在低的或高的浓度下对吞噬细胞的相对作用分别为 mu 和 δ 受体同等物所抵消。阿片样物质对白细胞趋化性的影响是特别有意思的。Mu 和 δ 阿片样物质受体的同等物对不同的白细胞群体起着有效的离体化学引诱剂作用（Grimm 等，1998a、b；Chadzinska 等，1999；Miyagi 等，2000；Chadzinska 和 Plytyez，2004），但同样的阿片样物质通过异性的脱敏作用过程亦能阻抑细胞向化学引诱剂移行（Grimm 等，1998a、b；Choi 等，1999；Miyagi 等，2000）。一般来说，一个类型 G-蛋白偶联受体（GPCR）的激活通过第二信使介导激酶如蛋白激酶 A（PKA）或 PKC，能够引起其他 GPCR C-端胞质尾部的磷酸化作用。一个磷酸化受体失去它和下游异源三聚 G 蛋白偶联的能力，就会对任何刺激失去敏感性（Rogers 等，2000）。

表 7.3 阿片样物质受体同等物对鱼类免疫参数的影响

同等物	处理	作用	种类	参考文献
吗啡	离体	$IL-1\beta$、$TNF-\alpha$、CXCa、CXCR1 和 iNOS 基因表达降低，抑制 LPS-刺激的吞噬细胞产生 NO	鲤鱼	Chadzinska 等，2009b
		白细胞向化学引诱物的趋化减弱	鲤鱼，野生金鱼	Chadzinska 和 Plytyez，2004；Chadzinska 等，2009b
	在体	在腹膜炎期间细胞流入和化学引诱剂水平降低	大西洋鲑，鲤鱼，野生金鱼	Chadzinska 等，1999，2000，2009 Chadzinska 等，1999 Chadzinska 等，2009b
		刺激渗出的白细胞呼吸爆发活性	大西洋鲑	
		TNF-、CXCa、CXCb、CXCR1、CXCR2、iNOS 基因表达降低，精氨酸酶 I 在渗出的白细胞内	鲤鱼	
β-END	离体	刺激超氧化阴离子产生和吞噬作用	虹鳟，鲤鱼	Watanuki 等，2000
		刺激吞噬作用和 NO 产生	纹鳢	Singh 和 Rai，2008
		刺激（在低剂量）和抑制（在高剂量）呼吸爆发活性		
		刺激白细胞的杀菌活性	虹鳟	Watanuki 等，2000
		刺激白细胞的细胞毒性和增殖	罗非鱼	Faisal 等，1989
	在体	刺激呼吸爆发作用和吞噬细胞的吞噬作用	虹鳟	Watanuki 等，1999
		刺激白细胞趋化性朝向调理的细菌		
内啡素	离体	降低吞噬朝向趋化因子的趋化性	鲤鱼	Chadzinska 等，2009a；Verburg-van Remenade 等，印刷中
		降低在 LPS-刺激吞噬细胞中 $TNF-\alpha$、CXCb 和 CXCR1 基因的表达		
δ 啡肽 II	离体	降低白细胞朝向化学引诱剂的趋化性	鲤鱼 野金鱼	Chadzinska 等，2009；Chadzinska 和 Plytyez，2004；Verburg-van Remenade 等，印刷中
		降低在 LPS-刺激吞噬细胞中 CXCa、CXCb、CXCR1 和 CXCR2 基因的表达	鲤鱼	Chadzinska 等，2009a；Verburg-van Remenade 等，印刷中
U50, 488H	离体	减弱吞噬细胞朝向趋化因子的趋化性，降低在 LPS-刺激吞噬细胞中 IL-10 和 CXCR1 的基因表达	鲤鱼	Chadzinska 等，2009a；Verburg-van Remenade 等，印刷中

IL，白细胞介素；$TNF-\alpha$，肿瘤坏死因子 α；CXC，趋化因子；CXCR，趋化因子受体；iNOS，诱导型-氧化氮合酶；NO，一氧化氮；LPS，脂多糖。

对金鱼的研究表明，用 mu 和 δ（不是 kappa）的同等物对白细胞进行预温育，能增强其不规则的移行，然而抑制朝向酵母聚糖-刺激的血清移行。这后一种情况能为特异的 mu 和 δ 阿片样物质受体的拮抗物所逆转（Chadzinska 和 Plytycz，2004）。还有，吗啡在体和离体都能下调趋化因子（CXCa 和 CXCb）以及趋化因子受体（CXCR1 和 CXCR2）在鲤鱼白细胞中的表达（Chadzinska 等，2009b）。用吗啡和其他 mu（不是 δ）阿片样物质受体同等物和哺乳类的巨噬细胞与单核细胞进行离体孵育能诱导细胞凋亡（Malik 等，2002；Bhat 等，2004）。我们的初步研究表明，用吗啡或者 δ 啡肽（deltorphine）Ⅱ和鲤鱼吞噬细胞进行预温育都不会引起细胞凋亡（Chadzinska 等，2009b；Verburg-van Kemenade 等，印刷中）。

由于阿片样物质对白细胞的影响被认为是炎症反应的一种重要变化，因而设计了使用促炎剂（酵母聚糖或巯基乙酸）的在体试验来证实这个观点。吗啡和酵母聚糖与巯基乙酸一起注射使鱼（金鱼、鲑鱼、鲤鱼）的炎症白细胞数量减少。这个有力的证据说明被吗啡处理的鱼腹膜白细胞数量减少，可以归因于化学引诱物和趋化因子受体水平的降低以及趋化因子受体异质性的脱敏作用（Chadzinska 等，1999，2000，2009a、b）。值得注意的是，Metz 和合作者的研究结果（Metz 等，2006）表明阿片样物质系统和炎症反应之间的交流沟通是双向的（见第 7.3 节）。

有关阿片样物质对免疫功能影响的许多研究都着重于研究天然杀伤和淋巴细胞功能的干扰（Shavit 等，1984；Yeager 等，1995）。对罗非鱼的研究表明，社会性对抗-诱导的免疫抑制能为阿片样物质受体拮抗物 naltrexone 所逆转（Faisal 等，1989）。这些研究结果间接地证明这种和社会性应激物相联系的免疫抑制至少部分是通过内源性阿片样物质系统介导的。然而，要注意的是 naltrexone 的作用只限于细胞毒素和 T-淋巴细胞。此外，外源性 β-END 诱导明显的细胞毒性以及增殖的抑制作用能为 naltrexone 所逆转（Faisal 等，1989，1992）。注射纳洛酮（阿片样物质拮抗物）使异源移植的大底鳉较长时间存活可以说明同样的情况（Nevid 和 Meier，1995）。

7.2.4　Ⅰ型细胞因子家庭成员：生长激素、催乳激素、瘦蛋白

7.2.4.1　Ⅰ型细胞因子及其与免疫系统的连锁

Ⅰ型细胞因子组成一个"内分泌"和"免疫"的信号家族，它们可能是从相同的祖先前体分子协同进化而来（Huising 等，2006b）。它们的特征是一串 4 个紧密包装的 α-螺旋，而它们的初级氨基酸序列只有很小的序列相似性。这个细胞因子家族的成员包括 GH 和 PRL，这两个多效性激素在它们的其他功能当中致力于恢复由皮质类固醇或者应激反应引起的细胞衰竭。例如，高催乳素血症（hyperprolactinemia）是和自身免疫病如风湿性关节炎和系统性红斑性狼疮（systemic lupus erythematosus）以及应激反应的激活阶段相联系的（Orbach 和 Shoenfeld，2007）。在健康的情况下，糖皮质激素起着抑制促炎细胞因子过度合成与释放的作用。炎症反应破坏脑或外周的细胞因子平衡，而身体反应是增强激素的释放以恢复这种平衡（Kelley 等，2007）。已确认硬骨鱼类有两

种类型的 PRL，即 PRL$_{177}$ 和 PRL$_{188}$（Swennen 等，1991；Flik 等，1994）。PRL 的浓度是和应激反应及感染的不同状况相关的（Wendelaar Bonga 等，1984）。PRL 和 GH 以及皮质醇参与渗透压调节，适应于淡水或者海水之后它们的表达和/或血浆蛋白质浓度会发生变化。所以，在这个转化过程期间免疫参数的缩减是和激素平衡的变化相联系的。

学者们已经充分了解Ⅰ型细胞因子瘦蛋白作为饱食因子的传统作用，并广泛进行研究以阐明体重状况调节及其相关的失调状况（Schwartz 等，2000）。此外，现在学者们的兴趣在于瘦蛋白作为Ⅰ型细胞因子在免疫中的潜在作用（Lago 等，2008）。最近阐述了河鲀和鲤鱼瘦蛋白的基因序列，开启了对变温动物在这一领域的研究。因而，我们将着重说明Ⅰ型细胞因子/激素对硬骨鱼类免疫防御相关的作用机理。

7.2.4.2　GH、PRL、SL 和瘦蛋白在免疫组织中的产生

GH、IGF-Ⅰ、PRL 和瘦蛋白的外周表达已经在哺乳类的淋巴组织和白细胞中被发现（Kelly 等，2007）。PRL 产生于脑垂体外的部位，包括神经元、前列腺、上皮和内皮（Ben-Jonathan 等，1996）。目前已检测到多种免疫反应的 PRL 分子（24kDa、21kDa 和 11kDa 类型在胸腺细胞和一个 27kDa 类型在单核细胞内），包括一个具有生物活性的脑垂体样 PRL（Montgomery，2001）。关于这些免疫细胞相关因子表达调控的资料还很少。PRL 在哺乳类 T-淋巴细胞的表达受到 IL-2、IL-4 和 IL-1β 的调节（Gerlo 等，2005）。此外，PRL 基因在淋巴组织的表达不受脑垂体转录因子 Pit-I 的调节（Berwaer，1994；Gellersen 等，1994），而现在认为 PRL 在人体淋巴细胞和脑垂体的表达受到一个不同的启动子调控（Gerlo 等，1994）。同样，GH 在哺乳类脑垂体或淋巴细胞的表达亦需要不同的转录因子（Kooijman 等，2000；Weigent 等，2000）。

在各种硬骨鱼类中，Ⅰ型细胞因子在白细胞和淋巴组织中表达（见表 7.4）。有意思的是 GH 在虹鳟白细胞中的合成受到皮质醇的刺激（Yada 等，2005），这证明由皮质醇与 GH/PRL 相对保持的促炎与抗炎信号是一种进化上保守的平衡。

表 7.4　Ⅰ型细胞因子在鱼类淋巴组织中的表达

细胞因子	表达部位	种类	参考文献
PRL	白细胞	虹鳟、罗非鱼	Yada 和 Azuma，2002
	头肾、脾脏、肠、PBL 和头肾白细胞	虹鳟、罗非鱼	Yada 等，2002
SL	脾脏	虹鳟	Yang 等，1997
GH	PBL	虹鳟、罗非鱼（不包括鲶鱼）	Yada 等，2001b，2002，2005
	头肾	金头鲷	Calduch-Giner 和 Perez-Sanchez，1999
	脾脏、肾脏、肠	大麻哈鱼	Mori 和 Devlin，1999
瘦蛋白	胸腺	鲤鱼	Huising 等，2006a

PRL，催乳激素；PBL，外周血液的白细胞；SL，生长乳素；GH，生长激素。

7.2.4.3　GH、PRL 和瘦蛋白在白细胞中的受体

GH、PRL、SL 和瘦蛋白受体属于一个大而异质性的受体家族，而 I 型细胞因子受体家族和它们是同源的，表明了它们和免疫系统在进化上的联系（Heinrich 等，2003）。和细胞因子结合后，由 JAK 家族酪氨酶介导酪氨酸磷酸化作用。接着，JAK 磷酸化细胞内酪氨酸残基处在受体链的细胞质区内。这个复合体由磷酸化作用激活 STAT。然后形成一个 STAT 二聚体转运到核并启动转录（Heinrick 等，2003）。转录哪个基因取决于形成二聚的 STAT 家族的不同成员。哺乳类 STAT 家族由 7 个成员组成，而硬骨鱼类曾被表明有几个成员（Lewis 和 Ward，2004）。许多细胞因子的生物学作用表现为多效性和丰余性，这可能是共同使用 GP130 信号链以及 JAK/STAT 通路成员的混杂性（promiscuity）造成的结果（Huising 等，2006b）。细胞因子信号阻抑剂（suppressor of cytokine signaling，SOCS）分子的诱导是启动 JAK/STAT 活化以确保一种负反馈的作用机理。这表现为一种和免疫功能活跃的相互作用，因为 SOCS 蛋白质通过各种细胞因子的活化亦能够上调（如 I 型细胞因子和 CXCL12 刺激 CXCR4 受体）（Redelman 等，2008）。

哺乳类有几种 PRL 受体（PRLR）的同种型，其中一种可能起诱饵受体（decoy receptor）的作用。和这些受体结合可能是不加区别的，如高浓度的 GH 可能结合并通过 PRLR 信号而影响人的嗜中性白细胞（Fu 等，1992；Soares，2004）。在配体与 PRLR 或 GHR 结合和同型二聚体化（homodimerization）后发生 JAK-STAT 信号传导途径。鱼类 PRLR 在构造上和哺乳类的配对物相似，虽然总的序列保守性很低（26%～37%）。在细胞内区域（ICD）的膜近侧区含有一个富于脯氨酸的区（盒1），对于哺乳类，它的作用是通过 JAK2-STAT 信号途径进行信号传递。鱼类 PRLR 的激活看来是启用同样的作用机理，因为尼罗罗非鱼的 PRLR 能够激活一个 STAT-5-效应的报道基因（reporter gene）的转录（Sohm 等，1998；Tse 等，2000）。I 型细胞因子受体在硬骨鱼类免疫细胞的表达归纳于表 7.5 中。

表 7.5　I 型细胞因子受体在鱼类淋巴组织中的表达

受体	表达或特异性结合	种类	参考文献
PRLR	头肾	金头鲷	Santos 等，2001
	淋巴细胞	罗非鱼	Sandra 等，2000；
			Yada 和 Azuma，2002
		虹鳟	Prunet 等，2000
GHR	淋巴细胞	海鲈	Calduch-Giner 等，1995
	脾脏	虹鳟	Very 等，2005
Leptin R	白细胞	鲤鱼	未发表的观察

PRLR，催乳激素受体；GHR，生长激素受体；Leptin R，瘦蛋白受体。

7.2.4.4 PRL、GH 和瘦蛋白对免疫反应的影响

对于切除脑垂体的侏儒型哺乳动物的免疫缺陷，学者们原先都将注意力集中于激素的免疫功能。同样，切除硬骨鱼类的脑垂体都导致先天性和适应性免疫功能下降（Hull 和 Harvey，1997；Yada 等，1999，2001a；Yada 和 Azuma，2002）。催乳激素的作用随后为哺乳类离体和在体的研究所确认（Kelley 等，2007）。GH 或 PRL 治疗亦能够逆转鱼类切除脑垂体后的效应。然而 PRL 对于免疫功能似乎并不是必要的，因为 PRL 和 PRLR 缺乏的小鼠具有良好的免疫防御能力（Horseman 等，1997；Bouchard 等，1999）。PRL 和 GH 对硬骨鱼类先天性和适应性免疫参数的特异性作用，与哺乳类一样，通常是刺激性的（Redelman 等，2008；见表 7.6）。

表 7.6 PRL 和 GH 对鱼类先天性和适应性免疫参数的影响

细胞因子	作用	种类	参考文献
PRL（不是 SL）			
离体	刺激白细胞吞噬作用	大西洋鲑鱼	Sakai 等，1996
	刺激白细胞增殖		Yada 等，2004
在体	刺激 Ig 产生	虹鳟	Yada 等，1999，2002
GH	刺激超氧化阴离子	虹鳟	Sakai 等，1995，1996；
在体和离体	产生和吞噬作用	罗非鱼	Yada 等，2001b
		海鲈	Yada 等，2002
			Munoz 等，1998
	脾脏、头肾和外周血液白细胞的 NCC 活性	虹鳟	Sakai 等，1995
	血清的溶血活性	虹鳟	Sakai 等，1995
	刺激 Ig 产生	虹鳟	Yada 等，1999，2002

PRL，催乳激素；SL，生长乳素；GH，生长激素。

可以设想肥胖人较常发生自身免疫疾病是和瘦蛋白相联系的（Harle 和 Straub，2006）。这是因为瘦蛋白对先天性免疫反应具有明显的刺激作用，如趋化性和氧自由基的分泌（La Cava 和 Matarese，2004），吞噬作用和急性期促炎介体的分泌（Sanchez-Margalet 等，2003）及 NCC 细胞毒的能力（Zhao 等，2003）。部分原因是最近才确定变温动物的瘦蛋白序列，至今还不清楚瘦素是否影响硬骨鱼类的先天性免疫。在哺乳类的适应性免疫中，瘦蛋白影响胸腺 T-细胞的增殖、成熟和成活（Howard 等，1999），增加幼稚 T-淋巴细胞发育与分泌 IL-2（Martin-Romero 和 Sanchez-Margalet，2001）。在记忆 T-细胞中，瘦蛋白通过增加 IFN-γ 和 TNF-γ 分泌而引起转向辅助性 1(Th1)-细胞的免疫反应以及 B-细胞产生 IgG2a。初步的研究结果表明，在免疫刺激时，瘦蛋白在 PBLs 和胸腺细胞中的表达增加，而在使用人的瘦蛋白时 INF-γ 的表达增加（Verburg-van Kemenade，未发表研究结果）。

GH、PRL 和瘦蛋白并不是和免疫系统交流沟通的唯一Ⅰ型细胞因子。在哺乳类相关研究中，目前对其他家族成员在这方面作用的证据正在不断增加，而这肯定是值得进一步研究的。IL-6 参与抗体产生、急性期反应和 T-细胞的刺激（Bird 等，2005）。对牙鲆（*Paralichthys olivaceus*）和东方红鳍鲀（*Takifugu rubripes*）的相关研究曾报道 IL-6。IL-6 具有保守的基因结构，表现出和哺乳类直向同源物相似的表达型式，说明基因功能具有保守性（Bird 等，2005；Nam 等，2007）。白血病抑制因子（LIF）和制癌蛋白 M（OSM）参与肿瘤细胞增殖的调控（Abe 等，2007）。最近，在对斑马鱼和黑青斑河鲀（*Tetraodon nigroviridis*）的研究中曾发现 LIF 的直向同源物。

7.2.5 下丘脑-脑垂体-甲状腺（HPT）轴

对于哺乳类，HPI 轴和免疫系统之间的相互作用在离体方面和以低的和高的甲状腺模型的在体方面都已经得到充分阐明（Bagriacik 和 Klein，2000；Dorshkind 和 Horseman，2001）。这些作用通常都被认为对适应性反应是刺激性的，特别是影响 B-细胞的发育（Dorshkind 和 Horseman，2000）。Halabe Bucay（2009）最近综述甲状腺激素对免疫参数的影响，包括激活增殖和细胞因子（如 IFN-γ）在白细胞内的产生。和适应性反应不同的是，Dorshkind 和 Horseman（2000）指出先天性反应的损害。此外，甲状腺刺激激素（TSH）亦可以在免疫系统内由白细胞产生（Weigent 和 Blalock，1995）。

硬骨鱼类的甲状腺功能亦可以由一个相当的 HPT 轴调控，尽管已经指出两者在结构和调控作用上有所不同。例如，已发现 CRF 而不是 TRH 成为主要的促甲状腺素释放激素（Flik 等，2006；Geven 等，2006）。有意思的是，有些鱼类在头肾的造血组织内发现甲状腺滤泡（Geven 等，2007）。此外，已表明 THs 对应激反应轴具有重要的调节作用，在鲤鱼诱导甲状腺功能亢进的试验中使血浆皮质醇水平急剧降低（Geven 等，2006）。早期在对甲状腺功能减退的鱼进行的研究中观察到血液循环中白细胞数量明显减少（Slicher，1961），而加入甲状腺素（T4）或哺乳类 TSH 后能使之恢复（Ball 和 Hawkins，1976）。Lam 等（2005）最先证明甲状腺和胸腺发育的关系从鱼类到较高等的脊椎动物都是保守的，因为胸腺发育和胸腺生成（胸腺大小，重组活化基因 RAG-I 正性，T-细胞抗原受体的表达）受到 T_4 的影响。特别有意思的是研究硬骨鱼类在一龄降海幼鲑准备向二龄鲑转变时期甲状腺-免疫的相互作用，研究结果表明其伴随着血浆中的 T_4 和皮质醇明显增加。用鳗弧菌（*Vibrio anguillarum*）抗原进行免疫接种后，外周血液的淋巴细胞和脾空斑形成细胞（splenic plaque forming cell）的数量减少，而同时小的淋巴细胞相对增加。然而，皮质醇对 T_4 的作用还是没有阐述清楚（Maule 等，1981）。

7.2.6 脑-脑垂体-性腺（BPG）轴

和识别应激反应对免疫的潜在影响一起的是应激反应状态经常伴随着性类固醇水平和性腺发育或性功能的变化。而且，哺乳类免疫细胞活性的变化和细胞凋亡能为雌激素和具有雌激素受体的免疫细胞所诱导（Ahmed，2000）。环境中的内分泌损害化合物（EDCs）亦表现出干扰一些鱼类的生殖活动和疾病抵抗能力。重要的 EDCs 是天然的与合成的雌激素和雄激素（Segner 等，2006），而水生生态系统处在天然雌激素如 17β-雌二醇（E_2）和合成雌激素如 17α-炔雌醇（17α-ethinylestradiol，EE_2）的影响之下（Cargouet 等，2004；Kolodziej 等，2004）。总而言之，关于性类固醇对硬骨鱼类免疫的影响，我们了解得还不完整。性别和种类的不同，以及发育阶段和产卵季节的不同都必须考虑到。此外，雌激素对 GH/IGF-I 系统的潜在干扰可能会造成复杂的后果（Riley 等，2004；Filby 等，2006）。

7.2.6.1 雌激素/雄激素受体和雌激素/雄激素对先天性免疫和适应性免疫的作用

在鲑鱼淋巴细胞中检测到雄激素受体，而曾报道在鲶鱼白细胞中有雌激素受体-α（ER-α）（Patino 等，2000），这些事实表明性类固醇参与硬骨鱼类的免疫作用。此外，金头鲷增强吞噬能力和对弧菌感染的抑制能力都和睾酮（不是 E_2）的水平有关（Deane 等，2001）。在体用 E_2 处理金鱼后引起对锥虫（*Trypanosoma donilewskyi*）的易感性，并降低白细胞的增殖（Wang 和 Belosevic，1994）。给鲤鱼注射 E_2、11-KT 和黄体酮能剂量依赖性地降低 NO 和吞噬作用（Watanuki 等，2002）。对鲤鱼头肾白细胞的离体分析表明毫微摩尔（nanomolar）浓度的 E_2、黄体酮和 11-KT 能损害吞噬作用，11-KT 和黄体酮能阻抑 NO，但任何一种类固醇都不会减弱呼吸爆发作用（Yamaguchi 等，2001）。Wang 和 Belosevic（1995）曾试验 E_2 对金鱼一个由巨噬细胞衍生的细胞系的影响，并测定对趋化和吞噬作用的阻抑，但和皮质醇不同的是，没有发现对 NO 产生影响。Saha 等（2003，2004）的研究表明，性腺类固醇对鲤鱼胸腺细胞的凋亡没有影响，亦不使 IgM 水平降低，虽然在产卵季节期间 IgM 通常都是高水平的。

7.3 细胞因子对神经内分泌的调节

由于研究技术的局限，对鱼类大多数细胞因子的研究都集中于细胞因子对免疫反应的作用，而只有很少的研究关注到细胞因子对神经内分泌的调节。然而，不断增加的研究结果表明存在着免疫-内分泌双向的相互作用。很明显，对于免疫和神经内分泌系统之间的生理学联系，必须有一种作用机理使信号释放到外周并能够诱导中枢的反应。科学家们确实经过二十多年的努力才建立起免疫系统和脑之间交流沟通的观点（Quan 和 Banks，2007）。在本节中，我们将综述和讨论当前对复杂的细胞因子信号在神经-免疫-内分泌相互作用中的认识，并尽可能涉及鱼类。

7.3.1 一个应激反应后的免疫反应

鱼类和哺乳类相似，身体感染或者炎症反应后出现的反应特征分为三期。首先，急性期反应（见第 7.1.2 节）的特点是引起先天性免疫反应。其次，激活 HPA 轴及脑的参与。然而，至今只有少数对鱼类的研究能够说明一个免疫的冲击对血浆皮质醇浓度的影响。对于哺乳类和鱼类，一个免疫的挑动能使神经内分泌通路的活性增强，进而使血浆皮质醇水平升高（Camp 等，2000；Haukenes 和 Barton，2004）。最后，糖皮质激素的强烈反馈作用，GH 和 PRL 的平衡作用，阻抑潜在的过多的细胞因子产生与促炎反应（见第 7.2.2 节）。

7.3.2　从外周到脑

如前所述，先天性免疫反应以 TLRs 对 PAMPs 的识别为基础，诱导促炎细胞因子增强表达。典型的是，IL-1β、IL-6 和 TNF-α 是最先表达的促炎细胞因子。除了它们"传统的"介导和协调对病原体局部的与身体的炎症反应作用之外，这些细胞因子还能诱导疾病综合症状，包括对感染的反应发生后所有内分泌的、自发的和行为的变化（Dantzer，2006）。例如，上述细胞因子诱导发热、食欲消失、退出社会性活动和疲劳。这些生理上的调整和积极的影响复原与存活一起是宿主对微生物感染的一种适应性反应（Dantzer，2006）。确实，疾病综合症状的干扰能导致如压抑和阿尔茨海默病神经错乱的发展。我们对于免疫系统-CNS 交流沟通了解的中心是要确定 CNS 如何获得有关在外周发生的免疫反应的"通知"。

7.3.3　外周免疫信号如何传达到脑

目前认为有四种作用机理将外周的炎症反应信号和 CNS 反应联系起来（Correa 等，2007）。第一个通路是在感染期间由迷走神经的传入神经纤维的活化而组成（Bluthe 等，1994；Goehler 等，1999）。这些神经纤维延伸到几个脑核，包括下丘脑的旁室核（paraventricular nucleus，PVN）。第二种作用机理是血液循环中的 PAMPS 引起 CVOs（血脑屏障缺损区，围绕脑室）的巨噬细胞产生与释放促炎细胞因子（Quan 等，1998）。这些细胞因子通过大量扩散而到达受体（Vitkovic 等，2000）。CVOs 神经元被发现在促炎反应细胞因子的受体表达，包括 IL-1R1、TNFR（Bette 等，2003；Nadjar 等，2003，2005）。这些神经元直接突入到下丘脑的 PVN（Rivest 等，2000）。第三种作用机理是由特异性载体主动运输以便通过血脑屏障（Banks，1989）。值得注意的是血脑屏障的通透性受到 IL-1β 的影响（Blamire 等，2000）。第四种作用机理是以促炎反应细胞因子在巨噬细胞和血管与脑界面的内皮细胞诱导前列腺素 E2 的局部合成为基础（Konsman 等，2004）。前列腺素 E2 能穿过血脑屏障，而它一旦在脑内就会引起发热（Engblom 等，2002）。

虽然还不清楚外周的免疫信号能否到达鱼类的脑部，但现有的研究结果表明和哺乳类相类似的作用机理亦会存在于鱼类。首先，最近已经在牙鲆、斑马鱼和鲶鱼的相关研究中确认由已经识别的 PAMPs 激活 TLRs（见第 7.1.2 节；Hirono 等，2004；Meijer 等，2004）。这些鱼类的 TLRs 是相当保守的，并且表明参与了感染的反应（Baoprasertkul 等，2007）。其次，鱼类具有 CVOs，如血管囊（saccus vasculosus）（Sueiro 等，2007）。鱼类的 CVOs 在解剖和功能方面都和哺乳类的 CVOs 相似，它们的血管具有和其他脑区相同的功能差别，包括相伴的血脑屏障的缺失（Jeong 等，2008）。再次，尽管还没有关于细胞因子通过鱼类血脑屏障的一项专门研究，但最近的研究结果表明斑马鱼的血脑屏障在分子结构和功能方面都和较高等脊椎动物的十分相似（Jeong 等，2008）。最后，腹腔注射 IL-1β 能促使环加氧酶-2（cyclooxygenase-2，COX-2）增加，而这种酶能促进前列腺素合成（Hong 等，2003）。综合这些研究结果清楚地表明鱼类和哺乳类一样，脑能够接近外周释放的因子或者对它们敏感。显然，这种接近能力和敏感性是否能够调控以及如何调控，还需进一步研究。

促炎反应细胞因子的重要功能是引起发热。初看起来这种功能似乎不适合于变温动物的鱼类。然而，在所有的脊椎动物类群包括鱼类的相关研究中都曾报道其调节体温的升高（Reynolds 等，1976）。虽然原先曾报道注射脂多糖（LPS）未能诱导驼背太阳鱼（*Lepomis gibbosus*）体温升高（Marx 等，1984），但后来的研究发现对金鱼注射 LPS 和 IL-2 都能引起发热（Cabanac 和 Laberge，1998）。

7.3.4 细胞因子在脑内产生

CNS 本身，包括下丘脑，神经元组成性地表达细胞因子并起着正常的脑功能作用，认识到这一点是重要的（Shintani 等，1995）。此外，星形胶质细胞和神经胶质细胞是具有免疫功能的细胞，在 CNS 内的反应和免疫细胞相似。据报道，细胞因子和趋化因子包括 IL-1（Giulian 等，1986）、IL-6（Frei 等，1989）和 TNF-α（Chung 和 Benveniste，1990）大量分泌与产生，对于自分泌的作用是明显足够的（Aloisi，2001）。促炎细胞因子亦在鱼脑中表达。例如，已证实 IL-1β 在鲤鱼视前核中表达（Metz 等，2006）。在虹鳟（Iliev 等，2007）和金头鲷（Castellana 等，2008）的相关研究中亦曾经报道 IL-1 和 IL-6 的中枢表达。

7.3.5 细胞因子诱导下丘脑-脑垂体-肾间腺轴的活化作用

在哺乳类关于细胞因子影响神经内分泌功能的第一批实例中发现，在初次免疫反应期间，小鼠的皮质酮水平明显增加（Shek 和 Sabiston，1983）。HPA 轴的细胞因子能的活化现在已成为免疫-神经内分泌调节的最稳定建立的类型（Berkenbosch 等，1987；Sapolsky 等，1987；Goshen 和 Yirmiya，2009）。在哺乳类中，IL-1、IL-2、IL-6、IL-10、IFN-β、IFN-γ、LIF、G-CSF 和 TNF-α 等都曾经影响 HPA 轴的活性（Hermus 和 Sweep，1990；Harbuz 等，1992；Mastorakos 等，1994；Shintani 等，1995；Kim 和 Melmed，1999；Smith 等，1999；Zylinska 等，1999；Dunn，2000）。大多数细胞因子是通过刺激下丘脑的 CRF 和 AVP 分泌而使哺乳类的 HPA 轴活化。确实是，IL-1R1 是在哺乳类的 PVN 内表达（Yabunchi 等，1994）。IL-1、IL-2、IL-6 和 IL-10 能刺激脑垂体细胞释放 ACTH（Woloski 等，1985；Prickett 等，2000）。有意思的是，IL-2 甚至比 CRF 更有能力刺激 ACTH 释放（Karanth 和 McCann，1991）。已有足够的证据表明 IL-1、IL-2、IL-6、G-CSF 和 TNF-α 能够离体和在体刺激肾上腺直接释放糖皮质激素（Turnbull 和 Rivier，1999）。

除了它们对 HPA 轴活性的作用之外，细胞因子还对其他几个内分泌过程起重要的作用。例如，已经表明许多细胞因子，包括 IL-1 能够影响脑垂体分泌促性腺激素以及精巢与卵巢的类固醇生成作用（Cannon，1998）。还有，IL-6 能减弱大鼠脑垂体分泌 GH 和 PRL（Tomida 等，2001）。

至今只有两个研究项目是研究细胞因子对鱼类 HPI 轴调节的潜在作用的，而这两项研究都聚焦于 IL-1β。和我们在哺乳类相关研究中所了解的相似，IL-1β 在体和离体都能激活 HPI 轴。对虹鳟注射 IL-1β 后血浆皮质醇水平升高（Holland 等，2002）。相反，对鲤鱼进行离体的灌注试验，IL-1β 并不影响皮质醇的产生，它刺激脑垂体释放 α-MSH 和 β-内啡肽（Metz 等，2006）。通过刺激 CRF 分泌所引起的一个可能的作用，至今尚未进行研究，但发现 IL-1 R1 mRNA 存在于鲤鱼脑的视前区（Metz 等，2006）。此外，如同在哺乳类相关研究中所观察到的，IL-1β 的表达受到 HPI 轴活性的影响。对鲤鱼的研究表明，监禁应激反应使 IL-1β 在视前区的表达上调 2.5 倍（Metz 等，2006）。因此，可以认为下丘脑的 IL-1β 对鱼类起着和哺乳类相似的功能，而这种作用机理在系统进化上是保守的。

7.4 结论和展望

硬骨鱼类神经内分泌和免疫系统之间的交流沟通现在已经明确阐明。尽管许多关键的作用因子目前已能确定，但是精确的调控细节，特别是它们双向的特征，以及所涉及的生理稳态和健康的问题都有待于进一步研究。

（1）对神经内分泌-免疫相互作用的研究还比较粗浅，但对这种相互作用的存在，目前已经普遍认同。

（2）这种沟通交流的生理重要性，从鱼类到人类在进化上是保守的。

（3）与哺乳类一样，应激反应的状态不仅抑制，而且能够增强特异性的免疫功能以确保适应性免疫。

（4）在应激反应轴的激素和 GH/PRL 释放之间，进化上保持的重要平衡对于适应性免疫反应是至关重要的。然而其他通路的潜在影响不应低估，因为直到现在，对于 HPT 和 BPG 轴在所有进化时期的全部种类当中都还没有得到足够的重视。

基因组序列研究的重大进展和检测手段的灵敏性不仅确保了对于发现一系列具有重要作用的调节因子以及它们各自的受体取得丰硕的成果，而且亦揭示了这些因子在不同的器官、组织和细胞中广泛延伸。所以，很明显的是这些因子的多效性和丰余性特征并不限于免疫系统。神经内分泌系统的调控因子亦可能具有多效性和丰余性。这种多效性特征肯定会增加分析解决问题的复杂程度，例如，来源于神经内分泌和免疫的配体必将受到多种来源的多个因子的调控。此外，信号会在不同的场合由不同的受体察觉，因而必须仔细进行检测和分析内分泌和副分泌型式的生理作用。所以，简单的应激反应或者感染模式是不存在的，甚至简单地施用一个因子都会得到多种结果。再者，我们必须注意到 3 亿年前硬骨鱼类曾经进行过额外的基因复制过程，许多复制的基因在进化过程中存活下来，其中多个单对的复制会获得不同的受体结合能力，或甚至完全不同的功能。还有，在鱼类和哺乳类中，我们发现许多剪接变异体和结合蛋白，它们能显著改变配体的生物活性或者配体的浓度，或者能明显地影响到可用受体的数量。

在这个复杂的系统中，免疫系统的复杂性还必须提到不同类型的细胞，它们持续地穿行于全身，在免疫刺激时急剧地发生反应，迅速增殖或者凋亡。在哺乳类和鱼类中，应激反应和感染具有广泛的影响并造成重大的再分配，它们在感染时必然会诱导一次强烈的局部性反应（Dhabhar，2003；Engelsma 等，2003；Huising 等，2003a）。

对于未来，非常重要的是我们不仅要聚焦于配体，受体和选择性受体调控作用的重要性亦不应忽视，因为它们可能成为一种重要因素以决定双向交流沟通的直接结果。存在着受体的多种亚型对于配体的特异性和灵敏性可能是一种完全不同的情况。越来越多的证据表明受体可以受到应激反应和感染的影响。在阐述阿片样物质受体和趋化性受体时，糖皮质激素和阿片样物质受体表达的调控或者异源二聚体化（heterodimerization）就是好的例证。最近已经阐明细胞因子、神经递质和氧自由基能够影响 G-蛋白偶联激酶的反应性。一种激酶的改变可以影响多个受体的灵敏度，因而会引起一个不同的"信号体"（signalosome）。Heijnen（2007）提出一个有意思的清楚表明是存在于哺乳类的假想。这个假想的意思是神经内分泌的或者免疫的配体活性，以及不同激酶的表达水平都参与到信号之中。G-蛋白偶联受体激酶 2（GRK2）在白细胞中是一个关键性激酶，它和重要的神经内分泌配体如 NA、DA、S-HT、阿片样物质和腺苷的受体偶联。受到感染的或者有自身免疫疾病的病人，GRK2 的表达水平会明显发生变化。由于这些神经内分泌配体和细胞因子会大量存在于脑内，它们一定会在这个水平起到明显的作用。Lombardi 等（2002）已证明在脑内，应激反应会引起 GRK2 表达水平下降。

虽然问题的复杂程度不断加深，我们对硬骨鱼类可进行的研究亦正在增加。斑马鱼和河鲀的全基因组已经迅速完成测序，而更多鱼类的全基因组序列研究亦在进行中。充满注释的生物芯片已经具备，能够帮助我们区分参与的各种不同的通路。在哺乳类，自身免疫病或肿瘤的模型已经产生有效的结果。对于鱼类，我们将选择感染的和应激反应的模型以及两者结合的模型。这种结合是非常重要的，因为我们通常所研究的神经内分泌因子对免疫系统的影响都没有免疫刺激的参与。在"正常的"生理状态下没有感染的严重威胁，研究的免疫系统表现低的活性，而它对于预防例如多种杀微生物因子（microbicidal factor）引起的组织损伤是必不可少的。正如曾经讨论过的，神经内分泌因子经常会操纵或者稳定一个免疫反应，但它们自己不会启动细胞的增殖或者增强活性。研究鱼类较之研究人类的好处是人们可以对试验刺激的应激反应或感染模型在不同的时间点从不同的器官检测细胞群体或者组织，这样就不仅能够研究参与的配体和受体，还可以阐述参与的启动子。采用已经可以提供的等基因系（isogenic line），可以使这些研究形式变得更为清晰。

从本文综述可以清楚看到，广泛的基因组研究对免疫系统和内分泌系统的协同进化已经提供并且将继续提供富于创意的结果。此外，在亿万年间进化的明显保守性为它们的生理影响提供证据。

最后，一个完整的蛋白质组学（proteomics）研究计划无疑是必不可少的。对硬骨鱼类的研究，其所提供的同源性蛋白质起着关键性作用。在人类和鱼类的直向同源细胞因子之间，序列的同一性通常是20%和30%。这意味着对抗体、受体或结合蛋白的结合或者亲和力存在着重大的差别。此外，当使用非同源性的蛋白质时，不能排除对这些蛋白质的抗免疫活性存在的危险性。

长期受到忽略的鱼类神经内分泌系统和免疫系统之间的双向交流沟通，现在已经普遍得到认可，而且这种相互作用在进化上的保守特性阐明了它们的生理重要性。然而，最大的挑战将是如何分析在中枢水平发生的作用过程，以及如何确定外周因子直接或间接的影响。直到现在，能够说明有关应激物的免疫冲击引起神经内分泌系统反应的证据还是很少的。这方面研究的实际意义是多方面的，对于富有成效的水产养殖生产实践亦有重大的应用价值。

B. M. V. -Van 凯门纳德

E. H. 斯托尔特

J. R. 梅特兹

M. 查德金斯卡

参考文献

Abe T, Mikekado T, Haga S, Kisara Y, Watanabe K, Kurokawa T, Suzuki T. 2007. Identification, cDNA cloning, and mRNA localization of a zebrafish ortholog of leukemia inhibitory factor. *Comp. Biochem. Physiol.* B147: 38-44.

Acerete L, Balasch J C, Castellana B, Redruello B, Roher N, Canario A V, Planas J V, MacKenzie S, Tort L. 2007. Cloning of the glucocorticoid receptor (GR) in gilthead seabream (*Sparus aurata*). Differential expression of GR and immune genes in gilthead seabream after an immune challenge. *Comp. Biochem. Physiol.* B148: 32-43.

Ader R, Cohen N, Felten D. 1995. Psychoneuroimmunology - Interactions between the nervous system and the immune system. *Lancet*, 345: 99-103.

Ahmed S R. 2000. The immune system as a potential target for environmental estrogens (endocrine disrupters): A new emerging field. *Toxicology*, 150: 191-206.

Aloisi F. 2001. Immune function of microglia. *Glia*, 36: 165-179.

Bagriacik E U, Klein J R. 2000. The thyrotropin (thyroid-stimulating hormone) receptor is expressed on murine dendritic cells and on a subset of CD45RB (high) lymph node T cells: Functional role for thyroid-stimulating hormone during immune activation. *J. Immunol*, 164: 6158-6165.

Ball J N, Hawkins E F. 1976. Adrenocortical (interrenal) responses to hypophysectomy and adenohypophyseal hormones in teleost Poecilia latipinna. *Gen. Comp. Endocrinol*, 28: 59-70.

Balm P H, van Lieshout E, Lokate J, Wendelaar Bonga S E. 1995. Bacterial lipopolysaccharide (LPS) and interleukin-1 (IL-1) exert multiple physiological effects in the tilapia Oreochromis mossambicus (Teleostei). *J. Comp. Physiol.* B165: 85-92.

Banks W A, Kastin A J, Durham D A. 1989. Bidirectional transport of interleukin-1-alpha across the blood-brain barrier. *Brain Research Bulletin*, 23: 433-437.

Baoprasertkul P, Peatman E, Abernathy J, Liu Z J. 2007. Structural characterisation and expression analysis of toll-like receptor 2 gene from catfish. *Fish Shellfish Immunol*, 22: 418-426.

Barrallo A, Gonzalez-Sarmiento R, Porteros A, Garcia-Isidoro M, Rodriguez R E. 1998. Cloning, molecular characterization, and distribution of a gene homologous to delta opioid receptor from zebrafish (Danio rerio). *Biochem. Biophys. Res. Commun*, 245: 544-548.

Barrallo A, Malvar F G, Gonzalez R, Rodriguez R E, Traynor J R. 1998. Cloning and characterization of a delta opioid receptor from zebrafish. *Biochem. Soc. Trans*, 26: S360.

Barrallo A, Gonzalez-Sarmiento R, Alvar F, Rodriguez R E. 2000. ZFOR2, a new opioid receptor-like gene from the teleost zebrafish (Danio rerio). *Brain Res. Mol. Brain Res*, 84: 1-6.

Bayne C J, Levy S. 1991. Modulation of the oxidative burst in trout myeloid cells by adrenocorticotropic hormone and catecholamines: Mechanisms of action. *J. Leukoc. Biol*, 50: 554-560.

Ben-Jonathan N, Mershon J L, Allen D L, Steinmetz R W. 1996. Extrapituitary prolactin: Distribution, regulation, functions, and clinical aspects. *Endocr. Rev*, 17: 639-669.

Berkenbosch F, Vanoers J, Delrey A, Tilders F, Besedovsky H. 1987. Corticotropin-releasing factor producing neurons in the rat activated by interleukin-1. *Science*, 238: 524-526.

Berwaer M, Martial J A, Davis J R. 1994. Characterization of an up-stream promoter directing extrapituitary expression of the human prolactin gene. *Mol. Endocrinol*, 8: 635-642.

Bette M, Kaut O, Schafer M K H, Weihe E. 2003. Constitutive expression of p55TNFR mRNA and mitogen-specific up-regulation of TNF alpha and p75TNFR mRNA in mouse brain. *J. Comp. Neurol*, 465: 417-430.

Bhat R S, Bhaskaran M, Mongia A, Hitosugi N, Singhal P C. 2004. Morphine-induced macrophage apoptosis: Oxidative stress and strategies for modulation. *J. Leu-*

koc. Biol, 75: 1131-1138.

Bird S, Zou J, Savan R, Kono T, Sakai M, Woo J, Secombes C. 2005. Characterisation and expression analysis of an interleukin-6 homologue in the Japanese pufferfish, *Fugu rubripes*. *Dev. Comp. Immunol*, 29: 775-789.

Blalock J E. 1984. The immune system as a sensory organ. *J. Immunol*, 132: 1067-1070.

Blamire A M, Anthony D C, Rajagopalan B, Sibson N R, Perry V H, Styles P. 2000. Interleukin-1 beta-induced changes in blood-brain barrier permeability, apparent diffusion coefficient, and cerebral blood volume in the rat brain: A magnetic resonance study. *J. Neurosci*, 20: 8153-8159.

Bluthé R M, Walter V, Parnet P, Laye S, Lestage J, Verrier D, Poole S, Stenning B E, Kelley K W, Dantzer R. 1994. Lipopolysaccharide induces sickness behavior in rats by a vagal mediated mechanism. *C. R. Acad. Sci. III*, 317: 499-503.

Bouchard B, Ormandy C J, Di Santo J P, Kelly P A. 1999. Immune system development and function in prolactin receptor-deficient mice. *J. Immunol*, 163: 576-582.

Boulay J L, O'Shea J J, Paul W E. 2003. Molecular phylogeny within type I cytokines and their cognate receptors. *Immunity*, 19: 159-163.

Bridgham J T, Carroll S M, Thornton J W. 2006. Evolution of hormone-receptor complexity by molecular exploitation. *Science*, 312: 97-101.

Brogden K A, Guthmiller J M, Salzet M, Zasloff M. 2005. The nervous system and innate immunity: The neuropeptide connection. *Nat. Immunol*, 6: 558-564.

Buonocore F, Mazzini M, Forlenza M, Randelli E, Secombes C J, Zou J, Scapigliati G. 2004. Expression in *Escherichia coli* and purification of sea bass (*Dicentrarchus labrax*) interleukin-1beta, a possible immunoadjuvant in aquaculture. *Mar. Biotechnol*, 6: 53-59.

Cabanac M, Laberge F. 1998. Fever in goldfish is induced by pyrogens but not by handling. *Physiol. Behav*, 63: 377-379.

Cain K D, Jones D R, Raison R L. 2002. Antibody-antigen kinetics following immunization of rainbow trout (*Oncorhynchus mykiss*) with a T-cell dependent antigen. *Dev. Comp. Immunol*, 26: 181-190.

Calduch-Giner J A, Perez-Sanchez J. 1999. Expression of growth hormone gene in the head kidney of gilthead sea bream (*Sparus aurata*). *J. Exp. Zool*, 283: 326-330.

Calduch-Giner J A, Sitja-Bobadilla A, Alvarez-Pellitero P, Perez-Sanchez J. 1995. Evidence for a direct action of GH on haemopoietic cells of a marine fish, the gilthead sea bream (*Sparus aurata*). *J. Endocrinol*, 146: 459-467.

Camp K L, Wolters W R, Rice C D. 2000. Survivability and immune responses after challenge with *Edwardsiella ictaluri* in susceptible and resistant families of channel catfish, *Ictalurus punctatus*. *Fish Shellfish Immunol*, 10: 475-487.

Cannon J G. 1998. Adaptive interactions between cytokines and the hypothalamic-pituitary-gonadal axis. *Ann. N. Y. Acad. Sci*, 856: 234-242.

Cargouet M, Perdiz D, Mouatassim-Souali A, Tamisier-Karolak S, Levi Y. 2004. Assessment of river contamination by estrogenic compounds in Paris area (France). *Sci. Total Environ*, 324: 55-66.

Castellana B, Iliev D B, Sepulcre M P, MacKenzie S, Goetz F W, Mulero V, Planas J V. 2008. Molecular characterization of interleukin-6 in the gilthead seabream (*Sparus aurata*). *Mol. Immunol*, 45: 3363-3370.

Chadzinska M, Plytycz B. 2004. Differential migratory properties of mouse, fish, and frog leukocytes treated with agonists of opioid receptors. *Dev. Comp. Immunol*, 28: 949-958.

Chadzinska M, Jozefowski S, Bigaj J, Plytycz B. 1997. Morphine modulation of thioglycollate-elicited peritoneal inflammation in the goldfish, *Carassius auratus*. *Arch. Immunol. Ther. Exp*, (*Warsz.*) 45: 321-327.

Chadzinska M, Kolaczkowska E, Seljelid R, Plytycz B. 1999. Morphine modulation of peritoneal inflammation in Atlantic salmon and CB6 mice. *J. Leukoc. Biol*, 65: 590-596.

Chadzinska M, Scislowska-Czarnecka A, Plytycz B. 2000. Inhibitory effects of morphine on some inflammation-related parameters in the goldfish *Carassius auratus* L. *Fish Shellfish Immunol*, 10: 531-542.

Chadzinska M, Leon-Kloosterziel K M, Plytycz B, Verburg-van Kemenade B M L. 2008. *In vivo* kinetics of cytokine expression during peritonitis in carp: Evidence for innate and alternative macrophage polarization. *Dev. Comp. Immunol*, 32: 509-518.

Chadzinska M, Hermsen T, Savelkoul H F, Verburg-van Kemenade B M. 2009. Cloning of opioid receptors in common carp (*Cyprinus carpio* L.) and their involvement in regulation of stress and immune response. *Brain Behav. Immun*, 23: 257-266.

Chadzinska M, Savelkoul H F J, Verburg-van Kemenade B M L. 2009. Morphine affects the inflammatory response in carp by impairment of leukocyte migration. *Dev. Comp. Immunol*, 33: 88-96.

Chen W H, Sun L T, Tsai C L, Song Y L, Chang C F. 2002. Cold-stress induced the modulation of catecholamines, cortisol, immunoglobulin M, and leukocyte phagocytosis in tilapia. *Gen. Comp. Endocrinol*, 126: 90-100.

Choi Y, Chuang L F, Lam K M, Kung H F, Wang J M, Osburn B I, Chuang R Y. 1999. Inhibition of chemokine-induced chemotaxis of monkey leukocytes by mu-opioid receptor agonists. *In Vivo*, 13: 389-396.

Chung I Y, Benveniste E N. 1990. Tumor necrosis factor-alpha production by astrocytes- induction by lipopolysaccharide, IFN-gamma, and IL-1-beta. *J. Immunol*, 144: 2999-3007.

Churchill L, Taishi P, Wang M F, Brandt J, Cearley C, Rehman A, Krueger J M. 2006. Brain distribution of cytokine mRNA induced by systemic administration of interleukin-1 beta or tumor necrosis factor alpha. *Brain Res*, 1120: 64-73.

Corbett A D, Henderson G, McKnight A T, Paterson S J. 2006. 75 years of opioid research: the exciting but vain quest for the Holy Grail. *Br. J. Pharmacol*, 147 (Suppl. 1): S153-S162.

Correa S G, Maccioni M, Rivero V E, Iribarren P, Sotomayor C E, Riera C M. 2007. Cytokines and the immune-neuroendocrine network: What did we learn from infection and autoimmunity?. *Cytokine Growth Factor Rev*, 18: 125-134.

Danielson P B, Dores R M. 1999. Molecular evolution of the opioid/orphanin gene family. *Gen. Comp. Endocrinol*, 113: 169-186.

Dantzer R. 2006. Cytokine, sickness behavior, and depression. *Neurol. Clin*, 24: 441-460.

Darlison M G, Greten F R, Harvey R J, Kreienkamp H J, Stuhmer T, Zwiers H, Lederis K, Richter D. 1997. Opioid receptors from a lower vertebrate (*Catostomus commersoni*): Sequence, pharmacology, coupling to a G-protein-gated inward-rectifying potassium channel (GIRK1), and evolution. *Proc. Natl. Acad. Sci. USA*, 94: 8214-8219.

Deane E E, Li J, Woo N Y S. 2001. Hormonal status and phagocytic activity in sea bream infected with vibriosis. *Comp. Biochem. Physiol.* B129: 687-693.

DeVries M E, Kelvin A A, Xu L L, Ran L S, Robinson J, Kelvin D J. 2006. Defining the origins and evolution of the chemokine/chemokine receptor system. *J. Immunol*, 176: 401-415.

Dhabhar F S. 2003. Stress, leukocyte trafficking, and the augmentation of skin immune function. *Ann. N. Y. Acad. Sci*, 992: 205-217.

Dorshkind K, Horseman N D. 2000. The roles of prolactin, growth hormone, insulin-like growth factor-I, and thyroid hormones in lymphocyte development and function: Insights from genetic models of hormone and hormone receptor deficiency. *Endocr. Rev*, 21: 292-312.

Dorshkind K, Horseman N D. 2001. Anterior pituitary hormones, stress, and immune system homeostasis. *Bioessays*, 23: 288-294.

Ducouret B, Tujague M, Ashraf J, Mouchel N, Servel N, Valotaire Y, Thompson E B. 1995. Cloning of a teleost fish glucocorticoid receptor shows that it contains a deoxyribonucleic acid-binding domain different from that of mammals. *Endocrinology*,

136: 3774-3783.

Dugan S G, Lortie M B, Nickerson J G, Moon T W. 2003. Regulation of the rainbow trout (*Oncorhynchus mykiss*) hepatic beta (2) -adrenoceptor by adrenergic agonists. *Comp. Biochem. Physiol. B*136: 331-342.

Dunn A J. 2000. Cytokine activation of the HPA axis. *Ann. N. Y. Acad. Sci*, 917: 608-617.

Edwards K M, Burns V E, Carroll D, Drayson M, Ring C. 2007. The acute stress-induced immunoenhancement hypothesis. *Exerc. Sport Sci. Rev*, 35: 150-155.

Eisenstein T K, Rahim R T, Feng P, Thingalaya N K, Meissler J J. 2006. Effects of opioid tolerance and withdrawal on the immune system. *J. Neuroimmune Pharmacol*, 1: 237-249.

Elenkov I J, Chrousos G P. 2006. Stress system - organization, physiology and immunoregulation. *Neuroimmunomodulation*, 13: 257-267.

Engblom D, Ek M, Saha S, Ericsson-Dahlstrand A, Jakobsson P J, Blomqvist A. 2002. Prostaglandins as inflammatory messengers across the blood-brain barrier. *J. Mol. Med*, 80: 5-15.

Engelsma M Y, Hougee S, Nap D, Hofenk M, Rombout J H, van Muiswinkel W B, Verburg-van Kemenade B M L. 2003. Multiple acute temperature stress affects leucocyte populations and antibody responses in common carp, *Cyprinus carpio* L. *Fish Shellfish Immunol*, 15: 397-410.

Esteban M A, Rodriguez A, Ayala A G, Meseguer J. 2004. Effects of high doses of cortisol on innate cellular immune response of seabream (*Sparus aurata* L.). *Gen. Comp. Endocrinol*, 137: 89-98.

Fabbri E, Chen X, Capuzzo A, Moon T W. 2008. Binding kinetics and sequencing of hepatic alpha (1) -adrenergic receptors in two marine teleosts, mackerel (*Scomber scombrus*) and anchovy (*Engraulis encrasicolus*). *J. Exp. Zool. A*, 309A: 157-165.

Faisal M, Chiappelli F, Ahmed I, Cooper E L, Weiner H. 1989. Social confrontation "stress" in aggressive fish is associated with an endogenous opioid-mediated suppression of proliferative response to mitogens and nonspecific cytotoxicity. *Brain Behav. Immun*, 3: 223-233.

Faisal M, Chiappelli F, II Ahmed, Cooper E L, Weiner H. 1992. The role of endogenous opioids in modulation of immunosuppression in fish. *Schriftenr. Ver. Wasser Boden. Lufthyg*, 89: 785-799.

Fast M D, Hosoya S, Johnson S C, Afonso L O B. 2008. Cortisol response and immune-related effects of Atlantic salmon (*Salmo salar* Linnaeus) subjected to short- and long-term stress. *Fish Shellfish Immunol*, 94: 194-204.

Felten D L, Felten S Y, Carlson S L, Olschowka J A, Livnat S. 1985. Noradrenergic and peptidergic innervation of lymphoid tissue. *J. Immunol*, 135: 755-765s.

Filby A L, Thorpe K L, Tyler C R. 2006. Multiple molecular effect pathways of an environmental oestrogen in fish. *J. Mol. Endocrinol*, 37: 121-134.

Finkenbine S S, Gettys T W, Burnett K G. 2002. Beta-adrenergic receptors on leukocytes of the channel catfish, *Ictalurus punctatus*. *Comp. Biochem. Physiol. C*131: 27-37.

Fischer U, Utke K, Somamoto T, Kollner B, Ototake M, Nakanishi T. 2006. Cytotoxic activities of fish leucocytes. *Fish Shellfish Immunol*, 20: 209-226.

Flik G, Rentier-Delrue F, Wendelaar Bonga S E. 1994. Calcitropic effects of recombinant prolactins in *Oreochromis mossambicus*. *Am. J. Physiol*, 266: R1302R1308.

Flik G, Klaren P H M, Van den Burg E H, Metz J R, Huising M O. 2006. CRF and stress in fish. *Gen. Comp. Endocrinol*, 146: 36-44.

Flory C M. 1989. Autonomic innervation of the spleen of the coho salmon, *Oncorhynchus kisutch*. A histochemical demonstration and preliminary assessment of its immunoregulatory role. *Brain Behav. Immun*, 3: 331-344.

Flory C M. 1990. Phylogeny of neuroimmunoregulation: effects of adrenergic and cholinergic agents on the in vitro antibody response of the rainbow trout, *Onchorynchus mykiss*. *Dev. Comp. Immunol*, 14: 283-294.

Flory C M, Bayne C J. 1991. The influence of adrenergic and cholinergic agents on the chemiluminescent and mitogenic responses of leukocytes from the rainbow trout, *Oncorhynchus mykiss*. *Dev. Comp. Immunol*, 15: 135-142.

Forlenza M, de Carvalho Dias J D, Vesely T, Pokorova D, Savelkoul H F, Wiegertjes G F. 2008. Transcription of signal-3 cytokines, IL-12 and IFN alpha beta, coincides with the timing of CD8 alpha beta up-regulation during viral infection of common carp (*Cyprinus carpio* L.). *Mol. Immunol*, 45: 1531-1547.

Frei K, Malipiero U V, Leist T P, Zinkernagel R M, Schwab M E, Fontana A. 1989. On the cellular source and function of interleukin-6 produced in the central nervous system in viral diseases. *Eur. J. Immunol*, 19: 689-694.

Fu Y K, Arkins S, Fuh G, Cunningham B C, Wells J A, Fong S, Cronin M J, Dantzer R, Kelley K W. 1992. Growth hormone augments superoxide anion secretion of human neutrophils by binding to the prolactin receptor. *J. Clin. Invest*, 89: 451-457.

Garden G A, Moller T. 2006. Microglia biology in health and disease. *J. Neuroimmune Pharmacol*, 1: 127-137.

Gellersen B, Kempf R, Telgmann R, DiMattia G E. 1994. Nonpituitary human prolactin gene transcription is independent of Pit-1 and differentially controlled in lymphocytes and in endometrial stroma. *Mol. Endocrinol*, 8: 356-373.

Gerlo S, Verdood P, Hooghe-Peters E L, Kooijman R. 2005. Modulation of prolactin expression in human T lymphocytes by cytokines. *J. Neuroimmunol*, 162: 190-193.

Gerlo S, Davis J R, Mager D L, Kooijman R. 2006. Prolactin in man: A tale of two promoters. *Bioessays*, 28: 1051-1055.

Geven E J W, Verkaar F, Flik G, Klaren P H M. 2006. Experimental hyperthyroidism and central mediators of stress axis and thyroid axis activity in common carp (*Cyprinus carpio* L.). *J. Mol. Endocrinol*, 37: 443-452.

Geven E J W, Nguyen N K, van den Boogaart M, Spanings F A T, Flik G, Klaren P H M. 2007. Comparative thyroidology: thyroid gland location and iodothyronine dynamics in Mozambique tilapia (*Oreochromis mossambicus* Peters) and common carp (*Cyprinus carpio* L.). *J. Exp. Biol*, 210: 4005-4015.

Giulian D, Baker T J, Shih L C N, Lachman L B. 1986. Interleukin-1 of the central nervous system is produced by ameboid microglia. *J. Exp. Med*, 164: 594-604.

Goehler L E, Gaykema R P A, Nguyen K T, Lee J E, Tilders F J H, Maier S F, Watkins L R. 1999. Interleukin-1 beta in immune cells of the abdominal vagus nerve: A link between the immune and nervous systems?. *J. Neurosci*, 19: 2799-2806.

Gomes I, Filipovska J, Devi L A. 2003. Opioid receptor oligomerization. Detection and functional characterization of interacting receptors. *Methods Mol. Med*, 84: 157-183.

Gomez G D, Balcazar J L. 2008. A review on the interactions between gut microbiota and innate immunity of fish. *FEMS Immunol. Med. Microbiol*, 52: 145-154.

Gonzalez-Nunez V, Gonzalez-Sarmiento R, Rodriguez R E. 2003. Characterization of zebrafish proenkephalin reveals novel opioid sequences. *Brain Res. Mol. Brain Res*, 114: 31-39.

Gonzalez-Nunez V, Gonzalez-Sarmiento R, Rodriguez R E. 2003. Cloning and characterization of a full-length pronociceptin in zebrafish: evidence of the existence of two different nociceptin sequences in the same precursor. *Biochim. Biophys. Acta*, 1629: 114-118.

Gonzalez-Nunez V, Gonzalez-Sarmiento R, Rodriguez R E. 2003. Identification of two proopiomelanocortin genes in zebrafish (*Danio rerio*). *Brain Res. Mol. Brain Res*, 120: 1-8.

Goshen I, Yirmiya R. 2009. Interleukin-1 (IL-1): A central regulator of stress responses. *Frontiers in Neuroendocrinol*, 30: 30-45.

Grether G F, Kasahara S, Kolluru G R, Cooper E L. 2004. Sex-specific effects of carotenoid intake on the immunological response to allografts in guppies (*Poecilia reticu*-

lata). *Proc. Biol. Sci*, 271: 45-49.

Grimm M C, Ben-Baruch A, Taub D D, Howard O M, Resau J H, Wang J M, Ali H, Richardson R, Snyderman R, Oppenheim J J. 1998. Opiates transdeactivate chemokine receptors: delta and mu opiate receptor-mediated heterologous desensitization. *J. Exp. Med*, 188: 317-325.

Grimm M C, Ben-Baruch A, Taub D D, Howard O M, Wang J M, Oppenheim J J. 1998. Opiate inhibition of chemokine-induced chemotaxis. *Ann. N. Y. Acad. Sci*, 840: 9-20.

Haddad G, Hanington P C, Wilson E C, Grayfer L, Belosevic M. 2008. Molecular and functional characterization of goldfish (*Carassius auratus* L.) transforming growth factor-beta. *Dev. Comp. Immunol*, 32: 654-663.

Halabe Bucay A. 2007. Clinical hypothesis: Application of AIDS vaccines together with thyroid hormones to increase their immunogenic effect. *Vaccine*, 25: 6292-6293.

Hansen J D, Strassburger P. 2000. Description of an ectothermic TCR coreceptor, CD8 alpha, in rainbow trout. *J. Immunol*, 164: 3132-3139.

Hansen J D, Landis E D, Phillips R B. 2005. Discovery of a unique Ig heavy-chain isotype (IgT) in rainbow trout: Implications for a distinctive B cell developmental pathway in teleost fish. *Proc. Natl. Acad. Sci. USA*, 102: 6919-6924.

Harbuz M S, Stephanou A, Sarlis N, Lightman S L. 1992. The effects of recombinant human interleukin (IL) -1-alpha, IL-1-beta or IL-6 on hypothalamo-pituitary-adrenal axis activation. *J. Endocrinol*, 133: 349-355.

Harle P, Straub R H. 2006. Leptin is a link between adipose tissue and inflammation. *Ann. N. Y. Acad. Sci*,. 1069: 454-462.

Harris J, Bird D J. 2000. Supernatants from leucocytes treated with melanin-concentrating hormone (MCH) and alpha-melanocyte stimulating hormone (alpha-MSH) have a stimulatory effect on rainbow trout (*Oncorhynchus mykiss*) phagocytes *in vitro*. *Vet. Immunol. Immunopathol*, 76: 117-124.

Haukenes A H, Barton B A. 2004. Characterization of the cortisol response following an acute challenge with lipopolysaccharide in yellow perch and the influence of rearing density. *J. Fish Biol*, 64: 851-862.

Heijnen C J. 2007. Receptor regulation in neuroendocrine-immune communication: Current knowledge and future perspectives. *Brain Behav. Immun*, 21: 1-8.

Heinrich P C, Behrmann I, Haan S, Hermanns H M, Muller-Newen G, Schaper F. 2003. Principles of interleukin (IL) -6-type cytokine signalling and its regulation. *Biochem. J*, 374: 1-20.

Hermus A R M M, Sweep C G J. 1990. Cytokines and the hypothalamic pituitary adrenal axis. *J. Ster. Biochem. Mol. Biol*, 37: 867-871.

Hirono I, Takami M, Miyata M, Miyazaki T, Han H J, Takano T, Endo M, Aoki T. 2004. Characterization of gene structure and expression of two toll-like receptors from Japanese flounder, *Paralichthys olivaceus*. *Immunogenetics*, 56: 38-46.

Holland J W, Pottinger T G, Secombes C J. 2002. Recombinant interleukin-1 beta activates the hypothalamic-pituitary-interrenal axis in rainbow trout, *Oncorhynchus mykiss*. *J. Endocrinol*, 175: 261-267.

Hong S H, Peddie S, Campos-Perez J J, Zou J, Secombes C J. 2003. The effect of intraperitoneally administered recombinant IL-1 beta on immune parameters and resistance to *Aeromonas salmonicida* in the rainbow trout (*Oncorhynchus mykiss*). *Dev. Comp. Immunol*, 27: 801-812.

Horseman N D, Zhao W, Montecino-Rodriguez E, Tanaka M, Nakashima K, Engle S J, Smith F, Markoff E, Dorshkind K. 1997. Defective mammopoiesis, but normal hematopoiesis, in mice with a targeted disruption of the prolactin gene. *EMBO J*, 16: 6926-6935.

Howard J K, Lord G M, Matarese G, Vendetti S, Ghatei M A, Ritter M A, Lechler R I, Bloom S R. 1999. Leptin protects mice from starvation-induced lymphoid atrophy and increases thymic cellularity in ob/ob mice. *J. Clin. Invest*, 104: 1051-1059.

Huising M O, Guichelaar T, Hoek C, Verburg-van Kemenade B M L, Flik G, Savelkoul H F J, Rombout J H W M. 2003. Increased efficacy of immersion vaccination in fish with hyperosmotic pretreatment. *Vaccine*, 21: 4178-4193.

Huising M O, Stet R J M, Kruiswijk C P, Savelkoul H F J, Verburg-van Kemenade B M L. 2003. Response to Shields: Molecular evolution of CXC chemokines and receptors. *Trends Immunol*, 24: 356-357.

Huising M O, Stolte E H, Flik G, Savelkoul H F J, Verburg-van Kemenade B M L. 2003. CXC chemokines and leukocyte chemotaxis in common carp (*Cyprinus carpio* L.). *Dev. Comp. Immunol*, 27: 875-888.

Huising M O, Metz J R, van Schooten C, Taverne-Thiele A J, Hermsen T, Verburg-van Kemenade B M L, Flik G. 2004. Structural characterisation of a cyprinid (*Cyprinus carpio* L.) CRH, CRH-BP and CRH-R1, and the role of these proteins in the acute stress response. *J. Mol. Endocrinol*, 32: 627-648.

Huising M O, Stet R J M, Savelkoul H F J, Verburg-van Kemenade B M L. 2004. The molecular evolution of the interleukin-1 family of cytokines; IL-18 in teleost fish. *Dev. Comp. Immunol*, 28: 395-413.

Huising M O, van der Meulen T, Flik G, Verburg-van Kemenade B M L. 2004. Three novel carp CXC chemokines are expressed early in ontogeny and at nonimmune sites. *Eur. J. Biochem*, 271: 4094-4106.

Huising M O, Kruiswijk C P, van Schijndel J E, Savelkoul H F J, Flik G, Verburg-van Kemenade B M L. 2005. Multiple and highly divergent IL-11 genes in teleost fish. *Immunogenetics*, 57: 432-443.

Huising M O, Geven E J W, Kruiswijk C P, Nabuurs S B, Stolte E H, Spanings F A T, Verburg-van Kemenade B M L, Flik G. 2006. Increased leptin expression in common carp (*Cyprinus carpio*) after food intake but not after fasting or feeding to satiation. *Endocrinology*, 147: 5786-5797.

Huising M O, Kruiswijk C P, Flik G. 2006. Phylogeny and evolution of class-I helical cytokines. *J. Endocrinol*, 189: 1-25.

Huising M O, van Schijndel J E, Kruiswijk C P, Nabuurs S B, Savelkoul H F J, Flik G, Verburg-van Kemenade B M L. 2006. The presence of multiple and differentially regulated interleukin-12p40 genes in bony fishes signifies an expansion of the vertebrate heterodimeric cytokine family. *Mol. Immunol*, 43: 1519-1533.

Hull K L, Harvey S. 1997. Growth hormone: An immune regulator in vertebrates. In: Kawashima S, Kikuyama S, Ed. *Advances in Comparative Endocrinology*. Monduzzi Editore: Bologna, 565-572.

Iliev D B, Castellana B, MacKenzie S, Planas J V, Goetz F W. 2007. Cloning and expression analysis of an IL-6 homolog in rainbow trout (*Oncorhynchus mykiss*). *Mol. Immunol*, 44: 1803-1807.

Jeong J Y, Kwon H B, Ahn J C, Kang D, Kwon S H, Park J A, Kim K W. 2008. Functional and developmental analysis of the blood-brain barrier in zebrafish. *Brain Res. Bull*, 75: 619-628.

Jordan B A, Trapaidze N, Gomes I, Nivarthi R, Devi L A. 2001. Oligomerization of opioid receptors with beta 2-adrenergic receptors: A role in trafficking and mitogen-activated protein kinase activation. *Proc. Natl. Acad. Sci. USA*, 98: 343-348.

Jozefowski S, Plytycz B. 1997. Characterization of opiate binding sites on the goldfish (*Carassius auratus* L.) pronephric leukocytes. *Pol. J. Pharmacol*, 49: 229-237.

Jozefowski S J, Plytycz B. 1998. Characterization of beta-adrenergic receptors in fish and amphibian lymphoid organs. *Dev. Comp. Immunol*, 22: 587-603.

Kaattari S L, Zhang H L, Khor I W, Kaattari I M, Shapiro D A. 2002. Affinity maturation in trout: Clonal dominance of high affinity antibodies late in the immune response. *Dev. Comp. Immunol*, 26: 191-200.

Karanth S, McCann S M. 1991. Anterior pituitary hormone control by interleukin-2. *Proc. Natl. Acad. Sci. USA*, 88: 2961-2965.

Kawashima K, Fujii T. 2003. The lymphocytic cholinergic system and its contribution to the regulation of immune activity. *Life Sci*, 74: 675-696.

Kelley K W, Bluthe R M, Dantzer R, Zhou J H, Shen W H, Johnson R W, Broussard S R. 2003. Cytokine-induced sickness behavior. *Brain Behav. Immun*, 17:

S112 - S118.

Kelley K W, Weigent D A, Kooijman R. 2007. Protein hormones and immunity. *Brain Behav. Immun*, 21: 384-392.

Kieffer B L, Gaveriaux-Ruff C. 2002. Exploring the opioid system by gene knockout. *Prog. Neurobiol*, 66: 285-306.

Kim D S, Melmed S. 1999. Stimulatory effect of leukemia inhibitory factor on ACTH secretion of dispersed rat pituitary cells. *Endocr. Res*, 25: 11-19.

Kolodziej E P, Harter T, Sedlak D L. 2004. Dairy wastewater, aquaculture, and spawning fish as sources of steroid hormones in the aquatic environment. *Environ. Sci. Technol*, 38: 6377-6384.

Konsman J P, Vigues S, Mackerlova L, Bristow A, Blomqvist A. 2004. Rat brain vascular distribution of interleukin-1 type-1 receptor immunoreactivity: Relationship to patterns of inducible cyclooxygenase expression by peripheral inflammatory stimuli. *J. Comp. Neurol*, 472: 113-129.

Kooijman R, Gerlo S, Coppens A, Hooghe-Peters E L. 2000. Growth hormone and prolactin expression in the immune system. *Ann. N. Y. Acad. Sci*, 917: 534-540.

La Cava A, Matarese G. 2004. The weight of leptin in immunity. *Nat. Rev. Immunol*, 4: 371-379.

Lago R, Gómez R, Lago F, Gómez-Reino J, Gualillo O. 2008. Leptin beyond body weight regulation - current concepts concerning its role in immune function and inflammation. *Cell. Immunol*, 252: 139-145.

Laing K J, Secombes C J. 2004. Chemokines. *Dev. Comp. Immunol*, 28: 443-460.

Lam S H, Sin Y M, Gong Z, Lam T J. 2005. Effects of thyroid hormone on the development of immune system in zebrafish. *Gen. Comp. Endocrinol*, 142: 325-335.

Laurent V, Jaubert-Miazza L, Desjardins R, Day R, Lindberg I. 2004. Biosynthesis of proopiomelanocortin-derived peptides in prohormone convertase 2 and 7B2 null mice. *Endocrinology*, 145: 519-528.

Law W Y, Chen W H, Song Y L, Dufour S, Chang C F. 2001. Differential *in vitro* suppressive effects of steroids on leukocyte phagocytosis in two teleosts, tilapia and common carp. *Gen. Comp. Endocrinol*, 121: 163-172.

Lewis R S, Ward A C. 2004. Conservation, duplication and divergence of the zebrafish stat5 genes. *Gene*, 338: 65-74.

Lombardi M S, Kavelaars A, Penela P, Scholtens E J, Roccio M, Schmidt R E, Schedlowski M, Mayor F, Heijnen C J. 2002. Oxidative stress decreases G protein-coupled receptor kinase 2 in lymphocytes via a calpain-dependent mechanism. *Mol. Pharmacol*, 62: 379-388.

Lutfalla G, Crollius H R, Stange-Thomann N, Jaillon O, Mogensen K, Monneron D. 2003. Comparative genomic analysis reveals independent expansion of a lineage-specific gene family in vertebrates: The class II cytokine receptors and their ligands in mammals and fish. *BMC Genomics*, 4: 29.

Ma Q, Jones D, Borghesani P R, Segal R A, Nagasawa T, Kishimoto T, Bronson R T, Springer T A. 1998. Impaired B-lymphopoiesis, myelopoiesis, and derailed cerebellar neuron migration in CXCR4- and SDF-1-deficient mice. *Proc. Natl. Acad. Sci. USA*, 95: 9448-9453.

Magnadottir B. 2006. Innate immunity of fish (overview). *Fish Shellfish Immunol*, 20: 137-151.

Magnadottir B, Lange S, Gudmundsdottir S, Bogwald J, Dalmo R A. 2005. Ontogeny of humoral immune parameters in fish. *Fish Shellfish Immunol*, 19: 429-439.

Malik A A, Radhakrishnan N, Reddy K, Smith A D, Singhal P C. 2002. Morphine-induced macrophage apoptosis modulates migration of macrophages: Use of *in vitro* model of urinary tract infection. *J. Endourol*, 16: 605-610.

Martin-Romero C, Sanchez-Margalet V. 2001. Human leptin activates PI3K and MAPK pathways in human peripheral blood mononuclear cells: Possible role of Sam68. *Cell. Immunol*, 212: 83-91.

Martin S A M, Zou J, Houlihan D F, Secombes C J. 2007. Directional responses following recombinant cytokine stimulation of rainbow trout (*Oncorhynchus mykiss*) RTS-11 macrophage cells as revealed by transcriptome profiling. *BMC Genomics*, 8: 150

Marx J, Hilbig R, Rahmann H. 1984. Endotoxin and prostaglandin E1 fail to induce fever in a teleost fish. *Comp. Biochem. Physiol*. A77: 483-487.

Mastorakos G, Weber J S, Magiakou M A, Gunn H, Chrousos G P. 1994. Hypothalamic pituitary adrenal axis activation and stimulation of systemic vasopressin secretion by recombinant interleukin-6 in humans - potential implications for the syndrome of inappropriate vasopressin secretion. *J. Clin. Endocrinol. Met*, 79: 934-939.

Maule A G, Schreck C B. 1991. Stress and cortisol treatment changed affinity and number of glucocorticoid receptors in leukocytes and gill of coho salmon. *Gen. Comp. Endocrinol*, 84: 83-93.

Maule A G, Schreck C B, Kaattari S L. 1987. Changes in the immune system of coho salmon (*Oncorhynchus kisutch*) during the parr to smolt transformation and after implantation of cortisol. *Can. J. Fish. Aquat. Sci*, 44: 161-166.

Mazon A F, Verburg-van Kemenade B M L, Flik G, Huising M O. 2006. Corticotropin-releasing hormone-receptor 1 (CRH-R1) and CRH-binding protein (CRH-BP) are expressed in the gills and skin of common carp *Cyprinus carpio* L. and respond to acute stress and infection. *J. Exp. Biol*, 209: 510-517.

McCarthy L, Wetzel M, Sliker J K, Eisenstein T K, Rogers T J. 2001. Opioids, opioid receptors, and the immune response. *Drug Alcohol Depend*, 62: 111-123.

Meijer A H, Gabby Krens S F, Medina Rodriguez I A, He S, Bitter W, Ewa Snaar-Jagalska B, Spaink H P. 2004. Expression analysis of the Toll-like receptor and TIR domain adaptor families of zebrafish. *Mol. Immunol*, 40: 773-783.

Mellon R D, Bayer B M. 1998. Evidence for central opioid receptors in the immunomodulatory effects of morphine: Review of potential mechanism (s) of action. *J. Neuroimmunol*, 83: 19-28.

Metz J R, Huising M O, Meek J, Taverne-Thiele A J, Wendelaar Bonga S E, Flik G. 2004. Localization, expression and control of adrenocorticotropic hormone in the nucleus preopticus and pituitary gland of common carp (*Cyprinus carpio* L.). *J. Endocrinol*, 182: 23-31.

Metz J R, Huising M O, Leon K, Verburg-van Kemenade B M L, Flik G. 2006. Central and peripheral interleukin-1beta and interleukin-1 receptor I expression and their role in the acute stress response of common carp, *Cyprinus carpio* L. *J. Endocrinol*, 191: 25-35.

Miyagi T, Chuang L F, Lam K M, Kung H, Wang J M, Osburn B I, Chuang R Y. 2000. Opioids suppress chemokine-mediated migration of monkey neutrophils and monocytes - an instant response. *Immunopharmacology*, 47: 53-62.

Mola L, Gambarelli A, Pederzoli A, Ottaviani E. 2005. ACTH response to LPS in the first stages of development of the fish *Dicentrarchus labrax* L. *Gen. Comp. Endocrinol*, 143: 99-103.

Montgomery D W. 2001. Prolactin production by immune cells. *Lupus*, 10: 665-675.

Mori T, Devlin R H. 1999. Transgene and host growth hormone gene expression in pituitary and nonpituitary tissues of normal and growth hormone transgenic salmon. *Mol. Cell. Endocrinol*, 149: 129-139.

Munoz P, Calduch-Giner J A, Sitja-Bobadilla A, Alvarez-Pellitero P, Perez-Sanchez J. 1998. Modulation of the respiratory burst activity of Mediterranean sea bass (*Dicentrarchus labrax* L.) phagocytes by growth hormone and parasitic status. *Fish Shellfish Immunol*, 8: 25-36.

Nadjar A, Combe C, Laye S, Tridon V, Dantzer R, Amedee T, Parnet P. 2003. Nuclear factor kappa B nuclear translocation as a crucial marker of brain response to interleukin-1. A study in rat and interleukin-1 type I deficient mouse. *J. Neurochem*, 87: 1024-1036.

Nadjar A, Combe C, Busquet P, Dantzer R, Parnet P. 2005: Signaling pathways of interleukin-1 actions in the brain: Anatomical distribution of phospho-ERK1/2 in the brain of rat treated systemically with interleukin-1 beta. *Neuroscience*, 134: 921-932.

Nakanishi T. 1986. Seasonal changes in the humoral immune response and the lymphoid tissues of the marine teleost, *Sebastiscus marmoratus*. *Vet. Immunol. Immunopathol*, 12: 213-221.

Nam B H, Byon J Y, Kim Y O, Park E M, Cho Y C, Cheong J. 2007. Molecular cloning and characterisation of the flounder (*Paralichthys olivaceus*) interleukin-6 gene. *Fish Shellfish Immunol*, 23: 231-236.

Nanaware Y K, Baker B I, Tomlinson M C. 1994. The effect of various stresses, corticosteroids and adrenergic agents on phagocytosis in the rainbow trout, *Oncorhynchus mykiss*. *Fish Physiol. Biochem*, 13: 31-40.

Nance D M, Sanders V M. 2007. Autonomic innervation and regulation of the immune system (1987-2007). *Brain Behav. Immun*, 21: 736-745.

Nascimento D S, do Vale A, Tomas A M, Zou J, Secombes C J, dos Santos N M. 2007. Cloning, promoter analysis and expression in response to bacterial exposure of sea bass (*Dicentrarchus labrax* L.) interleukin-12 p40 and p35 subunits. *Mol. Immunol*, 44: 2277-2291.

Nevid N J, Meier A H. 1995. Timed daily administrations of hormones and antagonists of neuroendocrine receptors alter day-night rhythms of allograft rejection in the gulf killifish, *Fundulus grandis*. *Gen. Comp. Endocrinol*,. 97: 327-339.

Orbach H, Shoenfeld Y. 2007. Hyperprolactinemia and autoimmune diseases. *Autoimmun. Rev*, 6: 537-542.

Ottaviani E, Franchini A, Franceschi C. 1995. Evidence for the presence of immunoreactive POMC-derived peptides and cytokines in the thymus of the goldfish (*Carassius auratus*). *Histochem. J*, 27: 597-601.

Pagniello K B, Bols N C, Lee L E. 2002. Effect of corticosteroids on viability and proliferation of the rainbow trout monocyte/macrophage cell line, RTS11. *Fish Shellfish Immunol*, 13: 199-214.

Palaksha K J, Shin G W, Kim Y R, Jung T S. 2008. Evaluation of non-specific immune components from the skin mucus of olive flounder (*Paralichthys olivaceus*). *Fish Shellfish Immunol*, 24: 479-488.

Patino R, Xia Z F, Gale W L, Wu C F, Maule A G, Chang X T. 2000. Novel transcripts of the estrogen receptor alpha gene in channel catfish. *Gen. Comp. Endocrinol*, 120: 314-325.

Peatman E, Liu Z J. 2007. Evolution of CC chemokines in teleost fish: A case study in gene duplication and implications for immune diversity. *Immunogenetics*, 59: 613-623.

Perez A R, Roggero E, Nicora A, Palazzi J, Besedovsky H O, del Rey A, Bottasso O A. 2007. Thymus atrophy during *Trypanosoma cruzi* infection is caused by an immunoendocrine imbalance. *Brain Behav. Immun*, 21: 890-900.

Peterson P K, Molitor T W, Chao C C. 1998. The opioid-cytokine connection. *J. Neuroimmunol*, 83: 63-69.

Pinal-Seoane N, Martin I R, Gonzalez-Nunez V, de Velasco E M, Alvarez F A, Sarmiento R G, Rodriguez R E. 2006. Characterization of a new duplicate delta-opioid receptor from zebrafish. *J. Mol. Endocrinol*, 37: 391-403.

Pinto R D, Nascimento D S, Reis M I R, do Vale A, dos Santos N M S. 2007. Molecular characterization, 3D modelling and expression analysis of sea bass (*Dicentrarchus labrax* L.) interleukin-10. *Mol. Immunol*, 44: 2056-2065.

Pratt W B, Toft D O. 1997. Steroid receptor interactions with heat shock protein and immunophilin chaperones. *Endocr. Rev*, 18: 306-360.

Prickett T C R, Inder W J, Evans M J, Donald R A. 2000. Interleukin-1 potentiates basal and AVP-stimulated ACTH secretion *in vitro*- The role of CRH pre-incubation. *Horm. Metab. Res*, 32: 350-354.

Prunet P, Sandra O, Le Rouzic P, Marchand O, Laudet V. 2000. Molecular characterization of the prolactin receptor in two fish species, tilapia *Oreochromis niloticus* and rainbow trout, *Oncorhynchus mykiss*: A comparative approach. *Can. J. Physiol. Pharmacol*, 78: 1086-1096.

Przewlocki R, Przewlocka B. 2005. Opioids in neuropathic pain. *Curr. Pharm. Des*, 11: 3013-3025.

Purcell M K, Smith K D, Aderem A, Hood L, Winton J R, Roach J C. 2006. Conservation of Toll-like receptor signaling pathways in teleost fish. *Comp. Biochem. Physiol*. D1: 77-88.

Quan N, Banks W A. 2007. Brain-immune communication pathways. *Brain Behav. Immun*, 21: 727-735.

Quan N, Whiteside M, Herkenham M. 1998. Time course and localization patterns of interleukin-1 beta messenger RNA expression in brain and pituitary after peripheral administration of lipopolysaccharide. *Neuroscience*, 83: 281-293.

Raida M K, Buchmann K. 2007. Temperature-dependent expression of immune-relevant genes in rainbow trout following *Yersinia ruckeri* vaccination. *Dis. Aquat. Organ*, 77: 41-52.

Redelman D, Welniak L A, Taub D, Murphy W J. 2008. Neuroendocrine hormones such as growth hormone and prolactin are integral members of the immunological cytokine network. *Cell. Immunol*, 252: 111-121.

Reynolds W W, Casterlin M E, Covert J B. 1976. Behavioural fever in teleost fishes. *Nature*, 259: 41-42.

Rijkers G T, Frederix-Wolters E M H, Van Muiswinkel W B. 1980. The immune system of cyprinid fish - kinetics and temperature dependence of antibody-producing cells in carp (*Cyprinus carpio*). *Immunology*, 41: 91-97.

Riley L G, Hirano T, Grau E G. 2004. Estradiol-17 beta and dihydrotestosterone differentially regulate vitellogenin and insulin-like growth factor-I production in primary hepatocytes of the tilapia *Oreochromis mossambicus*. *Comp. Biochem. Physiol. C*138: 177-186.

Rivest S, Lacroix S, Vallieres L, Nadeau S, Zhang J, Laflamme N. 2000. How the blood talks to the brain parenchyma and the paraventricular nucleus of the hypothalamus during systemic inflammatory and infectious stimuli. *Proc. Soc. Exp. Biol. Med*, 223: 22-38.

Robertsen B. 2006. The interferon system of teleost fish. *Fish Shellfish Immunol*, 20: 172-191.

Rogers T J, Steele A D, Howard O M, Oppenheim J J. 2000. Bidirectional heterologous desensitization of opioid and chemokine receptors. *Ann. N. Y. Acad. Sci*, 917: 19-28.

Roitt I, Delves P, Martin S, Burton D. 2006. *Roitt's Essential Immunology*. Blackwell Publishing: London.

Roy B, Rai U. 2008. Role of adrenoceptor-coupled second messenger system in sympatho-adrenomedullary modulation of splenic macrophage functions in live fish *Channa punctatus*. *Gen. Comp. Endocrinol*, 155: 298-306.

Saeij J P, Stet R J, de Vries B J, van Muiswinkel W B, Wiegertjes G F. 2003. Molecular and functional characterization of carp TNF: A link between TNF polymorphism and trypanotolerance?. *Dev. Comp. Immunol*, 27: 29-41.

Saeij J P, Verburg-van Kemenade B M L, van Muiswinkel W B, Wiegertjes G F. 2003. Daily handling stress reduces resistance of carp to *Trypanoplasma borreli*: In vitro modulatory effects of cortisol on leukocyte function and apoptosis. *Dev. Comp. Immunol*, 27: 233-245.

Saha N R, Usami T, Suzuki Y. 2003. A double staining flow cytometric assay for the detection of steroid induced apoptotic leucocytes in common carp (*Cyprinus carpio*). *Dev. Comp. Immunol*, 27: 351-363.

Saha N R, Usami T, Suzuki Y. 2004. *In vitro* effects of steroid hormones on IgM-secreting cells and IgM secretion in common carp (*Cyprinus carpio*). *Fish Shellfish Immunol*, 17: 149-158.

Sakai M, Kobayashi M, Kawauchi H. 1995. Enhancement of chemiluminescent responses of phagocytic cells from rainbow trout, *Oncorhynchus mykiss*, by injection of growth hormone. *Fish Shellfish Immunol*, 5: 375-379.

Sakai M, Kobayashi M, Kawauchi H. 1996. *In vitro* activation of fish phagocytic cells by GH, prolactin and somatolactin. *J. Endocrinol*, 151: 113-118.

Sakai M, Yamaguchi T, Watanuki H, Yasuda A, Takahashi A. 2001. Modulation of fish phagocytic cells by N-terminal peptides of proopiomelanocortin (NPP). *J. Exp. Zool*, 290: 341-346.

Salzet M, Tasiemski A, Cooper E. 2006. Innate immunity in lophotrochozoans: The annelids. *Curr. Pharm. Des*, 12: 3043-3050.

Sanchez-Margalet V, Martin-Romero C, Santos-Alvarez J, Goberna R, Najib S, Gonzalez-Yanes C. 2003. Role of leptin as an immunomodulator of blood mononuclear cells: mechanisms of action. *Clin. Exp. Immunol*, 133: 11-19.

Sandra O, Le Rouzic P, Cauty C, Edery M, Prunet P. 2000. Expression of the prolactin receptor (tiPRL-R) gene in tilapia *Oreochromis niloticus*: Tissue distribution and cellular localization in osmoregulatory organs. *J. Mol. Endocrinol*, 24: 215-224.

Sanmartin M L, Parama A, Castro R, Cabaleiro S, Leiro J, Lamas J, Barja J L. 2008. Vaccination of turbot, *Psetta maxima* (L.), against the protozoan parasite *Philasterides dicentrarchi*: Effects on antibody production and protection. *J. Fish Dis*, 31: 135-140.

Santos C R, Ingleton P M, Cavaco J E, Kelly P A, Edery M, Power D M. 2001. Cloning, characterization, and tissue distribution of prolactin receptor in the sea bream (*Sparus aurata*). *Gen. Comp. Endocrinol*, 121: 32-47.

Sapolsky R, Rivier C, Yamamoto G, Plotsky P, Vale W. 1987. Interleukin-1 stimulates the secretion of hypothalamic corticotropin-releasing factor. *Science*, 238: 522-524.

Savan R, Aman A, Nakao M, Watanuki H, Sakai M. 2005. Discovery of a novel immunoglobulin heavy chain gene chimera from common carp (*Cyprinus carpio* L.). *Immunogenetics*, 57: 458-463.

Schwartz M W, Woods S C, Porte D J, Seeley R J, Baskin D G. 2000. Central nervous system control of food intake. *Nature*, 6: 661-671.

Segner H, Eppler E, Reinecke M. 2006. The impact of environmental hormonally active substances on the endocrine and immune systems of fish. In: Reinecke M, Zaccone G, Kapoor B G, Ed. *Fish Endocrinology*. Science Publishers: Enfield, 809-865.

Seppola M, Larsen A N, Steiro K, Robertsen B, Jensen I. 2008. Characterisation and expression analysis of the interleukin genes, IL-1beta, IL-8 and IL-10, in Atlantic cod (*Gadus morhua* L.). *Mol. Immunol*, 45: 887-897.

Sharp B M. 2006. Multiple opioid receptors on immune cells modulate intracellular signaling. *Brain Behav. Immun*, 20: 9-14.

Shavit Y, Lewis J W, Terman G W, Gale R P, Liebeskind J C. 1984. Opioid peptides mediate the suppressive effect of stress on natural killer cell cytotoxicity. *Science*, 223: 188-190.

Shek P N, Sabiston B H. 1983. Neuroendocrine regulation of immune processes - change in circulating corticosterone levels induced by the primary antibody response in mice. *Int. J. Immunopharmacol*, 5: 23-33.

Shintani F, Nakaki T, Kanba S, Kato R, Asai M. 1995. Role of interleukin-1 in stress responses - a putative neurotransmitter. *Mol. Neurobiol*, 10: 47-71.

Singh R, Rai U. 2008. beta-Endorphin regulates diverse functions of splenic phagocytes through different opioid receptors in freshwater fish *Channa punctatus* (Bloch): An *in vitro* study. *Dev. Comp. Immunol*, 32: 330-338.

Slicher A M. 1961. Endocrinological and hematological studies in *Fundulus heteroclitus* (Linn.). *Bull. Bingham Oceanogr. Coll*, 17: 1-55.

Smith E M, Cadet P, Stefano G B, Opp M R, Hughes T K. 1999. IL-10 as a mediator in the HPA axis and brain. *J. Neuroimmunol*, 100: 140-148.

Soares M J. 2004. The prolactin and growth hormone families: Pregnancy-specific hormones/cytokines at the maternal-fetal interface. *Reprod. Biol. Endocrinol*, 2: 51.

Sohm F, Pezet A, Sandra O, Prunet P, de Luze A, Edery M. 1998. Activation of gene transcription by tilapia prolactin variants tiPRL188 and tiPRL177. *FEBS Lett*, 438: 119-123.

Stefano G B, Salzet B, Fricchione G L. 1998. Enkelytin and opioid peptide association in invertebrates and vertebrates: Immune activation and pain. *Immunol. Today*, 19: 265-268.

Stet R J M, Kruiswijk C P, Dixon B. 2003. Major histocompatibility lineages and immune gene function in teleost fishes: The road not taken. *Crit. Rev. Immunol*, 23: 441-471.

Stevens C W. 2004. Opioid research in amphibians: An alternative pain model yielding insights on the evolution of opioid receptors. *Brain Res. Brain Res. Rev*, 46: 204-215.

Stolte E H, Verburg-van Kemenade B M L, Savelkoul H F J, Flik G. 2006. Evolution of glucocorticoid receptors with different glucocorticoid sensitivity. *J. Endocrinol*, 190: 17-28.

Stolte E H, Chadzinska M, Przybylska D, Flik G, Savelkoul H F, Verburg-van Kemenade B M L. 2008. The immune response differentially regulates Hsp70 and glucocorticoid receptor expression *in vitro* and *in vivo* in common carp (*Cyprinus carpio*L.). *Fish Shellfish Immunol*, [Epub ahead of print].

Stolte E H, de Mazon A F, Leon-Kloosterziel K M, Jesiak M, Bury N R, Sturm A,

Savelkoul H F, Verburg-van Kemenade B M L, Flik G. 2008. Corticosteroid receptors involved in stress regulation in common carp, *Cyprinus carpio*. *J. Endocrinol*, 198: 403-417.

Stolte E H, Nabuurs S B, Bury N R, Sturm A, Flik G, Savelkoul H F, Verburg-van Kemenade B M L. 2008. Corticosteroid receptors and pro-inflammatory cytokines. *Mol. Immunol*, 46: 70-79.

Sueiro C, Carrera I, Ferreiro S, Molist P, Adrio F, Anadon R, Rodriguez-Moldes I. 2007. New insights on saccus vasculosus evolution: A developmental and immunohistochemical study in elasmobranchs. *Brain Behav. Evol*, 70: 187-204.

Suetake H, Araki K, Suzuki Y. 2004. Cloning, expression, and characterization of fugu CD4, the first ectothermic animal CD4. *Immunogenetics*, 56: 368-374.

Suzuki S, Chuang L F, Yau P, Doi R H, Chuang R Y. 2002. Interactions of opioid and chemokine receptors: Oligomerization of mu, kappa, and delta with CCR5 on immune cells. *Exp. Cell Res*, 280: 192-200.

Swennen D, Rentier-Delrue F, Auperin B, Prunet P, Flik G, Wendelaar Bonga S E, Lion M, Martial J A. 1991. Production and purification of biologically active recombinant tilapia (*Oreochromis niloticus*) prolactins. *J. Endocrinol*, 131: 219-227.

Tafalla C, Coll J, Secombes C J. 2005. Expression of genes related to the early immune response in rainbow trout (*Oncorhynchus mykiss*) after viral haemorrhagic septicemia virus (VHSV) infection. *Dev. Comp. Immunol*, 29: 615-626.

Terova G, Gornati R, Rimoldi S, Bernardini G, Saroglia M. 2005. Quantification of a glucocorticoid receptor in sea bass (*Dicentrarchus labrax*, L.) reared at high stocking density. *Gene*, 357: 144-151.

Tomida M, Yoshida U, Mogi C, Maruyama M, Goda H, Hatta Y, Inoue K. 2001. Leukaemia inhibitory factor and interleukin-6 inhibit secretion of prolactin and growth hormone by rat pituitary MtT/SM cells. *Cytokine*, 14: 202-207.

Tse D L, Chow B K, Chan C B, Lee L T, Cheng C H. 2000. Molecular cloning and expression studies of a prolactin receptor in goldfish (*Carassius auratus*). *Life Sci*, 66: 593-605.

Turnbull A V, Rivier C L. 1999. Regulation of the hypothalamic-pituitary-adrenal axis by cytokines: actions and mechanisms of action. *Physiol. Rev*, 79: 1-71.

van Muiswinkel W B. 1995. *The Piscine Immune System: Innate and Acquired Immunity* CAB International: Wallingford, UK.

Verburg-van Kemenade B M L, Nowak B, Engelsma M Y, Weyts F A A. 1999. Differential effects of cortisol on apoptosis and proliferation of carp B-lymphocytes from head kidney, spleen and blood. *Fish Shellfish Immunol*, 9: 405-415.

Verburg-van Kemenade B M L, Savelkoul H F, Chadzinska M. 2009. Function for the opioid system during inflammation in carp. *Ann. NY Acad Sci*, in press.

Very N M, Kittilson J D, Norbeck L A, Sheridan M A. 2005. Isolation, characterization, and distribution of two cDNAs encoding for growth hormone receptor in rainbow trout (*Oncorhynchus mykiss*). *Comp. Biochem. Physiol.* B140: 615-628.

Vilcek J. 2003. *The Cytokines*; *An Overview*. Academic Press: London.

Vitkovic L, Konsman J P, Bockaert J, Dantzer R, Homburger V, Jacque C. 2000. Cytokine signals propagate through the brain. *Mol. Psychiatry*, 5: 604-615.

Viveros-Paredes J M, Puebla-Perez A M, Gutierrez-Coronado O, Sandoval-Ramirez L, Villasenor-Garcia M M. 2006. Dysregulation of the Th1/Th2 cytokine profile is associated with immunosuppression induced by hypothalamic-pituitary-adrenal axis activation in mice. *Int. Immunopharmacol*, 6: 774-781.

Vizzini A, Vazzana M, Cammarata M, Parrinello N. 2007. Peritoneal cavity phagocytes from the teleost sea bass express a glucocorticoid receptor (cloned and sequenced) involved in genomic modulation of the *in vitro* chemiluminescence response to zymosan. *Gen. Comp. Endocrinol*, 150: 114-123.

Wang R, Belosevic M. 1994. Estradiol increases susceptibility of goldfish to *Trypanosoma danilewskyi*. *Dev. Comp. Immunol*, 18: 377-387.

Wang R, Belosevic M. 1995. The *in vitro* effects of estradiol and cortisol on the function of a long term goldfish macrophage cell line. *Dev. Comp. Immunol*, 19: 327-336.

Watanuki N, Takahashi A, Yasuda A, Sakai M. 1999. Kidney leucocytes of rainbow trout, *Oncorhynchus mykiss*, are activated by intraperitoneal injection of beta-endorphin. *Vet. Immunol. Immunopathol*, 71: 89-97.

Watanuki H, Gushiken Y, Takahashi A, Yasuda A, Sakai M. 2000. *In vitro* modulation of fish phagocytic cells by beta-endorphin. *Fish Shellfish Immunol*, 10: 203-212.

Watanuki H, Yamaguchi T, Sakai M. 2002. Suppression in function of phagocytic cells in common carp *Cyprinus carpio* L. injected with estradiol, progesterone or 11-ketotestosterone. *Comp. Biochem. Physiol. C. Toxicol. Pharmacol*, 132: 407-413.

Watanuki H, Sakai M, Takahashi A. 2003. Immunomodulatory effects of alpha melanocyte stimulating hormone on common carp (*Cyprinus carpio* L.). *Vet. Immunol. Immunopathol*, 91: 135-140.

Weigent D A, Blalock J E. 1995. Associations between the neuroendocrine and immune systems. *J. Leukoc. Biol*, 58: 137-150.

Weigent D A, Vines C R, Long J C, Blalock J E, Elton T S. 2000. Characterization of the promoter-directing expression of growth hormone in a monocyte cell line. *Neuro-*

immunomodulation, 7: 126-134.

Wendelaar Bonga S E. 1997. The stress response in fish. *Physiol. Rev*, 77: 591-625.

Wendelaar Bonga S E, van der Meij J C, Flik G. 1984. Prolactin and acid stress in the teleost *Oreochromis* (formerly *Sarotherodon*) *mossambicus*. *Gen. Comp. Endocrinol*, 55: 323-332.

Weyts F A, Verburg-van Kemenade B M, Flik G, Lambert J G, Wendelaar Bonga S E. 1997. Conservation of apoptosis as an immune regulatory mechanism: Effects of cortisol and cortisone on carp lymphocytes. *Brain Behav. Immun*, 11: 95-105.

Weyts F A, Flik G, Verburg-van Kemenade B M. 1998. Cortisol inhibits apoptosis in carp neutrophilic granulocytes. *Dev. Comp. Immunol*, 22: 563-572.

Weyts F A A, Flik G, Rombout J H, Verburg-van Kemenade B M L. 1998. Cortisol induces apoptosis in activated B cells, not in other lymphoid cells of the common carp, *Cyprinus carpio*L. *Dev. Comp. Immunol*, 22: 551-562.

Woloski B M M J, Smith E M, Meyer W J, Fuller G M, Blalock J E. 1985. Corticotropin-releasing activity of monokines. *Science*, 230: 1035-1037.

Yabuuchi K, Minami M, Katsumata S, Satoh M. 1994. Localization of type-I interleukin-1 receptor messenger RNA in the rat brain. *Mol. Brain Res*, 27: 27-36.

Yada T. 2007. Growth hormone and fish immune system. *Gen. Comp. Endocrinol*, 152: 353-358.

Yada T, Azuma T. 2002. Hypophysectomy depresses immune functions in rainbow trout. *Comp. Biochem. Physiol. C*131: 93-100.

Yada T, Nagae M, Moriyama S, Azuma T. 1999. Effects of prolactin and growth hormone on plasma immunoglobulin M levels of hypophysectomized rainbow trout, *Oncorhynchus mykiss*. *Gen. Comp. Endocrinol*, 115: 46-52.

Yada T, Azuma T, Hirano T, Grau E G. 2001. Effects of hypophysectomy on immune functions in channel catfish. In: Goos H J T, Rastogi R K, Vaudry H, Pierantoni R, Ed. *Perspective in Comparative Endocrinology: Unity and Diversity*. Monduzzi Editore: Bologna, 369-376.

Yada T, Azuma T, Takagi Y. 2001. Stimulation of non-specific immune functions in seawater-acclimated rainbow trout, *Oncorhynchus mykiss*, with reference to the role of growth hormone. *Comp. Biochem. Physiol. B*129: 695-701.

Yada T, Uchida K, Kajimura S, Azuma T, Hirano T, Grau E G. 2002. Immunomodulatory effects of prolactin and growth hormone in the tilapia, *Oreochromis mossambicus*. *J. Endocrinol*, 173: 483-492.

Yada T, Misumi I, Muto K, Azuma T, Schreck C B. 2004. Effects of prolactin and growth hormone on proliferation and survival of cultured trout leucocytes. *Gen. Comp.*

Endocrinol, 136: 298-306.

Yada T, Muto K, Azuma T, Hyodo S, Schreck C B. 2005. Cortisol stimulates growth hormone gene expression in rainbow trout leucocytes *in vitro*. *Gen. Comp. Endocrinol*, 142: 248-255.

Yada T, Hyodo S, Schreck C B. 2008. Effects of seawater acclimation on mRNA levels of corticosteroid receptor genes in osmoregulatory and immune systems in trout. *Gen. Comp. Endocrinol*, 156: 622-627.

Yamaguchi T, Watanuki H, Sakai M. 2001. Effects of estradiol, progesterone and testosterone on the function of carp, *Cyprinus carpio*, phagocytes *in vitro*. *Comp. Biochem. Physiol. C*129: 49-55.

Yang B Y, Arab M, Chen T T. 1997. Cloning and characterization of rainbow trout (*Oncorhynchus mykiss*) somatolactin cDNA and its expression in pituitary and nonpituitary tissues. *Gen. Comp. Endocrinol*, 106: 271-280.

Yeager M P, Colacchio T A, Yu C T, Hildebrandt L, Howell A L, Weiss J, Guyre P M. 1995. Morphine inhibits spontaneous and cytokine-enhanced natural killer cell cytotoxicity in volunteers. *Anesthesiology*, 83: 500-508.

Zhao Y R, Sun R, You L, Gao C Y, Tian Z G. 2003. Expression of leptin receptors and response to leptin stimulation of human natural killer cell lines. *Biochem. Biophys. Res. Commun*, 300: 247-252.

Zou J, Carrington A, Collet B, Dijkstra J M, Yoshiura Y, Bols N, Secombes C J. 2005. Identification and bioactivities of IFN-gamma in rainbow trout *Oncorhynchus mykiss*: The first Th1-type cytokine characterized functionally in fish. *J. Immunol*, 175: 2484-2494.

Zylinska K, Mucha S, Komorowski J, Korycka A, Pisarek H, Robak T, Stepien H. 1999. Influence of granulocyte-macrophage colony stimulating factor on pituitary-adrenal axis (PAA) in rats *in vivo*. *Pituitary*, 2: 211-216.

第8章 神经内分泌调控液体摄入和液体平衡

鱼类在脊椎动物当中是独特的，它们的体液通过很薄的呼吸上皮直接和环境中的水分接触。为了在不同盐度下保持体液的稳态（包括重量摩尔渗透压浓度和容积），通过食物摄入和渗透压调节器官的活性，以及包括许多来自脑的激素的调控与整合作用，离子及水分的丢失和积累保持平衡。对于海水硬骨鱼类，饮水对体液平衡起着必不可少的作用，因为这是补偿因渗透压失水的唯一途径。虽然一些神经肽类可以在中枢以旁分泌/自分泌的形式起着调节饮水的作用，而外周激素由于结构上不受血脑屏障影响亦起着很大的作用。鱼类调控饮水的作用机理和陆生动物有很大差别，这反映它们的水生生活方式。体液的平衡亦依靠于软骨鱼类内分泌和神经内分泌因子对鳃、消化道、肾脏和直肠腺的运输上皮的调控和整合作用，其中有些因子是由这些外周渗透压调节器官局部分泌和起作用的。

8.1 导　言

8.1.1 液体交换和平衡作用机理

鱼类体液只由一层薄的呼吸上皮和它们周围的介质隔离开。所有生活在淡水（FW）和海水（SW）环境中的主要类群的代表，都有一个从少量到1000 mOsm/kg体重的摩尔渗透压浓度（osmolality）的幅度。相应地，具有调节体液不受外界环境影响的能力对这些动物的生存是必不可少的。大多数种类只局限于一个或另一个介质中（狭盐性），但有少数种类的渗透压调节能力表现出极大的可塑性，能够在这些介质之间洄游（广盐性）。虽然只是少数的类群，并且可能不是鱼类渗透压调节生理学的真正代表，但这些种类都受到学者们的高度关注，因为广盐性（euryhalinity）明显地存在于圆口类（七鳃鳗）、软骨鱼类和硬骨鱼类当中。

对于生活在淡水中的鱼类，主要的挑战是面对持续不断从非常稀薄的介质渗透得到的水分，以及相偶联的主要体液离子如Na^+和Cl^-稳定扩散的丧失而必须保持住细胞外的液体成分和容量（见图8.1）。肾脏的肾小球提供了最重要的途径，它通过血液过滤以排除过多的水分，以及肾小管重吸收离子与其他溶质，从而产生大量稀释的尿液。鱼类膀胱的离子重吸收能力进一步帮助这一生理过程。离子的失去可以通过食物摄入和消化道吸收保持平衡，虽然这亦可以通过鳃瓣上富于线粒体的细胞［MRCs，通常称为氯细胞或电离细胞（ionocyte）］从水介质中主动摄入离子而得到补充（Evans等，2005）。H^+-ATP酶和Na/K-ATP酶的协同作用能促进对离子的主动摄入（Lin和Randall，1995）。

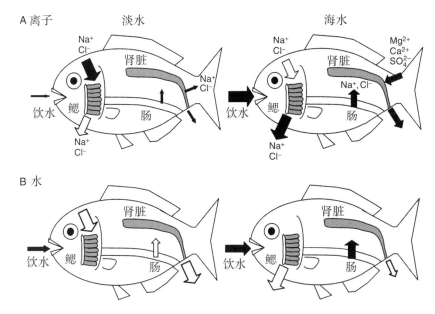

图8.1 在淡水中和海水中饥饿的硬骨鱼类身体和环境之间离子（A）和水分（B）的安排
空箭头表示被动运动，实箭头表示主动运动。

海水圆口类如盲鳗（*Mytine glutinosa*）血浆的重量摩尔渗透压浓度和海水相接近，体液的离子浓度亦和海水相似，所以和介质的水分与离子交换有限，很少进行渗透压调节。这些鱼类和渗透随变的（osmoconforming）海洋无脊椎动物相似，因为它们的体液重量摩尔渗透压浓度和周围介质的浓度一致（Foster 和 Foster，2007）。然而，存在于这些古老种类的肾脏肾小球可以维持身体的容量以及调节一些特异性离子，如 Mg^{2+} 和 SO_4^{2-}。

不同的是，海水硬骨鱼类保持体液的成分和淡水鱼类相似但渗透压要稍高一些，大约是 300～400 mOsm/kg。因此，对周围的海水（大约1000 mOsm/kg），它们显然是低渗透性的，所以就会持续不断地将渗透水分丢失到介质中（见图8.1）。它们通过调节吞饮海水以恢复水分丢失。在食道经过初步脱盐作用之后，余留的盐分随着水分在肠道内主动吸收。很明显，虽然通过吞饮海水恢复了总的液体平衡和体液容量，但亦给鱼体造成很大的盐分负担。许多研究已经证实在皮肤和鳃上皮的富于线粒体细胞起着将过多的 Na^+ 和 Cl^- 离子主动分泌到海水中的作用。这包括三种离子转运蛋白的协同作用：基底外侧位的 Na/K-ATP 酶、Na/K/2 Cl^- 协同转运蛋白（NKCC）和顶端囊性纤维化跨膜传导调节蛋白（CFTR）-Cl^- 型通道(Hiroi 和 McCormick，2007)。由于鱼类和介质之间现有的离子梯度（如海水中 Na^+ 和 Cl^- 的浓度要比硬骨鱼类血液高 2～3 倍）作用，通过扩散作用获得的盐分进一步加剧鱼体需要排出的盐分负担。因此，海水硬骨鱼类体液平衡的作用机理主要是吞饮海水和消化道吸收并偶联鳃的离子主动排出。海水硬骨鱼类的肾脏产生少量尿液以进一步保留体内水分，而尿液中含有许多二价离子，如 Mg^{2+}、SO_4^{2-}，它们是吸收海水后必然积累下来的（Beyenbach，1995）。和淡水鱼类不同，海水鱼类尿液的形成很少通过血液的过滤作用，而主要是由肾小管的分泌作用形成。许多海水硬骨鱼类的肾小球已经明显退化，有些种类如毒棘豹蟾鱼（*Opsanus tau*）的肾小球甚至完全消失（Beyenbach，2004）。

板鳃鱼类主要生活在海水中，其特点是在血浆中积累与保留尿素和三甲胺氧化物（trimethylamine oxide）。这些生理学上的尿毒物和 Na^+ 与 Cl^- 一起使血液浓度高于海水硬骨鱼类，并保持血液重量摩尔渗透压浓度比海水略高，因而，减少了它们对渗透压调节的需求。板鳃鱼类在脱水时亦会调节吞饮海水，虽然基础水平和海水硬骨鱼类相比要低些（Anderson 等，2000c）。在鱼体和周围海水之间的很低的离子梯度，使通过扩散作用获得而又必须主动排出的 Na^+ 和 Cl^- 亦减少很多。有证据表明鳃能分泌 NaCl，但板鳃鱼类还具有一个能进一步分泌盐分的组织——直肠腺，它能够迅速调节而排出浓的 NaCl，特别重要的是在摄食后将盐分清除（Anderson 等，2002a）。对环境盐度的试验性操作表明，当盐度降低时，血液的 Na^+、Cl^- 和尿素水平下调（Hazon 和 Henderson，1984）。保持渗透稳态的盐度阈值在不同种类中有所不同，它反映尿素、Na^+ 和 Cl^- 保留的不同能力（Hazon 等，2003）。全部或者部分在淡水中生活的少数鱼类持续地降低血液中 Na^+、Cl^- 和尿素的浓度，减少不可避免的渗透水分在体内的累积。然而，广盐性的低鳍真鲨（*Carcharhinus leucas*）生活在淡水中仍保持着相当于硬骨鱼类 2 倍的血浆重量摩尔渗透压浓度（650～700 mOsm/kg）（Pillans 和 Franklin，2004）。板鳃鱼类体液容量和盐分与尿素浓度的调控包括对肾排泄作用的调节（肾小球过滤，肾小管分泌和重吸收）、鳃的流出、吞饮和消化道吸收、直肠腺分泌和肝脏尿素的生物合成等作用的整合（Anderson 等，2006）。

鱼类所有的类群对外界盐度变化的顺应都包含鳃和其他渗透压调节上皮通透性的改变，并和特异性运转蛋白（transporter）活性的调节密切相关。越来越多的证据说明密蛋白（claudin）、闭合蛋白（occludin）和水通道蛋白（aquaporins，AQPs）在调节上皮通透性变化中的作用。在海鲈中已经分离和鉴别 cDNA 序列编码的哺乳类 AQP1 和 AQP3 的同源物（Giffard-Mena 等，2007）。变化的盐度影响水通道蛋白在鳃、肾脏和消化道的表达水平。这些学者表明 AQP1 的主要作用是适应于海水鱼类的肾脏和消化道内水分的运输，而 AQP3 参与适应于淡水鱼类鳃的水分运输。硬骨鱼类鳃的 AQP3 是在 MRC_3 中表达，而有些种类则是在其他的上皮细胞中表达。大多数对鱼类离子调节上皮的研究都聚焦于由主动运输引起的离子运动的跨细胞作用机理，而很少注意到由于限制旁细胞的丢失而增强离子保留的作用机理。目前对这种由紧密连接复合体控制的旁细胞通路还了解得很少。紧密连接由与跨膜和胞质的蛋白复合物组成，围绕上皮细胞顶区，在邻近的细胞之间相连接而形成一个半渗透性的旁细胞"封印"以限制溶质运动。已经确认 40 多个紧密连接和相关的蛋白质。密蛋白是紧密连接蛋白质的一个大家族，参与决定离子选择能力和上皮旁细胞通路的通透性。Tipsmark 和他的同事（2008b）在大西洋鲑的鳃表达系列标签（expresscd sequence tag，EST）文库中确定密蛋白家族的 5 个同种型（10e、27a、28a、28b 和 30）。驯化在海水中的鲑鱼是和密蛋白 27a 和 30 的表达减弱相联系的，密蛋白 28a 或 28b 没有明显的影响，而密蛋白 10e

的表达增加 4 倍。这些研究结果和大麻哈鱼一龄降海幼鲑向二龄鲑转变期间对海水的耐受力最强时密蛋白 10e 在鳃的表达处于高峰的观察结果是一致的。这些学者认为密蛋白 10e 对阳离子选择通道是重要的，而密蛋白 27a 和 30 的减少可以改变鳃的通透性以适合于海水中鳃的离子分泌方式。在罗非鱼的鳃中已经确定密蛋白 3-样免疫反应蛋白和密蛋白 4-样免疫反应蛋白（Tipsmark 等，2008a）。当鱼处在海水中时，鳃的密蛋白 3-样蛋白和 4-样蛋白减少，而转入淡水中后，这些蛋白的含量增加。在淡水中驯养期间，罗非鱼密蛋白 28a 和 30 mRNA 的表达亦增强。这些研究结果意味着密蛋白作用于和驯养盐度相关的鳃通透性的变化，可能是在淡水中时鳃形成较深的紧密连接。这就是所料想到的较低的离子通透性对于鱼在淡水中生存是必要的。驯养在淡水中的漠斑牙鲆，其鳃的密蛋白 3-样免疫反应蛋白和 4-样免疫反应蛋白亦比驯养在海水中时更高（Tigsmark 等，2008b）。闭合蛋白是最先确定的跨膜紧密连接蛋白质之一，最近曾对处在缺离子水中的金鱼研究它的动态。在突然处在缺离子的水中后，鳃组织的 Na/K ATP 酶活性和闭合蛋白表达增强。在缺离子水中 14 天和 28 天后，闭合蛋白表达亦提高，表明它在鱼驯养于缺离子的水中起着作用。对于了解鱼类体液和离子平衡所处困境的综合反应复杂性来说，这是一个重要的新兴研究领域。所以，迄今为止对于揭示密蛋白、闭合蛋白或水通道蛋白在鱼类运输上皮表达的内分泌调控取得的研究进展还很少。这将是一个非常富于成果的未来研究领域，而参与的许多因子都已经在下面讨论的特异性转运蛋白活性的调控中确定。

正如前面提到的，渗透压调节的需求（在淡水中血液稀释，在海水中脱水和盐分负荷），使得生理反应极大地受到环境盐度的影响。在淡水中生存的鱼类反应表现得较为一致，而生活在海水中的主要海洋鱼类类群之间表现出不同的适应策略（由低的对高的渗透压调节）。板鳃鱼类的尿毒素和分泌性的直肠腺明显不同于硬骨鱼类，虽然这两个类群都有肾脏的、鳃的和吞饮海水的作用机理。通过这些多样性的作用机理最终达到鱼类和它们的周围介质之间水分和溶质流入与排出的稳态平衡。这种动态的平衡是通过神经的和激素的综合输入，进而对这些作用机理进行整合调控而得以实现。本章中我们将注重内分泌调控的作用。这可以从两个主要的方面来认识：对现有的运输/流出作用机理（急性调控反应）做出迅速的调整，以及对运输／流出作用机理能够进行实质的改变，包括产生新的蛋白质／细胞（驯化反应）。后一种情况可能需要几小时或者几天，并且经常是对外界介质预期的（洄游）或者被迫改变的反应的一部分。要进一步了解这些原理，可以参考 McCormick 和 Bradshaw（2006）的综述。

8.1.2　体液调控的容量对渗透作用机理

对于体液的调控，血量（blood volume）和血浆容量摩尔渗透压浓度是相互紧密联系的，容量和盐分调节的相对重要性因动物的栖息地不同而有所不同。除了鱼类各种类群（圆口类、板鳃鱼类和硬骨鱼类）的多种体液调控作用之外，鱼类在体液调控的主要作用机理方面和陆生动物（哺乳类、鸟类和爬行类）相比是独特的。如上所述，硬骨鱼类在淡水中，经过鳃和大量尿液而失去离子，特别是 Na^+ 和 Cl^-，因为硬骨鱼类和哺乳类与鸟类相似，肾小球重吸收这些离子的能力是有限的（Brown 等，1993）。因此，通过鳃的主动摄取以获得离子的能力，对于硬骨鱼类在淡水中生存来说（如果它们不能从食物中摄入 Na^+ 和 Cl^-）是必不可少的。相反，生活在海水中的硬骨鱼类遇到过多的 Na^+ 和 Cl^-，因为它们可以通过鳃和肠道从摄入的海水中吸收这些离子进入体内，而海水鱼类需要吞饮大量海水以补偿因渗透压而失去的水分。所以，海水硬骨鱼类要在海水中生存就必须把进入体内过多的离子排出体外。由于这些原因，硬骨鱼类要适应在淡水或者海水中的生活，最重要的是离子的调控而不是水分的调控（Me Cormick 和 Bradshaw，2006；Takei，2008）。这和陆地动物不同，对它们来说水分的调控（保留）支配着离子的调控，例如，抗利尿激素（antidiuretic hormone）对体液的调控起着必要的作用，尽管容量的控制是由离子的保留来实现（Takei 等，2007）。这种差别源于鱼类的水栖生活方式，不管外界环境的盐度如何，它们都会接近水分。

虽然鱼类只要能够进行离子调控就能在多种环境盐度中生存，但当它们的体液受到干扰时，它们对保持血量似乎比保持血浆重量摩尔渗透压浓度更为严密（Takei，2000）。当广盐性的鳗鱼由淡水进入海水时，血浆 Na^+ 浓度和重量摩尔渗透压浓度立即增加，并且持续增加一个星期（Takei 等，1998）。进入海水后血量降低，但和通过体重变化（Oide 和 Utida，1968；Kirsch 和 Mayer-Gostan，1973）、血细胞比容的变化（Okawara 等，1987；Takei 等，1998）或染料稀释法（Takei，1988）所测定的重量摩尔渗透压浓度的变化相比，血量的变化要小得多，而且是短暂的。由于高渗性血液（hyperosmoremia），细胞液排出到细胞外的空间，血量的降低能够得到进一步的改善。鳗鱼进入海水后，血浆血管紧张素Ⅱ（angiotensinⅡ，ANG Ⅱ）的浓度只是短暂的微弱增加，使血量略为降低（Okawara 等，1987）。血容量减少（hypovolemia）对硬骨鱼类肾素（renin）的分泌活动是有力的刺激（Nishimura 等，1979）。

Ken Olson 和他的同事们通过对鳟鱼心血管调控作用的广泛研究（Olson，1992），提出鱼类调控的主要靶标是血量而不是盐分平衡的见解。他们在最近的实验中（Olson 和 Hoagland，2008），把鳟鱼分为三组不同的体液平衡，即淡水鱼（容量负荷，盐分减少）、海水鱼（容量减少，盐分负荷）和淡水鱼投喂以高盐分食物（容量负荷，盐分负荷），并测定血量和动脉与静脉血压，结果表明血压是和血量而不是和盐分平衡密切联系，虽然他们没有测定每个鱼群体液的离子浓度。这些结果说明血量是鳟鱼血压的主要决定因子。鳟鱼在每个实验条件下都驯养两个多星期，所以这些结果表明它们在不同的渗透情况下未能保持血量。鳗鱼在淡水和海水中都能保持血量（Nishimara 等，1976）和动脉血压（Nabata 等，2008）恒定，但血浆的重量摩尔渗透压浓度在海水鱼中要比淡水鱼更高些（Tsuchida 和 Takei，1998）。所以，当体液受到冲击时，鳗鱼对血量的调控要比鳟鱼更为严密些。鳟鱼基本上是淡水鱼类，而鳗鱼是海水鱼类（Tsukamoto 等，1998），虽然它们都是洄游性鱼类。在这两种鱼类之间，体液调节得虽小，而明显的差别似乎可能和它们在进化史上的差别有关。

陆生动物在丧失水分后保持血浆 Na^+ 浓度和重量摩尔渗透压浓度比保持血量更为重要，这和硬骨鱼类受到海水冲击的情况相似。当不给狗饮水时，在水分丧失的初期，虽然细胞外液容量降低，血浆重量摩尔渗透压浓度仍能保持（Elkinton 和 Taffel，1942）。因此，对于水分丧失的一系列生理反应是初期 Na^+ 在尿中丢失以保护血浆的张力（tonicity），稍后是对血量的有力防护，先是组织液耗费，接着是细胞液损耗，以保护循环系统不受损伤。缺水初期血浆张力的保护可以归因于增加血浆 Na^+ 浓度，使细胞超极化（hyperpolarize）并且减少神经元和肌肉的兴奋性。这种调控作用的类似型式，即初期保护血浆张力，接着保护容量，亦在鸟类丧失水分后观察到（Takei 等，1988a）。虽然在鱼类和四足类（哺乳类和鸟类）之间血量或重量摩尔渗透压浓度的调控存在明显的差别，但这两组参数是不可分开的，并且表明哺乳类主要是保持总的身体含钠量，从而保持体液容量（Reinhardt 和 Seeliger，2000）。

8.2 液体摄入的调控

动物通过口纳入水分，从食物中得到水分，还得到从碳水化合物、脂类和蛋白质的细胞代谢活动中产生的氧化水分（Schmidt-Nielsen，1997）。鱼生活在水中，它们对口中纳入水分的需求和离开水中生活的陆生动物有很大的不同。各种鱼类对口纳入水分的需求亦有差别，取决于水环境中的盐度，特别是血浆的血量摩尔渗透压浓度大约是海水 1/3 的硬骨鱼类。所以淡水鱼类通常很少饮水以避免体内水分过多（overhydration），因为它们通过渗透作用经过鳃能得到足够的水分，尽管淡水鱼类在幼鱼时期会饮水（Flik 等，2002），特别是因黏附食料而大量摄入水分（Ruohonen 等，1997）。另一方面，海水鱼类必须从口中吞饮水以补偿渗透作用的失水，这种情况和陆生动物相似。广盐性鱼类栖息环境的盐度经常变化，它们必须建立依赖环境盐度而调控饮水的作用机理。因此，它们是已经大量用来研究鱼类调控饮水作用机理的理想模型。

8.2.1 鱼类特有的调控作用机理

虽然海水硬骨鱼类从口中吞饮海水对于在高渗性海水中生存是必需的，如同陆生动物一样，但在这些动物之间，由于不同的栖息环境，调控饮水的作用机理有很大的差别。鱼类特有的作用机理主要是源自它们的水生生活方式。因为水分经常停留在鱼的口腔内，它们不必寻找水分而只要吞入就能饮入水分；而对陆生动物来说，饮入水分是一种专性的（obligatory）行为。此外，鱼类是处在有可能过度饮水的周围水环境中，抑制饮水可能比诱导饮水的作用机理更为重要；这和陆生动物的情况不同，对它们来说，渴感激发寻水行为是必不可少的。此外，在鱼类和陆生动物之间，体液调节的容量和渗透作用机理相对重要性的差别可能会影响到控制液体摄入作用机理出现一些不同之处。

8.2.1.1 反射性饮水与口渴激发的饮水

饮水行为由一系列过程组成而导致液体从环境中摄入到消化道内，由肠吸收而最终变为体液。陆生动物的饮水行为由一系列环节组成，即寻找水源、将水吸入口中、吞入。前两个行为由渴感激发，而吞入是反射性动作，当水分到达咽喉时，会自动地引起吞入（Doty，1968）。有学者对哺乳类渴感引起的作用机理曾进行了长期研究，并总结于一系列专题著作中（Fitzsimons，1979；Rolls 和 Rolls，1982；de Caro 等，1986；Grossman，1990；Ramsay 和 Booth，1991）。很明显，渴感对陆生动物饮水是必要的，但鱼类能够反射性地吞饮海水而无须渴感，因为水分经常存在于口内。所以还不知道鱼类饮水时是否有渴感。因为用导管在食道测定饮水速率时，海水鱼类几乎以相同的速率大量而不断地饮水，水分似乎是自动地吸入而没有调控。此外，致渴激素（dipsogenic hormones）甚至在切除前脑后对硬骨鱼类还是有作用的（Hirano 等，1972；Takei 等，1979），而哺乳类就是通过前脑的整合而引起渴感的（Fitzsimons，1998）。鱼类能够饮水而没有渴感的行为组成是鱼类特有的饮水活动的第一个特点。但是，必须清楚地确定鱼类到底有没有渴的感觉。以陆栖鱼类为例，当它们有渴感时就会进入水中。如果致渴刺激能够使它们进入水中并且饮水，"渴感激发的行为"似乎亦会由于这种刺激而产生。

8.2.1.2 促进饮水与抑制饮水

因为饮水的重要性，陆生动物必须建立促进饮水的作用机理，以保证能在缺水的环境中生存。现已阐明有多种作用机理协同作用以唤起渴感，这些作用机理能够互相补偿，甚至当一种作用机理受到损害时亦能保证饮水（Fitzsimons，1979）。相反，鱼类在任何时候都能摄入水分，它们只需要吞入水分。所以，鱼类经常处在过多饮水的危险之中，这就使得鱼类必须建立较为完善的作用机理以抑制饮水。正如在随后一节中所详细讨论的，这种看法和事实完全一致，即止渴激素（antidipsogenic hormone）在作用和数量上对于硬骨鱼类都要比对陆生动物更为重要，而致渴激素对于哺乳类与鸟类要比对鱼类起着更为明显的作用（Fitzsimons，1998；TaKei，2002；Kozaka 等，2003）。例如，缓激肽（bradykinin）对于哺乳类是致渴的，但对于鱼类来说是止渴的（TaKei 等，2001）。抑制性作用机理的支配作用亦是由于发现鳗鱼在切除前脑后 ANG Ⅱ 促进饮水活动的作用甚至有所加强（Takei 等，1979）。这项研究结果表明来自前脑的一些抑制性信号能够连续地传递到后脑以阻抑吞咽，而前脑切除后就会失去抑制作用，从而增强对致渴激素的敏感性。此外，渗透作用刺激或者细胞脱水对于哺乳类和鸟类是唤起渴感的强有力刺激（Fitzsimons，1979；Kaufman 和 Peters，1980；Takei 等，1988b），而对于硬骨鱼类（Takei 等，1988c）和板鳃鱼类（Anderson 等，2002b）的饮水是一种抑制性的刺激。这些研究结果亦说明鱼类饮水调控的抑制性作用机理占有明显优势。所以，止渴作用机理的优势是鱼类特有的饮水调控作用机理的第二个有意义的特点。

8.2.1.3 作为驱动刺激的渗透性因子与容量因子

和硬骨鱼类血量精确调控相一致的是通过抽取血液引起的容量刺激（血容量减少）能够强有力地促进鳗鱼饮水（Hirano，1974）。血容量减少伴随着血浆中 ANG Ⅱ（一种致渴激素）增加（fitzsimons），它可能在容量刺激诱导饮水增加的过程中起作用。相反，由注射高渗性的 NaCl、甘露醇或蔗糖溶液造成的渗透性刺激（高渗血）能抑制鳗鱼的饮水（Takei 等，1988c）。渗透性刺激后饮水受抑制并不是由经过鳃部流入水分增加引起的血容量过多（hypervolemia）所致，因为在对生活在海水中的鱼类进行的研究中亦观察到饮水受到抑制，而它们在同样的渗透性刺激后并不出现渗透的水分流入。值得注意的是，渗透性刺激后，鳗鱼血浆的 ANG Ⅱ 浓度增加。这和哺乳类和鸟类的情况不同，它们的肾远端小管致密斑（macula densa）的 Na^+ 和 Cl^- 负荷增加，抑制肾素释放而使 ANG Ⅱ 产生（Takei 等，1988b；Komlosi 等，2004）。由于硬骨鱼类肾单位（nephron）缺乏致密斑，这个结果表明渗透性刺激本身对于肾小球旁细胞分泌肾素是刺激性的。总的来说，即使血浆的 ANG Ⅱ 增加，渗透性刺激亦能抑制鳗鱼饮水。

对于板鳃鱼类，由出血引起的血容量减少对于促进饮水是一种有力的刺激，但是注射各种渗压剂（osmolyte）后使血浆重量摩尔渗透压浓度增加能抑制两种鲨鱼——小点猫鲨（*Scyliorhinus canicula*）和皱唇鲨（*Triakis scyllia*）的饮水活动（Anderson 等，2002b）。血容量减少伴随着血浆 ANG Ⅱ 增加，但是，大量注射高渗性 NaCl 溶液引起的高钠血（hypernatremia）并不能使 ANG Ⅱ 增加。高渗性蔗糖溶液能增加血浆的 ANG Ⅱ 浓度，而 NaCl 溶液不能，可能是因为板鳃鱼类肾脏存在着致密斑（Lacy 和 Reale，1990）。七鳃鳗（*Lampetra flaviatilis*）处在不同盐度的介质中，饮水率和血浆的重量摩尔渗透压浓度密切联系（Rankin，2002）。然而，还不能确定是渗透性刺激本身还是由介质重量摩尔渗透压浓度增加而引起的血容量减少起着增加饮水的作用。有必要检测高渗性溶液对七鳃鳗饮水的影响。所以，鱼类特有的饮水控制作用机理的第三个特点是容量比渗透性作用机理起着更大作用；而对于鱼类，比起调控血浆重量摩尔渗透性浓度，当然必须更加精确地调控血量。

确实，和鱼类不同，对于促进哺乳类和鸟类饮水，渗透性的刺激作用要比容量的刺激作用大得多（Fitzsimons，1979），即使这些作用能降低血浆 ANG Ⅱ 的含量。抽取血液或者腹腔注射高胶体渗透压（hyperoncotic）的聚乙二醇引起的血容量减少只能使哺乳类或鸟类少量地饮水，即使是同时增加血浆的 ANG Ⅱ 含量（Fitzsimons，1979；Takei 等，1989）。这些研究结果一致表明，对于陆生动物，比起血量的调控，必须更加精确地调控血浆重量摩尔渗透压浓度。

8.2.2 鱼类饮水的激素调控

由于饮水行为是在脑内整合的，神经作用机理对它的调控起着关键作用。然而，在 20 世纪 60 年代末期就已经证明内分泌系统参与诱导饮水行为，ANG Ⅱ 能诱使在正常水分平衡状态下的大鼠大量饮水（Epstein 等，1970）。这项研究报告揭开了内分泌学新的领域。这个新的研究领域，即"行为内分泌学"，由于它和传统的激素作用相比能创新性地证明激素作用于脑进行行为的调节，因而吸引许多内分泌学的学者们关注。激素调控饮水已经在一些鱼类包括硬骨鱼类（Takei，2002）、板鳃鱼类（Anderson 等，2002c）和圆口类（Rankin，2002）当中进行了研究。

8.2.2.1 诱导饮水的激素

1. 血管紧张素Ⅱ（AngiotensinⅡ，ANGⅡ）

最出名的致渴激素（dipogenic hormone）是 ANGⅡ，它是肾素-血管紧张素系统（RAS）中一个活跃的组分（见图8.2）。ANGⅡ能诱导迄今所研究的所有各纲脊椎动物大量饮水，虽然对生活在不同栖息环境或者对饮水需求不同的种类当中，其诱导的效能有很大差别（Kobayashi 等，1979，1983）。在鱼类中，ANGⅡ使生活在河口咸淡水中的广盐性鱼类增加饮水，当把它们移入高渗性水中时就开始饮水。然而，ANGⅡ并不能诱导一些限定生活在淡水或者海水中的狭盐性鱼类饮水。我们已经知道海水硬骨鱼类能大量地以稳定的速率饮水，表明水分通过松弛的上食道括约肌被动地进入食道。相反，ANGⅡ可能通过协调和吞咽有关的肌肉运动而引起突发性的饮水。所以，ANGⅡ并不参与海水鱼类经常性的饮水活动。事实上，通过灌注 ANGⅡ的抗血清将游离的 ANGⅡ从血浆中清除后并不能降低海水鳗鱼的饮水速率（Takei 和 Tsuchida，2000）。然而，硫甲丙脯酸（captopril）——一种转化酶抑制剂阻抑没有活性的 ANGⅠ转化为有活性的 ANGⅡ，能显著地抑制鳗鱼饮水（Tierney 等，1995）。硫甲丙脯酸的作用可能是非特异性的，因为抑制作用持续到灌注硫甲丙脯酸结束之后，当血浆 ANGⅡ浓度恢复到灌注前水平之上的时候（Takei 和 Tsuchida，2000）。饮水的抑制作用可能是由于经过硫甲丙脯酸处理后，血浆中增加了缓激肽，因为转化酶亦是一种缓激肽主要的降解酶，称为激肽酶Ⅱ（kininaseⅡ）（见图8.2），而缓激肽对鳗鱼是强有力的止渴剂（Takei 等，2001）。

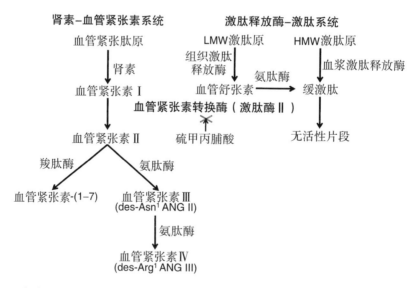

图8.2　肾素-血管紧张素和激肽释放酶-激肽级联以及它们通过血管紧张素-转化酶的关系
硫甲丙脯酸抑制血管紧张素-转化酶将 ANGⅠ转化为 ANGⅡ，但硫甲丙脯酸抑制这种酶亦减少缓激肽分解为没有活性片段的平截（Truncaition）。HMW，高分子质量；LMW，低分子质量。

外源的 ANGⅡ进入鳗鱼（Takei 等，1979）、底鳉（Malvin 等，1980）和川鲽（Carrick 和 Balment，1983）的外周组织中能诱导饮水。这些处理能激活内源的 RAS，如平滑肌松弛的罂粟碱（papaverine），亦使川鲽的饮水速率增加（Balment 和 Carrick，1985）。ANGⅡ是小点猫鲨和皱唇鲨的致渴剂，它激活 RAS 亦使饮水率提高（Anderson 等，2001b）。但是，外源的 ANGⅡ不能诱导洄游的七鳃鳗（*L. fluviatilis*）饮水（Rankin，2002）。在血液循环中增加 ANGⅡ似乎作用于缺乏血脑屏障（BBB）的脑靶标部位以及诱导饮水，例如，在哺乳类（Simpson 和 Routtenberg，1973；Nicolaidis 和 Fitzsimons，1975）和鸟类（Takei，1977）终板（lamina terminalis，OVLT）的穹窿下器（subfornical organ）和/或血管器（organon vasculosum）（见图 8.3）。在哺乳类和鸟类中，这些血管丰富的室周器（circumventricular organ）直接从血液中引起对 ANGⅡ的反应（McKinley 等，2003b），因为切除这些构造就会消除由外周注射 ANGⅡ引起的饮水活动。所以，它们被称为感觉室周器（Johnson 和 Thunhorst，1977）。这些前脑的结构可能传递由 ANGⅡ产生的致渴信息到相关的渴感中心，从而激活哺乳类和鸟类一系列饮水行为。但是，这些前脑的室周器至今未曾在鱼类的相关研究中确定（Mukuda 等，2005）。

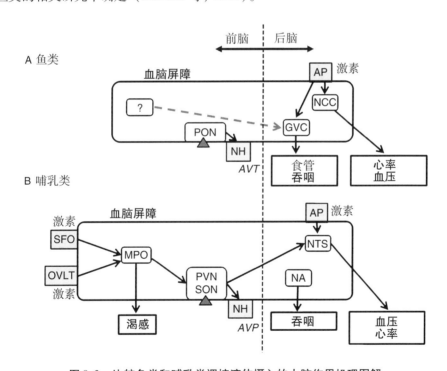

图 8.3　比较鱼类和哺乳类调控液体摄入的大脑作用机理图解

血液循环中的激素作用于血管屏障外的室周器，例如最后区（area postrema，AP）、终极（OVLT）的血管器（organon vasulosum）、穹窿下器（SFO）和神经脑垂体（NH）。产生神经脑垂体激素的视前核（PON）、室旁核（PVN）和视上核（SON）延伸树突到血脑屏障外侧，并从血液中直接接受信息。哺乳类调控渴感的激素，如血管紧张素Ⅱ，在 SFO、OVLT、中间视前核（MPO）和其他脑部位局部合成并调控液体摄入。AVP，精氨酸加压素；AVT，精氨酸催产素；GVC，舌咽-迷走运动复合体（glassopharyngeal-vagal motor complex）；NA，疑核（nucleus ambiguus）；NCC，Cajal 连合核；NTS，孤束核（nucleus tractus solitarius）。

和哺乳类与鸟类不同，硬骨鱼类的前脑似乎并不与 ANG Ⅱ 联系以启动饮水活动（Takei 等，2002）。如前所述，注射 ANG Ⅱ 进入血液循环能诱导鳗鱼大量饮水（Takei 等，1979）。但是，在"切除大脑"（"decerebration"），将整个前脑和大部分中脑切除后，ANG Ⅱ 还一直起作用或甚至有明显作用。"切除大脑"的鳗鱼可能没有渴感，因为渴感是在前脑中包括扁桃体（amygdala）整合的（Fitzsimons，1998）。所以，血液中的 ANG Ⅱ 可能作用于后脑，在延脑水平诱导反射性吞咽活动，饮入环境中的水分（见图 8.3）。事实是 ANG Ⅱ 灌注进入延脑的第四脑室诱导鳗鱼饮水（Kozaka 等，2003）。ANG Ⅱ 最可能作用的部位是最后区（AP），它是在后脑面对第四脑室的另一个室周器官（Mukuda 等，2005）。然而，ANG Ⅱ 确切的作用部位目前我们还不能确定。

除致渴作用外，中枢施给 ANG Ⅱ，主要通过交感神经激活使哺乳类的动脉血压和心率增加（Reid，1992）（见图 8.3）。在硬骨鱼类的相关研究中，Le Mevel 和他的同事曾经表明中枢的 ANG Ⅱ 能使动脉血压和心率增加，而将 ANG Ⅱ 注射到清醒的鳟鱼（*Oncorhynchus mykiss*）第三脑室内使心率的变动率和压力感受器反射（baroreflex）的敏感性降低（Le Mevel 等，1994，2002）。但在鳟鱼的研究中，其对心率的作用要明显强于对动脉血压的作用，表明心搏过速（tachycardia）介导血管升压（vasopressor）的作用（见图 8.3）。局部注射 ANG Ⅱ 到背迷走运动核是特别敏感的（Pamantung 等，1997），而给以阿托品处理（atropinized）的鱼，ANG Ⅱ 的心血管作用消失（Lancien 和 Le Mevel，2007）。所以，对于鳟鱼，ANG Ⅱ 的作用可能是由副交感神经的抑制而不是交感神经的激活所介导。在鳟鱼的相关研究中亦曾报道精氨酸催产素、硬骨鱼紧张肽 Ⅱ 和其他激素对中枢心血管的作用（Le Mevel 等，2008a、b），这将在下文叙述。

2. 肾上腺髓质素（Adrenomedulins，AMs）

AM 最先从肾上腺髓质来源的嗜铬细胞瘤（pheochromocytoma）细胞中分离出来（Kitamura 等，1993）。AM 起先被认为是降钙素基因相关肽（calcitonin gene-related peptide，CGRP）家族的成员，这个家族由 CGRP、AM 和支链淀粉（amylin）组成。然而，在硬骨鱼类中有 5 个 AMs 类型而成为独立的亚家族，命名为 AM1—AM5（Ogoshi 等，2003）。现在已经阐明哺乳类 AM 是硬骨鱼类 AM1 的直向同源物，而 AM1/4 和 AM2/3 只是在硬骨鱼类谱系发生的整个基因复制过程中产生的（Ogoshi 等，2006）。基于这些发现，AM2 和 AM5 最近已在哺乳类中确定（Takei 等，2004a，2008）。AM1 和降钙素受体类受体（calcitonin receptor-like receptor，CLR）结合，和硬骨鱼类与哺乳类的受体活性-修饰蛋白（receptor activity-modifying protein，RAMP）发生联系（Hay 等，2004；Nag 等，2006），与 CGRP 一样。但是，新的对 AM2 和/或 AM5 的特异性受体可能存在，因为 AM2 和 AM5 对中枢的作用要比 AM1 强得多，虽然它们对已知 AM 受体的亲和力低；还因为它们的作用只是部分地为大鼠 AM 受体特异性的拮抗物所阻抑（Hashimoto 等，2007b）。AM 基因在各种不同的组织中表达，所以它在哺乳类基本上是一个局部的旁分泌因子（Lopez 和 Martínezz，2001）。AM 的免疫原性（immunogenicity）特别低，因而难以获得良好的抗体；但是，在健康的志愿者中曾测定 AM 的血浆浓度

[（3.3±0.4）fmol/mL，$n=8$）]，而在高血压严重性不同的高血压病人中，AM 水平增加（Kitamura 等，1994）。血浆 AM 的来源目前还不清楚。AM 基因在河鲀（Ogoshi 等，2003）和鳗鱼（Nobata 等，2008）的各种不同组织中表达，除了 AM2 和 AM3 在河鲀中的表达之外，它们都特别地在脑中表达。

在动脉内大量注射 AM2 和 AM5 能诱导鳗鱼出现低血压（Nobata 等，2008）和大量饮水（Ogoshi 等，2008）。动脉血压在 1 nmol/kg 时降低超过 50%，这是在鱼类研究中所知道的最有效的降压剂。虽然血管降压物对于鱼类通常是致渴的（Hirano 和 Hasegawa，1984），AM 诱导饮水并不是由于低血压，因为以低而非压制的 AM2 和 AM5 剂量进行缓慢的灌注亦会致渴。AM2 的致渴作用是剂量依赖的，甚至比 ANGⅡ 的作用更为明显。但是，脑室内注射 AM2 和 AM5 并不会引起鳗鱼饮水，而 ANGⅡ 通过长期植入的导管注入同样的部位是致渴的。另外，在脑室内同一个部位注射 AM2 使动脉血压升高，和 ANGⅡ 相似，表明 AM2 对于致渴和血压升高的效应是在脑的不同部位起作用的。外周给以 AMs 应该作用于缺乏血脑屏障的脑靶标部位，如位于后脑的 AP。AMs 注射到脑室内亦作用于室周器的结构。由于 AMs 分子在相对分子质量方面比 ANGⅡ 大 5 倍，所以 AM 注射到脑室后可能不能通过脑脊液-脑屏障到达靶神经元（McKinley 等，2003b）。但亦有可能在外周和中枢施用的 AM 和 ANGⅡ 是在不同的靶部位起作用。

8.2.2.2 抑制饮水的激素

1. 利尿钠肽（NPs）

在硬骨鱼类中，NP 家族由 7 个成员组成：心房 NP、B-型 NP、心室 NP（ANP、BNP 和 VNP，主要在心脏合成），以及 4 个 C-型 NPs（CNP1、CNP2、CNP3、CNP4），它们共同的结构特征是缺少一个从分子内环延伸的 C 端"尾"序列，并且主要在脑内大量合成（Inoue 等，2003）。这 7 个成员都已经在软骨鱼类和一些原始硬骨鱼类如鳗鱼和鲑鳟鱼类中确定，但 ANP 和/或 VNP 在硬骨鱼类进化的过程中已经在一些鱼类中变得沉默了，这表明 BNP 是脊椎动物一个基本的心利尿钠肽（Inoue 等，2005）。

ANP 和 BNP 对鳗鱼是强有力的止渴激素（Tsukada 和 Takei 等，2006）。BNP 最近才在鳗鱼中确定，所以尚未检测它对饮水的作用。当灌注到海水鳗鱼的血液循环中时，ANP 能剂量依赖性地在 0.3～3 pmol/（kg·min）范围内降低饮水速率，而大量饮水的海水鳗鱼在 3 pmol/（kg·min）时几乎停止饮水。这个剂量还是非低血压的，可能是和生理学相关的作用（Tsuchida 和 Takei，1998）。事实上，渗透性刺激能使血浆内源的 ANP 浓度增加到灌注 3 pmol/（kg·min）期间的水平（Kaiya 和 Takei，1996b）。ANP 是哺乳类有效的利尿钠肽激素，对狗来说其排钠利尿作用接近 50 pmol/（kg·min），同时出现明显的低血压（Seymour 等，1986）。ANP 亦是哺乳类的止渴剂（Antunes-Rodrigues 等，1985），但是脑腔注射时，ANP 的最低有效剂量要比 ANGⅡ 引起致渴作用的剂量大得多。对比之下，ANP 的止渴作用要比鳗鱼在血液循环中注射 ANGⅡ 的致渴作用强两个数量级（Tsuchida 和 Takei，1998）。对板鳃鱼类亦是如此，它们只有 CNP 这种心激素，CNP 抑制饮水的作用要比 ANGⅡ 刺激饮水的作用强 50 倍（Anderson 等，

2001c）。鱼类和哺乳类致渴与止渴激素效能的差别可能和前面叙述的水栖鱼类对饮水抑制性作用机理的相对重要性有关。灌注 ANP 亦能剂量依赖地使海水鳗鱼血浆 ANG Ⅱ 浓度降低，并能增强止渴作用（Tsuchida 和 Takei，1998）。

ANP 的止渴作用对于适应海水生活是有利的（Takei 和 Hirose，2002）。如前所述，对于在高渗性海水环境中硬骨鱼类的生存来说，饮入海水是必不可少的。然而，鱼处在海水中时，它们很容易因过量饮水而导致高钠血。将鳗鱼由淡水移入海水中，由于对海水中高 Cl^- 浓度的反应，它在 1 min 内就会大量饮水（Hirano，1997；Takei 等，1998）。这种急性与过量的摄入水分在 15 min 后停止，而阻抑饮水于较低水平持续一两个小时，然后保持稳定的海水饮水率。这种暂时的抑制作用可能由一种抑制性信号引起，例如，在开始大量饮水后的胃膨胀（Hirano，1974），但这种抑制作用的时间进程和转移到海水后血浆 ANP 浓度的暂时增加完全一致（Kaiya 和 Takei，1996a）。所以进入海水后，ANP 能限制过量饮水，以改善血浆 Na^+ 浓度突然增加的状况，促进对海水的适应能力（Takei 和 Hirose，2002）。适应于海水和适应于淡水的鳗鱼之间血浆 ANP 浓度并没有差别（Kaiya 和 Takei，1996a）。然而，ANP 对适应海水生活的鳗鱼似乎能慢性地抑制过度饮水以保持血浆 Na^+ 浓度在一个比淡水稍高的水平，因为灌注 ANP 的抗血清以清除血液组织中的 ANP 后，能使海水鳗鱼饮水增加和血浆 Na^+ 浓度升高（Tsukada 和 Takei，2006）。

外周施用 ANP 最有可能作用于脑内缺乏血脑屏障（BBB）的构造，并减少渴感或抑制反射性吞咽活动。基于下列的证据，最有可能作用的部位是 AP：①将 ANP 注射到第四脑室靠近 AP 的部位能抑制海水鳗鱼饮水（Kozaka 等，2003）；②将伊文思蓝（Evnas blue）注射到血液循环中使 AP 和一些室周器结构染色，而其他的脑实质部分都没有染上色（Makuda，2005）；③ANP 受体（A-型利尿钠肽受体）采用免疫组织化学技术定位于 AP（Tsukuda 等，2007）；④对 AP 施行热凝固（heat coagulation）或者用 kinic 酸局部损伤神经元能明显减弱 ANP 抑制海水鳗鱼饮水的作用（Tsukada 等，2007）。然而，将 ANP 直接注射到脑室内和我们实验的鱼体外周施用相比较，是很少起止渴作用的，尽管脑室内注射 ANG Ⅱ 的效应要比外周注射 ANG Ⅱ 的效应大得多（Nobata S. 和 Takei Y. 未发表结果）。这可能是由于 ANP 较大的相对分子质量，和 ANG Ⅱ 相比，较难穿过室管膜屏障（ependymal barrier）（MeKinley，2003b）。

如同在第 8.1.2 节中所讲述的，在鳟鱼和鳗鱼之间容量对渗透调节具有显著意义的差别，对 ANP 的作用来说亦是这样。鳟鱼 ANP 的分泌活动主要由血量增加来调控，而 ANP 的作用是减少血量，如同在哺乳类中观察到的（Johnson 和 Olson，2008）。容量减少导致血压降低和心脏保护（cardioprotection）。但是，对于鳗鱼，重量摩尔渗透压浓度是 ANP 分泌的主要刺激因素，而 ANP 的作用是降低体内的钠和氯化物，从而有助于适应海水生活（Takei 和 Hirose，2002）。确实，ANP 的在体分泌发生在移入海水生活后不久，甚至是在血量降低时，而离体的分泌是分离出来的鳗鱼心房在受到介质重量摩尔渗透压浓度增加的刺激时发生的（Kaiya H. 和 Takei Y.，未发表结果）。虽然在鳟鱼的研究中曾经采用异源的 ANP 和放射免疫测定技术，但 ANP 的作用对鳟鱼容量的调控和对鳗鱼盐分的调控来说都是相对比较强的。这再次说明硬骨鱼类主要调控系统的多样性。

2. 生长素释放肽（ghrelin）

生长素释放肽最先是从胃里分离出来的，作为生长激素促分泌受体的配体（Kojima 等，1999）。生长素释放肽是一个直线肽，由 19～28 个氨基酸残基组成，在第三个丝氨酸或苏氨酸残基加上不同的中链脂肪酸如辛酰、葵酸或葵烯酸（Kaiya 等，2008）。生长素释放肽已经在 2 种板鳃鱼类（Kawakoshi 等，2007）和 11 种硬骨鱼类中鉴定，并且在虹鳟和斑点鲴中鉴定出两个不同的基因（Kaiya 等，2008）。在外周施用生长素释放肽能刺激生长激素释放，但更为广泛地引起学者们关注的是它在大鼠的脑内起着强有力的食欲激素作用（Nakazato 等，2001）。

在体液调控作用方面，将生长素释放肽注射到鳗鱼的脑和外周后出现强有力的止渴作用，这是在脊椎动物中首次进行的这类研究（Kozaka 等，2003）。基于这些发现，随后在大鼠（Hashimoto 等，2007a）和鸡（Tachibana，2006）的相关研究中都发现生长素释放肽是一个强有力的止渴激素。对于大鼠，生长素释放肽的止渴作用和增强食欲的作用一样强，所以，它甚至能在食物摄入增加的同时抑制饮水。一般认为动物取食时通常会饮水，即所谓的餐中饮水（Fitzsimons，1979）。迄今在所有检测过的种类包括硬骨鱼类、哺乳类和鸟类中都曾观察到生长素释放肽的止渴作用。然而，尽管生长素释放肽对哺乳类和硬骨鱼类是增强食欲的（Unniappan 等，2004；Riley 等，2005；Matsuda 等，2006），但对鸡来说是厌食性的（anorectic）（Saito 等，2005）。

当注射到适应海水生活的鳗鱼第四脑室内时，生长素释放肽的作用甚至比 ANP 更强（Kozaka 等，2003）。然而，注射或者灌注到血液循环中，这两种肽的止渴作用是相似的（Kaiya H.，Nobata S. 和 Takei Y.，未发表结果）。ANP 似乎是作用于 AP 以抑制吞咽活动（见前述；Tsukada 等，2007），但生长素释放肽在脑的作用部位完全不清楚。这两种肽在中枢和外周注射后相对效能的差别可能是从脑室易于进入实质神经元的差别导致的。鳗鱼生长素释放肽（21 个氨基酸）和鳗鱼 ANP（27 个氨基酸）具有相近的相对分子质量，但生长素释放肽的脂肪酸酰化，可能会影响其穿过心室表面管室膜层的能力。生长素释放肽从胃里分泌出来之后的作用部位应该是血脑屏障功能不全的室周器，例如 AP。所以，有必要阐明 AP 是否对在血液中产生的各种激素起着窗口的作用，并且成为在脑内局部合成的激素的靶标而整合鱼类饮水的调控作用。

对许多脊椎动物，包括鱼类，曾经着重研究生长素释放肽的食欲作用和刺激生长激素分泌的作用（Kaiya 等，2008）。因而测定了和禁食与能量代谢相关的血浆生长素释放肽浓度的变化。生长素释放肽有两种类型存在于血浆中，即酰化的有活性类型和非酰化的无活性类型。使用直接从 N-端得到的抗血清以放射免疫技术测定有活性的类型，而使用直接从 C-端得到的抗血清可以测定有活性的和无活性的类型（Hosoda 等，2000）。对大鼠的研究表明，血浆中酰化的有活性的生长素释放肽的浓度是 5～80 fmol/mL，大约低于生长素释放肽总浓度的 10 倍。对鳗鱼的研究表明，生长素释放肽有活性类型的血浆浓度是 50 fmol/mL，将鳗鱼从淡水移入海水后 6 h 该浓度升高（Kaiya 等。2006）。此外，同源性生长素释放肽在离体能增强罗非鱼生长激素的分泌活动并刺激生长激素/胰岛素样生长因子-I（IGF-I）轴（Fox 等，2007）。生长激素/IGF-I 轴参与对海水生活的适应（Sakamoto 和 McCormick，2006）。和很强的止渴作用一起，可以清楚地看到生长素释放肽参与了硬骨鱼类相当广泛的体液调控作用。

3. 缓激肽（Bradykinin）

缓激肽，一个直链的九肽激素，是激肽释放酶-激肽系统（kallikrein-kinin system，KKS）最后的有活性的产物，具有强有力的心血管和炎症作用（Regoli 和 Barabe，1980）。存在着两种 KKSs，即血浆的 KKS 和组织中的 KKS。血浆的激肽释放酶和高分子量的激肽原（kininogen）作用产生缓激肽，而组织中的激肽释放酶和低分子量的激肽原作用而产生胰激肽（kallidin），[赖^0]-缓激肽（见图 8.3）。缓激肽已经在从硬骨鱼类到哺乳类的各个类群脊椎动物中确定，但不存在于软骨鱼类和圆口类。在鱼类的相关研究中，用猪组织的激肽释放酶和鱼的血浆一起孵育通常产生胰激肽，它已经在辐鳍鱼类的所有类群（软骨硬鳞类、全骨鱼类、硬骨鱼类）和叶鳍鱼类（肺鱼）中测序（Conlon，1999）。还不知道缓激肽是否存在于鱼类的血液循环中，因为它的半寿期非常短。胰激肽([赖^0]-或[精^0]-缓激肽) 在鳕鱼（Platzack 和 Conlon，1997）、鲟鱼（Li 等，1998）、虹鳟（Olson 等，1997）、鳗鱼（Takei 等，2001）和肺鱼（Balment 等，2002）中显示强的心血管作用，而且胰激肽在鳗鱼中的作用要比同源的缓激肽强些。缓激肽受体（B2-类受体）已经在斑马鱼中克隆（Duner 等，2002）。

KKS 和 RAS 关系密切，因为这两个系统共用血管紧张素转化酶（angiotensin-converting enzyme，ACE），对系统分别进行激活或者失活（Skidgel 和 Erdos，2004）。ACE 亦称为激肽酶 II（kininase II），并能降解缓激肽（见图 8.3）。所以用硫甲丙脯酸（captopril）抑制 ACE 就会同时抑制缓激肽的降解。同源的 [精^0]-缓激肽给鳗鱼大量注射或者以不影响动脉血压的速率进行灌注时,是强有力的止渴激素（Takei 等，2001）。[精^0]-缓激肽注射到血液循环中起止渴作用甚至伴随着增加血浆中 ANG II 的浓度。血浆 ANG II 的增加可能是注射缓激肽后使 ACE 活性增强所引起。在止渴作用方面，[精^0]-缓激肽要比缓激肽或[精^0]-des-精^9-缓激肽强得多。还不清楚这些止渴作用是缓激肽直接的作用还是通过增强其他的止渴激素如 ANP 和生长素释放肽而介导的。所以，脑腔内注射同源的 [精^0]-缓激肽进入脑内对解决这个问题以及确定其作用的靶位置是必要的。由于缓激肽能诱导大鼠饮水（Fregly 和 Rowland，1991），它对鳗鱼的止渴作用亦支持这个观点；在水里的鱼类，抑制性的作用机理对饮水是起主导作用的。

8.2.2.3 调节饮水的其他激素

表 8.1 列出了对鱼类饮水起调节作用的激素。通常认为低血压的物质是致渴的，而高血压的物质是止渴的（Hirano 和 Hasegawa，1984）。这个观点的事实根据是血管加压剂 α-肾上腺素能同等物、肾上腺素或去甲肾上腺素、精氨酸催产素、催产素和尾垂体提取物（可能是硬骨鱼尾紧张肽 II）对鳗鱼都是止渴的，而血管减压剂 β-肾上激素

能同等物、异丙基肾上腺素（isoproterenol）、组胺（histamine）和乙酰胆碱对鳗鱼都是致渴的。在上述的激素当中，这个观点适用于 AM 和缓激肽，但 ANG Ⅱ 和 ANP 是例外；对鳗鱼，ANG Ⅱ 是血管加压的，但是致渴的，而 ANP 是血管减压的，但是止渴的。此外，哺乳类的 P 物质和 5-羟色胺使适应海水生活的鳗鱼增加水分摄入，而血管活性肠肽（vasoactive intestinal peptide）、鳗鱼肠五肽（intestinal pentapeptide）和缩胆囊肽抑制饮水（Ando 等，2000）。还有其他一些激素能略微影响鳗鱼的饮水率（见表 8.1），但还没有检测这些激素对心血管的作用。

表 8.1　至今检测的调控鱼类和哺乳类饮水的激素

鱼类		哺乳类	
刺激性的	抑制性的	刺激性的	抑制性的
血管紧张素 Ⅱ	去甲肾上腺素	血管紧张素 Ⅱ	心房钠尿肽
肾上腺髓质素 2	心房钠尿肽	去甲肾上腺素	肾上腺髓质素
P 物质	生长素释放肽	缓激肽	生长素释放肽
催乳激素（静脉内）	缓激肽	速激肽	速激肽（大鼠）
	精氨酸催产素	血管加压素	类胰高血糖素
	血管活性肠肽	PACAP	类铃蟾肽（大鼠）
	催乳激素（脑腔内）	催乳激素（脑腔内）	内皮缩血管肽
	鳗鱼肠肽	类铃蟾肽	阿片样物质（大鼠）
	硬骨鱼催产素	松弛素	
	缩胆囊肽	神经降压肽	
		神经肽 Y	
		阿片样物质	
		胰岛素	

原参考文献，见 Fitzsimons（1998），Ando 等（2000）和 Kozaka 等（2003）。
有些激素外周注射或中枢注射，或者在不同种类当中会出现相对立的功能。
icv，脑腔内；iv，静脉内；PACAP，脑垂体腺苷酸环化酶激活多肽。

当对海水鳗鱼进行中枢注射时，除 ANG Ⅱ 之外，乙酰胆碱、异丙基肾上腺素和 P 物质亦能增加饮水，而除 ANP 和生长素释放肽之外，5-羟色胺、α-氨基丁酸（GABA）、催乳激素、精氨酸催产素、血管活性肠肽和去甲肾上腺素等都能抑制饮水（Kozaka 等，2003）。还不清楚脑腔注射这些物质之后如何影响动脉血压，但脑腔注射 ANG Ⅱ 能提高动脉血压（Le Mevel 等，2008a）。皮质醇和生长激素对硬骨鱼类是适应海水的重要激素，它们对于鲑鳟鱼类洄游到高渗性的海水中提高饮水率起着随意的作用（Fuentes 等，1996；Nielsen 等，1999）。

综合至今所有关于饮水调控激素的资料，可以明显看到硬骨鱼类的止渴激素数量超过致渴激素的数量，这可能是由于前面已经提到的为了避免从周围环境中过量饮水的原因。今后，必须阐明这些外周的和中枢的调节剂如何在脑内对饮水调节进行互相作用，从而使硬骨鱼类得到最合适的水分摄入量。此外，还有一些其他的候选激素可能参与鱼类体液的调控，值得去研究，例如，已在哺乳类相关研究中提到的松弛素（relaxin）（Fitzsimons，1998；McKinley 等，2003b）。

8.2.2.4 中枢的对外周的激素

很明显的是外周产生的激素通过血脑屏障缺失的部位，例如，具有它们受体的 SFO、AP 和 OVLT，而作用于脑以调控鱼类的饮水活动（Ferguson 和 Bains，1996；Johnson 和 Thunhorst，1997；McKinley 等，2003b）。此外，大多数外周激素亦在脑内局部合成并以旁分泌或自分泌的方式作用于中枢的靶部位。例如，血液生成的 ANG Ⅱ 作用于终板（lamina terminalis）上的室周器诱导哺乳类（Johnson 和 Thunhorst，1997）和鸟类（Takei，1977）饮水；虽然 ANG Ⅱ 亦由脑 RAS 在脑内局部合成，并且作用于和外周 RAS 没有联系的 BBB 内的神经元（Bader 和 Ganten，2002；McKinley 等，2003a）。事实上，ANG Ⅱ 在血浆和脑的反应浓度在水分和电解质的平衡发生变化之后并不是平行的。关于中枢的和外周的激素为确保陆生或水生动物恰当的渴感和饮水反应而进行的互相作用还了解得很少。

对鱼类饮水调控激素的免疫组织化学定位还没有进行，包括 ANG Ⅱ 和 ANP。然而，最可能的是这些激素在鱼的脑内合成，以旁分泌/自分泌方式作用于调控的部位。至今得到的生理学资料表明 AP 是血液循环中激素调控鱼类吞咽活动的作用部位。但是，还有一个可能性是其他的室周器亦存在于前脑，它们亦对血液循环中的激素产生反应而把调控信号送到 AP。在大鼠中，SFO 和 OVLT 以神经纤维联系而互相沟通（Fitzsimons，1998；McKinley 等，2003b）。大脑作用机理的进一步分析可以阐明外周的和中枢的激素如何相互作用，以及在鱼脑内抑制性的作用机理如何支配刺激性的作用机理。虽然目前对鱼脑的结构和功能还了解得不多，但是，由于鱼脑的结构较为简单，它可以作为分析大脑作用机理的好模型。

8.2.3 引起鱼类饮水（吞饮）的神经作用机理

虽然饮水对海水适应是不可缺少的，但和陆生动物相比，对鱼类饮水调控作用机理的研究是相当少的。然而，考虑到鱼类非常容易接近水分，它们调控饮水行为的作用机理要比陆生动物简单一些。由于切除前脑和大部分中脑后并不影响鳗鱼大量饮水（Hirano 等，1972），表明鱼类的饮水基本上是由后脑调控。由于 ANG Ⅱ-诱导饮水的"去大脑"鳗鱼能一直存活，ANG Ⅱ 作用部位亦一定会在后脑（Takei 等，1979）。另外，第十对脑神经，迷走神经对饮水起重要作用，因为切断两侧迷走神经后能消除海水引起的饮水（Hirano 等，1972）和 ANG Ⅱ 诱导的饮水（Takei 等，1979）。

Mukuda 和 Ando（2003）已经表明不同的骨骼肌、胸骨舌骨肌（sternohyoid）、第三鳃弓的、第四鳃弓的、鳃盖的、咽的、上食管括约肌和食管体肌（aesphageal body muscles）等都由迷走神经和/或舌咽神经分布而引起一系列吞咽活动。亦曾发现每个肌肉都来自沿着延脑两侧由前向后延伸的舌咽-迷走运动复合体（GVC）的不同部分。在参与吞咽活动的肌肉当中，上食管括约肌（UES）由骨骼肌组成，起着吞入门户的作用，通常受胆碱能控制而收缩（Kozaka 和 Ando，2003）。他们亦证明参与吞咽活动的 GVC 神经元活性为儿茶酚胺所抑制，表明肾上腺素能的神经分布可能来自 Cajal 连合核和/或 AP，而前者和哺乳类的孤束核（nucleus tractus salitarius）相当。根据这些研究结果，可以推想紧张性的刺激信号通过迷走神经不断地使上食管括约肌（UES）收缩，阻止水分进入食管内；但是，当分布到 GVC 的肾上腺素能神经激活时，它抑制输送刺激性信号到 UES 的 GVC 神经元，引起 UES 的松弛而将水分摄入。然而，这个模式并不符合切断两侧迷走神经后消除海水和 ANG II 诱导鳗鱼饮水的研究结果，因为切断迷走神经会使 UES 从紧张性的收缩信号中得到放松。不过，一系列和吞咽活动相关的肌肉收缩都是由迷走神经控制的，因而协调活动是无能为力的，甚至在 UES 张开时亦不会引起饮水。

8.2.4 鱼类钠欲（偏爱）的调控

钠欲对体液平衡是一种重要的调节剂，在哺乳类中和渴感一起已经有很长的研究历史（Fitzsimons，1979；Denton，1982；de Caro 等，1986；Gnossman，1990；Ramsay 和 Booth，1991）。寻求盐分的行为是先天的，特别是对 NaCl 的寻求；在新生大鼠脑腔内给予肾素，其出生三天后就会渴求盐分（Leshem，1999）。所以，鱼类亦可能存在着和哺乳类相似的钠欲感觉。然而，有关可能诱导鱼类钠欲的激素，目前还不清楚。

鱼类在淡水和海水之间洄游是由许多因子调控的，包括性腺发育成熟等，但是钠偏爱亦可能参与最后朝向不同盐度水域游去的动机（motivation）。有一些报道谈到鲑鳟鱼类幼鱼开始往下游洄游的作用机理（Iwata，1995）。已经清楚阐明鱼类开始向海洄游之前鲑鳟鱼类发生二龄鲑完成降海洄游的转变（smoltification）（一龄降海幼鲑向二龄鲑转变）和鳗鱼由黄鳗向银鳗的转变（Folmer 和 Dickhoff，1980）。特别是随着鲑鳟鱼类二龄鲑完成降海洄游转变的进展，对盐度的耐受力（Parry，1960）和偏爱性（McInerney，1964）增加。几种激素，包括甲状腺激素、皮质醇、胰岛素、生长激素等，在二龄鲑完成降海洄游转变过程中增加，其中，甲状腺素 T_3 和 T_4 在二龄鲑完成降海洄游转变期间明显地大量释放（Hoar，1998）。

Iwata 等（1986）采用一种具有淡水和海水区室的选择水池，对孵化不久开始向下游洄游的大麻哈鱼（O. keta）幼鱼进行研究，检测其对海水偏爱变化的时间进程。他们发现幼鱼保持在淡水中会逐渐失去对海水的偏爱。已经证明血浆的甲状腺素水平和向海洄游之间有着密切的联系（Ojima 和 Iwata，2007），但还不清楚下海洄游是否由盐度偏爱激发。此外，还有待于研究解决的问题是：激素是否促进二龄鲑完成降海洄游转变和鱼类对海水的适应？哺乳类诱导钠欲的激素是否亦增强硬骨鱼类的盐分偏爱？有一项研究报告表明，应激反应妨碍大鳞大麻哈鱼（O. tshawytscha）幼鱼进入海水（Price 和 Schreck，2003），这说明皮质醇抑制这种鲑鳟鱼类的盐偏爱。洄游的软骨鱼类如低鳍真鲨（*Carcharhinus leucas*）、圆口类如海七鳃鳗（*Petromyzon marinus*）和河七鳃鳗（*L. fluviatilis*），目前对这些脊椎动物类群的钠偏爱还没有进行研究。

8.3 液体平衡的调控：肾的和肾外的作用机理

如同在第8.1节中提到的，鱼类通过食物摄取以及和外界环境交换，平衡离子与水分的损失与累积而保持体液。在这个过程中，离子平衡和身体水分含量紧密偶联，以保证体液的重量摩尔渗透性浓度和细胞内与细胞外液体分隔间总容量的稳定。除了前面讨论的液体摄入作用机理之外，液体平衡亦依靠鳃、消化道、肾脏和直肠腺上皮运输所起的作用。这些器官的活性为许多内分泌的和神经内分泌的因子所调控与整合，这就是下面要阐述的。

8.3.1 精氨酸催产素

两种神经脑垂体激素（哺乳类同源的血管升压素和催产素）存在于所有有颌类脊椎动物中。精氨酸催产素（arginine vasotocin，AVT）在所有的非哺乳类种类中通常是碱性肽，在哺乳类中为精氨酸升压素（AVP）所取代（Acher，1996）。在所有脊椎动物中，AVT/AVP系列都在密切地参与体液容量和组成的生理调控作用。细胞外脱水（容量降低）和细胞脱水（重量摩尔渗透压浓度升高）两者都是哺乳类AVP分泌的刺激因素，虽然重量摩尔渗透压浓度增加的作用要较强些（Balment，2002）。对广盐性川鲽（*Platichthys flesus*）的研究表明重量摩尔渗透压浓度升高对于AVT的分泌活动亦是比血量降低要强得多的刺激因素（Warne和Balment，1995）。川鲽不管是处在海水或者淡水中，其血浆AVT浓度都明显和血浆重量摩尔渗透压浓度处于一致的相互关系中（Balment等，2006）。AVT对于渗透压突然变化的急性反应以及较长时期的驯化反应起作用。鱼类在海水和淡水之间转移后，起初的几小时和几天，下丘脑AVT mRNA表达增加，脑垂体AVT含量降低，而血浆中AVT水平升高（Bond等，2002；Warne等，2005）。在进入淡水的起初数小时，血浆AVT浓度亦会降低；而当鱼转入海水中时，和血浆重量摩尔渗透压浓度增加相联系，血浆AVT浓度会急性上升。腹腔注射高渗的盐水能够模拟鱼类处在海水早期出现的盐分负荷和潜在的细胞脱水现象。给川鲽注射高渗盐水，60 min后血浆重量摩尔渗透压浓度急剧升高，与之相联系的是血浆AVT水平上升（Warne和Balment，1995）。Hyodo和同事最近对海水板鳃鱼类皱唇鲨（*T. scyllium*）的研究显示了细胞脱水刺激AVT分泌活动的相似结果（Hyodo等，2004）。将鱼放入高浓度海水（130%）中，2天后明显提高血浆AVT水平和下丘脑AVT mRNA表达水平。

对于哺乳类，AVP的主要靶器官是肾脏，特别是远端小管（收集管），它在那里调控水通道蛋白（aquaporin）的嵌入，以调节水的通透性以及最后尿液的容量和渗透压浓度。然而，必须注意的是，甚至在哺乳类中，AVP的生理相关水平不仅能影响水分排出，亦影响肾NaCl的丢失（Balment等，2006）。对哺乳类的研究已阐明AVP的三个主要受体类型：V1a 和血管平滑肌有联系，V1b 和脑垂体促肾上腺皮质激素细胞相关，V2 和肾小管功能调控相联系。到目前为止，在鱼类中只克隆了 V1 型加压催产素受体 (Mahlmann 等，1994；Warne，2001)，用细胞表达系统已证明它和磷脂酶 C-肌醇三磷酸信号通路偶联 (Warne，2001)。对硬骨鱼类注射 AVT 能产生全身的加压反应 (Le Mevel 等，1993)，并伴随着心输出量 (cardiac output) 和搏出量 (stroke volume) 的变化 (Oudit 和 Butler，1995)。典型的是背大动脉血压瞬时下降，这反映开始时鳃部血管的深度收缩 (Bennett 和 Rankin，1986)。鳃后身体血压升高的缓慢发展是较持续的外周血管收缩的结果。利用哺乳类限定的 V1 和 V2 受体的同等物，可以看到不同的受体类型能够介导 AVT 在鳃和外周血管的作用 (Warne 和 Balment，1997)。AVT 受体介导对 V2 同等物应答的鳃血管收缩，而它对外周的血管紧张度没有明显的作用。虽然这些研究结果表明 AVT 对一些心血管参数起着调控作用，但许多研究报道的作用都是 AVT 浓度超出生理学范围才得以实现的 (Warne 和 Balment，1997)。至今所研究的大多数硬骨鱼类，血液循环的 AVT 浓度下降到 $1 \sim 20$ fmol/mL，而最近报道的板鳃鱼类皱唇鲨 (*T. scyllium*) 的 AVT 浓度要稍高些，为 $50 \sim 100$ fmol/mL (Hyodo 等，2004)。所以，正如哺乳类的 AVP 一样，AVT 可能并不直接作用于鱼类维持身体血压或心血管功能，尽管它的作用是局部调节血流分布和重要组织中的血流速率。AVT 对液体和电解质保持平衡的作用亦十分重要，因为它间接地保证心血管功能和血压的调控。

肾的和肾外的靶组织都参与 AVT 对鱼类体液平衡所起的作用。肾的作用较少依靠肾小管的作用，如同哺乳类那样；而较多的是依靠调节肾小球的滤过速率。早期对淡水鳗鱼的研究曾报道对高剂量 AVT 的利尿 (diuresis) 和加压的反应 (Henderson 和 Wales，1974；Babiker 和 Rankin，1978)，而 AVT 较低的非加压的剂量引起尿流量明显下降或者抗利尿。我们现在了解到正是这些较低的剂量将会成为生理学上恰当的血浆浓度 (Warne 和 Balment，1997)。使用原位鳟鱼躯干部制品，能使身体压力保持在一个稳定的水平，Amer 和 Brown (1995) 能够证明 AVT 生理学上恰当的浓度的确是抗利尿的。这主要是由肾小球滤过率 (GFR) 降低所致，而不是肾小管的额外作用。AVT 使 GFR 降低是起过滤作用的肾单位 (nephron) 数量减少的结果，这是使用葡萄糖最大运送量作为滤过肾单位的数量标记物而测定的 (Henderson 和 Wales，1974；Amer 和 Brown，1995)。最近 Wells 等 (2002，2005) 使用小点猫鲨 (*S. canicula*) 原位灌注的肾脏制品进行试验，表明 AVT 亦能引起板鳃鱼类的肾小球抗利尿作用，这再次涉及肾小球的间歇性。采用川鲽同源的特异性抗体对 V1 受体进行免疫染色，展现了完整的肾血管分布 (Warne 等，2005)。免疫染色的受体明显出现在肾小球的入球和出球小动脉，并扩展到围绕收集管平滑肌层的小血管。很明显，AVT 诱导肾小球的入球和出球小动脉的收缩就会产生减弱或者增强 GFR 的作用机理。但是，AVT 诱导肾小球的间歇性至少包含两个非滤过性的肾单位群体，一个群体是血液灌注的肾小球，而另一个群体不是。这意味着 AVT 又多了一个难以辨认的作用，它可能是由围绕收集管的平滑肌内的受体提供的。有研究曾经表明这些平滑肌的收缩会增加收集管内的压力，因而降低滤过率 (Tsuneki 等，1984)。最近用定量 PCR 对总体的肾 V1 型受体 mRNA 表达进行评估，结果表明适应淡水的川鲽表达水平要低于驯化在海水中的川鲽。将川鲽从海水转移到淡水 24 h 后，相对的 AVT V1 受体 mRNA 表达亦明显降低（Balment 等，2006）。因此，AVT 在淡水和海水鱼中作用的部分变化很可能包含组织受体表达的变化以及对神经脑垂体分泌的肽类敏感性的变化。

虽然通常都认为 V2 受体出现在脊椎动物进化到四足类的时期，也就是出现在陆地上（Pang，1983），但有一些证据表明在鱼类肾脏存在一个和 cAMP 信号通路偶联的 AVT 受体（Perrott 等，1993；Warne 等，2002）。鳟鱼离体的肾小管制品表明积累的 cAMP 产生浓度依赖性的刺激作用。当把组织处在 10^{-12} M AVT 中时，亦就是所报道的血液循环水平相一致的浓度，可以观察到明显的 cAMP 产生。在四足类的 V2 受体，通过将水通道蛋白 2（AQP2）插入远端肾小管细胞顶膜内而和肾小管水分的重吸收相关联，而对鱼类则必须重新评估 AVT 是怎样调节尿量产生的。这可能是指肾小球和肾小管两种成分的结合，和在非哺乳类的四足类中看到的情况类似，根本上改变了我们当前关于 AVT 调控脊椎动物肾功能的进化发展的观点。所以，很明显，积极去确认和鉴定这个推定的 AVT V2 受体的特征是非常有意义的。

鱼类的鳃是肾脏以外的支持离子与水分平衡作用机理的一个组成部分，如上所述，AVT 受体存在于鳃内。碘化的加压催产素结合研究表明在鳗鱼鳃上存在着一个受体群体，其结合位点的数量在海水鱼类的细胞中要高于淡水鱼类的细胞（Guibbolini 等，1988）。加压催产素 V1 受体的 mRNA 在鳃中表达（Mahlmann 等，1994；Warne，2001），而驯养在淡水和海水中的川鲽，虽然它们的相对受体 mRNA 表达水平之间没有差别，但川鲽由海水转入淡水 24 h 后，相对受体 mRNA 表达水平明显下降（Balment 等，2006）。如上所述，AVT 能显著影响血流经过鳃部，降低鳃叶的灌注，间接影响离子主动和被动地穿过鳃上皮（Bennett 和 Rankin，1986；Olson，2002）。鳟鱼血管的受体在药理学上和哺乳类的 V1 受体与催产素受体相似（Conklin 等，1999）。

AVT 亦能直接影响鳃的离子和水分运输（Maetz 等，1994；Marshall，2003）。Guibbolini 和 Avella（2003）使用海鲈鳃呼吸细胞进行培育，表明 AVT 对单层短线路电流的直接作用。这可以解释为 V1-型受体介导的 AVT 刺激的 Cl^- 分泌，这是一种鱼类能在海水中存活的重要反应。值得注意的是，这些反应亦为其他的硬骨鱼神经脑垂体肽，如硬骨鱼催产素所共有。虽然这些研究着重于 AVT 和鳃呼吸细胞中的作用，但 Guibbolini 和 Avella（2003）并没有忽略 AVT 对鳃的 MRCs（富于线粒体的细胞）作用。确实，V1 受体在适应海水生活的川鲽鳃的免疫定位（immunolocalization）研究支持在推定的 MRCs 上是存在着受体的（Balment 等，2006）。总之，对 AVT 的零散的研究结果以及现今认可的鳃能够排出离子的作用，如果受体 mRNA 表达的变化是和蛋白质的改变有联系的话，那么，在海水环境中的重要性就要比在淡水环境中大些。

近年来增加了不少关于 AVT 参与各种行为，包括和生殖有关行为的证据（Balment 等，2006），所以，鱼类如大麻哈鱼（Hiraoka 等，1997）的产卵洄游和下丘脑 AVT mRNA 水平的变化相联系就不足为奇了。Kulczykowska（1995）曾认为褪黑激素和 AVT 相互作用以调控对季节的和每日的环境变化（光照、温度和盐度）的适应。鱼类和其他脊椎动物一样，血浆皮质醇浓度表现出昼夜节律。虹鳟下丘脑小细胞的 AVT mRNA 表达表现出日节律，而和皮质醇是相反的（Gilchriest 等，1998）。川鲽的血浆褪黑激素有非常清晰的昼夜变化，在黑暗期开始时达到高峰，这种型式甚至当鱼处在连续的黑暗中时亦能一直保持（Kulczykowska 等，2001）。血浆 AVT 浓度和褪黑激素是相反的情况，最高的 AVT 水平在白昼时间，这意味着在褪黑激素和 AVT 之间存在着某种功能关系，或许和褪黑激素与 AVT 在哺乳类中的情况相似，即 AVT/AVP 明显地抑制松果体分泌褪黑激素（Olcese 等，1993）。

脑垂体促肾上腺皮质激素细胞由 AVT 神经元的神经纤维分布（Batten 等，1990），使 AVT 能影响到促肾上腺皮质激素（ACTH）以及肾间腺皮质醇的分泌活动。Baker 等（1996）使用虹鳟离体的前脑垂体细胞孵育，表明 AVT 能剂量依存地刺激 ACTH 分泌，并且具有和促肾上腺皮质激素-释放激素（CRF）的协同刺激作用，与所报道的哺乳类 AVP 和 CRF 的作用一样（Rivier 和 Vale，1983）。对虹鳟视前神经元的原位杂交分析表明对禁闭的应激反应是小细胞的而不是大细胞的神经元的 AVT mRNA 表达增加（Gilchriest 等，2000）。这和川鲽的研究结果有所不同，川鲽在禁闭应激反应的 3 h、24 h 和 48 h 后，大细胞的而不是小细胞的 AVT mRNA 表达明显发生变化（Bond 等，2007）。这些神经元亦表达糖皮质激素受体，而这类受体能支持血浆皮质醇水平升高的负反馈作用。这些看来都是种类之间的差别，其中下丘脑 AVT 神经元的亚型能够调控下丘脑-脑垂体轴；然而，这就使得 AVT 在特异性应激物的作用下和在渗透压调节的作用下都能影响皮质醇的分泌活动。

8.3.2 肾素-血管紧张素系统

肾素-血管紧张素系统（renin-angiotensen system，RAS）是在整个脊椎动物中调控渗透压调节作用机理的主要内分泌系统（Kobayashi 和 Takei，1996）。肾脏是肾素的主要来源部位，鳃是转化酶的主要来源部位，它们不仅对 RAS 系统活性起着中枢的调控作用，而且它们本身就是 ANG Ⅱ 的靶组织。盐度转变或者出血和其他的冲击都能够产生激活 RAS 的低血压或者低血容量现象（Butler 和 Brown，2007）。一系列的研究表明，鱼处在高渗性介质中能激活 RAS，它的作用是使鱼在脱水的环境中维持血量和血压（Wong 等，2006）。除了前面提到 ANG Ⅱ 的加压和致渴的作用之外，ANG Ⅱ 进一步的直接和间接作用对于维持体液平衡是重要的。对于硬骨鱼类适应于在不同盐度的介质之间移动，由肾间腺产生的皮质醇对许多较长期驯化的调节是重要的，包括 MRC 的功能和消化道的运输能力。对川鲽使用 ANG Ⅱ 能刺激血浆皮质醇水平升高（Perrott 和 Balment，1990），而金头鲷适应于不同的盐度时，血浆皮质醇和 ANG Ⅱ 发生平行的变化（Wong 等，2006）。鳗鱼由淡水转移到海水中后皮质醇升高，而使用转化酶抑制剂卡托普利能抑制皮质醇的升高（Kenyon 等，1985）。因此，和 RAS 活性变化相关的一些作用看来是包括了调节皮质醇分泌活动所引起的间接作用，而皮质醇分泌的调节既可以在脑垂体 ACTH 分泌的水平（Weld 和 Fryer，1987），亦可以直接在肾间腺进行。最近的研究亦表明 RAS 对鳗鱼渗透压调节组织的 Na/K-ATP 酶活性具有直接的调控作用，包括鳃（Marsigliante 等，1997）、肾脏（Marsigliante 等，2000）和肠（Marsigliante 等，2001）。对金头鲷腹腔注射 ANG Ⅱ 能引起鳃 Na/K-ATP 酶活性的剂量依赖性增加（Wong 等，2006）。很明显，ANG Ⅱ 的这些作用能提高这些组织中潜在的离子运输活性。

和鱼类处在脱水状况使 RAS 明显激活相一致，ANG Ⅱ 对肾脏的直接作用是诱导肾微小血管收缩而使 GFR 降低与抗利尿（Gray 和 Brown，1985；Olson 等，1986）。为了排除由 ANG Ⅱ 在体引起的身体血压变化的混乱不清的影响，离体使用灌注的鳟鱼躯干部和原位肾脏制品已经证实 ANG Ⅱ 的这种抗利尿作用（Dunne 和 Rankin，1992；Brown 等，1993）。尽管通过改变 GFR 使肾脏的抗利尿反应占优势，但仍有一些证据表明 ANG Ⅱ 对肾小管可能有抗利尿作用（Brown 等，2000）。一些研究已经表明沿着硬骨鱼类的肾小管有 ANG Ⅱ 的受体（Cobb 和 Brown，1992；Marsigliante 等，1997）。ANG Ⅱ 在体的肾作用是 ANG Ⅱ 通过肾动脉的输送和 ANG Ⅱ 在肾内产生相结合的结果。不断增加的证据表明，与哺乳类一样，局部存在的肾内 RAS 系统能够精准地控制肾功能和肾小球的间歇性，而这些看来就是中枢调控鱼类尿的产生（Brown 等，2000）。

由于板鳃鱼类 ANG Ⅱ 的独特结构以及在哺乳类的生物检测中证明缺乏活性，起初学者们曾认为它们不具有一个 RAS 系统，但现在已经阐明它们的 RAS 存在着其他脊椎动物所具有的组成（Takei 等，2004b，Anderson 等，2001c）。肽的加压活性是明显的，RAS 的部分可能依靠于儿茶酚胺的刺激作用（Tierney 等，1997a）；而且，RAS 被认为在急性血量降低和/或血压下降期间起着血压调节作用。间接的渗透压调节作用包括板鳃鱼类特有的盐皮质激素，肾间分泌的 1α-羟基皮质酮的刺激作用（Armour 等，1993a、b）。在眼斑鳐（*Raja ocellata*）的直肠腺、肾脏和鳃中曾经确定这个类固醇的受体（Mood 和 Idler，1974；Idler 和 Kane，1980）。板鳃鱼类存在着肾小球旁细胞（juxtaglomerular cell）和致密斑（macula densa）（Lacy 和 Reale，1990），表明 RAS 和在硬骨鱼类中的作用一样，参与调节肾小球滤过率。确实是，ANG Ⅱ 能诱导小点猫鲨（*S. canicula*）游离躯干部制品的抗利尿作用（Anderson 等，2001c）。最近的研究结果证实对 ANG Ⅱ 的抗利尿反应是基于肾小球滤过率和滤过的肾小球比例下降。这和 CNP 在同一个制品中的利尿作用形成对照（Wells 等，2006）。板鳃鱼类的鳃具有 MRCs，但 Na/K-ATP 酶活性要比海水硬骨鱼类鳃的 Na/K-ATP 酶活性小 10~15 倍（Jampol 和 Epstein，1970）。特异性的 ANG Ⅱ 结合存在于板鳃鱼类的鳃细胞膜和鳃血管中（Tierney 等，1997b；Anderson 等，2001a）。ANG Ⅱ 在板鳃鱼类中的生理作用还有待阐明，但看来它对体液的离子与水分平衡起着作用。直肠腺是高度血管化的分泌组织，能间歇性地主动分泌多余的 NaCl（Shuttleworth，1998）。容量扩张能直接刺激直肠腺的分泌活动（Solomon 等，1985），在急性的容量/盐分负荷期间，如果在摄食情况下，这种作用就有助于清除鱼体内过多的盐分。和 ANG Ⅱ 的其他作用如支持盐分和液体的存留相一致的是 ANG Ⅱ 抑制直肠腺的分泌活动（Anderson 等，2001c）。对小点猫鲨（*S. canicula*）游离与灌注的直肠腺给予 ANG Ⅱ 能引起血管抗性增加，和 CNP 的松弛作用形成对照，表明 ANG Ⅱ 参与调控直肠腺的分泌活动，以达到可能的动态平衡。

在过去的五年，ANG Ⅱ 已经从两种圆口类，即海七鳃鳗（*P. marinus*）（Takei 等，2004b）和河七鳃鳗（*L. fluviatilis*）（Rankin 等，2004）中分离和鉴定。Rankin 等（2001）采用异源的分析技术表明在海水中驯养的河七鳃鳗的血浆 ANG Ⅱ 和 ANG Ⅲ 的浓度要高于在淡水中的河七鳃鳗。对于这个在江河与海水之间溯河洄游的种类，其 RAS 能参与体液的调控而和广盐类性硬骨鱼类的情况相似。最近，Brown 等（2005）表明血量降低对 RAS 激活是有效的刺激因素，而将鱼在淡水与海水和海水与淡水之间迅速地转移能分别使血浆中的 ANG 升高和降低。这些学者认为容量/压力受体和渗透受体相互作用以调控七鳃鳗 RAS 的活性。容量/压力的敏感性是和七鳃鳗同源的 ANG Ⅱ 的血管收缩作用相一致的（Rankin 等，2004），并且 RAS 维持血量与血压的基本作用在近 5 亿年的脊椎动物进化过程中是保守的。未来对鱼类这个奇特类群（即圆口类）有关 ANG 的其他直接与间接作用的研究将是令人期待的。

8.3.3 催乳激素、生长激素和皮质醇

催乳激素和生长激素属于脑垂体多肽激素的同一个家族，是和它们的受体一起通过一次基因复制以及随后在脊椎动物进化过程中的趋异而产生的（Forsyth 和 Wallis，2002；Kawauchi 和 Sower，2006）。这两种激素都在鱼类中为增强它们的离子与水分运输能力以满足新环境需求的适应过程中起重要作用。Grace Pickford 提出第一个有说服力的证据阐明催乳激素在淡水硬骨鱼类的离子摄取作用机理中起作用（Pickford 和 Phillips，1959）。此后，对许多鱼类研究累积的大量文献资料都证实催乳激素对支持硬骨鱼类在淡水中存活起着重要作用。鱼处在淡水中后，催乳激素的基因表达、合成、分泌和血浆的水平都增加（Manzon，2002）。血浆的重量摩尔渗透压浓度和皮质醇都直接影响催产激素的分泌（Seale 等，2006），而新增加的催乳激素释放肽（PrRP）最近亦已经在硬骨鱼类中确定。Sakamoto 等（2003）证明 PrRP 能特异性地促进催乳激素转录和分泌，而且 PrRP 神经末梢定位于靠近脑垂体的催乳激素细胞。值得注意的是，在不同种类的鱼类当中，没有脑垂体激素而在淡水中生存的能力存在着差别（可能是鲑鳟鱼类但不是底鳉或罗非鱼）。

催乳激素的功能主要是减弱渗透压调节上皮的离子和水分的通透性（Hirano，1986）。对于广盐性鱼类，催乳激素减少消化道水分和离子重吸收，尽管在一些鱼类中会有所差别（Manzon，2002）。这和皮质醇的作用形成对照，皮质醇倾向于增加消化道上皮离子和水分的通透性并主动摄取离子以增加渗透的水分摄入（Loretz，1995）。在一些淡水和广盐性硬骨鱼类中，催乳激素虽然和 Na 与 Cl 的存留相联系，但对准确的靶组织作用还了解得很少。在鱼的鳃内，催乳激素能影响 MRC 的发育，抑制海水硬骨鱼类 MRC 的出现（Herndon 等，1991），而促使淡水硬骨鱼类摄取离子 MRC 的形态发育（Pisam 等，1993）。皮质醇亦能促进一些鱼类摄取离子的 MRC 发育（Perry 和 Goss，1994），虽然皮质醇最被认可的作用是促使鱼类对海水的适应和海水鱼类 MRCs 的形成（Sakamoto 和 McCormick，2006）。有人提出在淡水适应过程中皮质醇和催乳激素之间可能存在相互作用，但支持这种想法的直接证据很少（McCormick，2001）。

GH 最先被认为能够改善褐鳟（Samlo trutta）处在海水中的耐受能力（Smith，1956）。随后的研究表明，由于 GH 能诱导鳃 MRCs 的数量与形状以及 NA/K-ATP 酶和 NKCC 的增加而提高鳃的盐分分泌能力（McCormick，2001；Pelis 和 McCormick，2001）。在这些作用中，GH 和皮质醇之间存在着重要的累加/协同作用（Madsen，1990）。GH 的一些作用是通过胰岛素样生长因子（IGF-I）介导的。用 IGF-I 处理能提高大西洋鲑、底鳉和虹鳟的盐耐受力（Mancera 和 McCormick，1998）。IGF 在硬骨鱼类的鳃内起着内分泌和自分泌/旁分泌的作用（Sakamoto 和 McCormick，2006）。在溯河洄游鲑科鱼类特有的由一龄降海幼鲑向二龄鲑转变的生理适应过程中，GH/IGF-I 轴显然起着重大的作用。

GH 和催乳激素的作用存在着种类的差异，例如，GH 对金头鲷（Sparus auratus）（Mancera 等，2002）或鳗鱼（Sakamoto 等，1993）的渗透压调节参数没有影响。催乳激素的作用对许多狭盐性海水硬骨鱼类亦未必是适合的，虽然我们还没有进行这方面的研究。另外要考虑的是，有报道称在硬骨鱼类中存在着高水平的脑垂体外的催产激素和 GH（Imaoka 等，2000；Sakamoto 等，2005a、b）。因此，可以好奇地考虑，如同 Sakamato 和 McCormick（2006）所认为的，随着四足类的出现，催乳激素的分泌会不会都集中到脑垂体里面去。

如上所述，皮质醇能够整合硬骨鱼类在淡水和海水中生存的复杂的渗透压调节过程。硬骨鱼类皮质醇从肾间腺分泌是通过 GRF 和 ACTH 轴在下丘脑-脑垂体的控制下进行的，和其他脊椎动物一样。然而，影响皮质醇分泌的深一层神经内分泌输入可能来自下丘脑的 AVT 神经元和尾神经内分泌系统（CNSS），它们通过 CRF 和硬骨鱼类紧张肽 I（UI）和 II（UII）的分泌而提供有效的刺激性输入（Lu 等，2004）。确实，最近对马苏大麻哈鱼的研究亦表明由下丘脑产生的 UI 对产卵期间调控皮质醇分泌活动的重要性比 CRF 还要大些（Westring 等，2008）。此外，ANP（Arnold-Reed 和 Balment，1991）和 ANG II（Perrott 和 Balment，1990）能够影响肾间腺，使皮质醇的分泌活动和渗透压与容量变化的内分泌反应的其他许多方面结合起来。这是特别重要的，因为皮质醇对鱼类的促进作用不仅在于渗透压调节，还包括其他重要的生理学方面，如生殖、免疫反应、生长和代谢（Mommsen 等，1999）。和四足类不同，硬骨鱼类并不分泌单独的盐皮质激素，如醛固酮（aldosterone），而这些类型的作用如上所述是由皮质醇介导的。皮质类固醇的作用依靠于细胞内的受体，它们作为配体-依赖的转录因子（ligand-dependent transcription factor）而起作用。在许多组织包括鳃和肠中都已经确定皮质醇高亲和力与低容量级别的结合部位。外界盐度的变化能影响结合部位的数量和亲和力（Prunet 等，2006）。糖皮质激素受体（GR）最先在虹鳟（Ducouret 等，1995）中克隆出来，随后在其他几种硬骨鱼类中克隆出来。但是，最近的研究表明，在鳟鱼中出现第二个 GR 的同种型（Burg 等，2003），亦出现一个盐皮质激素受体（MR）（Sturm 等，2005）。由于硬骨鱼类肾间腺并不分泌单独的盐皮质激素，这是令人费解的，虽然脱氧皮质酮（DOC）是 MR 的有效刺激剂，它在鱼类中的存在引起这样的可能性，即 DOC 成为硬骨鱼类 MR 的有生理功能的配体（Prunet 等，2006）。有关 DOC 在鱼类中的作用的资料很少，需要做进一步的研究以阐明这种设想的正确性。但是，最近在对大西洋鲑鱼的研究中比较了外源皮质醇、DOC 和醛固酮对盐度耐受力的影响，提供了有力的证据证明皮质醇可能通过和 GR 的作用，因而增强对盐度的耐受力和提高鳃的 Na/K-ATP 酶活性。DOC 或醛固酮并没有明确的作用，而 MR 的抑制剂螺甾内脂（spironolactone）并不能影响由皮质醇诱导而增强的盐度耐受力（McCormick，2008）。Shaw 及其同事（2007）对底鳉（*Fundulus heteroclitus*）的研究同样证明是 GR 而不是 MR 介导皮质类固醇促进对海水生活的适应性。用 RU-486 抑制 GR 后能阻止底鳉对盐度增加的适应能力，而螺甾内脂（MR 抑制剂）对海水的适应性没有影响。

8.3.4 硬骨鱼紧张肽

硬骨鱼紧张肽Ⅰ（UⅠ）是由 41 个氨基酸组成的多肽，和 CRF 的结构很相似。虽然 UⅠ 原先是在鱼类中确定的，而现在它作为神经递质/神经调节剂已存在于脊椎动物的各个类群中，并对应激反应起着显著的作用（见本书第 6 章）。UⅡ 是一个环肽，原先是从一种硬骨鱼类——长颌姬鰕虎鱼（*Gillichthys mirabilis*）的 CNSS 分离出来的（Pearson 等，1980），其基础是对平滑肌的刺激作用（见图 8.4）。最近已经阐明 UⅡ 是 GPR14 孤独受体的天然配体（Ames 等，1999），现定名为 UⅡ 受体或 UⅠ 受体。在哺乳类中，UⅡ 已表明具有非常强有力的心血管作用，并且明显和一些相关疾病如原发性高血压病相联系（Balment 等，2005）。

人	ETPDCFWKYCV	AA535545
川鲽	QFAGTTECFWKYCV	P21857
红鳍东方鲀	TGNNECFWKYCV	Scaf. 60
河鲀	HGNDECFWKYCV	Chr. 9
鰕虎鱼	AGTADCFWKYCV	P01147
鲤鱼-a	GGGADCFWKYCV	P04560
鲤鱼-b	GGNTECFWKYCV	P04561
斑马鱼-a	GGGADCFWKYCV	Chr. 7
斑马鱼-b	GSNTECFWKYCV	EH608865
白亚口鱼-a	GSGADCFWKYCV	P04558
白亚口鱼-b	GSNTECFWKYCV	P04559
青鳉	SGNTECFWKYCV	Chr. 7
三棘刺鱼	AGNSECFWKYCV	Contig 2571
匙吻鲟	GSTSECFWKYCV	P81022
鲨鱼	NNFSDCFWKYCV	P35490
河七鳃鳗	NNFSDCFWKYCV	Waugh et al. (1995)
海七鳃鳗	NNFSDCFWKYCV	Waugh et al. (1995)

图 8.4 至今在鱼类中确定的 UⅡ 序列

列出人的 UⅡ 序列作为参考。各个种类保守的氨基酸序列用阴影表示。蛋白质和 EST 数据库的检索号以及基因定位的染色体号列于左侧。

UⅠ和UⅡ都存在于鱼类和哺乳类的特别脑区，但鱼类血液循环中的UⅠ和UⅡ主要来源于CNSS。这个鱼类特有的神经内分泌结构存在于末端椎骨节段的脊髓内，由大的肽合成神经元，巨大细胞（dahlgren cell）组成；在硬骨鱼类中，巨大细胞伸出轴突到神经血器官，即尾垂体（urophysis）（Lu等，2004，2006；McCrohan等，2007）。板鳃鱼类亦有巨大细胞，但CNSS不发达，亦没有一个明确的尾垂体。最近对一些鱼类的研究提供了有关CNSS结构的详细资料（Parmentier等，2006）。这就显示了硬骨鱼类这个独特结构在神经内分泌基本作用机理研究中的重要价值（McCrohan等，2007）。不同于下丘脑-神经脑垂体系统，它们已有许多相似之处，而CNSS容易进行在体与离体的电生理学研究（Brierley等，2001；Ashworth等，2005）。这使得我们研究这个系统的神经生理学以及调节它们输出的因子能在近期取得良好进展（Lu等，2007；McCrohan等，2007；Marley等，2007）。

硬骨鱼紧张肽参与鱼类生理学的许多方面，包括渗透压调节、生殖和应激反应。已经证明UⅠ和UⅡ调节皮质醇从肾间腺分泌，这说明CNSS对皮质醇的分泌活动能提供应激反应特异性的刺激（Kelsall和Balment，1998）而不依赖于下丘脑-脑垂体的传入（Winter等，2000）。显然，如上所述，硬骨鱼紧张肽对皮质醇分泌活动的这种作用会影响到体液的平衡。的确，尾垂体肽的分泌活动对外界的盐度是敏感的。介质涨度（tonicity）降低使两种硬骨鱼类巨大细胞UⅠ的免疫反应增强（Minniti和Minniti，1995）。将长颌姬鰕虎鱼（*G. mirabilis*）移入淡水中，24 h后尾垂体的UⅠ含量要比保持在海水中时更高（Larson和Madani，1991）。最近对虹鳟的研究表明，将鱼由淡水转移到海水中后，前脑和CNSS CRF和UⅠ mRNA表达发生复杂的变化（Craig等，2005）。转移24 h后，血浆ACTH和皮质醇暂时升高，下丘脑和视前的CRF mRNA表达一起增加，而下丘脑UⅠ mRNA在CNSS中的表达亦同样增加。笔者认为前脑和CNSS特异性的神经元群体在协调体液受到干扰的急性和慢性反应中起着不同的作用。

已经阐明UⅠ对运输上皮的直接作用。对适应于淡水而不是海水的罗非鱼的分离肠片段，UⅠ能使水分和NaCl的吸收明显降低（Mainoya和Bern，1982）。UⅠ能刺激长颌姬鰕虎鱼鳃盖皮肤的MRCs主动分泌氯化物（Marshall和Bern，1981），这是海水鱼类保持离子平衡的重要作用。当河口鱼类如底鳉在稀释的和高渗性的微环境中移动时，UⅠ能和AVT一起启动迅速而复杂的鳃盐分分泌作用（Marshall，2003）。给未麻醉的鳟鱼脑腔注射UⅠ能引起背大动脉血压升高，而动脉内注射UⅠ能使血压暂时降低，和儿茶酚胺释放引起的高血压相反（Le Mevel等，2006）。对板鳃鱼类的小点猫鲨（*S. canicula*）大量注射UⅠ亦会使血压暂时降低然后持续升高（Platzack等，1998）。显然，这些作用看来都影响到肾的排泄作用，而对板鳃鱼类可能会影响到直肠腺的分泌作用，但这有待于进一步研究。

给鳟鱼脑腔内注射UⅡ亦能引起长时间的高血压反应，虽然使用的剂量比UⅠ要高些（Le Mevel等，2008b）。动脉内注射UⅡ引起明显的剂量依存的背大动脉血压升高和长时间的心率降低。UⅡ对肾和鳃组织的血流量和血压的影响亦使穿过这些上皮的离子流和水流发生变化，并且传递部分由使用UⅡ或者切除尾垂体对血浆组成所产生的影响（Lederis，1977）。所以，这是 UⅡ直接影响离子和水分运输的证据。为了排除 UⅡ对血流量的影响，Marshall 和 Bern（1979）使用长颌姬鰕虎鱼分离的鳃盖皮肤证明 UⅡ抑制短线路电流（SCC），这是一种测定氯化物穿过上皮主动分泌的方法。同样，Loretz 和 Bern（1981）使用长颌姬鰕虎鱼分离的膀胱制品，证明 UⅡ刺激 Na 的主动摄取。最近，使用适应淡水的鳗鱼（*A. anguilla*）分离的后肠制品，表明高剂量 UⅡ能刺激 SCC，有效地反映钠通过消化道壁的吸收增加（Baldisserotto 和 Mimura，1997）。这些在分离上皮中的作用证明鱼类在低渗的介质中能够提高保存 NaCl 的能力，而这是它们渗透压调节主要的挑战之一。鱼类血液循环中的 UⅡ主要来源于 CNSS，手术除去尾垂体后明显降低血浆的 UⅡ水平（Winter 等，1999；Lu 等，2006）。广盐性鱼类如川鲽（*P. flesus*）尾垂体中 UⅡ含量依外界的盐度而变化（Arnold-Reed 等，1991），而血浆 UⅡ水平亦由于鱼类在海水和淡水中的转移而迅速发生反应（Bond 等，2002）。鱼类转移到淡水中相对于保持在海水中，血浆 UⅡ水平降低而尾垂体 UⅡ贮存增加。

最近从川鲽中克隆了第一个鱼类 UⅡ的受体（Lu 等，2006），由于在组织中的广泛表达，它看来能介导 UⅡ的多种功能。但是，现在已经知道 UⅡ受体 mRNA 存在于所有重要的渗透压调节组织，如鳃、肾脏、膀胱和消化道，它们亦都表达 UⅡ mRNA，这表明 UⅡ可能通过旁分泌和内分泌调控离子和水分的运输。使用一种异源的 UⅡ受体的抗体，证明 UⅡ受体的免疫反应定位于肾脏和鳃的血管成分上（Lu 等，2006）。值得注意的是，UⅡ受体 mRNA 在适应淡水鱼的肾脏与鳃中的表达要低于适应海水中的鱼类，这可能是由于靶组织的敏感性发生变化。将川鲽直接从海水转移到淡水中，8 h 后 CNSS 的 UⅡ mRNA 表达增加；而由淡水到海水的相反转移，8 h 后相对于保持在淡水中的鱼，CNSS 的 UⅡ mRNA 表达下降，这突出表现 CNSS 神经内分泌系统对外界盐度变化迅速反应的能力（Lu 等，2006）。将鱼从海水移入淡水，8 h 和 24 h 后肾脏 UⅡ受体 mRNA 的表达降低，而将鱼做相反的转移 24 h 后，UⅡ受体 mRNA 水平明显增加。鳃 UⅡ受体 mRNA 水平亦明显出现类似型式的变化，表明在这些渗透压调节的冲击下，靶组织对 UⅡ的反应存在着明显的可塑性。

最近对哺乳类 UⅡ的广泛研究揭示了第二个基因编码 UⅡ种内类似物（paralogue）的前体，称为 UⅡ-相关肽（URP）。已知在人、小鼠和大鼠中确定的这个八肽（Sugo 等，2003），共有 UⅡ的环形六肽序列。UⅡ与 URP 和生长抑素Ⅰ（SSⅠ）的结构同一性有限，而 UⅡ和 SSⅠ共有一些功能的特点（Conlon 等，1997）。然而，前-UⅡ原和前-SSⅠ原的 cDNA 分析表明它们序列的同一性很少（Coulouarn 等，1999），它们可能不是由一个共同的祖先衍生而来。但使用比较基因组学的方法，Tostivint 等（2006）提供有说服力的证据表明 UⅡ和 URP 基因以及 SSⅠ与其相关肽的基因是通过两个互相联系的祖先基因的片段复制（segmental duplication）而产生。因此，UⅡ和生长抑素编码基因最后都属于同一个超家族。Tostivint 等（2006）未能在斑马鱼数据库检测到一个 URP-类基因，表明这个基因已经在斑马鱼中丢失。不过，一个 URP-类序列明显存在于河鲀（*Tetraodon*）和东方鲀（*Takifugu*）的基因组中，尽管所预测的肽和四足类 URP 相比较是非典型的。鱼类 URP 的系统进化和功能的重要性尚有待于进一步研究。

8.4 展　　望

在体液的调控方面，鱼类在脊椎动物当中处于独特的位置，因为它们的血液只通过单层的呼吸上皮和不同盐度的环境水体直接接触。所以，当鱼类在淡水和海水之间洄游时，水环境的盐度明显影响水分和离子的调控。海水硬骨鱼类通过渗透作用失去水分的速率甚至比陆生动物的还要大。因而，和陆生动物一样，从口摄入液体对它们的生存是必不可少的。此外，鱼类调节饮水的作用机理并不如陆生动物那样复杂，因为它们很容易接近水。所以，鱼类是分析研究这些作用机理最合适的模型。不过，和对哺乳类的研究相比较，鱼类在这方面的研究还是相当粗浅的。当鱼类饮水时，确定它们是否有口渴的感觉是重要的，因为它们简单地通过反射性吞咽活动就能从周围的水中摄入水分。青蛙虽然不从口中饮水，但当它们由腹部皮肤吸收水分时是有渴感的（Hillyard，1999）。口渴的感觉是由青蛙停留在水中持续的时间确定的，这是推想由于口渴而激发青蛙停留在水中以便皮肤吸收水分。类似的指标可以用来研究经常停留在陆地的两栖鱼类。

鱼类液体摄入的另一个重要研究领域是引起饮水行为的大脑作用机理。因为鱼类只要吞咽就可以饮入周围的水，由 GVC 通过迷走神经可以调控肌肉的协调活动。所以，在较高层次的神经元系统对 GVC 活性的调控将是第二个研究的目标。已有一些证据表明 AP 的神经元发出它们的轴突直接到达 GVC，或者通过其他脑核的神经元到达 GVC（Mukuda 等，2005）。已有一些详细的论据证明从前脑或者中枢发出抑制性的信号以防止过度饮水。还需要研究鱼类脑内调控饮水激素的存在和定位，以便分析内在的脑激素和外周激素的相互作用以获得最适宜的水分摄入。无论如何，鱼类并不那么复杂的脑结构对于这些课题都是非常合适的研究对象。

脑和 CNS 的神经内分泌活动对鳃、消化道和肾脏的离子和水分的流动起着直接和间接的调控作用。对 AVT 直接作用的观察研究还不多，下一步的研究应该包括评定 V2 受体介导的作用是否存在于鱼类当中。这个受体及与其相联系的反应看来未必就像目前所认为的那样只局限于陆生的四足类。最近对哺乳动物 UⅡ 迅速增长的研究使得我们亦必须考虑加强鱼类 UⅡ 作用的进一步深入研究。对于这些以及在鱼类和四足类的调控作用机理的认识之间存在着明显隔阂或者矛盾，两栖的鱼类如弹涂鱼，可以作为未来研究的很有意义的模型。最近，圆口类和板鳃鱼类在配体与受体的特性鉴定方面取得的进展为它们的功能研究奠定基础，而这些研究成果对于深入了解水生动物体液调控作用机理的进化是非常重要的。最后，越来越清楚的是调控通路能在脑内表现，而血液循环中的激素以及局部渗透压组织能在外周表达（配体和受体）。下一步的研究必须注重这些中枢的、内分泌的和自分泌/旁分泌的成分如何相互作用以达到适当调节液体和离子平衡的目的。

在本章中已经强调体液的稳态是以复杂而高度相互协调作用的调控通路为基础的。我们现在对这些通路当中一些相互作用还只有初步的了解，而随着研究手段的发展，可以采用多组分分析（multicomponent）和实时符合度量法（real-time coincident measure），我们预期在这一研究领域中将取得重大进展。由于鱼类经常受到液体平衡安全性的严重挑战，阐明这个复杂的问题是如何通过整合而实现对生理系统迅速而较长期间的调整是至关重要的。

竹井祥郎
R. J. 巴尔门特

参考文献

Acher R. 1996. Molecular evolution of fish neurohypophysial hormones: Neutral and selective evolutionary mechanisms. *Gen. Comp. Endocrinol*, 102: 157-172.

Amer S, Brown J A. 1995. Glomerular actions of arginine vasotocin in the *in situ* perfused trout kidney. *Am. J. Physiol*, 269: R775-R780.

Ames R S, Sarau H M, Chambers J K, Willette R N, Aiyar N V, Romanic A M, Louden C S, Foley J J, Sauermelch C F, Coatney R W, Ao Z, Disa J, et al. 1999. Human urotensin-II is a potent vasoconstrictor and agonist for the orphan receptor GPR14. *Nature*, 401: 282-286.

Anderson W G, Cerra M C, Wells A, Tierney M L, Tota B, Takei Y, Hazon N. 2001. Angiotensin and angiotensin receptors in cartilaginous fishes. *Comp. Biochem. Physiol*, 128A: 31-40.

Anderson W G, Takei Y, Hazon N. 2001. The dipsogenic effect of the renin-angiotensin system in elasmobranch fish. *Gen. Comp. Endocrinol*, 125: 300-307.

Anderson W G, Takei Y, Hazon N. 2001. Possible interaction between the renin-angiotensin system and natriuretic peptides on drinking in elasmobranch fish. In: Goos H J T, Rastoi R K, Vaudry H, Perantoni R, Ed. *Perspectives in Comparative Endocrinology*. Monduzzi Editore: Bologna, 753-758.

Anderson W G, Good J P, Hazon N. 2002. Changes in secretion rate and vascular perfusion in the rectal gland of the European lesser-spotted dogfish (*Scyliorhnus canicula*L.) in response to environmental and hormonal stimuli. *J. Fish Biol*, 60: 1580-1590.

Anderson W G, Takei Y, Hazon N. 2002. Osmotic and volemic effects on drinking rate in elasmobranch fish. *J. Exp. Biol*, 205: 1115-1122.

Anderson W G, Wells A, Takei Y, Hazon N. 2002. The control of drinking in elasmobranch fish with special reference to the renin-angiotensin system. In: Hazon N, Flik G, Ed. *Osmoregulation and Drinking in Vertebrates*. BIOS Scientific Publishers Ltd: Oxford, 19-30.

Anderson W G, Pillans R D, Hyodo S, Tsukada T, Good J P, Takei Y, Franklin C

E, Hazon N. 2006. The effects of freshwater to seawater transfer on circulating levels of angiotensin II, C-type natriuretic peptide and arginine vasotocin in the euryhaline elasmobranch, *Carchahinus leucas*. *Gen. Comp. Endocrinol*, 147: 39-46.

Ando M, Fujii Y, Kadota T, Kozaka T, Mukuda T, Takase I, Kawahara A. 2000. Some factors affecting drinking behavior and their interactions in seawater-acclimated eels, *Anguilla japonica*. *Zool. Sci*, 17: 171-178.

Antunes-Rodrigues J, McCann S M, Rogers L C, Samson W K. 1985. Atrial natriuretic factor inhibits dehydration- and angiotensin-induced water intake in the conscious unrestrained rat. *Proc. Natl. Acad. Sci. USA*, 82: 8720-8723.

Armour K J, O'Toole L B, Hazon N. 1993. The effect of dietary protein restriction on the secretory dynamics of 1α-hydroxycorticosterone and urea in the dogfish *Scyliorhinus canicula*: a possible role for 1α-hydroxycorticosterone in sodium retention. *J. Endocrinol*, 138: 275-282.

Armour K J, O'Toole L B, Hazon N. 1993. Mechanisms of ACTH- and angiotensin II-stimulated 1α-hydroxycorticosterone secretion in the dogfish, *Scyliorhinus canicula*. *J. Mol. Endocrinol*, 19: 235-242.

Arnold-Reed D E, Balment R J. 1991. Atrial naturetic factor stimulates in-vivo and in-vitro secretion of cortisol in teleosts. *J. Endocrinol*, 128: R17-R20.

Arnold-Reed D E, Balment R J, McCrohan C R, Hackney C M. 1991. The caudal neurosecretory system of *Platichthys flesus*: General morphology and responses to altered salinity. *Comp. Biochem. Physiol*, 99A: 137-143.

Ashworth A J, Banks J, Brierley M J, Balment R J, McCrohan C R. 2005. Electrical activity of caudal neurosecretory neurons in seawater and freshwater-adapted *Platichthys flesus*, in vivo. *J. Exp. Biol*, 208: 267-275.

Babiker M M, Rankin J C. 1978. Neurohypophysial hormone control of kidney function in the European eel (*Anguilla anguilla*) adapted to sea water or fresh water. *J. Endocrinol*, 78: 347-358.

Bader M, Ganten D. 2002. It's renin in the brain: Transgenic animals elucidate the brain renin-angiotensin system. *Circ. Res*, 90: 8-10.

Baker B I, Bird D J, Buckingham J C. 1996. In the trout, CRH and AVT synergize to stimulate ACTH release. *Regul. Pept*, 67: 207-210.

Baldisserotto B, Mimura O M. 1997. Changes in the electrophysiological parameters of the posterior intestine of *Anguilla anguilla* (Pisces) induced by oxytocin, urotensin II and aldosterone. *Braz. J. Med. Biol. Res*, 30: 35-39.

Balment R J. 2002. Control of water balance in mammals. In: Hazon N, Flik G, Ed. *Osmoregulation and Drinking in Vertebrates*. Bios Scientific Publishers Ltd: Oxford, 153-168.

Balment R J, Carrick S. 1985. Endogenous renin-angiotensin system and drinking behavior in flounder. *Am. J. Physiol*, 248: R157-R160.

Balment R J, Masini M A, Vallarino M, Conlon J M. 2002. Cardiovascular actions of lungfish bradykinin in the unanesthetised African lungfish, *Protopterus annectens*. *Comp. Biochem. Physiol*, 131A: 467-474.

Balment R J, Song W, Ashton N. 2005. Urotensin II - ancient hormone with new functions in vertebrate body fluid regulation. *Ann. N. Y. Acad. Sci*, 1040: 66-73.

Balment R J, Lu W, Weybourne E, Warne J M. 2006. Arginine vasotocin a key hormone in fish physiology and behaviour: A review with insights from mammalian models. *Gen. Comp. Endocrinol*, 147: 9-16.

Batten T F C, Cambre M L, Moons L, Vandesande F. 1990. Comparative distribution of neuropeptides-immunoreactive systems in the brain of the green molly (*Poecilia latipinna*). *J. Comp. Neurol*, 302: 893-919.

Bennett M B, Rankin J C. 1986. The effect of neurohypophysial hormones on the vascular resistance of the isolated perfused gill of the European eel *Anguilla anguilla*. *Gen. Comp. Endocrinol*, 64: 60-66.

Beyenbach K W. 1995. Secretory electrolyte transport in renal proximay tubules of fish. In: Wood C M, Shuttleworth T J, Ed. *Cellular and Molecular Approaches to Fish Ionic Regulation*. Academic Press: San Diego, 85-105.

Beyenbach K W. 2004. Kidneys sans glomeruli. *Am. J. Physiol*, 286: F811-F827.

Bond H, Winter M, Warne J M, Balment R J. 2002. Plasma concentrations of arginine vasotocin and urotensin II are reduced following transfer of flounder (*Platichthys flesus*) from sea water to fresh water. *Gen. Comp. Endocrinol*, 125: 113-120.

Bond H, Warne J M, Balment R J. 2007. Effect of acute restraint on hypothalamic provasotocin mRNA expression in flounder, *Platichthys flesus*. *Gen Comp. Endocrinol*, 153: 221-227.

Brierley M J, Ashworth A J, Banks R J, Balment R J, McCrohan C R. 2001. Bursting properties of caudal neurosecretory cells in the flounder, *Platichthys flesus*, in vitro. *J. Exp. Biol*, 204: 2733-2739.

Brown J A, Rankin J C, Yokota S D. 1993. Glomerular heamodynamics of filtration in single nephrons of non-mammalian vertebrates. In: Brown J A, Balment R J, Rankin J C, Ed. *New Insights in Vertebrate Kidney Function*. Cambridge University Press: Cambridge, 1-44.

Brown J A, Paley R K, Amer S, Aves S J. 2000. Evidence for an intrarenal renin-angiotensin system in the rainbow trout, *Oncorhynchus mykiss*. *Am. J. Physiol*, 278: R1685-R1691.

Brown J A, Cobb C S, Frankling S C, Rankin J C. 2005. Activation of the newly discovered cyclostome renin-angiotensin system in the river lamprey *Lampetra fluviatilis*. *J. Exp. Biol*, 208: 223-232.

Bury N R, Sturm A, Le Rouzic P, Lethimonier C, Ducouret B, Guiguen Y, Robinson-Rechavi M, Laudet V, Rafestin-Oblin M E, Prunet P. 2003. Evidence for two distinct functional glucocorticoid receptors in teleost fish. *J. Mol. Endocrinol*, 31: 141-156.

Butler D G, Brown J A. 2007. Stanniectomy attenuates the renin-angiotensin response to hypovolemic hypotension in freshwater eels (*Anguilla rostrata*) but not blood pressure. *J. Comp. Physiol*, 177B: 143-151.

Carrick S, Balment R J. 1983. The renin-angiotensin system and drinking in the euryhaline flounder, *Platichthys flesus*. *Gen. Comp. Endocrinol*, 51: 423-433.

Chasiotis H, Effendi J C, Kelly S P. 2009. Occludin expression in goldfish held in ion-poor water. *J. Comp. Physiol. B*, 179: 145-154.

Cobb C S, Brown J A. 1992. Localisation of angiotensin II binding to tissues of the rainbow trout, *Oncorhynchus mykiss*, adapted to freshwater and seawater: An autoradiographic study. *J. Comp. Physiol. B*162: 197-202.

Conklin D J, Smith M P, Olson K R. 1999. Pharmacological characterisation of arginine vasotocin vascular smooth muscle receptors in the trout (*Onorhynchus mykiss*) in vitro. *Gen. Comp. Endocrinol*, 114: 36-46.

Conlon J M. 1999. Bradykinin and its receptors in non-mammalian vertebrates. *Regul. Pept*, 79: 1-81.

Conlon J M, Tostivint H, Vaudry H. 1997. Somatostatin- and urotensin II-related peptides: Molecular diversity and evolutionary perspectives. *Regul. Pept*, 69: 95-103.

Coulouarn Y, Jégou S, Tostivint H, Vaudry H, Lihrmann I. 1999. Cloning, sequence analysis and tissue distribution of the mouse and rat urotensin II precursors. *FEBS Lett*, 457: 28-32.

Craig P M, Al-Timimi H, Bernier N J. 2005. Differential increase in forebrain and caudal neurosecretory system corticotrophin-releasing factor and urotensin I gene expression associated with seawater transfer in rainbow trout. *Endocrinology*, 146: 3851-3860.

de Caro G, Epstein A N, Massi M. 1986. *The Physiology of Thirst and Sodium Appetite*. Plenum Press: New York.

Denton D. 1982. *The Hunger for Salt*. Springer-Verlag: Berlin.

Doty R W. 1968. Neural organization of deglutition. In: Code C F, Ed. *Handbook of Physiology: Alimentary Canal*. American Physiological Society: Washington DC. 1861-1902.

Ducouret B, Tujague M, Ashraf J, Mouchel N, Servel N, Valotaire Y, Thompson E

B. 1995. Cloning of a teleost fish glucocorticoid receptor shows that it contains a deoxyribonucleic acid-binding domain different from that of mammals. *Endocrinology*, 136: 3774-3783.

Duner T, Conlon J M, Kukkonen J P, Akerman K E O, Yan Y, Postlethwait J H, Larhammer D. 2002. Cloning, structural characterization and functional expression of a zebrafish bradykinin B2-related receptor. *Biochem. J*, 364: 817-824.

Dunne J B, Rankin J C. 1992. Effects of atrial natriuretic peptide and angiotensin II on salt and water excretion by the perfused rainbow trout kidney. *J. Physiol. (Lond.)*, 446: 92P.

Elkinton J R, Taffel M. 1942. Prolonged water deprivation in the dog. *J. Clin. Invest*, 21: 787-794.

Epstein A N, Fitzsimons J T, Rolls B J. 1970. Drinking induced by injection of angiotensin into the brain of the rat. *J. Physiol. (London)*, 210: 457-474.

Evans D H, Piermarini P M, Choe K P. 2005. The multifunctional fish gill: Dominant site of gas exchange, osmoregulation, acid-base regulation and excretion of nitrogenous waste. *Physiol. Rev*, 85: 97-177.

Ferguson A V, Bains J S. 1996. Electrophysiology of the circumventricular organs. *Frontiers Neuroendocrinol*, 17: 440-475.

Fitzsimons J T. 1979. *The Physiology of Thirst and Sodium Appetite* Cambridge University Press: Cambridge.

Fitzsimons J T. 1998. Angiotensin, thirst, and sodium appetite. *Physiol. Rev*, 78: 583-686.

Flik G, Varsamos S, Guerreiro P M G, Fuentes X, Huising M O, Fenwick J C. 2002. Drinking in (very young) fish. In: Hazon N, Flik G, Ed. *Osmoregulation and Drinking in Vertebrates*. BIOS Scientific Publishers Ltd: Oxford, 31-47.

Folmar L C, Dickhoff W W. 1980. The Parr-smalt transformation (smoltification) and seawater adaptation in salmonids. *Aquaculture*, 21: 1-37.

Forsyth I A, Wallis M. 2002. Growth hormone and prolactin-molecular and functional evolution. *J. Mammary Gland Biol. Neoplasia*, 7: 291-312.

Foster J M, Forster M E. 2007. Effects of salinity manipulations on blood pressures in an osmoconforming chordate, the hagfish, *Eptatretus cirrhatus*. *J. Comp. Physiol. B*177: 31-39.

Fox B K, Riley L G, Dorough C, Kaiya H, Hirano T, Grau E G. 2007. Effects of homologous ghrelins on the growth hormone/ insulin-like growth factor-I axis in the tilapia, *Oreochromis mossambicus*. *Zool. Sci*, 24: 391-400.

Fregly M J, Rowland N E. 1991. Bradykinin-induced dipsogenesis in captopril-treated rats. *Brain Res. Bull*, 26: 169-172.

Fuentes J, Bury N R, Carroll S, Eddy F B. 1996. Drinking in Atlantic salmon presmolts (*Salmo salar* L.) and juvenile rainbow trout (*Oncorhynchus mykiss* Walbaum) in response to cortisol and seawater challenge. *Aquaculture*, 141: 129-137.

Giffard-Mena I, Boulo V, Aujoulat F, Fowden H, Castille R, Charmantier G, Cramb G. 2007. Aquaporin molecular characterization in the sea-bass (*Dicentrarchus labrax*): The effect of salinity on AQP1 and AQP3 expression. *Comp. Biochem. Physiol. A*, 148: 430-444.

Gilchriest B J, Tipping D R, Levy A, Baker B I. 1998. Diurnal changes in the expression of genes encoding for arginine vasotocin and pituitary proopiomelanocortin in the rainbow trout (*Oncorhynchus mykiss*): Correlation with changes in plasma hormones. *J. Neuroendocrinol*, 10: 937.

Gilchriest B J, Tipping D R, Hake L, Levy A, Baker B I. 2000. The effects of acute and chronic stresses on vasotocin gene transcripts in the brain of the rainbow trout (*Oncorhynchus mykiss*). *J. Neuroendocrinol*, 12: 795-801.

Gray J C, Brown J A. 1985. Renal and cardiovascular effects of angiotensin II in the rainbow trout, *Salmo gairdneri*. *Gen. Comp. Endocrinol*, 59: 375-381.

Grossman S P. 1990. *Thirst and Sodium Appetite. Physiological Basis*. Academic Press: San Diego.

Guibbolini M E, Avella M. 2003. Neurohypophysial hormone regulation of Cl- secretion: Physiological evidence for V1-type receptors in sea bass gill respiratory cells in culture. *J. Endocrinol*, 176: 111-119.

Guibbolini M E, Henderson I W, Mosley W, Lahlou B. 1988. Arginine vasotocin binding to isolated branchial cells of the eel: Effect of salinity. *J. Mol. Endocrinol*, 1: 125-130.

Hashimoto H, Fujihara H, Kawasaki M, Saito T, Shibata M, Takei Y, Ueta Y. 2007. Centrally and peripherally administered ghrelin potently inhibits water intake in rats. *Endocrinology*, 148: 1638-1647.

Hashimoto H, Hyodo S, Kawasaki M, Shibata M, Saito T, Suzuki H, Ohtsubo H, Yokoyama T, Fujihara H, Higuchi T, Takei Y, Ueta Y. 2007. Adrenomedullin 2 is a more potent activator of hypothalamic oxytocin-secreting neurons than adrenomedullin in rats, and its effects are only partially blocked by antagonists for adrenomedullin and calcitonin gene-related peptide receptors. *Peptides*, 28: 1104-1112.

Hay D L, Conner A C, Howitt S G, Smith D M, Poyner D R. 2004. The pharmacology of adrenomedullin receptors and their relationship to CGRP receptors. *J. Mol. Neurosci*, 22: 105-114.

Hazon N, Henderson I W. 1984. Secretory dynamics of 1α-hydroxycorticosterone in the elasmobranch fish, *Scyliorhinus canicula*. *J. Endocrinol*, 103: 205-211.

Hazon N, Wells A, Pillans R D, Good J P, Anderson W G, Franklin C E. 2003. Urea based osmoregulation and endocrine control in elasmobranch fish with special reference to euryhalinity. *Comp. Biochem. Physiol*, 136B: 685-700.

Henderson I W, Wales N A M. 1974. Renal diuresis and antidiuresis after injections of arginine vasotocin in the fresh water eel *Anguilla anguilla*. *J. Endocrinol*, 61: 487-500.

Herndon T M, McCormick S D, Bern H A. 1991. Effects of prolactin on chloride cells in opercular membrane of seawater-adapted tilapia. *Gen. Comp. Endocrinol*, 83: 283-289.

Hillyard S D. 1999. Behavioral, molecular and integrative mechanisms of amphibian osmoregulation. *J. Comp. Zool*, 283: 662-674.

Hirano T. 1974. Some factors regulating drinking by the eel, *Anguilla japonica*. *J. Exp. Biol*, 61: 737-747.

Hirano T. 1986. The spectrum of prolactin action in teleosts. *Prog. Clin. Biol. Res*, 205: 53-74.

Hirano T, Hasegawa S. 1984. Effects of angiotensins and other vasoactive substances on drinking in the eel *Anguilla japonica*. *Zool. Sci*, 1: 106-113.

Hirano T, Satou M, Utida S. 1972. Central nervous system control of osmoregulation in the eel (*Anguilla japonica*). *Comp. Biochem. Physiol*, 43A: 537-544.

Hiraoka S, Ando H, Ban M, Ueda H, Urano A. 1997. Changes in expression of neurohypophysial hormone genes during spawning migration in chum salmon, *Ocorhynchus keta*. *J. Mol. Endocrinol*, 18: 49-55.

Hiroi J, McCormick S D. 2007. Variation in salinity tolerance, gill $Na^+/K^+/2Cl^-$ cotransporter and mitochondrial-rich cell distribution in three salmonids *Salvelinus namaycush*, *Salvelinus fontinalis* and *Salmo salar*. *J. Exp. Biol*, 210: 1015-1024.

Hoar W S. 1988. The physiology of smolting salmonids. In: Hoar W S, Randall D J, Ed. *Fish Physiology*. Vol. 11B. Academic Press: San Diego, 275-344.

Hosoda H, Kojima M, Matsuo H, Kangawa K. 2000. Ghrelin and des-acyl ghrelin: Two major forms of rat ghrelin peptide in gastrointestinal tissues. *Biochem. Biophys. Res. Commun*, 279: 909-913.

Hyodo S, Tsukada T, Takei Y. 2004. Neurohypophysial hormones of dogfish, *Triakis scyllium*: Structures and salinity-dependent secretion. *Gen. Comp. Endocrinol*, 138: 97-104.

Idler D R, Kane K M. 1980. Cytosol receptor glycoprotein for 1α-hydroxycorticosterone in tissues of elasmobranch fish, *Raja ocellata*. *Gen. Comp. Endocrinol*, 42: 259-266.

Imaoka T, Matsuda M, Mori T. 2000. Extrapituitary expression of the prolactin gene in the goldfish, African clawed frog and mouse. *Zool. Sci*, 17: 791-796.

Inoue K, Naruse K, Yamagami S, Mitani H, Suzuki N, Takei Y. 2003. Four functionally distinct C-type natriuretic peptides found in fish reveal new evolutionary history of the natriuretic system. *Proc. Natl. Acad. Sci. USA*, 100: 10079-10084.

Inoue K, Sakamoto T, Yuge S, Iwatani H, Yamagami S, Tsutsumi M, Hori H, Cerra M C, Tota B, Suzuki N, Okamoto N, Takei Y. 2005. Structural and functional evolution of three cardiac natriuretic peptides. *Mol. Biol. Evol*, 22: 2428-2434.

Ito S, Mukuda T, Ando M. 2006. Catecholamines inhibit neuronal activity in the glossopharyngeal-vagal motor complex of the Japanese eel: Significance for controlling swallowing water. *J. Exp. Zool*, 305A: 499-506.

Iwata M. 1995. Downstream migratory behavior of salmonids and its relationship with cortisol and thyroid hormones: A review. *Aquaculture*, 135: 131-139.

Iwata M, Ogura H, Komatsu S, Suzuki K. 1986. Loss of seawater preference in chum salmon (*Oncorhynchus keta*) fry retained in fresh water after migration season. *J. Exp. Zool*, 240: 369-376.

Jampol L M, Epstein F M. 1970. Sodium-potassium-activated adenosine triphosphatase and osmotic regulation by fishes. *Am. J. Physiol*, 218: 607-611.

Johnson A K, Thunhorst R L. 1997. The neuroendocrinology of thirst and salt appetite: Visceral sensory signals and mechanisms of central integration. *Frontiers Neuroendocrinol*, 18: 292-353.

Johnson K R, Olson K R. 2008. Comparative physiology of the piscine natriuretic peptide system. *Gen. Comp. Endocrinol*, 157: 21-26.

Kaiya H, Takei Y. 1996. Changes in plasma atrial and ventricular natriuretic peptide concentrations after transfer of eels from fresh water and seawater or *vice versa*. *Gen. Comp. Endocrinol*, 104: 337-345.

Kaiya H, Takei Y. 1996. Osmotic and volaemic regulation of atrial and ventricular natriuretic peptide secretion in conscious eels. *J. Endocrinol*, 149: 441-447.

Kaiya H, Tsukada T, Yuge S, Mondo H, Kangawa K, Takei Y. 2006. Identification of eel ghrelin in plasma and stomach by raidoimmunoassay and histochemistry. *Gen. Comp. Endocrinol*, 148: 375-382.

Kaiya H, Miyazato M, Kangawa K, Peter R E, Unniappan S. 2008. Ghrelin: A multifunctional hormone in non-mammalian vertebrates. *Comp. Biochem. Physiol*, 149A: 109-128.

Kaufman S, Peters G. 1980. Regulatory drinking in the pigeon, *Columba livia*. *Am. J. Physiol*, 239: R219-R225.

Kawakoshi A, Kaiya H, Riley L G, Hirano T, Grau E G, Miyazato M, Hosoda H, Kangawa K. 2007. Identification of ghrelin-like peptide in two species of shark,

Sphyrna lewini and *Carcharhinus melanopterus*. *Gen. Comp. Endocrinol*, 151: 259-268.

Kawauchi H, Sower S A. 2006. The dawn and evolution of hormones in the adenohypophysis. *Gen. Comp. Endocrinol*, 148: 3-14.

Kelsall C J, Balment R J. 1998. Native urotensins influence cortisol secretion and plasma cortisol concentration in the euryhaline flounder, *Platichthys flesus*. *Gen. Comp. Endocrinol*, 112: 210-219.

Kenyon C J, McKeever A, Oliver J A, Henderson I W. 1985. Control of renal and adrenocortical function by the renin-angiotensin system in two euryhaline teleost fishes. *Gen. Comp. Endocrinol*, 58: 93-100.

Kirsch R, Mayer-Gostan N. 1973. Kinetics of water and chloride exchanges during adaptation of the European eel to sea water. *J. Exp. Biol*, 58: 105-121.

Kitamura K, Kangawa K, Kawamoto M, Ichiki Y, Nakamura S, Matsuo H, Eto T. 1993. Adrenomedullin: A novel hypotensive peptide isolated from human pheochromocytoma. *Biochem. Biophys. Res. Commun*, 192: 553-560.

Kitamura K, Ichiki Y, Tanaka M, Kawamoto M, Emura J, Sakakibara S, Kangawa K, Matsuo H, Eto T. 1994. Immunoreactive adrenomedullin in human plasma. *FEBS Lett*, 341: 288-290.

Kobayashi H, Takei Y. 1996. *The Renin-Angiotensin System: Comparative Aspects* Springer International: Berlin.

Kobayashi H, Uemura H, Wada M, Takei Y. 1979. Ecological adaptation of angiotensin II-induced thirst mechanism in tetrapods. *Gen. Comp. Endocrinol*, 38: 93-104.

Kobayashi H, Uemura H, Takei Y, Itazu N, Ozawa M, Ichinohe K. 1983. Drinking induced by angiotensin II in fishes. *Gen. Comp. Endocrinol*, 49: 295-306.

Kojima M, Hosoda H, Date Y, Nakazato M, Matsuo H, Kangawa K. 1999. Ghrelin is a growth-hormone-releasing acylated peptide from stomach. *Nature*, 402: 656-660.

Komlosi P, Fintha A, Bell P D. 2004. Current mechanisms of macula densa signaling. *Acta Physiol. Scand*, 181: 463-469.

Kozaka T, Ando M. 2003. Cholinergic innervations to the upper esophageal sphincter muscle in the eel, with special reference to drinking behavior. *J. Comp. Physiol. B*, 173: 135-140.

Kozaka T, Fujii Y, Ando M. 2003. Central effects of various ligands on drinking behavior in eels acclimated to seawater. *J. Exp. Biol*, 206: 687-692.

Kulczykowska E. 1995. Arginine vasotocin-melatonin interactions in fish: A hypothesis. *Fish Biol. Fish*, 5: 96-102.

Kulczykowska E, Warne J M, Balment R J. 2001. Day-night variations in plasma melatonin and arginine vasotocin concentrations in chronically cannulated flounder (*Platichthys flesus*). *Comp. Biochem. Physiol*, 130A: 827-834.

Lacy E R, Reale E. 1990. The presence of juxtaglomerular apparatus in elasmobranch fish. *Anat. Embryol*, 182: 249-262.

Lancien F, Le Mevel J C. 2007. Central actions of angiotensin II on spontaneous baroreflex sensitivity in the trout *Oncorhynchus mykiss*. *Regul. Pept*, 138: 94-102.

Larson B A, Madani Z. 1991. Increased urotensin I and II immunoreactivity in the urophysis of *Gillichthys mirabilis* transferred to low salinity. *Gen. Comp. Endocrinol*, 83: 379-387.

Lederis K. 1977. Chemical properties and the physiological and pharmacological actions of urophysial peptides. *Am. Zool*, 17: 823-832.

Le Mével J C, Pamantung T F, Mabin D, Vaudrey H. 1993. Effects of central and peripheral administration of arginine vasotocin and related neuropeptides on blood pressure and heart rate in the conscious trout. *Brain Res*, 610: 82-89.

Le Mével J C, Pamantung T F, Mabin D, Vaudry H. 1994. Intracerebroventricular administration of angiotensin II increases heart rate in the conscious trout. *Brain Res*, 654: 216-222.

Le Mével J C, Mimassi N, Lancien F, Mabin D, Boucher J M, Blanc J J. 2002. Heart rate variability, a target for the effects of angiotensin II in the brain of the trout *Oncorhynchus mykiss*. *Brain Res*, 947: 34-40.

Le Mével J C, Mimassi N, Lancien F, Mabin D, Conlon J M. 2006. Cardiovascular actions of the stress-related neurohormonal peptides corticotropin- releasing factor and urotensin-I in the trout *Oncorhynchus mykiss*. *Gen. Comp. Endocrinol*, 146: 56-61.

Le Mével J C, Lancien F, Mimassi N. 2008. Central cardiovascular actions of angiotensin in trout. *Gen. Comp. Endocrinol*, 157: 27-34.

Le Mével J C, Lancien F, Mimassi N, Leprince J, Conlon J M, Vaudry H. 2008. Central and peripheral cardiovascular, ventilatory, and motor effects of trout urotensin-II in the trout. *Peptides*, 29: 830-837.

Leshem M. 1999. The ontogeny of salt hunger in the rat. *Neurosci. Biobehav. Rev*, 23: 649-659.

Li Z, Smith M P, Duff D W, Barton B A, Olson K R, Conlon J M. 1998. Isolation and cardiovascular activity of [Met^1, Met^1] bradykinin from the plasma of a sturgeon (Acipenseriformes). *Peptides*, 19: 635-641.

Lin H, Randall D. 1995. Proton pump in fish gills. In: Wood C M, Shuttleworth T J, Ed. *Cellular and Molecular Approaches to Fish Ionic Regulation*. Academic Press: San Diego, 229-256.

López J, Martínez A. 2001. Cell and molecular biology of the multifunctional peptide, adrenomedullin. *Int. Rev. Cytol*, 221: 1-92.

Loretz C A. 1995. Electrophysiology of ion transport in teleost intestinal cells. In: Wood C M, Shuttleworth T J, Ed. *Cellular and Molecular Approaches to Fish Ionic Regulation*. Academic Press: San Diego, 25-56.

Loretz C A, Bern H A. 1981. Stimulation of sodium transport across the teleost urinary bladder by Urotensin II. *Gen. Comp. Endocrinol*, 43: 325-330.

Lu W, Gumusgoz S, Dow L, Bricrley M J, Warne J M, McCrohan C R, Balment R J, Riccardi D. 2004. Co-expression of corticotrophin-releasing hormone (CRH) and urotensin I (UI) precursor genes in the caudal neurosecretory system of the euryhaline flounder (*Platichthys flesus*): A possible shared role in phenotypic plasticity. *Endocrinology*, 145: 5786-5797.

Lu W, Greenwood M, Dow L, Yuill J, Worthington J, Brierley M J, McCrohan C R, Riccardi D, Balment R J. 2006. Molecular characterization and expression of urotensin II and its receptor in the flounder (*Platichthys flesus*): A hormone system supporting body fluid homeostasis in euryhaline fish. *Endocrinology*, 147: 3692-3708.

Lu W, Worthington J, Riccardi D, Balment R J, McCrohan C R. 2007. Seasonal changes in peptide, receptor and ion channel mRNA expression in the caudal neurosecretory system of the European flounder (*Platichthys flesus*). *Gen Comp. Endocrinol*, 153: 262-267.

Madsen S S. 1990. The role of cortisol and growth hormone in seawater adaptation and development of hypoosmoregulatory mechanisms in sea trout parr (*Salmo trutta trutta*). *Gen. Comp. Endocrinol*, 79: 1-11.

Maetz J, Bourguet J, Lahlou B, Hourdry J. 1964. Peptides neurohypophysiares et osmoregulation chez *Carassius auratus*. *Gen. Comp. Endocrinol*, 4: 508-522.

Mahlmann S, Meyerhof W, Hausmann H, Heierhorst J, Schönrock C, Zwiers H, Lederis K, Richter D. 1994. Structure, function, and phylogeny of [Arg^8] vasotocin receptors from teleost fish and toad. *Proc. Natl. Acad. Sci. USA*, 91: 1342-1345.

Mainoya J R, Bern H A. 1982. Effects of teleost urotensins on intestinal absorption of water and NaCl in Tilapia, *Saratherodon moassambicus*, adapted to fresh water or sea water. *Gen. Comp. Endocrinol*, 47: 54-58.

Malvin R L, Schiff D, Eiger S. 1980. Angiotensin and drinking rates in the euryhaline killifish. *Am. J. Physiol*, 239: R31-R34.

Mancera J M, McCormick S D. 1998. Osmoregulatory actions of the GH/IGF axis in non-salmonid teleosts. *Comp. Biochem. Physiol*, 121B: 43-48.

Mancera J M, Carrion R L, del Rio M D M. 2002. Osmoregulatory action of PRL, GH, and cortisol in the gilthead seabream (*Sparus aurata* L.). *Gen. Comp. Endocrinol*, 129: 95-103.

Manzon L A. 2002. The role of prolactin in fish osmoregulation: A review. *Gen. Comp. Endocrinol.* 125: 291-310.

Marley R, Lu, Balment R J, McCrohan C R. 2007. Evidence for nitric oxide role in the caudal neurosecretory system of the European flounder, *Platichthys flesus*. *Gen. Comp. Endocrinol*, 153: 251-261.

Marshall W S. 2003. Rapid regulation of NaCl secretion by estuarine teleost fish: Coping strategies for short-duration fresh water exposures. *Biochim. Biophys. Acta*, 1618: 95-105.

Marshall W S, Bern H A. 1979. Teleostean urophysis: Urotensin II and ion transport across the isolated skin of a marine teleost. *Science*, 204: 519-520.

Marshall W S, Bern H A. 1981. Active chloride transport by the skin of a marine teleost is stimulated by urotensin I and inhibited by urotensin II. *Gen. Comp. Endocrinol*, 43: 484-491.

Marsigliante S, Muscella A, Vinson G P, Storelli C. 1997. Angiotensin II receptors in the gill of sea water- and freshwater-adapted eel. *J. Mol. Endocrinol*, 18: 67-76.

Marsigliante S, Muscella A, Barker S, Storelli C. 2000. Angiotensin II modulates the activity of the Na^+/K^+ ATPase in eel kidney. *J. Endocrinol*, 165: 147-156.

Marsigliante S, Muscella A, Greco S, Elia M G, Vilella S, Storelli C. 2001. Na^+/K^+ ATPase activity inhibition and isoform-specific translocation of protein kinase C following angiotensin II administration in isolated eel enterocytes. *J. Endocrinol*, 168: 339-346.

Matsuda K, Miura T, Kaiya H, Maruyama K, Shimakura S, Uchiyama M, Kangawa K, Shioda S. 2006. Regulation of food intake by acyl and des-acyl ghrelins in the goldfish. *Peptides*, 27: 2321-2325.

McCormick S D. 2001. Endocrine control of osmoregulation in teleost fish. *Am. Zool*, 41: 781-794.

McCormick S D, Bradshaw D. 2006. Hormonal control of salt and water balance in vertebrates. *Gen. Comp. Endocrinol*, 147: 3-8.

McCormick S D, Regish A, O'Dea M F, Shrimpton J M. 2008. Are we missing a mineralocorticoid in teleost fish? Effects of cortisol, deoxycorticosterone and aldosterone on osmoregulation, gill Na^+, K^+-ATPase activity and isoform mRNA levels in Atlantic salmon. *Gen. Comp. Endocrinol*, 157: 35-41.

McCrohan C R, Lu W, Brierley M J, Dow L, Balment R J. 2007. Fish caudal neurosecretory system: A model for the study of neuroendocrine secretion. *Gen. Comp. Endocrinol*, 153: 243-250.

McInerney J E. 1964. Salinity preference: An orientation mechanism in salmon migration. *J. Fish. Res. Bd. Canada*, 21: 995-1018.

McKinley M J, Albiston A L, Allen A M, Mathai M L, May C N, McAllen R M, Oldfield B J, Mendelsohn F A O, Chai S Y. 2003. The brain renin-angiotensin system: Location and physiological roles. *Int. J. Biochem. Cell Biol*, 35: 901-918.

McKinley M J, McAllen R M, Davern P, Giles M E, Penschow J, Sunn N, Uschakov A, Oldfield B J. 2003. *The Sensory Circumventricular Organs of the Mammalian Brain. Advances in Anatomy, Embryology and Cell Biology* 172. Springer: Berlin.

Minniti F, Minniti G. 1995. Immunocytochemical and ultrastructural changes in the caudal neurosecretory system of a sea water fish *Boops boops* L (teleostei: Sparidae) in relation to the osmotic stress. *Eur. J. Morphol*, 33: 473-483.

Mommsen T P, Vijayan M M, Moon T W. 1999. Cortisol in teleosts: Dynamics, mechanisms of action, and metabolic regulation. *Rev. Fish Biol. Fish*, 9: 211-268.

Mood T W, Idler D R. 1974. The binding of 1α-hydroxycorticosterone to tissue soluble proteins in the skate, *Raja ocellata. Comp. Biochem. Physiol*, 48: 499-500.

Mukuda T, Ando M. 2003. Medullary motor neurons associated with drinking behavior of Japanese eels. *J. Fish Biol*, 62: 1-12.

Mukuda T, Matsunaga Y, Kawamoto K, Yamaguchi K, Ando M. 2005. "Blood-contacting neurons" in the brain of the Japanese eel, *Anguilla japonica. J. Exp. Zool*, 303A: 366-376.

Nag K, Kato A, Nakada T, Hoshijima K, Mistry A C, Takei Y, Hirose S. 2006. Molecular and functional characterization of adrenomedullin receptors in pufferfish. *Am. J. Physiol*, 290: R467-R478.

Nakazato M, Murakami N, Date Y, Kojima M, Matsuo H, Kangawa K, Matsukura S. 2001. A role for ghrelin in the central regulation of feeding. *Nature*, 409: 194-198.

Nicolaidis S, Fitzsimons J T. 1975. La dependence de la prise d'eau induite par l'angiotensine II envers la function vasomotrice cerebrale chez le rat. *C. R. Acad. Sci. Paris*, 281: 1417-1420.

Nielsen C, Madsen S S, Bjornsson T B. 1999. Changes in branchial and intestinal osmoregulatory mechanisms and growth hormone levels during smolting in hatchery-reared and wild brown trout. *J. Fish. Biol*, 54: 799-818.

Nishimura H, Sawyer W H, Nigrelli R F. 1976. Renin, cortisol and plasma volume in marine teleost fishes adapted to dilute media. *J. Endocrinol*, 70: 47-59.

Nishimura H, Lunde L G, Zucker A. 1979. Renin response to hemorrhage and hypotension in the aglomerular toadfish *Opsanus tau. Am. J. Physiol*, 237: H105-H111.

Nobata S, Ogoshi M, Takei Y. 2008. Potent cardiovascular actions of homologous adrenomedullins in eel. *Am. J. Physiol*, 294: R1544-R1553.

Ogoshi M, Inoue K, Takei Y. 2003. Identification of a novel adrenomedullin gene family in teleost fish. *Biochem. Biophys. Res. Commun*, 311: 1072-1077.

Ogoshi M, Inoue K, Naruse K, Takei Y. 2006. Evolutionary history of the calcitonin gene-related peptide family in vertebrates revealed by comparative genomic analyses. *Peptides*, 27: 3154-3164.

Ogoshi M, Nobata S, Takei Y. 2008. Potent osmoregulatory actions of peripherally and centrally administered homologous adrenomedullins in eels. *Am. J. Physiol*, 295: R2075-R2083.

Oide H, Utida S. 1968. Changes in intestinal absorption and renal excretion of water during adaptation to sea-water in the Japanese eel. *Marine Biol*, 1: 172-177.

Ojima D, Iwata M. 2007. The relationship between thyroxine surge and onset of downstream migration in chum salmon *Oncorhynchus keta* fry. *Aquaculture*, 273: 185-193.

Okawara Y, Karakida T, Aihara M, Yamaguchi K, Kobayashi H. 1987. Involvement of angiotensin II in water intake in the Japanese eel, *Anguilla japonica*. *Zool. Sci*, 4: 523-528.

Olcese J, Sinemus C, Ivell R. 1993. Vasopressinergic innervation of the bovine pineal gland: Is there a local source for arginine vasopressin?. *Mol. Cell. Neurosci*, 4: 47-54.

Olson K R. 1992. Blood and extracellular fluid volume regulation: Role of the renin-angiotensin kallikrein-kinin systems and atrial natriuretic peptides. In: Hoar W S, Randall D J, Farrell A P, Ed. *Fish Physiology*. Vol. XIIB. Academic Press: San Diego, 135-254.

Olson K R. 2002. Gill circulation: Regulation of perfusion distribution and metabolism of regulatory molecules. *J. Exp. Zool*, 293: 320-335.

Olson K R, Hoagland T M. 2008. Effects of freshwater/saltwater adaptation and dietary salt on fluid compartments, blood pressure and venous capacitance in trout. *Am. J. Physiol*, 294: R1061-R1067.

Olson K R, Kullman D, Nakartes A J, Oparil S. 1986. Angiotensin extraction by trout tissues *in vivo* and metabolism by the perfused gill. *Am. J. Physiol*, 250: R532-R538.

Olson K R, Conklin D J, WeaverJr L, Duff D W, Herman C A, Wang X, Conlon J M. 1997. Cardiovascular effects of homologous bradykinin in rainbow trout. *Am. J. Physiol*, 272: R1112-R1120.

Oudit G Y, Butler D G. 1995. Cardiovascular effects of arginine vasotocin, atrial natriuretic peptide, and epinephrine in freshwater eels. *Am. J. Physiol*, 268: R1273-R1280.

Pamantung T F, Leroy J P, Mabin D, Le Mével J C. 1997. Role of dorsal vagal motor nucleus in angiotensin II-mediated tachycardia in the conscious trout *Oncorhynchus mykiss*. *Brain Res*, 777: 167-175.

Pang P K T. 1983. Evolution of control of epithelial transport in vertebrates. *J. Exp. Biol*, 106: 549-556.

Parmentier C, Taxi J, Balment R J, Nicolas G, Calas A. 2006. Caudal neurosecretory system of the zebrafish: Ultrastructural organization and immunocytochemical detection of urotensins. *Cell Tiss. Res*, 325: 111-124.

Parry G. 1960. The development of salinity tolerance in the salmon, *Salmo salar* (L.) and some related species. *J. Exp. Biol*, 37: 425-434.

Pearson D, Shively J E, Clark B R, Geschwind I I, Barkley M, Nishioka R S, Bern H A. 1980. Urotensin II: A somatostatin-like peptide in the caudal neurosecretory system of fishes. *Proc. Natl. Acad. Sci. USA*, 77: 5021-5024.

Pelis R M, McCormick S D. 2001. Effects of growth hormone and cortisol on Na^+-K^+-$2Cl^-$ cotransporter localization and abundance in the gills of Atlantic salmon. *Gen. Comp. Endocrinol*, 124: 134-143.

Perrott M N, Balment R J. 1990. The renin-angiotensin system and the regulation of plasma cortisol in the flounder, *Platichthys flesus*. *Gen. Comp. Endcorinol*, 78: 414-420.

Perrott M N, Sainsbury R J, Balment R J. 1993. Peptide hormone stimulated second messenger production in the teleoston nephron. *Gen. Comp. Endocrinol*, 89: 387-395.

Perry R M, Goss G G. 1994. The effects of experimentally altered gill chloride cell surface area on acid-base regulation in rainbow trout during metabolic alkalosis. *J. Comp. Physiol. B*, 164: 327-336.

Pickford G E, Phillips J G. 1959. Prolactin, a factor promoting survival of hypophysectomized killifish in freshwater. *Science*, 130: 454-455.

Pillans R D, Franklin C E. 2004. Plasma osmolyte concentrations and rectal gland mass of bull sharks, *Carcharhinus leucas*, captured along a salinity gradient. *Comp. Biochem. Physiol*, 138A: 363-371.

Pisam M, Auperin B, Prunet P, Rentierdelrue F, Martial J, Rambourg A. 1993. Effects of prolactin on alpha and beta chloride cells in the gill epithelium of the saltwater adapted tilapia *Orecochromis niloticus*. *Anat. Rec*, 235: 275-284.

Platzack B, Conlon J M. 1997. Purification, structural characterization and cardiovascular activity of cod bradykinins. *Am. J. Physiol*, 272: R710-R717.

Platzack B, Schaffert C, Hazon N, Conlon J M. 1998. Cardiovascular actions of dogfish urotensin-I in the dogfish, *Scyliorhinus canicula*. *Gen. Comp. Endocrinol*, 109: 269-275.

Price C S, Schreck C B. 2003. Stress and saltwater-entry behavior of juvenile chinook salmon (*Oncorhynchus tshawytscha*): Conflicts in physiological motivation. *Can. J. Fish. Aquat. Sci*, 60: 910-918.

Prunet P, Sturm A, Milla S. 2006. Multiple corticosteroid receptors in fish: From old ideas to new concepts. *Gen. Comp. Endocrinol*, 147: 17-23.

Ramsay D J, Booth D A. 1991. *Thirst. Physiological and Psychological Aspects* Springer-Verlag: Berlin.

Rankin J C. 2002. Drinking in hagfishes and lampreys. In: Hazon N, Flik G, Ed. *Osmoregulation and Drinking in Vertebrates*. BIOS Scientific Publishers Ltd.: Oxford 1-18.

Rankin J C, Cobb C S, Frankling S C, Brown J A. 2001. Circulating angiotensins in the river lamprey, *Lampetra fluviatilis*, acclimated to freshwater and seawater: Possible involvement in the regulation of drinking. *Comp. Biochem. Physiol*, 129B: 311-318.

Rankin J C, Watanabe T X, Nakajima K, Broadhead C, Takei Y. 2004. Identification of angiotensin I in a cyclostome, *Lampetra fluviatilis*. *Zool. Sci*, 21: 173-179.

Regoli D, Barabe J. 1980. Pharmacology of bradykinin and related kinins. *Pharmacol. Rev*, 32: 1-46.

Reid I A. 1992. Interactions between ANG II, sympathetic nervous system, and baroreceptor reflexes in regulation of blood pressure. *Am. J. Physiol*, 262: E763-E778.

Reinhardt H W, Seeliger E. 2000. Toward an integrative concept of control of total body sodium. *News Physiol. Sci*, 15: 319-325.

Riley L G, Fox B K, Kaiya H, Hirano T, Grau E G. 2005. Long-term treatment of ghrelin stimulates feeding, fat deposition and alters the GH/IGF-I axis in the tilapia, *Oreochromis mossambicus*. *Gen. Comp. Endocrinol*, 142: 234-240.

Rivier C, Vale W. 1983. Modulation of stress-induced ACTH release by corticotrophin-releasing factor, catecholamines and vasopressin. *Nature*, 305: 325-327.

Rolls B J, Rolls E T. 1982. *Thirst*. Cambridge University Press: Cambridge.

Ruohonen K, Grove D J, McIloy J T. 1997. The amount of food ingested in a single meal by rainbow trout offered chopped herring, dry and wet diets. *J. Fish. Biol*, 51: 93-105.

Saito E, Kaiya H, Tachibana T, Tomonaga S, Denbow D M, Kangawa K, Furuse M. 2005. Inhibitory effect of ghrelin on food intake is mediated by the corticotropin-releasing factor system in neonatal chicks. *Regul. Pept*, 251: 201-208.

Sakamoto T, McCormick S D. 2006. Prolactin and growth hormone in fish osmoregulation. *Gen. Comp. Endocrinol*, 147: 24-30.

Sakamoto T, McCormick S D, Hirano T. 1993. Osmoregulatory actions of growth hormone and its mode of action in salmonids: A review. *Fish Physiol. Biochem*, 11: 155-164.

Sakamoto T, Fujimoto M, Ando M. 2003. Fishy tales of prolactin-releasing peptide. *Int. Rev. Cytol*, 225: 91-130.

Sakamoto T, Amano M, Hyodo S, Moriyama S, Takahashi A, Kawauchi H, Ando M. 2005. Expression of prolactin-releasing peptide and prolactin in the euryhaline mudskippers (*Periophthalmus modestus*): prolactin-releasing peptide as a primary regulator of prolactin. *J. Mol. Endocrinol*, 34: 825-834.

Sakamoto T, Oda A, Narita K, Takahashi H, Oda T, Fujiwara J, Godo W. 2005. Prolactin: Fishy tales of its primary regulator and function. *Ann. NY Acad. Sci*, 1040: 184-188.

Seale A P, Fiess J C, Hirano T, Cooke I M, Grau E G. 2006. Disparate release of prolactin and growth hormone from the tilapia pituitary in response to osmotic stimulation. *Gen. Comp. Endocrinol*, 145: 222-231.

Schmidt-Nielsen K. 1997. *Animal Physiology. Adaptation and Environment. Fifth Ed.* Cambridge University Press: Cambridge, 301-394.

Seymour A A, Smith S G, Mazack E K, Blaine E H. 1986. A comparison of synthetic rat and human atrial natriuretic factor in conscious dogs. *Hypertension*, 8: 211-216.

Shaw J R, Gabor K, Hand H, Lankowski A, Durant L, Thibodeau R, Stanton C R, Barnaby R, Coutermarsh B, Karlson K H, Sato J D, Hamilton J W, et al. 2007. Role of glucocorticoid receptor in acclimation of killifish (*Fundulus heteroclitus*) to seawater and effects of arsenic. *Am. J. Physiol*, 292: R1052-R1060.

Shuttleworth T J. 1988. Salt and water balance. In: Shuttleworth T J, Ed. *Physiology of Elasmobranch Fishes*. Springer Verlag: Berlin, 171-199.

Simpson J B, Routtenberg A. 1973. Subfornical organ: Site of drinking elicitation by angiotensin II. *Science*, 181: 1172-1175.

Skidgel R A, Erdos E G. 2004. Angiotensin converting enzyme (ACE) and neprilysin hydrolyze neuropeptides: A brief history, the beginning and follow-ups to early studies. *Peptides*, 25: 521-525.

Smith D C W. 1956. The role of the endocrine organs in the salinity tolerance of trout. *Mem. Soc. Endocrinol*, 5: 83-101.

Solomon R, Taylor M, Sheth S, Silva P, Epstein F H. 1985. Primary role of volume expansion in stimulation of rectal gland function. *Am. J. Physiol*, 248: R638-R640.

Sturm A, Bury N, Dengreville L, Fagart J, Flouriot G, Rafestin-Oblin M E, Prunet P. 2005. 11-deoxycorticosterone is a potent agonist of the rainbow trout (*Oncorhynchus mykiss*) mineralocorticoid receptor. *Endocrinology*, 146: 47-55.

Sugo T, Murakami Y, Shimomura Y, Harada M, Abe M, Ishibashi Y, Kitada C, Miyajima N, Suzuki N, Mori M, Fujino M. 2003. Identification of urotensin II-related peptide as the urotensin II-immunoreactive molecule in the rat brain. *Biochem. Biophys. Res. Commun*, 310: 860-868.

Tachibana T, Kaiya H, Denbow D M, Kangawa K, Furuse M. 2006. Central ghrelin acts as an anti-dipsogenic peptide in chicks. *Neurosci. Lett*, 405: 241-245.

Takei Y. 1977. The role of subfornical organ in drinking induced by angiotensin in the Japanese quail, *Coturnix coturnix japonica*. *Cell Tiss. Res*, 185: 175-185.

Takei Y. 1988. Changes in blood volume after alteration of hydromineral balance in conscious eels, *Anguilla japonica*. *Comp. Biochem. Physiol*, 91A: 293-297.

Takei Y. 2000. Comparative physiology of body fluid regulation in vertebrates with special reference to thirst regulation. *Jpn. J. Physiol*, 50: 171-186.

Takei Y. 2002. Hormonal control of drinking in the eel: An evolutionary approach. In: Hazon N, Flik G, Ed. *Osmoregulation and Drinking in Vertebrates*. BIOS Scientific Publishers Ltd.: Oxford, 61-82.

Takei Y. 2008. Exploring novel hormones essential for seawater adaptation in teleost fish. *Gen. Comp. Endocrinol*, 157: 3-13.

Takei Y, Hirose S. 2002. The natriuretic peptide system in eel: A key endocrine system for euryhalinity?. *Am. J. Physiol*, 282: R940-R951.

Takei Y, Tsuchida T. 2000. Role of the renin-angiotensin system in drinking of seawater-adapted eels *Anguilla japonica*. *Am. J. Physiol*, 279: R1105-R1111.

Takei Y, Hirano T, Kobayashi H. 1979. Angiotensin and water intake in the Japanese eel, *Anguilla japonica*. *Gen. Comp. Endocrinol*, 38: 446-475.

Takei Y, Okawara Y, Kobayashi H. 1988. Water intake caused by water deprivation in the quail, *Coturnix coturnix japonica*. *J. Comp. Physiol. B*, 158: 519-525.

Takei Y, Okawara Y, Kobayashi H. 1988. Drinking induced by cellular dehydration in the quail, *Coturnix coturnix japonica*. *Comp. Biochem. Physiol*, 90A: 291-296.

Takei Y, Okubo J, Yamaguchi K. 1988. Effect of cellular dehydration on drinking and plasma angiotensin II level in the eel, *Anguilla japonica*. *Zool. Sci*, 5: 43-51.

Takei Y, Okawara Y, Kobayashi H. 1989. Control of drinking in birds. In: Hughes M R, Chadwick A C, Ed. *Progress in Avian Osmoregulation*. Leeds Philosophical and Literary Society Ltd: Leeds, 1-12.

Takei Y, Tsuchida T, Tanakadate A. 1998. Evaluation of water intake in seawater adaptation in eels using a synchronized drop counter and pulse injector system. *Zool. Sci*, 15: 677-682.

Takei Y, Tsuchida T, Li Z, Conlon J M. 2001. Antidipsogenic effect of eel bradykinin in the eel, *Anguilla japonica*. *Am. J. Physiol*, 281: R1090-R1096.

Takei Y, Inoue K, Ogoshi M, Kawahara T, Bannai H, Miyano S. 2004. Mammalian homolog of fish adrenomedullin 2: Identification of a novel cardiovascular and renal regulator. *FEBS Lett*, 556: 53-58.

Takei Y, Joss J M P, Kloas W, Rankin J C. 2004. Identification of angiotensin I in several vertebrate species: Its structural and functional evolution. *Gen. Comp. Endocrinol*, 135: 286-292.

Takei Y, Ogoshi M, Inoue K. 2007. A "reverse" phylogenetic approach for identification of novel osmoregulatory and cardiovascular hormones in vertebrates. *Frontiers Neuroendocrinol*, 28: 143-160.

Takei Y, Hashimoto H, Inoue K, Osaki T, Yoshizawa-Kumagaye K, Watanabe T X, Minamino N, Ueta Y. 2008. Central and peripheral cardiovascular actions of adrenomedullin 5, a novel member of the calcitonin gene-related peptide family, in mammals. *J. Endocrinol*, 197: 391-400.

Tierney M L, Luke G, Cramb G, Hazon N. 1995. The role of the renin-angiotensin system in the control of blood pressure and drinking in the European eel, Anguilla anguilla. *Gen. Comp. Endocrinol*, 100: 39-48.

Tierney M L, Hamano K, Anderson G, Takei Y, Ashida K, Hazon N. 1997. Interactions between the renin-angiotensin system and catecholamines on the cardiovascular system of elasmobranchs. *Fish. Physiol. Biochem*, 17: 333-337.

Tierney M L, Takei Y, Hazon N. 1997. The presence of angiotensin II receptors in elasmobranchs. *Gen. Comp. Endocrinol*, 105: 9-17.

Tipsmark C K, Kiilerich P, Nilsen T O, Ebbesson L O, Stefansson S O, Madsen S S. 2008. Branchial expression patterns of claudin isoforms in Atlantic salmon during seawater acclimation and smoltification. *Am. J. Physiol*, 294: R1563-R1574.

Tipsmark C K, Luckenbach J A, Madsen S S, Kiilerich P, Borski R J. 2008. Osmoregulation and expression of ion transport proteins and putative claudins in the gill of southern flounder (Paralichthys lethostigma). *Comp. Biochem. Physiol. A*, 150: 265-273.

Tostivint H, Joly L, Lihrmann I, Parmentier C, Lebon A, Morisson M, Calas A, Ekker M, Vaudry H. 2006. Comparative genomics provides evidence for close evolutionary relationships between the urotensin II and somatostatin gene families. *Proc. Natl. Acad. Sci. USA*, 103: 2237-2242.

Tsuchida T, Takei Y. 1998. Effects of homologous atrial natriuretic peptide on drinking and plasma angiotensin II level in eels. *Am. J. Physiol*, 275: R1605-R1610.

Tsukada T, Takei Y. 2006. Integrative approach to osmoregulatory action of atrial natriuretic peptide in seawater eels. *Gen. Comp. Endocrinol*, 147: 31-38.

Tsukada T, Nobata S, Hyodo S, Takei Y. 2007. Area postrema, a brain circumventricular organ, is the site of antidipsogenic action of circulating atrial natriuretic peptide in eels. *J. Exp. Biol*, 210: 3970-3978.

Tsukamoto K, Nakai I, Tesch F W. 1998. Do all freshwater eels migrate?. *Nature*, 396: 635-636.

Tsuneki K, Kobayashi H, Pang P K T. 1984. Electron microscope study of innervation of smooth muscle cells surrounding collecting tubules of the fish kidney. *Cell Tiss. Res*, 238: 307-312.

Unniappan S, Canosa L F, Peter R E. 2004. Orexigenic actions of ghrelin in goldfish: feeding-induced changes in rain and gut mRNA expression and serum levels, and responses to central and peripheral injections. *Neuroendocrinology*, 79: 100-108.

Warne J M. 2001. Cloning and characterisation of an arginine vasotocin receptor from the euryhaline flounder *Platichthys flesus*. *Gen. Comp. Endocrinol*, 122: 312-319.

Warne J M, Balment R J. 1995. Effect of acute manipulation of blood volume and osmolality on plasma [AVT] in seawater flounder. *Am. J. Physiol*, 269: R1107-R1112.

Warne J M, Balment R J. 1997. Changes in plasma arginine vasotocin (AVT) concentration and dorsal aortic blood pressure following AVT injection in the teleost *Platichthys flesus*. *Gen. Comp. Endocrinol*, 105: 358-364.

Warne J M, Harding K E, Balment R J. 2002. Neurohypohysial hormones and renal function in fish and mammals. *Comp. Biochem. Physiol*, 132B: 231-237.

Warne J M, Bond H, Weybourne E, Sahajpal V, Lu W, Balment R J. 2005. Altered plasma and pituitary arginine vasotocin and hypothalamic provasotocin expression in flounder (*Platichthys flesus*) following hypertonic challenge and distribution of vasotocin receptors within the kidney. *Gen. Comp. Endocrinol*, 144: 240-247.

Weld M M, Fryer J N. 1987. Stimulation by angiotensin I and II of ACTH release from goldfish pituitary cell columns. *Gen. Comp. Endocrinol*, 68: 19-27.

Wells A, Anderson W G, Hazon N. 2002. Development of an *in situ* perfused kidney preparation for elasmobranch fish: Action of arginine vasotocin. *Am. J. Physiol*, 282: R1636-R1642.

Wells A, Anderson W G, Hazon N. 2005. Glomerular effects of AVT on the *in situ* perfused trunk preparation of the dogfish. *Ann. N. Y. Acad. Sci*, 1040: 515-517.

Wells A, Anderson W G, Cains J E, Cooper M W, Hazon N. 2006. Effects of angiotensin II and C-type natriuretic peptide on the *in situ* perfused trunk preparation of the dogfish, *Scyliorhinus canicula*. *Gen. Comp. Endocrinol*, 145: 109-115.

Westring C G, Ando H, Kitahashi T, Bhandari R K, Ueda H, Urano A, Dores R M, Sher A A, Danielson P B. 2008. Seasonal changes in CRF-1 and urotensin I transcript in masu salmon: Correlation with cortisol secretion during spawning. *Gen.*

Comp. Endocrinol, 155: 126-140.

Winter M J, Hubbard P C, McCrohan C R, Balment R J. 1999. A homologous radio-immunoassay for the measurement of urotensin II in the euryhaline flounder, *Platichthys flesus*. *Gen. Comp. Endocrinol*, 114: 249-256.

Winter M J, Ashworth A, Bond H, Brierley M J, McCrohan C R, Balment R J. 2000. The caudal neurosecretory system: control and function of a novel neuroendocrine system in fish. *Biochem. Cell Biol*, 78: 1-11.

Wong M K S, Takei Y, Woo N Y S. 2006. Differential status of the renin-angiotensin system of the silver bream (*Sparus sarba*) in different salinities. *Gen. Comp. Endocrinol*, 149: 81-89.

第9章 内分泌调控食物摄取

对于脊椎动物,食物摄取的调控是一个复杂的过程,包括许多中枢的与外周的内分泌因子。这些因子的作用受到内在的与外界的变化因素的调节,包括能量储备、代谢燃料使用、个体发育、生殖状态、外界环境,等等。至今,分子和行为的证据表明食物摄取的调控在脊椎动物谱系中是相当保守的,至少在鱼类和哺乳类当中是如此的。然而,在恒温动物的哺乳类和变温动物的鱼类之间生理学上的差异表明摄食行为的内分泌调控可能包含着两大动物类群各自不同的特异性作用机理。本章综述把重点放在我们现有的鱼类摄食内分泌调控的相关知识上,并且将注重如何把我们在鱼类系统中的知识整合到更为广阔的脊椎动物研究领域中。

9.1 导　　言

9.1.1 脊椎动物摄食的一般观察

对于所有的脊椎动物,能量平衡都是通过多种通路调控的,其中包括关键的刺激食欲的(食欲的)或者抑制食欲的(厌食的)因子。摄食的调控不仅包括中枢神经系统(CNS),亦包括外周器官如胃肠道(GI)和脂肪组织。脑,特别是下丘脑,对能量稳态的调节起着关键性作用(Valassi 等,2008)。起初设想是在下丘脑特异性的脑核影响下,而现在认为摄食是由一些神经元回路(neuronal circuit)所调控,它们整合反映动物能量状态的外周代谢的、内分泌的和神经元的信号。外周的激素通过迷走神经把信息传送到中枢的摄食中心,或者穿过血脑屏障直接作用于脑(Brightman 和 Broadwell,1976)。这些外周激素部分是由于食物引起的反应在消化道内释放(Holmgren 和 Olsson,本书第10章)。鱼类和哺乳类一样,一些脑区参与食物摄取的调控。这不仅包括下丘脑,还包括下丘脑以外的脑区,如腹端脑和嗅束(Peter,1979)。近年来,许多哺乳类的食欲调节肽已经在鱼类中鉴定出来(Volkoff 等,2005;Gorissen 等,2006)。这些肽类,包括中枢的和外周的因子,在鱼类中都具有和在哺乳类一样的食欲调控作用(见表9.1),表明虽然存在着群体特异性的差别,但在脊椎动物谱系中,食物摄取的调控是相当保守的。表9.1列出了至今所知的鱼类同源物(相应物)并归纳了它们对摄食的作用以及它们对营养状态的反应。

表9.1 鱼类食欲调控的内分泌因子，以及它们的组织分布、
对摄食的影响和对营养状态变化的 mRNA 表达的反应

包括提要：①编码食欲调节肽的 mRNA 在主要组织中的表达；②食欲调节因子通过中枢或外周施用后对摄食活动的影响；③禁食引起 mRNA 表达的变化；④餐后 mRNA 表达的变化。详细资料和更多参考资料可参阅正文中对特异性因子的小分段介绍。

内分泌因子	mRNA 表达的组织	食欲调节的作用和报道的最低有效剂量		禁食引起的 mRNA 表达的变化	餐后 mRNA 表达的变化
		脑腔内给药	腹腔内给药		
神经肽 Y（NPY）	脑 消化道	刺激，金鱼 0.5 ng/g gfNPY[1] 100 ng/g pNPY[2] 鳟鱼	没有作用，金鱼 0.1 或 0.33 μg/g pNPY[2]	是，+ve：消化道，脑[5-7]	是，+ve：脑[5-8]
	脑垂体	400 ng/g pNPY[3] 鲶鱼 50 ng/g pNPY[4]			
肽 Y	消化道 脑	ND	ND	是，+ve：消化道[9]	ND
食欲肽（OX）	脑 消化道	刺激，金鱼 1 ng/g hOX-A[10] 10 ng/g hOX-B[10]	ND	是，+ve：脑[11,12]	是，+ve：脑[13]
甘丙肽	脑 消化道	刺激，金鱼，丁鲹鱼 10 ng/g rGAL[14,15] 100 ng/g PGAL[14]	没有作用，金鱼，丁鲹鱼 100 ng/g PGAL[14]	是，+ve：脑[16]	没有，脑[16]
生长素释放肽（GHL）	消化道 脑	刺激，金鱼 10 ng/g gfGHL（[1-12][17]） 1 ng/g hGHL[17] ～0.5 ng/g gfAcylGHL[18]	刺激，金鱼 ～5 ng/g gfAcyl-GHL[17] 鳟鱼 ～3 ng/g GHL[19] 罗非鱼 ～1 ngg/h til GHL-C10[20]	是，+ve：消化道，脑[17,21,22]	是，-ve：脑，消化道[17,23]

续表 9.1

内分泌因子	mRNA表达的组织	食欲调节的作用和报道的最低有效剂量		禁食引起的mRNA表达的变化	餐后mRNA表达的变化
		脑腔内给药	腹腔内给药		
可卡因-安非他明-调节转录体（CART）	脑 消化道 性腺	抑制，金鱼 1 ng/g hCART 55-102[24] 10 ng/g CART62-76[24]	ND	是， −ve：脑[8,25,26]	是， +ve：脑[8] −ve：脑[25]
促肾上腺皮质激素释放因子（CRF）	脑	抑制，金鱼 2 ng/g rCRF[27] 丁鲹 ~200 ng/g oCRF[28]	ND	是， +ve：脑[29]	ND
硬骨鱼尾紧张肽（UI）	脑	抑制，金鱼 2 ng/g gfUI[27]	ND	是， +ve：脑[29]	ND
皮质醇	肾间腺	ND	抑制，金鱼 250 μg/g[30]	ND	ND
催乳激素释放肽（PrRP）	脑	抑制，金鱼 10 ng/g gfPrRP[31]	抑制，金鱼 25 ng/g gfPrRP[31]	是， +ve：脑[31]	是， +ve：脑[31]
速激肽	脑 消化道 神经元	ND	ND	ND	是， +ve：脑[32]
内大麻素（Endocannabinoids）	脑	ND	刺激/抑制，金鱼 +ve，1 pg/g 花生四烯酸乙醇酰胺[33] −ve，10 pg/g 花生四烯酸乙醇酰胺[33]	ND	ND

续表9.1

内分泌因子	mRNA表达的组织	食欲调节的作用和报道的最低有效剂量		禁食引起的mRNA表达的变化	餐后mRNA表达的变化
		脑腔内给药	腹腔内给药		
缩胆囊肽（CCK）	消化道 脑	抑制，金鱼 5ng/g CCK-8[34]	抑制，金鱼 50 ng/g CCK-8[34] 鲶鱼 12.5 ng/g CCK-8[4]	是，+ve：消化道[7,35]	是，+ve：消化道，脑[35,36]
胃泌素释放肽/铃蟾肽（BBS）	消化道 脑	抑制，金鱼 5 ng/g BBS[37]	抑制，金鱼 50 ng/g BBS[37]	是，+ve：消化道[38]	ND
支链淀粉	脑 性腺	抑制，金鱼 10 ng/g 支链淀粉[39]	抑制，金鱼 100 ng/g 支链淀粉[39]	ND	ND
降钙素基因相关肽（CGRP）	脑 脑垂体 消化道 性腺	抑制，金鱼 10 ng/g hCGRP[40]	ND	ND	ND
垂体中间叶激素（IM）	脑 脑垂体 消化道 性腺	抑制，金鱼 10 ng/g pfIM[40]	ND	ND	ND
神经调节肽U（NMU）	脑 消化道 性腺	抑制，金鱼 ～5 ng/g NMU-21[41]	ND	是 −ve：脑[41]	ND
瘦素	肝脏	抑制，金鱼 100 ng/g h瘦素[42]	抑制，金鱼 300 ng/g h瘦素[42] 鳟鱼 720 ng/g tr瘦素[43]	ND	是，+ve：肝脏[44]
胰高血糖素样肽-1（GLP-1）	胰腺 消化道	抑制，鲶鱼 0.25 ng/g cfGLP-1[45] 0.25 ng/g hGLP-1[45]	抑制，鲶鱼 150 ng/g cfGLP-1，iv[45]	ND	ND

续表9.1

内分泌因子	mRNA表达的组织	食欲调节的作用和报道的最低有效剂量		禁食引起的mRNA表达的变化	餐后mRNA表达的变化
		脑腔内给药	腹腔内给药		
胰岛素	胰腺	抑制/无,鳟鱼 −ve: 9 ng/g[46] 鲶鱼 无: 25 ng/g 和 50 ng/g[4]	抑制,鳟鱼 4 μg/g[46]	ND	ND
胰岛素样生长因子(IGFs)	肝脏 肌肉	ND	ND	是, −ve: 肌肉, 肝脏[47-49]	ND
生长激素(GH)	脑垂体	ND	刺激,鳟鱼 5 μg/g oGH[52] 25 μg/g oGH 埋植[51]	是, +ve: 脑垂体[47,49]	ND
脑垂体腺苷酸环化酶激活多肽(PACAP)	脑	抑制,金鱼 30 ng/g fPACAP[52]	抑制,金鱼 140 ng/g fPACAP[52]	无,脑[52,53]	ND
生长抑素	脑 消化道 胰腺	ND	没有作用,鳟鱼 ~1 ngg/h 20天, 埋植[54]	是, +ve: 脑, 胰脏[53,55]	ND
黑色素细胞刺激激素(MSH)	脑	抑制,金鱼 ~20 ng/gg NDP-α-MSH[56] ~15-20 ng/gg MTⅡ[57]	ND	ND	ND
灰蛋白相关蛋白	脑	ND	ND	是, +ve: 脑[58,59]	ND
黑色素浓集激素(MCH)	脑	抑制,金鱼 ~0.5 ng/gg flMCH[50] ~10 ng/gg hMCH[60]	ND	是, +ve: 脑[61]	ND

续表9.1

内分泌因子	mRNA表达的组织	食欲调节的作用和报道的最低有效剂量		禁食引起的mRNA表达的变化	餐后mRNA表达的变化
		脑腔内给药	腹腔内给药		
促性腺激素释放激素（GnRH）	脑	抑制，金鱼 0.1～0.5 ng/g cGnRH[62,63]	ND	ND	ND
睾酮	性腺	ND	抑制，海鲈 60 μg/g 埋植[64]	ND	ND
雌二醇	性腺	ND	抑制，海鲈 60 g/g 埋植[64]	ND	ND

+ve，mRNA 表达正变化（上调）；-ve，mRNA 表达负变化（下调）；ND，未确定；iv，静脉内。词头；h，人；p，猪；o，羊；r，大鼠；c，鸡；f，青蛙；gf，金鱼；til，罗非鱼；tr，鳟鱼；cf，鲶；fl，鲽鱼。

参考文献：[1] Narnaware et al. (2000)；[2] Lopez-Patino et al. (1999)；[3] Aldegunde and Mancebo (2006)；[4] Silverstein and Plisetskaya (2000)；[5] Narnaware and Peter (2001)；[6] Silverstein et al., 1999a；[7] MacDonald and Volkoff (2009)；[8] Kehoe and Volkoff (2007)；[9] Murashita et al. (2007)；[10] Volkoff et al. (1999)；[11] Nakamachi et al. (2006)；[12] Novak et al (2005)；[13] Xu and Volkoff (2007)；[14] de Pedro et al. (1995a)；[15] Volkoff and Peter (2001b)；[16] Unniappan et al. (2004b)；[17] Unniappan et al. (2004a)；[18] Matsuda et al. (2006a)；[19] Shepherd et al. (2007)；[20] Riley et al. (2005)；[21] Terova et al. (2008)；[22] Amole and Unniappan (2009)；[23] Peddu et al. (2009)；[24] Volkoff and Peter (2000)；[25] Volkoff and Peter (2001a)；[26] Kobayashi et al. (2008a)；[27] Bernier and Peter (2001)；[28] De Pedro et al. (1995)；[29] Bernier and Craig (2005)；[30] Gregory and Wood (1999)；[31] Kelly and Peter (2006)；[32] Peyon et al. (2000)；[33] Valenti et al. (2005)；[34] Himick and Peter (1994b)；[35] Murashita et al. (2006)；[36] Peyon et al. (1999)；[37] Himick and Peter (1994a)；[38] Xu and Volkoff (2009)；[39] Thavanathan and Volkoff (2006)；[40] Martinez-Alvarez et al. (2008)；[41] Maruyama et al. (2008)；[42] Volkoff et al. (2003)；[43] Murashita et al. (2008)；[44] Huising et al. (2006a)；[45] Silverstein et al. (2001)；[46] Soengas and Aldegunde (2004)；[47] Ayson et al. (2007)；[48] Terova et al. (2007)；[49] Pedroso et al. (2006)；[50] Johnsson and Bjornsson (1994)；[51] Johansson et al. (2005)；[52] Matsuda et al. (2005)；[53] Xu and Volkoff (2008)；[54] Very et al. (2001)；[55] Ehrman et al. (2002)；[56] Cerdá-Reverter et al. (2003)；[57] Matsuda et al. (2008b)；[58] Cerdá-Reverter et al. (2003)；[59] Song et al. (2003)；[60] Matsuda et al. (2006)；[61] Takahashi et al. (2004)；[62] Hoskins et al. (2008)；[63] Matsuda et al. (2008)；[64] Leal et al. (2009)。

9.1.2 鱼类作为摄食调控研究的模型

9.1.2.1 和哺乳类模型比较

越来越多的证据表明调控摄食行为的基本作用机理在脊椎动物中是相当保守的。作为试验的模型，鱼类是很有用的，因为它们通常可以忍受侵害性的试验操作（如注射）或者重复的取样而对摄食行为很少或者没有影响。此外，鱼类在形态、生态、行为和基因组方面都表现出非常明显的多样性（Volff，2004），使它们非常适合于研究脊椎动物食欲-调控系统的进化发展。这种多样性以及在恒温的哺乳类和变温的鱼类之间存在着较多的解剖学与生理学上的差别，表明内分泌调控摄食活动所包含的分子与作用机理可能对一定的种类或一定的类群都是特异性的。哺乳类和鱼类之间明显的解剖学差别包括脑和消化道的形态学差异以及鱼类具有的特异性器官而哺乳类是没有的（如尾神经内分泌器官）。在鱼类当中，亦存在解剖的和功能的差别。例如，鱼类摄食习性的范围有杂食性和肉食性，从而使胃肠道的形态构造和消化道的激素型式产生差异（Holmgren 和 Olsson，本书第 10 章）。在鱼类中，脑的神经肽类分布亦有所不同（Cerda-Reverter 和 Canosa，本书第 1 章）。鱼类的能量代谢虽然和哺乳类与鸟类相似，但鱼类无须花费能量去保持和外界环境不同的体温，而且只需要不多的能量就能将废弃的含氮产物排出体外。此外，和大多数有限生长的哺乳类不同，许多鱼类的生长是无限的，它们没有固定的大小，有的可以终生持续地生长。鱼类不仅需要摄食以保持它们的基础代谢，而且会在它们的生命过程中经历生殖与生长之间资源分配难以全面兼顾的状况（Heino 和 Kaitala，1999）。鱼类亦会遭受一系列环境的冲击，因而形成许多摄食的适应性，例如能应对长时间的禁食。

9.1.2.2 研究鱼类摄食的方法

虽然概念是简单的，但测定鱼类的食物摄取并不那么容易，并且需采用一系列途径。对野生鱼类可以通过检测胃或消化道内食物，或者使用模型估算捕获物分类群（taxa）比例的变量以评定食物的摄食情况（Godby 等，2007）。在实验室里，较精确的食物摄取可以通过定量计算每尾鱼取食的食物颗粒数量得到（Himick 和 Peter，1994a；Volkoff 等，1999）。通过计算饱满的和空的消化道之间的质量差别（Delicio 和 Vicentini-Paulino，1993）或者留在水池的食物和鱼吞食的食物之间的质量差别（Bernier 等，2004），通过定量计算放射性标志的食物（Ronnestad 等，2007）或者用 X-射线检测含有不透明微球的颗粒饲料（Craig 等，2005）等等，都可以评价鱼类的食物摄食情况。处理大量的鱼样本时，个别鱼的摄食量可以通过这群鱼的数量和平均鱼体质量除以整群鱼所取食的食物质量得到（Kehoe 和 Volkoff，2008），亦可以使用水下传感器检测没有吃完的食物（Noble 等，2007），或者计算经过自身摄食装置训练的鱼所需要的食物量获得（Millot 等，2008）。通过使用不同的激素处理，例如有些肽类由外周注射（腹腔注射，ip），或者脑注射（脑腔内注射，icv）（Silverstein 等，2001；Volkoff 等，2003），或者口服（Bernier，2006），通过腹腔埋植小丸（Johansson 等，2005），或者使用渗透性小泵（Riley 等，2003）等，都可以直接评定这些肽对鱼类摄食和代谢的影响。通过检测这些肽类在鱼类处于不同状况下（如饥饿、应激反应）的血液含量或者蛋白质/mRNA 表达水平，亦可以间接评估这些激素的作用。值得注意的是，鱼类基因组计划的进展和新的技术（如转基因和基因敲除）（Volff，2004）将会促进对鱼类摄食调控研究有学术价值的遗传模型的发展。

9.2 内分泌调控

食欲-调控因子可以分为饥饿信号与饱食信号、中枢信号和外周信号。本节将介绍所有已知的鱼类主要食欲调节剂（见图9.1和表9.1的总结）。

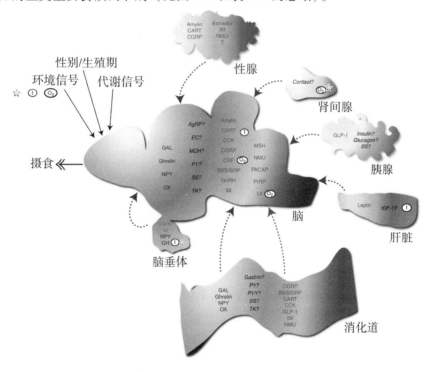

图9.1 目前所了解的激素和其他因子调控鱼类摄食活动的总结

注意：本图既不表示和种类特异性联系的许多激素以及它们的存在、定位和作用，亦不表示激素之间的相互作用。蓝色的肽是食欲性的，红色的肽是厌食性的，黑色斜体字的肽和带有"?"号的肽是对摄食的作用不了解或不清楚的。破折号弯曲的红箭头表示血液联系，而在脑与消化道之间的灰色破折号箭头表示神经（迷走）联系。激素附近圆圈内的"t"、"★"和"O₂"表示激素分别受到温度、光周期或氧气水平的影响。AgRP，野灰蛋白相关肽；BBS/GRP，铃蟾肽/胃泌素释放肽；CART，可卡因-安非他明调节转录体；CCK，缩胆囊肽；CGRP，降钙素基因-相关肽；CRF，促肾上腺素皮质激素释放因子；EC，内大麻素；GAL，甘丙肽；GH，生长激素；GLP-1，胰高血糖素样肽-1；GnRH，促性腺激素释放激素；IGF-I，胰岛素样生长因子-1；IM，垂体中间叶激素；MCH，黑色素浓集激素；MSH，黑色素细胞刺激激素；NMU，神经调节肽 U；NPY，神经肽 Y；OX，食欲肽；PACAP，脑垂体腺苷酸环化酶激活多肽；PY，肽 Y；PYY，肽 YY；SS，生长抑素；T，睾酮；TK，速激肽；UⅠ，硬骨鱼尾紧张肽Ⅰ。（见书后彩图）

9.2.1 饥饿信号

9.2.1.1 脑的信号

1. 神经肽 Y 家族肽

神经肽 Y（NPY）家族的肽类由 NPY 和肽 YY（PYY，发现于脊椎动物所有各纲）、胰多肽（PP，只在四足类的胰脏内发现）和肽 Y（PY，只在一些硬骨鱼类中发现）（Hoyle，1999）组成，它们有 5 个已知的受体，即 Y1、Y2、Y4、Y5 和 Y6（Kamiji 和 Inui，2007）。在哺乳类中，NPY 在 CNS 中是丰富的，特别是在参与摄食调控的下丘脑的核内，而且是已知的强有力的食欲肽之一，其作用由 Y1 和 Y5 两个受体亚型介导（Gao 和 Horvath，2007）。所有鱼类都产生 NPY 和 PYY，而只有一些硬骨鱼类产生 PY（Cerdá-Reverter 等，2000a）。NPY 序列已经在几种鱼类中测定，NPY 神经元广泛分布于鱼类的 CNS（Sundstrom 等，2005；Kehoe 和 Volkoff，2007；Sueiro 等，2007；MacDonald 和 Volkoff，2009）以及脑垂体和消化道（Cerdá-Reverter 等，2000a；Rodríguez-Gómez 等，2001）。至今，和 NPY 与 PYY 结合的 7 个 NPY 受体亚型已经在鱼类当中确定（Y1、Y2、Y4、Y5、Y6、Y7、Y8）（Salaneck 等，2008）。脑腔内而不是腹腔内注射哺乳类或鱼类的 NPY 能够增强鱼类的摄食活动（Lopez-Patino 等，1999；Narnaware 等，2000；Silver-Stein 和 Plisetskaya，2000；Aldegunde 和 Mancebo，2006），但还不清楚是哪一种 NPY 受体亚型介导这个反应（Salaneck 等，2008）。金鱼（Narnaware 和 Peter，2001）、眼斑鳐（*Raja ocellata*）（MacDonald 和 Volkoff，2009）、大麻哈鱼（*Oncorhynchus tshawytscha*）和银大麻哈鱼（*O. kisutch*）（Silverstein 等，1999a）等在禁食后，脑的 NPY mRNA 水平增加；金鱼（Narnaware 和 Peter，2001）、大西洋鳕鱼（*Gadus morhua*）（Kehoe 和 Volkoff，2007）和罗非鱼（*Oreochromis mossambicus*）（Peddu 等，2009）等在进食时，脑的 NPY mRNA 水平发生变化；而大量摄入营养物会影响金鱼（Narnaware 和 Peter，2002）脑的 NPY mRNA 水平。鱼类和哺乳类一样，NPY 对摄食的作用部分是由其他的食欲调节剂调节的，例如 CRF 和皮质醇（Bernier 等，2004）、CART（Volkoff 和 Peter，2000）、瘦素（Volkoff 等，2003）、MCH（Matsuda 等，2008b）、食欲肽（OXs）和甘丙肽（GAL）（Volkoff 和 Peter，2001b）、生长激素（GH）（Mazumdar 等，2006）和生长素释放肽（Miura 等，2006）。PY 和/或 PYY 虽然已在几种硬骨鱼类中鉴定（Cerdá-Reverter 等，2000a），它们对鱼类的生理作用还不清楚。PYY 基因在海鲈（*Dicentrarchus labrax*）（Cerdá-Reverter 等，2000b）和红鳍东方鲀（*Takifugu rubripes*）（Sundstrom 等，2005）的脑中表达，PY 基因在海鲈的脑内（Cerda-Reverter 等，2000b）和五条鰤（*Seriola quinqueradiata*）的肠内（Murashita 等，2007）表达，而且 mRNA 表达随着禁食而增加，表明 PY 是一个摄食的调节剂。

2. 食欲肽（Orexins, OX）

OXs（下丘泌素，hypocretins）由两个肽——食欲肽 A（OX-A）和食欲肽 B（OX-B）组成，它们是由一个前体——前食欲肽原（preproorexin）分裂而产生。在哺乳类中，OXs 主要在外侧下丘脑产生，通过和两个 G-蛋白偶联受体（OX_1 和 OX_2）作用而刺激摄食，控制胃分泌活动（Korczynski 等，2006）并调节睡眠和觉醒（Ohno 和 Sakurai，2007）。在鱼类的相关研究中，已经报道 6 种鱼类的 mRNA 编码前-OX（Faraco 等，2006；Xu 和 Volkoff，2007），它分布在几个脑区，包括下丘脑（Amiya 等，2007；Xu 和 Volkoff，2007）、脑垂体和外周组织如消化道（Xu 和 Volkoff，2007）。在花鲈（*Lateolabrax japonicus*）中，OX-A-免疫反应细胞在脑垂体和 GH-分泌细胞中共定位（Suzuki 等，2007），表明 OXs 在鱼类脑垂体分泌活动的调控中起作用。OXs 似乎能调控鱼类的摄食，因为脑腔注射 OXs 能刺激金鱼的食欲（Volkoff 等，1999；Nakamachi 等，2006）；在禁食的金鱼中，OXs 样免疫反应细胞在下丘脑增加，而注射葡萄糖的鱼则减少（Nakamachi 等，2006）；食物缺乏亦使金鱼（Nakamachi 等，2006）和斑马鱼（Novak 等，2005）脑的前-OX mRNA 水平升高。对大西洋鳕鱼的研究表明，前-OX mRNA 在脑的表达水平在正餐时出现变化，而在喂给低食粮时表达水平升高（Xu 和 Volkoff，2007）。给金鱼脑腔注射 OXs 使活动能力增强（Nakamachi 等，2006），缺乏 OX 受体的斑马鱼（Yokogawa 等，2007），或者 OX-过表达的斑马鱼幼鱼都表现出不正常的睡眠型式，表明 OXs 参与鱼类觉醒状态的调控，而这可以间接地影响摄食和代谢。鱼类和哺乳类一样（Saper，2006；Nishino，2007），OXs 和其他的食欲调节剂互相作用。在金鱼的相关研究中，阻断 OX 受体使 NPY-、GAL-（Volkoff 和 Peter，2001b）和生长素释放肽诱导的（Miura 等，2007）摄食活动降低；而阻断 NPY、GAL 或者生长素释放肽的受体会抑制 OX-诱导的摄食活动。此外，中枢给予 OX 能使 NPY 和生长素释放肽 mRNA 的表达增加（Volkoff 和 Peter，2001b；Miura 等，2007），而用生长素释放肽处理后能使 OX 在脑的表达增加（Miura 等，2007）。对金鱼的研究表明，CART（Volkoff 和 Peter，2000）和瘦素（Volkoff 等，2003）能抑制由 OX-诱导的摄食活动。在青鳉中，MCH-免疫反应纤维和 OX-生产细胞紧密接触（Amiya 等，2007）。这个证据表明 OX 和其他肽能系统之间存在着功能上的互相依赖，以调控金鱼的能量平衡。

3. 甘丙肽（Galanin，GAL）

在哺乳类中，甘丙肽广泛地在 CNS 和肠道表达，通过 GAL1R、GAL2R 和 GAL3R 受体而刺激食物摄取与体重增长（Walton 等，2006；Lang 等，2007）。GAL 曾经在几种鱼类中被分离出来（Volkoff 等，2005），而前甘丙肽原 mRNAs 编码 5 个预定的 GAL 肽已经在金鱼中确定（Unniappan 等，2003）。前甘丙肽原 mRNA（Unniappan 等，2004b）和甘丙肽样免疫反应活性（Jadhao 和 Meyer，2000；Jadhao 和 Pinelli，2001；Riley 等，2005）在鱼脑中广泛分布。脑腔注射而不是腹腔注射哺乳类的 GAL 刺激金鱼（De Pedro 等，1995a；Volkoff 和 Peter，2001b）和丁鲅（*Tinca tinca*）(Guijarro 等，1999）的食物摄取，而禁食使前甘丙肽原 mRNA 在金鱼脑中的表达增加（Unniappan 等，2004b），表明 GAL 在鱼类中起着增强食欲的作用。甘丙肽样肽（galanin-like peptide，GALP）是最近发现的 GAL 肽家族的第二个成员，其氨基酸组成和 GAL 相似，并且和 GAL 受体结合（Lang 等，2007）。学界对于 GALP 的食欲调节作用是有争论的，对啮齿类的相关研究曾报道其有刺激作用和抑制作用。鱼类 GALP 的结构和功能目前还不清楚。

9.2.1.2 外周的信号

生长素释放肽（Ghrelin，GRL）是 28 个氨基酸组成的酰基化肽，主要由胃分泌，亦由脑分泌。对于哺乳类，生长素释放肽刺激 GH 分泌和食欲，而这是仅有的已知 GI 激素证实具有增强食欲的特性（Olszewski 等，2008）。生长素释放肽已经在几种硬骨鱼类（Kaiya 等，2008；Manning 等，2008；Miura 等，2008；Olsson 等，2008；Terova 等，2008；Xu 和 Volkoff，2009）和板鳃鱼类（Kawakoshi 等，2007）中鉴定。鱼类和哺乳类一样，生长素释放肽 mRNA 主要在胃中表达，在脑中有低水平的表达（Kaiya 等，2008）。注射金鱼的或人的生长素释放肽能刺激金鱼的摄食活动（Unniappan 和 Peter，2005；Matsuda 等，2006a；Miura 等，2007，2008），而连续灌注罗非鱼生长素释放肽能增强罗非鱼的食物摄取活动（Riley 等，2005）。给鳟鱼腹腔注射生长素释放肽能使虹鳟的食物摄取增加（Shepherd 等，2007），但对另外一群鳟鱼，生长素释放肽对食物摄取却没有影响（Jonsson 等，2007）。在同一种鱼类中生长素释放肽调节食欲作用的差别可能是由于试验使用肽类的剂量和方法以及食物摄取的数量差别所致。金鱼（Unniappan 等，2004a）和罗非鱼（Peddu 等，2009）下丘脑和消化道前生长素释放肽原的 mRNA 表达以及血浆生长素释放肽水平在就餐时都出现变化。禁食诱导海鲈胃（Terova 等，2008）、金鱼下丘脑与消化道（Unniappan 等，2004a）和斑马鱼脑与消化道（Amole 和 Unniappan，2009）的生长素释放肽 mRNA 表达上调，但并不影响尼罗罗非鱼（*Oreochromis niloticus*）（Parhar 等，2003）和大西洋鳕鱼（Xu 和 Volkoff，2009）的消化道生长素释放肽 mRNA 水平。禁食引起鳟鱼（Jonsson 等，2007）和江鳕（*Lota lota*）（Nieminen 等，2003）类似的血浆生长素释放肽水平降低。对金鱼的研究曾证明生长素释放肽和其他的食欲相关肽相互作用，包括 NPY（Miura 等，2006）、OX（Miura 等，2007）和 GRP/BBS（Canosa 等，2005）。脑腔注射生长素释放肽使前-OX 原和 NPY mRNA 在脑中的表达增加。预先注射食欲肽受体 1A 的拮抗剂 SB334867 能清除生长素释放肽调节食欲的刺激作用，而 NPY Y1 受体拮抗剂能减弱生长素释放肽的食欲刺激作用。这些研究结果表明生长素释放肽的食欲作用是通过 NPY 和 OX-依赖的通路而介导的（Miura 等，2006，2007）。生长素释放肽的受体（GRLR）已在几种鱼类中鉴定（Kaiya 等，2008），它们在脑内高度表达（Chan 等，2004），表明生长素释放肽在中枢的直接作用。生长素释放肽受体在消化道和脑内的存在进一步证明生长素释放肽-GRLR 系统在鱼类中的食欲调节作用（Kaiya 等，2008）。罗非鱼在就餐时脑的 GRLR mRNA 水平明显升高（Peddu 等，2009）。生长素释放肽使离体的斑马鱼肠肌肉收缩，表明它具有调节消化道活动的作用（Olsson 等，2008），而这可能有助于证明生长素释放肽在鱼类中的食欲调控功能。

9.2.2 饱食信号

9.2.2.1 脑的信号

可卡因与安非他明-调节转录体（CART）肽原先是在急性使用心理动作（psychomotor）刺激剂后在大鼠脑内分离出来的一种上调 mRNA，随后能在脑、消化道和胰脏表达（Ekblad，2006；Vicentic 和 Jones，2007）。CART 能抑制哺乳类（Vicentic 和 Jones，2007）和鸟类（Tachibana 等，2003）的摄食，虽然亦曾报道给大鼠在特异性下丘脑区注射 CART 后引起促进食欲的作用（Abbott 等，2001）。在鱼类中，CART mRNA 序列已在金鱼（Volkoff 和 Peter，2001b）、斑点鮰（*Ictalurus punctatus*）（Kobayashi 等，2008a）、眼斑鳢（MacDonald 和 Volkoff，2009）和鳕鱼（Kehoe 和 Volkoff，2007）的相关研究中报道，在金鱼中确定有两个类型的 CART 肽前体，即 CART Ⅰ 和 CART Ⅱ（Volkoff 和 Peter，2001b）。在鱼类中，CART mRNA 存在于脑和外周组织，包括性腺和肾脏。CART 免疫反应亦曾在胡子鲶（*Clarias batrachus*）（Singra 等，2007）的脑和脑垂体中检测到，有意思的是在海蟾鱼（*Thalassophryne nattereri*）（Magalhaes 等，2006）的毒腺亦检测到。给金鱼脑腔注射人的 CART 能抑制摄食（Volkoff 和 Peter，2000），而金鱼（Volkoff 和 Peter，2001b）、斑点鮰（Kobayashi 等，2008a）和大西洋鳕鱼（Kehoe 和 Volkoff，2007）在禁食后，CART mRNA 水平降低，表明 CART 在鱼类中起着厌食因子的作用。金鱼餐后下丘脑 CART 的表达增加（Volkoff 和 Peter，2001a），但在鳕鱼中却是降低的（Kehoe 和 Volkoff，2007）。使用人的 CART 能抑制 NPY 和 OX-A 刺激的金鱼摄食活动，表明 CART 对 NPY 和 OX-A 起着抑制作用（Volkoff 和 Peter，2000）。此外，在鲶鱼脑中，CART 和 NPY 免疫反应轴突紧密联系（Singru 等，2008）。对金鱼的研究亦表明瘦素和 CART 之间存在协同的相互作用（Volkoff 等，2003）。

促肾上腺皮质激素释放因子（CRF）-相关神经肽包括 CRF、几种尾皮质素（urocortin，UCN）、硬骨鱼尾紧张肽（UⅠ）和两栖类蛙皮降压肽（sauvagine），它们对于和 CRF 受体两种亚型（CRF1 和 CRF2）结合的亲和力有所不同（Bernier，2006；Bernier 等，本书第 6 章）。编码 CRF、UⅠ 和 UCNs 以及 CRF1 和 CRF2 的转录体已经在几种硬骨鱼中报道（Alderman 和 Bernier，2007）。鱼类和其他脊椎动物一样，CRF 对应激反应起着重要的作用，但 CRF-表达神经元在脑内广泛分布，表明 CRF 具有多种生理和行为功能（Bernier 等，本书第 6 章），包括对摄食的调控。给金鱼腹腔注射 CRF 并不影响食物摄取（De Pedro 等，1993）。给丁鱥脑腔注射 CRF（De Pedro 等，1995b），以及给金鱼脑腔注射 CRF（De Pedro 等，1993；Bernier 和 Peter，2001）或 UⅠ（Bernier 和 Peter，2001）都抑制摄食活动。在腹腔埋植糖皮质激素受体拮抗物而引起厌食的金鱼，或者用皮质醇合成抑制剂处理的金鱼（Bernier 和 Peter，2001）以及受到低氧影响而厌食的鳟鱼（Bernier 和 Craig，2005）中，都观察到 UI 和 CRF 在脑 mRNA 水平的增高。UCNs 对鱼类食物摄取的调节作用还不清楚，皮质醇的作用亦不清楚。腹腔注射皮质醇对金鱼的食物摄取没有影响（De Pedro 等，1997），但对虹鳟能引起厌食作用（Gregory

和 Wood，1999）。对于斑点鮰（Peterson 和 Small，2005）和金鱼，血浆皮质醇的适度增加能刺激食物摄取与降低 CRF mRNA 的表达，而高剂量的皮质醇使 CRF mRNA 水平降低，但对食物摄取没有影响（Bernier 等，2004）。已经证明 CRF-相关肽在鱼类中能和食欲调控系统的其他成员如羟色胺（De Pedro 等，1998）和 NPY（Bernier 等，2004）相互作用。

RF 酰胺属于多样性的肽类群，其特征是有一个共同的 N-端序列。最初在帘蛤中发现，随后从鸡和哺乳类的脑中分离出来，还包括催乳激素释放肽（PrRP）、RF 酰胺相关肽（RFRPs）、metastin 和 kisspeptins（s）。还没有完全鉴定 RF 酰胺受体的特点（Dockray，2004）。至今所确定的脊椎动物 RF 酰胺，除了 kisspeptin 家族的成员之外，都已被证明参与调节哺乳类的食物摄取（Bechtold 和 Luckman，2007）。目前对于 RF 酰胺在鱼类中的结构和功能还了解得很少。RFRP-表达神经元定位于金鱼的中脑，表明它起着神经内分泌的作用（Sawada 等，2002），而哺乳类 PrRP 的直向同源物已经在一些鱼类中鉴定（Moriyama 等，2002；Seale 等，2002；Moriyama 等，2007）。在金鱼摄食后以及食物缺失 7 天后，下丘脑 PrRP mRNA 的表达明显增加（Kelly 和 Peter，2006）。在金鱼中，PrRP 表现厌食作用，但亦密切地参与水分与盐分平衡的调控（Kelly 和 Peter，2006），这可能是它在鱼类中的主要作用（Takei 和 Balment，本书第 8 章）。

速激肽（Tackykinins，TKs）组成一个小的神经肽家族，其结构特点是有一个共同的 C-端氨基酸序列，分布于脊索动物的神经系统。TK 家族由几个成员组成，包括 P 物质（哺乳类、鸟类和鱼类）、神经激肽（neurokinins）和 Carassin（金鱼）（Conlon 和 Larhammar，2005）。已证实几种鱼类的 TK 免疫反应活性出现在 CNS 和肠的神经元（enteric neuron）（Bermudez 等，2007；Pinuela 和 Northcutt，2007）。已证明 TKs 能引起几种鱼类消化道平滑肌的收缩（Liu 等，2002；Holmberg 等，2004；Kim 等，2005），而前 TK 原 mRNA 在下丘脑的表达在就餐时出现变化，表明 TKs 参与鱼类摄食和消化过程的调节作用（Holmgren 和 Olsson，本书第 10 章）。

内大麻素系统（endocannabinoid system）。内大麻素（ECs）是内源的磷脂衍生物，能激活两个大麻素受体亚型（CB1 和 CB2 受体）。在哺乳类中，ECs 通过下丘脑的 CB1 受体具有中枢的促食欲作用（Despres，2007）。哺乳类大麻素 CB1 和 CB2 受体的直向同源物最近已经在鱼类中鉴定（Elphick 和 Egertova，2001；McPartland 等，2007），但 EC 系统对鱼类食物摄取的调控作用还不清楚。在金鱼中，ECs［（花生四烯酸乙醇酰胺（anandamide）和花生四烯酸甘油（arachidonoylglycerol）］和 CB1-样免疫反应活性分布于整个脑内（Valenti 等，2005）。腹腔给以低剂量（1 pg/g 体重）和中等（10 pg/g 体重）剂量花生四烯酸乙醇酰胺时，分别增加或减少食物摄食，而食物缺失时使花生四烯酸乙醇酰胺（但不是 arachidonoylglycerol）mRNA 水平在端脑增加（Valenti 等，2005）。

其他肽类。神经调节肽（neuromedin U，NMU）最早从猪的脊髓中分离出来，在哺乳类中是一种厌食性神经肽。编码三种 NMU 直向同源物（NMU-21、NMU-25 和 NMU-38）的 cDNA 已经在金鱼中分离出来，NMU mRNAs 在脑、消化道和性腺中表达（Maruyama 等，2008）。在金鱼的相关研究中，禁食使脑原 NMU mRNA 水平在脑内降低，而脑腔注射 NMU-21 能抑制食物摄取（Maruyama 等，2008），表明 NMU 参与鱼类食欲的调节。

降钙素基因相关肽（CGRP）、肾上腺调节素（adrenomedullin，AM）、垂体中间叶激素（intermedin，IM）和支链淀粉一起是结构上相关的肽类，是降钙素/CGRP 肽家族的成员。这三个肽都参与哺乳类食物摄取的调控。在金鱼中，CGRP、IM 和 AM mRNAs 在脑、脑垂体和一些外周组织包括消化道和性腺中表达（Martínezz-Alvarez 等，2008）。脑腔注射 CGRP 或 IM，但不是 AM，能使金鱼的食物摄取活动明显降低（Martínezz-Alvarez 等，2009）。

9.2.2.2 外周信号

1. 缩胆囊肽/促胃液素

GI 肽类的缩胆囊肽（CCK）和促胃液素的特点是有一个共同的四肽序列。在哺乳类中，CCK 存在于脑和胃肠道，而促胃液素只由胃的内分泌细胞产生。这两种肽都有多种生物学功能并且和两种受体亚型（CCK1R 和 CCK2R）结合。CCK 通过迷走通路作用于外周以减弱胃排空，刺激胃分泌活动，减少食物摄食，但是 CCK 和脑的受体结合亦能导致饱食（Raybould，2007）。

CCK/促胃液素-样的免疫反应活动曾出现在一些鱼类的神经系统和消化道中（Aldman 等，1989；Himick 和 Peter，1994b；Bermudez 等，2007）。编码 CCK 的 mRNA 已经在一些鱼类中确定（Peyon 等，1998；Jensen 等，2001；Kurokawa 等，2003），而促胃液素 mRNAs 只在黑青斑河鲀（*Tetraodon nigroviridis*）、牙鲆（*Paralichthys olivaceus*）（Kurokawa 等，2003）和两种鲨鱼——白斑角鲨（*Squalus acanthias*）和鼠鲨（*Lamna cornubica*）（Johnsen 等，1997）中鉴定。CCK mRNA 表达（Peyon 等，1998；Jensen 等，2001；Kurokawa 等，2003；MacDonald 和 Volkoff，2009）和结合部位（Himick 等，1996；Oliver 和 Vigna，1996）出现于脑和肠，而促胃液素 mRNA 只在肠内观察到（Kurokawa 等，2003）。CCK-相关肽影响鱼类消化和摄食过程。当肠内有食物时，它们释放出来抑制胃排空与增强消化道活动能力（Forgan 和 Forster，2007；Holmgren 和 Olsson，本书第 10 章）。CCK 在鱼类中亦起着饱食因子的作用。给金鱼脑腔和腹腔注射 CCK（Himick 和 Peter，1994b；Thavanathan 和 Volkoff，2006）与给斑点鲖脑腔注射 CCK（Silverstein 和 Plisetskaya，2000）都能阻抑食物摄取，而用 CCK 的拮抗物处理后能引起虹鳟增强食物的摄取活动（Gelineau 和 Boujard，2001）。此外，在餐后金鱼脑（Peyon 等，1999）和五条鰤幽门盲囊（Murashita 等，2007）的 CCK mRNA 水平升高。禁食使五条鰤（Murashita 等，2006）和眼斑鳐（MacDonald 和 Volkoff，2009）消化道的 CCK mRNA 水平升高。给虹鳟投喂高脂肪饵料，其血浆的 CCK 水平要较高于投喂高蛋白质饵料的鱼（Jonsson 等，2006），表明鱼类的饲料能影响 CCK 的释放。CCK 能够部分介导瘦素（Volkoff 等，2003）和支链淀粉（Thavanathan 和 Volkoff，2006）对金鱼食物摄取的作用。亦已证明 CCK 和生长抑制（Eilerton 等，1996；Canosa 和 Peter，

2004）之间以及和 GH（Himick 等，1993）之间的相互作用。虽然促胃液素对摄食调控的作用还不清楚，但用促胃液素处理能使大麻哈鱼离体的消化道环收缩，表明它能对鱼类消化道的活动性起作用（Forgan 和 Forster，2007）。

2. 铃蟾肽/GRP

铃蟾肽（BBS）最初是从蛙皮肤分离出来的，和胃泌素释放肽（GRP）是结构上的相关肽，都有一个相似的 C-端序列，并具有相似的生物学功能。在脊椎动物中，BBS/GRP 肽广泛分布于胃肠道和 CNS（McCoy 和 Avery，1990）。BBS/GRP 样肽可以在鱼类的胃肠道和脑中检测到，包括板鳃鱼类（Bjenning 等，1991）和硬骨鱼类（Volkoff 等，2000；Bosi 等，2004；Xu 和 Volkoff，2009）。在鱼类中，BBS 样肽调控胃酸分泌和胃活动能力（Thorndyke 等，1990），并可能起饱食因子的作用，因为给金鱼注射 BBS 能抑制食物摄取（Himick 和 Peter，1994a）。对大西洋鳕鱼的研究表明，投喂高分量饲料的鱼，其 GRP mRNA 在消化道的表达水平要较高于投喂低分量饲料的鱼（Xu 和 Volkoff，2009）。然而在虹鳟的相关研究中，血浆 GRP 水平并不受摄食或饲料成分的影响（Jonsson 等，2006）。对金鱼用 BBS 处理能刺激 GH 释放并减弱前-SS 在前脑的表达（Canosa 和 Peter，2004；Canosa 等，2005），而这些作用部分是由于生长素释放肽受到了抑制，这表明 BBS、SS 和生长素释放肽之间存在着相互作用（Canosa 等，2005）。

3. 支链淀粉（Amylin）

哺乳类的支链淀粉是和胰岛素一起在就餐时由胰脏的 β-细胞分泌出来的。它和 CNS 中的特异性受体结合而抑制营养物-刺激的胰高血糖素分泌，减慢胃排空，减少食物摄取（Lutz，2006）。目前已经在斑马鱼、大西洋鲑、短角床杜父鱼（*Myoxocephahus scorpius*）（Westermark 等，2002）、金鱼（Martínezz-Alvarez 等，2008）和河鲀（Chang 等，2004）中鉴定支链淀粉，并且部分确定它们的推导氨基酸序列特点。支链淀粉的免疫反应活性已经在杜父鱼 Brockmann 小体和大西洋鲑内分泌胰脏的胰岛素细胞内观察到，表明鱼类和哺乳类一样，支链淀粉是在外周产生的（Westermark 等，2002）。在金鱼中，支链淀粉 mRNA 主要在脑表达，但亦出现在脑垂体、性腺、肾脏和肌肉中（Martínezz-Alvarez 等，2008）。尽管目前对支链淀粉在鱼类中的生理作用还了解得很少，给金鱼脑腔和腹腔注射人的支链淀粉都能降低食物摄取（Thavanathan 和 Volkoff，2006），表明它在鱼类起着厌食性作用。

4. 瘦素（Leptin）

在哺乳类中，瘦素由脂肪细胞产生，调控食物摄取、能量平衡和生殖活动。虽然瘦素样的免疫反应物质起初曾在几种鱼类的胃肠道（Muruzabal 等，2002；Bosi 等，2004）、血液（Mustonen 等，2002；Nieminen 等，2003；Nagasaka 等，2006）、脑、肝脏（Johnson 等，2000）和脂肪（Yaghoubian 等，2001；Vegusdal 等，2003）中检测到，但直到最近才鉴定鱼类的瘦素特征（Kurokawa 等，2005；Huising 等，2006a）。在鱼类的瘦素之间以及鱼类和哺乳类的瘦素之间只有低的氨基酸同一性（Kurokawa 等，2005）。和哺乳类瘦素主要在脂肪组织中表达形成对照的是鱼类瘦素的主要表达部位是肝脏（Kurokawa 等，2005；Huising 等，2006a、b）。最近报道编码瘦素受体的 cDNA 在青鳉（*Oryzias melastigma*）的脑和外周组织中大量表达（Wong 等，2007），以及在河鲀的脑垂体和卵巢中大量表达（Kurokawa 等，2008）。学界对于瘦素调节鱼类摄食的作用一直是有争论的。虽然用哺乳类瘦素处理对银大麻哈鱼（Baker 等，2000）或斑点鲷（Siluerstein 和 Plisetskaya，2000）没有明显的作用，但能使金鱼降低食物摄取（Volkoff 等，2003；De Pedro 等，2006）。对于金鱼，瘦素能增强 CART 和 CCK 的作用，而抑制 NPY 和 OX-A 的作用（Volkoff 和 Peter，2000；Volkoff 等，2003）。腹腔注射基因重组的鳟鱼瘦素能阻抑食物摄取，减弱 NPY mRNA 的表达，增加 POMC mRNA 水平（Murashita 等，2008）。对于产卵的香鱼，血液中高的瘦素免疫反应水平是和食欲的降低相联系的（Nagasaka 等，2006），禁食使江鳕（*Lota lota*）血浆中免疫反应的瘦素水平降低（Nieminen 等，2003），鲤鱼在喂食后肝脏瘦素 mRNA 的表达达到高峰（Huising 等，2006b），而使用哺乳类瘦素处理后使蓝太阳鱼（*Lepomis cyanellus*）的脂肪代谢增强（Londraville 和 Duvall，2002）。但是，禁食并不影响鲤鱼肝脏的瘦素表达（Huising 等，2006b）。

5. 胰高血糖素/胰高血糖素样肽-1 和胰岛素/胰岛素样生长因子（IGF）

胰高血糖素和胰高血糖素样肽（GLP）是肠促胰液素（secretin）家族的肽类，它还包括肠抑胃肽（gastric inhibitory peptide，GIP）、肠促胰液素、血管活性肠肽（VIP）和泌酸调节肽（oxyntomodulin）（Nelson 和 Sheridan，2006）。脊椎动物的前胰高血糖素基因编码 3 个胰高血糖样的序列［胰高血糖素、胰高血糖素样-肽 1（GLP-1）和胰高血糖素样-肽 2（GLP-2）］，它们对调控代谢活动有不同的功能（Irwin 和 Wong，2005）。在鱼类中，一个胰脏的胰高血糖素基因产生胰高血糖素和 GLP-1，而一个肠的基因编码所有 3 个胰高血糖素肽以及泌酸调节肽（Navarro 等，1990）。虽然没有证据表明胰高血糖素调控鱼类摄食的作用，但胰高血糖素是影响代谢活动的。用胰高血糖素处理能提高血浆脂肪酸水平（Albalat 等，2005），而高葡萄糖含量的饲料能使血液循环的胰高血糖素水平降低（del Sol Novoa 等，2004）。已经在几种鱼类中鉴定 GLP-1，它由胰脏和肠道产生（Nelson 和 Sheridan，2006）。在斑点鲷脑腔和腹腔注射 GLP-1 使食物摄取减少（Silverstein 等，2001），并通过糖原分解（glycogenolysis）和糖异生（gluconeogenesis）而影响代谢活动（Silverstein 等，2001）。和 GLP-1 在哺乳类中影响胃活动能力不同的是人的 GLP-1 对大麻哈鱼消化道环的活动能力并没有影响（Forgan 和 Forster，2007）。在肠促胰液素家族的成员中，只有 VIP 影响食物摄取，因为腹腔或脑腔注射 VIP 都使金鱼减少食物的摄取（Matsuda 等，2005a）。

在一些鱼类中已经鉴定胰岛素（Irwin，2004），并在胰岛中观察到胰岛素免疫反应细胞（Nelson 和 Sheridan，2006）。鱼类和其他脊椎动物一样，胰岛素起着合成代谢因子的作用，刺激糖原合成、脂肪生成和蛋白质合成（Nelson 和 Sheridan，2006）。但是，鱼类利用碳水化合物的能力低（Albalat 等，2007），而葡萄糖亦只能微弱地刺激胰岛素分泌，表明其他的化合物，可能是氨基酸成为主要的促胰岛素因子（insulinotropic factor）（Andoh，2007）。胰岛素对调节鱼类摄食的作用是不清晰的：禁食的鱼血浆的胰岛素和喂食的鱼相比通常处于较低的水平（Navarro 等，2006；Montserrat 等，2007），对虹鳟给予胰岛素能抑制食物的摄取（Soengas 和 Aldegunde，2004），但对斑点鮰没有影响（Silverstein 和 Plisetskaya，2000）。虽然曾经表明禁食使胰岛素样生长因子（IGFs）mRNA 在几种鱼类的肝脏和肌肉中的表达下降（Pedroso 等，2006；Ayson 等，2007；Terova 等，2007），但还没有证据说明 IGFs 对鱼类摄取的调控起作用。

9.2.3 生长激素系统

生长激素（GH）对所有脊椎动物（包括鱼类）的生长与代谢都起着关键性作用（Canosa 等，2007；Rousseau 和 Dufour，2007；Chang 和 Wong，本书第 4 章），用 GH-处理（Mclean 等，1997）和 GH 转基因鱼（Fu 等，2007；Hallennan 等，2007）都能促进生长。GH 亦参与鱼类摄食的调控，因为用 GH-处理硬骨鱼［腹腔注射（Johnsson 和 Bjornsson，1994）或者腹腔埋植小丸（Johansson 等，2005）］和 GH 转基因银大麻哈鱼（Stevens 和 Devlin，2005）都能刺激摄食行为，而饥饿会引起脑垂体 GH mRNA 和血浆 GH 水平发生变化（Small，2005；Pedroso 等，2006；Ayson 等，2007；Chang 和 Wong，本书第 4 章）。转基因动物较快地生长不仅是由于增加了食物的耗费，亦由于增强了食物的转化效率（Raven 等，2006）。

在脊椎动物中，GH 的分泌受到下丘脑激素，特别是生长激素释放激素（GHRH）/脑垂体腺苷酸环化酶激活多肽（PACAP）和生长抑素（SS）的调控。GHRH-样肽和 PACAP 已经在许多鱼类中鉴定（Chang 和 Wong，本书第 4 章）。鱼类和哺乳类一样，PACAP 刺激鱼类离体的脑垂体细胞释放 GH（Wong 等，2005；Sze 等，2007），并参与摄食的调控。中枢或外周注射 PACAP，能阻抑金鱼的食物摄食（Matsuda 等，2005b），而这些作用可能是由 POMC 和 CRF 介导的（Matsuda 和 Maruyama，2007）。对金鱼（Matsuda 等，2005a）和鳕鱼（Xu 和 Volkoff，2008）的研究表明，PACAP mRNA 的表达不受食物缺失的影响，但过度摄食或者再摄食能使 PACAP mRNA 的表达增加。PACAP 能增强餐食引起的胰岛素释放和胰岛素对脂肪细胞的脂类贮存作用（Nakata 和 Yada，2007），表明鱼类的 PACAP 在几个组织中调控能量代谢。

在哺乳类中，生长抑素（SS）存在着两个生物活性类型，即 SS-14 和 SS-28，它们是由同一个前体（前生长抑素原 I，PSSI）的可变裂解（alternative cleavage）而产生，并且抑制 GH 分泌（Patel，1999）。SS 蛋白质或者编码 PPSs 的 mRNA 已经在 20 多种鱼类当中分离出来。3 个 PPS 基因已经鉴定：PPS I 编码 SS14，而 PPS II 和 PPS III 产生 SS14 变异体。已证明 SS14 能抑制一些鱼类在体与离体的 GH 分泌（Klein 和 Sheridan，2008；Chang 和 Wong，本书第 4 章）。目前对 SS 对鱼类食物摄取的作用还了解得很少（Klein 和 Sheridan，2008）。给虹鳟埋植 SS-14-I 能抑制生长，但对食物摄取没有影响（Very 等，2001）。然而，已证明 SS 能影响代谢和能量稳态。SS 不仅能促进脂类转移和高血糖（Eilertson 和 Sheridan，1993），而且 SS 的生物合成和分泌活动都受到营养物如脂类与葡萄糖（Sheridan 和 Kittilson，2004）和食欲调节剂如 CCK、NPY 与 GAL

(Eilertson 等，1996) 的调控。此外，禁食能引起鳕鱼的脑 PPS Ⅰ 表达水平升高（Xu 和 Volkoff，2008）以及虹鳟胰脏 PPS Ⅰ mRNA 增加（Ehrman 等，2002）。

9.2.4 黑皮质素系统和黑色素浓集激素

黑皮质素（melanocortin，MC）系统对调控体重和色素形成起着关键的作用（Millington，2007）。它由黑色素细胞刺激激素（MSH）、促肾上腺皮质激素（ACTH）、黑皮质素受体（MCRs）和两个内源性拮抗物——野灰蛋白相关蛋白（AgRP）和野灰蛋白信号传递蛋白（ASP）组成。ACTH 和 MSH 是由一个共同的前体——阿黑皮素原（POMC）产生，其表达的主要部位是脑垂体的促肾上腺皮质激素细胞和促黑色素激素细胞。促肾上腺皮质激素细胞主要产生 ACTH，而促黑色素激素细胞产生 MSH（Metz 等，2006）。MCRs 已经在硬骨鱼类、无颌类和板鳃鱼类中确定（Haitina 等，2007；Selz 等，2007），它至少由 6 个亚型组成（Schioth 等，2005）。在哺乳类和鱼类中，皮肤颜色由 MC1R 介导，而 MC4R 对食物摄取和能量平衡的中枢调控起着关键性作用（Metz 等，2006）。尽管 MC4R 在哺乳类中主要是在 CNS 表达，但它亦在鱼类的外周表达（Metz 等，2006）。在中枢给予 MC4R 同等物能抑制金鱼摄取食物，而用 MC4R 的拮抗物处理后能刺激金鱼摄食（Cerdá-Reverter 等，2003）；而食物缺失时能使川鲽的肝脏（但不是脑）MC4R 转录体的数量增加（Kobayashi 等，2008b），这表明通过中枢和外周，MC4R 都能够部分参与鱼类摄食的调控。亦有证据表明 AgRP 和 MSH 对鱼类摄食和代谢的调控起作用。斑马鱼孵化 5 天后，当幼鱼开始主动摄食时，α-MSH-和 AgRP-免疫反应细胞显著分布在腹室周下丘脑区，表明这些肽对摄食行为的调控起着作用（Forlano 和 Cone，2007）。禁食能上调金鱼和斑马鱼下丘脑 AgRP mRNA 水平（Cerdá-Reverter 等，2003；Song 等，2003），而过表达 AgRP 的转基因斑马鱼呈现肥胖症、生长加快、脂肪细胞过度增生（Song 和 Cone，2007），表明 AgRP 起着促食欲作用。对金鱼中枢注射 NDP-α-MSH（一种 MC4R 同等物）（Cerdá-Reverter 等，2003）或者 MT Ⅱ（一种 MSH 同等物）（Matsuda 等，2008a）能抑制摄食活动。MSH 的厌食作用是由 CRF-信号通路介导的（Matsuda 等，2008a）。对鲑鳟鱼类的研究表明，α-MSH 刺激肝脏的脂肪酶活性，增加血液循环中的脂肪酸水平，而 α-MSH 合成有缺陷的鱼是摄食过度的（hyperphagic），具有增大的肝脏，腹部积累脂肪（Yada 等，2002）。这些研究结果表明 α-MSH 抑制鱼类的摄食，影响能量平衡。对金鱼禁食并不明显地影响其下丘脑的 POMC mRNA 水平（Cerdá-Reverter 等，2003），表明 POMC 并不直接参与摄食的调控。ASP 已在鱼类中确定，起着 MC1R 和 MC4R 的黑皮质素拮抗物和黑变作用（melanization）抑制因子的作用（Cerdá-Reverter 等，2005；Klovins 和 Schioth，2005），但它对鱼类摄食的调控作用还不清楚。

黑色素浓集激素（MCH）对哺乳类摄食和能量稳态的调控起着重要作用。在鱼类中，MCH 通过拮抗 α-MSH 的作用而调节体色（Pissios 等，2006）。两种鱼类的 MCH 基因，即 MCH1 和 MCH2，以及鱼类的 MCH 受体都已经鉴定，它们对鱼类食物摄取的调节作用还是有争议的。在金鱼脑腔内注射哺乳类或川鲽的 MCH 能使金鱼摄食减少

（Matsuda 等，2006b；Shimakura 等，2008），而禁食使下丘脑 MCH-样免疫反应神经元的数量减少（Matsuda 等，2007）。在金鱼中，MCH 的厌食作用是由 NPY 和 MSH 而不是由 CRF、PACAP 或者 CCK 介导的（Matsuda 等，2008b；Shimakura 等，2008）。但是，禁食使条斑星鲽（*Verasper moseri*）下丘脑 MCH 表达增加，而过度表达 MCH 基因的转基因青鳉体色发生变化，但保持正常的生长和摄食行为（Kinoshita 等，2001）。据我们所知，到目前为止还没有研究 ACTH 对鱼类摄食的作用。

9.3 内在因子的影响

9.3.1 代谢信号/能量储备

血液循环的代谢物水平影响鱼类的食物摄取。使用葡萄糖能引起鳟鱼高血糖和减少食物摄取（Banos 等，1998）。相反，使用 2-脱氧葡萄糖（一种非-可代谢的葡萄糖同等物）能刺激食物摄取（Delicio 和 Vicentini-Paulino，1993；Soengas 和 Aldegunde，2004）。高血糖亦能增加鱼类摄食的潜伏时间（Kuz'mina 和 Garina，2001；Kuz'mina 等，2002；Kuz'mina，2005）。除碳水化合物之外，蛋白质和脂类的代谢物亦已被证明能影响鱼类的摄食活动。腹腔注射单个的或者混合的氨基酸能减弱鲤鱼的食物摄取（Kuz'mina，2005）。在大麻哈鱼和斑点鲖，肥胖的鱼比瘦鱼吃得少，表明食物摄取受到脂肪的抑制性（lipostatic）控制（Shearer 等，1997；Silverstein 和 Plisetskaya，2000）。然而，对金头鲷喂食富含亚油酸的饲料后能使其食物摄取减少以及鱼体总脂肪含量降低（Diez 等，2007）。

9.3.2 个体发育

已研究几个鱼类食欲调控因子在发育过程中的变化，其结果是在分布的型式和出现的定时方面都表现出调控因子特异性和种类特异性的差别。通常，食欲调控因子在发育早期就已出现，表明它们在胚胎发育过程（孵化前）以及鱼类幼苗的营养物吸收/获得期（孵化前和孵化后）起作用。稍后出现的食欲调控因子，特别是和变态期或者混合摄食期（如开始外源性摄食）相联系的，可以反映摄食在个体发育过程中的变化。由于 GI 激素在鱼类幼鱼消化生理中的重要作用而被广泛研究（Ronnestad 等，2007）。从目前已经进行的鱼类内分泌因子的研究中可以推想 CCK 和生长素释放肽在幼鱼发育的早期出现，至少在一些鱼类中是这样的（Kamisaka 等，2003；Parhar 等，2003；Kamisaka 等，2005；Manning 等，2008）。GI CCK 的免疫反应在鲱鱼（*Clupea harengus*）（Kamisaka 等，2005）刚孵化时就可检测到，而生长素释放肽在斑马鱼（Pauls 等，2007）孵化后 48 h 检测到。但是，促胃液素和 GLR1 在发育的稍后时期才检测到（Kurokawa 等，2003；Navarro 等，2006）。参与鱼类摄食调控的一些其他信号都是在孵化之前出现，如 PACAP（Krueckl 等，2003；Xu 和 Volkoff，2008）、生长抑素（Xing 等，2005；Xu 和 Volkoff，2008）、NPY（Xu 和 Volkoff，未发表研究结果）、OX（Xu 和

Volkoff, 2007)、α-MSH 和 AgRP (Forlano 和 Cone, 2007)。

9.3.3 性别和生殖状态

曾经报道几种食欲调控剂的分布和表达水平的性别特异性差别，包括 TKs (Peyon 等, 2000)、GAL (Rao 等, 1996; Jadhao 和 Meyer, 2000) 和生长素释放肽 (Parhar 等, 2003)。这些研究结果表明性别或者更为特定的是性类固醇能影响摄食。在自然条件下，当鱼进行繁殖或者准备进行繁殖时 (如洄游)，这些影响就变得比较明显。在这方面已有的文献说明在产卵洄游或发生其他生殖行为 (如交配、产卵、占有领地、保护巢区) 时摄食活动降低或停止 (van Ginneken 和 Maes, 2005)。有关食欲调控因子参与产卵引起的厌食证据还有待研究。然而，食欲调控因子和生殖之间明显的相互作用已经在金鱼的相关研究中展现。给金鱼脑腔注射促性腺激素释放激素 (GnRH) 引起食物摄取降低 (Hoskins 等, 2008; Matsuda 等, 2008c)，而这种作用部分是由于下调了脑 OX mRNA 的表达 (Hoskins 等, 2008)。相反，脑腔注射 OX-A 使产卵行为减弱以及 GnRH mRNA 在脑的表达水平下降 (Hoskins 等, 2008)。此外，和性腺发育周期相联系的是香鱼 (Chiba 等, 1996) 和鲶鱼 (Mazumdar 等, 2007) 的 NPY 以及马苏大麻哈鱼的 (Westring 等, 2008) 的 UI mRNA 与血液循环中皮质醇水平出现季节变化。至今，性类固醇对鱼类食物摄取的影响还不清楚。睾酮 (T) 处理使雄鲈鱼 (Mandiki 等, 2005) 的食物摄取减弱，而使雌性和雄性金鱼 MCH mRNA 在下丘脑的表达升高 (Cerdá-Reverter 等, 2006)，在阉割后使罗非鱼前脑 NPY 免疫反应纤维的密度降低 (Sakharkar 等, 2005)，这些都表明 T 具有厌食性作用。然而，T 处理曾表明能增加真鲷的摄食 (Woo 等, 1993)。用雌二醇处理能刺激雌性鲈鱼 (但不是雄鱼) 摄食 (Mandiki 等, 2005)，而对真鲷的食物摄取没有影响 (Woo 等, 1993)，还能提高金鱼下丘脑 MCH (一种推定的摄食抑制剂) mRNA 表达水平 (Cerdá-Reverter 等, 2006)。在海鲈腹腔埋植 17-雌二醇或睾酮都使食物摄食减少，而 T 的作用可能是它在中枢芳香化为雌二醇后才介导的 (Leal 等, 2009)。

9.3.4 遗传的影响

在遗传背景不同的鱼类品系当中曾表现出食物摄取的差异。在被监禁的鳟鱼中，不同的品系在摄食活动与生长 (Mambrini 等, 2004)、摄食的型式 (Boujard 等, 2007) 和营养物利用 (Quillet 等, 2007) 等方面都出现差别。来自不同表型的个体处在特异性的环境中或者给以特异性的处理亦会出现不同的摄食与生长型式。例如，对于斑点鮰，厌食性化合物抑制摄食的作用在一些品系中要比另外一些品系更明显些，而处在低温下对不同品系摄食效率的影响亦有所不同 (Silverstein 等, 1996b)。在斑点鮰和虹鳟 (Silverstein, 2002) 的 NPY 基因中曾经检测到微卫星 (串联重复) 的差异，表明食欲调节肽的遗传差异可能部分是来自这些趋异的表型。

9.4 外在因子的影响

9.4.1 温度

在一些鱼类中已经证明温度和食物摄取之间的关系。鱼处在极端的温度条件下会降低食物摄取,但通常在温度升高到"可忍受的"范围内时会增加食物消耗和提高生长率(Russell 等,1996;Guijarro 等,1999;Bendiksen 等,2002;Sunuma 等,2007;Kehoe 和 Volkoff,2008)。目前对于温度调控引起摄食变化的内分泌作用机理还了解得很少。温度升高通常会使血浆的 GH 和 IGF-I 水平升高(Gabillard 等,2005)。大西洋鳕鱼脑的 NPY mRNA 表达似乎不受温度的影响,但脑的 CART mRNA 表达的水平处于 2℃中的鱼要较高于处在 11℃或 15℃中的鱼。这表明是 CART 而不是 NPY 对温度引起鱼食欲的变化起着作用(Kehoe 和 Volkoff,2008)。

9.4.2 光周期

对一些鱼类的研究已经表明光周期和光的作用能影响摄食活动(Noble 等,2005;Tucker 等,2006;Sunuma 等,2007),尽管亦有例外的报道(Canavate 等,2006)。已经表明光周期可以通过增加食物摄取和/或通过运动增加肌肉质量而间接影响生长(Boeuf 和 Le Bail,1999)。引起这些变化的特异性内分泌作用机理还不清楚,有待进一步研究。对于虹鳟,光的作用能影响血浆的 GH、甲状腺素和皮质醇的浓度(Reddy 和 Leatherland,2003),而大西洋鲑对增强的光周期反应是血浆 GH 水平增高(Nordgarden 等,2007)。相反,光周期并不影响阳光鲈鱼血液循环中的 IGF-I 水平(Davis 和 McEntire,2006)。这些研究结果表明至少在鲑鳟鱼类中存在着刺激 GH 相关通路的食欲调控因子的作用。松果体合成的褪黑激素起着协调环境和节律性生理过程的神经内分泌信号的作用。给金鱼(Pinillos 等,2001;De Pedro 等,2008)和丁鱥(Lopez-Olmeda 等,2006)腹腔注射以及给海鲈(Rubio 等,2004)口服褪黑激素能使摄食减少和体重降低。对金鱼腹腔注射褪黑激素并不会影响血液瘦素、生长素释放肽以及下丘脑 NPY 的水平,表明这些摄食调节剂并不参与褪黑激素对鱼类能量稳态的影响(De Pedro 等,2008)。

9.4.3 盐度

对一些鱼类的研究已经表明盐度能影响摄食(De Boeck 等,2000),但还不清楚有关食欲调控因子可能参与的情况。在鲑鳟鱼类中,促生长轴是生长和海水耐受力的主要调节剂,这个系统亦可能起着和盐度相关的影响摄食的作用(Boeuf 和 Payan,2001)。对金鱼进行食物摄取和水盐平衡相联系的研究,结果表明 PrRP 起着调控食物摄取和离子调节稳态的作用,但是要按照全身的需求来参加两者当中的一个生理过程(Kelly 和 Peter,2006)。最后,在急性渗透压调节受到干扰的情况下,鱼类通常会暂时减少食物摄取，并伴随鳟鱼脑 CRF 和 UI mRNA 表达水平的提高（Craig 等，2005），表明这些因子参与了调控作用。

9.4.4 低氧

至今对一些鱼类的研究表明低氧具有独特的抑制食欲的作用（Buentello 等，2000；Ripley 和 Foran，2007）。虹鳟处于低氧水平时，前脑 CRF 和 UI mRNA 水平以及血浆皮质醇水平提高，表明 CRF 相关肽类起着生理作用，介导至少一部分低氧状态对减少食物摄取的影响（Bernier 和 Craig，2005）。

9.4.5 污染物和健康状态

污染物和疾病亦能影响摄食，但依不同的种类以及不同的疾病和接触的方式而有所不同。例如，虹鳟在水源金属浓度升高的情况下摄食减少（Todd 等，2007），而对银大麻哈鱼投喂高锌含量的饵料能增加其摄食率（Bowen 等，2006）。用 LPS 处理产生的免疫冲击使金鱼减少摄食，脑的 CART、NPY 和 CRF 基因表达发生变化（Volkoff 和 Peter，2004），亦使罗非鱼脑的 CRF 免疫反应发生变化（Pepels 等，2004）。湖鳟（*Salvelinus namaycush*）和低剂量的杀虫剂 tebufenozide 接触后，虽然食物摄取以及脑的 NPY 和 CRF mRNA 不受影响，但脑的 CART mRNA 表达水平明显高于对照组鱼类（Volkoff 等，2007）。CART 以及其他食欲调节剂的这些变化可能和 tebufenozide 引起对湖鳟的免疫刺激相联系（Hamoutene 等，2008）。

9.5　参与鱼类摄食的内分泌线路模型和结束语

我们目前对鱼类摄食的内分泌调控以及内因与外因影响的了解只能使我们建立一个简单而相对不完整的鱼类食欲调控模型（见表 9.1、图 9.1）。然而，鱼类摄食调控的一般模式显然和其他脊椎动物相似，由中枢的摄食中心调控食欲，而中枢的摄食中心受到来自脑和外周的激素因子影响。和哺乳类一样，鱼类的模型是丰余的（redundancy），存在着一些功能明显相似的食欲性和厌食性因子以及在内分泌系统之间高度的相互作用。至今得到的证据表明鱼类食欲调控的神经肽类是平行地密切相互作用，而不是阶梯级的作用型式。在哺乳类中已经界定了长期的和短期的调节剂，短期的主要由饱食信号组成，它们在餐后立即起作用；而长期调节剂感受身体的能量状态，根据脂肪质量/能量储备来调控摄食。至今对鱼类食欲调控激素所进行的大多数研究都是采用急性处理的相对较短时期的研究，所以，在短期和长期摄食相关因子之间的差别还难以确定。

哺乳类和鱼类之间存在的差别看来是存在于一些内分泌食欲调节剂的合成部位、结构和功能方面。正在增加的证据表明脑区产生的摄食相关激素以及它们作用的靶标在鱼类和哺乳类之间以及鱼类自身之间都有所不同，因为在鱼类的不同类群之间脑的结构都有很大差别（Cerdá-Reverter 和 Canosa，本书第 1 章）。例如，OXs 主要在哺乳类的外侧下丘脑产生，而采用 mRNA 表达的研究证明 OXs 广泛分布在鱼类脑内。关于结构，虽然激素的氨基酸组成及其编码 cDNA 可能在哺乳类与鱼类之间有很大差别，它们可能一直存在着保守的功能。例如，虽然哺乳类和鱼类的瘦素在结构上差别很大（蛋白质水平的同一性少于 10%），并且由不同的组织（脂肪组织和肝脏）产生，但它们在这两类动物中都起着调控摄食和代谢的作用。相反，有些激素可能具有保守的结构却表现出不同的功能。例如，GLPs 在鱼类和哺乳类中有相似的结构，并且都影响摄食活动，但 GLP 还影响哺乳类的胃活动能力，而在鱼类中并没有这种作用。NPY 在鱼类和哺乳类之间结构上亦相当保守，它在哺乳类中是主要的食欲肽，但它的表达并不受有些鱼类（如鳕鱼）禁食的影响，表明它的相对重要性可能是种类特异性的。

鱼类存在着哺乳类所没有的特别组织（如尾神经内分泌器官）以及辐鳍鱼类在进化发展中经历的额外的基因复制过程（导致鱼类比陆栖脊椎动物有更多的基因，从而相对于哺乳类的一个肽类型，在鱼类中会出现多个肽类型），可能预示着鱼类会出现一些具有不同生理功能的附加的内分泌作用物（因子）。鱼类模型的进一步复杂性还在于，鱼类在栖息地和环境条件方面通常会比哺乳类展现出更高程度的异质性（heterogeneity），从而形成种类特异性的生理适应，包括摄食行为的变化。至今对调控这些生理适应的内分泌作用机理还了解得很少。

总而言之，尽管在近年来对鱼类摄食和能量平衡的内分泌调控研究已经取得明显的进展，但研究的深度还不够，存在争议的问题还不少，而整个研究领域的难题还远未能解决。对哺乳类激素的鱼类相应物做进一步的分析鉴定将会揭示在结构方面的类群-特异性差别，进而导致发现哺乳类相应物在鱼类中的新功能。鱼类食欲调控因子的目录亦必定会不断增加，并且还可能包括在哺乳类尚未分离出来的鱼类特异性激素。

H. 沃尔科夫

S. 安尼阿潘

S. P. 凯利

参考文献

Abbott C R, Rossi M, Wren A M, Murphy K G, Kennedy A R, Stanley S A, Zollner A N, Morgan D G, Morgan I, Ghatei M A, Small C J, Bloom S R. 2001. Evidence of an orexigenic role for cocaine- and amphetamine-regulated transcript after administration into discrete hypothalamic nuclei. *Endocrinology*, 142: 3457-3463.

Albalat A, Gutierrez J, Navarro I. 2005. Regulation of lipolysis in isolated adipocytes of rainbow trout (*Oncorhynchus mykiss*): The role of insulin and glucagon. *Comp. Biochem. Physiol. A*, 142: 347-354.

Albalat A, Saera-Vila A, Capilla E, Gutierrez J, Perez-Sanchez J, Navarro I. 2007. Insulin regulation of lipoprotein lipase (LPL) activity and expression in gilthead sea bream (*Sparus aurata*). *Comp. Biochem. Physiol. B*, 148: 151-159.

Aldegunde M, Mancebo M. 2006. Effects of neuropeptide Y on food intake and brain biogenic amines in the rainbow trout (*Oncorhynchus mykiss*). *Peptides*, 27: 719-727.

Alderman S L, Bernier N J. 2007. Localization of corticotropin-releasing factor, urotensin I, and CRF-binding protein gene expression in the brain of the zebrafish, *Danio rerio*. *J. Comp. Neurol*, 502: 783-793.

Aldman G, Jonsson A C, Jensen J, Holmgren S. 1989. Gastrin/CCK-like peptides in the spiny dogfish, *Squalus acanthias*; concentrations and actions in the gut. *Comp. Biochem. Physiol. C*, 92: 103-108.

Amiya N, Amano M, Oka Y, Iigo M, Takahashi A, Yamamori K. 2007. Immunohistochemical localization of orexin/hypocretin-like immunoreactive peptides and melanin-concentrating hormone in the brain and pituitary of medaka. *Neurosci. Lett*, 427: 16-21.

Amole N, Unniappan S. 2009. Fasting induces preproghrelin mRNA expression in the brain and gut of zebrafish. *Danio rerio. Gen. Comp. Endocrinol*, 161 (1): 133-137.

Andoh T. 2007. Amino acids are more important insulinotropins than glucose in a teleost fish, barfin flounder (*Verasper moseri*). *Gen. Comp. Endocrinol*, 151: 308-317.

Ayson F G, de Jesus-Ayson E G T, Takemura A. 2007. mRNA expression patterns for GH, PRL, SL, IGF-I and IGF-II during altered feeding status in rabbitfish, *Siganus guttatus. Gen. Comp. Endocrinol*, 150: 196-204.

Baker D M, Larsen D A, Swanson P, Dickhoff W W. 2000. Long-term peripheral treatment of immature coho salmon (*Oncorhynchus kisutch*) with human leptin has no clear physiologic effect. *Gen. Comp. Endocrinol*, 118: 134-138.

Banos N, Baro J, Castejon C, Navarro I, Gutierrez J. 1998. Influence of high-carbohydrate enriched diets on plasma insulin levels and insulin and IGF-I receptors in trout. *Regul. Pept*, 77: 55-62.

Bechtold D A, Luckman S M. 2007. The role of RFamide peptides in feeding. *J. Endocrinol*, 192: 3-15.

Bendiksen E, Jobling M, Arnesen A. 2002. Feed intake of Atlantic salmon parr *Salmo salar*L. in relation to temperature and feed composition. *Aquac. Res*, 33: 525-532.

Bermudez R, Vigliano F, Quiroga M I, Nieto J M, Bosi G, Domeneghini C. 2007. Immunohistochemical study on the neuroendocrine system of the digestive tract of turbot, *Scophthalmus maximus* (L.), infected by *Enteromyxum scophthalmi* (Myxozoa). *Fish Shellfish Immunol*, 22: 252-263.

Bernier N J. 2006. The corticotropin-releasing factor system as a mediator of the appetite-suppressing effects of stress in fish. *Gen. Comp. Endocrinol*, 146: 45-55.

Bernier N J, Craig P M. 2005. CRF-related peptides contribute to stress response and regulation of appetite in hypoxic rainbow trout. *Am. J. Physiol. Regul. Integr. Comp. Physiol*, 289: R982-R990.

Bernier N J, Peter R E. 2001. Appetite-suppressing effects of urotensin I and corticotropin-releasing hormone in goldfish (*Carassius auratus*). *Neuroendocrinology*, 73: 248-260.

Bernier N J, Bedard N, Peter R E. 2004. Effects of cortisol on food intake, growth, and forebrain neuropeptide Y and corticotropin-releasing factor gene expression in goldfish. *Gen. Comp. Endocrinol*, 135: 230-240.

Bjenning C, Farrell A, Holmgren S. 1991. Bombesin-like immunoreactivity in skates and the *in vitro* effect of bombesin on coronary vessels from the longnose skate, *Raja rhina*. *Regul. Pept*, 35: 207-219.

Boeuf G, Le Bail P Y. 1999. Does light have an influence on fish growth?. *Aquaculture*, 177: 129-152.

Boeuf G, Payan P. 2001. How should salinity influence fish growth?. *Comp. Biochem. Physiol. C*, 130: 411-423.

Bosi G, Di Giancamillo A, Arrighi S, Domeneghini C. 2004. An immunohistochemical study on the neuroendocrine system in the alimentary canal of the brown trout, *Salmo trutta*, L., 1758. *Gen. Comp. Endocrinol*, 138: 166-181.

Boujard T, Ramezi J, Vandeputte M, Labbe L, Mambrini M. 2007. Group feeding behavior of brown trout is a correlated response to selection for growth shaped by the environment. *Behav. Genet*, 37: 525-534.

Bowen L, Werner I, Johnson M L. 2006. Physiological and behavioral effects of zinc and temperature on coho salmon (*Oncorhynchus kisutch*). *Hydrobiologia*, 559: 161-168.

Brightman M W, Broadwell R D. 1976. The morphological approach to the study of normal and abnormal brain permeability. *Adv. Exp. Med. Biol*, 69: 41-54.

Buentello J A, Gatlin D M, Neill W H. 2000. Effects of water temperature and dissolved oxygen on daily feed consumption, feed utilization and growth of channel catfish (*Ictalurus punctatus*). *Aquaculture*, 182: 339-352.

Canavate J P, Zerolo R, Fernandez-Diaz C. 2006. Feeding and development of Senegal sole (*Solea senegalensis*) larvae reared in different photoperiods. *Aquaculture*, 258: 368-377.

Canosa L F, Peter R E. 2004. Effects of cholecystokinin and bombesin on the expression of preprosomatostatin-encoding genes in goldfish forebrain. *Regul. Pept*, 121: 99-105.

Canosa L F, Unniappan S, Peter R E. 2005. Periprandial changes in growth hormone release in goldfish: Role of somatostatin, ghrelin, and gastrin-releasing peptide. *Am. J. Physiol. Regul. Integr. Comp. Physiol*, 289: R125-133.

Canosa L F, Chang J P, Peter R E. 2007. Neuroendocrine control of growth hormone in fish. *Gen. Comp. Endocrinol*, 151: 1-26.

Cerdá-Reverter J, Martinez-Rodriguez G, Zanuy S, Carrillo M, Larhammar D. 2000. Molecular evolution of the neuropeptide Y (NPY) family of peptides: Cloning of three NPY-related peptides from the sea bass (*Dicentrarchus labrax*). *Regul. Pept*, 95: 25-34.

Cerdá-Reverter J M, Martinez-Rodriguez G, Anglade I, Kah O, Zanuy S. 2000. Peptide YY (PYY) and fish pancreatic peptide Y (PY) expression in the brain of the sea bass (*Dicentrarchus labrax*) as revealed by *in situ* hybridization. *J. Comp. Neurol*, 426: 197-208.

Cerdá-Reverter J, Schioth H, Peter R. 2003. The central melanocortin system regulates food intake in goldfish. *Regul. Pept*, 115: 101-113.

Cerdá-Reverter J M, Haitina T, Schioth H B, Peter R E. 2005. Gene structure of the goldfish agouti-signaling protein: A putative role in the dorsal-ventral pigment pattern of fish. *Endocrinology*, 146: 1597-1610.

Cerdá-Reverter J M, Canosa L F, Peter R E. 2006. Regulation of the hypothalamic melanin-concentrating hormone neurons by sex steroids in the goldfish: Possible role in the modulation of luteinizing hormone secretion. *Neuroendocrinol*, 84: 364-377.

Chan C B, Leung P K, Wise H, Cheng C H. 2004. Signal transduction mechanism of the seabream growth hormone secretagogue receptor. *FEBS Lett*, 577: 147-153.

Chang C L, Roh J, Hsu S Y. 2004. Intermedin, a novel calcitonin family peptide that exists in teleosts as well as in mammals: A comparison with other calcitonin/intermedin family peptides in vertebrates. *Peptides*, 25: 1633-1642.

Chiba A, Sohn Y C, Honma Y. 1996. Distribution of neuropeptide Y and gonadotropin-releasing hormone immunoreactivities in the brain and hypophysis of the ayu, *Plecoglossus altivelis* (Teleostei). *Arch. Histol. Cytol*, 59: 137-148.

Conlon J M, Larhammar D. 2005. The evolution of neuroendocrine peptides. *Gen. Comp. Endocrinol*, 142: 53-59.

Craig P M, Al-Timimi H, Bernier N J. 2005. Differential increase in forebrain and caudal neurosecretory system corticotropin-releasing factor and urotensin I gene expression associated with seawater transfer in rainbow trout. *Endocrinology*, 146: 3851-3860.

Davis K B, McEntire M. 2006. Effect of photoperiod on feeding, intraperitoneal fat, and insulin-like growth factor-I in sunshine bass. *J. World Aquac. Soc*, 37:

431-436.

De Boeck G, Vlaeminck A, Van der Linden A, Blust R. 2000. The energy metabolism of common carp (*Cyprinus carpio*) when exposed to salt stress: An increase in energy expenditure or effects of starvation?. *Physio. Bioch. Zool*, 73: 102-111.

De Pedro N, Alonso-Gomez A, Gancedo B, Delgado M, Alonso-Bedate M. 1993. Role of corticotropin-releasing factor (CRF) as a food intake regulator in goldfish. *Physiol. Behav*, 53: 517-520.

De Pedro N, Cespedes M V, Delgado M J, Alonsobedate M. 1995. The galanin-induced feeding stimulation is mediated via alpha (2) -adrenergic receptors in goldfish. *Regul. Pept*, 57: 77-84.

De Pedro N, Gancedo B, Alonsogomez A L, Delgado M J, Alonsobedate M. 1995. Alterations in food-intake and thyroid-tissue content by corticotropin-releasing factor in *Tinca-tinca*. *Rev. Esp. Fisiol*, 51: 71-75.

De Pedro N, Alonso-Gomez A L, Gancedo B, Valenciano A I, Delgado M J, Alonso-Bedate M. 1997. Effect of alpha-helical-CRF [9-41] on feeding in goldfish: Involvement of cortisol and catecholamines. *Behav. Neurosci*, 111: 398-403.

De Pedro N, Pinillos M L, Valenciano A I, Alonso-Bedate M, Delgado M J. 1998. Inhibitory effect of serotonin on feeding behavior in goldfish: Involvement of CRF. *Peptides*, 19: 505-511.

De Pedro N, Martinez-Alvarez R, Delgado M J. 2006. Acute and chronic leptin reduces food intake and body weight in goldfish (*Carassius auratus*). *J. Endocrinol*, 188: 513-520.

De Pedro N, Martinez-Alvarez R M, Delgado M J. 2008. Melatonin reduces body weight in goldfish (*Carassius auratus*): Effects on metabolic resources and some feeding regulators. *J. Pineal Res*, 45: 32-39.

del Sol Novoa M, Capilla E, Rojas P, Baro J, Gutierrez J, Navarro I. 2004. Glucagon and insulin response to dietary carbohydrate in rainbow trout (*Oncorhynchus mykiss*). *Gen. Comp. Endocrinol*, 139: 48-54.

Delicio H, Vicentini-Paulino M. 1993. 2-deoxyglucose-induced food-intake by Nile tilapia, *Oreochromis-niloticus* (L). *Braz. J. Med. Biol. Res*, 26: 327-331.

Despres J P. 2007. The endocannabinoid system: A new target for the regulation of energy balance and metabolism. *Crit. Pathw. Cardiol*, 6: 46-50.

Diez A, Menoyo D, Perez-Benavente S, Calduch-Giner J A, Vega-Rubin de Celis S, Obach A, Favre-Krey L, Boukouvala E, Leaver M J, Tocher D R, Perez-Sanchez J, Krey G, et al. 2007. Conjugated linoleic acid affects lipid composition, metabolism, and gene expression in gilthead sea bream (*Sparus aurata*L). *J. Nutr*, 137: 1363-1369.

Dockray G J. 2004. The expanding family of -RFamide peptides and their effects on feeding behaviour. *Exp. Physiol*, 89: 229-235.

Ehrman M, Melroe G, Moore C, Kittilson J D, Sheridan M. 2002. Nutritional regulation of somatostatin expression in rainbow trout, Oncorhynchus mykiss. *Fish Physiol. Biochem*, 26: 309-314.

Eilertson C D, Carneiro N M, Kittilson J D, Comley C, Sheridan M A. 1996. Cholecystokinin, neuropeptide Y and galanin modulate the release of pancreatic somatostatin-25 and somatostatin-14 *in vitro*. *Regul. Pept*, 63: 105-112.

Eilertson C D, Sheridan M A. 1993. Differential effects of somatostatin-14 and somatostatin-25 on carbohydrate and lipid metabolism in rainbow trout Oncorhynchus mykiss. *Gen. Comp. Endocrinol*, 92: 62-70.

Ekblad E. 2006. CART in the enteric nervous system. *Peptides*, 27: 2024-2030.

Elphick M R, Egertova M. 2001. The neurobiology and evolution of cannabinoid signalling. *Philos. Trans. R. Soc. Lond. B. Biol. Sci*, 356: 381-408.

Faraco J H, Appelbaum L, Marin W, Gaus S E, Mourrain P, Mignot E. 2006. Regulation of hypocretin (orexin) expression in embryonic zebrafish. *J. Biol. Chem*, 281: 29753-29761.

Forgan L G, Forster M E. 2007. Effects of potential mediators of an intestinal brake mechanism on gut motility in Chinook salmon (Oncorhynchus tshawytscha). *Comp. Biochem. Physiol. C*, 146: 343-347.

Forlano P M, Cone R D. 2007. Conserved neurochemical pathways involved in hypothalamic control of energy homeostasis. *J. Comp. Neurol*, 505: 235-248.

Fu C, Li D, Hu W, Wang Y, Zhu Z. 2007. Fast-growing transgenic common carp mounting compensatory growth. *J. Fish Biol*, 71: 174-185.

Gabillard J C, Weil C, Rescan P Y, Navarro I, Gutierrez J, Le Bail P Y. 2005. Does the GH/IGF system mediate the effect of water temperature on fish growth? A review. *Cybium*, 29: 107-117.

Gao Q, Horvath T L. 2007. Neurobiology of feeding and energy expenditure. *Annu. Rev. Neurosci*, 30: 367-398.

Gelineau A, Boujard T. 2001. Oral administration of cholecystokinin receptor antagonists increase feed intake in rainbow trout. *J. Fish Biol*, 58: 716-724.

Godby N A, Rutherford E S, Mason D M. 2007. Diet, feeding rate, growth, mortality, and production of juvenile steelhead in a Lake Michigan tributary. *N. Am. J. Fish. Man*, 27: 578-592.

Gorissen M H A G, Flik G, Huising M O. 2006. Peptides and proteins regulating food intake: A comparative view. *Animal Biol*, 56: 447-473.

Gregory T R, Wood C M. 1999. The effects of chronic plasma cortisol elevation on the feeding behaviour, growth, competitive ability, and swimming performance of juvenile rainbow trout. *Physiol. Biochem. Zool*, 72: 286-295.

Guijarro A I, Delgado M J, Pinillos M L, Lopez-Patino M A, Alonso-Bedate M, De Pedro N. 1999. Galanin and beta-endorphin as feeding regulators in cyprinids: Effect of temperature. *Aquac. Res*, 30: 483-489.

Haitina T, Klovins J, Takahashi A, Lowgren M, Ringholm A, Enberg J, Kawauchi H, Larson E T, Fredriksson R, Schioth H B. 2007. Functional characterization of two melanocortin (MC) receptors in lamprey showing orthology to the MC1 and MC4 receptor subtypes. *BMC Evol. Biol*, 7: 101.

Hallennan E M, McLean E, Fleming I A. 2007. Effects of growth hormone transgenes on the behavior and welfare of aquacultured fishes: A review identifying research needs. *Appl. Anim. Behav. Sci*, 104: 265-294.

Hamoutene D, Payne J F, Volkoff H. 2008. Effects of tebufenozide on some aspects of lake trout (*Salvelinus namaycush*) immune response. *Ecotoxicol. Environ. Saf*, 69: 173-179.

Heino M, Kaitala V. 1999. Evolution of resource allocation between growth and reproduction in animals with indeterminate growth. *J. Evol. Biol*, 12: 423-429.

Himick B A, Peter R E. 1994. Bombesin acts to suppress feeding behavior and alter serum growth hormone in goldfish. *Physiol. Behav*, 55: 65-72.

Himick B A, Peter R E. 1994. CCK/gastrin-like immunoreactivity in brain and gut, and CCK suppression of feeding in goldfish. *Am. J. Physiol*, 267: R841-851.

Himick B A, Golosinski A A, Jonsson A C, Peter R E. 1993. CCK/gastrin-like immunoreactivity in the goldfish pituitary: Regulation of pituitary hormone secretion by CCK-like peptides *in vitro*. *Gen. Comp. Endocrinol*, 92: 88-103.

Himick B A, Vigna S R, Peter R E. 1996. Characterization of cholecystokinin binding sites in goldfish brain and pituitary. *Am. J. Physiol*, 271: R137-143.

Holmberg A, Schwerte T, Pelster B, Holmgren S. 2004. Ontogeny of the gut motility control system in zebrafish *Danio rerio* embryos and larvae. *J. Exp. Biol*, 207: 4085-4094.

Hoskins L J, Xu M, Volkoff H. 2008. Interactions between gonadotropin-releasing hormone (GnRH) and orexin in the regulation of feeding and reproduction in goldfish (*Carassius auratus*). *Horm. Behav*, 54: 379-385.

Hoyle C H V. 1999. Neuropeptide families and their receptors: Evolutionary perspectives. *Brain Res*, 848: 1-25.

Huising M O, Geven E J, Kruiswijk C P, Nabuurs S B, Stolte E H, Spanings F A, Verburg-van Kemenade B M, Flik G. 2006. Increased leptin expression in common carp (*Cyprinus carpio*) after food intake but not after fasting or feeding to satiation.

Endocrinology, 147: 5786-5797.

Huising M O, Kruiswijk C P, Flik G. 2006. Phylogeny and evolution of class-I helical cytokines. *J. Endocrinol*, 189: 1-25.

Irwin D. 2004. A second insulin gene in fish genomes. *Gen. Comp. Endocrinol*, 135: 150-158.

Irwin D M, Wong K. 2005. Evolution of new hormone function: Loss and gain of a receptor. *J. Heredity*, 96: 205-211.

Jadhao A G, Meyer D L. 2000. Sexually dimorphic distribution of galanin in the preoptic area of red salmon, *Oncorhynchus nerka*. *Cell Tissue Res*, 302: 199-203.

Jadhao A, Pinelli C. 2001. Galanin-like immunoreactivity in the brain and pituitary of the "four-eyed" fish, *Anableps anableps*. *Cell Tissue Res*, 306: 309-318.

Jensen H, Rourke I J, Moller M, Jonson L, Johnsen A H. 2001. Identification and distribution of CCK-related peptides and mRNAs in the rainbow trout, *Oncorhynchus mykiss*. *Biochim. Biophys. Acta*, 1517: 190-201.

Johansson V, Winberg S, Bjornsson B T. 2005. Growth hormone-induced stimulation of swimming and feeding behaviour of rainbow trout is abolished by the D-1 dopamine antagonist SCH23390. *Gen. Comp. Endocrinol*, 141: 58-65.

Johnsen A H, Jonson L, Rourke I J, Rehfeld J F. 1997. Elasmobranchs express separate cholecystokinin and gastrin genes. *Proc. Natl. Acad. Sci. USA*, 94: 10221-10226.

Johnson R M, Johnson T M, Londraville R L. 2000. Evidence for leptin expression in fishes. *J. Exp. Zool*, 286: 718-724.

Johnsson J I, Bjornsson B T. 1994. Growth hormone increases growth rate, appetite and dominance in juvenile rainbow trout, *Oncorhynchus mykiss*. *Anim. Behav*, 48: 177-186.

Jonsson E, Forsman A, Einarsdottir I E, Egner B, Ruohonen K, Bjornsson B T. 2006. Circulating levels of cholecystokinin and gastrin-releasing peptide in rainbow trout fed different diets. *Gen. Comp. Endocrinol*, 148: 187-194.

Jonsson E, Forsman A, Einarsdottir I E, Kaiya H, Ruohonen K, Bjornsson B T. 2007. Plasma ghrelin levels in rainbow trout in response to fasting, feeding and food composition, and effects of ghrelin on voluntary food intake. *Comp. Biochem. Physiol. A*, 147: 1116-1124.

Kaiya H, Miyazato M, Kangawa K, Peter R E, Unniappan S. 2008. Ghrelin: A multifunctional hormone in non-mammalian vertebrates. *Comp. Biochem. Physiol. A*, 149: 109-128.

Kamiji M M, Inui A. 2007. Neuropeptide Y receptor selective ligands in the treatment of obesity. *Endocr. Rev*, 28: 664-684.

Kamisaka Y, Fujii Y, Yamamoto S, Kurokawa T, Ronnestad I, Totland G K, Tagawa M, Tanaka M. 2003. Distribution of cholecystokinin-immunoreactive cells in the digestive tract of the larval teleost, ayu, *Plecoglossus altivelis*. *Gen. Comp. Endocrinol*, 134: 116-121.

Kamisaka Y, Drivenes O, Kurokawa T, Tagawa M, Ronnestad I, Tanaka M, Helvik J V. 2005. Cholecystokinin mRNA in Atlantic herring, *Clupea harengus*- molecular cloning, characterization, and distribution in the digestive tract during the early life stages. *Peptides*, 26: 385-393.

Kawakoshi A, Kaiya H, Riley L G, Hirano T, Grau E G, Miyazato M, Hosoda H, Kangawa K. 2007. Identification of a ghrelin-like peptide in two species of shark, *Sphyrna lewini* and *Carcharhinus melanopterus*. *Gen. Comp. Endocrinol*, 151: 259-268.

Kehoe A S, Volkoff H. 2007. Cloning and characterization of neuropeptide Y (NPY) and cocaine and amphetamine regulated transcript (CART) in Atlantic cod (*Gadus morhua*). *Comp. Biochem. Physiol. A*, 146: 451-461.

Kehoe A S, Volkoff H. 2008. The effects of temperature on feeding and expression of two appetite-related factors, neuropeptide Y and cocaine- and amphetamine-regulated transcript, in Atlantic cod, *Gadus morhua*. *J. World Aquac. Soc*, 39: 790-796.

Kelly S P, Peter R E. 2006. Prolactin-releasing peptide, food intake, and hydromineral balance in goldfish. *Am. J. Physiol. Regul. Integr. Comp. Physiol*, 291: R1474-1481.

Kim E J, Kim C H, Seo J K, Go H J, Lee S, Takano Y, Chung J K, Hong Y K, Park N G. 2005. Structure-activity relationship of neuropeptide gamma derived from mammalian and fish. *J. Pept. Res*, 66: 395-403.

Kinoshita M, Morita T, Toyohara H, Hirata T, Sakaguchi M, Ono M, Inoue K, Wakamatsu Y, Ozato K. 2001. Transgenic medaka overexpressing a melanin-concentrating hormone exhibit lightened body color but no remarkable abnormality. *Marine Biotechnology*, 3: 536-543.

Klein S E, Sheridan M A. 2008. Somatostatin signaling and the regulation of growth and metabolism in fish. *Mol. Cell Endocrinol*, 286: 148-154.

Klovins J, Schioth H B. 2005. Agouti-related proteins (AGRPs) and agouti-signaling peptide (ASIP) in fish and chicken. *Ann. N. Y. Acad. Sci*, 1040: 363-367.

Kobayashi Y, Peterson B C, Waldbieser G C. 2008. Association of cocaine- and amphetamine-regulated transcript (CART) messenger RNA level, food intake, and growth in channel catfish. *Comp. Biochem. Physiol. A*, 151: 219-225.

Kobayashi Y, Tsuchiya K, Yamanome T, Schioth H B, Kawauchi H, Takahashi A. 2008. Food deprivation increases the expression of melanocortin-4 receptor in the liver of barfin flounder, *Verasper moseri*. *Gen. Comp. Endocrinol*, 155: 280-287.

Korczynski W, Ceregrzyn M, Matyjek R, Kato I, Kuwahara A, Wolinski J, Zabielski R. 2006. Central and local (enteric) action of orexins. *J. Physiol. Pharmacol*, 57 (Suppl 6): 17-42.

Krueckl S L, Fradinger E A, Sherwood N M. 2003. Developmental changes in the expression of growth hormone-releasing hormone and pituitary adenylate cyclase-activating polypeptide in zebrafish. *J. Comp. Neurol*, 455: 396-405.

Kurokawa T, Suzuki T, Hashimoto H. 2003. Identification of gastrin and multiple cholecystokinin genes in teleost. *Peptides*, 24: 227-235.

Kurokawa T, Uji S, Suzuki T. 2005. Identification of cDNA coding for a homologue to mammalian leptin from pufferfish, *Takifugu rubripes*. *Peptides*, 26: 745-750.

Kurokawa T, Murashita K, Suzuki T, Uji S. 2008. Genomic characterization and tissue distribution of leptin receptor and leptin receptor overlapping transcript genes in the pufferfish, *Takifugu rubripes*. *Gen. Comp. Endocrinol*, 158: 108-114.

Kuz'mina V. 200. Regulation of the fish alimentary behavior: Role of humoral component. *J. Evol. Biochem. Physiol*, 41: 282-295.

Kuz'mina V, Garina D. 2001. Glucose, insulin, and adrenaline effects on some aspects of fish feeding behavior. *J. Evol. Biochem. Physiol*, 37: 154-160.

Kuz'mina V V, Garina D V, Gerasimov Y V. 2002. The role of glucose in regulation of feeding behavior of fish. *J. Ichthyol*, 42: 210-215.

Lang R, Gundlach A L, Kofler B. 2007. The galanin peptide family: Receptor pharmacology, pleiotropic biological actions, and implications in health and disease. *Pharmacol. Ther*, 115: 177-207.

Leal E, Sanchez E, Muriach B, Cerdá-Reverter J M. 2009. Sex steroid-induced inhibition of food intake in sea bass (*Dicentrarchus labrax*). *J. Comp. Physiol. B*, 179: 77-86.

Liu L, Conlon J M, Joss J M, Burcher E. 2002. Purification, characterization, and biological activity of a substance P-related peptide from the gut of the Australian lungfish, *Neoceratodus forsteri*. *Gen. Comp. Endocrinol*, 125: 104-112.

Londraville R L, Duvall C S. 2002. Murine leptin injections increase intracellular fatty acid-binding protein in green sunfish (*Lepomis cyanellus*). *Gen. Comp. Endocrinol*, 129: 56-62.

Lopez-Olmeda J F, Madrid J A, Sanchez-Vazquez F J. 2006. Melatonin effects on food intake and activity rhythms in two fish species with different activity patterns: Diurnal (goldfish) and nocturnal (tench). *Comp. Biochem. Physiol. A*, 144: 180-187.

Lopez-Patino M A, Guijarro A I, Isorna E, Delgado M J, Alonso-Bedate M, De Pedro N. 1999. Neuropeptide Y has a stimulatory action on feeding behavior in goldfish

(*Carassius auratus*). *Eur. J. Pharmacol*, 377: 147-153.

Lutz T A. 2006. Amylinergic control of food intake. *Physiol. Behav*, 89: 465-471.

MacDonald E, Volkoff H. 2009. Neuropeptide Y (NPY), cocaine- and amphetamine-regulated transcript (CART) and cholecystokinin (CCK) in winter skate (*Raja ocellata*): cDNA cloning, tissue distribution and mRNA expression responses to fasting. *Gen. Comp. Endocrinol*, 161: 252-261.

Magalhaes G S, Junqueira-de-Azevedo I L, Lopes-Ferreira M, Lorenzini D M, Ho P L, Moura-da-Silva A M. 2006. Transcriptome analysis of expressed sequence tags from the venom glands of the fish *Thalassophryne nattereri*. *Biochimie*, 88: 693-699.

Mambrini M, Medale F, Sanchez M P, Recalde B, Chevassus B, Labbe L, Quillet E, Boujard T. 2004. Selection for growth in brown trout increases feed intake capacity without affecting maintenance and growth requirements. *J. Anim. Sci*, 82: 2865-2875.

Mandiki S N M, Babiak I, Bopopi J M, Leprieur F, Kestemont P. 2005. Effects of sex steroids and their inhibitors on endocrine parameters and gender growth differences in Eurasian perch (*Perca fluviatilis*) juveniles. *Steroids*, 70: 85-94.

Manning A J, Murray H M, Gallant J W, Matsuoka M P, Radford E, Douglas S E. 2008. Ontogenetic and tissue-specific expression of preproghrelin in the Atlantic halibut, *Hippoglossus hippoglossus* L. *J. Endocrinol*, 196: 181-192.

Martinez-Alvarez R M, Volkoff H, Cueto J A, Delgado M J. 2008. Molecular characterization of calcitonin gene-related peptide (CGRP) related peptides (CGRP, amylin, adrenomedullin and adrenomedullin-2/intermedin) in goldfish (*Carassius auratus*): Cloning and distribution. *Peptides*, 29: 1534-1543.

Martinez-Alvarez R M, Volkoff H, Munoz-Cueto J A, Delgado M J. 2009. Effect of calcitonin gene-related peptide (CGRP), adrenomedullin and adrenomedullin-2/intermedin on food intake in goldfish (*Carassius auratus*). *Peptides*, 30: 803-807.

Maruyama K, Konno N, Ishiguro K, Wakasugi T, Uchiyama M, Shioda S, Matsuda K. 2008. Isolation and characterisation of four cDNAs encoding neuromedin U (NMU) from the brain and gut of goldfish, and the inhibitory effect of a deduced NMU on food intake and locomotor activity. *J. Neuroendocrinol*, 20: 71-78.

Matsuda K, Maruyama K. 2007. Regulation of feeding behavior by pituitary adenylate cyclase-activating polypeptide (PACAP) and vasoactive intestinal polypeptide (VIP) in vertebrates. *Peptides*, 28: 1761-1766.

Matsuda K, Maruyama K, Miura T, Uchiyama M, Shioda S. 2005. Anorexigenic action of pituitary adenylate cyclase-activating polypeptide (PACAP) in the goldfish: feeding-induced changes in the expression of mRNAs for PACAP and its receptors in the brain, and locomotor response to central injection. *Neurosci. Lett*, 386: 9-13.

Matsuda K, Nagano Y, Uchiyama M, Takahashi A, Kawauchi H. 2005. Immunohistochemical observation of pituitary adenylate cyclase-activating polypeptide (PACAP) and adenohypophysial hormones in the pituitary of a teleost, *Uranoscopus japonicus*. *Zool. Sci*, 22: 71-76.

Matsuda K, Miura T, Kaiya H, Maruyama K, Shimakura S, Uchiyama M, Kangawa K, Shioda S. 2006. Regulation of food intake by acyl and des-acyl ghrelins in the goldfish. *Peptides*, 27: 2321-2325.

Matsuda K, Shimakura S, Maruyama K, Miura T, Uchiyama M, Kawauchi H, Shioda S, Takahashi A. 2006. Central administration of melanin-concentrating hormone (MCH) suppresses food intake, but not locomotor activity, in the goldfish, *Carassius auratus*. *Neurosci. Lett*, 399: 259-263.

Matsuda K, Shimakura S, Miura T, Maruyama K, Uchiyama M, Kawauchi H, Shioda S, Takahashi A. 2007. Feeding-induced changes of melanin-concentrating hormone (MCH) -like immunoreactivity in goldfish brain. *Cell Tissue Res*, 328: 375-382.

Matsuda K, Kojima K, Shimakura S, Wada K, Maruyama K, Uchiyama M, Kikuyama S, Shioda S. 2008. Corticotropin-releasing hormone mediates alpha-melanocyte-stimulating hormone-induced anorexigenic action in goldfish. *Peptides*, 29: 1930-1936.

Matsuda K, Kojima K, Shimakura S I, Miura T, Uchiyama M, Shioda S, Ando H, Takahashi A. 2008. Relationship between melanin-concentrating hormone- and neuropeptide Y-containing neurons in the goldfish hypothalamus. *Comp. Biochem. Physiol. A. Mol. Integr. Physiol*, In press.

Matsuda K, Nakamura K, Shimakura S, Miura T, Kageyama H, Uchiyama M, Shioda S, Ando H. 2008. Inhibitory effect of chicken gonadotropin-releasing hormone II on food intake in the goldfish, *Carassius auratus*. *Horm. Behav*, 54: 83-89.

Mazumdar M, Lal B, Sakharkar A J, Deshmukh M, Singru P S, Subhedar N. 2006. Involvement of neuropeptide Y Y1 receptors in the regulation of LH and GH cells in the pituitary of the catfish, *Clarias batrachus*: An immunocytochemical study. *Gen. Comp. Endocrinol*, 149: 190-196.

Mazumdar M, Sakharkar A J, Singru P S, Subhedar N. 2007. Reproduction phase-related variations in neuropeptide Y immunoreactivity in the olfactory system, forebrain, and pituitary of the female catfish, *Clarias batrachus* (Linn.). *J. Comp. Neurol*, 504: 450-469.

McCoy J G, Avery D D. 1990. Bombesin: Potential integrative peptide for feeding and satiety. *Peptides*, 11: 595-607.

Mclean E, Devlin R H, Byatt J C, Clarke W C, Donaldson E M. 1997. Impact of a controlled release formulation of recombinant bovine growth hormone upon growth and seawater adaptation in coho (*Oncorhynchus kisutch*) and chinook (*Oncorhynchus tshawytscha*) salmon. *Aquaculture*, 156: 113-128.

McPartland J M, Glass M, Matias I, Norris R W, Kilpatrick C W. 2007. A shifted repertoire of endocannabinoid genes in the zebrafish (*Danio rerio*). *Mol. Genet. Genomics*, 277: 555-570.

Metz J R, Peters J J, Flik G. 2006. Molecular biology and physiology of the melanocortin system in fish: A review. *Gen. Comp. Endocrinol*, 148: 150-162.

Millington G W. 2007. The role of proopiomelanocortin (POMC) neurones in feeding behaviour. *Nutr. Metab. (Lond)*, 4: 18.

Millot S, Begout M L, Ruyet J P L, Breuil G, Di-Poi C, Fievet J, Pineau P, Roue M, Severe A. 2008. Feed demand behavior in sea bass juveniles: Effects on individual specific growth rate variation and health (inter-individual and inter-group variation). *Aquaculture*, 274: 87-95.

Miura T, Maruyama K, Shimakura S, Kaiya H, Uchiyama M, Kangawa K, Shioda S, Matsuda K. 2006. Neuropeptide Y mediates ghrelin-induced feeding in the goldfish, *Carassius auratus*. *Neurosci. Lett*, 407: 279-283.

Miura T, Maruyama K, Shimakura S, Kaiya H, Uchiyama M, Kangawa K, Shioda S, Matsuda. 2007. Regulation of food intake in the goldfish by interaction between ghrelin and orexin. *Peptides*, 28: 1207-1213.

Miura T, Maruyama K, Kaiya H, Miyazato M, Kangawa K, Uchiyama M, Shioda S, Matsuda K. 2008. Purification and properties of ghrelin from the intestine of the goldfish. *Carassius auratus*. *Peptides*, 30: 758-765.

Montserrat N, Gabillard J C, Capilla E, Navarro M I, Gutierrez J. 2007. Role of insulin, insulin-like growth factors, and muscle regulatory factors in the compensatory growth of the trout (*Oncorhynchus mykiss*). *Gen. Comp. Endocrinol*, 150: 462-472.

Moriyama S, Ito T, Takahashi A, Amano M, Sower S A, Hirano T, Yamamori K, Kawauchi H. 2002. A homolog of mammalian PRL-releasing peptide (fish arginyl-phenylalanyl-amide peptide) is a major hypothalamic peptide of PRL release in teleost fish. *Endocrinology*, 143: 2071-2079.

Moriyama S, Kasahara M, Amiya N, Takahashi A, Amano M, Sower S A, Yamamori K, Kawauchi H. 2007. RFamide peptides inhibit the expression of melanotropin and growth hormone genes in the pituitary of an Agnathan, the sea lamprey, *Petromyzon marinus*. *Endocrinology*, 148: 3740-3749.

Murashita K, Fukada H, Hosokawa H, Masumoto T. 2006. Cholecystokinin and peptide Y in yellowtail (*Seriola quinqueradiata*): Molecular cloning, real-time quantitative RT-PCR, and response to feeding and fasting. *Gen. Comp. Endocrinol*, 145:

287-297.

Murashita K, Fukada H, Hosokawa H, Masumoto T. 2007. Changes in cholecystokinin and peptide Y gene expression with feeding in yellowtail (*Seriola quinqueradiata*): Relation to pancreatic exocrine regulation. *Comp. Biochem. Physiol. B*, 146: 318-325.

Murashita K, Uji S, Yamamoto T, Ronnestad I, Kurokawa T. 2008. Production of recombinant leptin and its effects on food intake in rainbow trout (*Oncorhynchus mykiss*). *Comp. Biochem. Physiol. B*, 150: 377-384.

Muruzabal F J, Fruhbeck G, Gomez-Ambrosi J, Archanco M, Burrell M A. 2002. Immunocytochemical detection of leptin in non-mammalian vertebrate stomach. *Gen. Comp. Endocrinol*, 128: 149-152.

Mustonen A M, Nieminen P, Hyvarinen H. 2002. Leptin, ghrelin, and energy metabolism of the spawning burbot (*Lota lota*, L.). *J. Exp. Zool*, 293: 119-126.

Nagasaka R, Okamoto N, Ushio H. 2006. Increased leptin may be involved in the short life span of ayu (*Plecoglossus altivelis*). *J. Exp. Zool*, 305: 507-512.

Nakamachi T, Matsuda K, Maruyama K, Miura T, Uchiyama M, Funahashi H, Sakurai T, Shioda S. 2006. Regulation by orexin of feeding behaviour and locomotor activity in the goldfish. *J. Neuroendocrinol*, 18: 290-297.

Nakata M, Yada T. 2007. PACAP in the glucose and energy homeostasis: Physiological role and therapeutic potential. *Curr. Pharm. Des*, 13: 1105-1112.

Narnaware Y K, Peter R E. 2001. Effects of food deprivation and refeeding on neuropeptide Y (NPY) mRNA levels in goldfish. *Comp. Biochem. Physiol. B*, 129: 633-637.

Narnaware Y K, Peter R E. 2002. Influence of diet composition on food intake and neuropeptide Y (NPY) gene expression in goldfish brain. *Regul. Pept*, 103: 75-83.

Narnaware Y K, Peyon P P, Lin X, Peter R E. 2000. Regulation of food intake by neuropeptide Y in goldfish. *Am. J. Physiol. Regul. Integr. Comp. Physiol*, 279: R1025-1034.

Navarro I, Leibush B, Moon T, Plisetskaya E, Banos N, Mendez E, Planas J, Gutierrez J. 1999. Insulin, insulin-like growth factor-I (IGF-I) and glucagon: The evolution of their receptors. *Com. Biochem. Physiol. B*, 122: 137-153.

Navarro M H, Lozano M T, Agulleiro B. 2006. Ontogeny of the endocrine pancreatic cells of the gilthead sea bream, *Sparus aurata* (Teleost). *Gen. Comp. Endocrinol*, 148: 213-226.

Nelson L E, Sheridan M A. 2006. Gastroenteropancreatic hormones and metabolism in fish. *Gen. Comp. Endocrinol*, 148: 116-124.

Nieminen P, Mustonen A M, Hyvarinen H. 2003. Fasting reduces plasma leptin-and ghrelin-immunoreactive peptide concentrations of the burbot (*Lota lota*) at 2 degrees C but not at 10 degrees C. *Zoolog. Sci*, 20: 1109-1115.

Nishino S. 2007. The hypothalamic peptidergic system, hypocretin/orexin and vigilance control. *Neuropeptides*, 41: 117-133.

Noble C, Mizusawa K, Tabata M. 2005. Does light intensity affect self-feeding and food wastage in group-held rainbow trout and white-spotted charr?. *J. Fish Biol*, 66: 1387-1399.

Noble C, Kadri S, Mitchell D F, Huntingford F A. 2007. The impact of environmental variables on the feeding rhythms and daily feed intake of cage-held 1 + Atlantic salmon parr (*Salmo salar* L.). *Aquaculture*, 269: 290-298.

Nordgarden U, Bjornsson B T, Hansen T. 2007. Developmental stage of Atlantic salmon parr regulates pituitary GH secretion and parr-smolt transformation. *Aquaculture*, 264: 441-448.

Novak C M, Jiang X, Wang C, Teske J A, Kotz C M, Levine J A. 2005. Caloric restriction and physical activity in zebrafish (*Danio rerio*). *Neurosci. Lett*, 383: 99-104.

Ohno K, Sakurai T. 2007. Orexin neuronal circuitry: Role in the regulation of sleep and wakefulness. *Front. Neuroendocrinol*, 29: 70-87.

Oliver A S, Vigna S R. 1996. CCK-X receptors in the endothermic mako shark (*Isurus oxyrinchus*). *Gen. Comp. Endocrinol*, 102: 61-73.

Olsson C, Holbrook J D, Bompadre G, Jonsson E, Hoyle C H, Sanger G J, Holmgren S, Andrews P L. 2008. Identification of genes for the ghrelin and motilin receptors and a novel related gene in fish, and stimulation of intestinal motility in zebrafish (*Danio rerio*) by ghrelin and motilin. *Gen. Comp. Endocrinol*, 155: 217-226.

Olszewski P K, Schioth H B, Levine A S. 2008. Ghrelin in the CNS: From hunger to a rewarding and memorable meal?. *Brain Res. Rev*, 58: 160-170.

Parhar I S, Sato H, Sakuma Y. 2003. Ghrelin gene in cichlid fish is modulated by sex and development. *Biochem. Biophys. Res. Commun*, 305: 169-175.

Patel Y C. 1999. Somatostatin and its receptor family. *Front. Neuroendocrinol*, 20: 157-198.

Pauls S, Zecchin E, Tiso N, Bortolussi M, Argenton F. 2007. Function and regulation of zebrafish nkx2. 2a during development of pancreatic islet and ducts. *Develop. Biol*, 304: 875-890.

Peddu S C, Breves J P, Kaiya H, Gordon Grau E, RileyJr L G. 2009. Pre- and postprandial effects on ghrelin signaling in the brain and on the GH/IGF-I axis in the mozambique tilapia (*Oreochromis mossambicus*). *Gen. Comp. Endocrinol*, 161: 412-418.

Pedroso F L, de Jesus-Ayson E G T, Cortado H H, Hyodo S, Ayson F G. 2006. Changes in mRNA expression of grouper (*Epinephelus coioides*) growth hormone and insulin-like growth factor I in response to nutritional status. *Gen. Comp. Endocrinol*, 145: 237-246.

Pepels P P L M, Bonga S E W, Balm P H M. 2004. Bacterial lipopolysaccharide (LPS) modulates corticotropin-releasing hormone (CRH) content and release in the brain of juvenile and adult tilapia (*Oreochromis mossambicus*; Teleostei). *J. Exp. Biol*, 207: 4479-4488.

Peter R E. 1979. The brain and feeding behavior. In: Hoar W S, Randall D J, Brett J R, Ed. *Fish Physiology*. Vol. VIII. Academic Press: New York, NY, 121-159.

Peterson B C, Small B C. 2005. Effects of exogenous cortisol on the GH/IGF-I/IGFBP network in channel catfish. *Domest. Anim. Endocrinol*, 28: 391-404.

Peyon P, Lin X W, Himick B A, Peter R E. 1998. Molecular cloning and expression of cDNA encoding brain preprocholecystokinin in goldfish. *Peptides*, 19: 199-210.

Peyon P, Saied H, Lin X, Peter R E. 1999. Postprandial, seasonal and sexual variations in cholecystokinin gene expression in goldfish brain. *Brain Res. Mol. Brain Res*, 74: 190-196.

Peyon P, Saied H, Lin X, Peter R E. 2000. Preprotachykinin gene expression in goldfish brain: Sexual, seasonal, and postprandial variations. *Peptides*, 21: 225-231.

Pinillos M L, De Pedro N, Alonso-Gomez A L, Alonso-Bedate M, Delgado M J. 2001. Food intake inhibition by melatonin in goldfish (*Carassius auratus*). *Physiol. Behav*, 72: 629-634.

Pinuela C, Northcutt R G. 2007. Immunohistochemical organization of the forebrain in the white sturgeon, *Acipenser transmontanus*. *Brain Behav. Evol*, 69: 229-253.

Pissios P, Bradley R L, Maratos-Flier E. 2006. Expanding the scales: The multiple roles of MCH in regulating energy balance and other biological functions. *Endocr. Rev*, 27: 606-620.

Prober D A, Rihel J, Onah A A, Sung R J, Schier A F. 2006. Hypocretin/orexin overexpression induces an insomnia-like phenotype in zebrafish. *J. Neurosci*, 26: 13400-13410.

Quillet E, Le Guillou S, Aubin J, Labbe L, Fauconneau B, Medale F. 2007. Response of a lean muscle and a fat muscle rainbow trout (*Oncorhynchus mykiss*) line on growth, nutrient utilization, body composition and carcass traits when fed two different diets. *Aquaculture*, 269: 220-231.

Rao P D, Murthy C K, Cook H, Peter R E. 1996. Sexual dimorphism of galanin-like immunoreactivity in the brain and pituitary of goldfish, *Carassius auratus*. *J. Chem. Neuroanat*, 10: 119-135.

Raven P A, Devlin R H, Higgs D A. 2006. Influence of dietary digestible energy content on growth, protein and energy utilization and body composition of growth hormone transgenic and non-transgenic coho salmon (*Oncorhynchus kisutch*). *Aquaculture*, 254: 730-747.

Raybould H E. 2007. Mechanisms of CCK signaling from gut to brain. *Curr. Opin. Pharmacol*, 7: 570-574.

Reddy P K, Leatherland J F. 2003. Influences of photoperiod and alternate days of feeding on plasma growth hormone and thyroid hormone levels in juvenile rainbow trout. *J. Fish Biol*, 63: 197-212.

Riley L G, Fox B K, Kaiya H, Hirano T, Grau E G. 2005. Long-term treatment of ghrelin stimulates feeding, fat deposition, and alters the GH/IGF-I axis in the tilapia, *Oreochromis mossambicus*. *Gen. Comp. Endocrinol*, 142: 234-240.

Ripley J L, Foran C M. 2007. Influence of estuarine hypoxia on feeding and sound production by two sympatric pipefish species (Syngnathidae). *Mar. Environ. Res*, 63: 350-367.

Rodriguez-Gomez F J, Rendon-Unceta C, Sarasquete C, Munoz-Cueto J A. 2001. Distribution of neuropeptide Y-like immunoreactivity in the brain of the Senegalese sole (*Solea senegalensis*). *Anat. Rec*, 262: 227-237.

Ronnestad I, Kamisaka Y, Conceicao L E C, Morais S, Tonheim S K. 2007. Digestive physiology of marine fish larvae: Hormonal control and processing capacity for proteins, peptides and amino acids. *Aquaculture*, 268: 82-97.

Rousseau K, Dufour S. 2007. Comparative aspects of GH and metabolic regulation in lower vertebrates. *Neuroendocrinol*, 86: 165-174.

Rubio V C, Sanchez-Vazquez F J, Madrid J A. 2004. Oral administration of melatonin reduces food intake and modifies macronutrient selection in European sea bass (*Dicentrarchus labrax*, L.). *J. Pineal Res*, 37: 42-47.

Russell N R, Fish J D, Wootton R J. 1996. Feeding and growth of juvenile sea bass: The effect of ration and temperature on growth rate and efficiency. *J. Fish Biol*, 49: 206-220.

Sakharkar A J, Singru P S, Sarkar K, Subhedar N K. 2005. Neuropeptide Y in the forebrain of the adult male cichlid fish *Oreochromis mossambicus*: Distribution, effects of castration and testosterone replacement. *J. Comp. Neurol*, 489: 148-165.

Salaneck E, Larsson T A, Larson E T, Larhammar D. 2008. Birth and death of neuropeptide Y receptor genes in relation to the teleost fish tetraploidization. *Gene*, 409: 61-71.

Saper C B. 2006. Staying awake for dinner: Hypothalamic integration of sleep, feeding, and circadian rhythms. *Hypoth. Integ. Energy Metabol*, 153: 243-252.

Sawada K, Ukena K, Satake H, Iwakoshi E, Minakata H, Tsutsui K. 2002. Novel fish hypothalamic neuropeptide. *Eur. J. Biochem*, 269: 6000-6008.

Schioth H B, Haitina T, Fridmanis D, Klovins J. 2005. Unusual genomic structure: Melanocortin receptors in Fugu. *Ann. N. Y. Acad. Sci*, 1040: 460-463.

Seale A P, Itoh T, Moriyama S, Takahashi A, Kawauchi H, Sakamoto T, Fujimoto M, Riley L G, Hirano T, Grau E G. 2002. Isolation and characterization of a homologue of mammalian prolactin-releasing peptide from the tilapia brain and its effect on prolactin release from the tilapia pituitary. *Gen. Comp. Endocrinol*, 125: 328-339.

Selz Y, Braasch I, Hoffmann C, Schmidt C, Schultheis C, Schartl M, Volff J N. 2007. Evolution of melanocortin receptors in teleost fish: The melanocortin type 1 receptor. *Gene*, 401: 114-122.

Shearer K D, Silverstein J T, Plisetskaya E M. 1997. Role of adiposity in food intake control of juvenile chinook salmon (*Oncorhynchus tshawytscha*). *Comp. Biochem. Physiol. A*, 118: 1209-1215.

Shepherd B S, Johnson J K, Silverstein J T, Parhar I S, Vijayan M M, McGuire A, Weber G M. 2007. Endocrine and orexigenic actions of growth hormone secretagogues in rainbow trout (*Oncorhynchus mykiss*). *Comp. Biochem. Physiol. A*, 146: 390-399.

Sheridan M A, Kittilson J D. 2004. The role of somatostatins in the regulation of metabolism in fish. *Comp. Biochem. Physiol. B*, 138: 323-330.

Shimakura S, Kojima K, Nakamachi T, Kageyama H, Uchiyama M, Shioda S, Takahashi A, Matsuda K. 2008. Neuronal interaction between melanin-concentrating hormone- and alpha-melanocyte-stimulating hormone-containing neurons in the goldfish hypothalamus. *Peptides*, 29: 1432-1440.

Silverstein J T. 2002. Using genetic variation to understand control of feed intake in fish. *Fish Physiol. Biochem*, 27: 173-178.

Silverstein J T, Plisetskaya E M. 2000. The effects of NPY and insulin on food intake regulation in fish. *Amer. Zool*, 40: 296-308.

Silverstein J T, Shearer K D, Dickhoff W W, Plisetskaya E M. 1999. Regulation of nutrient intake and energy balance in salmon. *Aquaculture*, 177: 161-169.

Silverstein J T, Wolters W R, Holland M. 1999. Evidence of differences in growth and food intake regulation in different genetic strains of channel catfish. *J. Fish Biol*, 54: 607-615.

Silverstein J T, Bondareva V M, Leonard J B, Plisetskaya E M. 2001. Neuropeptide regulation of feeding in catfish, *Ictalurus punctatus*: A role for glucagon-like peptide-1 (GLP-1)?. *Comp. Biochem. Physiol. B*, 129: 623-631.

Singru P S, Mazumdar M, Sakharkar A J, Lechan R M, Thim L, Clausen J T, Subhedar N K. 2007. Immunohistochemical localization of cocaine- and amphetamine-regulated transcript peptide in the brain of the catfish, *Clarias batrachus* (Linn.). *J. Comp. Neurol*, 502: 215-235.

Singru P S, Mazumdar M, Barsagade V, Lechan R M, Thim L, Clausen J T, Subhedar N. 2008. Association of cocaine- and amphetamine-regulated transcript and neuropeptide Y in the forebrain and pituitary of the catfish, *Clarias batrachus*: A double immunofluorescent labeling study. *J. Chem. Neuroanat*, 36: 239-250.

Small B C. 2005. Effect of fasting on nychthemeral concentrations of plasma growth hormone (GH), insulin-like growth factor I (IGF-1), and cortisol in channel catfish (*Ictalurus punctatus*). *Comp. Biochem. Physiol. B*, 142: 217-223.

Soengas J, Aldegunde M. 2004. Brain glucose and insulin: Effects on food intake and brain biogenic amines of rainbow trout. *J. Comp. Physiol. A*, 190: 641-649.

Song Y, Cone R D. 2007. Creation of a genetic model of obesity in a teleost. *Faseb J*, 21: 2042-2049.

Song Y, Golling G, Thacker T L, Cone R D. 2003. Agouti-related protein (AGRP) is conserved and regulated by metabolic state in the zebrafish, *Danio rerio*. *Endocrine*, 22: 257-265.

Stevens E D, Devlin R H. 2005. Gut size in GH-transgenic coho salmon is enhanced by both the GH transgene and increased food intake. *J. Fish Biol*, 66: 1633-1648.

Sueiro C, Carrera I, Ferreiro S, Molist P, Adrio F, Anadon R, Rodriguez-Moldes I. 2007. New insights on Saccus vasculosus evolution: A developmental and immunohistochemical study in elasmobranchs. *Brain Behav. Evol*, 70: 187-204.

Sundstrom G, Larsson T A, Brenner S, Venkatesh B, Larhammar D. 2005. Ray-fin fish tetraploidization gave rise to pufferfish duplicates of NPY and PYY, but zebrafish NPY duplicate was lost. *Ann. N. Y. Acad. Sci*, 1040: 476-478.

Sunuma T, Amano M, Yamanome T, Furukawa K, Yamamori K. 2007. Self-feeding activity of a pleuronectiform fish, the barfin flounder. *Aquaculture*, 270: 566-569.

Suzuki H, Miyoshi Y, Yamamoto T. 2007. Orexin-A (hypocretin 1) -like immunoreactivity in growth hormone-containing cells of the Japanese seaperch (*Lateolabrax japonicus*) pituitary. *Gen. Comp. Endocrinol*, 150: 205-211.

Sze K H, Zhou H, Yang Y, He M, Jiang Y, Wong A O. 2007. Pituitary adenylate cyclase-activating polypeptide (PACAP) as a growth hormone (GH) -releasing factor in grass carp: II. Solution structure of a brain-specific PACAP by nuclear magnetic resonance spectroscopy and functional studies on GH release and gene expression. *Endocrinology*, 148: 5042-5059.

Tachibana T, Takagi T, Tomonaga S, Ohgushi A, Ando R, Denbow D M, Furuse M. 2003. Central administration of cocaine- and amphetamine-regulated transcript inhibits food intake in chicks. *Neurosci. Lett*, 337: 131-134.

Takahashi A, Tsuchiya K, Yamanome T, Amano M, Yasuda A, Yamamori K, Kawauchi H. 2004. Possible involvement of melanin-concentrating hormone in food intake in a teleost fish, barfin flounder. *Peptides*, 25: 1613-1622.

Terova G, Rimoldi S, Chini V, Gornati R, Bernardini G, Saroglia M. 2007. Cloning and expression analysis of insulin-like growth factor I and II in liver and muscle of sea bass (*Dicentrarchus labrax*, L.) during long-term fasting and refeeding. *J. Fish Biol*, 70: 219-233.

Terova G, Rimoldi S, Bernardini G, Gornati R, Saroglia M. 2008. Sea bass ghrelin: Molecular cloning and mRNA quantification during fasting and refeeding. *Gen. Comp. Endocrinol*, 155: 341-351.

Thavanathan R, Volkoff H. 2006. Effects of amylin on feeding of goldfish: Interactions with CCK. *Regul. Pept*, 133: 90-96.

Thorndyke M C, ReeveJr J R, Vigna S R. 1990. Biological activity of a bombesin-like peptide extracted from the intestine of the ratfish, *Hydrolagus colliei*. *Comp. Biochem. Physiol. C*, 96: 135-140.

Todd A S, McKnight D M, Jaros C L, Marchitto T M. 2007. Effects of acid rock drainage on stocked rainbow trout (*Oncorhynchus mykiss*): An in-situ, caged fish experiment. *Environmental Monitoring and Assessment*, 130: 111-127.

Tucker B J, Booth M A, Allan G L, Booth D, Fielder D S. 2006. Effects of photoperiod and feeding frequency on performance of newly weaned Australian snapper *Pagrus auratus*. *Aquaculture*, 258: 514-520.

Unniappan S, Peter R E. 2005. Structure, distribution and physiological functions of ghrelin in fish. *Comp. Biochem. Physiol. A*, 140: 396-408.

Unniappan S, Lin X, Peter R E. 2003. Characterization of complementary deoxyribonucleic acids encoding preprogalanin and its alternative splice variants in the goldfish. *Mol. Cell Endocrinol*, 200: 177-187.

Unniappan S, Canosa L F, Peter R E. 2004. Orexigenic actions of ghrelin in goldfish: Feeding-induced changes in brain and gut mRNA expression and serum levels, and responses to central and peripheral injections. *Neuroendocrinology*, 79: 100-108.

Unniappan S, Cerdá-Reverter J M, Peter R E. 2004. *In situ* localization of preprogalanin mRNA in the goldfish brain and changes in its expression during feeding and starvation. *Gen. Comp. Endocrinol*, 136: 200-207.

Valassi E, Scacchi M, Cavagnini F. 2008. Neuroendocrine control of food intake. *Nutr. Metab. Cardiovasc. Dis*, 18: 158-168.

Valenti M, Cottone E, Martinez R, De Pedro N, Rubio M, Viveros M P, Franzoni M F, Delgado M J, Di Marzo V. 2005. The endocannabinoid system in the brain of *Carassius auratus* and its possible role in the control of food intake. *J. Neurochem*, 95: 662-672.

van Ginneken V J T, Maes G E. 2005. The european eel (*Anguilla anguilla*, Linnaeus), its lifecycle, evolution and reproduction: A literature review. *Rev. Fish. Biol. Fisheries*, 15: 367-398.

Vegusdal A, Sundvold H, Gjoen T, Ruyter B. 2003. An *in vitro* method for studying the proliferation and differentiation of Atlantic salmon preadipocytes. *Lipids*, 38: 289-296.

Very N M, Knutson D, Kittilson J D, Sheridan M A. 2001. Somatostatin inhibits growth of rainbow trout. *J. Fish Biol*, 59: 157-165.

Vicentic A, Jones D C. 2007. The CART (cocaine- and amphetamine-regulated transcript) system in appetite and drug addiction. *J. Pharmacol. Exp. Ther*, 320: 499-506.

Volff J N. 2004. Genome evolution and biodiversity in teleost fish. *Heredity*, 94: 280-294.

Volkoff H, Peter R E. 2000. Effects of CART peptides on food consumption, feeding and associated behaviors in the goldfish, *Carassius auratus*: Actions on neuropeptide Y- and orexin A-induced feeding. *Brain Res*, 887: 125-133.

Volkoff H, Peter R E. 2001. Characterization of two forms of cocaine- and amphetamine-regulated transcript (CART) peptide precursors in goldfish: Molecular cloning and distribution, modulation of expression by nutritional status, and interactions with leptin. *Endocrinology*, 142: 5076-5088.

Volkoff H, Peter R E. 2001. Interactions between orexin A, NPY and galanin in the control of food intake of the goldfish, *Carassius auratus*. *Regul. Pept*, 101: 59-72.

Volkoff H, Peter R E. 2004. Effects of lipopolysaccharide treatment on feeding of goldfish: Role of appetite-regulating peptides. *Brain Res*, 998: 139-147.

Volkoff H, Bjorklund J M, Peter R E. 1999. Stimulation of feeding behavior and food consumption in the goldfish, *Carassius auratus*, by orexin-A and orexin-B. *Brain Res*, 846: 204-209.

Volkoff H, Peyon P, Lin X, Peter R. 2000. Molecular cloning and expression of cDNA encoding a brain bombesin/gastrin-releasing peptide-like peptide in goldfish. *Peptides*, 21: 639-648.

Volkoff H, Eykelbosh A J, Peter R E. 2003. Role of leptin in the control of feeding of goldfish *Carassius auratus*: Interactions with cholecystokinin, neuropeptide Y and orexin A, and modulation by fasting. *Brain Res*, 972: 90-109.

Volkoff H, Canosa L F, Unniappan S, Cerdá-Reverter J M, Bernier N J, Kelly S P, Peter R E. 2005. Neuropeptides and the control of food intake in fish. *Gen. Comp. Endocrinol*, 142: 3-19.

Volkoff H, Hamoutene D, Payne J F. 2007. Potential effects of tebufenozide on feeding and metabolism of lake trout (*Salvelinus namaycush*). *Can. Tech. Rep. Fish. Aquat. Sci*, 2777: iv + 19.

Walton K M, Chin J E, Duplantier A J, Mather R J. 2006. Galanin function in the central nervous system. *Curr. Opin. Drug Discov. Devel*, 9: 560-570.

Westermark G T, Falkmer S, Steiner D F, Chan S J, Engstrom U, Westermark P. 2002. Islet amyloid polypeptide is expressed in the pancreatic islet parenchyma of the teleostean fish, *Myoxocephalus* (*cottus*) *scorpius*. *Comp. Biochem. Physiol. B*, 133: 119-125.

Westring C G, Ando H, Kitahashi T, Bhandari R K, Ueda H, Urano A, Dores R M, Sher A A, Danielson P B. 2008. Seasonal changes in CRF-I and urotensin I transcript levels in masu salmon: Correlation with cortisol secretion during spawning. *Gen. Comp. Endocrinol*, 155: 126-140.

Wong A O, Li W, Leung C Y, Huo L, Zhou H. 2005. Pituitary adenylate cyclase-activating polypeptide (PACAP) as a growth hormone (GH) -releasing factor in grass carp. I. Functional coupling of cyclic adenosine 3', 5'-monophosphate and Ca^{2+}/calmodulin-dependent signaling pathways in PACAP-induced GH secretion and GH gene expression in grass carp pituitary cells. *Endocrinology*, 146: 5407-5424.

Wong M M, Yu R M, Ng P K, Law S H, Tsang A K, Kong R Y. 2007. Characterization of a hypoxia-responsive leptin receptor (omLepR (L)) cDNA from the marine medaka (*Oryzias melastigma*). *Mar. Pollut. Bull*, 54: 797-803.

Woo N Y S, Chung A S B, Ng T B. 1993. Influence of oral-administration of estradiol-17-beta and testosterone on growth, digestion, food conversion and metabolism in the underyearling red-sea bream, *Chrysophrys major*. *Fish Physiol. Biochem*, 10: 377-387.

Xing Y, Wensheng L, Haoran L. 2005. Polygenic expression of somatostatin in orange-spotted grouper (*Epinephelus coioides*): Molecular cloning and distribution of the mRNAs encoding three somatostatin precursors. *Mol. Cell Endocrinol*, 241: 62-72.

Xu M, Volkoff H. 2007. Molecular characterization of prepro-orexin in Atlantic cod (*Gadus morhua*): Cloning, localization, developmental profile and role in food intake regulation. *Mol. Cell Endocrinol*, 271: 28-37.

Xu M, Volkoff H. 2008. Cloning, tissue distribution and effects of food deprivation on pituitary adenylate cyclase activating polypeptide (PACAP) /PACAP-related peptide (PRP) and preprosomatostatin 1 (PPSS 1) in Atlantic cod (*Gadus morhua*). *Peptides*, 30: 766-776.

Xu M, Volkoff H. 2009. Molecular characterization of ghrelin and gastrin-releasing peptide in Atlantic cod (*Gadus morhua*): Cloning, localization, developmental profile and role in food intake regulation. *Gen. Comp. Endocrinol*, 160: 250-258.

Yada T, Moriyama S, Suzuki Y, Azuma T, Takahashi A, Hirose S, Naito N. 2002. Relationships between obesity and metabolic hormones in the "cobalt" variant of rainbow trout. *Gen. Comp. Endocrinol*, 128: 36-43.

Yaghoubian S, Filosa M F, Youson J H. 2001. Proteins immunoreactive with antibody against a human leptin fragment are found in serum and tissues of the sea lamprey, *Petromyzon marinus* L. *Comp. Biochem. Physiol. B*, 129: 777-785.

Yokogawa T, Marin W, Faraco J, Pezeron G, Appelbaum L, Zhang J, Rosa F, Mourrain P, Mignot E. 2007. Characterization of sleep in zebrafish and insomnia in hypocretin receptor mutants. *PLoS Biol*, 5: e277.

第 10 章　神经和内分泌调控消化道功能

本章归纳与总结目前对鱼类消化道功能包括活动、分泌和吸收的神经与内分泌调控的研究进展。大多数研究成果是在板鳃鱼类和硬骨鱼类中取得的，但相应地亦包括从其他类群中得到的不多的资料。概括介绍消化道神经分布的解剖学和消化道内分泌系统。许多研究都涉及神经递质和消化道激素的鉴定和分布，这将在一节中概述。大多数功能研究都涉及活动性调控的各个方面，这就组成本章的一部分主要内容，但同时亦包括一些消化道血液循环、分泌、水分和离子运输、吸收过程等较为难得的研究结果。此外，还报道了最近涉及鱼类消化道神经与内分泌调控系统发育的研究进展，并对鱼类各个种类与类群之间，以及相应的鱼类和脊椎动物之间进行比较。

10.1　引　言

本章将着重概述神经内分泌调控参与食物消化与吸收的各种生理过程。消化道的其他功能，例如局部免疫防御的控制，不在本章的范围之内。分泌过程、活动型式和通过消化道的血流是食物加工过程的中枢作用机理，主要由作用于精确反射通路的神经与内分泌细胞以及肌肉细胞，Cajal（ICCs）的间质细胞与分泌细胞一起调控。反射既是肠道的，即局部的到达消化道，亦有外在的通路参与。两种类型都会对一定的刺激产生综合反应。

在许多情况下，神经和内分泌系统在基础的刺激和消化道功能调整之间只是一个分界面。食物本身就是最重要的基础刺激。对哺乳类的研究表明食物的组成和性质能调节消化道活动以便进行最适的消化与吸收（Schneeman，2002）。感觉的刺激，甚至对食物的渴望，通过中枢神经系统（CNS）和迷走神经而作用于消化道。消化道壁的膨胀、食物的成分、消化道腔内 pH 的变化都可以激发局部的反射和激素的释放。

10.2　消化道神经和内分泌系统的解剖学

鱼类消化道的神经与内分泌系统依照着脊椎动物的一般模式，但有些种类偏离这个模式而表现解剖的特异性。例如，圆口类、全头类和一些硬骨鱼类没有胃，即缺乏酸的分泌，胆管的入口出现在消化道紧密靠近食道的位置。在有幽门盲囊的情况下，其大小和数量在不同种类之间是不同的。板鳃鱼类、全头鱼类和软骨硬鳞鱼类通常具有螺旋状的肠，即短而宽的中肠，内有精致的螺旋状的黏膜皱褶（Stevens 和 Hume，

1995）。鱼类的胰脏通常是弥散状的，不同种类的结构差异明显（见第 10.2.2 节）。所有这些都反映在肠道的神经与内分泌细胞的分布方面。

10.2.1 消化道的神经分布

胃肠（GI）道是一个神经分布稠密的器官。消化道的平滑肌、血管、内分泌细胞等都由自主神经控制。这些神经很多都定位在消化道壁内，形成肠神经系统（ENS）。消化道分别通过迷走神经和内脏神经亦接受外在的脑和脊髓的自主神经分布。有些无胃的鱼类，如斑马鱼的迷走神经分布于从食道开始的整个消化道（Olsson 等，2008b）。大多数的其他鱼类的迷走神经分布于食道一部分、胃和近端的肠；而有些鱼类的迷走神经纤维只分布于食道（Burnstock，1969；Nilsson，1983）。迷走神经发出感觉的和运动的神经纤维，但一些研究曾提到鱼类感觉的和运动的纤维比例问题。对大西洋鳕鱼（*Gadus morhua*）进行的追踪实验表明外周感觉神经末梢分布在消化道壁内（Karila，1997）。

圆口类的脊髓神经主要分布在消化道后部。在板鳃鱼类中，一对前内脏神经分布于胃的后部和前肠，中与后内脏神经分布于螺旋肠和直肠。硬骨鱼类的前内脏神经分布到胃的后部和肠的大部分。在板鳃鱼类和硬骨鱼类中，最前的脊神经分支经常加入迷走神经，形成迷走-交感神经干，分布到消化道（Nilsson，1983）。

在所有的脊椎动物，ENS 都含有大量神经细胞（Nilsson，1983）。在鱼类中，神经细胞的细胞体主要位于肠肌丛内，在纵肌和环肌层之间。在圆口类中，发达地散布着神经细胞体的肠丛沿着直的肠管分布（Kirtisinghe，1940；Baumgarten 等，1973）。同样，在硬骨鱼类中，肠肌的神经元沿着消化道均匀地散布，只有少数形成多于 2～3 个细胞的微小神经结（Kirtisinghe，1940；Olsson 和 Karila，1995；见图 10.1A）。但实际的分布情况，在不同种类之间以及不同区域之间都有所不同。此外，硬骨鱼类肠肌神经细胞体的总体密度和小型哺乳动物相似。在板鳃鱼类中，神经细胞体通常在肠肌丛内形成结节（Kirtisinghe，1940；Olsson 和 Karila，1995）。

其他的细胞类型，如 ICCs 和可能对消化道活动的调控起部分作用的肠胶质细胞，在鱼类中还研究得不多。Kirtisinghe（1940）描述了鱼类消化道的 ICCs，而最近采用哺乳类 ICC 标记进行的研究，所得结果是与之互相矛盾的（Mellgren 和 Johnson，2005；Rich 等，2007）。

10.2.2 消化道作为一个内分泌器官

消化道具有广泛的内分泌功能。大量的内分泌细胞分布于消化道的黏膜层，而胰脏形成所谓的"胃肠胰（GEP）内分泌系统"。有些信号物质对内分泌细胞是特有的，而其他的则和神经递质一样或者相似（详见第 10.3 节）。在肠黏膜层的一群内分泌细胞和释放酶的外分泌细胞，能释放和胰脏细胞相同的物质，推想是相当于胰脏进化的原始早期阶段的情况（Van Noorden 和 Falkmer，1980；Youson 和 Al-Mahrouki，1999）。

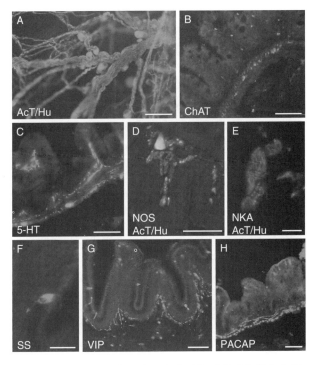

图 10.1　鱼类消化道免疫反应信号物质的形态学和实例

（A）杜父鱼（*Myoxocephalus scorpius*）消化道的整体制片（whole-mount preparation）；（B、C、G、H）斑马鱼（*Danio rerio*）消化道切片；（D-F）杜父鱼消化道的切片。（A）贲门胃肠肌丛内的神经细胞体和神经束。（B）肠球内的神经纤维和神经细胞体的 ChAT 免疫反应，表示胆碱能的神经分布。（C）中肠的神经纤维和神经细胞体的 5-HT 免疫反应。（D）贲门胃的抗 NOS（绿色）和 AT/Hu（红色）的双重标志，表示在肠肌丛内的氧化氮能（nitrergic）神经细胞体。（E）贲门胃的抗 NKA（红色）和 AT/Hu（绿色）的双重标志，表示在肠肌丛内的神经细胞体的 NKA 免疫反应。（F）贲门胃黏膜层内分泌细胞的 SS-免疫反应。（G）幽门胃黏膜层神经纤维和分泌细胞的 VIP 免疫反应。（H）中肠肠丛和环肌层神经纤维的 PACAP-免疫反应。AcT，乙酰微管蛋白；ChAT，胆碱乙酰转移酶；5-HT，5-羟色胺；Hu，人神经元蛋白质 C/D；NKA，神经激肽 A；NOS，一氧化氮合成酶；PACAP，脑垂体腺苷酸环化酶激活多肽；SS，生长抑素；VIP，血管活性肠多肽。标尺：50 μm（A-D，G、H），25 μm（E、F）。（见书后彩图）

在鱼类中，胰脏的解剖学差异明显。在圆口类中，外分泌的胰腺组织出现在近端的肠壁上，而散布的内分泌胰岛器官则存在于肠系膜（Van Noorden，1990）。在板鳃鱼类和肺鱼类中，胰脏是一个结实的器官，明显分为内分泌和外分泌两部分（Van Noorden 和 Falkmer，1980）。在硬骨鱼类中，较小或较大的胰岛散布在靠近肠和幽门盲囊的系膜中。外分泌的细胞组成一薄层围绕着内分泌组织，如同虹鳟的 brockmann 小体，或者形成一个与内分泌组织分开的叶状器官。在板鳃鱼类和硬骨鱼类中，存在着 4 个内分泌细胞的主要类型，即胰岛素、胰高血糖素、生长抑素（SS）和胰多肽（PP），以及分泌其他肽类的小群细胞，但在不同的种类之间，会出现一些差别（Youson 和 Al-Mahrouki，1999）。自主神经分布于板鳃鱼类和硬骨鱼类胰脏的外分泌和内分泌组织中（Van Noorden 和 Patent，1980；Jonsson，1991，1993）。

10.3 消化道的神经递质和激素

信号物质可以在内分泌细胞和神经中合成。许多信号物质和一个以上的受体互相作用，即神经递质的实际作用取决于受体的亚型。有关鱼类受体分布的资料很少，但一些和哺乳类受体亚型有联系的基因已经在鱼类中确定。对于下面介绍的已经确定基因的详细资料，可参阅国家生物信息中心（NCBI；www.ncbi.nlm.nih.gov）网站信息。

下面概括介绍鱼类消化道中已被详细研究的信号物质。它们的鉴别和存在由选用的参考文献举例说明，并将分析讨论总的型式以及这些型式的例外情况。表10.1列出短角床杜父鱼（*Myoxocephalus scorpius*）消化道的概况作为一些神经内分泌物质在消化道分布的详细例子。更多的表格和早期文献的参考目录可参阅 Nilsson 和 Holmgren（1994）的著作。在神经或者内分泌细胞中可能共存的信号物质通常都不予以列出与说明，因为它们在鱼类中还了解得很少。只有个别经过抗血清测试的混合体，将在总的化学编码中做简略的介绍。一些因子能影响消化道神经递质与激素的分布与丰度，如摄食状态、食物类型和季节性差异等，但这方面的研究才刚刚开始。此外，感染和炎症亦能影响到表达的情况（Bosi 等，2005）。

表10.1 采用免疫组织化学技术测定杜父鱼（*Myoxocephalus scorpius*）消化道的神经纤维相对密度（+）、神经细胞体（X）和黏膜内分泌细胞（O）

	CCK/促胃液素[a]	CGRP	GAL	GRL	GRP[a]	NPY	SS	TK	VIP	PACAP	5-HT	NOS
CS	O	−	+++OOO	−	−	++	OOO	+++XOOO+++	+++	OOO	+++X
PS	O	+	+OOO	O	−	++	OOO	+++XOOO+++	+++OO	OOO	+++
PI	OOO	+++	(+)OOO	(0)	OOO	+++OOO	−	+++OOO+++O+++OOO			+++X	+++
MI	OOO	弱	(+)OOO	(0)	OO	+++OOO	−	+++OOO+++O+++OO			+++X	+++X
DI	OOO	弱	(+)OOO	(0)	OO	+++OOO	−	+++OOO+++OO+++OO			+++	+++X
R	?	弱	(+)OOO	(0)	OO	?	−	+++OOO+++OO+++OO			+++	弱

CS，贲门胃；PS，幽门胃；PI，近端肠；MI，中肠；DI，远端肠；R，直肠；CCK，缩胆囊肽；CGRP，降钙素相关肽；GAL，甘丙肽；GRL，生长素释放肽；GRP，促胃液素释放肽；NPY，神经肽Y；SS，生长抑素；TK，速激肽（P物质/神经激肽A）；VIP，血管活性肠多肽；PACAP，脑垂体腺苷酸环化酶激活多肽；5-HT，5-羟色胺；NOS，一氧化氮合成酶。

[a] Bjenning 和 Holmgren（1988）在胃和肠内发现促胃液素/CCK 样免疫反应神经纤维；在胃内发现铃蟾肽样-免疫反应（＝GRP 样-免疫反应）神经纤维。

10.3.1 乙酰胆碱

药理学和生理学的研究表明乙酰胆碱（ACh）对调控鱼类和其他脊椎动物消化道的功能起着重要作用（见第10.5.2节）。组织化学研究的发现支持这个论点，在消化道的神经中，例如鳕鱼肠肌的神经细胞中，出现不多的合成酶，如胆碱乙酰转移酶（ChAT），或者小泡乙酰胆碱转运蛋白（VAChT）（Karila 等，1998）。已经表明鳕鱼的胆碱能神经包括下行的（向肛门方向延伸）中间神经元和上行的（向口腔方向延伸）运动神经元。对斑马鱼的研究亦证明存在着 ChAT-免疫反应神经（见图10.1B；Olsson 等，2008b）。大多数硬骨鱼都具有迷走的胆碱能神经分布，但有些硬骨鱼类和板鳃鱼类是没有的（Campbell 和 Burnstock，1968；Campbell，1975；Holmgren 和 Nilsson，1981；Nilsson，1983）。毒蝇碱性（muscarinic）M3-样受体是调控虹鳟消化道活动性最重要的胆碱能受体亚型（Aronsson 和 Holmgren，2000）。

10.3.2 胺类、氨基酸和嘌呤

成为鱼类消化道信号物质的胺类包括儿茶酚胺、组胺和5-羟色胺（5-HT）。儿茶酚胺［肾上腺素（AD）、去甲肾上腺素（NA）和多巴胺（DA）］由肾上腺素能神经和嗜铬组织释放。鱼类大多数类群的消化道都有稠密的肾上腺素能神经分布，它们围绕肠肌神经细胞体，并且分布神经到消化道的血管（Baumgarten 等，1973；Anderson，1983；Jensen 和 Halmgren，1985；Karila 等，1997）。这些神经纤维的来源主要是外源的（脊髓的自主神经）。但是，我们未发表的研究结果表明在一些硬骨鱼类的肠神经细胞体内存在着合成酶——酪氨酸羟化酶（TH）。TH 催化儿茶酚胺合成的第一步，而TH-阳性细胞可能成为肾上腺素能的、去甲肾上腺素能的或者多巴胺能的细胞。已经表明在河七鳃鳗（*Lampetra fluviatilis*）（Baumgarten 等，1973）的消化道内存在着多巴胺能的和肾上腺素能的神经细胞体，而在硬骨鱼类肠内的一些荧光的神经可能含有 DA（Nilsson，1983）。板鳃鱼类和哺乳类、鸟类与爬行类一样，NA 是消化道肾上腺素能神经中的主要儿茶酚胺，而 DA 是硬骨鱼类和全骨鱼类主要的儿茶酚胺类型（Von Euier 和 Fange，1961；Burnstock，1969；Abrahamsson 和 Nilsson，1961）。

组胺出现在鱼类胃和肠黏膜层的内分泌细胞内（Reite，1972；Hakanson 等，1986；D'Este 和 Rende，1995；本人未发表研究结果）。它们在消化道神经和肥大细胞中的存在是很不确定的。深入的研究表明，组胺只存在于肺鱼类和鲈形目这些进化到最高等的硬骨鱼类的肥大细胞中，并且推想这种表达首先是出现在爬行类中，而在稍后时期独立地出现在这两个类群的鱼类中（Reite，1972；Mulero 等，2007）。药理学研究表明 HI-样受体介导消化道活动能力的调控（见第10.5.2.1节），而 H2 受体参与调控胃酸分泌活动（见第10.6.1节）。

5-HT 是一个有意思的事例。在圆口类，它只出现于神经细胞之中（Baumgarten 等，1973；Goodrich 等，1980）。在板鳃鱼类中，神经稀少，内分泌细胞通常含有5-HT。鲟鱼类（软骨硬鳞类）的内分泌细胞和神经细胞都有5-HT（Salimova 和 Feher，1982）。

硬骨鱼类在神经细胞和内分泌细胞中含有 5-HT 的相对比例方面表现出明显的种间差别（Anderson，1983）。在鲤科鱼类中，5-HT 只存在于神经细胞内（Pan 和 Fang，1993；Pederzoli 等，2004；Olsson 等，2008b；见图 10.1C）。在欧洲鳗鱼（*Anguilla anguilla*）中，5-HT 只在内分泌细胞中发现（Domeneghini 等，2000），而一些相近的种类则在内分泌细胞和神经细胞中都含有 5-HT（Anderson 和 Campbell，1988）。但神经元的染色微弱。Anderson 和 Campbell（1988）坚持认为，如果没有 5-HT 神经纤维到达黏膜层，内分泌细胞就只表达 5-HT。至少有七类 5-HT 受体，它们的成员都已经从不同的鱼类中分离出来，但目前对这些受体在消化道的分布还研究得很少。

γ-氨基丁酸（GABA）存在于 CNS，但至今在鱼类消化道中还很少报道。Gabriel 等（1990）证明鲤鱼（*Cyprinus carpio*）出现 GABA 免疫反应的神经细胞体。在虹鳟中有时会发现 GABA 免疫反应的肠肌神经细胞（我们未发表研究结果）。GABA 和三种类型的受体结合，其中两种（A 和 B）已经在硬骨鱼类中测定序列，但还不清楚它们在消化道的分布情况。

腺苷和 ATP 是嘌呤衍生物，当它们从神经末梢的突触囊泡释放达到足够高的浓度时具有递质的功能。它们的功能由许多受体亚型介导，这些受体属于由腺苷刺激的 P1-受体群或者和 ATP 及其他核苷酸相互作用的 P2-受体。P1-受体和 P2-受体都存在于鱼类的消化道中（Lennard 和 Huddart，1989；Knight 和 Burnstock，1993）。

10.3.3 一氧化氮

一氧化氮（NO）由一氧化氮合成酶（NOS）合成，其神经元的同等型已经从一些鱼类，如斑马鱼、虹鳟和河鲀（NCBI，2008）等中分离和测定序列。NOS 的存在通常用来作为氧化氮能神经的指示物（见图 10.1D）。Li 和 Furness（1993）最先在虹鳟整个消化道的肠肌丛中发现含有 NOS 的神经细胞体。这些细胞主要发送神经纤维到肌肉层。此后证实在一些鱼类中存在着氧化氮能神经元。大西洋鳕鱼和白斑角鲨（*Squalas acanthias*）接近或者超过半数的肠肌神经细胞含有 NOS，而在盲鳗（*Myxine glutinosa*）的肠肌内则没有这些细胞（Olsson 和 Karila，1995）。在金鱼消化道的所有括约肌内都发现密集的氧化氮能神经元（Bruning 等，1996）。

10.3.4 肽类

消化道内的信号肽类由只有几个到大约 40 个氨基酸组成。它们按照结构和遗传相似性而形成家族。肽的氨基酸序列变化反映种类之间的进化距离。硬骨鱼类经历过额外的基因复制过程，增加了特有信号物质进化的可能性（Conlon 和 Larhammar，2005）。下面介绍曾经深入研究的一些参与神经内分泌调控消化道的肽类。但是，这并不是一个存在于胃肠消化道的信号肽类的完整目录。

缩胆囊肽（CCK）/促胃液素家族包括几个长度不同的肽，其特征是 4 个共同的 C-端氨基酸（Johnsen，1998；Kurokawa 等，2003）。在大多数种类中，CCK/促胃液素样物质出现在胃和肠的内分泌细胞中，而且通常具有附加的免疫反应神经纤维（Van

Noorden 和 Pearse，1974；Holmgren 和 Nilsson，1983a；Burkhardt-Holm 和 Holmgren，1989；Yui 等，1990；Holmgren 等，1994）。个别的免疫组织化学研究曾明确地区分 CCK 和促胃液素，但按照脊椎动物一般的情况，最常见的是胃内分泌细胞含有促胃液素，而肠内分泌细胞和消化道神经含有 CCK 肽。已证明在鱼类中有一个 CCK/促胃液素受体类型（Oliver 和 Vigna，1996）。这个 CCK-X 受体可能是哺乳类 CCK-A 和 CCK-B 受体的祖先类型。Ronnestad 等（2000）曾详细综述 CCK 对鱼类（幼鱼）消化活动的中枢调控功能。

降钙素基因相关肽（CGRP）是一个由 37 个氨基酸组成的肽，由降钙素/CGRP 基因的可变剪接（alternative splicing）而产生（Wimalawansa，1997）。在硬骨鱼类（Ogoshi 等，2006）和哺乳类中存在着两个相似的同类型。在大西洋鲑鱼中，CGRP 在肠的神经细胞体和纤维中表达，但在胃的神经细胞体中没有出现表达（Karila，1997；Shahbazi 等，1998）。在斑马鱼中，迷走的 CGRP 神经纤维分布于近侧的肠，而外在的和内在的神经纤维分布在远端的肠（Olsson 等，2008b）。CGRP 神经亦存在于北极七鳃鳗（*Lampetra japonica*）（Yui 等，1988）和澳洲肺鱼（*Neoceratodus forsteri*）（Holmgren 等，1994）。对鳕鱼和肺鱼的研究曾报道在黏膜层有内分泌细胞。CGRP 受体在牙鲆（*Paralichthys olivaceus*）的消化道中表达（Suzuki 等，2000）。

甘丙肽（GAL）是 29 个氨基酸组成的肽，从古老的鱼类群体以及硬骨鱼类与板鳃鱼类中分离与测序（Habu 等，1994；Wang 和 Conlon，1994；Wang 等，1999）。甘丙肽的初级结构高度保守，特别是在 N-端。在七鳃鳗的内分泌细胞出现 GAL 是一个例外（Bosi 等，2004），所有报道几乎全部表明 GAL 存在于肠的神经元和周维管纤维（perivascular fiber）（Karila 等，1993；Holmgren 等，1994；Preston 等，1995）。大多数的肠纤维出现在黏膜层和黏膜下层，但亦有一些存在于肌肉层，表明它们具有分泌和运动的作用。神经细胞体存在于肠肌丛内，表明至少有一部分神经分布是消化道内在的（Bosi 等，2007）；而在迷走神经，既延伸到 CNS，亦伸展到消化道（Karila 等，1993）。由胃到直肠，神经分布的密度通常逐渐降低。

生长素释放肽（GRL）最近才加入消化道激素的目录中，已在一些鱼类中鉴定和测序。GRL 的长度在不同种类中有差别，从金鱼的 19 个氨基酸到鲨鱼的 25 个氨基酸。有些鱼类表达多于一个同类型（Kaiya 等，2008）。GRL 存在于虹鳟的内分泌细胞（Sakata 等，2004）和鳗鱼的胃内（Kaiya 等，2006）。然而，GRL-免疫反应的内分泌细胞亦存在于一些无胃鱼类如斑马鱼和鲤鱼的肠内（Kono 等，2008；Olsson 等，2008a）。GRL 和促胃动素受体相关联的受体结合。除 GRL 受体和促胃动素受体之外，在河鲀基因组中曾发现第三个受体（Olsson 等，2008a）。GRL 对这个受体的亲和力还不清楚。

促胃液素-释放肽（GRP）和铃蟾肽（BBS）属于具有 8 个相同 C-端氨基酸的一些肽类家族。GRP 已经从鲨鱼（Conlon 等，1987）和虹鳟（Jenson 和 Conlon，1992a）的胃肠消化道内分离出来，而相关的 cDNA 已经从斑马鱼和金鱼中测定序列（NCBI，2008；Volkoff 等，2000）。一个和神经调节肽 B 一样的序列亦存在于斑马鱼中（NCBI，2008）。在虹鳟中存在着一个较短的 GRP，而提取物中并不含 BBS（Jensen 和 Conlon，

1992a)。此外，免疫组织化学研究表明在一种板鳃鱼类和一种全头鱼类银鲛的消化道内存在着一种比 GRP 更像 BBS 的肽（Cimini 等，1985；Thorndyke 等，1990）。

GRP/BBS-样免疫反应神经存在于七鳃鳗的消化道内（Yui 等，1988）。在板鳃鱼类和硬骨鱼类的一些种类中，神经和内分泌细胞沿着整个胃肠消化道分布，其密度在不同种类和消化道的不同区中有所差异（Cimini 等，1985；Bjenning 和 Holmgren，1988；Tagliafierro 等，1988）。神经细胞体出现在白斑角鲨和短角床杜父鱼的肠肌丛内，表明神经分布是肠内源的特征（Holmgrem 和 Nilsson，1993a；Bjenning 和 Holmgrem，1988）。至少在板鳃鱼类中，消化道壁血管的周维管神经纤维以及消化道的主要血管表达一种 GRP/BBS-样肽（Holmgrem 和 Nilsson，1983a；Tagliafierro 等，1988；Bjenning 等，1990）。

多年来，神经肽 Y（NPY）肽类家族的进化受到明显的关注。Sundstrom 等（2008）经过深入研究，阐述经过额外的基因复制后鱼类（硬骨鱼类）的基因组中已经演化出两种类型的 NPY 和两种类型的肽 YY（PYY），它们在不同种类中的表达略微不同。真正的 PP 已经消失。一些研究曾报道 NPY/PYY-样物质出现在鱼的消化道和胰脏，但对这些肽的确切鉴定还很少阐明。PYY 通常是全身的包括消化道的神经和内分泌细胞中主要的类型（Sundstron 等，2008）。

在圆口类中，NPY/PYY-样物质只出现在内分泌细胞内（Yui 等，1988），而在板鳃鱼类和硬骨鱼类的消化道，神经纤维存在于肠肌内和肌肉层，有些围绕着血管（Bjenning 等，1989；Burkhardt-Holm 和 Holmgren，1989；Preston 等，1998；Shahbazi 等，2002）。PP-样物质一直存在于内分泌胰脏内，只有圆口类例外（Van Noorden 和 Patent，1980），它们亦存在于肠的内分泌细胞，在个别例子中亦存在于胃内（Langer 等，1979；Cimini 等，1985；Yui 等，1990）。在脊椎动物中至少有 7 个 NPY-受体类型。在鱼类当中，种类-依赖性的差别是常见的（Salaneck 等，2008），但对鱼类消化道存在的各个受体类型目前还知道得不多。

生长抑素（SS）有几个类型，在鱼类中，通常的长度是 14 个或约 28 个氨基酸。SS-14(1)是高度保守的,在盲鳗、肺鱼、鳟鱼和哺乳类中都是一样的。其他的SS-14s以及较长的类型如 SS-28 在不同种类之间变化较大，而有些类型是鱼类特有的（Nelson 和 Sheridan，2005）。除了通常出现在胰脏细胞之外，SS 还出现在所有研究过的鱼类消化道的内分泌细胞内（见图 10.1F）。神经元的 SS 在板鳃鱼类中是常见的，但在硬骨鱼类中则很少见。SS 的分布在不同种类中有所不同：在角鲨中，神经纤维在直肠中最为常见，而在较大的斑点猫鲨（*Scyliorhinus stellaris*）中，它们只出现在胃内（Holmgren 和 Nilsson，1983a；Cimini 等，1985），在硬骨鱼类的玫瑰鲃（*Barbas conchonius*）中，它们只出现在肠内（Rombout 等，1986）。已经鉴定 5 个类型的 SS 受体（SSR_{1-5}），除 SSR_4 外，其他的受体类型都已经在鱼类中发现（Nelson 和 Sheridan，2005）。亚型 1A、1B 和 2 只存在于肝脏和胰脏内。受体表达受到激素水平和摄食状态的影响。例如，在虹鳟中，胰岛素样生长因子-1（IGF-I）下调而禁食上调所有 3 个亚型的受体（Slagter 等，2005）。

速激肽（TK）组成一个大的肽类家族，其中 P 物质和神经激肽 A（NKA）是研究得最多的。P 物质含有 11 个氨基酸，板鳃鱼类与硬骨鱼类的序列和哺乳类相比是在第 3~4 位的氨基酸不同。NKA 由 10 个氨基酸组成，其最后的 5 个氨基酸通常和 P 物质 C-端的一致（Holmgren 和 Jensen，2001；Severini 等，2002）。此外，曾分离出几个和鱼类相关的肽，如从鲨鱼分离猫鲨激肽（scyliorhinin）Ⅰ和Ⅱ（Colon 等，1986）。速激肽能的神经经常分布在所有鱼类的消化道各区，只有圆口类例外。神经纤维最主要分布在肠肌丛内，但亦存在于消化道壁的其他层内（Jensen 和 Holmgren，1985，1991；Jensen 等，1993a）。肠的神经细胞体很常见，表明至少有一部分的神经纤维是内源的（见图 10.1E；Karila 等，1998）。采用抗鱼类 TK 的抗血清确定虹鳟和角鲨的神经纤维分布到消化道的血管（Kagstrom 等，1996a；Kagstrom 和 Holmgren，1998）。此外，在所有的消化道区都发现内分泌细胞表达 TK（Jensen 和 Holmgren，1991）。TK 至少通过 3 个不同的受体起作用，而神经激肽（NK）1-3 受体都出现在鱼类。功能的研究结果表明鱼类的消化道有 NK1 受体（Jensen 等，1993b）。

血管活性肠多肽（VIP）和脑垂体腺苷酸环化酶激活多肽（PACAP）是两个密切相联系的多肽，分别含有 28 个氨基酸和 38 个或 27 个氨基酸。VIP 和 PACAP 已经在几种鱼类中测定序列，并且表现出和哺乳类的相关肽类有相当大的相似性（Hoyle，1998，Holmgren 和 Jensen，2001）。除了在圆口类中只出现在内分泌细胞之外（Reinecke 等，1981；Van Noorden，1990），神经元的 VIP 出现在大多数曾经检测过的鱼类中。VIP 神经纤维分布在消化道壁的所有各层中（见图 10.1G）。虽然发现不多的标志神经细胞体，但大部分的 VIP 神经分布仍然是内源性的。在消化道的大动脉中只有相当稀少的 VIP 神经分布（Langer 等，1979；Holmgren 和 Nilsson，1983a、b；Jensen 和 Holmgren，1985；Yui 等，1990；Holmgren，1994；Kagstrom 和 Holmgren，1997）。VIP-样物质亦经常出现在许多鱼类的内分泌细胞内（Falkmer 等，1980；Reinecke 等，1981；Rajjo 等，1989；Olsson 和 Holmgren，1994）。但是，有些鱼类可能没有这类细胞（Domeneghini 等，2000；Lee 等，2004）。这种差异可能是来源于不同种类之间真正的差别，但亦可能是因为使用的抗血清产生的不同特异性。

已证明 PACAP 存在于一些鱼类的消化道神经内（Olsson 和 Holmgren，1994；Holmgrem 等，2004；见图 10.1H）。在鳕鱼、虹鳟和角鲨中，肠神经元和 VIP 100% 共定位（Olsson 和 Holmgren，1994；我们未发表的研究结果）。在内分泌细胞当中，通常有 PACAP 细胞的亚群体是 VIP-阴性的。PACAP 和 VIP 通过 3 个主要的受体类型——VPAC1、VPAC2 和 PAC1 而互相作用，有些种类在受体中分布方面不同。3 种鲀形目鱼类在消化道内表达 VPAC2 和 PAC1（Cardoso 等，2004）。在金鱼肠内只有 VAPC1（Chow，1997），而在斑马鱼肠内存在着一个 PAC1 的同类型（Fradinger 等，2005）。

10.4 消化道神经分布和神经内分泌系统的发育

鱼类神经系统早期发育的大部分知识都来自对斑马鱼的研究（Riable 等，1992；Shepherd 等，2001；Holmberg 等，2003，2004，2006，2007），虽然亦包括对大菱鲆

（Scophthalmus maximus）和绿剑尾鱼（Xiphophorus helleri）的一些研究结果（Sadaghiani 和 Vielkind, 1990; Reinecke 等, 1997）。消化道的自主神经分布由神经嵴（neural crest）衍生而来（Lamers 等, 1981; Sadaghiani 和 Vielkind, 1990; Raibe 等, 1992）。在各种化学信息的影响下, 迷走神经嵴细胞迁移到消化道。在斑马鱼孵化后 48 h 内, 消化道出现第一个神经细胞（Bisgrove 等, 1997; Holmberg 等, 2003）; 孵化后 4～5 天开始外源摄食, 但在孵化后 3 天肠道已经有丰富的神经分布, 并且神经开始表达各种不同的信号物质（Holmberg 等, 2004, 2006; Olsson 等, 2008b）, 但是, 在最近端的部分（肠球）并没有很多的神经分布; 直到稍后, 孵化后 11～13 天, 大部分消化道具有稠密的神经纤维网络（Olsson 等, 2008b）。消化道的内分泌细胞由原始消化道的内胚层衍生（Wallace 和 Park, 2003）。斑马鱼内分泌细胞的分化与成熟比较迟, 出现在孵化后 4 天左右（Ng 等, 2005）。

许多递质和激素在外源性摄食开始之前就已经表达。信息物质第一次出现的准确时间和顺序在不同种类之间有差别。此外, 一种物质能否在神经和内分泌细胞中表达, 以及首次出现在一个或另一个细胞类型亦是不同的。有些种类的神经出现早于内分泌细胞, 另一些种类的神经和内分泌细胞亦会在同一个发育阶段出现。表 10.2 对一些鱼类做了较详细的比较。此外, 在大西洋庸鲽（Hippoglossus hippoglossus）的幼鱼中, GRL 在外源性摄食开始之前表达, 而在变态前达到高峰（Manning 等, 2008）。相反, CCK 在第一次摄食之后出现在大西洋庸鲽的内分泌细胞内（Kamisaka 等, 2001）, 但在第一次摄食的前一天出现在牙鲆的内分泌细胞内（Kurokawa 等, 2000）。不同的分泌物质亦表现不同的区域性发育。在海鲈中, TK 最早是在直肠内表达, 随后在胃内表达, 而 VIP 则是相反的型式（Pederzoli 等, 2004）。在虎纹猫鲨（Scyliorhinus torazame）中, 5-HT 最先在消化道中部的内分泌细胞内表达, 然后扩散到胃, 有些细胞亦分布在后消化道中（Chiba, 1998）。

表 10.2　在发育过程中一些递质和激素在不同鱼类消化道内的第一次出现（采用免疫组织化学技术测定）

	海鲈[1]		大菱鲆[2]		斑马鱼[3]		鲨鱼[4]		鲨鱼[5]	
	ec	nf	ec	nf	ec	nf	ec	nf	ec	nf
CCK			11				4～5	—		
NPY			8	24			2.5	4		
TK	4	12	11	8	3	3				
VIP/PACAP	18	4	24	5	4	3			4	7～8
5-HT			10	15	—	4	3		4～5	—
NOS	—	8[a]			—	3				
	dph		dph		dpf		月[b]		月	

ec，内分泌细胞；nf，神经纤维；CCK，缩胆囊肽；NPY，神经肽 Y；TK，速激肽（P 物质/神经激肽 A）；VIP，血管活性肠多肽；PACAP，脑垂体腺苷酸环化酶激活多肽；5-HT，5-羟色胺；NOS，一氧化氮合成酶。

dpf，受精后天数；dph，孵化后天数。

1. Pederzoli 等，2004，2007（*Dientrarchus labrax*）。
2. Reinecke 等，1997（*Scophthalmus maximus*）。
3. Holmberg 等，2004，2006；Olsson 等，2008b（*Danio rerio*）。
4. Chiba，1998（*Scyliorhinus torazame*）。
5. Tagliafierro 等，（*Scyliorhinus stellaris*）。
a. 在较早的阶段没有测定；b. 孵化后 8~9 月。

有些分泌物质在发育期间可以暂时性表达。在七鳃鳗的幼鱼中，SS-或 CCK/促胃液素免疫反应的内分泌细胞出现在肠内，但它们都不存在于成鱼中（Yui 等，1988）。在较大的斑点猫鲨中，胰高血糖-样物质只在胚胎发育的中期在幽门胃的内分泌细胞中表达（但在整个胚胎发育期和出生时，它们都存在于胃和螺旋肠内）（Tagliafierro 等，1989）。

10.5 消化道活动性的调控

消化道活动的目的是混合和推动消化道的内含物，以便沿着胃肠道得到最合适的消化和吸收。主要的作用机理是消化道的平滑肌在摄入食物的化学与机械刺激下进行收缩与舒张的活动。消化道活动亦发生在禁食的动物中。消化道的不同区有不同的活动型式，在不同的状态下调控的通路亦不同。

10.5.1 消化道肌肉和自发性活动

胃肠道平滑肌的电活动表现为环状型式，即所谓的"慢波"（基本电节律），它依赖于自发的去极化和复极化。慢波开始于 ICCs，然后被动地扩散到整个 ICC 网络以及同步作用的肌肉细胞。ICC 设置慢波的频率并存在于整个胃肠道的几个起搏点区。神经元的和激素的输入转换慢波活动到肌肉的收缩中。这方面的情况在鱼类中还很少研究。

在实验方面，如果是发生在制品处理之前，如毒物接触或者电刺激之前，收缩活动通常都归类为"自发性的"活动，尽管它们大多数是由 ICCs 和/或神经或激素启动的。鱼类消化道制品的自发性收缩活动早期在板鳃鱼类和硬骨鱼类中已得到证实（Young，1933；Burnstock，1958a、b）。例如，在大西洋鳕鱼的肠道中，这些收缩活动是神经依赖性的，因为它们为河鲀毒素（tetrodotoxin，TTX）所减弱或消除。胆碱能的和羟色胺能的神经能够保持一种刺激的紧张性，而由抑制性的氧化氮能紧张性使它均衡（Jensen 和 Holmgren，1985；Karila 和 Holmgren，1995；Olsson 和 Holmgren，2000）。其他一些制品对神经（兴奋）阻滞可能是不敏感的，或者是在阻滞之后改变了活动的型式（我们未发表的研究结果）。

10.5.2 个别信号物质的作用

消化道活动在每一种情况下都是依赖于神经和激素的相互作用。我们所了解的这种整合作用的基础是阐明每种物质所起的作用。对鱼类消化道活动药理性研究的数量有限，而大多数研究采用的是离体的分离制品。下面总结大部分详细研究的递质对消化道活动的作用（见图10.2）。虽然时常有例外，但 Ach、组胺、5-HT 和 TK 通常是兴奋性的，NO 和 VIP/PACAP 是抑制性的，而儿茶酚胺、嘌呤能物质和 CCK 的作用可能在不同的种类甚至在消化道的不同部分是不同的。

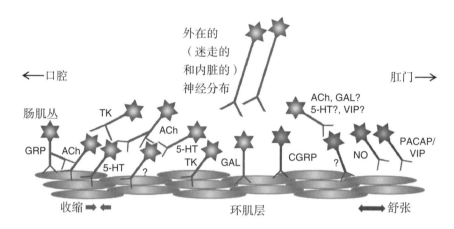

图10.2 在硬骨鱼类肠肌丛内的神经元类型（由免疫组织化学方法确证），包括在消化道平滑肌的主要延伸和作用

本图为检测的不同鱼类在不同消化道区的试验性总结。从神经或者内分泌细胞释放的附加的信号物质都包括在内。神经接触到环状肌肉层，表示为已证实的对平滑肌的直接作用。如果神经分布的特征未证实，表示为神经没有和其他细胞接触。垂直的神经表示没有进行分支延伸的研究。红色＝兴奋性，蓝色＝抑制性。5-HT, 5-羟色胺；ACh, 乙酰胆碱；CGRP, 降钙素基因相关肽；GAL, 生长素释放肽；GRP, 促胃液素释放肽；NO, 一氧化氮；PACAP, 脑垂体腺苷酸环化酶激活多肽；TK, 速激肽；VIP, 血管活性肠多肽。（见书后彩图）

10.5.2.1 刺激性的信号物质

乙酰胆碱（ACh）是大多数硬骨鱼类消化道的强有力的兴奋性递质（Nilsson, 1983；Jensen 和 Holmgren, 1994）。胆碱性神经可以是迷走外在的，或者内源的（见第10.3.1节）。鳕鱼肠道的胆碱能收缩能为阿托品所减弱，但不受 TTX 的影响，表明平滑肌毒蕈碱性受体（muscarinic receptor）的直接作用（Jensen 和 Holmgren, 1985）。此外，对游离肠道的离体研究表明存在着胆碱能内源的运动神经元向口腔方向延伸（Karila 和 Holmgren, 1995）。阿托品亦能减弱虹鳟对胃膨胀反应的反射性收缩活动（Grove 和 Holmgren, 1992a）。

外源的 ACh 能引起板鳃鱼类和盲鳗类消化道的收缩（Von Euler 和 Ostlund, 1957；

Holmgren 和 Fange，1981；Nilsson 和 Holmgren，1983），但对在体的胆碱能调控作用了解得很少。板鳃鱼类可能缺乏一个外源的兴奋性（胆碱能的）神经分布（Nilsson，1993），表明 ACh 主要是由肠神经元释放。此外，刺激盲鳗的迷走神经能产生微弱而无说服力的反应（Patterson 和 Fair，1933）。

组胺对许多鱼类消化道平滑肌没有反应或者只有微弱的反应（Reite，1972），但能引起其他许多鱼类肠道肌肉明显而连贯的收缩（Von Euler 和 Ostlund，1957；Valette 和 Augeraux，1958）。Burnstock（1958a）发现低浓度的组胺没有作用，但高浓度（非生理性作用的？）能引起收缩活动。组胺能使鲈形目的金头鲷（*Sparas aurata*）肠的平滑肌收缩，这种鱼的组胺包含在肥大细胞内。特异性的同等物和拮抗物的作用表明这种作用是由 H2 样受体介导的，但亦是需要非常高的浓度（$10^{-5} \sim 10^{-2}$ M）（Mulero 等，2007）。相反，对虹鳟游离的胃制品进行试验，用 pD_2（即-Log EC_{50}）5.65 能产生传统的收缩反应；而明显的是，在这个例子中，特异性药物证明这是一种通过 HI 受体的反应（Manera 等，2008）。

5-HT 能刺激鱼类消化道的收缩活动（Young，1980a、b，1983；Nilsson 和 Holmgren，1983；Jensen 和 Holmgren，1994）。如上所述，5-HT 的来源是内分泌细胞或者肠神经。例如，在巴斯鲉（*Platycephalus bassensis*）中缺乏内分泌的羟色胺能细胞，但能从肠神经自发地释放 5-HT，形成一种羟色胺能的紧张（Anderson 等，1991）。

5-HT 对虹鳟消化道的收缩作用不受阿托品或 TTX 的影响，这排除了和胆碱能的联系，但表明是对平滑肌的直接作用（Holmgren 等，1985；Burka 等，1989；Grove 和 Holmgren，1992a）。在鳕鱼的肠内，TTX 减弱 5-HT 诱导的收缩活动，表明这至少部分是间接的作用，最主要的是 ACh 释放的参与（Jensen 和 Holmgren，1985），而在其他几种硬骨鱼类中亦有相似的情况（Grove 和 Campbell，1979b；Kiliaan 等，1989）。除了它们对肌肉的刺激作用之外，5-HT 已被表明能从鳕鱼肠下行的中间神经元释放（Karila，1997）。

速激肽（TK）通常刺激收缩活动。有些学者研究了 P 物质和 NKA 对鱼类消化道活动物的影响，有些研究使用了内源的肽类（Jensen 等，1993b；Karila 等，1998）。一项对大多数鱼类类群的代表种类进行的游离带状制品的研究表明，P 物质和 NKA 能诱导或者增强星斑鳐、南美肺鱼（*Lepidosiren paradoxa*）、多鳍鱼（*Polypterus senegalensis*）和虹鳟消化道的收缩，但对盲鳗和河七鳃鳗没有影响（Jensen 和 Holmgren，1991）。P 物质和 NKA 亦能刺激鳕鱼灌注的胃和肠的活动（Jensen 和 Holmgren，1985；Jensen 等，1987，1993a）。对鳕鱼和虹鳟的肠施用天然的肽，P 物质和 NKA 的效能是相同的，但 P 物质对虹鳟胃的效能要强得多（Jensen 等，1993b；Karila 等，1998），表明胃和肠内存在不同的受体。

TK 既能对消化道肌肉直接起作用，如证明施用 TTX 没有影响的试验那样，亦能间接起作用。在大多数鱼类中，P 物质起直接的作用（Andrews 和 Young，1988a；Jensen 和 Holmgren，1991；Jensen 等，1993a）。此外，在硬骨鱼类的肠内，通过胆碱能神经元而起间接作用（Kitazawa 等，1986a；Jensen 和 Holmgren，1991）。还有，通过羟色胺神经元对虹鳟的胃与肠和鳕鱼的肠起间接的作用（Holmgren 等，1985；Jensen 和 Holmgren，1991）。对虹鳟和鳕鱼肠的三种不同作用模式的生理意义，还没有阐明它们是否会涉及不同的反射作用。

生长素释放肽（GRL）是一个食欲的信号，从哺乳类的胃壁释放出来，能对消化道的活动起作用。在斑马鱼中，GRL 能使肠的带状制品产生小而明显增强的基础张力（basal tension）（Olsson 等，2008a）。

甘丙肽（GAL）对离体的大西洋鳕鱼胃制品能引起微弱的直接兴奋作用（Karila 等，1993）。根据它的分布以及它在消化道神经内和胆碱能酶 ChAT 的共存，表明 GAL 在鱼类消化道中可能起着胆碱能活性调节剂的作用（Bosi 等，2007）。

鳕鱼的 NPY（Jensen 和 Conlon，1992b）能使鳕鱼肠的环状制品收缩，但只在高浓度下才起作用，并且在使用免疫组织化学技术后发现这种内源物质是和 PYY 而不是和 NPY 相似（Shahbazi 等，2002）。

10.5.2.2 抑制性的信号物质

一氧化氮（NO）抑制胃肠道的活动。NO 对鱼类作用的研究是用鳕鱼的肠进行的。使用 NO 供体硝普钠（sodium nitroprusside）能使游离的带状制品自发的收缩活动消除（Olsson 和 Holmgren，2000）。此外，NOS 抑制剂如 L-NAME（NG-氮-L-精氨酸甲基酯）能增大自发收缩的幅度，有时亦增大基础张力，表明 NO 在制品中内源的紧张性释放。肠道经 L-NAME 处理使舒张活动减弱，刺激朝向肛门方向发展，表明 NO 参与下行的抑制性通路，可能是由运动神经元释放的（Karila 和 Holmgren，1995）。

NO 亦使电诱导虹鳟胃的收缩活动减弱（Green 和 Campbell，1994）。此外，迷走刺激的抑制性组分能明显被 NOS 抑制剂减弱。在鳟鱼胃内似乎没有内源的 NO 紧张性释放，因为 L-NAME 并不影响基础的活性（Olsson 等，1999）。进一步的研究表明，NO 亦抑制鳕鱼胃与虹鳟肠（我们未发表的研究结果）以及斑马鱼肠（Holmberg 等，2007）的收缩活动。

VIP 和 PACAP 对鱼类消化道平滑肌具有明显的抑制作用。在大西洋鳕鱼的胃内，VIP 能消除在体的自发性收缩活动（Jensen 等，1991）。同样，VIP 能使灌注的虹鳟胃膨胀引起的收缩活动减弱或者使虹鳟胃环状制品的自发性收缩活动减退（Holmgren，1983；Grove 和 Holmgren，1992a）。VIP 对虹鳟胃的作用能被二甲麦角新碱（methysergide）阻抑，表明这是一种通过羟色胺能神经元抑制性的作用（Grove 和 Holmgren，1992a）。相反，哺乳类的和鳕鱼的 VIP 对鳕鱼的肠都没有作用（Jensen 和 Holmgren，1985；Olsson 和 Holmgren，2000）。另一方面，PACAP 能使鳕鱼和虹鳟肠的环状制品自发的收缩活动减弱（Olsson 和 Holmgren，2000；我们未发表的研究结果）。同样，PACAP 抑制斑马鱼肠和日本䲢（*Uranoscopus japonicas*）直肠的活动（Matsuda 等，2000；Holmberg 等，2004）。对鳐鱼的研究表明，VIP 对胃肠的每个部分都没有作用（Andrews 和 Young，1988a），但 VIP 抑制角鲨直肠的活动（Lundin 等，1984）。

CGRP 抑制游离的鳕鱼肠的自发收缩活动（Shahbazi 等，1998）。这种作用不受

TTX 的影响，表明主要是对平滑肌的直接作用。CGRP 已证明对下行的抑制性通路起作用。在哺乳类中，CGRP 通常是感觉神经元的递质，但 CGRP 是否存在于鱼类消化道的外源神经中并参与传入的信号，目前还不清楚。

生长抑素（SS）抑制鳕鱼和虹鳟由于胃的膨胀而引起的活动，可能和哺乳类一样是通过胆碱能神经元的抑制作用实现的（Grove 和 Holmgren，1992a、b）。对鳕鱼的肠用 SS 处理后，开始是舒张，而有时接着是收缩的活动（Jensen 和 Holmgren，1985）。

10.5.2.3 变动的作用

儿茶酚胺。儿茶酚胺的作用在不同种类以及消化道的不同部分有所不同（Nilsson，1983；Jensen 和 Holmgren，1994）。在盲鳗中，其反应是微弱而不规则的（Holmgren 和 Fange，1981）。在板鳃鱼类中，AD 通常和 α 肾上腺素受体（AR）作用而使胃和肠收缩，但却使直肠舒张（Young，1980b，1983，1988；Nilsson 和 Holmgren，1983）。在有些硬骨鱼类，如鲑鱼（*Salmo* spp.）和鳗鱼中，对内脏神经的刺激以及对 AD 和 NA 的反应而使胃收缩；而在另一些鱼类，如鳕鱼、鮟鱇（*Lophius piscatorius*）和日本鰧中，则表现为混合的反应或者舒张活动。由于儿茶酚胺的释放，肠对内脏神经刺激的明显反应是一种松弛现象（Grove 和 Campbell，1979b）。收缩活动由 α-AR 介导，而舒张活动由 α-AR 或 β-AR 介导，或者两种共同介导（Burnstock，1958a、b；Grove 和 Campbell，1979a、b；Young，1936，1980b）。鲤鱼肠球的舒张活动由突触前 α_2-AR 对胆碱能神经元的刺激而引起，因为它抑制 ACh 的释放（Kitazawa 等，1986b）。虽然药理学的证据还不清晰，AR 看起来在虹鳟的胃中亦是 α_2 的变异体（Kitazawa 等，1986a，1988）。一般都认为鱼类肾上腺素能的调控作用主要是由外在的内脏神经实行的。

DA 使辐鳐（*Raja radiata*）和鳕鱼胃壁的电活动增强，而药理学的证据表明存在着一种特异性 DA 受体的作用（Groisman 和 Shparkovskii，1989）。

腺苷和 ATP。有关鱼类消化道内嘌呤能作用机理的零星资料表明，不同种类和消化道的不同部位，其作用是多样性的，经常是混合的。ATP 刺激背棘鳐（*Raja clavata*）（Young，1983）的胃环，但抑制直肠的纵走肌，这是对平滑肌的一种直接作用（Young，1988）。同样，对于硬骨鱼类的鮟鱇，ATP 刺激胃的活动，但抑制肠道有节奏的活动，而这两种情况都和迷走的刺激相似（Young，1980a）。对虹鳟的胃环，腺苷和 ATP 都能使纵肌收缩，而使环肌舒张（Holmgren，1983）。这两种物质都使灌流的鳕鱼肠舒张（Jensen 和 Holmgren，1985），这些结果表明在鱼类消化道内嘌呤能受体亦有许多变种。在川鲽（*Platichthys flesus*）的肠内存在着两个嘌呤能受体的主要类型，即 P1 和 P2。P1 受体可能有两个不同的亚型，介导腺苷在肠内的抑制性作用和在直肠的兴奋性作用。P2 受体介导 ATP 在肠内的兴奋性作用（Lennard 和 Huddart，1989）。

CCK 对虹鳟的胃排空有抑制作用，在 20 h 后明显增加停留在胃内的食物量（Olsson 等，1999）。CCK 使肠内自发收缩的幅度降低，但同时增加静息张力（resting tension）（Olsson 等，1999）。相反，CCK 对大西洋鳕鱼游离的消化道环状制品具有刺激作用（Jonsson 等，1987）。对在体的胃活动作用是浓度依赖性的，在低浓度时降低频率和幅度，而在较高的浓度时引起收缩活动（Olsson 等，1999）。CCK 和相关肽对鳐鱼和鲨鱼的消化道亦起兴奋性作用，但反应微弱而不规则（Andrews 和 Young，1988a；Aldman 等，1989）。

GRP 或 BBS 刺激鱼类胃的活动（Holmgren，1983；Holmgren 和 Jonsson，1988）。同样，BBS 刺激角鲨直肠平滑肌的基本状态的和节律性的活动（Lundin 等，1984）。相反，BBS 微弱地抑制鳕鱼肠的活动（Jensen 和 Holmgren，1985；Holmgren 和 Jonsson，1988）。TTX 并不影响鳕鱼或虹鳟的胃对 BBS 的反应，表明它是一种直接的作用。此外，BBS 还能增强 ACh 的作用（Thorndyke 和 Holmgren，1990）。

在鳕鱼的胃内，BBS 的作用要比哺乳类的 GRP 大得多（Bjenning 和 Holmgren，1988）。据我们所知，还没有一项研究是使用种类特异性的内源肽的，因为 GRP-样肽在鱼类在体的确切效能还没有确定，亦没有检测参与调控消化道活动的相关受体类型。

10.5.3 传播性活动的控制

典型的传播性活动是使消化道内含物沿着消化道移动，其速度、频率和距离依食物的类型、种类以及消化道的不同部位而有所差异。Burnstock（1958a）曾报道褐鳟胃和肠在消化道膨胀时引起传播性收缩活动。平均频率是每 2 min 一次。收缩活动是神经依赖性的，因为它为己烷双胺（hexamethonium）所阻断，而且是外在的组分参与的。

蠕动包含上行和下行反射通路的混合动作，亦包括兴奋性和抑制性的肠神经作用。感觉神经元为食物的机械刺激和化学刺激所激活，其作用是直接地或者由内分泌细胞释放的 5-HT 激活。这种内分泌细胞存在于许多种鱼类中，但亦不是全部鱼类（见第 10.3.2 节）。感觉神经同时激活上行和下行的中间神经元，进而分别激活兴奋性的或者抑制性的运动神经元。鱼类蠕动活动的在体研究相对比较少，所以，我们在这方面的见解都来自哺乳类以外推法（ectrapolation）进行的离体的或者原位的试验。关于鱼类消化道内源的感觉神经的分布、递质含量和功能亦了解得不多，但可以认为其总体模式和哺乳类是相似的（Olsson 和 Holmgren，2001）。对大西洋鳕鱼的研究已证明存在着一个上行的兴奋性通路和一个下行的抑制性通路（Karila 和 Holmgren，1995，1997；Karila 等，1998）。上行的通路包括胆碱能的运动神经元和一个羟色胺的作用机理，它可能是由羟色胺能的中间神经元作用于胆碱能的神经元（Karila 和 Holmgren，1995）。TK 亦参与兴奋期，它直接作用于平滑肌，或者间接通过其他神经元而起作用（Jensen 等，1987；Jensen 和 Holmgren，1994），而 NO 对上行的反射活动引起一种抑制性的紧张状态（Karila 和 Holmgren，1995）。下行通路明显地包括 NO 以及可能包括 VIP（Karila 和 Holmgren，1995，1997；Olsson 和 Karila，1995；Olsson 和 Holmgren，2000）；而羟色胺能的和胆碱能的神经元在下行的抑制性通路中起着兴奋性中间神经元的作用（Karila 等，1998）。

迁移运动复合体（migrating motor complexes，MMCs）出现在消化间期的状态中，组成不同类型的传播性收缩活动。它们比餐后的收缩波更为缓慢，通常传播较长的距离。在哺乳类中，MMCs 包含三个不同的时相，时相Ⅲ最明显的是有节律的收缩活动。

时相Ⅱ较不规则，而时相Ⅰ或多或少是静止的。在鱼类中，至今尚未证明包含所有三个时相的完整的 MMC，但禁食鱼的传播性收缩活动表明和 MMC 的时相Ⅲ有联系。禁食的虹鳟在体的胃出现收缩活动，频率约为每分钟 0.6 次（Olsson 等，1999），即和褐鳟的频率相似。同样，对于鳕鱼的胃和肠，离体记录的收缩活动平均频率约为每分钟 0.5 次（Jensen 等，1991；Karila 和 Holmgren，1995）。鳕鱼的肠收缩活动传播的速度和哺乳类的 MMCs 相似（约 3.5 cm/min）（Karila 和 Holmgren，1995）。

逆行的蠕动，即收缩波由肛门朝口腔方向传播，已经在鱼类中观察到。在发育的斑马鱼中，视频影片证明在肠的上部和直肠区频繁出现逆行蠕动波（Holmberg 等，2003）。这可能是幼鱼开始摄食后，这种收缩活动有助于混合食糜。同样，在庸鲽的幼鱼中亦观察到逆行蠕动，表明它在幽门区对将食糜和消化酶混合起重要作用（Ronnestad 等，2000）。在成年的小点猫鲨（*Scyliorhinus canicula*）中曾观察到逆行蠕动和呕吐相联系。刺激内脏神经可以引起蠕动波，这可能是由于 5-HT 的释放（Andrews 和 Young，1993）。

10.5.4 非传播性活动的调控

非传播性活动的型式包括胃调节（gastric accommodation）、胃排空和混合食物的分节收缩活动。胃调节是对胃壁膨胀的舒适反应，可以包含中枢和局部的反射性通路。在鱼类中，这种胃调节甚至在迷走和内脏神经输入被切断后亦能发生，表明它主要是局部胃肠调控的作用机理（Grove 和 Holmgren，1992a、b）。

胃排空是在胃的压力超过近端的肠道时发生的。蠕动的收缩活动将食糜推向处于紧张性收缩因而正常关闭的幽门括约肌。曾采用各种不同的方法研究鱼类的胃排空，从鱼死后检查消化道的内含物到使用 X-射线（Fange 和 Grove，1979）。虹鳟胃半排空的时间大约是 24 h（Olsson 等，1999）。肠排空之前的滞后期通常在 0～5 h 之间，这主要取决于食物的干燥程度（Bucking 和 Wood，2006）。食糜成分进入肠道对于排空率起着反馈信号的作用。特别是脂肪和酸能刺激近端肠的内分泌细胞释放 CCK，它通过作用于外源的和/或内源的神经而减弱胃排空，即所谓的"肠刹车"。在虹鳟中，CCK 抑制胃排空（Olsson 等，1999）。

胃和肠的局部持续收缩使消化道分节运动而将消化道内含物混合起来。除了前面报道个别递质的作用之外（见第 10.5.2 节），对于这种运动的综合调控作用还研究得很少。胃的膨胀，除了引起胃调节之外，亦会引起收缩活动。对于虹鳟，胃膨胀能激活刺激性的胆碱能和羟色胺能通路。这些可能是由于局部内分泌细胞释放 SS 和 VIP 神经元而引起调节（抑制）作用（Grove 和 Holmgren，1992a）。在鳕鱼中，VIP/胆碱能的联系显然已经消失（Grove 和 Holmgren，1992b）。

10.5.5 胆囊活动的调控

盲鳗胆囊由迷走胆碱能神经纤维分布，并通过毒蕈碱性受体而使胆囊收缩（Holmgren 和 Fange，1981），而药理学的证据表明在虹鳟中亦是类似的情况（Aldman

和 Holmgren，1987）。CCK 可能由十二指肠的内分泌细胞释放，以收缩硬骨鱼类、板鳃鱼类和全骨鱼类的胆囊（Aldman 和 Holmgren，1987；Andrews 和 Young，1988b；Rajjo 等，1988）。十二指肠酸化或者脂肪与氨基酸灌注都能增强鱼类胆囊的活动，而在哺乳类中，这是由 CCK 介导的（Aldman 等，1992；Aldman 和 Holmgren，1995）。深入的在体药理学研究表明，CCK 在低浓度时通过刺激胆碱能神经元而起作用，在高浓度时对胆囊肌起直接作用（Aldman 和 Holmgren，1995）。AD 或者 VIP 通过 β-肾上腺素能通路的作用而抑制虹鳟胆囊的活动（Aldman 和 Holmgren，1987，1992）。

10.5.6 消化道活动性调控的发育

斑马鱼在受精后 3 天，消化道出现不协调的收缩活动（Holmberg 等，2003）；受精后 4 天，传播性的收缩活动波从肠前端和中肠之间的区域向肛门和口腔方向扩散；受精后 5～6 天，斑马鱼开始外源性摄食；而在受精后 4～7 天之间，神经元的调控作用增强（Holmberg 等，2001）。在庸鲽刚变态的幼鱼中亦观察到传播性的收缩活动（Rennestad 等，2000）。在幼鱼表达的一系列信号物质都可能影响消化道的收缩活动（Holmberg 等，2004，2006）。在受精后 4 天，ACh 开始释放并作用于毒蕈碱性受体，从而增加消化道收缩活动波从基础活性开始的频率。内源性 NO 的释放能产生抑制性的紧张状态。从受精后 5 天起，PACAP、NKA 等神经肽开始发挥作用。

10.6 分泌作用和消化作用的调控

胃酸分泌是消化道仅有的分泌活动作用机理，在鱼类中已经进行了一定程度的机械动力学方面的研究。有关胃蛋白酶原分泌或胰液与胆汁的分泌和/或释放的神经元和内分泌调控的研究资料还是零散的，而关于黏液分泌调控的资料会较多些。消化道分泌活动亦包括消化道黏膜层内分泌细胞的活动，在这种情况下，既包括对内分泌细胞的调控，亦包括对内分泌细胞引起作用的其他相关细胞的调控。

10.6.1 胃酸分泌

鱼类和许多其他非哺乳类脊椎动物一样，胃酸是由所谓的"泌酸消化"（oxynticopeptic）细胞分泌的。泌酸消化细胞通常聚集在胃黏膜层的胃腺内，它们分泌酸和胃蛋白酶原（Bishop 和 Odense，1966；Mattisson 和 Holstein，1980；Ezeasor，1981）。对美洲拟鲽（*Pseudopleuronectes americanus*）的原位杂交研究证实存在着表达这种鱼的胃蛋白酶原和质子泵（proton pump）的细胞（Gawlicka 等，2001）。然而，泌酸消化细胞并不是所有鱼类的特征，甚至在关系密切的种类之间亦常出现种类的差别（Rebolledo 和 Vial，1979；Michelangeli 等，1988）。在有些种类中，和泌酸消化细胞一起还出现单纯的泌酸的和/或消化的细胞（Smolka 等，1994）。有意思的是，在美洲拟鲽的黏液细胞中发现了质子泵，表明这些细胞亦分泌酸（Gawlicka 等，2001）。

胃腺的分布表现明显的种类差别。通常，它们都分布在胃的贲门部（Darias 等，

2005),但有时亦出现在胃的各个部位(Arellano 等,2001)。有些鱼类如虹鳟,分泌酸的黏膜延续到食道的末端(Ezeasor,1984)。无胃的鱼一般都缺乏分泌酸的黏膜(Koelz,1992)。鱼类胃酸的基础性分泌活动,甚至在胃没有消化作用时亦出现。大西洋鳕鱼在体的胃酸基础分泌活动主要是由迷走神经的紧张性调控的,因为切断迷走神经后分泌活动几乎完全消失(Holstein 和 Cederberg,1980)。食物摄入能增加胃酸分泌活动。这早在1905年就已经在板鳃鱼类的相关研究中观察到(Sullivan,1905),并且在一些鱼类的相关研究中被重复地证实(Norris 等,1973;Maier 和 Tullis,1984)。至少一部分刺激是通过胃壁膨胀引起的(Smit,1968)。

一些神经元的或内分泌的信号物质参与泌酸消化细胞的调控(见图10.3)。正如在哺乳类中,促胃液素、组胺、ACh 和 SS 起着酸分泌活动调控的中枢作用。组胺和 ACh 刺激一些板鳃鱼类和硬骨鱼类的酸分泌活动(Holstein,1976;Smit,1968)。在大西洋鳕鱼中,组胺的作用可能是由 H2 介导的,因为一些 H2 的同等物刺激而另一些 H2 受体的拮抗物阻抑或消除酸的分泌活动(Bomgren 和 Jonssen,1996;Holstein,1976,1986)。此外,VIP 和 SS 抑制组胺诱导鳕鱼的酸分泌活动(Holmgren 等,1986;Holstein,1983)。当胃腔的 pH 降低时,反馈作用机理使胃酸分泌速率减退,但这种介导的作用机理还没有被阐明(Bomgren 和 Jonssen,1996)。ACh 很可能是从鳕鱼的迷走通路释放出来,并作用于毒蕈碱性受体,因为切断迷走神经几乎完全阻止胃酸分泌活动,而阿托品能阻断胆碱能药物的作用(Holstein,1977;Holstein 和 Cederberg,1980)。此外,一种 H2 拮抗物能阻断 ACh 的作用,表明这是通过组胺释放而引起的反应(Holstein,1976)。

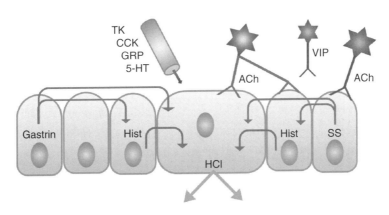

图10.3 概括总结鱼类胃酸分泌活动的旁分泌、内分泌和神经元调控作用(以 Bomgren,2001 为基础)

红色=刺激作用;蓝色=抑制作用。5-HT,5-羟色胺;ACh,乙酰胆碱;CCK,缩胆囊肽;GRP,促胃液素释放肽;HCl,盐酸;Hist,组胺;SS,生长抑素;TK,速激肽;VIP,血管活性肠多肽。(见书后彩图)

促胃液素在大西洋鳕鱼中具有一种不寻常的作用，即对酸分泌活动的抑制作用（Holstein，1982）。在赤虹（*Dasyatis pastinaca*）中，五肽促胃液素刺激胃的分泌活动，但胃腔内的 pH 并没有改变（Zaks 等，1975）。这可能是在进化过程中，鱼类和四足类相比，引起了不同作用机理的改变（Vigna，1983）。然而，在这些研究中使用了哺乳类的促胃液素，可能是和鱼类的促胃液素相比氨基酸序列发生了微小的变化，就会使它们对鱼类的受体产生拮抗性作用。对大西洋鳕鱼的研究表明，低剂量 BBS 刺激胃酸的分泌活动（Holstein 和 Humphrey，1980）。在哺乳类中，低剂量的 BBS（或者天然的 GRP）通过促胃液素的释放而起作用。但是，如果哺乳类促胃液素的抑制作用在鳕鱼的相关研究中是反映了鱼类天然的促胃液素的作用，那么，由 BBS 引起的刺激作用可能并不包括促胃液素。使用 BBS 处理后并没有使血浆的促胃液素水平发生变化，这一研究结果支持上述看法（Holstein 和 Humphrey，1980）。不同的是，阻抑一种抑制性 VIP 紧张状态，表明至少介导了一部分 BBS 的作用，因为 BBS 能使血浆 VIP 的水平降低（Holstein 和 Humphrey，1980；Holstein，1983）。

10.6.2 胃蛋白酶原/胃蛋白酶的分泌作用

泌酸消化细胞除分泌胃酸外，还合成、贮存和分泌胃蛋白酶原。胃蛋白酶原在分泌出来后就转变为有活性的胃蛋白酶（Smit，1968）。在硬骨鱼类和板鳃鱼类中都发现胃蛋白酶原免疫反应细胞（Yasugi 等，1988）。在大西洋鳕鱼中，TK 和 5-HT 强有力地刺激胃蛋白酶原分泌。和哺乳类相比较，不同的是鱼类的 ACh 和组胺只有微弱的作用。胃酸和胃蛋白酶原的分泌活动对不同的刺激因子表现出不同的敏感性。例如，TK 对胃蛋白酶原的分泌有很强的刺激作用，但对胃酸的分泌只有微弱的作用，而组胺和卡巴胆碱（carbachol）的作用正好相反。这意味着胃酸和胃蛋白酶原是从泌酸消化细胞中分开地释放出来的（Holstein 和 Cederberg，1986）。

10.6.3 胰脏的分泌作用

鱼类的内分泌胰脏已经被广泛地阐述（Youson 和 Al-Mahrouki，1999；Youson 等，2006），但对鱼类胰脏激素释放的神经元和内分泌的调控还了解得非常少。胰岛组织的神经分布包括 CCK-、GAL-、NPY-、催产素-和 VIP-的免疫反应纤维（Noe 等，1986；McDonald 等，1987；Jonsson，1991）。CCK、GAL 和 NPY 影响 SSs 从虹鳟胰岛的 Brockmann 小体中释放出来，其中对 SST-14 和 SST-35 的作用是分开的（Eilertson 等，1996）。

外分泌的胰脏组织分泌碳酸氢盐和各种消化营养物质所需要的酶类。食物的摄取和食物组分刺激胰脏酶类的分泌活动与活性（Krogdahl 和 Bakke-McKellep，2005；Murashita 等，2007）。一些研究表明 CCK 是这些作用的重要介体。在虹鳟（Jonsson 等，2006）和五条鰤（*Seriola quinqueradiata*）（Murashita 等，2007）摄食后，曾测定 CCK 水平或其 mRNA 明显增加。给大西洋鲑鱼（*Salmo salar*）在体腹腔注射哺乳类 CCK，模拟激素的作用，能刺激胰蛋白酶和胰凝乳蛋白酶的分泌（Einarsson 等，1997）。Murashita 等（2007）亦曾证明肽 Y（PY）的 mRNA 水平和 CCK 相比较是一种相反的趋势，即摄食后它暂时地减少，表明 PY 对 CCK 调控鱼类胰脏消化酶的分泌活动起着拮抗性作用。鱼类或许缺少促胰液素（Cardoso 等，2004）。

10.6.4 板鳃鱼类直肠腺的分泌作用

板鳃鱼类的直肠腺是一个分泌盐（NaCl）的外分泌腺体，其管道开口于直肠。这个腺体使板鳃鱼类具有额外的能力以对付由于尿素存在于它们的血液中引起的高渗透摩尔浓度。Olson（1999）已经全面综述直肠腺的功能及其神经元的与内分泌的调控。简言之，cAMP 活性增加能够增强分泌活动。最强有力的刺激剂是猫鲨激肽 II（scyliorhinin II，它在未完全测定序列之前曾被称为直肠肽，rectin）（Shuttleworth 和 Thorndyke，1984；Anderson 等，1995）。在角鲨中，VIP 亦能刺激直肠腺的分泌活动（Stoff 等，1979），但是，这个作用可能是由于 VIP 引起经过直肠腺的血液流量增加所致（Thorndyke 等，1989）。腺苷通过作用于腺苷 AI 受体而抑制角鲨直肠腺的分泌作用（Kelley 等，1991）。同样，NPY 与 BBS 和 SS 的互相作用亦抑制直肠腺分泌活动（Silva 等，1990，1993）。免疫组织化学技术支持生理学的研究（Holmgren 和 Nilsson，1983a；Stoff 等，1988）。此外，还曾推测和心房钠尿肽（ANP）相关的肽类对直肠腺分泌活动可能起着激素调控作用（Olson，1999）。

10.7 营养物吸收的调控

在黏膜肠细胞的细胞膜顶部与基底外侧的转运蛋白将氨基酸和单糖从食糜移送到血液内。适宜的转运蛋白数量与活性的增加或减少能调节吸收速率。在鱼类中，食物成分能影响葡萄糖和氨基酸的相对吸收，从而影响转运蛋白的群体（Buddington 等，1987）。一些激素能影响转运蛋白，但是，肠神经虽然分布于消化道的黏膜层，对于它们是否影响吸收率以及如何影响吸收率，目前还了解得很少。

有关激素影响鱼类吸收的报道亦很少，而且有时候是依环境情况而变的。降钙素存在于金鱼肠黏膜的内分泌细胞内。食物影响这些细胞的数量，可以推测降钙素释放通过旁分泌的作用能抑制营养物的吸收（Okuda 等，1999）。哺乳类中已知的影响营养物吸收的局部消化道激素，如肠高血糖素（enteroglucagon），亦在鱼类中发现（Rombout，1977）。对血液循环中激素研究得到的最完整证据表明它们会影响吸收率。例如，甲基睾酮能快速刺激葡萄糖吸收，这可能是由于将贮存的葡萄糖转运蛋白更多地插入到细胞膜内所致（Hazzard 和 Ahearn，1992）。在银大麻哈鱼中，皮质醇和生长激素增加而 AD 减少脯氨酸的吸收（Collie 和 Stevens，1985）。性类固醇能增加虹鳟的肠对亮氨酸的吸收（Habibi 和 Ince，1984）。

10.8 水分和离子运输的调控

局部的和血液循环中的激素,以及黏膜神经(可能在一定程度上)都参与消化道黏膜层的离子运输调控作用。这些作用机理是和稳态保持以及渗透压调节紧密整合在一起的。皮质醇和催乳激素在水分和离子的流动中起着重要的作用,特别是在鱼类适应于盐度变化的情况下(Santos 等,2001;Takahashi 等,2006)。SS 调控鳗鱼消化道 Na^+、Cl^- 和水分的流动(Uesaka 等,1994)。ACh 和 NA(来自神经)降低跨细胞的离子运输,而 5-HT 增强这种运输。亦可以推想 NA 降低而 5-HT 增强对阴离子的旁细胞传导性(Trischitta 等,1999)。NO 通过改变紧密连接(tight junction)的通透性而直接调节旁细胞通路,又通过抑制碳酸氢盐离子进入内皮细胞而间接调节跨细胞运输(Trischitta 等,2007)。斯丹尼亚调钙素(Stanniocalcin,由斯丹尼亚氏器官释放)能调控硬骨鱼类的肠对钙的吸收(Sundell 等,1992)。详细阐述鱼类消化道的渗透压调节不在本章的范围之内,读者可以参考这个领域的许多精彩综述,例如 McCormick(2001)。

10.9 内脏血液循环的调控

鱼类消化道通过背大动脉分支的动脉供给血液,形成内脏的血液循环(Farrell 等,2001)。为使消化道功能处于最佳状态,血液能从身体的其他部分再分配到消化道中。进入消化道的血液决定于整体的血压以及胃肠道血管床阻力和其余(即身体的)血管床阻力的相对比例。局部的阻力取决于一系列因素,包括血管的神经分布、激素的紧张性、局部的氧水平、从消化道组织释放的代谢物,以及围绕消化道壁的平滑肌引起的加压水平,等等(Farrell 等,2001)。本章阐述的范围只限于由神经元和激素所实行的调控作用,但必须记住的是在整体的心血管活动中,血液循环的血管活性激素以及其他的局部因素对消化道的血流量都产生明显的影响。

分布到消化道动脉和微动脉的神经通带在血管外膜形成曲张的纤维网络,浓集于中部血管外膜的边缘,只有少量纤维插入到血管中膜。激素由附近的内分泌/旁分泌细胞释放,而神经递质主要直接作用于血管的平滑肌。血液循环的激素或者血管活性物质(血管紧张肽、缓激肽等)从内侧以两种方式影响血管平滑肌:直接起作用,或者较为常见的是通过内皮因子的释放而间接起作用。

在正常情况下,食物摄入都和增加血流量进入消化道相联系,即所谓的"餐后充血"。对一些鱼类进行直接的测定证实了这一点(Axelsson 等,1989,2000;Axelsson 和 Fritsche,1991),但在其他一些鱼类中得到的数据不是很确定。运动和/或应激反应使进入消化道的血流量减少。

增加血液流量可以依赖于胃肠道血管床阻力的降低,和/或身体血管床阻力的同时增加,两者都使得血液分流到消化道。一系列神经元的和内分泌的物质都能使血管松弛/扩张,或者增加血液流向鱼类的消化道,并且或多或少成为餐后充血的重要介体

（见图 10.4）。内脏血管阻力的增加主要是由于肾上腺素能的活性，但其他的物质如利尿钠肽和内皮缩血管肽（endothelin）能起辅助作用（Evans，2001）。

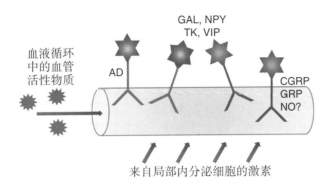

图 10.4 激素和神经元调控鱼类消化道血液流量

红色 = 减少血流；蓝色 = 增加血流。AD，腺苷；GAL，甘丙肽；CGRP，降钙素基因相关肽；GRP，促胃液素释放肽；NO，一氧化氮；NPY，神经肽 Y；TK，速激肽；VIP，血管活性肠多肽。（见书后彩图）

10.9.1 血流的减弱——儿茶酚胺

肾上腺素能的作用机理支配着消化道血管收缩的调控作用。肾上腺素能的神经分布于一些板鳃鱼类和硬骨鱼类消化道壁的血管并延伸到消化道的主要血管（Anderson，1983；Holmgren 和 Nilsson，1983a），而且在随意运动和应激反应期间血液循环的儿茶酚胺水平增加（Abrahamsson，1979；Axelsson，1988）。α-AR 介导消化道主要血管的肾上腺素能收缩活动（Holmgren 和 Nilsson，1974；Nilsson 等，1975），而在体的研究表明未摄食的鱼存在着 α-肾上腺素能的紧张状态（Holmgren 等，1992a；Axelsson 等，2000）。由于这种紧张状态的释放，摄食后消化道增加血液流量的多少存在着种类的差别（Axelsson 等，1989；Axelsson 和 Fritsche，1991）。

在体注射儿茶酚胺可以模拟由于运动和惊吓引起的消化道血流量降低。在角鲨中，AD 和 NA 能增加腹腔动脉血管床的阻力，从而减少进入消化道的血流（Holmgren 等，1992a；Axelsson 等，2000）。目前对于肾上腺素能神经在血液循环中儿茶酚胺的相对重要性还不清楚，但血液循环中儿茶酚胺的作用可能处于支配地位。

10.9.2 血流的增加

CGRP 通过平滑肌细胞上的受体引起虹鳟游离的腹腔动脉血管舒张（Kagstrom 和 Holmgren，1998）。同样，曾报道在全头鱼类的米氏叶吻银鲛（*Callarhinchus milii*）的背大动脉，存在着非内皮依赖的由 CGRP 诱导的血管舒张作用机理（Jennings 等，2007）。

当食物摄入时，消化道的 GRP/BBS 释放增加（Jonsson 等，2006），其来源可能是消化道的内分泌细胞和肠神经元。BBS 和/或 GRP 可能通过扩张消化道血管而增加进入消化道的血流量，就如同灌注角鲨的消化道那样（Bjenning 等，1990），但起主导作用的似乎是身体血流阻力的整体增加，将血液分流到消化道（Holmgren 等，1992b）。

NO 只存在一些鱼类控制消化道的血管中。因为加入底物（L-精氨酸）或者 NOS 拮抗物对虹鳟消化道血管没有作用，因此可以认为这些血管并不受到氧化氮能作用机能的调控，尽管 NO 供体硝普钠能诱导血管舒张（Olsson 和 Villa，1991）。但是，随后对其他鱼类的研究表明从血管周围的神经衍生的 NO 能使血管舒张（Donald 和 Broughton，2005；Jennings 等，2007）。值得注意的是，鱼类能够利用包含在血管周围神经纤维内的神经元 NOS 来调控血管，而不是像哺乳类那样利用血液循环中因子的刺激而产生的内皮 NOS。

10.9.3　可变的作用

甘丙肽（GAL）能使游离的大西洋鳟鱼腹腔动脉和肠系膜动脉以及板鳃鱼类胰脏-肠系膜动脉收缩（Karila 等，1993；Preston 等，1995），但使肺鱼在体的腹腔动脉阻力降低（Holmgren 等，1994）。鳕鱼的血管收缩不受肾上腺素能作用机理的影响。肺鱼血管阻力的减弱可能和氧化氮能作用相联系，正如 GAL-诱导鳟鱼整体血压降低的情况那样（Le Mevel 等，1998）。

和 NA 一起从交感神经释放的 NPY 是调控哺乳类血管床的重要因子。NPY 通过减少局部血管的阻力使角鲨进入消化道的血液流量增加（Holmgren 等，1992b）。这和 NPY 在哺乳类中最常见的作用形成对照。然而，在对另外三种板鳃鱼类的研究中，Preston 等（1998）发现 NPY 的作用使游离的消化道动脉收缩。现在不清楚这是否是真正的种类之间的差异，抑或是受到实验设置与操作的影响。和角鲨的动脉相似，大西洋鳕鱼未经处理的或者预先收缩的腹腔动脉，用鳕鱼 NPY 处理后都能使它们舒张。这个作用部分是由于对血管平滑肌的直接作用，部分是由于前列腺素的释放，但没有通过 NO 或其他内皮衍生因子的释放（Shahbazi 等，2002）。NPY 存在于延伸到肠血管和黏膜下层的大约 10% 的肾上腺素能神经元中（Karila 等，1997）。但是，进一步的研究必须阐明同时释放一种刺激性和一种抑制性物质的生理学意义。

速激肽（TKs）对于哺乳类的大多数血管床是血管舒张的，通过内皮依赖的作用机理而起作用，但是血管收缩是对一些血管平滑肌直接起作用的（Severini 等，2002）。在鱼类中的作用依不同种类而有所不同，但亦有在一个种类内因为和不同类型受体起作用而出现差异的情况。所以，哺乳类的 P 物质使角鲨消化道的血液流量增加，而使澳洲肺鱼的消化道血流量减少（Holmgren 等，1992a、1994）。哺乳类的 P 物质使鳕鱼在体产生一种混合的反应：开始时较大的反应是增加血液流量，而瞬时地被小而重叠的胆碱能减弱反应所阻断（Jensen 等，1991）。随后使用一些天然的鱼类 TKs 对角鲨和虹鳟进行的研究表明 TKs 能使血液分流到胃内，而且存在着几种 TK 的受体类型，有的介导兴奋性反应，有的介导抑制性反应（Kagstrom 等，1996a、b）。例如，天然的鳟鱼 P 物质和 NKA 偶尔会诱导虹鳟分离的血管收缩，但是，它们会诱导虹鳟在体早期增加腹腔动脉血液流量，随后又使血液流量减少（Kagstrom 等，1996b；Kagstrom 和 Holmgren，1998）。

Holder 等（1983）使用从鳟鱼和黑鲖鱼（*Ictalurus meles*）消化道粗提的 VIP 提取物证明灌注鲶鱼消化道的肠环产生血管舒张作用。随后，在大西洋鳕鱼在体发现 VIP 能使腹腔动脉和肠系膜动脉的血流增加；但是，腹腔动脉的血流增加依赖于血管阻力降低和心输出量增加，而在神经元分布较不稠密的肠系膜动脉，血流增加只是由于心输出量增加（Jensen 等，1991）。Kagstrom 和 Holmgren（1997）使用从虹鳟肠取得的小血管，确定 VIP 诱导的血管壁舒张不依赖于内皮和氧化氮能的联系（和许多哺乳类的血管不同），但有前列腺素合成的参与。在角鲨中，VIP 增加血管阻力，因而减少进入消化道的血液流量（Holmgren 等，1992a），这是一种不寻常的作用，和同一种类的直肠腺的腺体分泌增加（如同哺乳类的外分泌腺）相结合的血管舒张（Thorndyke 等，1989）有所不同。

10.10 结 束 语

鱼类基本的消化道解剖学和脊椎动物整体是一样的，但也有一些明显的不同，如有些科的鱼类没有胃，许多板鳃鱼类有螺旋瓣的肠。同样，在神经元和内分泌系统方面表现出相同的趋势，和脊椎动物整体相似而有一些明显例外。一系列越来越精确可靠的免疫组织化学研究和完成序列测定的物质有效性的增强是同时发生的。它们向我们展示鱼类消化道神经元和内分泌系统信号物质很明显的是和其他脊椎动物一样发生着变化，如果不是更多的话。同时，还有少数信号物质已经发现在鱼类中完全是独特的。

目前我们对于个别神经激素物质的功能已有所了解，但是，对不同的鱼类类群状况进行系统的研究还是欠缺的。总之，对鱼类消化道神经激素作用机理的综合功能还研究得很少，而这一研究领域随着气候和其他环境变化影响的增加，其重要性正在增强。虽然鱼类的许多作用机理一般是按照脊椎动物的模式，但在细节上存在着许多的变异，例如在不同的信号物质之间、不同的鱼类类群之间，或者和哺乳类相比较，都会出现相对立的作用或者不同的相互作用。必须进行许多比较性的分析研究，以便确定这些例外是进化重合（evolutionary coincidence）的结果，还是由进化压力（evolutionary pressure）所致。

S. 霍尔姆格伦
C. 奥尔逊

参考文献

Abrahamsson T. 1979. Phenylethanolamine-N-methyl transferase (PNMT) activity and catecholamine storage and release from chromaffin tissue of the spiny dogfish, *Squalus acanthias*. *Comp. Biochem. Physiol*, 64C: 169-172.

Abrahamsson T, Nilsson S. 1976. Phenylethanolamine-N-methyl transferase (PNMT) activity and catecholamine content in chromaffin tissue and sympathetic neurons in the cod, *Gadus morhua*. *Acta Physiol. Scand*, 96: 94-99.

Aldman G, Holmgren S. 1987. Control of gallbladder motility in the rainbow trout, *Salmo gairdneri*. *Fish Physiol. Biochem*, 4: 143-155.

Aldman G, Holmgren S. 1992. VIP inhibits CCK-induced gallbladder contraction involving a β-adrenoceptor mediated pathway in the rainbow trout, *Oncorhynchus mykiss*, in vivo. *Gen. Comp. Endocrinol*,. 88: 287-291.

Aldman G, Holmgren S. 1995. Intraduodenal fat and amino acids activate gallbladder motility in the rainbow trout, *Oncorhynchus mykiss*. *Gen. Comp. Endocrinol*, 100: 27-32.

Aldman G, Jönsson A C, Jensen J, Holmgren S. 1989. Gastrin/CCK-like peptides in the spiny dogfish, *Squalus acanthias*; concentrations and actions in the gut. *Comp. Biochem. Physiol. C*, 92: 103-108.

Aldman G, Grove D J, Holmgren S. 1992. Duodenal acidification and intra-arterial injection of CCK8 increase gallbladder motility in the rainbow trout *Oncorhynchus mykiss*. *Gen. Comp. Endocrinol*, 86: 20-25.

Anderson C. 1983. Evidence for 5-HT-containing intrinsic neurons in the teleost intestine. *Cell Tissue Res*, 230: 377-386.

Anderson C, Campbell G. 1988. Immunohistochemical study of 5-HT-containing neurons in the teleost intestine: Relationship to the presence of enterochromaffin cells. *Cell Tissue Res*, 254: 553-559.

Anderson C R, Campbell G, O'Shea F, Payne M. 1991. The release of neuronal 5-HT from the intestine of a teleost fish, *Platycephalus bassensis*. *J. Auton. Nerv. Syst*, 33: 239-246.

Anderson W G, Conlon J M, Hazon N. 1995. Characterization of the endogenous intestinal peptide that stimulates the rectal gland of *Scyliorhinus canicula*. *Am. J. Physiol*, 268: R1359-R1364.

Andrews P L, Young J Z. 1988. The effect of peptides on the motility of the stomach, intestine and rectum in the skate (*Raja*). *Comp. Biochem. Physiol. C*, 89: 343-348.

Andrews P L, Young J Z. 1988. A pharmacological study of the control of motility in the gallbladder of the skate. *Comp. Biochem. Physiol. C*, 89: 349-354.

Andrews P L R, Young J Z. 1993. Gastric motility patterns for digestion and vomiting evoked by sympathetic nerve stimulation and 5-hydroxytryptamine in the dogfish *Scyliorhinus canicula*. *Phil. Trans. R Soc. Lond*, 342: 363-380.

Arellano J M, Storch V, Sarasquete C. 2001. Histological and histochemical observations in the stomach of the Senegal sole, *Solea senegalensis*. *Histol. Histopathol*, 16: 511-521.

Aronsson U, Holmgren S. 2000. Muscarinic M3-like receptors, cyclic AMP and L-type calcium channels are involved in the contractile response to cholinergic agents in gut smooth muscle of the rainbow trout, *Oncorhynchus mykiss*. *Fish Physiol. Biochem*, 23: 353-361.

Axelsson M. 1988. The importance of nervous and humoral mechanisms in the control of cardiac performance in the Atlantic cod *Gadus morhua* at rest and during non-exhaustive exercise. *J. Exp. Biol*, 137: 287-302.

Axelsson M, Fritsche R. 1991. Effects of exercise, hypoxia and feeding on the gastrointestinal blood flow in the Atlantic cod *Gadus morhua*. *J. Exp. Biol*, 158: 181-198.

Axelsson M, Driedzic W R, Farrell A P, Nilsson S. 1989. Regulation of cardiac output and gut flow in the sea raven, *Hemitripterus americanus*. *Fish Physiol. Biochem*, 196: 2-12.

Axelsson M, Thorarensen H, Nilsson S, Farrell A P. 2000. Gastrointestinal blood flow in the red Irish lord, *Hemilepidotus hemilepidotus*: long-term effects of feeding and adrenergic control. *J. Comp. Physiol. B*, 170: 145-152.

Baumgarten H G, Björklund A, Lachenmayer L, Nobin A, Rosengren E. 1973. Evidence for the existence of serotonin-, dopamine-, and noradrenaline-containing neurons in the gut of *Lampetra fluviatilis*. *Z. Zellforsch. Mikrosk. Anat*, 141: 33-54.

Bisgrove B W, Raible D W, Walter V, Eisen J S, Grunwald D J. 1997. Expression of c-ret in the zebrafish embryo: Potential roles in motoneuronal development. *J. Neurobiol*, 33: 749-768.

Bishop C, Odense P H. 1966. Morphology of the digestive tract of the Atlantic cod, *Gadus morhua*. *J. Fish. Res. Bd. Canada*, 23: 1607-1615.

Bjenning C, Holmgren S. 1988. Neuropeptides in the fish gut. An immunohistochemical study of evolutionary patterns. *Histochemistry*, 88: 155-163.

Bjenning C, Driedzic W R, Holmgren S. 1989. Neuropeptide Y-like immunoreactivity in the cardiovascular nerve plexus of the elasmobranchs *Raja erinacea* and *Raja radiata*. *Cell Tiss. Res*, 255: 481-486.

Bjenning C, Jönsson A C, Holmgren S. 1990. Bombesin-like immunoreactive material in the gut, and the effect of bombesin on the stomach circulatory system of an elasmobranch fish, *Squalus acanthias*. *Regul. Pept*, 28: 57-69.

Bomgren P. 2001. *Gastrin and histamine in amphibians and fish*. University of Gothenburg: Göteborg, Sweden PhD thesis.

Bomgren P, Jönsson A C. 1996. Basal, H2-receptor stimulated and pH-dependent gastric acid secretion from an isolated stomach mucosa preparation of the Atlantic cod, *Gadus morhua*, studied using a modified pH-static titration method. *Fish Physiol. Biochem*, 15: 275-285.

Bosi G, Shinn A P, Giari L, Arrighi S, Domeneghini C. 2004. The presence of a galanin-like peptide in the gut neuroendocrine system of *Lampetra fluviatilis* and *Acipenser transmontanus*: an immunohistochemical study. *Tissue Cell*, 36: 283-292.

Bosi G, Shinn A P, Giari L, Simoni E, Pironi F, Dezfuli B S. 2005. Changes in the neuromodulators of the diffuse endocrine system of the alimentary canal of farmed rainbow trout, *Oncorhynchus mykiss* (Walbaum), naturally infected with *Eubothrium crassum* (*Cestoda*). *J. Fish. Dis*, 28: 703-711.

Bosi G, Bermudez R, Domeneghini C. 2007. The galaninergic enteric nervous system of *Pleuronectiformes* (*Pisces*, *Osteichthyes*): An immunohistochemical and confocal laser scanning immunofluorescence study. *Gen. Comp. Endocrinol*, 152: 22-29.

Brüning G, Hattwig K, Mayer B. 1996. Nitric oxide synthase in the peripheral nervous system of the goldfish, *Carassius auratus*. *Cell Tissue Res*, 284: 87-98.

Bucking C, Wood C M. 2006. Water dynamics in the digestive tract of the freshwater rainbow trout during the processing of a single meal. *J. Exp. Biol*, 209: 1883-1893.

Buddington R K, Chen J W, Diamond J. 1987. Genetic and phenotypic adaptation of intestinal nutrient transport to diet in fish. *J. Physiol*, 393: 261-281.

Burka J F, Blair R M J, Hogan J E. 1989. Characterization of the muscarinic and serotonergic receptors of the intestine of the rainbow trout, *Salmo gairdneri*. *Can. J. Physiol. Pharmacol*, 67: 477-482.

Burkhardt-Holm P, Holmgren S. 1989. A comparative study of neuropeptides in the intestine of two stomachless teleosts (*Poecilia reticulata*, *Leuciscus idus melanotus*) under conditions of feeding and starvation. *Cell Tissue Res*, 255: 245-254.

Burnstock G. 1958. The effect of drugs on spontaneous motility and on response to stimulation of the extrinsic nerves of the gut of a teleostean fish. *Br. J. Pharmacol. Chemother*, 13: 216-226.

Burnstock G. 1958. Reversible inactivation of nervous activity in a fish gut. *J. Physiol*, 141: 35-45.

Burnstock G. 1969. Evolution of the autonomic innervation of visceral and cardiovascular systems in vertebrates. *Pharmacol. Rev*, 21: 247-324.

Campbell G. 1975. Inhibitory vagal innervation of the stomach in fish. *Comp. Biochem. Physiol*, 50C: 169-170.

Campbell G, Burnstock G. 1968. Comparative physiology of gastrointestinal motility. In: Code C F, Ed. *Handbook of Physiology*. American Physiological Society: Washington DC, 2213-2266.

Cardoso J C, Power D M, Elgar G, Clark M S. 2004. Duplicated receptors for VIP and PACAP (VPAC1R and PAC1R) in a teleost fish, *Fugu rubripes*. *J. Mol. Endocrinol*, 33: 411-428.

Chiba A. 1998. Ontogeny of serotonin-immunoreactive cells in the gut epithelium of the cloudy dogfish, *Scyliorhinus torazame*, with reference to coexistence of serotonin and neuropeptide Y. *Gen. Comp. Endocrinol*, 111: 290-298.

Chow B K C. 1997. The goldfish vasoactive intestinal polypeptide receptor: Functional studies and tissue distribution. *Fish Physiol. Biochem*, 17: 213-222.

Cimini V, Van Noorden S, Giordano-Lanza G, Nardini V, McGregor G P, Bloom S R, Polak J M. 1985. Neuropeptides and 5-HT immunoreactivity in the gastric nerves of the dogfish (*Scyliorhinus stellaris*). *Peptides* 6 (Suppl 3), 373-377.

Collie N L, Stevens J J. 1985. Hormonal effects on L-proline transport in coho salmon (*Oncorhynchus kisutch*) intestine. *Gen. Comp. Endocrinol*, 59: 399-409.

Conlon J M, Larhammar D. 2005. The evolution of neuroendocrine peptides. *Gen. Comp. Endocrinol*, 142: 53-59.

Conlon J M, Deacon C F, O'Toole L, Thim L. 1986. Scyliorhinin I and II: two novel tachykinins from dogfish gut. *FEBS Lett*, 200: 111-116.

Conlon J M, Henderson I W, Thim L. 1987. Gastrin-releasing peptide from the intestine of the elasmobranch fish, *Scyliorhinus canicula* (common dogfish). *Gen. Comp. Endocrinol*, 68: 415-420.

D'Este L, Renda T. 1995. Phylogenetic study on distribution and chromogranin/secretogranin content of histamine immunoreactive elements in the gut. *Ital. J. Anat. Embryol*, 100 (Suppl 1): 403-410.

Darias M J, Murray H M, Martínez-Rodrígueza G, Cárdenasc S, Yúfera M. 2005. Gene expression of pepsinogen during the larval development of red porgy (*Pagrus pagrus*). *Aquaculture*, 248: 245-252.

Domeneghini C, Radaelli G, Arrighi S, Mascarello F, Veggetti A. 2000. Neurotransmitters and putative neuromodulators in the gut of *Anguilla anguilla* (L.). Localizations in the enteric nervous and endocrine systems. *Eur. J. Histochem*, 44: 295-306.

Donald J A, Broughton B R. 2005. Nitric oxide control of lower vertebrate blood vessels by vasomotor nerves. *Comp. Biochem. Physiol*, 142A: 188-197.

Eilertson C D, Carneiro N M, Kittilson J D, Comley C, Sheridan M A. 1996. Cholecystokinin, neuropeptide Y and galanin modulate the release of pancreatic somatostatin-25 and somatostatin-14 *in vitro*. *Regul. Pept*, 63: 105-112.

Einarsson S, Davies P S, Talbot C. 1997. Effect of exogenous cholecystokinin on the discharge of the gallbladder and the secretion of trypsin and chymotrypsin from the pancreas of the Atlantic salmon, *Salmo salar*L. *Comp. Biochem. Physiol*, 117C: 63-67.

Evans D H. 2001. Vasoactive receptors in abdominal blood vessels of the dogfish shark, *Squalus acanthias*. *Physiol. Biochem. Zool*, 74: 120-126.

Ezeasor D N. 1981. The fine structure of the gastric epithelium on the rainbow trout, *Salmo gairdneri*, Richardson. *J. Fish Biol*, 19: 611-627.

Ezeasor D N. 1984. Light and electron microscopic studies on the oesophageal epithelium of the rainbow trout, *Salmo gairdneri*. *Anat. Anz*, 155: 71-83.

Falkmer S, Fahrenkrug J, Alumets J, Håkanson R, Sundler F. 1980. Vasoactive intestinal polypeptide (VIP) in epithelial cells of the gut mucosa of an elasmobranchian cartilaginous fish, the ray. *Endocrinol. Japon*, 1: 31-35.

Fänge R, Grove D J. 1979. Digestion. In: Hoar W S, Randall, Ed. *Fish Physiology*. Academic Press: London, 161-260.

Farrell A P, Thorarensen H, Axelsson M, Crocker C E, Gamperl A K, CechJr J J. 2001. Gut blood flow in fish during exercise and severe hypercapnia. *Comp. Biochem. Physiol. A Mol. Integr. Physiol*, 128: 551-563.

Fradinger E A, Tello J A, Rivier J E, Sherwood N M. 2005. Characterization of four receptor cDNAs: PAC1, VPAC1, a novel PAC1 and a partial GHRH in zebrafish. *Mol. Cell. Endocrinol*, 231: 49-63.

Gábriel R, Halasy K, Fekete É, Eckert M, Benedeczki I. 1990. Distribution of GABA-like immunoreactivity in myenteric plexus of carp, frog and chicken. *Histochemistry*, 94: 323-328.

Gawlicka A, Leggiadro C T, Gallant J W, Douglas S E. 2001. Cellular expression of the pepsinogen and the gastric proton pump genes in the stomach of winter flounder as determined by *in situ* hybridization. *J. Fish. Biol*, 58: 529-536.

Goodrich J T, Bernd P, Sherman D, Gershon M D. 1980. Phylogeny of enteric serotonergic neurons. *J. Comp. Neurol*, 190: 15-28.

Green K, Campbell G. 1994. Nitric oxide formation is involved in vagal inhibition of the stomach of the trout (*Salmo gairdneri*). *J. Auton. Nerv. Syst*, 50: 221-229.

Groisman S D, Shparkovskii I A. 1989. Effect of dopamine and DOPA on electrical activity of stomach muscles in the skate *Raja radiata* and cod *Gadus morhua*. *Zh. Evoly-*

ut. Biokhim. Fiziol, 25: 505-511.

Grove D J, Campbell G. 1979. Effects of extrinsic nerve stimulation on the stomach of the flathead, *Platychephalus bassensis* Cuvier and Valenciennes. *Comp. Biochem. Physiol. C*, 63C: 373-380.

Grove D J, Campbell G. 1979. The role of extrinsic and intrinsic nerves in the coordination of gut motility in the stomachless flatfish *Rhombosolea tapirina* and *Ammotretis rostrata* Guenther. *Comp. Biochem. Physiol*, 63C: 143-159.

Grove D J, Holmgren S. 1992. Intrinsic mechanisms controlling cardiac stomach volume of the rainbow trout (*Oncorhynchus mykiss*) following gastric distension. *J. Exp. Biol*, 163: 33-48.

Grove D J, Holmgren S. 1992. Mechanisms controlling stomach volume of the Atlantic cod *Gadus morhua* following gastric distension. *J. Exp. Biol*, 163: 49-63.

Habibi H R, Ince B W. 1984. A study of androgen-stimulated L-leucine transport by the intestine of rainbow trout (*Salmo gairdneri* Richardson) *in vitro*. *Comp. Biochem. Physiol. A*, 79: 143-149.

Habu A, Ohishi T, Mihara S, Ohkubo R, Hong Y M, Mochizuki T, Yanaihara N. 1994. Isolation and sequenc determination of galanin from the pituitary of yellowfin tuna. *Biomed Res - Tokyo*, 15: 357-362.

Håkanson R, Böttcher G, Ekblad E, Panula P, Simonsson M, Dohlsten M, Hallberg T, Sundler F. 1986. Histamine in endocrine cells in the stomach. A survey of several species using a panel of histamine antibodies. *Histochemistry*, 86: 5-17.

Hazzard C E, Ahearn G A. 1992. Rapid stimulation of intestinal D-glucose transport in teleosts by 17 alpha-methyltestosterone. *Am. J. Physiol*, 262: R412-R418.

Holder F C, Vincent B, Ristori M T, Laurent P. 1983. Vascular perfusion of an intestinal segment in the catfish (*Ictalurus melas*, R): demonstration of the vasoactive effects of mammalian VIP and of gastrointestinal extracts from teleost fish. *C. R. Seances Acad. Sci. III*, 296: 783-788.

Holmberg A, Schwerte T, Fritsche R, Pelster B, Holmgren S. 2003. Ontogeny of intestinal motility in correlation to neuronal development in zebrafish embryos and larvae. *J. Fish Biol*, 63: 318-331.

Holmberg A, Schwerte T, Pelster B, Holmgren S. 2004. Ontogeny of the gut motility control system in zebrafish *Danio rerio* embryos and larvae. *J. Exp. Biol*, 207: 4085-4094.

Holmberg A, Olsson C, Holmgren S. 2006. The effects of endogenous and exogenous nitric oxide on gut motility in zebrafish *Danio rerio* embryos and larvae. *J. Exp. Biol*, 209: 2472-2479.

Holmberg A, Olsson C, Hennig G W. 2007. TTX-sensitive and TTX-insensitive control of spontaneous gut motility in the developing zebrafish (*Danio rerio*) larvae. *J. Exp. Biol*, 210: 1084-1091.

Holmgren S. 1983. The effects of putative non-adrenergic, non-cholinergic autonomic transmitters on isolated strips from the stomach of the rainbow trout *Salmo gairdneri*. *Comp. Biochem. Physiol. C*, 74: 229-238.

Holmgren S, Fänge R. 1981. Effects of cholinergic drugs on the intestine and gallbladder of the hagfish, *Myxine glutinosa* L., with a report on the inconsistent effects of catecholamines. *Mar. Biol. Lett*, 2: 265-277.

Holmgren S, Jensen J. 2001. Evolution of vertebrate neuropeptides. *Brain Res. Bull*, 55: 723-735.

Holmgren S, Jönsson A C. 1988. Occurrence and effects on motility of bombesin related peptides in the gastrointestinal tract of the Atlantic cod, *Gadus morhua*. *Comp. Biochem. Physiol*, 89C: 249-256.

Holmgren S, Nilsson S. 1974. Drug effects on isolated artery strips from two teleosts, *Gadus morhua* and *Salmo gairdneri*. *Acta Physiol. Scand*, 90: 431-437.

Holmgren S, Nilsson S. 1981. On the non-adrenergic, non-cholinergic innervation of the rainbow trout stomach. *Comp. Biochem. Physiol*, 70C: 65-69.

Holmgren S, Nilsson S. 1983. Bombesin-, gastrin/CCK-, 5-hydroxytryptamine-, neurotensin-, somatostatin-, and VIP-like immunoreactivity and catecholamine fluorescence in the gut of the elasmobranch, *Squalus acanthias*. *Cell Tissue Res*, 234: 595-618.

Holmgren S, Nilsson S. 1983. VIP-, bombesin- and neurotensin-like immunoreactivity in neurons of the gut of the holostean fish, *Lepisosteus platyrhincus*. *Acta Zool. (Stockh.)*, 64: 25-32.

Holmgren S, Grove D J, Nilsson S. 1985. Substance P acts by releasing 5-hydroxytryptamine from enteric neurons in the stomach of the rainbow trout, *Salmo gairdneri*. *Neuroscience*, 14: 683-693.

Holmgren S, Jönsson A C, Holstein B. 1986. Gastrointestinal peptides in fish. In: Nilsson S, Holmgren S, Ed. *Fish Physiology - Recent Advances*. Croom Helm Ltd: London, 119-139.

Holmgren S, Axelsson M, Farrell A P. 1992. The effect of catecholamines, substance P and vasoactive intestinal polypeptide on blood flow to the gut in the dogfish *Squalus acanthias*. *J. Exp. Biol*, 168: 161-175.

Holmgren S, Axelsson M, Farrell A P. 1992. The effects of neuropeptide Y and bombesin on blood flow to the gut in dogfish *Squalus acanthias*. *Regul. Pept*, 40: 169.

Holmgren S, Fritsche R, Karila P, Gibbins I, Axelsson M, Franklin C, Grigg G, Nilsson S. 1994. Neuropeptides in the Australian lungfish *Neoceratodus forsteri*: effects *in* vivo and presence in autonomic nerves. *Am. J. Physiol*, 266: R1568-R1577.

Holstein B. 1976. Effect of the H2-receptor antagonist metiamide on carbachol- and histamine-induced gastric acid secretion in the Atlantic cod, *Gadus morhua*. *Acta Physiol. Scand*, 97: 189-195.

Holstein B. 1977. Effect of atropine and SC-15396 on stimulated gastric acid secretion in the Atlantic cod, *Gadus morhua*. *Acta Physiol. Scand*, 101: 185-193.

Holstein B. 1982. Inhibition of gastric acid secretion inthe Atlantic cod, *Gadus morhua*, by sulphated and desulphated gastrin, caerulein, and CCK-octapeptide. *Acta Physiol. Scand*, 114: 453-459.

Holstein B. 1983. Effect of vasoactive intestinal polypeptide on gastric acid secretion and mucosal blood flow in the Atlantic cod, *Gadus morhua*. *Gen. Comp. Endocrinol*, 52: 471-473.

Holstein B. 1986. Characterization with agonists of the histamine receptors mediating stimulation of gastric acid secretion in the Atlantic cod, *Gadus morhua*. *Agents Actions*, 19: 42-47.

Holstein B, Cederberg C. 1980. Effect of vagotomy and glucose administration on gastric acid secretion in the Atlantic cod, *Gadus morhua*. *Acta Physiol. Scand*, 109: 37-44.

Holstein B, Cederberg C. 1986. Effects of tachykinins on gastric acid and pepsin secretion and on gastric outflow in the Atlantic cod, *Gadus morhua*. *Am. J. Physiol*, 250: G309-G315.

Holstein B, Humphrey C S. 1980. Stimulation of gastric acid secretion and suppression of VIP-like immunoreactivity by bombesin in the Atlantic codfish, *Gadus morhua*. *Acta Physiol. Scand*, 109: 217-223.

Hoyle C H. 1998. Neuropeptide families: evolutionary perspectives. *Regul. Pept*, 73: 1-33.

Jennings B L, Bell J D, Hyodo S, Toop T, Donald J A. 2007. Mechanisms of vasodilation in the dorsal aorta of the elephant fish, *Callorhinchus milii* (Chimaeriformes: Holocephali). *J. Comp. Physiol. B*, 177: 557-567.

Jensen J, Conlon J M. 1992. Isolation and primary structure of gastrin-releasing peptide from a teleost fish, the trout (*Oncorhynchus mykiss*). *Peptides*, 13: 995-999.

Jensen J, Conlon J M. 1992. Substance-P-related and neurokinin-A-related peptides from the brain of the cod and trout. *Eur. J. Biochem*, 206: 659-664.

Jensen J, Holmgren S. 1985. Neurotransmitters in the intestine of the Atlantic cod, *Gadus morhua*. *Comp. Biochem. Physiol. C*, 82: 81-89.

Jensen J, Holmgren S. 1991. Tachykinins and intestinal motility in different fish groups. *Gen. Comp. Endocrinol*, 83: 388-396.

Jensen J, Holmgren S. 1994. The gastrointestinal canal. In: Nilsson S, Holmgren S, Ed. *Comparative Physiology and Evolution of the Autonomic Nervous System*. Harwood Academic Publishers: Chur, Switzerland, 119-167.

Jensen J, Holmgren S, Jönsson A C. 1987. Substance P-like immunoreactivity and the effects of tachykinins in the intestine of the Atlantic cod, *Gadus morhua*. *J. Auton. Nerv. Syst*, 20: 25-33.

Jensen J, Axelsson M, Holmgren S. 1991. Effects of substance P and VIP on the gastrointestinal blood flow in the Atlantic cod, *Gadus morhua*. *J. Exp. Biol*, 156: 361-373.

Jensen J, Karila P, Jönsson A C, Aldman G, Holmgren S. 1993. Effects of substance P and distribution of substance P-like immunoreactivity in nerves supplying the stomach of the cod, *Gadus morhua*. *Fish Physiol. Biochem*, 12: 237-247.

Jensen J, Olson K R, Conlon J M. 1993. Primary structures and effects on gastrointestinal motility of tachykinins from the rainbow trout. *Am. J. Physiol*, 265: R804-R810.

Johnsen A H. 1998. Phylogeny of the cholecystokinin/gastrin family. *Front. Neuroendocrinol*, 19: 73-99.

Jönsson A C. 1991. Regulatory peptides in the pancreas of two species of elasmobranchs and in the Brockmann bodies of four teleost species. *Cell Tissue Res*, 266: 163-172.

Jönsson A C. 1993. Co-localization of peptides in the Brockmann bodies of the cod (*Gadus morhua*) and the rainbow trout (*Oncorhynchus mykiss*). *Cell Tissue Res*, 273: 547-555.

Jönsson A C, Holmgren S, Holstein B. 1987. Gastrin/CCK-like immunoreactivity in endocrine cells and nerves in the gastrointestinal tract of the cod, *Gadus morhua*, and the effect of peptides of the gastrin/CCK family on cod gastrointestinal smooth muscle. *Gen. Comp. Endocrinol*, 66: 190-202.

Jönsson E, Forsman A, Einarsdottir I E, Egnér B, Ruohonen K, Björnsson B T. 2006. Circulating levels of cholecystokinin and gastrin-releasing peptide in rainbow trout fed different diets. *Gen. Comp. Endocrinol*, 148: 187-194.

Kågström J, Holmgren S. 1997. VIP-induced relaxation of small arteries of the rainbow trout, *Oncorhynchus mykiss*, involves prostaglandin synthesis but not nitric oxide. *J. Auton. Nerv. Syst*, 63: 68-76.

Kågström J, Holmgren S. 1998. Calcitonin gene-related peptide (CGRP), but not tachykinins, causes relaxation of small arteries from the rainbow trout gut. *Peptides*, 19: 577-584.

Kågström J, Axelsson M, Jensen J, Farrell A P, Holmgren S. 1996. Vasoactivity and immunoreactivity of fish tachykinins in the vascular system of the spiny dogfish. *Am.*

J. Physiol. - Reg. Int. Comp. Phys, 39: R585-R593.

Kågström J, Holmgren S, Olson K R, Conlon J M, Jensen J. 1996. Vasoconstrictive effects of native tachykinins in the rainbow trout, *Oncorhynchus mykiss*. *Peptides*, 17: 39-45.

Kaiya H, Tsukada T, Yuge S, Mondo H, Kangawa K, Takei Y. 2006. Identification of eel ghrelin in plasma and stomach by radioimmunoassay and histochemistry. *Gen. Comp. Endocrinol*, 148: 375-382.

Kaiya H, Miyazato M, Kangawa K, Peter R E, Unniappan S. 2008. Ghrelin: A multifunctional hormone in non-mammalian vertebrates. *Comp. Biochem. Physiol. A Mol. Integr. Physiol*, 149: 109-128.

Kamisaka Y, Totland G K, Tagawa M, Kurokawa T, Suzuki T, Tanaka M, RØnnestad I. 2001. Ontogeny of cholecystokinin-immunoreactive cells in the digestive tract of Atlantic halibut, *Hippoglossus hippoglossus*, larvae. *Gen. Comp. Endocrinol*, 123: 31-37.

Karila P. 1997. *Nervous Control of Gastrointestinal Motility in the Atlantic Cod, Gadus morhua*. University of Gothenburg: Göteborg, Sweden PhD thesis.

Karila P, Holmgren S. 1995. Enteric reflexes and nitric oxide in the fish intestine. *J. Exp. Biol*, 198: 2405-2411.

Karila P, Holmgren S. 1997. Anally projecting neurons exhibiting immunoreactivity to galanin, nitric oxide synthase and vasoactive intestinal peptide, detected by confocal laser scanning microscopy, in the intestine of the Atlantic cod, *Gadus morhua*. *Cell Tissue Res*, 287: 525-533.

Karila P, Jönsson A C, Jensen J, Holmgren S. 1993. Galanin-like immunoreactivity in extrinsic and intrinsic nerves to the gut of the Atlantic cod, *Gadus morhua*, and the effect of galanin on the smooth muscle of the gut. *Cell Tissue Res*, 271: 537-544.

Karila P, Messenger J, Holmgren S. 1997. Nitric oxide synthase- and neuropeptide Y-containing subpopulations of sympathetic neurons in the coeliac ganglion of the Atlantic cod, *Gadus morhua*, revealed by immunohistochemistry and retrograde tracing from the stomach. *J. Auton. Nerv. Syst*, 66: 35-45.

Karila P, Shahbazi F, Jensen J, Holmgren S. 1998. Projections and actions of tachykininergic, cholinergic, and serotonergic neurones in the intestine of the Atlantic cod. *Cell Tissue Res*, 291: 403-413.

Kelley G G, Aassar O S, Forrest J N J. 1991. Endogenous adenosine is an autacoid feedback inhibitor of chloride transport in the shark rectal gland. *J. Clin. Invest*, 88: 1933-1939.

Kiliaan A J, Joosten H W J, Bakker R, Dekker K, Grooth J A. 1989. Serotonergic neurons in the intestine of two teleosts, *Carassius auratus* and *Oreochromis mossambi-*

cus, and the effect of serotonin on trans-epithelial ion-selectivity and muscle tension. *Neuroscience*, 31: 817-824.

Kirtisinghe P. 1940. The myenteric nerve plexus in some lower chordates. *Quart. J. Microscop. Sci*, 81: 521-539.

Kitazawa T, Kondo H, Temma K. 1986. Alpha 2-adrenoceptor-mediated contractile response to catecholamines in smooth muscle strips isolated from rainbow trout stomach (*Salmo gairdneri*). *Br. J. Pharmacol*, 89: 259-266.

Kitazawa T, Temma K, Kondo H. 1986. Presynaptic alpha-adrenoceptor mediated inhibition of the neurogenic cholinergic contraction of the isolated intestinal bulb of the carp (*Cyprinus carpio*). *Comp. Biochem. Physiol*, 83C: 271-277.

Kitazawa T, Miyashita N, Chugun A, Temma K, Kondo H. 1988. Antagonist-like action of synthetic alpha2-adrenoceptor agonist on contractile response to catecholamines in smooth muscle strips isolated from rainbow trout stomach (*Salmo gairdneri*). *Comp. Biochem. Physiol*, 91C: 585-588.

Knight G E, Burnstock G. 1993. Identification of purinoceptors in the isolated stomach and intestine of the three-spined stickleback *Gasterosteus aculeatus*L. *Comp. Biochem. Physiol. C*, 106: 71-78.

Koelz H R. 1992. Gastric acid in vertebrates. *Scand. J. Gastroenterol. Suppl*, 193: 2-6.

Kono T, Kitao Y, Sonoda K, Nomoto R, Mekata T, Sakai M. 2008. Identification and expression analysis of ghrelin gene in common carp *Cyprinus carpio*. *Fisheries Science*, 74: 603-612.

Krogdahl A, Bakke-McKellep A M. 2005. Fasting and refeeding cause rapid changes in intestinal tissue mass and digestive enzyme capacities of Atlantic salmon (*Salmo salar*L.). *Comp. Biochem. Physiol. A Mol. Integr. Physiol*, 141: 450-460.

Kurokawa T, Suzuki T, Andoh T. 2000. Development of cholecystokinin and pancreatic polypeptide endocrine systems during the larval stage of Japanese flounder, *Paralichthys olivaceus*. *Gen. Comp. Endocrinol*, 120: 8-16.

Kurokawa T, Suzuki T, Hashimoto H. 2003. Identification of gastrin and multiple cholecystokinin genes in teleost. *Peptides*, 24: 227-235.

Lamers C H, Rombout J W, Timmermans L P. 1981. An experimental study on neural crest migration in *Barbus conchonius* (*Cyprinidae*, *Teleostei*), with special reference to the origin of the enteroendocrine cells. *J. Embryol. Exp. Morphol*, 62: 309-323.

Langer M, Van Noorden S, Polak J M, Pearse A G. 1979. Peptide hormone-like immunoreactivity in the gastrointestinal tract and endocrine pancreas of eleven teleost species. *Cell Tissue Res*, 199: 493-508.

Le Mevel J C, Mabin D, Hanley A M, Conlon J M. 1998. Contrasting cardiovascular effects following central and peripheral injections of trout galanin in trout. *Am. J. Physiol*, 275: R1118-R1126.

Lee J H, Ku S K, Park K D, Lee H S. 2004. Immunohistochemical study of the gastrointestinal endocrine cells in the Korean aucha perch. *J. Fish Biol*, 65: 170-181.

Lennard R, Huddart H. 1989. Purinergic modulation in the flounder gut. *Gen. Pharmacol*, 20: 849-853.

Li Z S, Furness J B. 1993. Nitric oxide synthase in the enteric nervous system of the rainbow trout, *Salmo gairdneri*. *Arch. Histol. Cytol*, 56: 185-193.

Lundin K, Holmgren S, Nilsson S. 1984. Peptidergic functions in the dogfish rectum. *Acta Physiol. Scand*, 121: 46A.

Maier K J, Tullis R E. 1984. The effects of diet a digestive cycle on the gastrointestinal tract pH values in the goldfish, *Carassius auratus* L., Mozambique tilapia, *Oreochromis mossambicus* (Peters), and channel catfish, *Ictalurus punctatus* (Rafinesque). *J. Fish Biol*, 25: 151-165.

Manera M, Giammarino A, Perugini M, Amorena M. 2008. *In vitro* evaluation of gut contractile response to histamine in rainbow trout (*Oncorhynchus mykiss* Walbaum, 1792). *Res. Vet. Sci*, 84: 126-131.

Manning A J, Murray H M, Gallant J W, Matsuoka M P, Radford E, Douglas S E. 2008. Ontogenetic and tissue-specific expression of preproghrelin in the Atlantic halibut, *Hippoglossus hippoglossus* L. *J. Endocrinol*, 196: 181-192.

Matsuda K, Kashimoto K, Higuchi T, Yoshida T, Uchiyama M, Shioda S, Arimura A, Okamura T. 2000. Presence of pituitary adenylate cyclase-activating polypeptide (PACAP) and its relaxant activity in the rectum of a teleost, the stargazer, *Uranoscopus japonicus*. *Peptides*, 21: 821-827.

Mattisson A, Holstein B. 1980. The ultrastructure of the gastric glands and its relation to induced secretory activity of cod, *Gadus morhua* (Day). *Acta Physiol. Scand*, 109: 51-59.

McCormick S. 2001. Endocrine control of osmoregulation in teleost fish. *Amer. Zool*, 41: 781-794.

McDonald J K, Greiner F, Bauer G E, Elde R P, Noe B D. 1987. Separate cell types that express two different forms of somatostatin in anglerfish islets can be immunocytochemically differentiated. *J. Histochem. Cytochem*, 35: 155-162.

Mellgren E M, Johnson S L. 2005. kitb, a second zebrafish ortholog of mouse Kit. *Dev. Genes Evol*, 215: 470-477.

Michelangeli F, Ruiz M C, Dominguez M G, Parthe V. 1988. Mammalian-like differentiation of gastric cells in the shark *Hexanchus griseus*. *Cell Tissue Res*, 251:

225-227.

Mulero I, Sepulcre M P, Meseguer J, Garcia-Ayala A, Mulero V. 2007. Histamine is stored in mast cells of most evolutionarily advanced fish and regulates the fish inflammatory response. *Proc. Natl. Acad. Sci. USA*, 104: 19434-19439.

Murashita K, Fukada H, Hosokawa H, Masumoto T. 2007. Changes in cholecystokinin and peptide Y gene expression with feeding in yellowtail (*Seriola quinqueradiata*): Relation to pancreatic exocrine regulation. *Comp. Biochem. Physiol. B Biochem. Mol. Biol*, 146: 318-325.

NCBI (National Center for Biotechnology Information) www. ncbi. nlm. nih. gov.

Nelson L E, Sheridan M A. 2005. Regulation of somatostatins and their receptors in fish. *Gen. Comp. Endocrinol*, 142: 117-133.

Ng A N, de Jong-Curtain T A, Mawdsley D J, White S J, Shin J, Appel B, Dong P D, Stainier D Y, Heath J K. 2005. Formation of the digestive system in zebrafish: III. Intestinal epithelium morphogenesis. *Dev. Biol*, 286: 114-135.

Nilsson S. 1983. *Autonomic Nerve Function in the Vertebrates.* Springer Verlag: Berlin.

Nilsson S, Holmgren S. 1983. Splanchnic nervous control of the stomach of the spiny dogfish, *Squalus acanthias. Comp. Biochem. Physiol. C*, 76: 271-276.

Nilsson S, Holmgren S. 1994. *Comparative Physiology and Evolution of the Autonomic Nervous System.* Harwood Academic Publishers: Chur, Switzerland.

Nilsson S, Holmgren S, Grove D J. 1975. Effects of drugs and nerve stimulation on the spleen and arteries of two species of dogfish, *Scyliorhinus canicula*and *Squalus acanthias. Acta Physiol. Scand*, 95: 219-230.

Noe B D, McDonald J K, Greiner F, Wood J D. 1986. Anglerfish islets contain NPY immunoreactive nerves and produce the NPY analog aPY. *Peptides*, 7: 147-154.

Norris J S, Norris D O, Windell J T. 1973. Effect of simulated meal size of gastric acid and pepsin secretory rates in bluegill (*Lepomis macrochirus*). *J. Fisheries. Res. Board Canada*, 30: 201-204.

Ogoshi M, Inoue K, Naruse K, Takei Y. 2006. Evolutionary history of the calcitonin gene-related peptide family in vertebrates revealed by comparative genomic analyses. *Peptides*, 27: 3154-3164.

Okuda R, Sasayama Y, Suzuki N, Kambegawa A, Srivastav A K. 1999. Calcitonin cells in the intestine of goldfish and a comparison of the number of cells among saline-fed, soup-fed, or high Ca soup-fed fishes. *Gen. Comp. Endocrinol*, 113: 267-273.

Oliver A S, Vigna S R. 1996. CCK-X receptors in the endothermic mako shark (*Isurus oxyrinchus*). *Gen. Comp. Endocrinol*, 102: 61-73.

Olson K R. 1999. Chapter 13. Rectal gland and volume homeostasis. In: W. C.

Hamlett, Ed. *Sharks, Skates and Rays: The Biology of Elasmobranch Fishes* The Johns Hopkins University Press: Baltimore, Maryland, 329-352.

Olson K R, Villa J. 1991. Evidence against nonprostanoid endothelium-derived relaxing factor (s) in trout vessels. *Am. J. Physiol*, 260: R925-R933.

Olsson C, Holmgren S. 1994. Distribution of PACAP (pituitary adenylate cyclase-activating polypeptide) -like and helospectin-like peptides in the teleost gut. *Cell Tissue Res*, 277: 539-547.

Olsson C, Holmgren S. 2000. PACAP and nitric oxide inhibit contractions in proximal intestine of the Atlantic cod, *Gadus morhua*. *J. Exp. Biol*, 203: 575-583.

Olsson C, Holmgren S. 2001. The control of gut motility. *Comp. Biochem. Physiol. A Mol. Integr. Physiol*, 128: 481-503.

Olsson C, Karila P. 1995. Coexistence of NADPH-diaphorase and vasoactive intestinal polypeptide in the enteric nervous system of the Atlantic cod (*Gadus morhua*) and the spiny dogfish (*Squalus acanthias*). *Cell Tissue Res*, 280: 297-305.

Olsson C, Aldman G, Larsson A, Holmgren S. 1999. Cholecystokinin affects gastric emptying and stomach motility in the rainbow trout *Oncorhynchus mykiss*. *J. Exp. Biol*, 202: 161-170.

Olsson C, Holbrook J D, Bompadre G, Jönsson E, Hoyle C H, Sanger G J, Holmgren S, Andrews P L. 2008. Identification of genes for the ghrelin and motilin receptors and a novel related gene in fish, and stimulation of intestinal motility in zebrafish (*Danio rerio*) by ghrelin and motilin. *Gen. Comp. Endocrinol*, 155: 217-226.

Olsson C, Holmberg A, Holmgren S. 2008. Development of enteric and vagal innervation of the zebrafish (*Danio rerio*) gut. *J. Comp. Neurol*, 508: 756-770.

Pan Q S, Fang Z P. 1993. An immunocytochemical study of endocrine cells in the gut of a stomachless teleost fish, grass carp, *Cyprinidae*. *Cell Transplant*, 2: 419-427.

Patterson T L, Fair E. 1933. The action of the vagus on the stomach-intestine of the hagfish. Comparative studies. VIII. *J. Cell. Comp. Physiol*, 3: 113-199.

Pederzoli A, Bertacchi I, Gambarelli A, Mola L. 2004. Immunolocalisation of vasoactive intestinal peptide and substance P in the developing gut of *Dicentrarchus labrax* (L.). *Eur. J. Histochem*, 48: 179-184.

Pederzoli A, Conte A, Tagliazucchi D, Gambarelli A, Mola L. 2007. Occurrence of two NOS isoforms in the developing gut of sea bass *Dicentrarchus labrax* (L.). *Histol. Histopathol*, 22: 1057-1064.

Preston E, McManus C D, Jönsson A C, Courtice G P. 1995. Vasoconstrictor effects of galanin and distribution of galanin containing fibres in three species of elasmobranch fish. *Regul. Pept*, 58: 123-134.

Preston E, Jonsson A C, McManus C D, Conlon J M, Courtice G P. 1998. Comparative vascular responses in elasmobranchs to different structures of neuropeptide Y and peptide YY. *Regul. Pept*, 78: 57-67.

Raible D W, Wood A, Hodsdon W, Henion P D, Weston J A, Eisen J S. 1992. Segregation and early dispersal of neural crest cells in the embryonic zebrafish. *Dev. Dyn*, 195: 29-42.

Rajjo I M, Vigna S R, Crim J W. 1988. Actions of cholecystokinin-related peptides on the gallbladder of bony fishes *in vitro*. *Comp. Biochem. Physiol. C*, 90: 267-273.

Rajjo I M, Vigna S R, Crim J W. 1989. Immunocytochemical localization of vasoactive intestinal polypeptide in the digestive tracts of a holostean and a teleostean fish. *Comp. Biochem. Physiol*, 94C: 411-418.

Rebolledo I M, Vial J D. 1979. Fine structure of the oxynticopeptic cell in the gastric glands of an elasmobranch species (*Halaelurus chilensis*). *Anat. Rec*, 193: 805-822.

Reinecke M, Schluter P, Yanaihara N, Forssmann W G. 1981. VIP immunoreactivity in enteric nerves and endocrine cells of the vertebrate gut. *Peptides* 2 (Suppl 2), 149-156.

Reinecke M, Muller C, Segner H. 1997. An immunohistochemical analysis of the ontogeny, distribution and coexistence of 12 regulatory peptides and serotonin in endocrine cells and nerve fibers of the digestive tract of the turbot, *Scophthalmus maximus* (*Teleostei*). *Anat. Embryol. (Berl)*, 195: 87-101.

Reite O B. 1972. Comparative physiology of histamine. *Physiol. Rev*, 52: 778-819.

Rich A, Leddon S A, Hess S L, Gibbons S J, Miller S, Xu X, Farrugia G. 2007. Kit-like immunoreactivity in the zebrafish gastrointestinal tract reveals putative ICC. *Dev. Dyn*, 236: 903-911.

Rombout J H. 1977. Enteroendocrine cells in the digestive tract of *Barbus conchonius* (*Teleostei, Cyprinidae*). *Cell Tissue Res*, 185: 435-450.

Rombout J H, van der Grinten C P, Binkhorst F M, Taverne-Thiele J J, Schooneveld H. 1986. Immunocytochemical identification and localization of peptide hormones in the gastro-entero-pancreatic (GEP) endocrine system of the mouse and a stomachless fish, *Barbus conchonius*. *Histochemistry*, 84: 471-483.

RØnnestad I, Rojas-Garcia C R, Skadal J. 2000. Retrograde peristalsis, a possible mechanism for filling the pyloric cecae?. *J. Fish. Biol*, 56: 216-218.

Sadaghiani B, Vielkind J R. 1990. Distribution and migration pathways of HNK-1-immunoreactive neural crest cells in teleost fish embryos. *Development*, 110: 197-209.

Sakata I, Mori T, Kaiya H, Yamazaki M, Kangawa K, Inoue K, Sakai T. 2004. Localization of ghrelin-producing cells in the stomach of the rainbow trout (*Oncorhyn-*

chus mykiss). *Zoolog. Sci*, 21: 757-762.

Salaneck E, Larsson T A, Larson E T, Larhammar D. 2008. Birth and death of neuropeptide Y receptor genes in relation to the teleost fish tetraploidization. *Gene*, 409: 61-71.

Salimova N, Fehér E. 1982. Innervation of the alimentary tract in Chondrostean fish (*Acipenseridae*). A histochemical, microspectro-fluorimetric and ultrastructural study. *Acta Morphol. Acad. Sci. Hung*, 30: 213-222.

Santos C R, Ingleton P M, Cavaco J E, Kelly P A, Edery M, Power D M. 2001. Cloning, characterization, and tissue distribution of prolactin receptor in the sea bream (*Sparus aurata*). *Gen. Comp. Endocrinol*, 121: 32-47.

Schneeman B O. 2002. Gastrointestinal physiology and functions. *Br. J. Nutr.* 88 (Suppl 2), S159-S163.

Severini C, Improta G, Falconieri-Erspamer G, Salvadori S, Erspamer V. 2002. The tachykinin peptide family. *Pharmacol. Rev*, 54: 285-322.

Shahbazi F, Karila P, Olsson C, Holmgren S, Jensen J. 1998. Primary structure, distribution, and effects on motility of CGRP in the intestine of the cod *Gadus morhua*. *Am. J. Physiol*, 275: R19-R28.

Shahbazi F, Holmgren S, Larhammar D, Jensen J. 2002. Neuropeptide Y effects on vasorelaxation and intestinal contraction in the Atlantic cod *Gadus morhua*. *Am. J. Physiol. Regul. Integr. Comp. Physiol*, 282: R1414-R1421.

Shepherd I T, Beattie C E, Raible D W. 2001. Functional analysis of zebrafish GDNF. *Dev. Biol*, 231: 420-435.

Shuttleworth T J, Thorndyke M C. 1984. An endogenous peptide stimulates secretory activity in the elasmobranch rectal gland. *Science*, 225: 319-321.

Silva P, Lear S, Reichlin S, Epstein F H. 1990. Somatostatin mediates bombesin inhibition of chloride secretion by rectal gland. *Am. J. Physiol*, 258: R1459-1463.

Silva P, Epstein F H, KarnakyJr K J, Reichlin S, ForrestJr J N. 1993. Neuropeptide Y inhibits chloride secretion in the shark rectal gland. *Am. J. Physiol*, 265: R439-R446.

Slagter B J, Kittilson J, Sheridan M A. 2005. Expression of somatostatin receptor mRNAs is regulated *in vivo* by growth hormone, insulin, and insulin-like growth factor-I in rainbow trout (*Oncorhynchus mykiss*). *Regul. Pept*, 128: 27-32.

Smit H. 1968. Gastric secretion in the lower vertebrates and birds. In: Code C F, Ed. *Handbook of Physiology*, sect 6: *Alimentary canal*, volume V: *Bile, digestion, ruminal physiology* American Physiological Society: Washington DC, 2791-2805.

Smolka A J, Lacy E R, Luciano L, Reale E. 1994. Identification of gastric H, K-ATPase in an early vertebrate, the Atlantic stingray *Dasyatis sabina*. *J. Histochem.*

Cytochem, 42: 1323-1332.

Stevens C E, Hume I D. 1995. *Comparative Physiology of the Vertebrate Digestive System*. Cambridge University Press: Cambridge.

Stoff J S, Rosa R, Hallac R, Silva P, Epstein F H. 1979. Hormonal regulation of active chloride transport in the dogfish rectal gland. *Am. J. Physiol*, 237: F138-F144.

Stoff J S, Silva P, Lechan R, Solomon R, Epstein F H. 1988. Neural control of shark rectal gland. *Am. J. Physiol*, 255: R212-R216.

Sullivan M X. 1905—1906. The physiology of the digestive tract of elasmobranchs. *Am. J. Physiol*, 15: 42-45.

Sundell K, Björnsson B T, Itoh H, Kawauchi H. 1992. Chum salmon (*Oncorhynchus keta*) stanniocalcin inhibits *in vitro* intestinal calcium uptake in Atlantic cod (*Gadus morhua*). *J. Comp. Physiol. B*, 162: 489-495.

Sundström G, Larsson T A, Brenner S, Venkatesh B, Larhammar D. 2008. Evolution of the neuropeptide Y family: New genes by chromosome duplications in early vertebrates and in teleost fishes. *Gen. Comp. Endocrinol*, 155: 705-716.

Suzuki N, Suzuki T, Kurokawa T. 2000. Cloning of a calcitonin gene-related peptide receptor and a novel calcitonin receptor-like receptor from the gill of flounder, *Paralichthys olivaceus*. *Gene*, 244: 81-88.

Tagliafierro G, Zaccone G, Bonini E, Faraldi G, Farina L, Fasulo S, Rossi G G. 1988. Bombesin-like immunoreactivity in the gastrointestinal tract of some lower vertebrates. *Ann. NY Acad. Sci*, 547: 458-460.

Tagliafierro G, Rossi G G, Bonini E, Faraldi G, Farina L. 1989. Ontogeny and differentiation of regulatory peptide- and serotonin-immunoreactivity in the gastrointestinal tract of an elasmobranch. *J. Exp. Zool*, 252: 165-174.

Takahashi H, Sakamoto T, Hyodo S, Shepherd B S, Kaneko T, Grau E G. 2006. Expression of glucocorticoid receptor in the intestine of a euryhaline teleost, the Mozambique tilapia (*Oreochromis mossambicus*): Effect of seawater exposure and cortisol treatment. *Life Sciences*, 78: 2329-2335.

Thorndyke M, Holmgren S. 1990. Bombesin potentiates the effect of acetylcholine on isolated strips of fish stomach. *Regul. Pept*, 30: 125-135.

Thorndyke M C, Riddell J H, Thwaites D T, Dimaline R. 1989. Vasoactive intestinal polypeptide and its relatives - biochemistry, distributions and functions. *Biol. Bull*, 177: 183-186.

Thorndyke M C, ReeveJr J R, Vigna S R. 1990. Biological activity of a bombesin-like peptide extracted from the intestine of the ratfish, *Hydrolagus colliei*. *Comp. Biochem. Physiol. C*, 96: 135-140.

Trischitta F, Denaro M G, Faggio C. 1999. Effects of acetylcholine, serotonin and noradrenaline on ion transport in the middle and posterior part of *Anguilla anguilla* intestine. *J. Comp. Physiol. B*, 169: 370-376.

Trischitta F, Pidalà P, Faggio C. 2007. Nitric oxide modulates ionic transport in the isolated intestine of the eel, *Anguilla anguilla*. *Comp. Biochem. Physiol*, 148A: 368-373.

Uesaka T, Yano K, Yamasaki M, Nagashima K, Ando M. 1994. Somatostatin-related peptides isolated from the eel gut: Effects on ion and water absorption across the intestine of the seawater eel. *J. Exp. Biol*, 188: 205-216.

Valette G, Augeraux P. 1958. Réactivité des muscles lisses des poissons a l'histamine et a d'autres agents contracturants (5-hydroxytryptamine, acetylcholine et chlorure de baryum). *J. Physiol. (Paris)*, 50: 1067-1074.

Van Noorden S. 1990. Gut hormones in cyclostomes. *Fish Physiol. Biochem*, 8: 399-408.

Van Noorden S, Falkmer S. 1980. Gut-islet endocrinology-some evolutionary aspects. *Invest. Cell. Pathol*, 3: 21-35.

Van Noorden S, Patent G J. 1980. Vasoactive intestinal polypeptide-like immunoreactivity in nerves of the pancreatic islet of the teleost fish, *Gillichthys mirabilis*. *Cell Tissue Res*, 212: 139-146.

Van Noorden S, Pearse A G. 1974. Immunoreactive polypeptide hormones in the pancreas and gut of the lamprey. *Gen. Comp. Endocrinol*, 23: 311-324.

Vigna S R. 1983. Evolution of endocrine regulation of gastrointestinal function in lower vertebrates. *Amer. Zool*, 23: 512-520.

Volkoff H, Peyon P, Lin X, Peter R E. 2000. Molecular cloning and expression of cDNA encoding a brain bombesin/gastrin-releasing peptide-like peptide in goldfish. *Peptides*, 21: 639-648.

von Euler U S, Fänge R. 1961. Catecholamines in nerves and organs of *Myxine glutinosa*, *Squalus acanthias* and *Gadus callarias*. *Gen. Comp. Endocrinol*, 1: 191-194.

von Euler U S, Östlund E. 1957. Effects of certain biologically occurring substances on the isolated intestine of fish. *Acta Physiol. Scand*, 38: 364-372.

Wallace K N, Pack M. 2003. Unique and conserved aspects of gut development in zebrafish. *Dev. Biol*, 255: 12-29.

Wang Y, Conlon J M. 1994. Purification and characterization of galanin from the phylogenetically ancient fish, the bowfin (*Amia calva*) and dogfish (*Scyliorhinus canicula*). *Peptides*, 15: 981-986.

Wang Y, Barton B A, Thim L, Nielsen P F, Conlon J M. 1999. Purification and characterization of galanin and scyliorhinin I from the hybrid sturgeon, *Scaphirhynchus platorynchus x Scaphirhynchus albus* (Acipenseriformes). *Gen. Comp. Endocrinol*,

113: 38-45.

Wimalawansa S J. 1997. Amylin, calcitonin gene-related peptide, calcitonin, and adrenomedullin: A peptide superfamily. *Crit. Rev. Neurobiol*, 11: 167-239.

Yasugi S, Matsunaga T, Mizuno T. 1988. Presence of pepsinogens immunoreactive to anti-embryonic chicken pepsinogen antiserum in fish stomachs: Possible ancestor molecules of chymosin of higher vertebrates. *Comp. Biochem. Physiol. A*, 91: 565-569.

Young J Z. 1933. The autonomic nervous system of selachians. *Q. J. Microsc. Sci*, 75: 571-624.

Young J Z. 1936. The innervation and reactions to drugs of the viscera of teleostean fish. *Proc. R. Soc. London, Ser. B, Biol. Sci*, 120: 303-318.

Young J Z. 1980. Nervous control of gut movements in *Lophius*. *J. Mar. Biol. Assoc. UK*, 60: 19-30.

Young J Z. 1980. Nervous control of stomach movements in dogfishes and rays. *J. Mar. Biol. Assoc. UK*, 60: 1-17.

Young J Z. 1983. Control of movements of the stomach and spiral intestine of *Raja* and *Scyliorinus*. *J. Mar. Biol. Assoc. UK*, 63: 557-574.

Young J Z. 1988. Sympathetic innervation of the rectum and bladder of the skate and parallel effects of ATP and adrenaline. *Comp. Biochem. Physiol. C*, 89: 101-107.

Youson J H, Al-Mahrouki A A. 1999. Ontogenetic and phylogenetic development of the endocrine pancreas (islet organ) in fish. *Gen. Comp. Endocrinol*, 116: 303-335.

Youson J H, Al-Mahrouki A A, Amemiya Y, Graham L C, Montpetit C J, Irwin D M. 2006. The fish endocrine pancreas: review, new data, and future research directions in ontogeny and phylogeny. *Gen. Comp. Endocrinol*, 148: 105-115.

Yui R, Nagata Y, Fujita T. 1988. Immunocytochemical studies on the islet and the gut of the arctic lamprey, *Lampetra japonica*. *Arch. Histol. Cytol*, 51: 109-119.

Yui R, Shimada M, Fujita T. 1990. Immunohistochemical studies on peptide- and amine-containing endocrine cells and nerves in the gut and rectal gland of the ratfish *Chimaera monstrosa*. *Cell Tissue Res*, 260: 193-201.

Zaks M G, Gazhala E M, Gzgzian D M, Kuzina M M, Pesennikova D I. 1975. Comparative physiological characteristics of the action of pentagastrin on gastric secretion in fish, frogs, turtles and chickens. *Zh. Evol. Biokhim. Fiziol*, 11: 594-600.

索 引

A

Acetylated β-endorphins 乙酰化 β-内啡肽 242
Acetylcholine (Ach) 乙酰胆碱 296，352，435-436，443，450
Activator protein 1 (AP-1) 激活蛋白 1 140
Adaptive immune response 适应性免疫反应 289，292，312
Adenohypophysis, fish 腺垂体，鱼类
 hormone synthesis 激素合成 4-6
 regionalization, nomenclatures 区域化，命名 4-5
Adrenaline (AD) 肾上腺素 437，456
Adrenergic receptor agonists, on immune parameters 肾上腺素能受体激动剂，对免疫参数 297
Adrenocorticotropic hormone (ACTH) 促肾上腺皮质激素 6-7，181，187，219-225，239，242，302
 contribution of MCH in regulation of MCH 作用的调控 256
 releasing activity in fish 鱼类的释放活性 248
 releasing efficacy 释放效能 247
 secretion in fish 鱼类的分泌作用 224
 secretion in teleost fish, factors affecting 硬骨鱼类的分泌作用 220-222
Adrenomedullins (AMs) 肾上腺髓质素 348，353
African catfish (*Clarias gariepinus*) 非洲胡鲇 109，166
African cichlid fish (*Haplochromis burtoni*) 伯氏朴丽鱼 245-246，255
AGGT (C/A) A consensus sequence AGGT (C/A) A 共有系列 250
Agouti-related protein (AgRP) 野灰蛋白相关蛋白 32-33，405
Agouti-signaling peptide (ASP) 野灰蛋白信号肽 242，405
Aldosterone 醛固酮 81，245，362-363
Alpha-2-macroglobulin (α-2-M) α-2-巨球蛋白 291
Amago Salmon (*Oncorhynchus rhodurus*) 玫瑰大麻哈鱼 244
Androgen receptors (AR), in fish brain 雄激素受体，在鱼脑 74，79-80

Angiotensin-converting enzyme（ACE） 血管紧张肽转变酶 352
Angiotensin Ⅱ（ANG Ⅱ） 血管紧张素Ⅱ 346-347，353
　　dipsogenic effect 致渴作用 348-350，359
　　induced drinking 诱导饮水 354
　　in mammals 在哺乳类 354
　　for sustaining body fluid balance 保持体液平衡 359
Angiotensins 血管紧张素 223
ANP and VNP, antidipsogenic hormones in eel ANP 和 VNP, 鳗鱼的止渴作用激素 349
Anterior tuberal nucleus（NAT） 前结节核 12-13，16-17，
Anteroventral periventricular nucleus（AVPV） 前腹室周核 78
Appetite-regulating factors 食欲调节因子 389
Appetite regulators 食欲调节剂 395-397，404，409
　　brain signals 脑信号
　　　　galanin 甘丙肽 397，439，446
　　　　neuropeptide Y family of peptides 神经肽 Y 家族肽 396
　　　　orexins 食欲肽 396，398，410
　　gender and reproductive status 性别和生殖状态 407
　　genetic influence 遗传影响 407
　　ontogeny 个体发育 406
　　peripheral signals, ghrelin 外周信号, 生长素释放肽 395，401
　　satiation signals 饱食信号 409
　　　　amylin 支链淀粉 401-402
　　　　bombesin/GRP 铃蟾肽/GRP 402
　　　　calcitonin gene-related peptide 降钙素基因相关肽 401，439
　　　　cholecystokinin/gastrin 缩胆囊肽/促胃液素 401
　　　　cocaine-and amphetamine-regulated transcript 可卡因-和安非他明-调节转录体 399
　　　　corticotropin-releasing factor-related neuropeptides 促肾上腺皮质激素释放因子相关神经肽 399
　　　　endocannabinoids 内大麻素 390
　　　　glucagon/glucagon-like peptide-1 胰高血糖素/胰高血糖素样肽 403
　　　　insulin/insulin-like growth factors 胰岛素/胰岛素样生长因子 403
　　　　leptin 瘦蛋白（瘦素） 309-309
　　　　neuromedin U 神经调节肽 U 391，395
　　　　RFamides RF 酰胺 400
　　　　tachykinins 速激肽 400，440-441
Appetite regulatory endocrine factors 食欲调节内分泌因子 389-393
Arctic charr（*Salvelinus alpinus*） 北极红点鲑 232，253

Arctic grayling (*Thymallus arcticus*)　北极茴鱼　152
Arginine-vasopressin (AVP)　精氨酸加压素　347
Arginine-vasotocin (AVT)　精氨酸加压催产素　220，222-223，227
Aromatase B expression, in fish brain　芳化酶B在鱼脑表达　72-74
Atlantic cod (*Gadus morhua*)　大西洋鳕鱼　194，396-399，402，408，434，438，443，446-448，450-452，457
Atlantic croaker (*Micropogonias undulatus*)　大西洋绒须石首鱼　41，76，80，112，113，115-116，147，195-196
Atlantic halibut (*Hippoglossus hippoglossus*)　大西洋庸鲽　143，152，195，442
Atlantic salmon (*Salmo salar*)　大西洋鲑鱼　17，122，143，157，183，188-189，243，363，439，452
Atrial natriuretic peptide (ANP)　心房钠尿肽　154，453
Atrial NP (ANP)　心房钠尿肽　193
Autocrine/paracrine signals in pituitary, in GH release　脑垂体自分泌/旁分泌信号，对GH的释放　139
AVT and IST expression, in teleost brain　AVT和IST在硬骨鱼脑的表达　223
AVT-ir neurons　AVT-免疫反应神经元　223
AVT-stimulated ACTH release, in rainbow trout pituitary　AVT-刺激ACTH在虹鳟脑垂体的释放　223
AVT V2 receptor　AVT V2受体　358
Ayu (*Plecoglossus altivelis*)　香鱼　34-35

B

Barfin flounder (*Verasper moseri*)　条斑星鲽　33，190，229，406
Black porgy (*Acanthopagrus schlegeli*)　黑棘鲷　72，156，161，163，183
Black sea bream (*Acanthopagrus schlegeli*)　黑棘鲷　72，156，161，163，183
Blood-brain barrier (BBB)　血脑屏障　347，349-351
Bluehead wrasse (*Thalassoma bifasciatum*)　双带锦鱼　72，118
B-lymphocyte　B-淋巴细胞　293，297，300
Body fluid regulation　体液调节　342-343
Body fluid, fish　体液，鱼类　338
Bombesin (BBS)　铃蟾肽　28，153，391，395，402，439
Bradykinin　缓激肽　344，346，352-353
Brain-pituitary-gonadal (BPG) axis　脑-脑垂体-性腺轴　122

Brown ghost knifefish (*Apteronotus leptorhynchus*) 电鳗 10, 16, 24-26, 38, 219
Brown trout (*Salmo trutta*) 褐鳟 86, 155, 248, 254, 362, 448
Brown trout (*Salmo trutta fario*) 河鳟 27-28
B-type natriuretic peptide (BNP) B-型利尿钠肽 154
B-type NP (BNP) B-型利尿钠肽 154
Budget of ions and water 离子和水分的预算 339
Bullshark (*Carcharhinus leucas*) 低鳍真鲨 340, 356
Burton's mouthbrooder (*Haplochromis burtoni*) 伯氏朴丽鱼 245-246, 255

C

Calcitonin gene-related peptide (CGRP) 降钙素基因相关肽 348, 391, 401, 439, 444, 455
cAMP response element (CRE) cAMP 反应元件 140
cAMP signaling cascade cAMP 信号级联 239
CART mRNA expression, in goldfish CART mRNA 在金鱼表达 24
Catecholamines 儿茶酚胺
    effect on innate and adaptive immunity 对先天的和适应性免疫的影响 297
    in gut motility 对消化道活动 447
    and receptors, in immune system 对免疫系统受体 296
Catecholamines, definition 儿茶酚胺，定义 186
Catfish (*Clarias batrachus*) 胡子鲶 10, 118
Catfish (*Clarias gariepinus*) 非洲胡鲶 34, 109, 147
Catfish (*Ictalurus melas*) 黑鮰 456
Catfish (*Ictalurus punctatus*) 斑点鮰 141, 194, 297, 399
Caudal neurosecretory system (CNSS) 尾神经内分泌系统 246, 362
Caudal pars distalis (CPD) 尾远侧部 232
CCK/gastrin-ir perikarya and fibers, in goldfish CCK/促胃液素免疫反应核周体和纤维，在金鱼 23-24
Central nervous system (CNS) 中枢神经系统 7, 22, 39, 69, 87, 141, 190, 241, 388, 433
Central nucleus of the inferior lobe (NCLI) 下叶的中央核 14
ChAT-immunoreactive (ir) nerves ChAT-免疫反应神经 437
Chemokines and chemokine receptors 趋化因子和趋化因子受体 295
Chemotaxis 趋化因子 291-292, 294-295
Cherry salmon (*Oncorhynchus masou*) 马苏大麻哈鱼 86, 110, 120, 149, 151, 158, 191, 194, 255, 362, 407

Chinook salmon (*Oncorhynchus tshawytscha*) 大鳞大麻哈鱼 355
Cholecystokinin (CCK) 缩胆囊素 21
Choline acetyl transferase (ChAT) 胆碱乙酰转移酶 435，437
Chum salmon (*Oncorhynchus keta*) 大麻哈鱼 22，26，36，39，143，151，181，188，192
Cichlid (*Astatotilapia burtoni*) 妊丽鱼 80-81，84，110
Clawed toad (*Xenopus laevis*) 爪蟾 188，224，225，247
Climbing perch (*Anabas testudineys*) 攀鲈 184
Cloudy dogfish (*Scyliorhinus torazame*) 虎纹猫鲨 442
Cobia (*Rachycentron canadun*) 军曹鱼 117
Cocaine-and amphetamine-regulated transcript (CART) 可卡因-和安非他明-相关转录体（CART） 19，21，24，399
Cod (*Gadus morhua*) 大西洋鳕鱼 194，396-399，402，408，434，438，443，446-448，450-452，456-457
co-gonadotropin (co-GTH) 共-促性腺激素 141，143
Coho salmon (*Oncorhynchus kisutch*) 银大麻哈鱼 110，112，154，158，195-196，236-238，244，248，297，396，403-404，409，453
Colony-stimulating factors (CSFs) 集落刺激因子 295
Common carp (*Cyprinus carpio*) 鲤鱼 82，140，181，221，226，228，233，237
Common stingray (*Dasyatis pastinaca*) 蓝纹虹 451
Copeptine peptide 混合肽素 193
Corticoid receptors 肾上腺皮质激素受体 81
    glucocorticoid receptors 糖皮质激素受体
        chemically identified 化学鉴定 84
        distribution of 分布 82-83
    mineralocorticoid receptors 盐皮质激素受体 82
Corticosteroid receptor (CR), in fish brain 糖皮质激素受体（CR），在鱼脑内 82，245
Corticotropes 促肾上腺皮质激素细胞
    hypothalamic regulation 下丘脑调控 219
        inhibitory factors 抑制因子 224
        stimulatory factors 刺激因子 219，222-224
    secretion, targets and functions 分泌、靶标和功能 239
        ACTH and non-acetylatedβEND ACTH 和非乙酰化βEND 241，254
Corticotropic axis 促肾上腺皮质激素细胞轴 228
Corticotropin-like intermediate peptide (CLIP) 促肾上腺皮质激素样中间肽

（CLIP） 32, 239

Corticotropin-releasing factor（CRF） 促肾上腺皮质激素释放因子（CRF） 19, 21, 83, 220, 222, 225, 228, 231, 238, 390, 399

Corticotropin-releasing factor（CRF）-related neuropeptides 促肾上腺皮质激素释放因子（CRF）-相关神经肽 399

Cortisol 皮质醇 220, 224
  DOC and DOC 和 245
  effect on immune responses 对免疫反应的作用 299-300
  functions 功能 246-247
  in GH release 对 GH 释放 157
  negative feedback 负反馈作用 247-248
  P450 c11, role P450 c11, 作用 240
  receptor types in immune cells 在免疫细胞的受体类型 298-299
  roles in control of water and ion fluxes 水分和离子流出的调控作用 453
  vs. T4 和 T4 309

CRF-BP gene expression CRF-BP 基因表达 255

CRF-expressing neurons, in zebrafish preoptic area CRF-表达神经元，在斑马鱼视前区 25

CRF gene expression CRF-基因表达 247-248

CRF-related peptides CRF-相关肽 255, 257, 400

CRF/UI, in thyrotropic axis regulation CRF/UI, 在促甲状腺素轴的调控 255

CRH-induced ACTH release, in rainbow trout CRH-诱导 ACTH 释放，在虹鳟 229

C-terminal peptide of isotocin precursor（cNpCp） 硬骨鱼类催产素前体的 C-端肽 193

C-type natriuretic peptide（CNP） C-利尿钠肽（CNP） 154-155

C type NPs（CNP） C 利尿钠肽（CNP） 193, 349

CXC chemokine CXC 趋化因子 296

CXCL12/CXCR4 signaling CXCL12/CXCR4 信号 296

Cyp19a1b gene Cyp19a1b 基因 71-72

Cyprinid（*Spinibarbus denticulatus*） 锯齿倒刺鲃 81

Cystic fibrosis transmembrane conductance regulator（CFTR） 囊性纤维化跨膜传导调节蛋白（CFTR） 247, 339

Cytochrome P450 细胞色素 P450 240

Cytochrome P450c17（CYP17） 细胞色素 P450c17 71

Cytokines 细胞因子 289, 292
  classification and receptors 分类和受体 295
  family members, in teleost fish species 家族成员，在硬骨鱼类 292
  and phylogenetic relationship 和系统进化的关系 295

D

5'-Deiodinases D1 and D2 5'-脱碘酶 D1 和 D2 248
Deiodination 脱碘作用
    T4 50-deiodination activities T4 50-脱碘作用活性 248
    of T4's phenolic ring T4 的酚环 252
11-Deoxycorticosterone（DOC） 11-脱氧皮质酮（DOC） 245
    role in spermiation 在精子排放中的作用 247
11-deoxycorticosterone（DOC） 11-脱氧皮质酮（DOC） 81
Diffuse nucleus of inferior lobe（NDLI） 下叶的弥散核 14
Dihydroxyphenylacetic acid（DOPAC） 二羟苯丙乙酸 151
Dogfish（*Scyliorhinus canicula*） 小点猫鲨 154，345，347，357，360，364，449
Dogfish（*Triakis scyllia*） 皱唇鲨 194，345，347，356-357
Dopamine（DA） 多巴胺 186-187，437
    functions of 功能 107
    in GH release 对 GH 释放 151
    inhibiting basal ACTH secretion 抑制基础的 ACTH 分泌 224
    inhibitory effects and endocrine stress response 抑制作用和内分泌应激反应 256
    neurotransmitters 神经递质 40
    role as gonadotropin release inhibitory factor 对促性腺激素释放的抑制因子作用 113
    stimulatory role in stressed fish 对应激反应鱼的刺激作用 230
Dopaminergic neurons, in fish brain 多巴胺能神经元，在鱼脑内 79，84
Drinking in fish, hormonal regulation 鱼类的饮水活动，激素调控 346
    central vs. peripheral hormones 中枢和外周激素 354
    hormones inducing drinking 激素诱导饮水 346
        adrenomedullins 肾上腺髓质素 348
        angiotensin II 血管紧张素 II 346-347
    hormones inhibiting drinking 激素抑制饮水 349
        bradykinin 缓激肽 352
        ghrelin 生长素释放肽 351
        natriuretic peptides 利尿钠肽 349
    hormones regulating, in fishes and mammals 激素调控，在鱼类和哺乳类 353
    neural mechanisms to elicit drinking 引起饮水的神经作用机理 354-355
    sodium appetite, regulation of 钠欲，调控作用 355

E

Eel (*Anguilla anguilla*)　欧洲鳗鲡　5, 31
Eel (*Anguilla japonica*)　日本鳗鲡　227, 232
Eelpout (*Zoarces viviparous*)　绵鳚　79
Elasmobranchs fish, neurohypophysis　板鳃鱼类, 神经脑垂体　4
Elephant fish (*Callorhinchus milii*)　米氏叶吻银鲛　455
Endocannabinoids (ECs)　内大麻素　390, 395, 400
Endocrine drisrupting compounds (EDCs)　内分泌破坏化合物　310
Endocrine-immune interactions, in teleosts　内分泌-免疫相互作用, 在硬骨鱼类　184
Endocrine pancreas, fish　内分泌胰脏, 鱼类　402, 440, 452
Endogenous UI in fish　内源的 UI 在鱼类　223
β-Endorphin (β-END)　β-内啡肽 (β-END)　219, 241
　　secretion in teleost fish, factors affecting　在硬骨鱼类的分泌, 因子影响　228
Endorphins, role　内啡肽, 作用　241
English sole (*Parophrys vetulus*)　侧枝鲽　195
Enteric netvous system (ENS)　肠神经系统 (ENS)　434
Erythropoietin　红细胞生成素　292, 295
17β-Estradiol (E2)　17β-雌二醇 (E2)　112, 310
Estrogen/androgen receptors　雌激素/雄激素受体　310
Estrogen receptors (ERs)　雌激素受体 (ERs)　75-79
Estrogen responsive element (ERE)　雌激素反应元件 (ERE)　75
European eel (*Anguilla anguilla*)　欧洲鳗鲡　31
European plaice (*Pleuronectes platessa*)　欧洲鲽鱼　21
European sea bass (*Dicentrarchus labrax*)　舌齿鲈　7, 9-11, 13, 109, 198
Extant teleosts, stress responses　现存的硬骨鱼类, 应激反应　39
Extracellular signal-regulated kinase (ERK)　胞外信号调节激酶　151
Extrathyroidal deiodination　甲状腺外脱碘作用　248-249

F

Fathead minnow (*Pimephales promelas*)　胖头鲅　109, 155
Feeding activity and effect of photoperiod　摄食活动和光周期效应　408
Feeding, in vertebrates　摄食, 在脊椎动物　388
Feeding regulation　摄食调控
　　fishes as models for study　鱼类作为研究的模型　393

mammalian models, comparison 和哺乳类模式对比 393
methods for studying 研究的方法 394
Fish brain, estrogen receptors distribution 鱼类脑，雌激素受体分布 76-78
Fish, growth hormone 鱼类，生长激素
autocrine/paracrine signals in pituitary, GH release 在脑垂体的自分泌/旁分泌信号，GH 释放 159-160
biological actions 生物作用 140-142
and growth hormone receptors 和生长激素受体 139-140
at hypothalamic and pituitary levels 在下丘脑和脑垂体水平 160-163
hypothalamic signals from CNS, GH release 来自 CNS 的下丘脑信号，GH 释放
inhibitors 抑制剂 143-147
neuroendocrine regulators, GH GH 的神经内分泌调节剂 154-155
stimulators 刺激剂 147-154
peripheral organs/tissues signals, in GH release GH 释放的外周器官/组织信号
inhibitors 抑制剂 155-156
stimulators 刺激剂 156-159
secretion and synthesis regulation 分泌与合成的调节 142-143
Flounder (*Platichthys flesus*) 川鲽 18，255
Fluid balance, regulation 液体平衡，调节
renal and extrarenal mechanisms 肾的和肾外的作用机理 356-359
prolactin, growth hormone and cortisol 催乳激素、生长激素和皮质醇 361-363
renin-angiotensin system 肾素-血管紧张系统 359-361
urotensin Ⅰ 鱼类尾紧张素 Ⅰ 363-365
Fluid exchange and balancing mechanisms 液体交换和平衡的作用机理 338-341
elasmobranches 板鳃鱼类 340
for fish in FW, major challenges 淡水鱼类，主要挑战 338
for marine cyclostome and teleosts 海水圆口类和硬骨鱼类 339
roles for claudins, occluding and aquaporins 密蛋白、闭合蛋白和水通道蛋白的作用 340
Fluid intake, regulation 液体摄入，调节 343
regulatory mechanism in fish 鱼类调控的作用机理 343
acceleration vs. inhibition of drinking 饮水的促进作用对抑制作用 344
reflex drinking vs. thirst-motivated drinking 反射性饮水对口渴激发的饮水 343-344
volemic vs. osmotic factors 容量对渗透因子 344-345
fOat protein 一种有机阴离子转运蛋白 251
Follicle-stimulating hormone (FSH), in teleost fish 促卵泡激素（FSH），在硬

骨鱼类 5, 239
 dopamine in release of 多巴胺的释放 114
 gonadal steroids regulation 性类固醇激素调节作用 114-115
  negative feedback effects 负反馈作用 115-116
  positive feedback effdcts 正反馈作用 116-117
Four eyed fish (*Anableps anableps*) 四眼鱼 27
Free T4 in Senegal sole 塞内加尔鳎的游离 T4 254
Freshwater, fish in 淡水鱼类 338
FSHβmRNA expression, measurement of FSHβmRNA 表达、测定 110

G

Galanin (GAL) 甘丙肽 397, 439, 446, 456
 in GH release 对 GH 释放 154
Galanin-ir cell populations, in teleost fish 硬骨鱼类甘丙肽免疫反应细胞群 27
Galanin peptide 甘丙肽 27-28
Gamma-amino butyric acid (GABA) γ-氨基丁酸(GABA) 39, 79, 116, 438
 in GH release GH 释放 147
 neurons, in fish brain 鱼脑的神经元 79
 neurotransmitter, in LH and FSH neuroendocrine regulation 神经递质，对 LH 和 FSH 的神经内分泌调控 116
Gar (*Lepisosteus oculatus*) 雀鳝 34
Gastric acid 胃酸 402, 437, 450-452
Gastric inhibitory peptide (GIP) 肠抑胃肽 403
Gastrin, effect in Atlantic cod 促胃液素对大西洋鳕鱼的作用 451
Gastrin-releasing peptide (GRP) 胃泌素释放肽 28-29, 153-154, 402, 439
Gastroenteropancreatic (GEP) endocrine system 胃肠胰内分泌系统 434
Gastrointestinal (GI) tract 胃肠道 434
GH/IGF-I axis GH/IGF-I 轴 362
Ghrelin (GRL) 生长素释放肽 (GRL) 192, 351-353, 445
 in fish, structure and function of 在鱼类，结构和功能 156
GHRH and PACAP fibers, in pituitary zone GHRH 和 PACAP 纤维，在脑垂体区 31
GHRH-like peptides GHRH-样肽 404
Gill claudin 3-and 4-like immunoreactive proteins 鳃密蛋白 3-和 4-样免疫反应蛋白质 341
Gilthead sea bream (*Sparus auratus*) 金头鲷 181, 362
Glandular nerve cells, role 腺神经细胞，作用 3
Glossopharyngeal-vagal motor complex (GVC) 吞咽-迷走运动复合体 355

Glucagon-like peptide-1（GLP-1） 胰高血糖素-样肽-1　403
Glucagon treatment　胰高血糖素处理　403
Glucocorticoid receptor（GR）　糖皮质激素受体　245，362-363
Glucocorticoid-responsive element（GRE）　糖皮质激素反应元件　82
Glucocorticoids receptors，in fish brain　糖皮质激素受体，在鱼脑　81
　　chemically identified　化学鉴定　84
　　distribution of　分布　82
Glutamate，in GH release　谷氨酸，GH 释放　154
Glutamate neurotransmitters　谷氨酸神经递质　39
Glutamic acid decarboxylase（GAD）　谷氨酸脱羧酶　39-40
Glutamic acid decarboxylase 67（GAD67）　谷氨酸脱羧酶 67　114
*gnrh*1 gene, role　*gnrh*1 基因, 作用　111
GnRH-induced LH synthesis, DA inhibitory control　GnRH 诱导 LH 合成, DA 抑制性控制　113-114
GnRH neurons, in fish brain　GnRH 神经元, 在鱼脑内　78
Goldfish（*Carassius auratus*）　金鱼, 鲫鱼　10，76，109，140，438
Gonadal growth and reproduction seasonality, GnRH and GnRH receptors in　性腺生长与生殖季节性, GnRH 和 GnRH 受体　111
Gonadal steroid hormones　性类固醇激素　70-71
　　steroid production in fish brain　类固醇激素在鱼脑内产生　71-74
　　steroid receptors　类固醇受体　74-75
　　　　androgen receptors　雄激素受体　79-81
　　　　estrogen receptors　雌激素受体　75-79
　　　　progesterone receptors　黄体酮受体　81
Gonadectomy, in LH and FSH secrete evaluation　性腺切除, LH 和 FSH 分泌评价　112
Gonadotropin-releasing hormone（GnRH）　促性腺激素释放激素　29，72，108，149，190，407
　　actions of　作用　110-111
　　in GH release　对 GH 的释放　149-151
　　gonadal steroids regulation　性类固醇激素调节　111-112
　　　　negative feedback effects　负反馈作用　112-113
　　　　positive feedback effects　正反馈作用　113-114
　　multiplicity　多样性　108-109
　　receptors　受体　109-110
Gonadotropins hormone（GTH）　促性腺激素　5，71
G-protein coupled receptors（GPCR）　G-蛋白偶联受体　37，181
Grass carp（*Ctenopharyngodon idellus*）　草鱼　142

Green molly (*Poecilia latipina*)　绿花鳉　22
Green swordtail (*Xiphophorus helleri*)　剑尾鱼　441
Grey mullet (*Mugil cephalus*)　鲻鱼　31, 111, 113-114, 117
Growth hormone factor-1 (GHF-1)　生长激素因子-1 (GHF-1)　140
Growth hormone (GH)　生长激素 (GH)　5, 69, 139, 396, 404
　　autocrine/paracrine signals in pituitary, in GH release　脑垂体自分泌/旁分泌信号，GH 释放　159-160
　　biological actions　生物学作用　140-142
　　and growth hormone receptors　和生长激素受体　139-140
　　at hypothalamic and pituitary levels　在下丘脑和脑垂体水平　160-163
　　hypothalamic signals from CNS, in GH release　CNS 的下丘脑信号，对 GH 释放
　　　　inhibitors　抑制剂　143-147
　　　　neuroendocrine regulators of GH　GH 的神经内分泌调节剂　154-155
　　　　stimulators　刺激因子　147-154
　　peripheral organs/tissues signals, in GH release　外周器官/组织器官，对 GH 释放
　　　　inhibitors　抑制剂　155-156
　　　　stimulators　刺激剂　156-159
　　in regulation of feeding in fishes　鱼类摄食的调节作用　404
　　secretion and synthesis regulation　分泌和合成的调节　142-143
Growth hormone-releasing hormone (GHRH)　生长激素释放激素 (GHRH)　30, 142, 147, 404
Gulf toadfish (*Opsanus beta*)　海湾豹蟾鱼　21
Guppy (*Poecilia reticulata*)　网纹花鳉　234
Gut as endocrine organ　消化道作为内分泌器官　436
Gut blood flow, in teleosts　消化道血流，在硬骨鱼类　455
Gut innervation　消化道神经分布　434
　　development, neuroendocrine system and　发育、神经内分泌系统　441-443
　　in elasmobranchs, nerve cell bodies　在板鳃鱼类，神经细胞体　434
　　In fish, cell bodies　在鱼类，细胞体　434
　　ICCs in fish gut　ICCs 在鱼类消化道　435
　　spinal innervation in cyclostomes　圆口类脊髓神经分布　434
Gut motility control　消化道活动控制　443
　　Development　发育　450
　　gallbladder motility, control　胆囊活动，调控　449
　　gut muscles and spontaneous activity　消化道肌肉和自发活动　443-444
　　individual signal substances, effects　个别信号物质，作用　444
　　　　acetylcholine　乙酰胆碱　445

adenosine and ATP 腺苷和 ATP 447
catecholamines 儿茶酚胺 447
CCK CCK 447
CGRP CGRP 446
Galanin 甘丙肽 446
Ghrelin 生长素释放肽 445
GRP/BBS GRP/BBS 447
Histamine 组胺 444-445
5-HT 5-羟色胺 445
nitric oxide 一氧化氮 446
NPY NPY 446
Somatostatin 生长抑素 446
Tachykinins 速激肽 445
VIP and PACAP VIP 和 PACAP 446
non-propagating activity, control 非传播性活动，调控 449
propagating activity, control 传播性活动，调控 448
migrating motor complexes 迁移运动复合体 448
peristalsis 蠕动 448-449
retrograde peristalsis 逆行蠕动 449-450
Gut meuronal and endocrine systems in fish 鱼类消化道神经元和内分泌系统 433

H

Hagfish (*Myxine glutinosa*) 盲鳗 339, 438, 444, 447, 449
Hawaiian parrotfish (*Scarus dubius*) 夏威夷鹦嘴鱼 244
Hematopoietic growth factors 造血生长因子 295
Herring (*Clupea harengus*) 太平洋鲱鱼 109
Histamine 组胺 437, 444-445, 451-452
Horseradish peroxidase (HRP) 辣根过氧化物酶（HRP） 15
HPI axis in fish 鱼类 HPI 轴 224
5-Hydroxyindoleacetic acid (5-HIAA) 5-羟基吲哚乙酸（5-HIAA） 41
3β-Hydroxysteroid dehydrogenase/D4-D5 isomerase (3βHSD) 3β-羟基类固醇脱氢酶/D4-D5 异构酶（3βHSD） 71
5-Hydroxytryptamine (5-HT) 5-羟色胺（5-HT） 41, 115, 147, 437
Hypophysiotropic factors to stress response 下丘脑调控脑垂体因子对应激反应 254-257
Hypophysiotropic peptides, in teleost fish 硬骨鱼类下丘脑调控脑垂体的肽类

cholecystokinin 缩胆囊肽 23-24
cocaine-and amphetamine-regulated transcript 可卡因-和安非他明-调节转录体 24-25
corticotropin-releasing factor and related peptides 促肾上腺皮质激素释放因子和相
　　关肽 25-27
galanin 甘丙肽 27-28
gastrin-releasing peptide 胃泌素释放肽 28-29
GHRH and PACAP GHRH 和 PACAP 30-32
gonadotropin-releasing hormone 促性腺激素释放激素 29-30
melanocortin system 黑皮质素系统 32-33
neuropeptide tyrosine family of peptides 神经肽酪氨酸家族肽 33-35
orexins 食欲肽 35-36
RF-amide peptides RF 酰胺肽 36-38
somatostatin peptides 生长抑素肽 38-39
thyrotropin-releasing hormone 促甲状腺素释放激素 39-40
Hypophysiotropic territories 下丘脑调控脑垂体区域
　cytoarchitecture of 细胞结构
　　hypothalamus 下丘脑 11-14
　　preoptic area 视前区 10-11
　　telencephalon 端脑 8-10
　　in fish brain 在鱼脑 15，17
Hypothalamic MCH, orexigenic role 下丘脑的 MCH，食欲性作用 229
Hypothalamic neurotransmitters 下丘脑的神经递质
　amino acid neurotransmitters 氨基酸神经递质 39-40
　dopamine 多巴胺 40
　serotonin 5-羟色胺 41
Hypothalamic peptides, prolactin secretion 下丘脑肽，催乳激素分泌 185-186
　inhibiting factors 抑制因子 186-188
　stimulating peptides 刺激肽类
　　GnRH GnRH 190-191
　　GRL GRL 192-193
　　NpCp and NPs NpCp 和 NPs 193
　　PACAP PACAP 192
　　PrRP PrRP 188-189
　　TRH TRH 191-192
　　VIP VIP 192
Hypothalamic-pituitary-adrenal（HPA） 下丘脑-脑垂体-肾上腺（HPA）轴 141
Hypothalamic-pituitary-interrenal（HPI）axis 下丘脑-脑垂体-肾间腺（HPI）轴 218

Hypothalamic-pituitary-thyroid（HPT）axis　下丘脑-脑垂体-甲状腺（HPT）轴　218，219，309
Hyporthalamic signals from CNS, in GH release　来自 CNS 的下丘脑信号，对 GH 释放
 inhibitors　抑制剂　143
  5-HT and GABA in　5-HT 和 GABA　147
  somatostatin　生长抑素　143
 stimulators　刺激剂　147
  dopamine　多巴胺　151-152
  GHRH and GnRHs　GHRH 和 GnRHs　147-151
  GRP and Galanin　GRP 和甘丙肽　153-154
  NPY　NPY　152-153
  PACAP　PACAP　151-152
  TRH, CRF and CCK　TRH、CRF 和 CCK　153
Hypothalamo-hypophysial system　下丘脑-脑垂体系统　3
Hypothalamo-pituitary complex, steroid effects　下丘脑-脑垂体复合物，类固醇作用　71-74
Hypothalamus, cytoarchitecture　下丘脑，细胞结构　11-14
Hypoxia, appetite suppressive effects　低氧，食欲抑制作用　409

I

icv OX-A injections　脑腔 OX-A 注射　407
IL-1 family and receptors　IL-1 家族和受体　295
Immune modulation, in teleost fish　免疫调节，在硬骨鱼类
 BPG axis　BPG 轴　309-310
 by cytokines　由细胞因子　310
  cytokine-induced activation　细胞因子诱导活化　312-313
  peripheral immune signals　外周免疫信号　311-312
  from periphery to brain　由外周到脑　310
  production in brain　在脑内产生　312
  stress response　应激反应　310
 HPA axis　HPA 轴　298
 HPT axis　HPT 轴　309
 opioid system　阿片样物质系统　301
 sympathetic innervation　交感神经分布　297
 type I cytokine family　I 型细胞因子家族　295

  effects of PRL, GH and leptin PRL、GH 和瘦素的作用 308-309
  expression in lymphoid tissues of fish 在鱼类淋巴组织中的表达 307
  production ofGH, PRL, SL and leptin GH、PRL、SL 和瘦素的产生 306
  receptors in lymphoid tissues of fish 在鱼类淋巴组织中的受体 308
  receptors on leukocytes 白细胞的受体 307-308

Immunity 免疫性 290
  adaptive immune response 适应性免疫反应 293
  cell types for innate immune system 先天性免疫系统细胞类型 291
  immune organs and communication of 免疫器官和交流沟通 290
  immune response, termination of 免疫反应，终止 293-294
  innate immune response 先天性免疫反应 292
  and interaction with neuroendocrine system 和神经内分泌系统相互作用 294
  and pathogen recognition 和病原体识别 291

Immunocytochemical staining 免疫细胞化学染色 37
  in rainbow trout 在虹鳟 37
  in sea bass 在海鲈 27

Immunoprotective effects, of GH GH 的免疫保护作用 141

Immunostaining studies, in plainfin midshipman 在光蟾鱼的免疫染色反应 17

Indian catfish (*Clarias batrachus*) 蟾胡鲶 118

Indian catfish (*Heteropneustes fossilis*) 印度鲶鱼 112, 147

Innate immune response 先天性免疫反应 292

Inositol triphosphate (IP3) 肌醇三磷酸 191

Insulin and insulin-like growth factor 胰岛素和胰岛素生长因子 85-86

Insulin-like growth factor (IGF) 胰岛素生长因子（IGF） 69, 85-86, 155, 180

Insulin-like growth factor I (IGF-I) 胰岛素生长因子 I（IGF-I） 119-120, 362

Interferon gamma (IFN-γ) 干扰素-γ 292

Interleukin 1 beta (IL-1β) 白细胞介素-1β 292

Interleukin 6 (IF-6) 白细胞介素 6 292

Interleukin 12 (IF-12) 白细胞介素 12 292

Intermedin (IM) 垂体中间叶激素 401

Internal cell layer (ICL) 内细胞层（ICL） 34

Interstitial cells of Cajal (ICCs) Cajal 间质细胞（ICCs） 433-434, 443

Isotocin peptide (IST) 硬骨鱼催产素（IST） 19, 21, 220

J

Janus kinase（JAK） 詹纳斯激酶（JAK） 180，290
Janus kinase 2（JAK 2） 詹纳斯激酶 2（JAK2） 140
Japanese conger eel（*Conger myriaster*） 星康吉鳗 87
Japanese crucian carp（*Carassius auratus langsdorfi*） 兰氏鲫 188
Japanese crucian carp（*Carassius cuvieri*） 高鲫 36
Japanese flounder（*Paralichthys olivaceus*） 牙鲆 183，309，401，439
Japanese stargazer（*Uranoscopus japonicas*） 日本䲢 446

K

Kallikrein-kinin system（KKS） 激肽释放酶-激肽素（KKS） 352
Ketanserin, role Ketanserin，作用 115
Killifish（*Fundulus heteroclitus*） 底鳉 10，235，363
*kiss*1 gene *Kiss* 1 轴 117
KISSneurons, in fish brain KISS 神经元，在鱼脑内 78
Kisspeptins, fish reproductive function of 吻素，鱼的生殖功能 117-118

L

Lake trout（*Salvelinus namaycush*） 大湖红点鲑 409
Lamprey（*Lampetra fluviatilis*） 河七鳃鳗 154，437
Lateral forebrain bundle（LFB） 外侧前脑束（LFB） 10
Lateral hypothalamic nucleus（LH） 下丘脑外侧核（LH） 12
Lateral recess（LR） 外侧隐窝（LR） 12，14
Lateral tuberal nucleus（NLT） 外侧结节核（NLT） 14
Leptin 瘦素
    in GH release 在 GH 释放 158-159
    and leptin receptors 和瘦素受体 85
Ligand-binding domain（LBD） 配体结合区 74
Lipopolysaccharide（LPS） 脂多糖类 291
β-Lipotrophic hormone（β-LPH） β-促脂解素（β-LPH） 239
Lungfish（*Protopterus aethiopicus*） 东非肺鱼 181
Lungfish（*Protopterus annectens*） 非洲肺鱼 122
Luteinizing hormone（LH）, in teleost fish 硬骨鱼类促黄体激素 5，239

gonadal steroids regulation 性类固醇激素调节作用 111-112
negative feedback effects 负反馈作用 112
positive feedback effects 正反馈作用 112-113

M

Major histocompatibility complex（MHC） 主要组织相容性复合体（MHC） 293
Mammalian genes, positively regulated by T3 哺乳类基因，由 T3 正的调节 251
Marine cyclostome species, plasma osmolality 海水圆口类，血浆重量摩尔渗透压浓度 338
Marine goby（*Trimma okinawae*） 黄点拟肌塘鳢 18
Masu salmon（*Oncorhynchus masu*） 马苏大麻哈鱼 86, 110, 149, 151, 191, 194, 255, 407
MCH genes MCH 基因 406
MC4R-and MC5R-expressing neurons MC4R-和 MC5R-表达神经元 224
MCT proteins MCT 蛋白 251
Medaka（*Oryzias latipes*） 青鳉 117, 195
Medaka（*Oryzias melastigma*） 青鳉 78, 403
Melanin-concentrating hormone（MCH） 黑色素浓集激素（MCH） 14, 22, 79, 119, 224, 405
Melanocortin（MC）system 黑皮质素系统（MC） 405
Melanocortin system, in teleost fish 硬骨鱼类黑皮质素系统 32-33
α-melanocyte-stimulating hormone（α-MSH） α-促黑色素细胞激素（α-MSH） 86, 224, 241-243, 254
    contribution of MCH in regulation MCH 调控的作用 256
    secretion in teleost fish, factors affecting 硬骨鱼类的分泌，因子作用 226-228
Melanocyte-stimulating hormone（MSH） 促黑色素细胞激素（MSH） 6
Melanotropes 促黑色素细胞激素
    hypothalamic regulation of 下丘脑调节作用 224-225
        hypothalamic factors TRH and CRF 下丘脑因子 TRH 和 CRF 228-229
        MCH and DA MCH 和 DA 229-230
    secretion, targets and functions of 分泌、靶标和功能 239
        MSH and N-acetylated β-ENDs MSH 和 N-乙酰 β-END 241-243
Melanotropic axis 促黑色素细胞激素轴 218, 253-254
    CRF/UIcontribute to regulation of CRF/UI 对调节的作用 255
    factors affecting activity, in teleosts 在硬骨鱼类，因子作用活性 225
Melatonin, in LH and FSH neuroendocrine regulation 褪黑激素，对 LH 和 FSH 的神

经内分泌调节作用 115-116
Metabolic hormones 代谢激素 84
insulin and insulin-like growth factor 胰岛素和胰岛素生长因子 85-86
 leptin and leptin receptors 瘦素和瘦素受体 85
 thyroid hormone receptors 甲状腺素受体 86-87
Metabolic signals/energy reserves 代谢信号/能量保存 406
 hyperglycemia and feeding latency time 高血糖和摄食潜伏时间 406
 linoleic acid and food intake 亚油酸和食物摄取 406
 lipostatic control of food intake 食物摄取的脂质静态调控 406
Main brain and hormone relationships, in fish hormonal milieu maintenance 主要的脑与激素关系，鱼类激素的环境保持 70
Mineralocorticoid receptors (MR) 盐皮质激素受体 81，245
Mitogen-activated protein kinase (MAPK) 促分裂原活化蛋白激酶 (MAPK) 76，155，161，251
Molly (*Poecilia latipinna*) 帆鳍花鳉 187
Monoamine and amino acid neurotransmitters, fish reproductive function 单胺和氨基酸神经递质，鱼类生殖功能
 dopamine 多巴胺 113-115
 γ-aminobutyric acid γ-氨基丁酸 116
 noradrenaline and melatonin 去甲肾上腺素和褪黑激素 115-116
 serotonin 5-羟色胺 115
Monocarboxylate transporters (MCT) 单羧化转运蛋白 (MCT) 251
Mozambique tilapia (*Oreochromis mossambicus*) 莫桑比克罗非鱼 188，229，235，242，248
Mudskipper (*Periophthalmus modestus*) 弹涂鱼 154，187，189，366
Mudsucker (*Gillichthys mirabilis*) 长颌姬鰕虎鱼 363
MuscarinicM3-like receptors 蝇蕈碱M3-样受体 437

N

Na and Cl retention Na和Cl保存 361
N-AcetylatedβEND N-乙酰化βEND 228，241
Na/K-ATPase Na/K-ATP酶 338-339，363
Na/K/2Clcotransporter (NKCC) Na/K/2Cl共转运蛋白 (NKCC) 339，362
Natriuretic peptides (NPs) 利尿钠肽 (NPs) 154，165，193，349-350
Natural killer (NK) cells 天然杀伤 (NK) 细胞 291
Neuroendocrine control of GH release GH释放的神经内分泌调控 142-143

Neuroendocrine-immune interaction　神经内分泌-免疫反应　289
Neuroendocrine pathways, in LH and FSH regulation　神经内分泌通路，LH 和 FSH 调控　107-108，121-123
　　future development　未来的发展　123-124
　　gonadotropin-releasing hormone　促性腺激素释放激素
　　　　actions of　作用　110-111
　　　　gonadal steroids regulation of　性类固醇激素的调控作用　111-113
　　　　multiplicity of　多样性的　108-109
　　　　receptors of　受体的　109-110
　　monoamine and amino acid neurotransmitters　单胺和氨基酸神经递质
　　　　dopamine　多巴胺　113-115
　　　　γ-aminobutyric acid　γ-氨基丁酸　116-117
　　　　noradrenaline and melatonin　去甲肾上腺素和褪黑激素　115-116
　　　　serotonin　5-羟色胺　115
　　neuropeptides　神经肽类
　　　　kisspeptins　kisspeptins　117-118
　　　　neuropeptides and neuropeptide Y　神经肽和神经肽 Y　118-119
　　　　pituitary adenylate cyclase-activating polypeptide　脑垂体腺苷酸环化酶激活多肽　118
　　protein hormones　蛋白质激素　119-121
Neuroendocrinology, definition　神经内分泌定义　69
Neurohypophysial hormones　神经脑垂体激素
　　arginine-vasotocin peptide　精氨酸加压催产素　17-21
　　isotocin　硬骨鱼催产素　21-22
　　melanin-concentrating hormone　黑色素浓集激素　22-23
Neurohypophysis structure　神经脑垂体结构　3-4
Neurointermediate lobe (NIL)　神经中叶 (NIL)　4, 25
Neurokinin A (NKA)　神经激肽 A (NKA)　440, 445, 450, 456
Neurokinin (NK) 1-3 receptors　神经激肽 (NK) 1-3 受体　441
Neuromedin U (NMU)　神经调节肽 U (NMU)　400
Neuronal cell bodies, in producing MCH　神经元细胞体，产生 MCH　23
Neuronal tract tracing and AVT immunohistochemical studies, in Atlantic salmon　神经元神经束追踪和 AVT 免疫组织化学研究，在大西洋鲑鱼　17-18
Neuropeptides, fish reproductive function of　神经肽，鱼类生殖功能
　　kisspeptins　kisspeptins　117-118
　　neuropeptides and neuropeptide Y　神经肽和神经肽 Y　118-119
　　pituitary adenylate cyclase-activating polypeptide　脑垂体腺苷酸环化酶激活多肽　118

Neuropeptide tyrosine peptides（NPY peptides） 神经肽酪氨酸肽（NPY 肽） 33-35，114，118，152-153，185
Neuropeptide Y（NPY） 神经肽 Y（NPY） 396，440，446，456
Neurotransmitters and hormones of gut 神经递质和消化道激素 436
 acetylcholine（Ach），for gut functions in fish 乙酰胆碱，对鱼类的消化道功能 437
 amines，as signal substances in fish gut 胺类在鱼类消化道作为信号物质 437-438
 adenosine and ATP 腺苷和 ATP 438
 histamine，5-HT and GABA 组胺、5-HT 和 GABA 437
 nitric oxide 一氧化氮 438
 signal peptides in gut 消化道内的信号肽 438
  calcitonin gene-related peptide 降钙素基因-相关肽 439
  cholecystokinin/gastrin family 缩胆囊肽/促胃液素家族 438-439
  forms of somatostatin 生长抑素的类型 440
  galanin 甘丙肽 439
  gastrin-releasing peptide and bombesin 胃泌素释放肽和铃蟾肽 439
  ghrelin 生长素释放肽 439
  GRP/BBS-like-ir nerves GRP/BBS-样免疫反应神经 439
  neuropeptide Y family 神经肽 Y 家族 440
  NPY/PYY-like material NPY/PYY-样物质 440
  pituitary adenylate cyclaseactivating polypeptide 脑垂体腺苷酸环化酶激活多肽 441
  tachykinins 速激肽 440-441
  vasoactive intestinal polypeptide 血管活性肠多肽 441
Nile perch（*Lates niloticus*） 尼罗尖吻鲈 29，197
Nile tilapia（*Oreochromis niloticus*） 尼罗罗非鱼 110，113，183，244，398
Nissl's staining，in CNS cytoarchitecture investigation 尼氏染色，在 CNS 细胞结构的研究中 7
N-methyl-D-aspartate（NMDA） N-甲基-D-天冬氨酸 154
Non-acetylatedβEND 非-乙酰化 βEND 241
Non-specific cytotoxic cells（NCC） 非特异性细胞毒性细胞（NCC） 291
Non-teleost bony ray-finned fish，neurohypophysis 非硬骨鱼类辐鳍鱼类，神经脑垂体 3
Noradrenaline（NA） 去甲肾上腺素 115，437
NOS-containing in Atlantic cod and spiny dogfish 在大西洋鳕鱼和白斑角鲨含有NOS 438
NP family，in Telsost fish 硬骨鱼类，NP 家族 349
NPO CRF gene expression NPO CRF 基因表达 247

NP receptors (NPRs)　NP 受体（NPRs）　193
NPY genes　NPY 基因　407
Nuclear estrogen receptors　雌激素核受体　76
Nucleus lateralis tuberis (NLT)　外界结节核（NLT）　218
Nucleus of the saccus vasculosus (NSV)　血管囊核（NSV）　12
Nucleus posterioris periventricularis (NPPv)　后室周核（NPPv）　117
Nucleus preopticus (NPO)　视前核（NPO）　218-219，222-223，312
Nucleus preopticus periventricularis (NPP)　视前室周核（NPP）　113
Nucleus ventral tuberis (NVT)　腹结节核（NVT）　117
Nutrient absorption, control　营养物吸收的调控　453

O

OATP family, mediating thyroid hormone　OATP 家族，介导甲状腺激素　251
Occludin expression　闭合蛋白表达　341
Opioid　阿片样物质　301
　　opioid receptor agonists, on immune parameters　阿片样物质受体同等物，免疫参数　304
　　peptide synthesis, in immune cells　在免疫细胞内的肽合成　301
　　prohormones, for opioid peptides and hormones　阿片样物质的肽类与激素的前激素　302
　　receptors and function, in immue system　免疫系统的受体和功能　302-305
　　receptor types and ligands in mammals　哺乳类的受体类型和配体　303
Orexin A (OX-A)　食欲肽 A（OX-A）　396-397
Orexins peptide　食欲肽　35-36
Organic anion transporter polypeptides (OATP)　有机阴离子转运多肽（OATP）　251
Organon vasculosum of the lamina terminalis (OVLT)　终板的血管器（OVLT）　347
Ostariophysan fish, GH of　骨鳔鱼类，GH　139
Oxyntomodulin　泌酸调节肽　403
Oxytocin　催产素　223
Oyater toadfish (*Opsanus tau*)　毒棘豹蟾鱼　80

P

PACAP-related peptide (PRP)　PACAP 相关肽（PRP）　30-31
Pancreas, fish　鱼类胰脏　434
Pancreatic polypeptide (PP)　胰脏多肽（PP）　435

Paraventricular organ (PVO)　室旁器（PVO）　14
Pars distalis (PD)　远侧部（PD）　4
Pars intermedia (PI)　中间部（PI）　4，154，180，196
Pars nervosa (PN)　神经部（PN）　4
Pathogen-associated molecular patterns (PAMPs)　病原体相联系的分子型式（PAMPs）　291
Pejerrey (*Odontesthes bonariensis*)　银汉鱼　72，149
Pepsinogen/pepsin secretion　胃蛋白酶原/胃蛋白酶分泌　452
Periodic acid-Schiff (PAS)　过碘酸希夫反应（PAS）　194
Peripheral organs/tissues, signals from　外周器官/组织，信号来自
　　inhibitors　抑制剂　155
　　　　insulin-like growth factor　胰岛素样生长因子　155
　　　　noradrenaline　去甲肾上腺素　155-156
　　stimulators　刺激剂　156-159
Phosphatidylinositol 3-kinase (PI3K)　磷脂酰肌酸 3-激酶（PI3K）　76
Phosphoenolpyruvate carboxykinase (PEPCK)　烯醇丙酮酸磷酸羧激酶（PEPCK）　251
Phospholipase C (PLC)　磷脂酶 C（PLC）　152，191
Photoperiod and feeding activity in fish　光周期与鱼类摄食活动　408
Pituitary adenylate cyclase-activating polypeptide (PACAP)　脑垂体腺苷酸环化酶激活多肽　30，118，142，151，192，404，441
Pituitary cells, hypothalamic regulation　脑垂体细胞，下丘脑调节
　　corticotropes in　促肾上腺皮质激素细胞　219
　　　　inhibitory factors　抑制因子　224
　　　　stimulatory factors　刺激因子　219，222-224
　　melanotropes in　黑色素细胞　224
　　　　hypothalamic factors TRH and CRF　下丘脑因子 TRH 和 CRF　228-229
　　　　MCH and DA　MCH 和 DA　229-230
　　thyrotropes in　促甲状腺激素细胞　230
　　　　hypothalamic factors　下丘脑因子　236，238-239
　　　　thyrotropin-releasing hormone　促甲状腺激素释放激素　231，324-236
Pituitary　脑垂体
　　estrogen receptor in　雌激素受体　79
　　structure　结构　4
Plainfin midshipman (*Porichthys notatus*)　光蟾鱼　3，78
Plasma Na$^+$ concentration　血浆 Na$^+$ 浓度　342
Platyfish (*Xhiphophorus maculates*)　斑剑尾鱼　27
Pollutants and health status　污染物和健康状态　409

Posterior recess nucleus (NRP)　后隐窝核 (NRP)　12
Posterior tuberal nucleus (NPT)　后结节核 (NPT)　12
Postprandial hyperemia　餐后充血　454
Preoptic area　视前区
　　cytoarchitecture of　细胞结构　10
　　in teleost pituitary　在硬骨鱼类脑垂体　17
Preoptic area (POA)　视前区 (POA)　223
Prepro-SS peptides (PSSs)　前-SS 原肽 (PSSs)　145
PRL$_{188}$ gene transcript, in tilapia larvae　PRL$_{188}$ 基因转录体，在罗非鱼幼鱼　184
Proconvertase 1 (PC1)　前转变酶 1 (PC1)　32
Progesterone receptors (PRs), in fish brain　黄体酮受体 (PRs)，在鱼脑内　81
Prohormone convertases (PCs)　前激素转变酶 (PCs)　239
Prolactin (PRL) hormone　催乳激素　5, 295, 361, 453
　　neuroendocrine regulation of prolactin secretion　催乳激素分泌的神经内分泌调控
　　　　functions of　功能　183-185
　　　　　　hypothalamic peptides　下丘脑肽　185-193
　　　　　　PRL and receptor　PRL 和受体　181-183
Prolactin-releasing peptide (PrRP)　催乳激素释放肽 (PrRP)　36, 188, 361, 400
Pro-opiomelanocortin (POMC)　阿黑皮素原 (POMC)　6, 219, 300, 405
Pro-opiomelanocortin-related peptides　阿黑皮素原相关肽　300-301
Protein hormones, fish reproductive function　蛋白质激素，鱼类生殖功能　119-121
Protein kinase C (PKC)　蛋白质激酶 C (PKC)　76
Proximal pars distalis (PPD)　近端远侧部 (PPD)　4, 180
PrRP-immunoreactive (tPrRP-ir)　PrRP-免疫反应 (tPrRP-ir)　188
PSS-I and PSS-III, in preoptic area　PSS-I 和 PSS-III，在视前区　38
Pufferfish (*Fugu ruprides*)　东方红鳍鲀　110
Pufferfish (*Takifugu rubripes*)　红鳍东方鲀　24, 140, 183, 240, 309, 396
Pufferfish (*Teraodon nigrovidis*)　黑青斑河鲀　110

R

Rabbitfish (*Siganus guttatus*)　点蓝子鱼　142
Radial cells, in embryonic neurogenesis　径向细胞，在胚胎神经发生中　72
Rainbow trout (*Oncorhynchus mykiss*)　虹鳟　18, 72, 116, 140, 181, 223, 401, 434
Ray-finned fish, telencephalon topology　辐鳍鱼类，端脑局部解剖学　7

Reactive oxygen species (ROS)　活性氧类别（ROS）　291
Reciprocal interactions　交互作用　69-70
Rectal gland ofelasmobranches　板鳃鱼类直肠腺　452-453
Red drum (*Sciaenops ocellatus*)　红拟石首鱼　195
Red sea bream (*Pagrus major*)　真鲷　110
Renin-angiotensin system (RAS)　肾素-血管紧张素系统（RAS）　346，359
Reproduction　生殖
　　FSH and LH in　FSH 和 LH　107
　　reproductive functions, GH receptors　生殖功能，GH 受体　140-141
Reverse transcriptase polymerase chain reaction (RT-PCR)　逆转录酶-多聚酶链式反应（RT-PCR）　182
RF-amide peptides　RF-酰胺肽　36，400
　　in GH release　对 GH 释放　154
　　in teleost fish　在硬骨鱼类　36-37
RF-amide-related peptide (RFRP)　RF 酰胺相关肽（RFRP）　36
RNA polymerase　RNA 多聚酶　250
Roach (*Rutilus rutilus*)　拟鲤　109
Rostral pars distalis (RPD)　吻端远侧部　4，84，180，224

S

Saddleback wrasse (*Thalassoma duperrey*)　锦鱼　112
Salinity and feeding　盐度和摄食　408
Sculpin (*Myoxocephalus octodecimspinosus*)　多刺床杜父鱼　31
Sculpin (*Myoxocephalus scorpius*)　短角床杜父鱼　435-436
Sea bass (*Dicentrarchus labrax*)　舌齿鲈　7，10，198，396
Sea bream (*Sparus aurata*)　金头鲷　25，140
Sea lamprey (*Petromyzon marinus*)　海七鳃鳗　356
Sea perch (*Lateolabrax japonicas*)　花鲈　36，397
Secretion and digestion, control　分泌与消化，调控
　　gastric acid secretion　胃酸分泌　450-452
　　pancreatic secretion　胰脏分泌　452
　　pepsinogen/pepsin secretion　胃蛋白酶原/胃蛋白酶分泌　452
　　rectal gland of elasmobranchs, secretion　板鳃鱼类直肠腺，分泌　452-453
Secretogranin-II (Sg-II), goldfish LH release regulation　分泌粒蛋白 II（Sg-II），金鱼 LH 分泌调控　119
Secretoneurin (SN), goldfish LH release regulation　分泌神经碱（SN），金鱼 LH 释

放调控 119

Senegalese sole (*Solea senegalensis*) 塞内加尔鳎 28，254

Serotonin 5-羟色胺

    neurotransmitters 神经递质 41

    in teleost fish 在硬骨鱼类 115

Serum amyloid A (SAA) 血清淀粉状蛋白A (SAA) 291，300

Sex steroids, in GH release 性类固醇，对GH释放 158

Signal transducer and activator of transcription (STAT) 信号转导物和转录激活剂 (STAT) 290

Signal transducers and activators of transcription (STAT) 信号转导物和转录激活剂 (STAT) 180

Sockeye salmon (*Oncorhynchus nerka*) 红大麻哈鱼 119，151，198

Sodium appetite, (preference) 钠欲 (偏爱) 355

Somatolactin receptor (SLR) 生长乳素受体 194-195

Somatolactin (SL) hormone 生长乳素激素 5，180

    neuroendocrine regulation of somatolactin secretion 神经内分泌调控生长乳素分泌

        functions of 功能 195-197

        hypothalamic control of 下丘脑调控 197-198

        somatolactin and receptor 生长乳素和受体 194-195

Somatostatin gene 生长抑素基因 186，188

Somatostatin (SS) peptide, in teleosts 生长抑素 (SS) 肽，在硬骨鱼类 37-38，143-147

South American cichlid (*Cichlasoma dimerus*) 南美丽体鱼 30

Spadefish (*Chaetodipterus faber*) 大西洋棘白鲳 122

Spiny dogfish (*Squalus acanthias*) 白斑角鲨 242，401

Splanchnic circulation, control 内脏血液循环，调控 454

    increase in flow 增加流量 455

    reduction in flow 减少流量 455

    variable effects 可变作用 456-457

Spotted dogfish (*Scyliorhinus stellaris*) 斑点猫鲨 440

Spotted sea trout (*Cynoscion nebulosus*) 云纹犬牙石首鱼 81

Spotted snakehead (*Channa punctatus*) 翠鳢 241

SS-or CCK/gastrin-ir endocrine cells SS-或CCK-/促胃液素-免疫反应内分泌细胞 443

SS receptors (SST) SS受体 (SST) 187

Stargazer (*Uranoscopus japonicas*) 日本䲢 446

Steroid effects, on hypothalamo-pituitary omplex 类固醇作用，对下丘脑-脑垂体复合体 72-74

Steroid hormone replacement, in LH and FSH secrete evaluation 类固醇激素取代，LH 和 FSH 分泌评价 112
Steroid membrane receptors 类固醇膜受体 74-75
 androgen receptors 雄激素受体 79-81
 estrogen receptors 雌激素受体 75-79
 progesterone receptors 黄体酮受体 81
Stingray (*Dasyatis akajei*) 赤魟 152
Striped bass (*Morone saxatilus*) 条纹狼鲈 109
Sturgeon (*Acipenser gueldenstaedtii*) 俄罗斯鲟 181
SW-adapting hormones in teleosts 硬骨鱼类海水适应激素 352
SW-induced drinking 海水诱导饮水 354
Swordtail (*Xyphophorus helleii*) 剑尾鱼 36

T

Tachykinins (TKs) 速激肽（TKs） 400, 407, 456
T3 and T4 levels, in catfish 鲶鱼的 T3 和 T4 水平 254
Target genes, for T3 in teleost species 靶基因，硬骨鱼类的 T3 251
T-cell receptor (TCR) T-细胞受体（TCR） 293
Telencephalon, development of 端脑，发育 8-10
Teleost CXC cytokines 硬骨鱼类 CXC 细胞因子 295-296
Teleostean growth hormone 硬骨鱼类生长激素 139
Teleostean hypothalamus, cytoarchitecture 硬骨鱼类下丘脑，细胞结构 11-14
Temperature and food intake, relationship 温度和食物摄取的关系 408
Testosterone, in LH and FSH synthesis 睾酮，LH 和 FSh 合成 112
Tetrodotoxin (TTX) 河鲀毒素（TTX） 443-446, 448
TGF beta (TGF-β) TGF-β 294
T4 hormone T4 激素 235, 243, 248
 levels, in rainbow trout 在虹鳟的水平 254
Three-spined sticklebacks (*Gasterosteus aculeatus*) 三棘刺鱼 185
Thyroid hormone 甲状腺激素
 biologically active metabolites 生物活性代谢产物 251-252
 in GH release 对 GH 释放 157
 metabolism 代谢产物 249
 negative feedback 负反馈作用 252
 signaling 信号 250
 targets and functions 靶标和功能 251

Thyroid hormone receptors（TRs） 甲状腺激素受体（TRs） 86-87，250
Thyroid peroxidase（TPO） 甲状腺过氧化物酶（TPO） 244
Thyroid response element（TRE） 甲状腺反应元件（TRE） 250
Thyroid-stimulating hormone（TSH） 甲状腺刺激激素（TSH） 5，234-235，256-257
 production within immune system 在免疫系统内产生 309
 targets and functions of 靶标和功能 243-244
  receptor 受体 244-245
  structure and functions 结构和功能 243-244
Thyrotropes 促甲状腺细胞
 hypothalamic regulation of 下丘脑调节 230
  hypothalamic factors 下丘脑因子 236，238-239
  thyrotropin-releasing hormone 促甲状腺素释放激素 231，324-236
  secretion, targets and functions of 分泌、靶标和功能 239
Thyrotropic axis 促甲状腺素细胞轴 254
Thyrotropin-releasing hormone（TRH） 促甲状腺素释放激素（TRH） 38-39，191，223-224，231，234
 effects on thyrotropin secretion, in teleost fish 促甲状腺素分泌的作用，在硬骨鱼类 232-234
 in GH release 对 GH 的释放 153
Thyrotropin secretion in teleost fish, factors affecting 硬骨鱼类的促甲状腺素分泌，因子的影响 237-238
Thyroxine 甲状腺素 248-249
 and extrathyroidal deiodination 甲状腺外的脱碘作用 248-249
 factors affecting secretion 影响分泌活动的因子 231
Tilapia（*Astatotilapia burtoni*） 妊丽鱼 80，84
Tilapia larvae, PRL_{188} gene transcript in 罗非鱼幼鱼，PRL_{188} 基因转录 184
Tilapia（*Oreochromis mossambicus*） 莫桑比克罗非鱼 188，229，235，242
TLR activation TLR 激活作用 291
T-lymphocytes T-淋巴细胞 292-294
Toadfish（*Opsanus tau*） 毒棘豹蟾鱼 80，340
Toll-like receptors（TLRs） Toll-样受体（TLRs） 291
Torpedo fish（*Torpedo marmorata*） 纹电鳐 147
T3-responsive genes T3-反应基因 250
TRH-ir fibers TRH-免疫反应纤维 223
3,5,3'-Triiodo-L-thyronine（T3） 3,5,3'-三碘-L-甲状腺原氨酸（T3） 86，184，230，248，251

TR-RXR-TRE complex　TR-RXR-TRE 复合体　250
TSH receptor (TSHR)　TSH 受体（TSHR）　244
TSHR gene　TSHR 基因　244
T3-target genes　T3 靶标基因　250
Tuberal hypothalamus, in teleost pituitary　结节下丘脑，在硬骨鱼类脑垂体　17
Tumor necrosis factor alpha (TNF-α)　肿瘤坏死因子 α（TNF-α）　292
Turbot (*Psetta maxima*)　瘤棘鲆　143
Turbot (*Scophthalmus maximus*)　大菱鲆　145，441
TypeⅠcytokines and linkage to immune system　Ⅰ型细胞因子和免疫系统连接　305-306
TypeⅡ helical cytokines (HCⅡ)　Ⅱ型螺旋细胞因子（HCⅡ）　295
Tyrosine hydroxylase (TH)　酪氨酸羟化酶（TH）　74，437-438

U

UⅠ-ir nerve fibers　UⅠ-免疫反应神经纤维　223
Ultrastructural studies　超显微结构研究
　　in goldfish　在金鱼　17
　　in green molly　在绿花鳉　22
Upper esophageal sphincter muscle (UES)　上食道括约肌（UES）　346
UrotensinⅡ sequences　UⅡ序列　363
UrotensinsⅠ(UⅠ) andⅡ(UⅡ)　硬骨鱼类紧张肽Ⅰ(UⅠ) 和Ⅱ(UⅡ)　362

V

Valley pupfish (*Cyprinodon nevadensis*)　内华达鳉　17
Vasoactive intestinal peptide (VIP)　血管活性肠肽（VIP）　30，192，353，403
Vasotocin V1 receptor　加压催产素 V1 受体　358
Ventricular NP (VNP)　心室 NP（VNP）　193，349
Vertebrates, feeding　脊椎动物，摄食　388
Vesicular Ach transporter (VAChT)　小泡 ACh 转运蛋白（VAChT）　437
Volemic vs. osmotic mechanisms　容量对渗透作用机理　341-343
　　blood volume, determinant of blood pressure in trout　血量，鳟鱼血压的决定因素　342
　　euryhaline eels　广盐性鳗鱼　342
　　SW fish　海水鱼类　342
　　teleost fish in FW　淡水硬骨鱼类　341
　　terrestrial animals, to maintain plasma Na$^+$　陆生动物，保持血浆 Na$^+$　343

Voltage-sensitive Ca^{2+} channels (VSCC)　电压敏感 Ca^{2+} 通道（VSCC）　149，151-152，156

V1-type vasopressin receptor　V1-型血管升压素　223

W

Water and ion transport, control　水分和离子运输，调控　453-454

Whitefish (*Coregonus clupeaformis*)　鲱形白鲑　152

White sea bream (*Diplodus sargus*)　异带重牙鲷　21，25-26

White sturgeon (*Ascipenser transmontanus*)　高首鲟　152

White sucker (*Catostomus commersoni*)　白亚口鱼　25，143

Winter flounder (*Pseudopleuronectes americanus*)　美洲拟鲽　196，251，450

Y

Yellow snapper (*Lutjanus argentiventris*)　纹眼笛鲷　41

Yellowtail flounder (*Pleuronectes ferrugineus*)　大西洋鲽　152

Yellowtail (*Seriola quinqueradiata*)　五条鰤　396，452

Z

Zebrafish (*Danio rerio*)　斑马鱼　12，71，109，140，183

译 后 记

鱼类内分泌学，包括鱼类神经内分泌学，是鱼类生理学的重要分支学科之一，近数十年来发展迅速，具有重大的科学意义和应用前景。我自 1980 年在加拿大师从阿尔伯塔大学 R. E. Peter 教授开展鱼类促性腺激素分泌活动受到神经内分泌双重调节作用机理的研究以来，就一直从事鱼类神经内分泌学的研究。1982 年回国后，一方面，和 R. E. Peter 教授以及其他一些国外学者持续开展科研合作与学术交流；另一方面，组织和带领科研团队，设立学科点和重点实验室，培养硕士和博士研究生。30 多年来，紧密结合我国鱼类养殖生产对苗种繁育的需求，陆续开展神经内分泌对鱼类繁殖、生长、摄食、免疫等方面调控作用机理的研究，发表了一系列学术论文，解决了一些生产实际问题；同时，亦培养和造就了一批从事鱼类神经内分泌学教学与科学研究的中青年学术骨干。

为满足鱼类内分泌学科的发展、青年科技队伍的成长以及教学与科学研究进展的需求，几年前，我们曾经试图组织一批相关领域的学者编写一部关于"鱼类内分泌学"的专著，总结国际特别是国内学者们在近 20 多年来取得的研究进展，以促进该学科在我国的进一步发展，亦为青年人才培养提供学习参考用书。但是，经过一番努力之后却未能完成编写成书的预定计划。我想主要原因有二：一是参加编写的学者们都是处于鱼类内分泌学科领域第一线的科研骨干，他们承受着大量教学与科学研究任务的重担，主要精力都集中于亟待完成的创新性研究课题和青年人才的培养，难以抽出足够的时间去归纳、总结、整合前人取得的研究成果；二是我国学者们在这一学科领域的学术积累还不够充实深厚，一些研究成果还比较分散，难以整理成为系统而新颖的学术论述。

在搁置编书计划之后，我仔细阅读了美国科学出版社"鱼类生理学"系列专著第 28 卷的《鱼类神经内分泌学》。这是鱼类神经内分泌学一部比较系统与新颖的专门参考书。接着，到加拿大访问期间，我得到"鱼类生理学"系列专著的资深编辑、加拿大著名鱼类生理学家 D. J. Randall 教授的支持和帮助，他为我提供该书的原版和相关信息。于是，我于 2016—2017 年翻译这部专著。我的主要目的是通过翻译出版这部国外学者编写的专著，向我国从事这一学科相关领域的广大科学研究工作者和师生介绍鱼类神经内分泌学的系统学科内容和研究进展，借此带动我国鱼类神经内分泌学学习与研究的深入开展，增强学科研究基础的积累，包括研究平台的建设和专门人才的培养，推进学科研究水平的提高与国际学术交流的拓展，以期在几年后具备充足的条件之时，由我国的学者们编写一部具有我们自己特色的《鱼类内分泌学》（包括《鱼类神经内分泌学》）专著。我深信在中青年同行们的共同努力下，这个目标一定能够达到。

  这本专著的内容系统全面、新颖充实，涉及的研究领域较为广泛，我只用了一年多的时间翻译，时间比较紧迫，加之本人能力和水平所限，翻译过程中难免有失误之处，恳请读者赐正为盼！

林浩然
2017 年 11 月

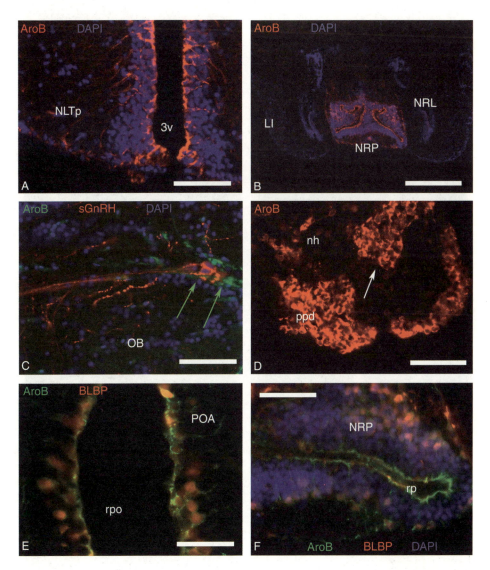

图 2.2

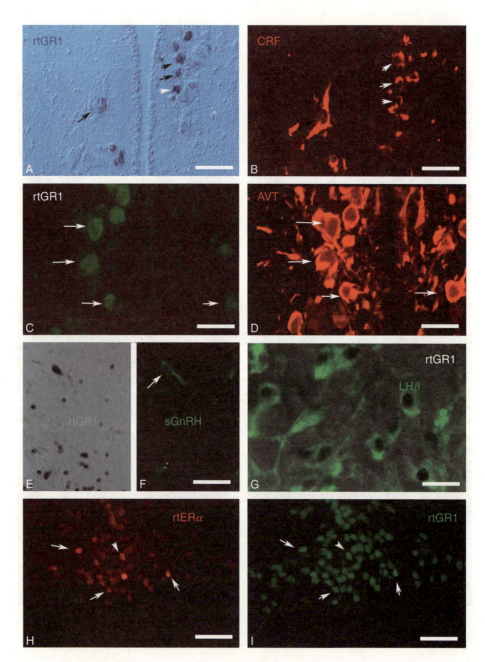

图 2.4

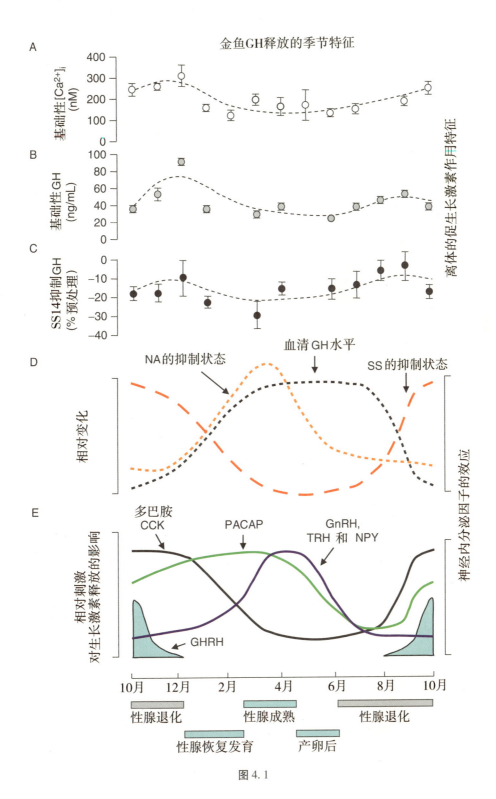

图 4.1

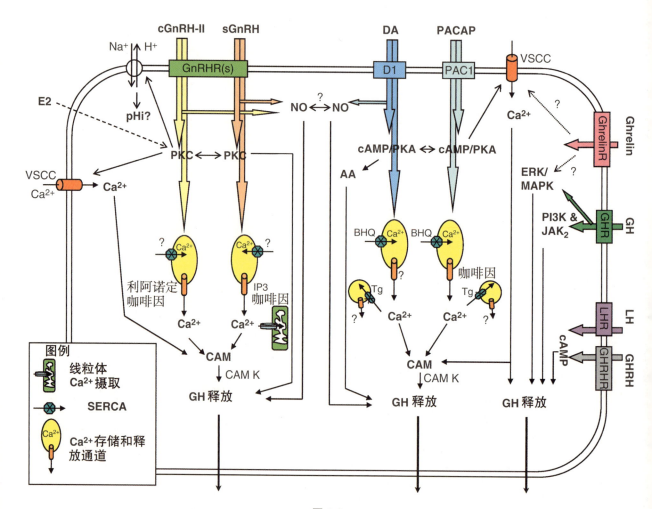

图 4.2

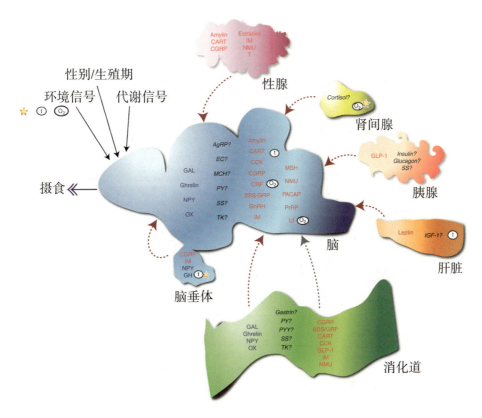

图 9.1

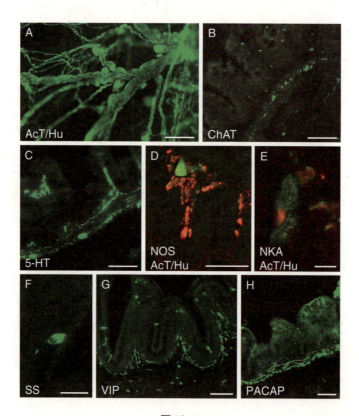

图 10.1

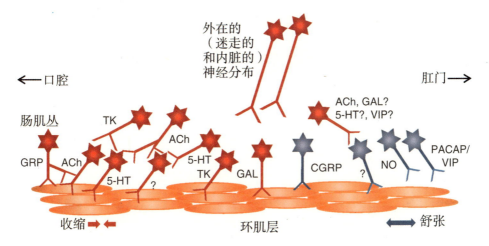

图 10.2

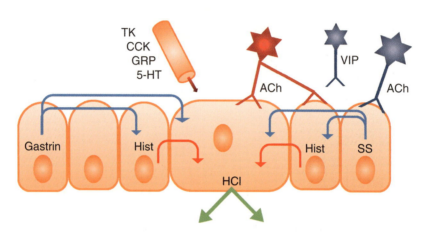

图 10.3

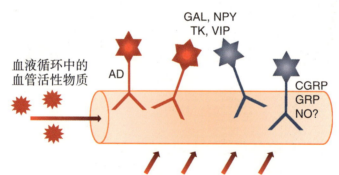

图 10.4